AF686848

Franz Rothe

Old and New Topics in Geometry

Volume I:
Projective, Neutral and Basic Euclidean Geometry

LitPrime Solutions
21250 Hawthorne Blvd
Suite 500, Torrance, CA 90503
www.litprime.com
Phone: 1-800-981-9893

© 2023 Franz Rothe. All rights reserved.

No part of this book may be reproduced, stored in a retrieval system, or transmitted by any means without the written permission of the author.

Published by LitPrime Solutions 05/08/2023

ISBN: 979-8-88703-249-8(sc)
ISBN: 979-8-88703-152-1(hc)
ISBN: 979-8-88703-252-8(e)

Library of Congress Control Number: 2022915982

Any people depicted in stock imagery provided by iStock are models, and such images are being used for illustrative purposes only.

Certain stock imagery © iStock.

Because of the dynamic nature of the Internet, any web addresses or links contained in this book may have changed since publication and may no longer be valid. The views expressed in this work are solely those of the author and do not necessarily reflect the views of the publisher, and the publisher hereby disclaims any responsibility for them.

LitPrime
"Your story is our priority"

Preface

The *front cover* shows the figure as on page 430 from the geometric proof that the segment arithmetic is commutative. The complete proof is given in the section "Arithmetic of Segments—Hilbert's Road from Geometry to Algebra" starting page 427. This line of reasoning does not presuppose Pappus' theorem, instead we use properties of circles in a Pythagorean plane.

A decade long experience of teaching the course *Fundamental of Geometry* at the University of North Carolina at Charlotte, many notes for exercises, and hours of extra reading are the bases for this bulky work. A rather complete manuscript of ca. 1100 pages, even online for some time, has been built over the years. This manuscript consists of three parts about neutral geometry, Euclidean geometry and hyperbolic geometry. It has grown from a translation of Hilbert's foundation of geometry and other German works on hyperbolic geometry, which in this way can become more easily available. The online manuscript nearly includes all topics from the fundamentals of geometry with many exercises including solutions and hundreds a elaborate computer generated drawings.

The first volume begins with Hilbert's axioms from the *Foundations of Geometry.* After some discussion of logic and axioms in general, we go on and treat in some detail the incidence geometries, especially the finite ones, and affine and projective geometry in two and three dimensions. As in Hilbert's system, there follow detailed sections about the axioms of order, next about congruence in neutral geometry,—which means without assumption of the axiom of parallelism. I deal with the axioms of measurement and of completeness, and deviating from Hilbert, with axioms about circles.

A large part of the material in the next sections originates from the century long efforts to prove the parallel axiom, from Proclus in antiquity to Legendre and his contemporaries. Lateron, the insight of independence of the parallel axiom leaves many open roads to pursue, but the desire to develop a natural as well as completely axiomatic system remains nevertheless. In this context, the classification of Hilbert planes into three types,—as semi-euclidean, semi-elliptic or semi-hyperbolic, known as the uniformity theorem is a important step beyond Euclid. A further step is the introduction of the axiom of the unbounded opening of an angle. In the spirit of Hilbert, I have set up a completely axiomatic theory of area in neutral geometry, before branching into the cases of Euclidean, or semi-hyperbolic, or semi-elliptic geometry.

Finally, the attention has to focus of Euclidean geometry, and the text gets a more educational and even more elementary flavour. Here I start with Thales theorem about the angle in a semicircle, and continue with Euclid's related theorems about angles in a circle. The most simple parts of the Euclidean geometry are given in detail, as well as the later parts about similarity, and finally area in Euclidean geometry with the theorem of Pythagoras and trigonometry, and the measurement of the circle. Too, the lens equation from geometrical optics is treated. In the table of contents the reader can get a first impression what else I have gathered and has been included into the present first volume.

The second volume deals with many more advanced topics form Euclidean geometry and contains a thorough treatise about hyperbolic geometry. Retired because of my poor eyesight, I could find the time and leisure to prepare these two volumes. My heartfelt thanks go to everyone involved, this means famous authors as well as my former students. Charlotte, August 2022

CLASSIFICATION: 51-01; 51A05; 51A30; 51Exx; 51Gxx; 51Fxx;

KEY WORDS: Instructional exposition, General theory and projective geometries, Desarguesian and Pappian geometries, Finite geometry and special incidence structures, Ordered geometries, Metric geometry,

Relation of geometry and algebra, Axiomatic theory of area in neutral and Euclidean geometry, Elementary problems in Euclidean geometry

Remark. Recently, I have observed that my efforts are not the end of the surprising developments in geometry. In his thesis *On the Formalization of Foundations of Geometry,* Pierre Boutry from the group of Julien Narboux at the university of Strasbourg in France, is going further with the formalization of geometry by means of automatic reasoning. Here is the abstract:

> In this thesis, we investigate how a proof assistant can be used to study the foundations of geometry. We start by focusing on ways to axiomatize Euclidean geometry and their relationship to each other. Then, we expose a new proof that Euclid's parallel postulate is not derivable from the other axioms of first-order Euclidean geometry. This leads us to refine Pejas' classification of parallel postulates. We do so by considering decidability properties when classifying the postulates. However, our intuition often guides us to overlook uses of such properties. A proof assistant allows us to use a perfect tool which possesses no intuition: a computer. Moreover, proof assistants let us leverage the computational capabilities of computers. We demonstrate how we enable the use of algebraic automated deduction methods thanks to the arithmetization of geometry. Finally, we present a specific procedure designed to automate proofs of incidence properties.

About the author:

Franz Rothe graduated from high school in Karlsruhe and studied mathematics, physics and music there. Graduated with a diploma in mathematics from the E T H Zürich, a doctorate in Tübingen. He got his Habilitation and venia legendi from the university at Tübingen (1984) and the Ludwig Maximilian university of Munich (1988). He has published about 40 articles, and meanwhile three books about mathematics. Since 1990 a professorship at the University of North Carolina at Charlotte, USA. His is now retired because of health problems.

In addition, Rothe and pianist Thomas Turner have developed a repertoire of classical music for flute and piano, and have recorded and released three CDs. This collection also contains several of their own transcriptions.

Some recent information about activity during the years 2017-2021, in all fields, is gathered on the websites

www.franzrothe.com

www.rothegeometry.com

Contents

Part I

Incidence, Affine and Projective Geometry

I.2 Hilbert's Axioms of Geometry

I.2.1 Logic

This subsection is too short to do justice to its topic. A comprehensive introduction into mathematical logic may be found in the book *"A Tour through Mathematical Logic"* by Robert S. Wolf, see [35].

By a *statement* is meant any sentence for which it is meaningful to consider its truth or falsehood. [1] Thus any statement can be either true or false,— but a statement is not automatically assumed to be true. Often, I shall put quotation marks around statements in order to remind the reader to this important remark. Let A and B denote any statements. The *negation* of a statement is true if and only if the statement itself is false. The negation of formula A is denoted by ¬A or not(A) or simply not A.

By $A \equiv B$, we denote a true equivalence. Thus A ≡ B means that it is true that either formulas A and B are both true, or formulas A and B are both false. [2]

A *formal implication* A ⇒ B is defined by a truth table—without assuming or presupposing any causal relation between the statements A and B. According to such a definition, the implication A ⇒ B becomes false if and only if statement A is true and statement B is false. Thus

$$\begin{aligned} \neg(\mathrm{A} \Rightarrow \mathrm{B}) &\equiv \mathrm{A}\,\mathrm{and}\,\neg\mathrm{B} \\ \mathrm{A} \Rightarrow \mathrm{B} &\equiv \neg\mathrm{A}\,\mathrm{or}\,\mathrm{B} \end{aligned}$$

The statements A ⇒ B and A ⇒ ¬B can both be true simultaneously. They are not negations of each other. Indeed

$$(\mathrm{A} \Rightarrow \mathrm{B}) \text{ and } (\mathrm{A} \Rightarrow \neg\mathrm{B}) \equiv \mathrm{A} \text{ is false}$$

The *converse* of the statement A ⇒ B is the statement B ⇒ A. Remember: The converse of a true statement may be either true or false.

The *contrapositive* of the statement A ⇒ B is the statement ¬B ⇒ ¬A.

Lemma I.2.1. *Any statement is equivalent to its contrapositive. As a formula:*

$$\mathrm{A} \Rightarrow \mathrm{B} \equiv \neg\mathrm{B} \Rightarrow \neg\mathrm{A}$$

Further considerations of such a handling of true and false statements is the topic of *propositional logic.*

Problem I.2.1. *Give the converse of the following sentences:*

"If Peter drives fast, he does not read the traffic signs."

Answer. "If Peter does not read the traffic signs, then he drives fast."

"If Tom reads the traffic signs, then he drives fast."

---

[1] This is a rather ad hoc definition, but I cannot find any better one.

[2] It would be good to have a different notation for an arbitrary equivalence, which may be true or not. I do not suggest to give the symbol ⇔ such a meaning. Above all, the reader should remember that it is unusual and dangerous to write down false statements!

Answer. "If Tom drives fast, then he reads the traffic signs."

Problem I.2.2. *Give the negation of the following sentences in a clear and simple form. You are not supposed to simply say that the statement is false!*

"Bill drives slowly and he reads the traffic signs."

Answer. "Either Bill drives fast or he does not read the traffic signs."

A second possible answer. "If Bill drives slowly, he does not read the traffic signs."

"If Peter drives fast, he does not read the traffic signs."

Answer. "Peter drives fast and he reads the traffic signs."

"If Tom reads the traffic signs, then he drives fast."

Answer. "Tom reads the traffic signs, but he does not drive fast."

"If Paul drives fast, then he does not read the traffic signs."

Answer. "Paul drives fast and he reads the traffic signs."

For the formulation of the major part of substantial mathematics, one needs more: the predicate logic which deals with *variables* and *propositional functions*. Indeed the *predicate logic*, also called *quantified logic*, is the Swiss pocket knife of mathematical logic. In principle, it is not necessary to specify the meaning or kind of a variable. The important assumption is an infinite supply of values for any variable.

The propositional functions are properties which may be true or false depending on the value of the *variable* x. They are denoted by $\mathrm{P}(x)$. Such a sentence $\mathrm{P}(x)$ is also called a *predicate*, because in the English language the property is grammatically a predicate.

The *universal quantification* of a predicate $\mathrm{P}(x)$ is the statement

"For all values of x, the predicate $\mathrm{P}(x)$ is true."

This universal quantification of $\mathrm{P}(x)$ is denoted by $\forall x \mathrm{P}(x)$. The symbol $\forall$ is called the *universal quantifier.*

The *existential quantification* of a predicate $\mathrm{P}(x)$ is the statement

"There exist values of x for which the predicate $\mathrm{P}(x)$ is true."

This existential quantification of $\mathrm{P}(x)$ is denoted by $\exists x \mathrm{P}(x)$. The symbol $\exists$ is called the *existential quantifier.*

Problem I.2.3. *Find the pair of equivalent statements. Find all pairs of statement and its negation and mark them with matching color. How many pairs (colors) are there? Find the statements, of which the negation is not listed and encircle them with a closed line.*

1. *"If a triangle is isosceles, the base angles are congruent."*
2. *"All men like to drive."*
3. *"If a triangle is isosceles, one base angle is larger than the other one."*

4. *"Some men do not like to drive."*

5. *"It rains and the streets are dry."*

6. *"No woman likes to drive."*

7. *"Some women like to drive."*

8. *"There exists a triangle with sum of its angles equal to two right angles."*

9. *"Every triangle has sum of angles either less or more than two right angles."*

10. *"No triangle is equilateral."*

11. *"All triangles are equilateral."*

12. *"No point has three or more lines passing through it."*

13. *"There exists a point through which at most two lines pass."*

14. *"There exists a point through which three or more lines pass."*

15. *"Every point has at most two lines passing through it."*

16. *"If it rains, the streets are wet."*

Answer. Statements (12) and (15) are equivalent. Five pairs of statement and negation are [(2)(4)] [(5)(16)] [(6)(7)] [(8)(9)] [(12)(14)], or equivalently [(14)(15)]. Statements (1)(3)(10)(11)(13) are not negated.

Remark. It helps to write some statements in the symbols from *predicate logic.*
Let Pn denote the predicate
"Through point P pass exactly n lines."

Here are four statements in symbolic logic:

12. "No point has three or more lines passing through it."

$$\neg\exists P\exists n\,(n \geq 3 \wedge\ Pn)$$

13. "There exists a point through which at most two lines pass."

$$\exists P\exists n\,(n \leq 2 \wedge\ Pn)$$

14. "There exists a point through which three or more lines pass."

$$\exists P\exists n\,(n \geq 3 \wedge\ Pn)$$

15. "Every point has at most two lines passing through it."

$$\forall P\exists n\,(n \leq 2 \wedge\ Pn)$$

The term *"axiomatic method"* was coined by Hilbert to describe part of his formalist program. The essentials were already developed in ancient times in Euclid's Elements, and refined in Hilbert's foundations of geometry of 1899.

Euclid's Elements is the oldest surviving work in which mathematical subjects were developed from scratch in a thorough, rigorous and axiomatic way. He puts his principles at the beginning of the Elements and names them common notions and postulates. In place of the common notion, today are put the **logical axioms**. In place of the postulates, one has the **proper axioms**, which are specific to the first-order language and the subject under consideration.

The *formalist program* goes beyond the classical axiomatic approach by explicitly defining not only the language and axioms to be used, but also the **rules of inference**. In the usual approach to first-order logic, two features of the rules of inference are worth noting:

- first, they are based entirely on logic;
- second, they are the only way of generating new steps and and proving any theorem.

Without rules of inference no mathematics can be done—any axiomatic system would be useless.

Classically, postulates were supposed to be evident truths. ***But truth does not enter into the formalist viewpoint.*** Nevertheless, even today, to build any meaningful mathematical theory, the formulas taken to be axioms should be very few, and be based on as simple as possible principles. They should be justified on intrinsic and extrinsic reasons, in other words ***well motivated and needed for the consequential development.***

I.2.2 David Hilbert's axiomatization of Euclidean geometry

Already in 1899, on the occasion of the inauguration of the Gauss-Weber monument in Göttingen, David Hilbert had published his first edition of *Foundations of Geometry.* This work is considered to be the first version of Euclidean geometry that is *truly axiomatic*, in the sense that there were no *hidden* appeals to spatial intuition. Too, Hilbert's work has much contributed to a deeper understanding of the relation between geometric and algebraic structures. This work has been leading for the clear axiomatic way of doing mathematics in the twentieth century.

Introduction from Hilbert's Foundations of Geometry

> We shall be lead to several, apparently very simple, but nevertheless very deep and difficult problems. We shall be challenged by very new and—as I believe fruitful—problems, and see remarkable connections between the elements of arithmetic and geometry, gaining another insight into the unity of mathematics.
>
> From the preface of Hilbert's lecture notes *"Foundations of Euclidean Geometry"* (1898/99)

> Thus all human knowledge begins with intuition, proceeds to notions, and ends with ideas.
>
> Hilbert's preamble, citing Kant's *"Critique of Pure Reason"*(1781)

> Geometry needs—similar to arithmetic—only a few simple basic principles for its consequential development. These basic principles are called the *axioms* of geometry. Starting with Euclid, the setup of the axioms of geometry and the investigation of their mutual connections has been the subject of many excellent treatises in the mathematical literature. The problem in question is basically a logical analysis of our spatial imagination.
>
> The present—meaning Hilbert's—investigation is a new attempt to set up a *complete* and *as simple as possible* system of axioms. And, furthermore, to deduct from them the most important geometric theorems, such that the meaning and importance of the different axioms and their consequences become clear.
>
> Hilbert's introduction to *"Foundations of Geometry"* (1899)

Hilbert's axioms

As suggested in this paragraph, Hilbert introduces the modern habit to develop the consequences of the different groups of axioms immediately after introducing them. But—because I guess it is more convenient for the reader—we shall now state the complete axiomatic system at once, at the beginning. I have included all axioms, even those only needed for three dimensional geometry.

0. Undefined elements and relations

Elements:

- A class of undefined objects called *points*, denoted by $A, B, C, \ldots$.
- A class of undefined objects called *lines*, denoted by $a, b, c, \ldots$.
- A class of undefined objects called *planes*, denoted by $\alpha, \beta, \gamma, \ldots$.

Relations:

- *Incidence* (being *incident, lying on, containing*)
- *Order* (lying *between*) (for points on a line)
- *Congruence* (for segments and angles)

Remark. Planes are only needed to include three dimensional geometry.

I. Axioms of incidence

I.1 For two points A and B there exists a line that contains each of the points A, B.

I.2 For two [different] points A and B there exists no more than one line that contains each of the points A, B.

I.3a On any line, there exist at least two points.

I.3b There exist at least three points that do not lie on a line.

Remark. We did separate axiom I.3 into I.3a and I.3b, in order to stress there is no direct logical connection between the two sentences intended.

Remark. Additional axioms I.4 through I.8 are only needed for three dimensional geometry.

I.4 For any three points A, B, C [that do not lie on the same line], there exists a plane α that contains each of the points A, B, C.

I.4a For every plane there exists a point which it contains.

I.5 For any three points A, B, C that do not lie on the same line, there exists no more than one plane that contains each of the points A, B, C.

I.6 If two different points $A \neq B$ of a line a lie on a plane α, then every point of line a lies in the same plane α.

Remark. In this case, we say that the line a lies in the plane α.

I.7 If two planes α, β have a point A in common, then they have at least one more point B in common.

I.8 There exist at least four points which do not lie in a plane.

II. Axioms of order

II.1 If a point B lies between a point A and a point C, then the points A, B, C are three distinct points of a line, and B lies between C and A.

II.2 For two points A and C, there also exists at least one point D on the line AC such that C lies between A and D.

II.3 Of any three points on a line there exists no more than one that lies between the other two.

DEFINITIONS: Segment, point of segment, interior and exterior of segment, ray, half plane, triangle. [3]

Remark. A segment AB is assumed to have two different endpoints A and B.

II.4 (Pasch' Axiom) Let A, B, C be three points that do not lie on a line and let a be a line in the plane ABC which does not meet any of the points A, B, C. If the line a passes through a point of the segment AB, it also passes through a point of the segment AC, or through a point of the segment BC.

[3]lateron: quadrilateral, polygon, side of polygon, vertex of polygon, closed and open polygon, simple polygon.

III. Axioms of congruence

III.1 If A, B are two points on a line a, and A' is a point on the same or another line a', then it is always possible to find a point B' on a given side of the line a' through A' such that the segment AB is congruent to the segment $A'B'$. In symbols $AB \cong A'B'$.

III.1b For any two different points A and B, the segments AB and BA are congruent. [4]

III.2 If a segment $A'B'$ and a segment $A''B''$ are congruent to the same segment AB, then segment $A'B'$ is also congruent to segment $A''B''$.

III.3 On a line, let AB and BC be two segments which except for B have no point in common. Furthermore, on the same or another line a', let $A'B'$ and $B'C'$ be two segments which except for B' also have no point in common. In that case,

$$\text{if} \quad AB \cong A'B' \text{ and } \quad BC \cong B'C' \,, \quad \text{then} \quad AC \cong A'C'$$

DEFINITION: Angle.

III.4a Let $\angle(h, k)$ be an angle in a plane α and h' a ray in a plane α' that emanates from the point O'. Then there exists in the plane α' one (and only one) ray k' such that the angle $\angle(h, k)$ is congruent to the angle $\angle(h', k')$ and at the same time all interior points of the angle $\angle(h', k')$ lie on the given side of h'. This means that

$$\angle(h, k) \cong \angle(h', k')$$

III.4b The ray k' in (III.4a) is unique.

III.4c Every angle is congruent to itself

$$\angle(h, k) \cong \angle(h, k)$$

III.4d Every angle is congruent to the angle with the legs switched

$$\angle(h, k) \cong \angle(k, h)$$

III.5 If for two triangles $\triangle ABC$ and $\triangle A'B'C'$ the congruences

$$AB \cong A'B' \,, \; AC \cong A'C' \,, \;\; \angle BAC \cong \angle B'A'C'$$

hold, then the congruence $\angle ABC \cong \angle A'B'C'$ is also satisfied.

IV. Axiom of parallelism

IV.1 Let a be any line and A a point not on a. Then there exists at most one line in the plane determined by line a and point A that passes through A and does not intersect a.

[4] I have added the item (III.1b) to Hilbert's axioms. This is necessary in my opinion.

V. Axioms of continuity

V.1 (Axiom of Archimedes) If AB and CD are any segments, then there exists a number n such that n segments congruent to CD constructed contiguously from A, along a ray from A through B, will pass beyond B.

V.2 (Axiom of completeness) An extension of a set of points on a line, with its order and congruence relations existing among the original elements as well as the fundamental properties of line order and congruence that follow from Axioms I-III and from (V.1), is impossible.

Definition I.2.1 (Hilbert plane). A *Hilbert plane* is any model for two-dimensional geometry where Hilbert's axioms of incidence (I.1)(I.2)(I.3a)(I.3b), order (II.1) through (II.4), and congruence (III.1) through (III.5) hold.

Neither the axioms of continuity—Archimedean axiom and the axiom of completeness—nor the parallel axiom need to hold for an arbitrary Hilbert plane.

Definition I.2.2 (Pythagorean plane). A *Pythagorean plane* is a Hilbert plane for which the axiom of parallelism (IV.1) holds.

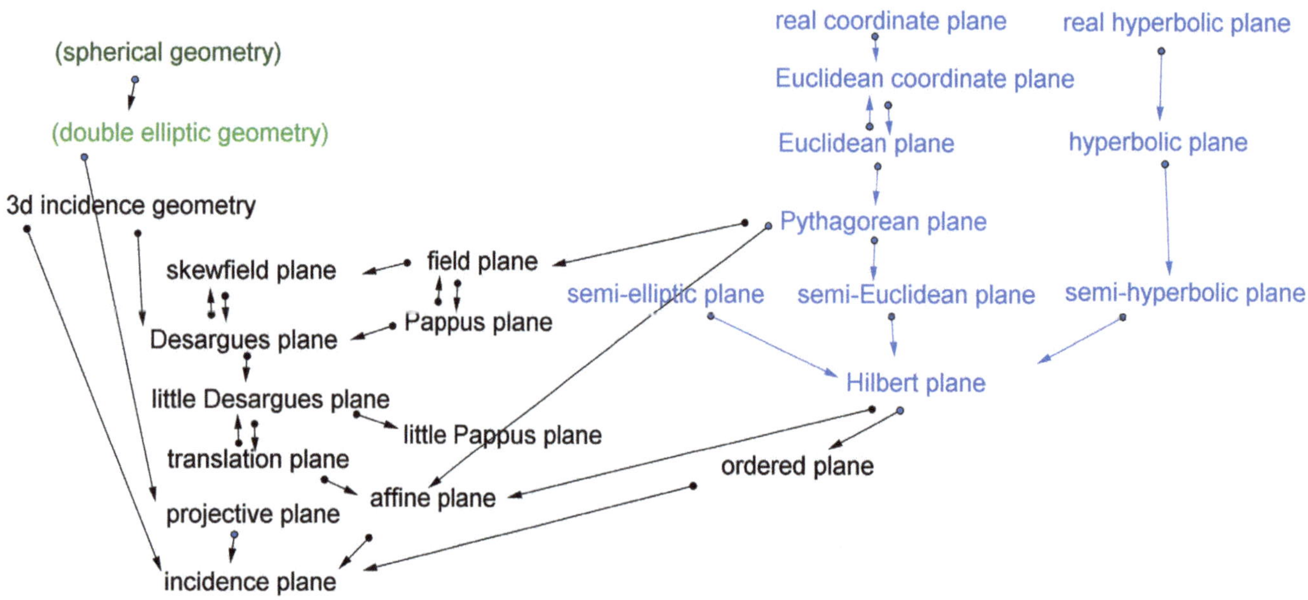

Figure I.2.1: Logical relations of geometries

I.2.3 Importance and Impact of Hilbert's Foundations of Geometry

In this essay, we address the following questions:

- Which features in Hilbert's *Foundations of Geometry* are new and different from Euclid's Elements?
- Which topics are still not included, and what are the results achieved elsewhere about these topics. Of course looking at first in this manuscript, but more important beyond at the worldwide research?
- Which important questions are still open?
- Which benefits for mathematics have been achieved by Hilbert's *Foundations of Geometry*?

Hilbert's Foundations in Comparison with Euclid's Elements

As already stressed in his introduction, Hilbert's goal is the *consequential development* of geometry. This is already an important justification for the axioms themselves, especially since no claim of absolute truth of the content of the axioms is intended. Secondly the style of presentation is determined by this goal. The axioms are presented in five groups: axioms of incidence, axioms of order, axioms of congruence, the axiom of parallelism, and the axioms of continuity. For each group, the theorems spelling out their consequences are proved, and examples of different structure are given. Such a study is done on purpose separately for the different groups. This approach is now commonplace in algebra or topology, but it is different from the style of classical geometry texts.

The role of the primary elements and relations is seen differently from Euclid. For Euclid, these were abstract entities given by nature. They are given without question, nevertheless they are still explained by phrases like "A point has no parts." Modern mathematicians point out that such a sentence poses more questions than it answers. What is the way out to avoid questioning every and even the most basic notions and how can one break the infinite chain of regress? The key point is the use of *primary elements and relations*. These entities cannot, may not, and need not to be defined. They get their meaning only via the way they are used in the axioms, proofs and theorems. To start this process, not only primary elements, but moreover primary relations, too, have to been postulated, in order to get the connections between the abstract objects. In the *Foundations of Geometry* the points, lines and planes are used as primary objects. They are connected, by the relations of incidence, order, congruence of segments and congruence of angles. Furthermore, equality is a relation from mathematical logic.

Their exist quite a few different equivalence relation important for geometry: equality, congruence of segments, congruence of angles, similarity for triangles and other figures, having same area for figures, having same volume for three dimensional polyhedra. In sweeping simplicity, Euclid used the same word "equal" for all these relations,—and afterwards even seemed to have justified the properties of an equivalence relation simply by the use of the word "equal". On the contrary, Hilbert and his followers clearly distinguish these and still several further relations, use different words and symbols for them, and prove their properties.

Among Hilbert's five groups of axioms (incidence, order, congruence, parallelism, continuity), only the axioms of congruence and parallelism have a clear-cut counterpart in Euclid. Only Euclid's first postulate "to draw a line between two points" refers to incidence. Hilbert clearly separates the two questions of *existence* and *uniqueness*, by postulating them in the two different axioms (I.1) and (I.2). The axioms of incidence referring to three dimensional geometry have no correspondence in Euclid. There are hints to the Archimedean axiom, but the axioms of order are totally omitted.

The axioms of order are a striking innovation based on the work of Pasch of 1880. They were totally omitted in Euclid's Elements. Axiom (II.4) is still named Pasch's axiom. Put into colloquial language, it tells that a line which intersects one side of triangle, intersect a second side, too. The axioms of order have been simplified in later editions of the *Foundations of Geometry*, taking advantage of work of E.H.Moore and Veblen.

Hilbert introduces two axioms of continuity: (V.1) is the Archimedean axiom and (V.2) the axiom of completeness. The axioms of continuity do not appear in Euclid's postulates. But the definition of same ratios $a : b$ and $c : d$ from Euclid's book V (credited to Eudoxus) makes only sense, if one assumes the Archimedean axiom.

The Archimedean axiom allows the *measurement* of segments and angles using real numbers. During the measurement process, a real number giving the length of a segment is produced, digit by digit in the form of a binary fraction. Since Hilbert, this axiom is also known as the *axiom of measurement.*

The clear-cut understanding of continuity was only achieved by Dedekind and Cantor in the late nineteenth century. There are several axioms for completeness, with very similar implications, which nevertheless have slight but deep differences. It is hard to say which one of these alternatives is the most natural axiom. Even Hilbert has suggested different axioms of continuity in different editions of his foundations of geometry. My favorite is Cantor's axiom, which occurs in the very first edition of Hilbert and in *Nichteuklidische Geometrie, Hyperbolische Geometrie der Ebene* by Baldus and Löbell [7], p.43.

The axioms of congruence resemble more to Euclid's Elements than the other groups, but even here we find important differences and innovations. Nowhere in Hilbert's *Foundations of Geometry*, any circle appears at all, indeed they are not even defined. Instead of Euclid's straightedge and compass, the *transfer of segments and angles* become the basic tools for geometric constructions. These tools turn out to be a bid weaker than straightedge and compass, but suffice for a few fundamental constructions. More important, the SAS congruence is introduced as an axiom. Even more, Hilbert proves the independence of the SAS axiom.

Euclid has tried to justify the SAS congruence by his principle of superposition. Because of the independence of the SAS axiom, the principle of superposition turns out to be at best a physical thought experiment, but cannot replace the SAS axiom. In a totally different approach, it is possible to use the motion of figures as a building block of geometry. But in this approach, extra work is needed to clarify what kind of motions are allowed. An (not totally rigorous) attempt in this direction is Hadamard's *Leçons de Géometrie Elémentaire* of 1901-1906.

So far, we have seen that Hilbert has achieved to make the foundations of geometry rigorous, without any hidden appeal to intuition, but kept the spirit of Euclid's Elements as much as possible. The investigations about the nature of axioms, are topics totally different from Euclid. In Hilbert's *Foundations of Geometry*, the questions of *consistency, categorial nature*, and *independence* of his axioms are addressed. I think that only a person of Hilbert's optimism could address such questions at that time. Now we know from the work of Gödel and Tarski, that consistency can only be proved for a too small part of mathematics, but not for any topic involving the natural numbers.

The most accessible topic is independence. Hilbert proves the independence of the SAS-axiom, the parallel axiom, and the Archimedean axiom. The independence of the parallel axiom is rather informally justified via the spherical geometry. In an appendix to the foundations, Hilbert gives a detailed axiomatic approach to hyperbolic geometry. Legendre's theorems the angle sum of triangles in neutral geometry, as exposed in detail.

Relative consistency of the entire system of axioms is proved, once consistency of the real number system is taken for granted—which turned out to be the really deep unsolvable problem!

Hilbert proves that his axiom system is categorial, once his axiom (V.2) of completeness is assumed, but states clearly that the system without this axiom is not categorial. Here are his own words:

> As one realizes, there are infinitely many geometries which satisfy the axiom groups I through IV and (V.1). On the other hand, there is only one—namely the Cartesian geometry—which satisfies the completeness axiom (V.2), too.

The Impact of Hilbert's Foundations of Geometry

Hilbert's work is considered to be the first version of Euclidean geometry that is *truly axiomatic*, in the sense that there were no *hidden* appeals to spatial intuition. But the *Foundations of Geometry* are much more than just a clarification of Euclid's Elements. Clearly this is one goal of Hilbert. A second goal of equal importance is a deeper understanding of the relation between geometric and algebraic structures.

Already in the introduction, Hilbert says: "We shall be challenged by very new and—as I believe fruitful—problems, and see remarkable connections between the elements of arithmetic and geometry, gaining another insight into the unity of mathematics." Such a claim is well justified. It was Hilbert who first established a clear correlation between geometric and algebraic structures. These investigation came out of projective geometry, which is a historic predecessor and Hilbert's starting point for the Foundations. It turns out that in coordinate geometry

- the Theorem of Pappus is equivalent to commutative multiplication of the coordinate field,
- the Theorem of Desargues is equivalent to associative multiplication of the coordinate field

Further results were obtained and are included in the latest edition of Hilbert's foundations. Here are two examples:

Hessenberg gave in 1904 a purely geometric proof that the Theorem of Pappus implies the Theorem of Desargues (see Theorem of Hessenberg I.4.8).

A simple example for a non-Desarguean projective plane was introduced by E. R. Moulton in the article [12] *A simple non-desarguesian plane geometry*, Trans. Math. Soc. (1902). The Moulton plane is useful to clarify the logical relations between different geometric structures.

The separate investigations about parts of the axioms have become more and more detailed and refined. Further research has extended the correlations of algebra and geometry to more exotic structures. The article of Hubert Kiechle, Alexander Kreuzer and Heinrich Wefelscheid in the fourteenth edition [19] of Hilbert's foundations from 1999 contains some relevant information.

For some of these ideas, an accessible account with examples are given by John Stillwell [31] in his exposition *The Four Pillars of Geometry*, Springer, 2005.

A totally new topic is *finite incidence geometry*. The connections to scheduling problems in computer science, large scale computation, and to sophisticated algebraic structures has lead to new research. Some results are indicated in the section on Finite Affine and Projective Incidence Planes and Latin Squares.

Finally, we all know that the axiomatic method is now commonplace in modern mathematics. Where does the word "complete" for existence of limits of Cauchy sequences come from? Many mathematicians may not even realize that it comes from the axiom of completeness in Hilbert's *Foundations of Geometry.* Here the axiomatic method is introduced in such a satisfactory way that it has been exemplary for the modern style of research and presentation in pure mathematics. Let me finish with this citation:

> We shall be lead to several, apparently very simple, but nevertheless very deep and difficult problems. (from Hilbert's preamble)
>
> Thus all human knowledge begins with intuition, proceeds to notions, and ends with ideas. (from Kant's *"Critique of Pure Reason"*(1781))

Drawbacks and Lacuna of Hilbert's Foundations

I want to ask, which topics are still not included, and what are the results achieved elsewhere about these topics. Of course I look at first at the present manuscript, after that beyond at the worldwide research.

The first subject I miss in Hilbert's foundation, are the axioms and theorems dealing with circles. Researchers like Greenberg and Hartshorne have meanwhile introduced the *line-circle intersection property* II.10.5 and the *Circle-circle intersection property* II.10.6 as the relevant axioms about the intersection of circles with lines, and of circles with circles. Several interesting theorems can be investigated already in neutral geometry. There is the remarkable three-circle theorem about of the threefold intersection of their three common chords, which holds even in neutral geometry. By the way, the figure is depicted on the cover of Robin Hartshorne's book [16] *Geometry: Euclid and Beyond.* Here, it would be nice to have a proof totally in neutral geometry. Presently, I know only a proof done at first in Euclidean geometry, and in a second step extended to hyperbolic geometry by means of the Poincaré disk model.

The *Uniformity Theorem* 17 allows to classify all Hilbert planes into one of three types, depending solely on the angle sum of triangles. This is a sharpening of Legendre's Second Theorem. Hilbert includes the two Legendre Theorems, but only mentions the more general Uniformity Theorem, giving credit to his student Max Dehn. Actually, the result may have been known earlier, A complete proof is contained in Robin Hartshorne's book *Geometry: Euclid and Beyond* [16] and in the present manuscript.

Greenberg has recently stressed the equivalence of the Euclidean parallel axiom with the angle sum of the triangle being two right angles together with *Aristole's Angle Unboundedness Axiom* II.11.2. Proclus' Theorem 18 clarifies and sharpens very old ideas of Proclus about the parallel postulate.

A corresponding result in hyperbolic geometry is much deeper and harder to prove. Here it would be a progress to have a proof totally in neutral geometry. The current proof mentioned by Greenberg [15] in his article *Old and new results in the foundations of elementary plane Euclidean and Non-Euclidean geometries*, is based on rather sophisticated elaborations of the Klein disk model of hyperbolic geometry.

I.2.4 Frege's Critique and Hilbert's answer

The book *Foundations of Geometry* was discussed and critized by Frege in an exchange of letters between the two scholars. Between October 1895 and November 1903, we know of nine letters having been exchanged. In the letter of December 1899 Hilbert writes:

> I still have to talk about one objection. You tell me, my notions, for example "point", "between" have not been fixed uniquely—for example the notion "between" is grasped (gefasst) differently on p.20 and there a point is a pair of numbers—.
>
> Yes, it is indeed self evident that every theory is only a framework or scheme of notions together with the necessary relations between them, and the basic elements (Grundelemente) can be thought of in any arbitrary manner. When I imagine as my points [their relation of order, my lines] any arbitrary system of things, for example the system: love, law, chimney sweeper...,
>
> I imagine [this system] and then I think of all my axioms as connections between these things; henceforth my theorems, for example the theorem of Pythagoras, are valid about these things, too.
>
> In other words: every theory can always be applied to infinitely many systems of basic elements (Grundelemente). One has indeed only to apply a one-to-one invertible transformation and agree [to postulate] the same corresponding axioms for the transformed basic elements. In fact this state of affairs is used frequently, for example in projective geometry as principle of duality, and by me in proofs of independence.

On another occasion, Hilbert made the point that his proofs should stay completely correct even if the words "point", "line", and "plane" were replaced throughout by "table", "chair", and "beer mug." In the two last letters of September 1900 and November 1903, Hilbert formulates his point of view concisely. In September 1900 Hilbert answers to Frege:

> It is really my opinion that a notion can only be logically clarified by its relations with other notions. These relations, formulated in distinct assertions, I call axioms. Hence I arrive at considering the axioms (together with the names for the notions) as the definitions of the notions. These convictions did not arise for me out of pure fancy, but I see myself pushed to them as a necessity for the rigorous and logical building of a theory. I have arrived at the conviction that subtle facts in mathematics and the natural sciences can be treated with certainty only in this way—otherwise one only turns oneself in cycles.

I.2.5 About the consistency proof for geometry

Around 1900, Hilbert originally wanted to obtain a consistency proof for geometry based on the well known construction of the Cartesian geometry with real-number valued coordinates. Hence he called for a proof of consistency of the real number system, among his 23 Paris problems. Because of Gödel's incompleteness theorem, it turns out to be provably impossible to prove consistency of the real number system. Today, it is clear to everybody that a consistent proof of geometry cannot be based on the coordinate geometry with real numbers as coordinates.

The consistency of the axioms of geometry, <u>without</u> the axioms of continuity, has been proved by Alfred Tarski. A rather popular and readable account is contained in Feferman's biography [10] of Tarski. Tarski's consistency proof uses his earlier result that the theory of real-closed fields [5] is consistent. One needs an infinity number of axioms to characterize a real-closed field, but one can still use a language of first-order logic. Secondly, a model of geometry is constructed: this is simply the analytic geometry with a real-closed field as coordinates. Both the axioms of a real-closed field, and the axioms of geometry (I),(II),(III),(IV) are formulated in a language of first-order logic.

Question. Why is there no contradiction between Tarski's proof of consistency of geometry and Gödel's incompleteness theorem?

Answer. The axiomatic of the Tarski geometry does not have enough power to define the Peano arithmetic. Indeed, even with any infinitely many axioms formulated in first-order arithmetic one can never define Peano arithmetic. Hence one cannot introduce a Gödel numbering for theorems and proofs nor obtain incompleteness via a diagonalization argument. Following Tarski's approach, we stay within first-order logic, and cannot define the Peano arithmetic. No contradiction to Gödel's incompleteness theorem arises.

Question. Why is it impossible to prove consistency of geometry including the axioms of continuity?

Answer. The situation becomes different, and indeed any hope for a consistency proof is doomed, once we introduce either one of the axioms of continuity (V.1) or (V.2). These axioms cannot be formulated in any language of first-order logic. Even more, they either need for their formulation the natural real numbers, or enough theory of sets as is needed for the construction of the real numbers. Hence Gödel's incompleteness theorem applies. No consistency proof is possible neither for the axiomatic system of the real numbers, nor for the Cartesian geometry with real-valued coordinates, nor for Hilbert's system of axioms (I),(I),III),(IV),(V).

I.2.6 General remark about models in mathematics

The meaning and usage of the notion "model" in the natural sciences on the one side, and mathematics on the other side, are different. For example, the astronomer says that Newton's theory of gravitation is a model for the solar system. The mathematician says that the solar system is a model of Newton's theory of gravitation. [6]

In the natural sciences, one has to begin with a complicated reality. One wants to extract some salient features, and describe or explain them in hopefully more simple mathematical terms. The scientist calls such a mathematical description a model.

On the other hand, the mathematician already begins with a theory, which often can be pinpointed by the appropriate axioms. In a second step, one looks for other examples, problems or contexts to which the theory fits. To achieve such a fit, the abstract objects and relations of the theory have to be interpreted as specific notions occurring in the example at hand. Being interpreted in the context of the example, one has to ask and check whether the objects and relations from the abstract theory actually satisfy the axioms originally postulated. That can turn out to be true or not, depending on the properties of the example chosen. Furthermore, it can turn out that the same theory fits for quite different contexts.

In building a model, the mathematician has to think on two levels. The basic level is given by the specific example at hand and its accepted properties. In favorite circumstances, one needs

[5] A field is called real-closed if and only if every odd-order polynomial has a zero

[6] Another consequence of these matters: A scientific model can only be falsified. A mathematical model can be verified

only rudiments of set theory and the number system as accepted properties. One can think of this level as the "background ontology". The secondary level brings in the originally undefined objects and relations of an abstract, and often more modern, theory, which now get interpreted in terms of the basic level. If the axioms can be proved to hold for this specific example, we say that we have constructed a model for the abstract theory. In this way, one obtains a proof of relative consistency of the abstract theory.

I.2.7 What is completeness?

As currently used in the foundations of mathematics, in mathematical research, and by the overall scientific community, the word completeness has several—I am sure at least five—different meanings.

- In the foundations of mathematics "completeness" of a theory can mean "validity implies provability", which has been confirmed, in the case of first-order logic by Gödel's completeness theorem of first-order logic. But, on the other hand, the term "completeness" of a theory has the second, much more powerful, meaning that "either a statement or its contrary is always provable". This has been rejected for theories including the natural numbers by Gödel's famous incompleteness theorem. The confusion is even made worse by the fact that the same famous logician proved these two important theorems very shortly one after the other.

- Beyond its usage in model theory and mathematical logic, further meanings are given to the same word "completeness". Actually, we are no longer talking about lack or availability or proofs, but about lack or availability of elements occurring as limits. It is customary in mathematical research that the name "completeness" can refer to either axiom (V.2) from Hilbert's foundations of geometry, and the corresponding axiom for the real number system,—or refer to the postulate stating convergence of every Cauchy sequence in a Hilbert space, Banach space or metric space. The latter is the notion of completeness occurring in real analysis and functional analysis. Even in that more narrow context, there are different axioms running under the name of completeness. For a final clarification, one needs to use names as Cauchy-completeness, or Dedekind-completeness.

  Remark. It is worthwhile to realize that either Cauchy or Dedekind completeness property of the reals is a fundamentally more complex statement than the properties that define real-closed fields, and cannot be replaced by any list of first-order properties. No countable list of first-order properties can characterize the real numbers uniquely.

- Furthermore, in the overall scientific community, the term "completeness of a theory" means that a theory includes all relevant notions and their logical connections—and thus does no longer refer to notions from outside a single given theory. In this sense, the term is used by Hilbert in his introduction to the foundations of geometry, where he announces "...a new attempt to set up a *complete* and *as simple as possible* system of axioms." Too, in this sense, Einstein talks about his belief that quantum mechanics is incomplete. Abraham Pais [26] states in his Einstein biography on p.449: "From 1931 on, the issue for Einstein was no longer the consistency of quantum mechanics but rather its completeness."

Here is my briefest attempt for a clarification: In the first case, completeness is related to lack versus availability of proofs In the second case, completeness is related to lack versus availability of elements occurring as limits. In the third case, completeness asks for availability of an entire theory.

I.2.8 More metamathematical considerations

Remark (About the two meanings of completeness). A sentence from a theory is called *valid* if and only if the sentence holds in every structure which is a model of the theory. This definition has a precise meaning in mathematical logic, once one has gives precise definitions for the notions of—: language, sentence, theory, structure, model. Following this approach, Tarski has introduced into mathematical logic the notion of truth—also called validity.

Mathematical logic provides an exact definition of the notion of a proof, and hence provability of a sentence, too. It is one of the basic problems of mathematical logic to investigate the relation of validity and provability. Is every provable sentence valid? Is every valid sentence provable?

The first question turns out to be much easier to answer. Indeed, the *Theorem of Soundness* states that a sentence which can be proved is valid. The second question cannot be answered in general. A theory in which every valid sentence can be proved is called "complete". (I prefer the term model-complete). Gödel's completeness theorem states that every first-order theory is complete. Thus he has given us a positive answer to the second question, at least in the context of first-order logic. This achievement was actually the topic of his Ph.D. thesis. By Gödel theorem any valid sentence from a first-order theory has a correct logical proof.

A theory is called negation-complete if and only if for every sentence, either the sentence or its negation, but not both, can be proved in the theory. The most common meaning of completeness is indeed negation-completeness. This is a very strong requirement, so strong indeed that most of the time it turns out to be just wishful thinking.

Here is a statement of the Turing-Church Theorem: *The set of laws of logic in the language of Peano arithmetic is undecidable. The set of laws of logic in the language of any first-order language with at least one relation or function symbol of two or more variables is undecidable.* The Turing-Church Theorem provides a negative answer to Hilbert's "Entscheidungsproblem" (decision problem).

The Gödel-Rosser Incompleteness Theorem is a consequence to the negative answer of the decision problem: There does not exist a negation-complete, axiomatizable extension of Peano arithmetic.

Question. Why does there not arise a contradiction Gödel's completeness theorem of first-order logic and the Gödel-Rosser Incompleteness Theorem?

Answer. Here is my (perhaps awkward) attempt of a brief clarification or "enlightenment", following mainly Robert Wolf [35]. It is certainly not true that every sentence or its negation is provable in first-order logic. But Gödel's completeness theorem [7] does not yield negation completeness of first-order logic. We have instead only the <u>weaker</u> result that every valid sentence is provable.

Indeed *validity* is a rather strong model-theoretic assumption. Only under this strong assumption of validity of any statement does Gödel's completeness theorem assure the existence of a proof for this statement. There does not occur a contradiction between Gödel's completeness theorem—for which I would prefer the name "Realization Theorem"—and the Turing-Church Theorem.

Remark (About second-order logic). The caveat in Gödel's completeness theorem is its restriction to first-order logic. A lot of mathematics can actually be expressed in first-order logic, but first-order logic is not powerful enough to express all mathematics and even less to express all scientific theory. By definition, first-order logic includes the propositional logic with connectives and, or, if...then, if and only if. First-order logic does include quantified logic and thus allows statements of the form

[7] I would prefer the name "Realization Theorem"

"For all x a statement $S(x)$ holds" and "There exists an x for which a statement $S(x)$ holds. Here x is allowed to be any primary (undefined) element occurring in the respective theory, and $S(x)$ is any statement in this theory in which may occur the element x. Too, first-order logic does include the notion of equality as a primary relation.

Scientific theory and mathematics as a whole needs to include the natural numbers and even set theory. In terms of mathematical logic, this means that we have to include (at least) Peano arithmetic into the relevant theory and its language. Indeed Peano arithmetic is only the most barren way for an abstract theory about the natural numbers. It has been proved that Peano arithmetic cannot be stated purely within the restriction to first-order logic.

Second-order logic allows quantifier to have as their range no longer single primary objects, but allows quantifier *ranging over subsets* of the primary objects, properties, or even axioms. In Peano arithmetics, the induction axiom—in each of its different forms—needs quantifiers of one of these types. All the different axioms of continuity need such quantifiers. Leibniz' definition of equality as "Two objects are equal iff they behave equally in all circumstances" depends on second-order logic. Indeed, every really powerful axiomatic system contains somewhere second-order logic.

I would not call these axioms "a flaw in the ointment" as Stefan Mykytiuk and Abe Shenitzer [24] seem to suggest in their nice article *"Four significant axiomatic systems and some of the issues associated with them"*. Look the other way: the language to first-order logic has a real caveat—a serious restriction one cannot live with forever. Let me use a metapher used by Hilbert in another context: I think of this restriction being similarly embarrassing as taking away the telescope from an astronomer, or the microscope from a biologist—.

So, in the end, practicable mathematics and actual mathematical research uses Peano arithmetic, continuity, and even more than that, and hence will need second-order logic.

I.3 Incidence Geometry

I.3.1 Elementary propositions about incidence planes

Definition I.3.1 (Incidence plane). Any model for which Hilbert's axioms of incidence (I.1)(I.2)(I.3a)(I. are valid is called an *incidence plane.*

Lemma I.3.1. *Any two different lines of an incidence plane have either no or exactly one point in common.*

Problem I.3.1. *Give a short explanation to convince yourself that Lemma I.3.1 does hold.*

Answer. Take any two different lines $l \neq m$. Assume towards a contradiction they intersect at two different points $A \neq B$.

By axiom (I.2), there exists at most one line through these two points. Hence $l = m$, contradicting the assumption that l and m are two different lines. Hence two different lines intersect at most in one point.

Definition I.3.2 (Parallel lines). Two different lines which lie in one plane and do not intersect are called *parallel.*

Question. Give at least two further useful formulations of Lemma I.3.1.

Answer. Here are three possible answers:

- Any two different lines lying in a plane and not parallel, have a unique point of intersection.
- If two lines have two or more points in common, they are equal.
- Two different lines have at most one point in common.

Lemma I.3.2. *In any incidence plane, there exist two different lines through every point.*

Proof. By axiom (I.3b), there exist at least three points that do not lie on a line. We call them A, B and C. Let any point P be given. The point P may be among the three points A, B, C or not. Thus the proof involves four points.

- In the case that point P is one of the three points A, B, C, we draw the three lines AB, BC and CA. These are three different lines, and two of them go through the given point P.
- In the case that point P is different from all three points A, B, C, we draw the three lines PA, PB and PC. At least two of them are different since A, B, C do not lie on a line.

In both cases we have obtained two different lines through the arbitrary point P. □

For two dimensional geometry, all lines and points are assumed to lie in the same plane. Referring to the parallel postulates, we need not state that the parallel to a given line through a given point actually lies in this plane. This leads to the following a bid simpler formulation:

Definition I.3.3 (Euclidean Parallel Postulate). For every line l and for every point P lying not on l, there *exists* a *unique* parallel m to l through point P.

But Hilbert's parallel axiom (IV) is a weaker assertion, it postulates only the uniqueness of the parallel to a given line through a given point.

Definition I.3.4 (Hilbert's Parallel Postulate). For every line l and for every point P lying not on l, there exists *at most* one parallel m to l through point P.

Proposition II.7.43 implies existence of a parallel once the axioms of order and congruence are assumed. Hence Euclid's and Hilbert's parallel postulate turn out to be equivalent in Hilbert planes. But in the case of incidence planes, such an equivalence does not hold.

Definition I.3.5. We say that a two-dimensional incidence geometry for which the Euclidean Parallel Postulate holds, has the *Euclidean parallel property.*

Definition I.3.6. We say that a two-dimensional incidence geometry has the *elliptic parallel property* iff any two lines do intersect—necessarily at a unique point.

Definition I.3.7. We say that a two-dimensional incidence geometry has the *hyperbolic parallel property* iff for every line l and point P not on line l, there exist *at least* two parallels to line l through point P.

Lemma I.3.3 (Proclus' Lemma). *In any incidence plane where Hilbert's Parallel Postulate for plane geometry I.3.4 holds, the following statements are valid:*

- *a third line intersecting one of two parallel lines intersects the other one, too.*
- *a third line parallel to one of two parallel lines, is parallel to the other one, too.*

Problem I.3.2. *Explain why Proclus' Lemma is an easy consequence of the uniqueness of parallels. Convince yourself that, conversely, Proclus' Lemma implies the uniqueness of parallels.*

Solution. Suppose towards a contradiction that the transversal t intersects one of the parallel lines l and m, but not the other one.

We may assume that P is the intersection point of lines t and m. If lines t and l would not intersect, then t and m would be two different parallels of line l through point P. This contradicts the uniqueness of parallels.

Conversely, we now assume Proclus' Lemma to be true and check the uniqueness of the parallel to a given line l through a given point P not on line l. Let m and t be two parallels to line l through point P—these may be equal or different lines. The line t is a transversal intersecting one of the two parallel lines $m \parallel l$ at point P. Hence it intersects the second line l, too, contrary to the assumption. The only possibility left is that $m = t$. Hence the parallel to a given line through a given point is unique. □

Remark. Don't fool yourself to think that all these words amount to a proof of the parallel postulate! All we have shown is: either both uniqueness of parallels and Proclus' Lemma hold or neither uniqueness of parallels nor Proclus' Lemma do hold.

Problem I.3.3. *Given any incidence plane where Hilbert's Parallel Postulate for plane geometry I.3.4 holds. Convince yourself that the relation that two lines are either equal or parallel is an equivalence relation.*

Answer. For any two lines, let $l \sim m$ mean $l = m$ or $l \parallel m$. The relation $\sim$ has the three defining properties of an equivalence relation. Indeed this relation is

reflexive: since $l = l$ is a logical axiom.

symmetric: since $l = m$ implies $m = l$, and $l \parallel m$ implies $m \parallel l$.

transitive: assume that $m \sim l$ and $l \sim k$. If two of the three lines are equal, we substitute equals to get $m \sim k$. We now assume that all three lines are different. The two lines m and k are either parallel or they intersect at one point, by Hilbert's first Proposition I.3.6. If m and k would intersect at point P, the line l would have two different parallels through P, which is impossible in an affine plane. Therefore m and k are parallel, as to be shown.

A relation that is reflexive, symmetric and transitive is called an equivalence relation.

Corollary 1 (About uniqueness of parallels). *For any incidence plane the following statements are equivalent:*

(a) *uniqueness of parallels,—stated in Hilbert's Parallel Postulate I.3.4 for plane geometry,—holds;*

(b) *any third line intersecting one of any two parallel lines intersects the other one, too;*

(c) *any third line parallel to one of any two parallel lines, is parallel to the other one, too;*

(d) *being equal or parallel defines an equivalence relation among the lines.*

I.3.2 Finite incidence geometries

Problem I.3.4 (The four-point incidence geometrics). *Find all non isomorphic incidence geometries with four points. Which parallel property (elliptic, Euclidean, hyperbolic, or neither) does hold? Which one is the smallest affine plane?*

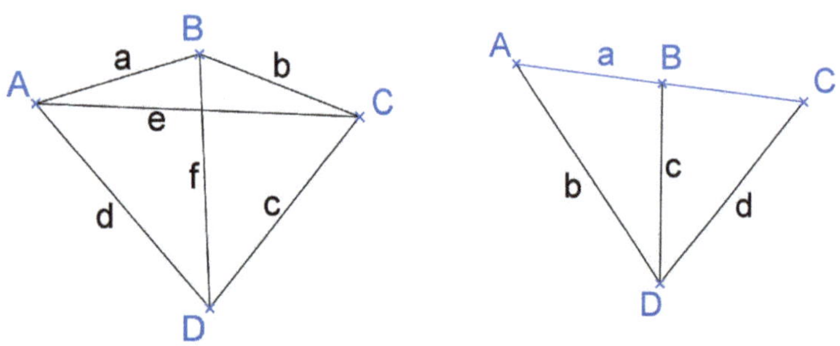

Figure I.3.1: There are two four-point incidence geometries.

Answer. There exist two non-isomorphic four-point geometries.

(a) Six lines with each one two points. It has the Euclidean parallel property, and is the smallest affine plane.

(b) There are four lines, one of which has three points. It has the elliptic parallel property.

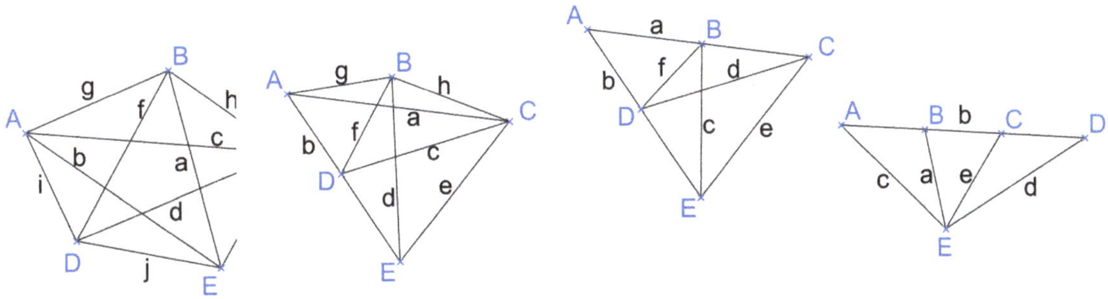

Figure I.3.2: There are four five-point incidence geometries.

Problem I.3.5 (The five-point incidence geometries). *Find all non isomorphic incidence geometries with five points. Describe the properties of their points and lines. Which parallel property (elliptic, Euclidean, hyperbolic, or neither) does hold?*

Answer. There exist four different five-point geometries.

(a) Ten lines have each two points. This model has the hyperbolic parallel property.

(b) Exactly one line with three points. Altogether, there are eight lines. Neither parallel property holds.

(c) Two intersecting lines with three points. Altogether, there are six lines. Neither parallel property holds.

(d) There are five lines, one of which has four points. This model has the elliptic parallel property.

Problem I.3.6. (i) *Which ones of the four- or five-point incidence geometries satisfy the statement:*

> *On every line there exist at least two points, and furthermore, there exist three points not on this line.*

(ii) *Why is this statement different from Hilbert's axioms?*

Answer. **(i)** The statement of the problem postulates existence of at least two points on any given line, additionally three extra points lying not this line. Thus the statement implies existence of five different points. Hence it does not hold for any four point incidence geometry. Moreover, it can only hold for a five point geometry, where every line has exactly two points. Indeed it holds only for the five-point handshake model.

(ii) Hilbert's axiom of incidence for plane geometry are (I.1)(I.2) and (I.3). Axiom (I.3) consists of two sentences:

I.3a There exist at least two points on a line.

I.3b There exist at least three points that do not lie on a line.

Here part (I.3a) and (I.3b) do not refer to each other. Part (I.3a) is a universal statement about any line. But part (I.3b) is a purely existential statement about three points.

On the other hand, the statement given in the problem is a different universal statement about a line. The first part repeats Hilbert's (I.3a). The second part goes on requiring additionally existence three points which are not lying on this same line.

Problem I.3.7. *Which ones of the incidence axioms* (I.1), (I.2), (I.3), *and which parallel property hold in the first model drawn on page 24.*

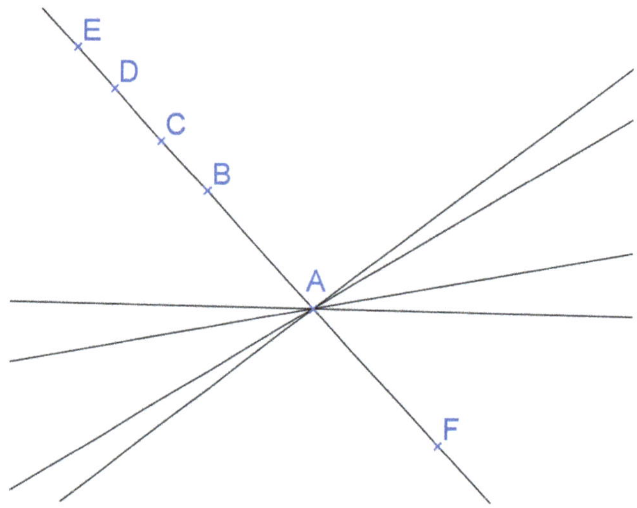

Figure I.3.3: Too simple, this is not even an incidence geometry.

Answer. Every two points lie on a unique line, and every two lines intersect at a unique point. Hence axioms (I.1) and (I.2), and the elliptic parallel property hold. Axiom (I.3) does not hold.

Before we proceed to six or more points, we take notice of two simple models of finite incidence geometry—the extreme ones in a sense—which exist for all numbers $n \geq 4$ of points.

Definition I.3.8 (Hand-shake model). A *hand shake model* is an incidence geometry for which every line has exactly two points.

Definition I.3.9 (Straight fan). A *straight fan* is an incidence geometry with all but one point lying on one line.

Problem I.3.8. *How many lines does the hand-shake incidence geometry with n points have. How many lines does the straight fan with n points have. Why do there exist two or more non-isomorphic incidence geometries for any number $n \geq 4$ of points.*

Answer. The hand-shake incidence geometry with n points has

$$(n-1) + (n-2) + \cdots + 1 = \frac{(n-1)n}{2}$$

lines. We see this as follows. We can connect the first point to the other $(n-1)$ points. Disregarding this point, we connect the second point to $(n-2)$ different points, and so on. The last line to be drawn is between the $(n-1)$-th and the n-th point.

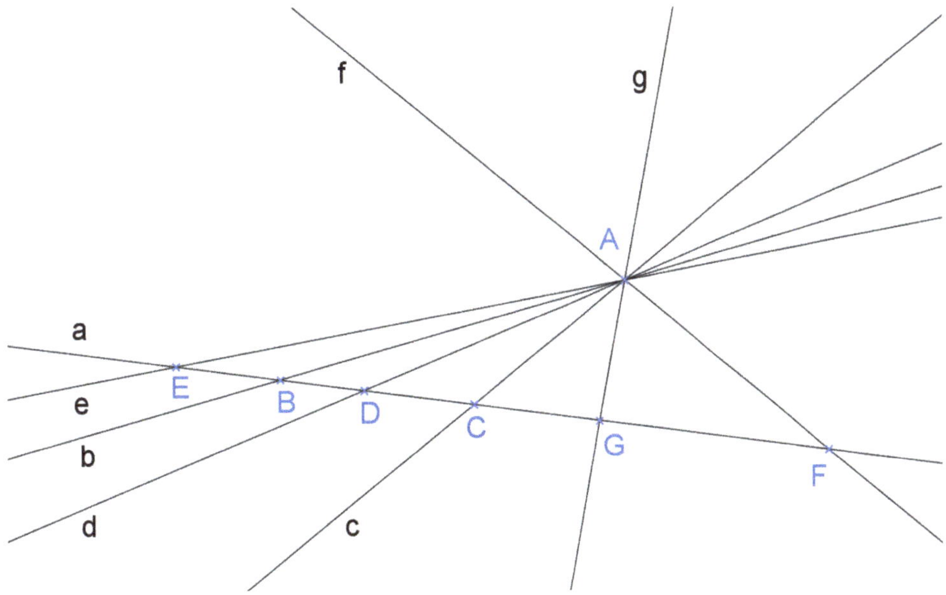

Figure I.3.4: A straight fan is an incidence geometry—but still no projective plane.

The straight fan with n points has n lines. There is one long line with $n-1$ points, and only one point P does not lie on this line. There are $n-1$ lines with two points each of which connects P to a different point on the long line.

We see that for $n \geq 4$, the hand-shake model with n points has more lines than the straight fan with n points. Hence there exist two or more non-isomorphic incidence geometries for any number $n \geq 4$ of points.

Problem I.3.9. *Which parallel property holds for the hand-shake model with 4 points. Which parallel property holds for the hand-shake model with $n \geq 5$ points. Which parallel property holds for a straight fan.*

Answer. The hand-shake model with 4 points has the Euclidean parallel property. All the hand-shake models with $n \geq 5$ points have the hyperbolic parallel property. For all straight fans, the elliptic parallel property holds.

Problem I.3.10 (Optional)**.** *Show that every finite incidence geometry has at least as many lines as points.*

`http://en.wikipedia.org/wiki/De_Bruijn`
`%E2%80%93Erd%C5%91s_theorem_%28incidence_geometry%29`

`http://en.wikipedia.org/wiki/Sylvester%E2%80%93Gallai_theorem`

Definition I.3.10. The *dual* of an incidence plane with elliptic parallel property is the plane where the points are the lines of the old one, and vice versa.

Question. What is the dual of a straight fan with n points.

Answer. The dual is isomorphic to the primal, with n points (old lines), and n lines (old points) having the corresponding incidence relations.

Problem I.3.11 (The six-point incidence geometries). *Find all non-isomorphic incidence geometries with six points. Count how many non-isomorphic models do exist. For the models different from the handshake and straight fan, mark all three-point lines with different blue shades, and any four-point line in red.*

Describe the properties of their points and lines. Count the lines. Which parallel property (elliptic, Euclidean, hyperbolic, or neither) does hold?

Answer. There are nine non-isomorphic six-point incidence planes:

1. The handshake model with 15 short lines.
2. The model with only one three-point line has 13 lines.
3. The model with only one four-point line has $1 + 1 + 2 \cdot 4 = 10$ lines.
4. The model with two parallel three-point lines has $2 + 3 \cdot 3 = 11$ lines.
5. The model with two intersecting three-point lines has $2 + 2 \cdot 2 + 5 = 11$ lines.
6. The model with a intersecting three-point line and four-point line has $2 + 2 \cdot 3 = 8$ lines.
7. The model with three intersecting three-point lines forming a "triangle" has $3+3\cdot 2 = 9$ lines.
8. The model with three intersecting three-point lines forming a "triangle" and the in-circle a fourth three-point line has $4 + 3 = 7$ lines.
9. The straight fan with $1 + 5 = 6$ lines.

Only the straight fan is elliptic. There is no Euclidean model. There are several mixed models, with are not hyperbolic neither.

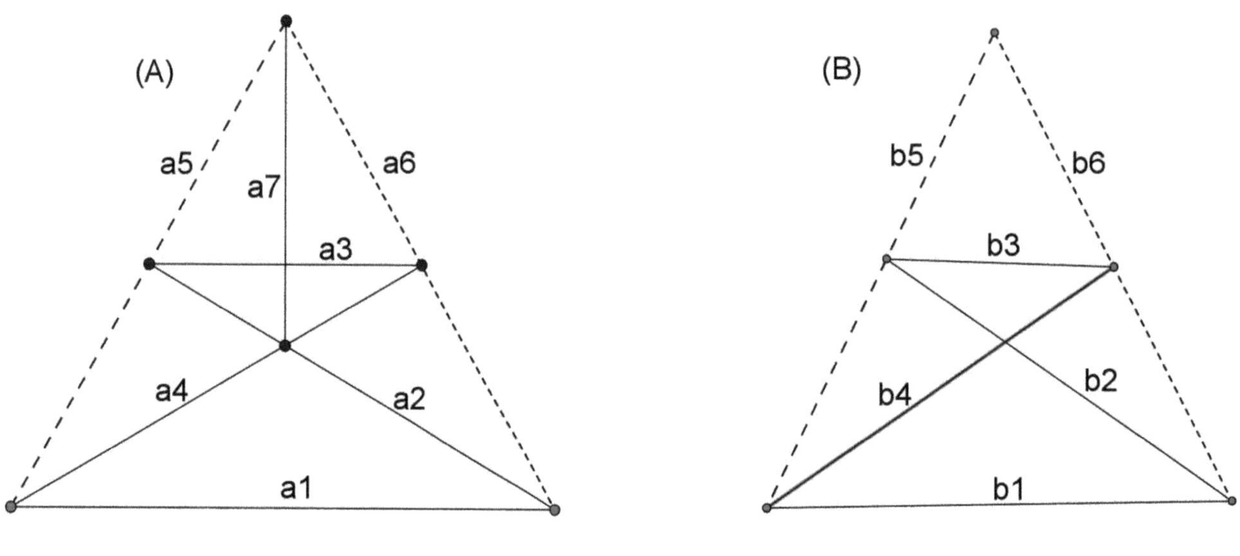

Figure I.3.5: Enumerate the pencils for the two models (A) and (B).

Definition I.3.11 (Pencil). An equivalence class of (equal or) parallel lines is called a *pencil.*

Problem I.3.12. *Given an incidence plane for which Hilbert's Parallel Postulate is valid. As mentioned in the lecture, and stated in Corollary 1: being equal or parallel defines an equivalence relation among the lines. Each one of the equivalence classes of equal or parallel lines is still called a* pencil. *But, contrary to the situation for an affine plane, not all pencils need to consist of the same number of lines.*

Determine and enumerate the pencils for the two models (A) *and* (B) *drawn in the figure on page 26. Put curely brackets around each pencil!*

Answer. Pencils of parallels for model (A): $\{a1, a3, a7\}$, $\{a2\}$, $\{a4\}$, $\{a5\}$, $\{a6\}$.
Pencils of parallels for model (B): $\{b1, b3\}$, $\{b2, b4\}$, $\{b5\}$, $\{b6\}$.

Definition I.3.12 (Isomorphism of incidence planes). Two models for incidence planes are called *isomorphic* if and only if there exists a bijection ϕ between the points of the two planes, and a bijection φ between the lines of the two planes such that incidence is preserved.

Remark. For incidence to be preserved, the two bijections are required to satisfy the assumption

$$\phi(PQ) = \phi(P)\phi(Q) \tag{I.3.1}$$

for all points P, Q from the first incidence plane. In other words, if the bijection of the points maps $P \mapsto \phi(P)$ and $Q \mapsto \phi(Q)$, then the bijection of the lines maps $PQ \mapsto \phi(P)\phi(Q)$.

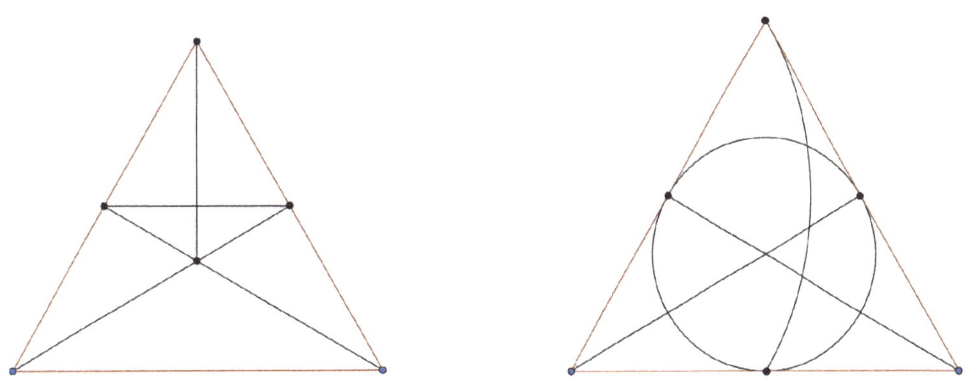

Figure I.3.6: Two isomorphic six-point incidence geometries

Problem I.3.13. *Given two incidence geometries, it is not obvious whether they are isomorphic. By corresponding labelling of the points in both geometries, show an isomorphism between the two six-point incidence geometries in the figure on page 27.*

Answer. The isomorphism is indicated in the figure on page 28.

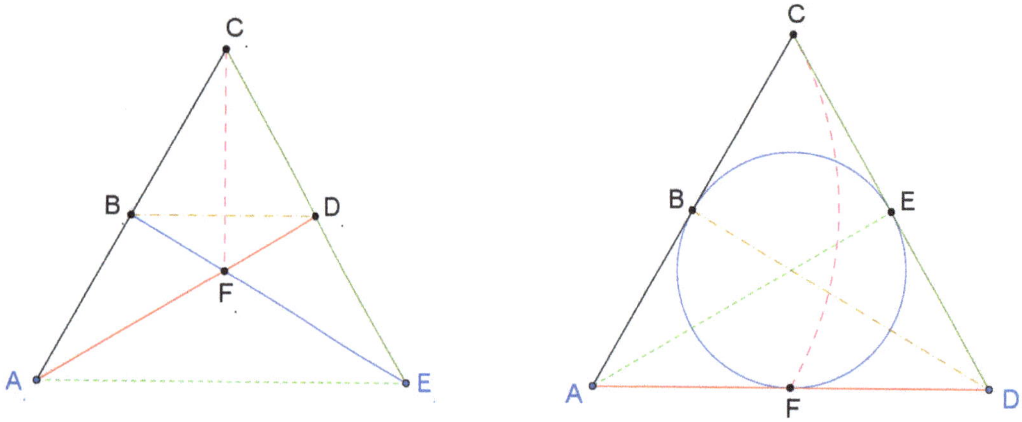

Figure I.3.7: The labelling of the points shows the isomorphism.

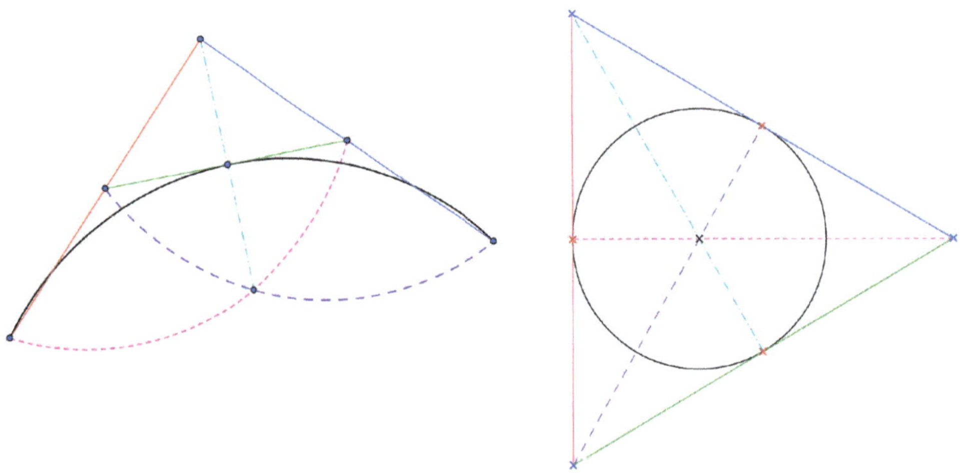

Figure I.3.8: Name the points as to show the isomorphism.

Problem I.3.14. *Given two incidence geometries, it is not obvious whether they are isomorphic. By corresponding labelling of the points in both geometries, show an isomorphism between the two seven-point incidence geometries in the figure on page 28.*

Answer. The isomorphism is indicated in the figure on page 29.

Problem I.3.15 (A nine point incidence geometry). *Find a model of incidence geometry with nine points, which satisfies*

> *(I3+) "Every line contains exactly three points."*

and for which the Euclidean parallel axiom does hold:

> *"For every line l and every point P not on l, there exists exactly one parallel to the line l through P."*

It is enough to provide a drawing to explain the model. How many lines does the model have?

Use colors for the lines. Choose clearly different colors for the lines in different directions, but give each set of three parallel lines different shades of nearby color.

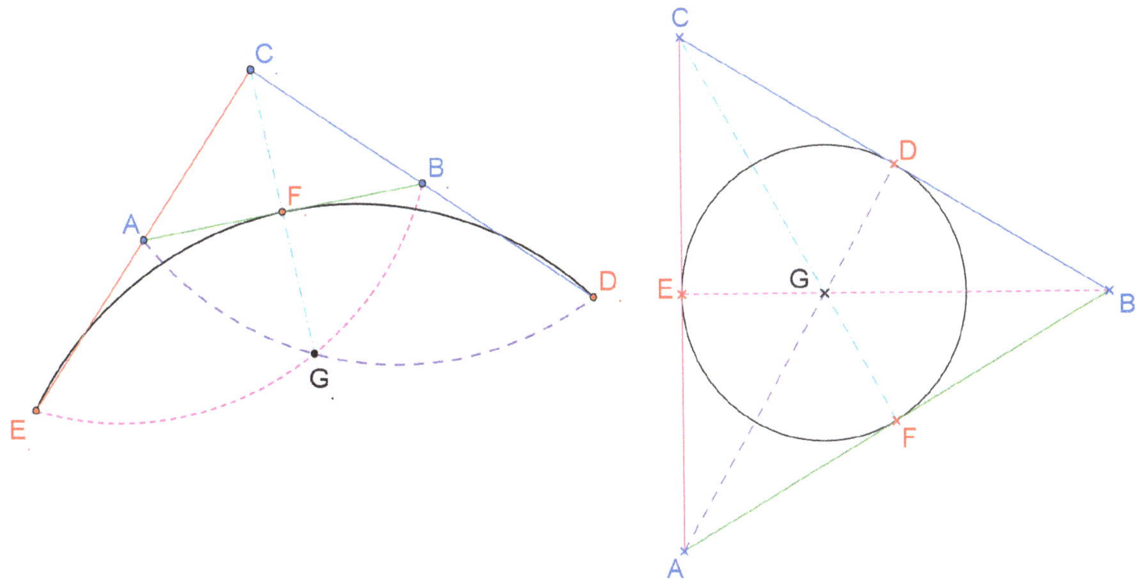

Figure I.3.9: The names of the points show the two drawings of the Fano-plane are really isomorphic.

Answer. There is exactly one such model. It has 12 lines.

Name any point A and let two lines through it consist of the points $a = \{A, B, C\}$ and $d = \{A, D, G\}$ (lines a and d are drawn horizontally and vertically). Let $\{D, E, F\}$ be the parallel to line $\{A, B, C\}$ through D. Two different parallels to a given line cannot intersect, because of the uniqueness of parallels. Hence the parallel to a through point G contains the remaining points and necessarily is $\{G, H, I\}$. The parallel to line $\{A, D, G\}$ through B contains one of the points E, F and one of the points G, H. Possibly by exchanging names of those four points, we can assume that $\{B, E, H\}$ is the parallel to line $\{A, D, G\}$ through B. Finally, $\{C, F, I\}$ is the parallel to line $\{A, D, G\}$ through C.

Up to now, we have mentioned and drawn six lines. Because of incidence axiom (I.1), there is a unique line through any two points. Hence there need to exist further lines in the model. As an example, we find the line g through points A and E. It cannot pass through any of the points B, C, D, F, H, I, because otherwise we get a line with four or more points on it. Hence $g = \{A, E, I\}$ is a line. Similarly, one finds lines $\{H, E, C\}$, and finally $\{B, D, I\}$,$\{A, F, H\}$,$\{B, F, G\}$,$\{C, D, I\}$.

It does not matter that the last four lines cannot be drawn as straight Euclidean lines. Neither does it matter that they have more intersection points—those are not included as points of the model. The model contains 12 lines. It is the unique model satisfying all requirements.

Remark. The model can be constructed using analytic geometry and arithmetic modulo 3. Take as "points" the ordered pairs (a, b) with $a, b \in \mathbf{Z}_3$. As in analytic geometry, lines are given by linear equations $ax + by + c = 0$ with $a, b, c \in \mathbf{Z}_3$ and a, b are not both zero. One defines the slope of a line in the usual way. Lines are parallel if and only if they have the same slope. To find the parallel to a given line through a given point, one uses the point slope equation of a straight line. This procedure shows that the Euclidean parallel property holds.

I.3.3 Affine incidence planes

Definition I.3.13 (Affine plane). An *affine plane* is a set of points, and a set of lines, satisfying the axioms:

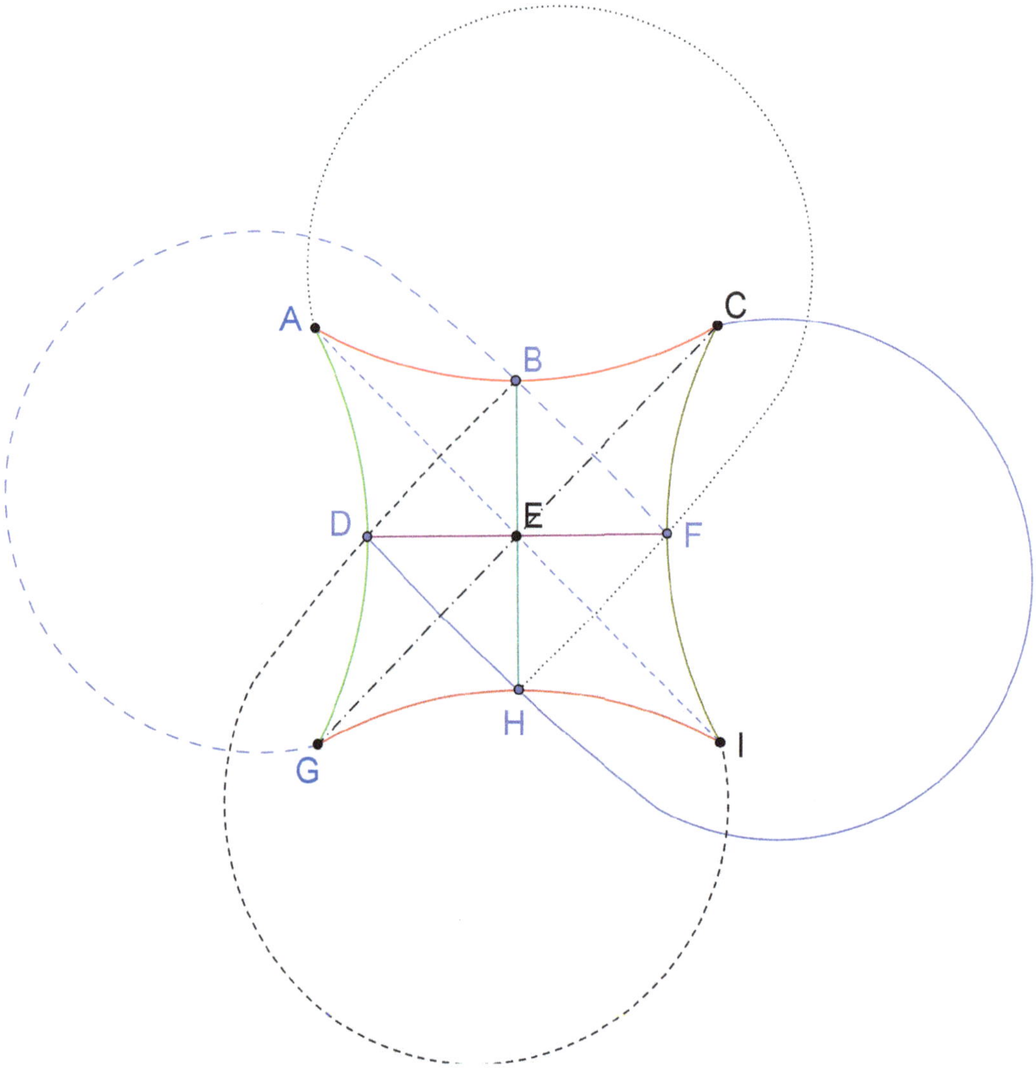

Figure I.3.10: A nine-point incidence geometry

A.1 Every two different points lie on exactly one line.

A.2 If point P does not lie on a line l, there exists exactly one line m through the point P that does not intersect l.

A.3 There exist three points that do not lie on a line.

Problem I.3.16. *The reader should convince himself that in an affine plane*

1. *on every line lie at least two points;*
2. *in every point intersect at least two lines.*

Proof. Let A, B, C be the three points not on a line which exist according to axiom (A.3).

1. Given is any line l. At most one of the three lines AB, BC and AC is parallel or equal to the given line l. We may assume that neither line AB nor line AC is parallel or equal to the given line l. Hence they intersect the given line in the points $P = AB \cap l$ and $Q = AC \cap l$.

- If these two points $P \neq Q$ are different, we are ready.
- We get a more involved argument in the special case that $P = Q$. In this case, we get $A = P = Q$. If the line l intersects the line BC, we get a second intersection point $R = BC \cap l$ on the line l and are ready again.
- Still open is only the case that lines BC and l are parallel and the point $A = P$ lies on the line l. Let m be the parallel to line $c = AB$ through point C. Since $m \neq BC \parallel l$, the line m is not parallel to l. Hence these two lines intersect at point $R = m \cap l \neq A$, which is a second point on the line l.

2. Given is any point P. Assume that $P \neq A, B, C$. Since these three points do not lie on a line, we get from lines PA, PB and PC at least two different ones through point P.

   In the special case that $P = A$, we get two different lines AB and AC through point P. We argue similarly in the special cases where $P = B$ or $P = C$.

□

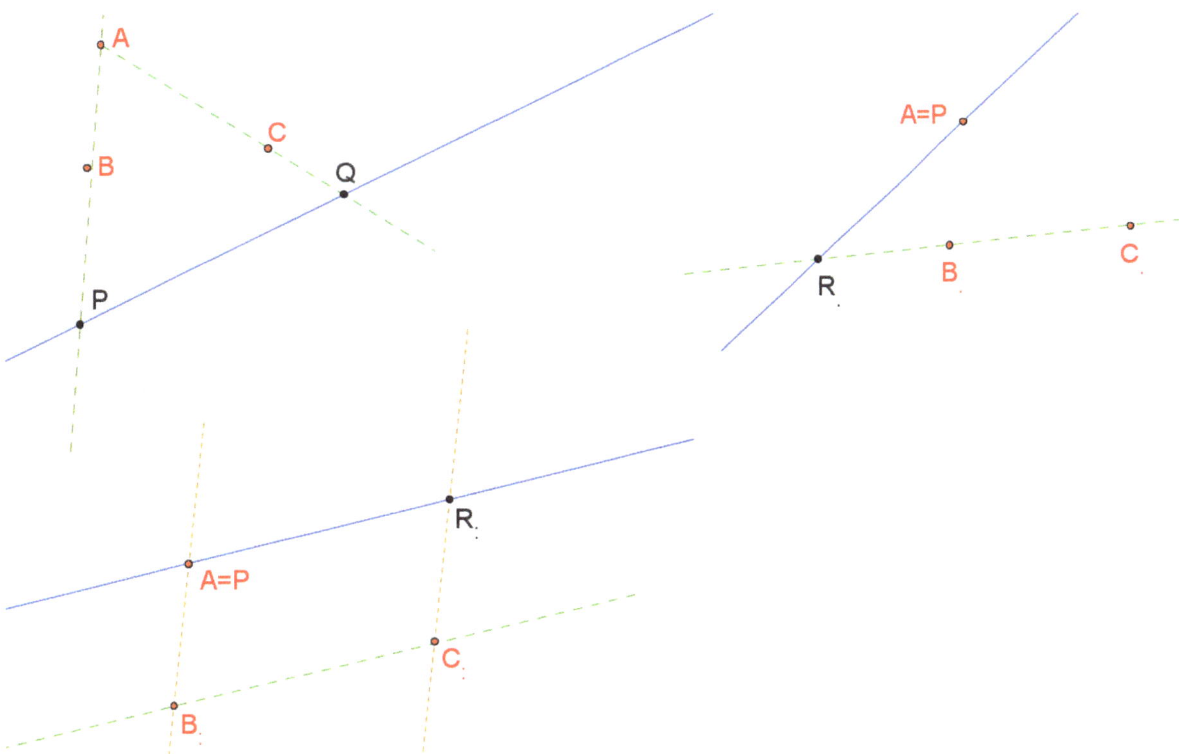

Figure I.3.11: In an affine plane, we find a second point on any given line.

Axiom (A.1) repeats Hilbert's incidence axioms (I.1) and (I.2). Axiom (A.2) is the Euclidean parallel property. In problem I.3.16, it is proved that on every line of an affine plane lie at least two points. Together with (A.3), we see that Hilbert's incidence axiom (I.3) holds for an affine plane, too. Hence an affine plane is just an incidence plane for which the Euclidean parallel property holds.

Corollary 2 (Characterizations of an affine plane). *For any incidence plane the following statements are equivalent:*

(a) *The incidence plane is an affine plane;*

(b) *for every line l and for every point P lying not on l, there exists a unique parallel m to l through point P. Thus the Euclidean parallel postulate I.3.3 holds.*

I.3.4 Introduction of coordinates

Definition I.3.14 (The Cartesian plane). The `"points"` of the Cartesian plane are ordered pairs (x, y) of real numbers $x, y, \in \mathbf{R}$. The `"lines"` of the Cartesian plane are given by equations

$$ax + by + c = 0 \tag{I.3.2}$$

with reals coefficients a, b, c, and a and b not both equal to zero. A `"point lies on a line"` if and only if the coordinate pair (x, y) satisfies the equation of the line. `Two lines are equal` if and only if they contain the same points.

Remark. Because coefficients (a, b, c) and $(\lambda a, \lambda b, \lambda c)$ with $\lambda \neq 0$ give rise to the same line, the triples (a, b, c) are *homogeneous coordinates.*

Theorem I.3.1. *The Cartesian plane is an affine plane.*

Because the construction of the affine plane from the real numbers uses only addition, subtraction, multiplication and division, replacing the real numbers by any finite or infinite field $\mathbf{F}$ and doing the same construction once more, leads to an affine plane, too. Thus one gets the following general definition.

Definition I.3.15 (The Cartesian plane over field F). The `"points"` of the Cartesian plane are ordered pairs (x, y) of elements $x, y \in \mathbf{F}$. The `"lines"` of the Cartesian plane of the field $\mathbf{F}$ are equations

$$ax + by + c = 0$$

with coefficients a, b, c from the field, of which a and b are not both zero. A `"point lies on a line"` if and only if the coordinate pair (x, y) satisfies the equation of the line. `Two lines are equal` if and only if they contain the same points.

Theorem I.3.2 (The Cartesian plane over an arbitrary field). *In a Cartesian plane over any field, there exist a unique line between any two points. There exists a unique parallel to a line through a given point. Hilbert's axioms* (I.1)(I.2)(I.3a)(I.3b) *and the Euclidean parallel postulate* (IV*) *hold. Hence a Cartesian plane over a field is an affine plane.*

Remark. Any axioms may only be used without further justification, in the context that they refer to primary (undefined) elements and relations. The situation is different for the Cartesian plane introduced above, which is another important example for a model or structure. For a *model* or *structure*, the primary elements and relations are replaced by elements and relations which are *explicitly defined.* For a model one has to check whether the *axioms are valid* for these (new) elements and relations.

Thus to check theorem I.3.2, one has to give a justification, why the incidence axioms (I.1), (I,2), (I.3a), (I.3b) and the Euclidean parallel postulate hold for the `"points"`, the `"lines"` and the `"incidences"` of the Cartesian plane specified above.

Problem I.3.17. *Use your knowledge of college algebra, and give in a few words an explanation why Theorem I.3.2 is valid.*

Answer. We begin by checking postulates (I.1) and (I.2). Given are two distinct points (x_1, y_1) and (x_2, y_2) through which we need a unique line. In the ususal way, one defines the slope

$$m = \frac{y_2 - y_1}{x_2 - x_1}$$

and puts $m = \infty$ in the special case that $x_1 = x_2$. The line through point (x_1, y_1) with slope m has the equation

$$y = y_1 + m(x - x_1) \tag{I.3.3}$$

in the usual case of finite slope, and $x = x_1 = x_2$ for $m = \infty$. This equation of a line in this form is uniquely determined by two points $(x_1, y_1) \neq (x_2, y_2)$, as the reader should check. Hence two lines through two distinct points contain the same points and are equal.

To check axiom (I.3a), we use that each field contains the two different elements $0 \neq 1$. On any given line with finite slope, we get two points on a line by putting $x = 0, x = 1$. We put $y = 0, y = 1$ for a vertical line. To check axiom (I.3b), we may for example use the three points $(0,0)$, $(1,0)$ and $(1,1)$. Hence we see that the Cartesian plane over any field is an incidence plane.

To get an affine plane, we need still to check that the Euclidean parallel postulate (IV*) does hold. We see that two distinct lines

$$y = mx + k \text{ and } y = m'x + k'$$

with different slopes $m \neq m'$ do intersect. But two distinct lines with the same slope have $m = m'$ and $k \neq k'$,— they do not intersect. Too, one may include vertical line in this statement. Given any line $y = mx + k$ and point (x_1, y_1), the parallel through the point is obtained by the point-slope formula (I.3.3). An easy reasoning shows that Euclidean parallel postulate (IV*) does hold in the case of a vertical line, too.

I.3.5 Finite coordinate planes

Recall that the number of points on any line is called the *order* of a finite affine plane.

Problem I.3.18 (A nine point incidence geometry). *Find an affine plane of order three.*

Answer. As a simple example, we see that the affine plane given by the figure on page 30 can be constructed as the Cartesian product $\mathbf{Z}_3 \times \mathbf{Z}_3$, using the field $\mathbf{Z}_3$, where the arithmetic is calculation modulo 3. This becomes even more visible by repeating copies of the same figure, shifted by three units along the x and y axes in a double periodic way.

Obviously, the model has order 3, since there are $n^2 = 9$ points. Hence there are $n^2 + n = 12$ lines. There is exactly one such model.

The finite fields are well- known from modern algebra. For any order $n = p^r$ where p is a prime number, and $r \geq 1$ any natural number, there exists exactly one finite field. It is the *Galois field*, denoted by $\mathbf{F}_n$.

Corollary 3 (Affine planes from finite fields). *If the order $n = p^r$ is a prime power, there exists exactly an affine incidence plane of order p^r. It has $n^2 = p^{2r}$ points and $n^2 + n = p^{2r} + p^r$ lines.*

The points are ordered pairs (x, y) with arbitrary $x, y \in \mathbf{F}_n$. In other words, the set of points is the Cartesian product $\mathbf{F}_n \times \mathbf{F}_n$. Hence there are $n^2 = p^{2r}$ points.

The lines are the n vertical lines $x = c$ with any $c \in \mathbf{F}_n$, as well as the other lines with an equation of the form $y = mx + k$ with arbitrary $m, k \in \mathbf{F}_n$. Hence there are $n^2 + n = p^{2r} + p^r$ lines.

Problem I.3.19 (Scheduling problem I). *Make a 6 day schedule for a school with 25 students. Each day the students are divided in a different way into 5 groups of 5 students. Never are two students in the same group more than one time during the week.*

Answer. We use the coordinate plane $\mathbf{Z}_5 \times \mathbf{Z}_5$. With addition and multiplication modulo 5, the set $\mathbf{Z}_5$ is a field, because 5 is a prime number. Hence the coordinate plane $\mathbf{Z}_5 \times \mathbf{Z}_5$ is an affine plane with order 5. As explained by Proposition I.3.4, the number of points on any line in a affine plane is given by the order, here $n = 5$. The lines can be partitioned into $n + 1 = 6$ pencils, each containing 5 parallel lines.

Hence for each pencil, the set of parallel lines contained in the pencil, determines the groups of students for one weekday. For each of the 6 weekdays, we use parallel lines with a different slope $0, 1, 2, 3, 4$, and finally on one day the vertical parallel lines.

One can make a picture (see page 35) of the schedule by drawing the 5×5 pattern of dots separately for every day. The five parallel lines in every pattern are indicated by different symbols for their points. One needs curved lines to connect all five points of a line. [8] Clearly such a picture contains more insight than a bare-bone list. □

Problem I.3.20 (Scheduling problem II). *Make a 5 day schedule for a school with 15 students. Each day the students are divided in a different way into 5 groups of 3 students. Never are two students in the same group more than one time during the week.*

Answer. We cut down the solution of the last problem and retain just the 15 students from three groups on the sixth weekday. For the remaining five days, we obtain five groups with three students, as required. □

Problem I.3.21 (Scheduling problem III). *Make a 7 day schedule for a school with 35 students. Each day the students are divided in a different way into 7 groups of 5 students. Never are two students in the same group more than one time during the week.*

I.3.6 Projective incidence planes

Projective planes are incidence planes with elliptic parallel property. But an additional axiom is needed, in order to rule out uninteresting cases. A straight fan (see definition I.3.9) is too simple for what one wants to be a projective plane. In a projective plane, one wants to build perspective and the corresponding mappings and measurements. As natural starting point of such an endeavor, one needs four points in a convenient configuration. I like to think of them, as the standpoint of the observer, a point some distance away from him, and two points on the horizon to the left and right.

[8] The figure leaves out these curve parts, the reader should put them in.

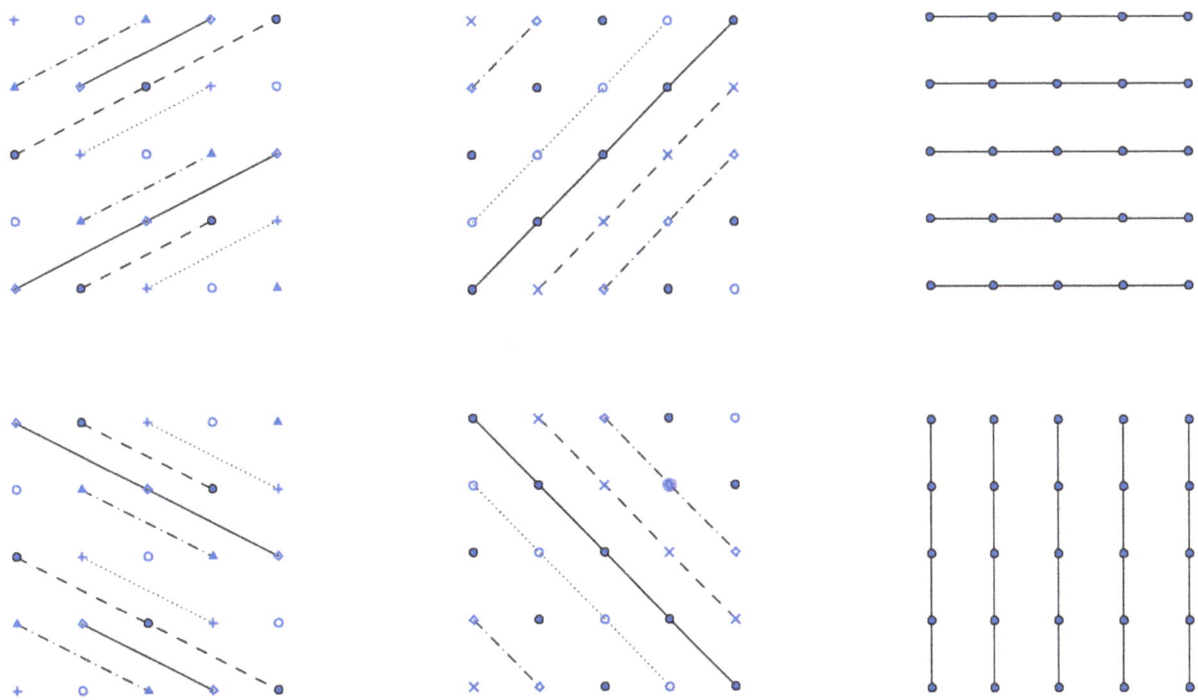

Figure I.3.12: A picture to explain the schedule for five groups of five students each, on six days.

Definition I.3.16 (Quadrangle and quadrilateral). Four points, no three of which lie on a line, are said to be a *quadrangle*. These four points are necessarily distinct.

Four lines, no three of which intersect at a point, are said to be a *quadrilateral*. Again these four lines are necessarily distinct.

Definition I.3.17 (Projective plane). A *projective plane* is a class of points, and a class of lines satisfying the axioms:

P.1 Every two distinct points lie on exactly one line.

P.2 Every two distinct lines intersect at exactly one point.

P.3 There exist four points of which no three lie on a line.

Remark. The lines of a projective plane are not necessarily sets of points.

Problem I.3.22. *Convince yourself that in every point of a projective plane intersect at least three lines.*

Secondly, convince yourself that on every line of a projective plane lie at least three points. Finally check that a quadrilateral exists.

Answer. By axiom (P.3), there exist at least four points A, B, C, D of which no three lie on a line. Let any point P be given. In the case that point P is different from all four points A, B, C, D, we draw the lines PA, PB, PC and PD. At least three of them are different since no three points among A, B, C, D lie on a line. In the case that point P is one of the four points A, B, C, D, we get three different lines through the given point P among the six lines connecting A, B, C, D. In both cases we have obtained three different lines through the arbitrary point P.

The proof confirming that three points lie on every line is done quite similarly, interchanging the roles of points and lines.

Proposition I.3.1 (The straight fan is the only exception). *An incidence geometry with elliptic parallel property is either a straight fan (see definition I.3.9), or a quadrangle exists. Hence, in the latter case, it is a projective plane.*

Proof. We assume the elliptic parallel property holds, but no quadrangle exists. We have to check that we get a straight fan. The three-point incidence geometry is a straight fan, as claimed. Hence we now assume at least four points to exist.

Take any four points of the given incidence plane. Since no quadrangle exists, three among them lie on a line. We can assume that the three points A, B, C lie on the line l. By the axiom (I.3) for the incidence plane, there exists a further point P not on the line l. We claim there cannot exist two points $Q \neq P$ neither of which lies on the line l.

Assume towards a contradiction that the two points $Q \neq P$ both do not lie on l. The lines PQ and l do intersect, either in one of the three points A, B, C, or still another point. We can assume the intersection point A' is different from both B and C. Of the four points B, C, P, Q at least three lie on a line. This could be points P, Q, B or P, Q, C. Both cases are impossible, since the lines PQ and $l = BC$ intersect only in point $A' \neq B, C$.

Hence there exists exactly one point P not on the line l. Thus the incidence geometry is a straight fan. □

Recall that an equivalence classes of equal or parallel lines is called a *pencil*.

Definition I.3.18 (Improper elements and projective completion of an affine plane). The points and lines from the original affine plane are called *proper points* and *proper lines*. To each pencil corresponds an *improper point*. [9] The line through all improper points is called the *improper line*. [10] Both the proper and the improper points (lines) of an affine plane are the points (lines) of a projective plane. By adding the improper points and the improper line, one may construct from any affine plane a projective plane, called the *projective completion*.

Problem I.3.23. *Convince yourself that the projective completion of an affine plane is indeed a projective plane. Starting from the axioms* (A.1)(A.2)(A.3) *for the affine plane, one needs to check the validity of axioms* (P.1)(P.2)(P.3).

Check of validity of axiom (P.1). Given are two distinct points $A \neq B$ of the projective plane. We distinguish three cases, depending on whether these are proper or improper points.

- In case that both A and B are proper points, existence and uniqueness of the line through A and B is postulated by axiom (A.1).
- Assume now that only one of the two points is proper. We may assume that point A is proper, and B_∞ is an improper point. By definition, the point B_∞ is a pencil of parallel lines. Take any line b in that pencil. By axiom (A.2) there exists a unique parallel m to line b through point A. In the projective completion, the line m is the unique line through points A and B_∞.

[9] This name is given for visual reasons. We want to catch the vague idea of a "point very far away" in the direction pointed to by a pencil, and make it precise in terms of rudimental set theory.

[10] I reserve the names *points at infinity* and *line at infinity* to hyperbolic geometry.

- Finally, we assume that both A_∞ and B_∞ are improper points. In the projective completion, the improper line l_∞ is a line through points A_∞ and B_∞. Any other line cannot go through both A_∞ and B_∞, since the extension of each proper line contains only one improper point.

We see that in all possible cases, there exists a unique line through points A and B. Hence axiom (P.1) is valid. □

Check of validity of axiom (P.2). Given are two distinct lines $a \neq b$ of the projective plane. We distinguish three cases, depending on whether these are proper or improper lines.

- Let both a and b be distinct proper lines, and assume they intersect at point P in the affine plane. They cannot have at second proper intersection point since by axiom (A.1) the line through two distinct points is unique. They cannot have any further improper intersection point, otherwise lines a and b would have to be parallel to two distinct pencils, which is impossible.

- Let both a and b be distinct proper lines, and assume they do not intersect in the affine plane. Hence they belong to the same pencil P_∞ of parallel lines. Extended to the projective plane, the two lines intersect at the improper point P_∞. They cannot have any further improper intersection point, otherwise lines a and b could not be parallel.

- Assume now that only line a is proper, and b_∞ is the improper line. The line a is contained in exactly one pencil A_∞ of parallel lines. This pencil is the unique intersection of lines a and b_∞ in the projective completion.

We see that in all possible cases, there exists a unique intersection point of lines a and b. Hence axiom (P.2) is valid. □

Check of validity of axiom (P.3). Finally, to confirm (A.3), we obtain four points no three on which lie on a line. By axiom (A.3), there exist, in the affine plane, three points A, B, C not lying on a line. We draw the parallel m to line AB through point C. Let C_∞ be the improper point for that direction. We draw the parallel n to line AC through point B. Let B_∞ be the improper point for that direction.

The lines m and n are not parallel since the respectively parallel lines $AB \parallel m$ and $AC \parallel n$ intersect at A. Hence the lines m and n intersect, in the affine plane, at a point $D = m \cap n$. Of the four points A, D, B_∞, C_∞ no three lie on a (projective) line, confirming axiom (P.3). □

Remark. Here is more simple of axiom (P.3): Finally, to confirm (P.3), we show existence of four points no three on which lie on a line. By axiom (A.3), there exist, in the affine plane, three points A, B, C not lying on a line. We draw the parallel m to line AB through point C, and the parallel n to line AC through point B. The lines m and n are not parallel since the respectively parallel lines AB and AC intersect at A. Hence the lines m and n intersect, in the affine plane, at a point $D = m \cap n$.

Of the four points A, B, C, D no three lie on a (projective) line, thus confirming axiom (P.3).

Proposition I.3.2 (Cutdown of a projective plane to an affine plane, and reconstruction of the projective plane). *Given is a projective plane. Any line—and all points on it—are marked as improper.*

(a) *The proper points and lines yield an affine plane.*

(b) *Concatination of improper elements leads back to the original projective plane.*

Check of validity of axiom (A.1). We begin to check the validity of the axioms for an affine plane. Existence and uniqueness of a line through any two distinct points is postulated by axiom (P.1), confirming axiom (A.1). □

Check of validity of axiom (A.2). Next we check existence and uniqueness of a parallel to a given proper line l through the proper point P not lying on l. Let l_∞ be the line marked as improper. Since $l_\infty \neq l$, axiom (P.2) yields, in the projective plane, existence of a unique intersection point $Q_\infty = l_\infty \cap l$. The line $m = PQ_\infty$ is retained in the affine plane. Since $Q_\infty = l \cap m$ is an improper point, as seen in the affine plane, the lines l and m are parallel. Hence we have obtained the parallel m to the given line l through point P, as required. Any other parallel $m' \parallel l$ through point P extended, goes through point Q_∞, too. By the uniqueness of the line through P and Q_∞, we conclude $m = m'$. Hence uniqueness of parallels holds, too. □

Check of validity of axiom (I.3a). Let l_∞ be the line marked as improper. The set of points $\mathcal{P}$ of the original projective plane is cut down to the set of points $\mathcal{A} = \mathcal{P} \setminus l_\infty$ of the affine plane. Each line of the projective plane, at least three points, as has been shown in the solution of Problem I.3.22. For a line $l \neq l_\infty$, only one of them, the point $l \cap l_\infty$, is improper. Hence there are at least two proper points on the line l. □

Check of validity of axiom (I.3b). Finally, to confirm (A.3), we obtain three points not lying on a line. By axiom (P.3), there exist, in the projective plane, four points A, B, C, D of which no three lie on a line. In the cases that none, or only one of them lies on the improper line l_∞, we are ready. Now assume that two of them, say C and D lie on the improper line. Hence A and B are proper points. The intersection $P = AC \cap BD$ exists.

- Moreover, point P is a proper point. Indeed if $P \neq C$ would be two improper points, then point A would be improper. If $P \neq D$ would be two improper points, then point B would be improper.
- Points $A \neq P$ are distinct, otherwise A, B, D would lie on a line. Points $B \neq P$ are distinct, otherwise A, B, C would lie on a line.
- Finally, point P cannot lie on the line AB. Otherwise all points A, B, P, C, D would lie on this single line.

Hence we have found three proper points A, B and P not lying on a line, thus confirming axiom (A.3). □

Proof of item (b). The set of points $\mathcal{P}$ of the original projective plane is cut down to the set of points $\mathcal{A} = \mathcal{P} \setminus l_\infty$ of the affine plane. All the parallel lines in one pencil, extended, have the same intersection point with the improper line l_∞. There is a bijection between these pencils and the improper points on l_∞. Hence the reconstruction of the improper points via pencils brings back the original projective plane. □

Definition I.3.19 (Dual projective plane). The *dual* of a projective plane is the projective plane where the points are the lines of the original one, and vice versa. The `"points"` of the dual projective plane are the lines of the primal projective plane. The `"lines"` of the dual projective plane are the points of the primal projective plane.

I.3.7 The Fano Plane

Problem I.3.24. *The smallest projective plane is obtained by projective completion of the affine plane of order 2. Find and describe this projective plane. How many points, and how many lines does it have. How are they situated.*

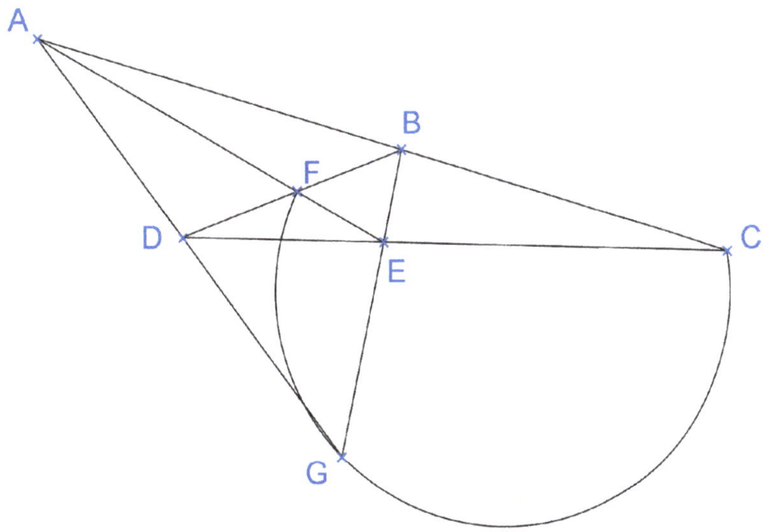

Figure I.3.13: Fano's seven-point projective plane

Answer. We begin with the affine plane with the four points A, B, D and E and six lines. The pencil of parallels $AB \parallel DE$ produces the improper point C. Similarly, the pencil of parallels $AE \parallel DB$ produces the improper point F and the pencil $AD \parallel BE$ produces the improper point G. Finally, we draw the improper line through the points C, F and G.

We get a projective plane with seven points and seven lines. Each line has exactly three points. In each point intersect exact three lines.

Problem I.3.25. *Give a highly symmetric illustration for the Fano plane based on an equilateral triangle. Your symmetric illustration is really isomorphic to the projective plane from page 39 above, which was obtained in problem I.3.24 by completion.*

Denote the seven points with the same names in both drawings, consistently in a way to show the isomorphism. After you have obtained the isomorphism, color the lines with seven different colors. Give the corresponding lines in the other model the same colors.

Answer. The figure on page 40 give an illustration based on an equilateral triangle. To check that this symmetric illustration is isomorphic to the illustration I have given on the left side, one needs to names the points in both illustrations in a way that the incidence relations hold for the same names. Thus the isomorphism is given by the correspondence of names.

To find such an isomorphism, the key observation is that a triangle can be mapped to any triangle, but afterwards the correspondence of the remaining points is uniquely determined.

Problem I.3.26 (Self-duality of the Fano geometry). *The dual of a projective plane is obtained by making the originally given lines the new points, and the originally given points the new lines (recall definition I.3.19). Find and describe the dual of the Fano plane.*

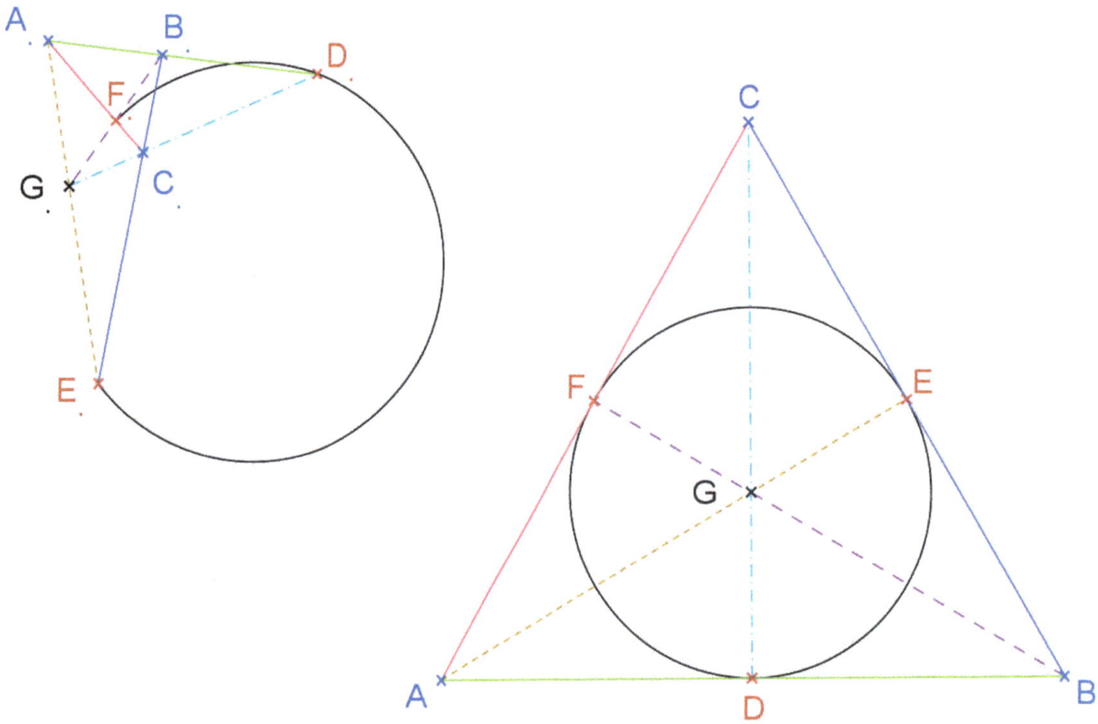

Figure I.3.14: The symmetric drawing of the Fano-plane is really isomorphic to the projective completion of the affine plane of order 2.

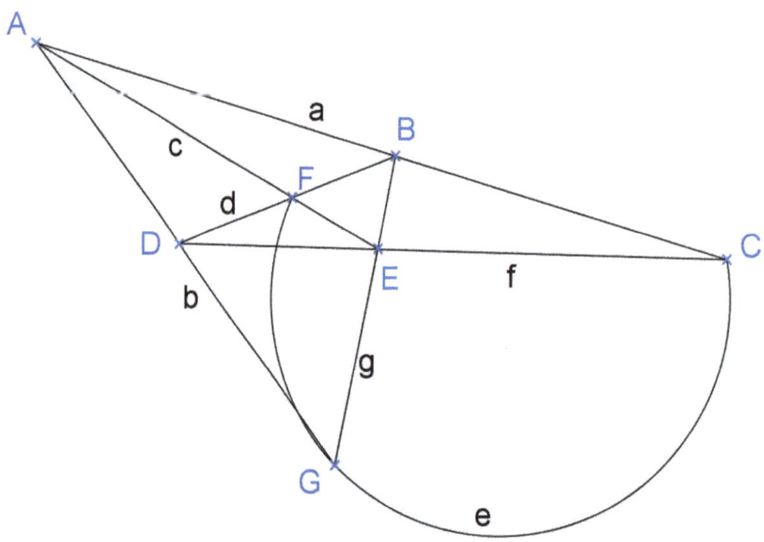

Figure I.3.15: Fano's seven-point incidence geometry is selfdual

Answer. For the Fano plane, one can name lines in a way such that the incidence relations of the dual are exactly the incidence relations of the primal with capital and small letters switched. It take a while to rename lines in a way to show the self-duality directly in this manner.

This way of choosing names gives an isomorphism of the primal and dual, and hence confirms self-duality. In the figure on page 40, the names have already been chosen to confirm self-duality.

For example, the three points A, B, D lie on line c— and the lines a, d, b intersect at the point C.

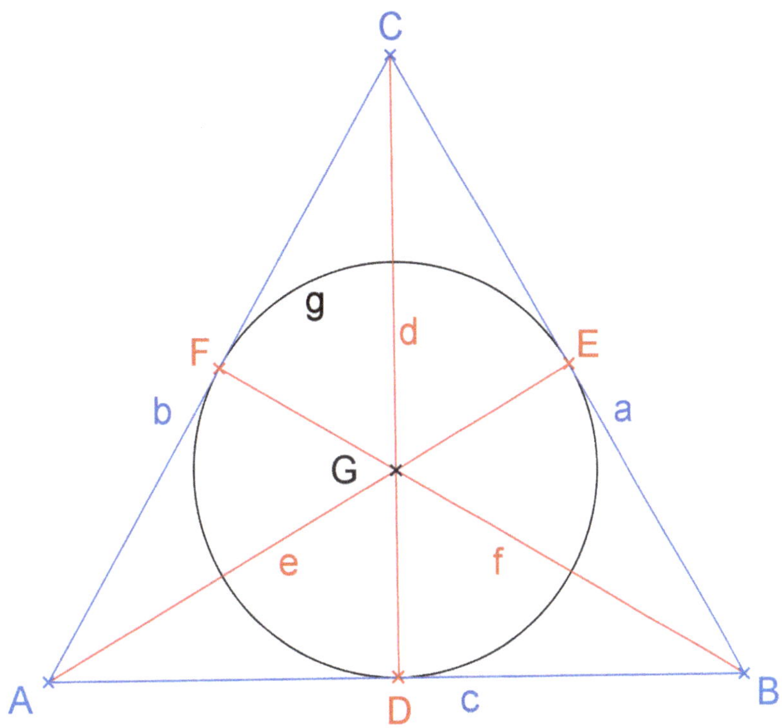

Figure I.3.16: The points and lines of the Fano-plane named in a way to confirm self-duality.

Problem I.3.27. *Here is, once more, a highly symmetric illustration for the Fano plane based on an equilateral triangle. Find the 7×7 incidence-matrix giving the incidence between the lines and points.*

What does it means geometrically that the matrix is symmetric? Would the matrix be always symmetric for any arbitrary choice of the names of points and lines?

Answer. Here is the incidence matrix:

	A	B	C	D	E	F	G
a	0	1	1	0	1	0	0
b	1	0	1	0	0	1	0
c	1	1	0	1	0	0	0
d	0	0	1	1	0	0	1
e	1	0	0	0	1	0	1
f	0	1	0	0	0	1	1
g	0	0	0	1	1	1	0

Since the incidence matrix is symmetric, we see once more that the Fano plane is isomorphic to its dual. For any arbitrary choice of the names of points and lines, a symmetric matrix would only arise after a suitable permutation of the rows.

I.3.8 Projective plane with coordinates

Definition I.3.20 (The projective plane over a field F). The "`points`" of the projective plane $\mathbf{PF}^2$ are the sets of equivalent triples $(x\lambda, y\lambda, z\lambda)$, where $x, y, z \in \mathbf{F}$ and λ runs over the nonzero elements of $\mathbf{F}$. The "`lines`" of the projective plane are equations

$$ax + by + cz = 0$$

with coefficients $a, b, c \in \mathbf{F}$, not all three equal to zero.

A "`point lies on a line`" if and only if the coordinate triple (x, y, z) satisfies the equation of the line.

The triples $(x, y, z) \in \mathbf{F}^3 \backslash (0, 0, 0)$ are the *homogeneous coordinates* for the points of the projective plane. Similarly, the triples $(a, b, c) \in \mathbf{F}^3 \setminus (0, 0, 0)$ are the *homogeneous coordinates* for the lines of the projective plane.

Again, it is easy to check, but important to confirm:

Main Theorem 1. *The projective plane over any field is a projective plane.*

Proof. One needs to check the incidence axioms (P.1), (P.2) and (P.3) from the definition (I.3.20) of a projective plane—which is not standard College algebra, but almost as easy:

P.1 We confirm that every two points lie on exactly one line. Take two points with homogeneous coordinates $(x_1, y_1, z_1) \neq (0, 0, 0)$ and (x_2, y_2, z_2). Because these points are different, one gets $(x_2, y_2, z_2) \neq (x_1\lambda, y_1\lambda, z_1\lambda)$ for all $\lambda \in \mathbf{F}$.

The homogeneous coordinates (a, b, c) of a line through the two points are a solution of the system

$$\begin{aligned} ax_1 + by_1 + cz_1 &= 0 \\ ax_2 + by_2 + cz_2 &= 0 \end{aligned} \tag{I.3.4}$$

These are two equations with three unknowns, hence there exists a nontrivial solution $(a, b, c) \neq (0, 0, 0)$. The system (I.3.4) has rank 2, because (x_1, y_1, z_1) and (x_2, y_2, z_2) are linearly independent. Hence the solution space is one dimensional, and hence (a, b, c) are the homogeneous coordinates for a unique line.

P.2 We confirm that every two lines intersect at exactly one point. In the proof, the roles of points and lines interchanged: Take two lines with homogeneous coordinates $(a_1, b_1, c_1) \neq (0, 0, 0)$ and (a_2, b_2, c_2). Because they are different, one gets $(a_2, b_2, c_2) \neq (\lambda a_1, \lambda b_1, \lambda c_1)$ for all $\lambda \in \mathbf{F}$.

The homogeneous coordinates (x, y, z) of the intersection point are a solution of the system

$$\begin{aligned} a_1x + b_1y + c_1z &= 0 \\ a_2x + b_2y + c_2z &= 0 \end{aligned} \tag{I.3.5}$$

Again, the solution space of system (I.3.5) is one dimensional, and hence (x, y, z) are the homogeneous coordinates for a unique point.

P.3 Four points of which no three lie on a line are given by the homogeneous coordinates $(1, 0, 0)$, $(0, 1, 0)$, $(0, 0, 1)$ and $(1, 1, 1)$.

□

Question. Definition I.3.20 tells that the points of the projective plane $l \in \mathbf{PF}^2$ are defined to be the lines through the origin in the three dimension space $\mathbf{F}^3$. On the other hand, one obtains a visual picture of the projective plane by adjoining to the set of points (x, y) of the affine plane the set of equivalence classes of parallel lines in this plane. How can both points of view be valid?

Answer. To obtain the isomorphism between these two models of the projective plane, we take intersections with the plane $\mathcal{P} = \{(x, y, 1)\} : x, y \in \mathbf{F}\}$. Let be given any point $l \in \mathbf{PF}^2$. In coordinates this means

$$l = \{(a\lambda, b\lambda, c\lambda) : \lambda \in \mathbf{F}\}$$

We take the intersection $l \cap \mathcal{P}$ and distinguish two cases:

- In case that $c \neq 0$, we get a unique intersection point $(\frac{a}{c}, \frac{b}{c}, 1)$ which determines a proper point corresponding to the point $(\frac{a}{c}, \frac{b}{c})$ of the affine plane.
- In case that $c = 0$, we get no intersection point. But we see that the given line l is parallel to the line $\{(a\lambda, b\lambda, 1) : \lambda \in \mathbf{F}\}$ lying in the intersection plane and determining the equivalence class of parallel lines for an improper point.

These two cases yield the proper and improper points of the projective plane. The isomorphism between the abstract $\mathbf{PF}^2$ definition, and the visual definition of the projective plane using improper points is provided.

Main Theorem 2. *Given are four points $x_1, x_2, x_3, x_4 \in \mathbf{PR}^2$ with no three of them lying on a line. Similarly, there are given any four image points $x'_1, x'_2, x'_3, x'_4 \in \mathbf{PR}^2$. Once more, it is assumed that no three of them lie on a line. Then there exists exactly one projective mapping which takes x_i to x'_i for $i = 1, 2, 3, 4$.*

Proof. Let $x_1, x_2, x_3, x_4 \in \mathbf{R}^3 \backslash \{0\}$ be any homogenous coordinates of the four given points $x_i \in \mathbf{PR}^2$. Since any four vectors in $\mathbf{R}^3$ are linearly dependent, there exists $\lambda_1, \lambda_2, \lambda_3, \lambda_4 \in \mathbf{R}$ such that

$$\lambda_1 x_1 + \lambda_2 x_2 + \lambda_3 x_3 + \lambda_4 x_4 = 0$$

Since by assumption no three of the four points x_i lie on a line, *all four* $\lambda_i \neq 0$ are nonzero. The matrix $A = [\lambda_1 x_1, \lambda_2 x_2, \lambda_3 x_3]$ maps

$$p_1 = (1, 0, 0) \mapsto \lambda_1 x_1 \ , \ p_2 = (0, 1, 0) \mapsto \lambda_2 x_2 \ , \ p_3 = (0, 0, 1) \mapsto \lambda_3 x_3 \ ,$$
$$p_4 = (1, 1, 1) \mapsto -\lambda_4 x_4 \ ,$$

and is nonsingular. The induced projective mapping $\phi_A : x \in \mathbf{PR}^2 \mapsto x' = Ax \in \mathbf{PR}^2$, takes the points $f_i \mapsto x_i$ for $i = 1, 2, 3, 4$.

By means of composition $A' \circ A^{-1}$, we find a projective mapping such that

$$x_i \in \mathbf{PR}^2 \mapsto x'_i \in \mathbf{PR}^2$$

where the four preimages x_i and images x'_i can be arbitrarily prescribed. □

The points of the affine plane are the proper points of the projective plane, whereas the equivalence classes of parallel lines in the affine plane are the improper points of the projective plane. The set of the points of the projective plane with the homogeneous coordinates $(x, y, 0)$. is the *improper line.* From the Cartesian plane over a field, the projective plane is obtained by adjoining the improper elements. One gets back the Cartesian plane after deleting the improper elements.

Problem I.3.28 (Three points on a line). *In a projective coordinate plane three points with the homogeneous coordinates $(x_1, y_1, z_1), (x_2, y_2, z_2)$ and (x_3, y_3, z_3) are given. Check that the three points lie on a line if and only if the determinant*

$$\begin{vmatrix} x_1 & x_2 & x_3 \\ y_1 & y_2 & y_3 \\ z_1 & z_2 & z_3 \end{vmatrix} = 0$$

Simple solution. The three points of the projective plane with the homogeneous coordinates $(x_1, y_1, z_1), (x_2$ and (x_3, y_3, z_3) lie on a line if and only if the three lines through the origin

$$\{(\lambda x_1, \lambda y_1, \lambda z_1) : \lambda \in \mathbf{F}\} \quad \{(\lambda x_2, \lambda y_2, \lambda z_2) : \lambda \in \mathbf{F}\} \quad \{(\lambda x_3, \lambda y_3, \lambda z_3) : \lambda \in \mathbf{F}\}$$

lie in a plane in $\mathbf{F}^3$. It is know that this happens if and only if the determinant of their coordinates from above is zero. □

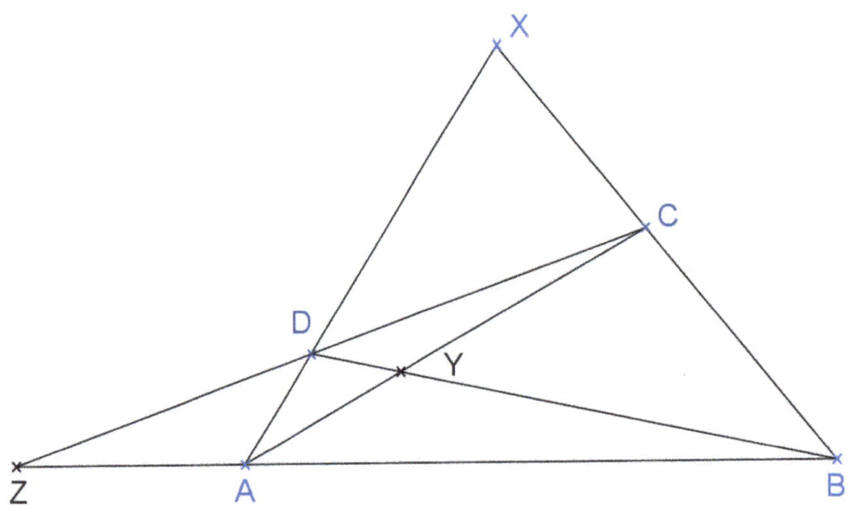

Figure I.3.17: The complete quadrilateral.

Theorem I.3.3. *Given is any projective coordinate plane $\mathbf{PF}^2$. A quadrangle $\square ABCD$ has its three intersections $X = AD \cap BC$, $Y = AC \cap BD$ and $Z = AB \cap CD$ lying on a line if and only if field $\mathbf{F}$ has characteristic two.*

Proof. For any two given quadrangles $x_1, x_2, x_3, x_4 \in \mathbf{PF}^2$ and $x_1', x_2', x_3', x_4' \in \mathbf{PF}^2$, there exists a projective mapping which takes $x_i \mapsto x_i'$ for $i = 1, 2, 3, 4$. This is shown in Theorem 2. Hence it is enough to prove the theorem above for just one special quadrangle. Conveniently, we choose

$$A = (1, 0, 0),\ B = (0, 1, 0),\ C = (0, 0, 1),\ D = (1, 1, 1)$$

In this special case, the points X, Y, Z turn out to have the homogeneous coordinates

$$X = (0, 1, 1),\ Y = (1, 0, 1),\ Z = (1, 1, 0)$$

By the previous problem, they lie on a line if and only if

$$\begin{vmatrix} 0 & 1 & 1 \\ 1 & 0 & 1 \\ 1 & 1 & 0 \end{vmatrix}$$

is zero. Elementary arithmetic yields the value 2 for the determinant. Hence the points X, Y and Z lie on a line if and only if $2 = 0$ which is true if and only if the field $\mathbf{F}$ has characteristic two. □

Proposition I.3.3. *The* dual *of a projective coordinate plane of a field is isomorphic to the plane.*

Proof. Once commutativity holds, the statement is easy to check since both points and lines use the same triples of homogeneous coordinates.

A point P with homogeneous coordinates (x, y, z) lies on a line l with homogeneous coordinates (a, b, c) if and only if

$$ax + by + cz = 0$$

The points are left vector spaces identifying

$$(x, y, z) \sim (x\lambda, y\lambda, z\lambda)$$

with arbitrary $\lambda \neq 0$. But the lines are right vector spaces, in which one has to identify,

$$(a, b, c) \sim (\mu a, \mu b, \mu c)$$

with arbitrary $\mu \neq 0$ as homogeneous coordinates of the same projective line. Without assuming commutativity, a difference arises. For the dual geometry, exchange of points with lines means that the dual points become right vector spaces, whereas the dual lines become left vector spaces. The involution mapping σ

$$\sigma : \mathcal{P}oints \mapsto \mathcal{L}ines$$

between left and right vector spaces does not exist in the non commutative case. Only after having assumed commutativity, this mapping is provided simply by the identity. We get an incidence preserving bijection, and hence are able confirm that the geometry is self dual. `Point` P `lies on line` l `if and only if line` $\sigma(P)$ `goes through point` $\sigma^{-1}(l)$. □

Remark. For a non-Desargues projective plane, the claim of Proposition I.3.3 need not to be true.

Problem I.3.29 (Self-duality of the Fano plane by means of coordinates). *Explain a systematic procedure to get the dual of the Fano plane, using homogeneous coordinates. Give an illustration based on the equilateral triangle.*

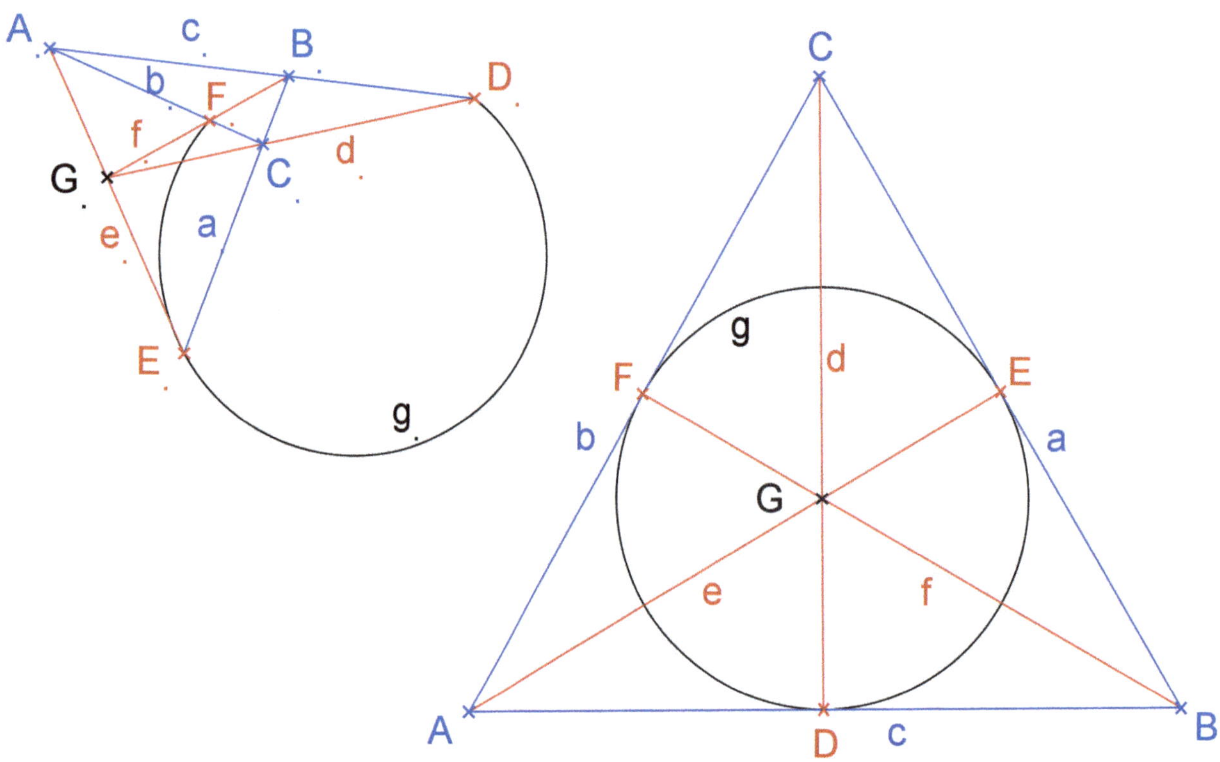

Figure I.3.18: Both illustrations confirm the self-duality

Answer. The order of the Fano geometry is two. The homogeneous coordinates are the seven nonzero elements of $\mathbf{Z}_2 \times \mathbf{Z}_2 \times \mathbf{Z}_2$. Three points lie on a line if and only if the determinant of their homogeneous coordinates is zero modulo two. Corresponding to the symmetric illustration, one can set:

$$\begin{aligned} A &:= (1,0,0),\ B := (0,1,0),\ C := (0,0,1), \\ D &:= (1,1,0),\ E := (1,0,1),\ F := (0,1,1),\ G := (1,1,1) \end{aligned}$$

In homogeneous coordinates, the dual of a line is the point with homogeneous coordinates orthogonal (dot product zero) to the coordinates of all the points on the line.

For all lines, I have calculated their dual point, and have chosen names for the lines according to homogeneous coordinates of the dual point. I have indicated this bijection by choosing the same capital and small letters for points and lines. The preservation of the incidence relations is now easy to check directly.

Problem I.3.30 (The group of symmetries of the Fano plane)**.** *The permutations of the points that carry collinear points to collinear points are called* collineations, automorphisms *or* symmetries *of the plane. Determine the order and describe the group of the automorphisms for the Fano plane.*

Answer. The group of automorphisms of the Fano plane is a *simple group of order* 168. To find and count the automorphisms, fix one triangle. As already mentioned in problem I.3.24, a triangle can be mapped to any triangle, but afterwards the correspondence of the remaining points is uniquely determined. One has 7 choices to map the first vertex, 6 choices to map the second, and only 4 choices to map the third vertex, since the third vertex cannot be mapped onto the line given by the images of the first two vertices. Hence there are $7 \times 6 \times 4 = 168$ automorphisms.

There are subgroups which fix all three points of a given lines. They have order 4 and consist of the identity and pairs of two transpositions of the remaining four points.

Furthermore, there are larger subgroups which fix a given line 123 as one object. They can be understood by the action on the four remaining points $4, 5, 6, 7$. Since all permutations of these four points turn out to be poosible, they have order 24 and consist of the identity, pairs of two transpositions of the remaining points $4, 5, 6, 7$, and permutations $(123)(456)$, and of type $(12)(4567)$. These four cases given permutations belonging to different conjugacy classes.

Finally, a permutation of one 7-cycle can be an automorphism. If the permutation (1234567) is an automorphism, either point 4 or point 6 is the third point on the line 12. Both cases do occur. They belong to different conjugacy classes.

From the above, we get 6 conjugacy classes:

the identity

21 permutation of type $(45)(67)$, with the three points $1, 2, 3$ lying on one line

56 permutations of type $(123)(456)$ with the three points $1, 2, 3$ lying on one line, but the three points $4, 5, 6$ not on a line

42 permutations of type $(12)(4567)$ with points $1, 2, 3$ collinear

24 permutations of type (1234567) with $1, 2, 4$ collinear

24 permutations of type (1234567) with $1, 2, 6$ collinear

Since these numbers add up to 168, we have found all conjugacy classes. Since 1 and 168 are the only divisors of 168 which are sums of sizes of conjugacy classes including the identity, the automorphism group has no normal subgroups. I leave it to reader to find all other subgroups.

Problem I.3.31 (Character table). *Find the character table for the group of the automorphisms of the Fano plane.*

I.3.9 Finite affine and projective incidence planes

Proposition I.3.4 (The finite affine plane of order n). *For every affine plane*

(a) *Each line has the same number n of points.*

(b) *The lines can be partitioned into $n+1$ classes, each containing n parallel lines.*

(c) *In every point intersect $n+1$ different lines.*

(d) *There are n^2 points and n^2+n lines.*

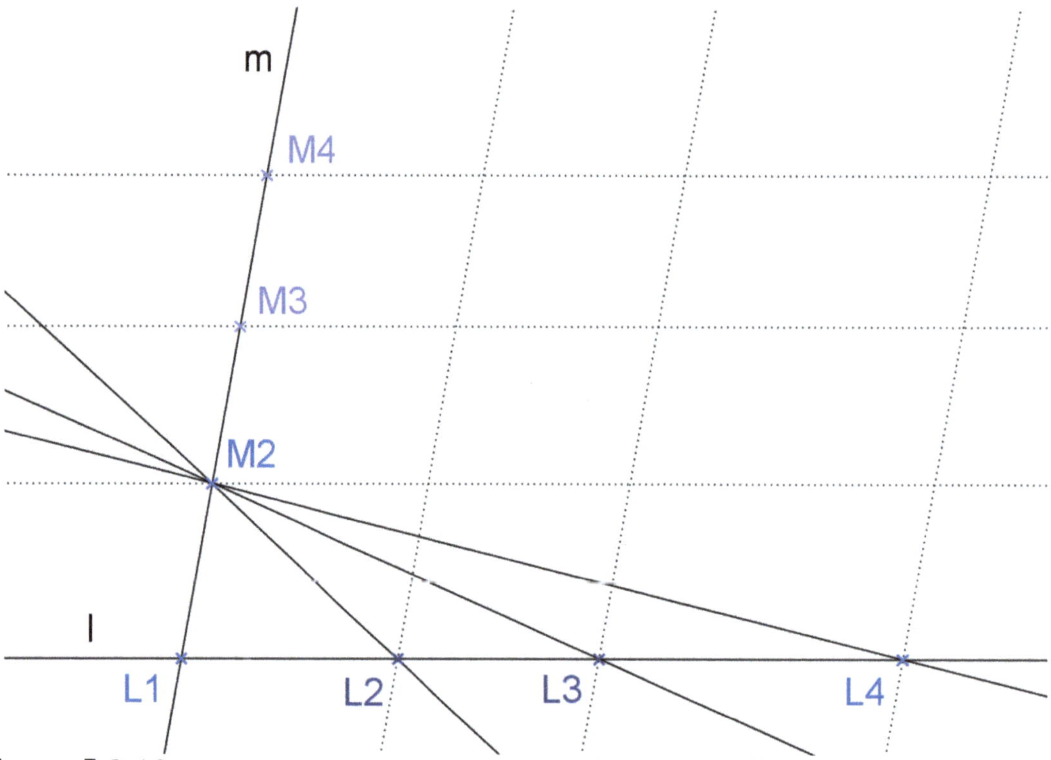

Figure I.3.19: $n+1=5$ pairwise intersecting lines in an affine plane of order $n=4$.

Proof. Let n be the number of points on any line. Using the unique parallels, one shows that every other line (intersecting or parallel) has the same number of points. Take any two lines l and m intersecting in point L_1. No third line can be parallel to both of them. Line m, and the parallels to line m through the $n-1$ points $L_2, \dots, L_n$ on line l yield a pencil of n parallel lines. There cannot exist more parallels, because that would give more intersection points on line l. Hence the lines occur in pencils of n parallels.

Take a second point M_2 on line m. One gets the $n+1$ lines

(I.3.6) $$l,\ m = L_1M_2,\ L_2M_2,\ L_3M_2, \dots, L_nM_2$$

which are pairwise intersecting. If a line intersects <u>all</u> these $n+1$ lines, it is equal to one of them. Hence any line is parallel to one of the lines in the list (I.3.6). Thus we get $n+1$ pencils of n parallel lines, which are n^2+n lines altogether.

Because of the $n+1$ pairwise intersecting lines, in every point there intersect $n+1$ lines, each one of which is either one of the lines (I.3.6) or a parallel to one of them.

The intersection points of the parallels to the intersecting lines l and m yield n^2 points. There can be no other point that this square grid, because there would be more than $n+1$ lines through any outside hypothetical point. □

Definition I.3.21 (Order of an affine plane). The number of points on any line is called the *order* of the affine plane.

Proposition I.3.5 (The projective completion of a finite affine plane). *From an affine plane of order n, one gets a projective plane with n^2+n+1 points and the same number of lines. Each line has the same number $n+1$ of points. Through each point go $n+1$ lines.*

Conversely, after delating any line from any given projective plane, one is left with an affine plane. In the case above with are left with n^2 points.

Reason. Discarding any line and its points as improper leave an affine plane. By problem (I.3.4), we are left with n^2 proper points and n^2+n proper lines. Each line has n points.

By adjoining the improper elements, we get $n+1$ extra improper points, which together yields n^2+n+1 points for the original projective plane. Too, we get one extra line, and hence n^2+n+1 lines in the projective plane. Each proper line gets one extra improper point—in its direction at infinity—and thus has $n+1$ points. The same number of points lie on the improper line.

Through each proper point go $n+1$ lines, as before. Through each improper point go the n parallel proper lines and the improper line, hence $n+1$ lines together. □

Definition I.3.22 (Order of a projective plane). The number of points on any line minus one, or, equivalently, the number of lines through a given point minus one, is called the *order* of the projective plane.

Thus the affine plane, and the projective plane corresponding to it by concatenating and deleting improper elements, have the same order.

Problem I.3.32. *Use homogeneous coordinates and the shorthand formula*

$$\mathbf{PF}^2 = \frac{\mathbf{F}^3 \setminus \{(0,0,0)\}}{\mathbf{F} \setminus \{0\}}$$

to count the points of a projective plane over a finite field with n points. Convince yourself that you get the same result as explained above.

Answer. Since $(\mathbf{F}^3 \setminus \{(0,0,0)\})$ has n^3-1 points, and we take equivalence classes over $(\mathbf{F} \setminus \{0\})$—which has $n-1$ points—we can count the points of the projective plane

$$|\mathbf{PF}^2| = \frac{n^3-1}{n-1} = n^2+n+1$$

This is the same result as obtained earlier. We need just one extra line—the improper line—additionally to the n^2+n lines from the affine plane of order n.

I.3.10 Elementary propositions for three-dimensional incidence spaces

We go back to Hilbert's system of axioms, and investigate now the three-dimensional incidence spaces. Among the consequences of the axioms of incidence, Hilbert spells out two propositions.

Proposition I.3.6 (Hilbert's Proposition 1).

(a) *Any two different lines have either no or exactly one point in common.*

(b) *Any two different planes have either no point in common, or they have one line and no further points in common.*

(c) *A plane and a line not lying in this plane have either no or exactly one point in common.*

Problem I.3.33. *Prove statement* (b). *Use definite names for the objects, explain the steps, and state which axioms are involved.*

Prove statement (c). *Write a short clear proof.*

Answer. **(b)** Take any two different planes $\alpha \neq \beta$. If they do not intersect, we are ready. Assume that the two planes have point of intersection A in common. By axiom (I.7), the two planes have a further point $B \neq A$ in common. By axioms (I.1) and (I.2), there exists a unique line a through points A and B. By axiom (I.6), every point of line a lies in the plane α.

Reasoning about the same points A and B and line a, axiom (I.6) yields that every point of line a lies in the plane β, too. In other words, we get the inclusion

$$a \subseteq \alpha \cap \beta$$

We need still to check that the two planes α and β do not have any point outside the line a in common.

Take any point C lying in both planes α and β. We assume towards a contradiction that point C does not lie on the line a. Both planes α and β contain the three points A, B and C. By axiom (I.5), there exists at most one plane through the three points A, B and C. Hence these two planes α and β would be equal, contradicting the assumption $\alpha \neq \beta$.

Hence the two planes α and β have no point outside the line a in common. Together we get the equality

$$a = \alpha \cap \beta$$

We have shown that two planes intersecting at one point have a line through this point as their intersection.

(c) Take a plane α and a line a not lying in α. We assume towards a contradiction they intersect at two different points $A \neq B$. By axiom (I.6), every point of a lies in the plane α, contradicting the assumption that the line a is not lying in α.

Hence the plane α and line a have at most one point in common.

Proposition I.3.7 (Hilbert's Proposition 2).

(a) *Given a line and a point not lying on it, there exists a unique plane containing the line and the point.*

(b) *Given two different intersecting lines, there exists a unique plane containing these two lines.*

As far as two dimensional geometry is concerned, Proposition I.3.6 reduces to the one simple statement that any two different lines either intersect at one point, or are parallel. But the present section is concerned with the three-dimensional case.

Proof. **(a)** Given is the line a and a point C not lying on it. By axiom (I.3a), there exist two points $A \neq B$ on the line a. The three points A, B, C do not lie on a line. Hence by axioms (I.4) and (I.5), there exists a unique plane α through these three points. By axiom (I.6), all points of line a lie in the plane α. Hence this plane contains both the given line and the given point. Uniqueness of the plane follows by axiom (I.5).

(b) Given are two different lines a and b, intersecting in point C. By axiom (I.3a), there exists a second point $A \neq C$ on line a, and a point $B \neq C$ on the line b. The three points A, B and C do not lie on one line, since the given lines $a \neq b$ are different.

Hence, again by axioms (I.4) and (I.5), there exists a unique plane α through these three points. By axiom (I.6), all points of line a, as well as of line b lie in the plane α. Hence this plane contains both given lines. Uniqueness of the plane follows by axiom (I.5).

□

Lemma I.3.4. *Any three lines in three-dimensional incidence space that go through one point but do not lie in a plane are different from each other.*

Proof. Let the three lines g, h, l all go through point P but not lie in a plane. The assumption that two of them are equal leads to a contradiction. Indeed, under the assumption that two of them would be the same line $g = h$, Hilbert's proposition I.3.7 part (b) would imply existence of a plane through line g and the third line l, since they intersect at point P. Hence all three lines would lie in a plane, a contradiction. Because of this contradiction, we see all three lines are different from each other. □

Proposition I.3.8 (Three lines through any point). *Through any point of a three-dimensional incidence space, there exist three lines not lying in a plane. Especially, these three lines are different from each other.*

A second variant. We use a modification of the proof for lemma I.3.2. Let any point P be given. By axiom (I.8), there exist at least four points not lying in a plane. We call them A, B, C and D.

In the case that point P is among the four points A, B, C, D, we may assume $P = A$. We draw the three lines AB, AC and AD. We check that these three lines do not lie in a plane,—otherwise the four points A, B, C, D would lie in a plane, contrary to the assumption.

In the case that point P is different from all four points A, B, C, D, we draw all four lines PA, PB, PC and PD.

- In the case that either lines PA, PB, PC or lines PA, PB, PD do not lie in a plane, we are ready.

- It may happen that both lines PA, PB, PC lie in the plane γ, and lines PA, PB, PC lie in a plane δ.

  But it cannot happen that any three lines, say e.g. PA, PB, PC, are equal. Otherwise Hilbert's proposition I.3.7 part (a) would imply existence of a plane through line $l = PA = PB = PC$ and the fourth point D. Thus the four points A, B, C, D would lie in a plane, contrary to the assumption.

  Hence lines PA, PB, PC determine the plane γ uniquely. The lines PA, PB, PD determine the plane δ uniquely. Moreover, these are two different planes,—otherwise all four lines PA, PB, PC, PD and all the four points A, B, C, D would lie in a plane, contrary to the assumption.

  Now Hilbert's proposition I.3.6 part (b) implies that the intersection $\gamma \cap \delta$ is a unique line, which is line $PA = PB$.

  We conclude that the lines PB, PC, PD cannot lie in a plane,—otherwise all four lines PA, PB, PC, PD and all the four points A, B, C, D would lie in a plane, contrary to the assumption. Hence we have obtained three lines PB, PC and PD not lying in a plane.

In all cases we have obtained three different lines through the arbitrary point P.

□

Proposition I.3.9 (Every plane is spanned by three points). *In every plane, there exist three points not lying on a line. Hence any plane can be uniquely specified by three points lying on it.*

Proof. By axiom (I.4a), there exists a point P in the given plane π. By proposition I.3.8, there exist three lines g, h and l through point P, which do not lie in a plane.

By Hilbert's proposition I.3.7 part (b), there exists a plane spanned by the lines g and h, which we denote by $\alpha := \text{span}(P, g, h)$. Similarly, there exist the planes $\beta := \text{span}(P, g, l)$ and $\gamma := \text{span}(P, h, l)$. All three planes are different from each other,—otherwise the three lines g, h, l would lie in a plane.

By Hilbert's Proposition I.3.6 part (b), one gets the intersection lines $\alpha \cap \pi$, $\beta \cap \pi$ and $\gamma \cap \pi$. Two of them may be equal. Nevertheless, it is impossible that they are all three equal,—otherwise all three line $g, h,$ and l would be equal.

We may assume the two lines $\alpha \cap \pi \neq \beta \cap \pi$ to be different. By axiom (I.3a), a second point $Q \neq P$ on the line $\alpha \cap \pi$ exists. Similarly, a point $R \neq P$ exists on the second line $\beta \cap \pi$, again by axiom (I.3a). All three points P, Q, R lie in the given plane π, but not on one line. By axiom (I.5), they determine the plane π uniquely. Hence we say they span the plane π. □

I.3.11 Three-dimensional Euclidean incidence geometry

For three dimensional incidence geometry, but no congruence assumed, one needs a strict version of the axiom of parallelism, including existence and uniqueness.

(IV*) Strict axiom of parallelism Given is a line a and a point P not lying on a. In the plane determined by the point P and the line a, there *exists a unique* parallel m to a through point P

Definition I.3.23 (Parallel lines and planes in three dimensions).

(a) Two different lines which <u>lie in one plane</u> and do not intersect are called *parallel.*

(b) Two different planes which do not intersect are called *parallel.*

(c) A line and a plane that do not intersect are called *parallel.*

In the following, we assume at first the axioms of incidence (I.1) through (I.8), and Hilbert's parallel axiom (IV) in order to assure the uniqueness of a parallel to a given line through a given point. The strict axiom of parallelism (IV*) is required later at the point where the existence of the parallel lines and planes is needed.

By generalizing Proclus' Lemma I.3.3 and problem I.3.3 to three dimensions, we are led to the following results.

Proposition I.3.10. *Given is a plane and two parallel lines. We assume that the plane intersects one of them. Then either one of the three cases occurs:*

(i) *both lines lie in the plane;*

(ii) *one of the lines lies in the plane, and the other one is parallel to the plane;*

(iii) *both lines intersect the plane, each one <u>at a different single point</u>.*

Proof. Let the plane α and the line a intersect at point A. The given lines a and b are parallel, and both lie in the plane β. The two planes α and β can be either equal or different. The special case $\alpha = \beta$ leads to alternative (i).

In the other case $\alpha \neq \beta$, the two planes intersect at point A. By Hilbert's Proposition 1(b) (see Proposition I.3.6), they intersect in a line $c := \alpha \cap \beta$. Either one of the following cases occurs:

- $c = a$. Hence line a lies in the plane α and line b is parallel to the plane α, leading to case (ii);
- $c \neq a$. Now we use Proclus' Lemma in the plane β, and conclude that line c intersects the parallel $b \parallel a$, too. The intersection point $B := b \cap c$ lies in the plane α, too. Hence the second parallel b intersects the given plane, as to be shown. This leads to case (iii).

□

Proposition I.3.11.

(a) *a plane intersecting one of two parallel lines <u>at only one point</u> intersects the other line, too.*

(b) *a line intersecting one of two parallel planes and not lying entirely in the first plane, intersects the other plane, too.*

(c) *a third plane intersecting one of two parallel planes intersects the other plane, too.*

Proof of item (b). Let the line c and the plane α intersect at point A. Let $\beta \parallel \alpha$ be second parallel plane.

We choose any point Q in the plane β, which is possible by axiom (I.4a). Let $\gamma := \operatorname{span}(c, Q)$ be the plane containing line c and point Q, which exists because of Proposition I.3.7 part (a). Let

$$p := \alpha \cap \gamma \text{ and } q := \beta \cap \gamma$$

be the respective intersection lines, which exist according to Proposition I.3.7 part (b). The lines $p \parallel q$ are parallel since they lie in the parallel planes $\alpha \parallel \beta$. Now we use Proclus' Lemma for these two parallels in the plane γ. The given line c intersects the first parallel p in point A, hence it intersects the second parallel q in a point $B := q \cap c \subset \beta \cap a$. This is the intersection of the second parallel plane β with line a to be obtained. □

Proof of item (c). Let the third plane $\gamma \neq \alpha, \beta$ and the plane α intersect, say in the line a. Let $\beta \parallel \alpha$ be second parallel plane.

In the plane γ, we choose a second line c which intersects line a, say in a point A. The line a lies entirely in the plane α, but the line c does not lie entirely in the plane α. We can now apply part (b) and get that the line c intersects the plane β in some point $B = c \cap \beta \subset \gamma \cap \beta$.

Hence the given plane γ intersects the second parallel plane $\beta \parallel \alpha$, as to be shown. □

The contrapositive of Proposition I.3.11 part (a) and (b) can be formulated as follows:

Proposition I.3.12.

(a') *If one of two parallel lines is parallel to a plane, then the second line either lies entire in this plane, or is parallel to the plane, too.*

(b') *a line which is parallel to one of two parallel planes, either lies entirely in the second plane or is parallel to the other plane, too.*

Proposition I.3.11 part (c) has further easy consequences:

Lemma I.3.5 (Uniqueness of a parallel plane). *For any plane α and point P not lying on it, there exists at most one parallel plane through this point.*

Proof. Assume towards a contradiction that $\beta \neq \gamma$ are two distinct planes through point P, and both $\beta \parallel \alpha$ and $\gamma \parallel \alpha$.

Now plane γ is the third plane, intersecting one of two parallel planes $\beta \parallel \alpha$, namely plane β in the point P. Hence by part (c) of Proposition I.3.11, the plane γ intersects the other plane α, too. This contradicts the assumption $\gamma \parallel \alpha$.

From this contradiction, we conclude there cannot exist two distinct planes through point P which are both parallel to plane α. Thus the uniqueness of the parallel plane; parallel to the plane α through point P not lying on α; has been confirmed. □

Proposition I.3.13. *Two planes which are both parallel to a third plane are either equal or parallel planes. Consequently, being equal or parallel is an equivalence relation among planes.*

Proof. We assume $\beta \parallel \alpha$ and $\gamma \parallel \alpha$: both planes β and γ are parallel to plane α.

If $\beta \neq \gamma$, the planes β and γ cannot intersect;—otherwise there would be two distinct planes through the intersection point P, and both are parallel to plane α. By Lemma I.3.5 the parallel plane is unique, and consequently $\beta = \gamma$ contrary to the assumption.

Hence either $\beta = \gamma$, or $\beta \parallel \gamma$. □

More important, being equal or parallel defines an equivalence relation among lines, too. This follows from the following a bid more difficult result. Effectively, we have to construct the third plane of a prism with the three lines as edges.

Proposition I.3.14. *Two lines which are both parallel to a third line are either equal or parallel lines. Consequently, being equal or parallel is an equivalence relation among lines.*

Proof. Let the third line $b \neq a, c$ be parallel to both lines $a \parallel b$ and $c \parallel b$. Let α be the plane in which a and b lie and γ be the plane in which c and b lie. In the case that these two planes are equal, we are back to the two dimensional version of problem I.3.3. We conclude that the lines a and c are either parallel or equal.

We now assume that the planes $\alpha \neq \gamma$ are different. We choose any point C on the line c and let $\beta = \operatorname{span}(a, C)$ be the plane in which line a and point C lie. The planes $\beta \neq \gamma$ are different,—otherwise all three planes α, β and γ would be equal.

We define the intersection line $d = \beta \cap \gamma$ and have to confirm that $c = d$. Thus we get the third plane at the surface of a prism with the three lines a, b and c as edges. The lines b, c, d all lie in the plane γ. Lines $b \parallel c$ are parallel by assumption. Since

$$b \cap d = (\alpha \cap \gamma) \cap (\beta \cap \gamma) = (\alpha \cap \beta) \cap (\alpha \cap \gamma) = a \cap b = \emptyset$$

we see that the lines $b \parallel d$ are parallel, too. We now use Proclus' Lemma in the plane β. The line d intersects the parallel $c \parallel b$ in point C. If lines d and c would be different, the line d would intersect the parallel b, too. That is a contradiction to the last formula above.

Hence the lines d and c are equal. Thus we conclude that both lines: line a and line $c = d = \beta \cap \gamma$ lie in the plane $\beta = \operatorname{span}(a, C)$. Moreover, the lines a and c do not intersect since

$$a \cap c = a \cap d = (\alpha \cap \beta) \cap (\beta \cap \gamma) = \emptyset$$

Hence the lines a and c are parallel, as to be shown. □

As a consequence of Proposition I.3.12, part (a') and (b'), the two equivalence relations for parallelism of lines and planes are compatible.

In the following, we shall use existence of the parallel lines and planes. Because there is no direct construction of parallels, we need not only Hilbert's parallel axiom, but the strict axiom of parallelism (IV*).

Proposition I.3.15 (Existence and uniqueness of a parallel plane). *We assume the axioms of incidence* (I.1) *through* (I.8) *and the Euclidean parallel postulate* (IV*). *For any plane α and point P not lying on it, there exists a unique parallel plane through this point.*

Proof. The easier proof of uniqueness is left to the reader. To show existence, we use two intersecting lines in the given plane, and their parallels to the given point.

As shown in Proposition I.3.9, we can choose three points A, B and C spanning the given plane α. In the plane spanned by points A, B and the given point P, we get the parallel m through P to the line AB. In the plane spanned by points A, C and the given point P, we get the parallel l through P to the line AC.

By Hilbert's Proposition 2 (see proposition I.3.7), there exists a unique plane π containing the intersecting lines m and l. Clearly this plane contains point P. We need still to check that the planes α and π are parallel.

Assume towards a contradiction that the planes $\alpha \neq \pi$ intersect, say at point X. By Hilbert's Proposition 1b (see proposition I.3.6), they would intersect in a line x through point X. All three

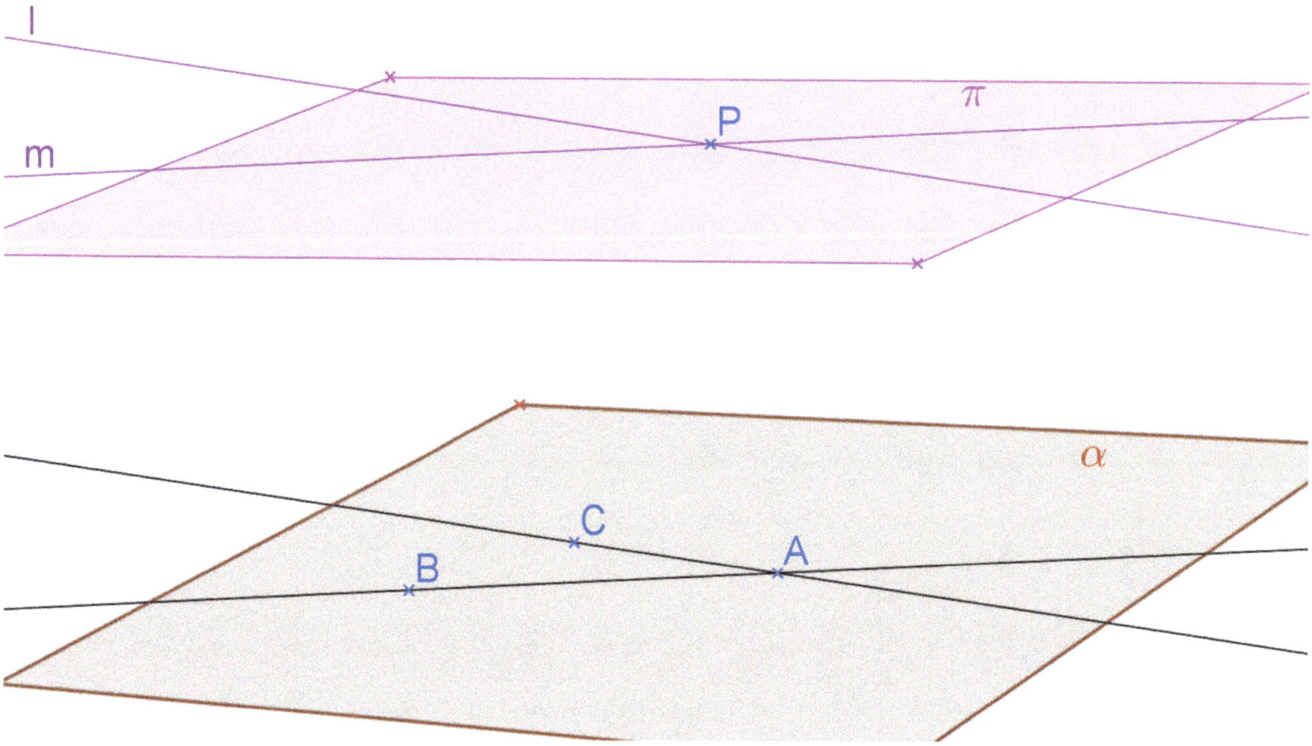

Figure I.3.20: Construction of the plane through a given point parallel to a given plane.

lines m, l and x lie in the plane π. The lines m and x do not intersect, hence lines $m \parallel x$ are parallel. Similarly, we see that lines $l \parallel x$ are parallel. The line x would have two different intersecting parallel m and l, contradicting uniqueness of the parallel.

The only way to avoid the contradiction is planes α and π being parallel. □

I.4 The Theorems of Desargues and Pappus

I.4.1 Desargues' Theorem

For a *triangle* $\triangle ABC$, it is assumed that the three vertices A,B and C do not lie on a line. The order of the three vertices A, B and C matters. The lines $a = BC$, $b = AC$, and $c = AB$ are the *sides* of the triangle. See also definition II.14.3 above. [11]

Proposition I.4.1 (Theorem of Desargues in parallel setting).

(i) *If two triangles in a plane have three pairs of parallel corresponding sides, then the three lines through their corresponding vertices either meet in one point, or all three are parallel.*

(ii) *If the three lines through the corresponding vertices of two triangles in a plane either meet in one point, or all three are parallel, and moreover, two pairs of corresponding sides are parallel, then the third pair of corresponding sides are parallel, too.*

[11] We cannot and need not at this point use the axioms of order.

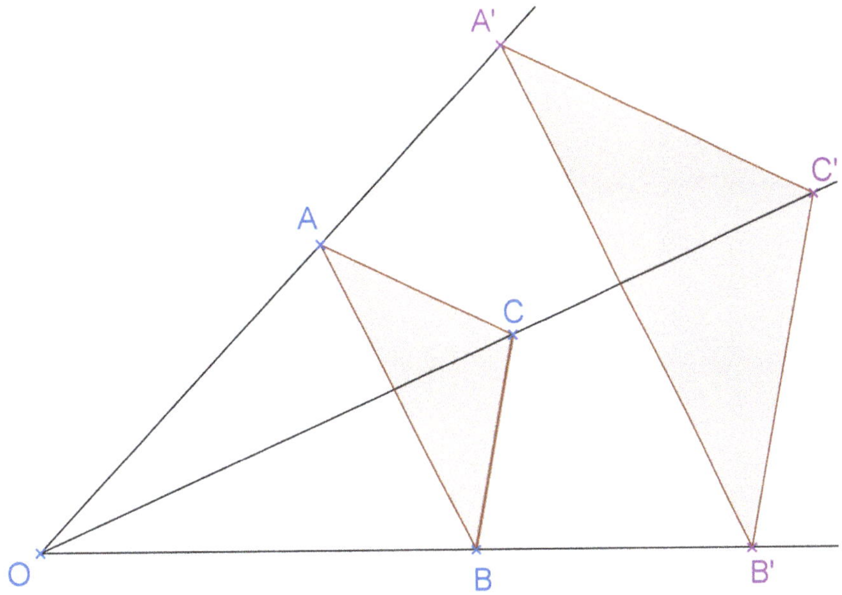

Figure I.4.1: Desargues' configuration in parallel setting.

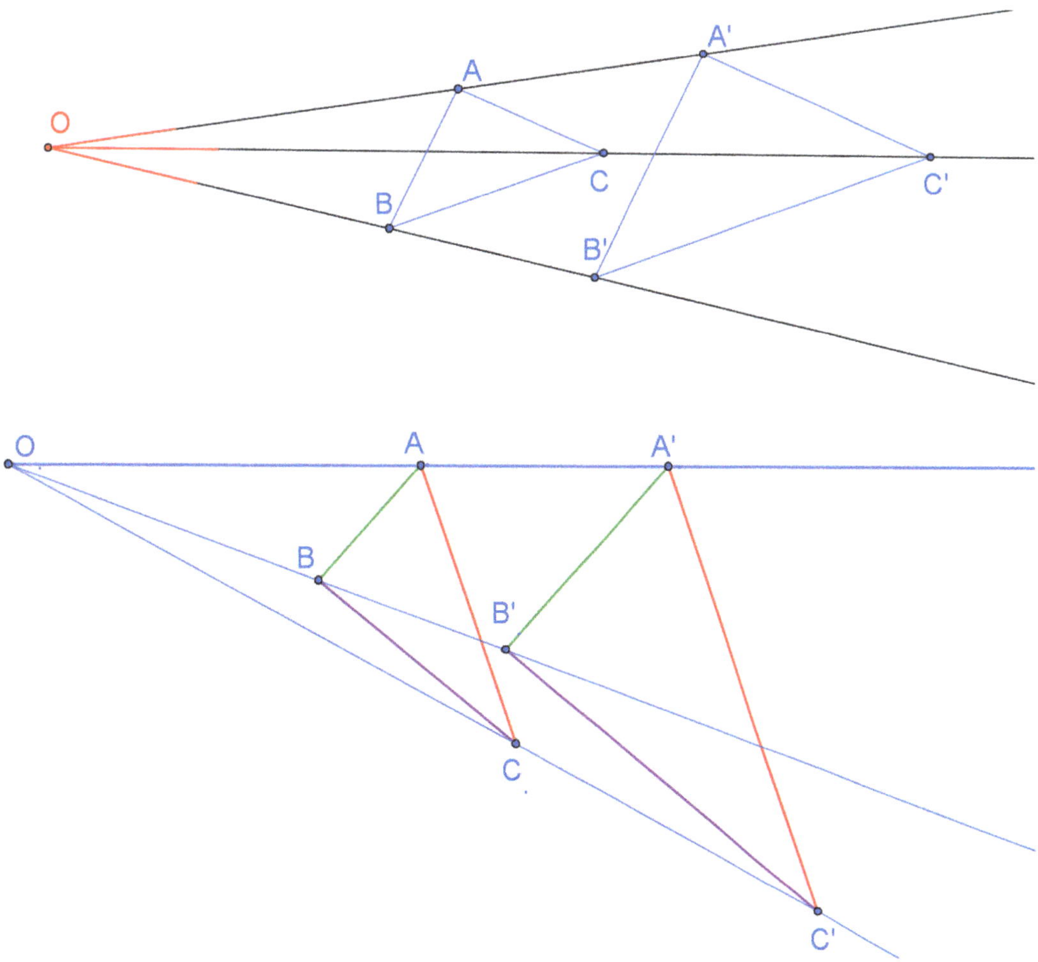

Figure I.4.2: The Theorem of Desargues in parallel setting.

This theorem is really quite easy to understand in two dimensions. While attempting to prove it, one sees that one has either to use the entire axioms of plane Euclidean geometry, or as an alternative use only incidence geometry but in three dimensions. In this first small section, I postpone the proofs, and begin with some easy observations.

The following definition strictly speaking refers to the projective completion of the Euclidean space or plane.

Definition I.4.1 (Triangles in perspective). The vertices of two triangles $\triangle ABC$ and $\triangle A'B'C'$ *lie in perspective from a point* O iff the three lines AA', BB' and CC' through corresponding vertices go through one point O, or are all three parallel. In either case, we may either say simply that the two triangles *lie in perspective.*

Definition I.4.2. The sides of two triangles *lie in perspective* iff the three intersection points $a \cap a'$, $b \cap b'$ and $c \cap c'$ of corresponding sides lie on a line. Again, we have to assert that these intersections exist either as proper or improper points.

Remark. The colorings of the first and the second part of the figure on page 57 are chosen in a way that color red always refers to the conclusion. In the upper figure, red colored parts stress the conclusion of converse Lemma I.4.2. In the lower figure red colored parts stress the conclusion of the Lemma I.4.1.

Lemma I.4.1 (First part of Desargues' Theorem). *If two triangles are in perspective, and, furthermore, two pairs of corresponding sides are parallel, then the third pair of sides are parallel, too.*

Lemma I.4.2 (Converse of the first part of Desargues' Theorem). *If the sides of two triangles are pairwise parallel, then the two triangles are either in perspective from a point, or the three lines through pairs of corresponding vertices are parallel.*

Question. Convince yourself that the first part I.4.1 of Desargues' Theorem implies the converse I.4.2.

Question. Convince yourself that the converse I.4.2 implies the first part I.4.1 of Desargues Theorem.

I.4.2 Theorem of Desargues and related theorems in projective setting

Proposition I.4.2 (Theorem of Desargues in projective setting). *We assume the axioms of incidence* (I.1) *through* (I.8) *and the strict axiom of parallelism* (IV*).
The vertices of two triangles lie in perspective if and only if its sides lie in perspective.

For any two triangles, for the assertions occurring above, I use the short hands

(Verts) "the vertices of two triangles lie in perspective"

(Sides) "the sides of two triangles lie in perspective"

Problem I.4.1. *Assume that the triangles lie in different planes. Convince yourself that* (Verts) $\Rightarrow$ (Sides).

Problem I.4.2. *Assume that the triangles lie in different planes. Convince yourself that* (Sides) $\Rightarrow$ (Verts).

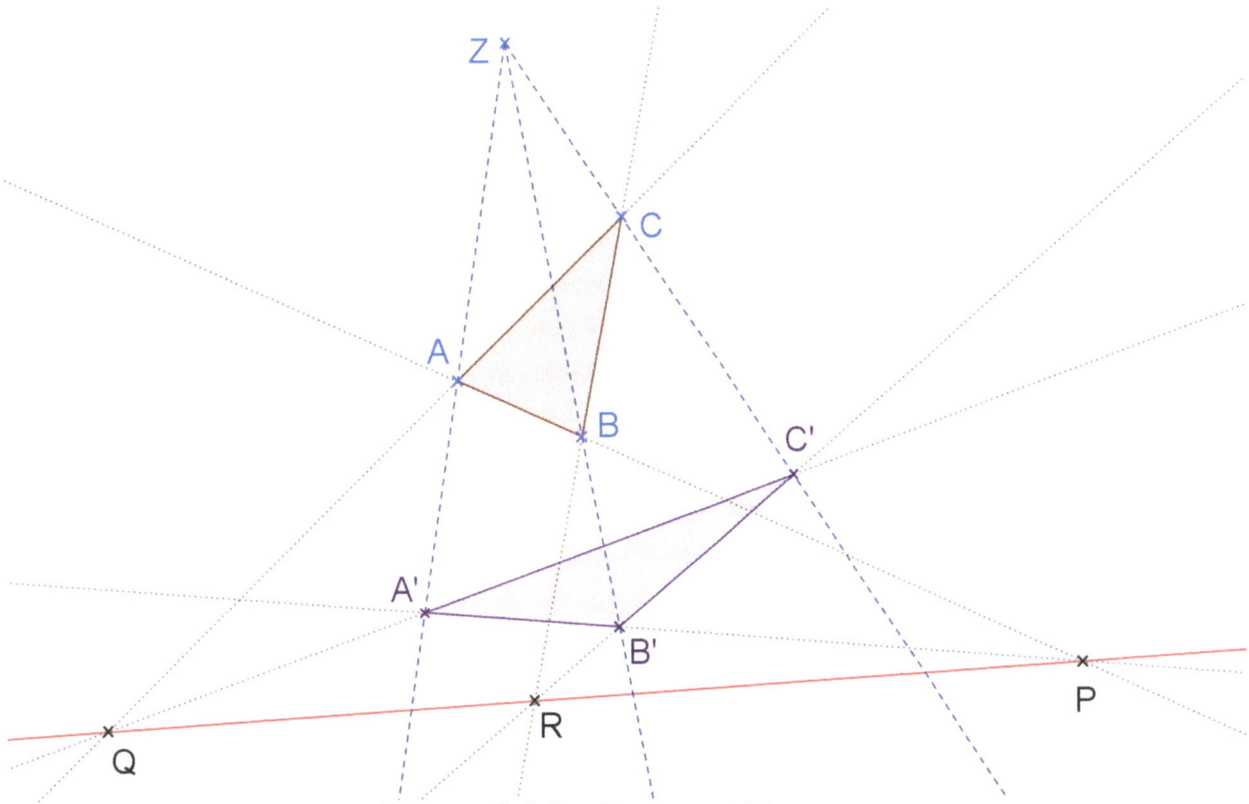

Figure I.4.3: Desargues' Theorem.

Lemma I.4.3. *If* (Verts) $\Rightarrow$ (Sides) *holds for any two triangles lying in different planes, it holds for any two different triangles in the same plane, too.*

Reason. Given are two triangles $\triangle ABC$ and $\triangle A'B'C'$ in the same plane π and its vertices lie in perspective, say at center Z. At least one pair of corresponding sides are different. We can assume that the sides AC and $A'C'$ are different and intersect at point Q.

Let X be a point not lying in the plane π. The plane spanned by points Z, B and X contains at least three points on every line. [12] Hence there exists a further point $D \neq X, D \neq B$ on the line XB. Let D' be the intersection of lines ZD and XB'. Indeed, if $B \neq B'$, all seven points X, Z, D, D', B, B' are different (why?).

We use point X as center of projection to produce a new three dimensional Desargues configuration, consisting of the triangles $\triangle ADC$ and $\triangle A'D'C'$. They lie in different planes (why?). By construction, their vertices lie in perspective. Hence, by the three-dimensional Desargues theorem, their sides lie in perspective. The three intersection points

$$Q = AC \cap A'C' \ , \ S := AD \cap A'D' \ , \ T := CD \cap C'D'$$

exist and lie on a line. A central projection with center X onto the plane π maps $Q \mapsto Q, S \mapsto P, T \mapsto R$ where

$$P := AB \cap A'B' \ , \ Q = AC \cap A'C' \ , \ R := BC \cap B'C'$$

Hence these three points lie on a line, as to be shown. □

[12]At this point, we have used axiom [P.3] "There exist four points of which no three lie on a line", from the definition I.3.20 of a projective plane.

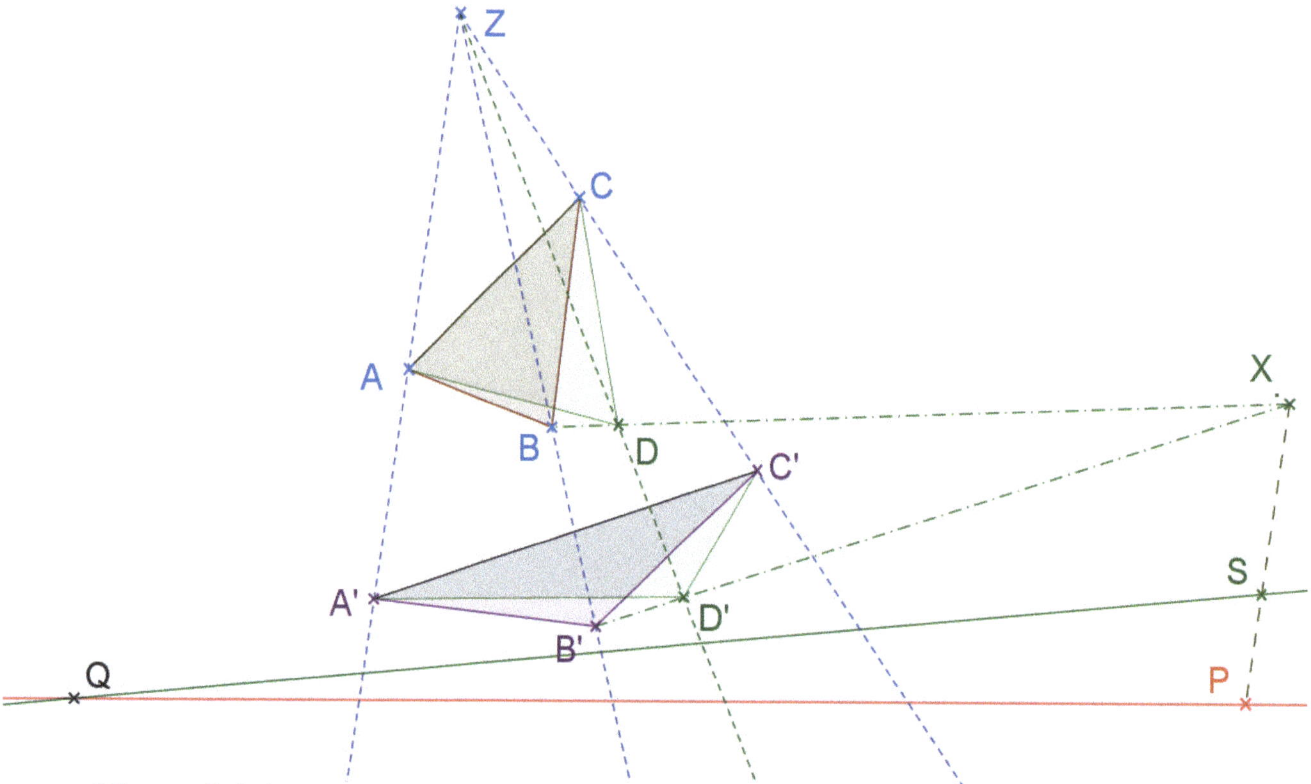

Figure I.4.4: Projecting the three-dimensional Desargues to the two-dimensional one.

Remark. Work in the completion of the Euclidean space. Convince yourself that including cases were several of the vertices or sides turn out to be equal do not lead to contradictions.

Lemma I.4.4. *If* (Verts) $\Rightarrow$ (Sides) *holds for any two triangles lying in the same plane, the converse* (Sides) $\Rightarrow$ (Verts) *holds for any two triangles in the same plane, too.*

Reason. Given are two triangles $\triangle ABC$ and $\triangle A'B'C'$ in the same plane, the sides of which lie in perspective. The three points

$$P := AB \cap A'B' \ , \ Q = AC \cap A'C' \ , \ R := BC \cap B'C'$$

lie on a line. Let $Z = BB' \cap CC'$ be the intersection of the lines through two pairs of corresponding vertices.

We apply the first part of Desargues theorem to the triangles $\triangle PBB'$ and $\triangle QCC'$. Their vertices lie in perspective with center R. Hence their sides lie in perspective. Thus the three intersections points

$$A = PB \cap QC \ , \ A' := PB' \cap QC' \ , \ Z := BB' \cap CC'$$

lie on a line. Hence the vertices of triangles $\triangle ABC$ and $\triangle A'B'C'$ lie in perspective with center Z, as to be shown. $\square$

Remark. Convince yourself that including cases were several of the vertices or sides turn out to be equal do not lead to contradictions.

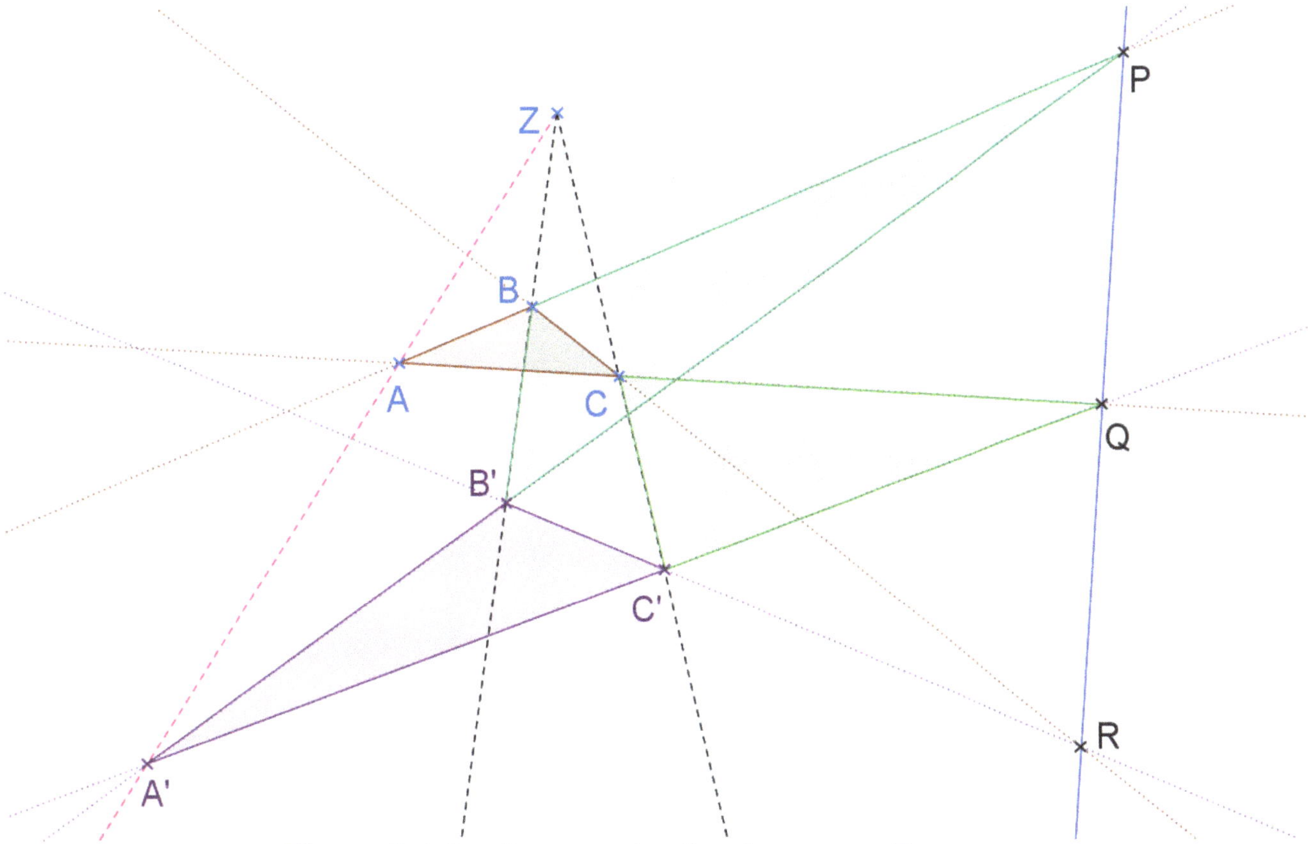

Figure I.4.5: Desargues implies the converse Desargues.

Problem I.4.3. *Convince yourself: If* (Verts) $\Rightarrow$ (Sides) *holds for any two triangles lying in different planes, the converse* (Sides) $\Rightarrow$ (Verts) *holds for any two triangles in different planes, too.*

Proposition I.4.3. *If an affine plane is embedded into a three dimensional incidence geometry, where the axioms of incidence* (I.1) *through* (I.8) *and the Euclidean parallel postulate* (IV*) *hold, then the Theorem of Desargues holds in this affine plane, respectively its completion.*

Proof. This proof puts together the lemmas above and is left as an exercise for the reader. □

The Theorem of Desargues I.4.1 in parallel setting is a special case of the general theorem of Desargues I.4.2 in projective setting, This special case is produced by the extra assumption that the *line* witnessing the sides of the triangles are in perspective, is an improper line. A different special case may be produced by the assumption that the *point* witnessing the vertices of the triangles are lying in perspective, is an improper point. In that manner, we get the following less common case worth to be stated:

Lemma I.4.5. *Given is an affine plane having a projective completion for which Desargues' Theorem holds. Assume the sides of the triangles $\triangle ABC$ and $\triangle A'B'C'$ lie in perspective, and the lines AA' and BB' are parallel. Then the line CC' is parallel to these two lines.*

Problem I.4.4. *Given is an affine plane of order $n \geq 4$ having a projective completion for which Desargues' Theorem holds. Given are* <u>*two different parallel lines*</u> *$AA' \parallel BB'$ and the point C. Use Lemma I.4.5 to construct the parallel to line AA' through the point C.*

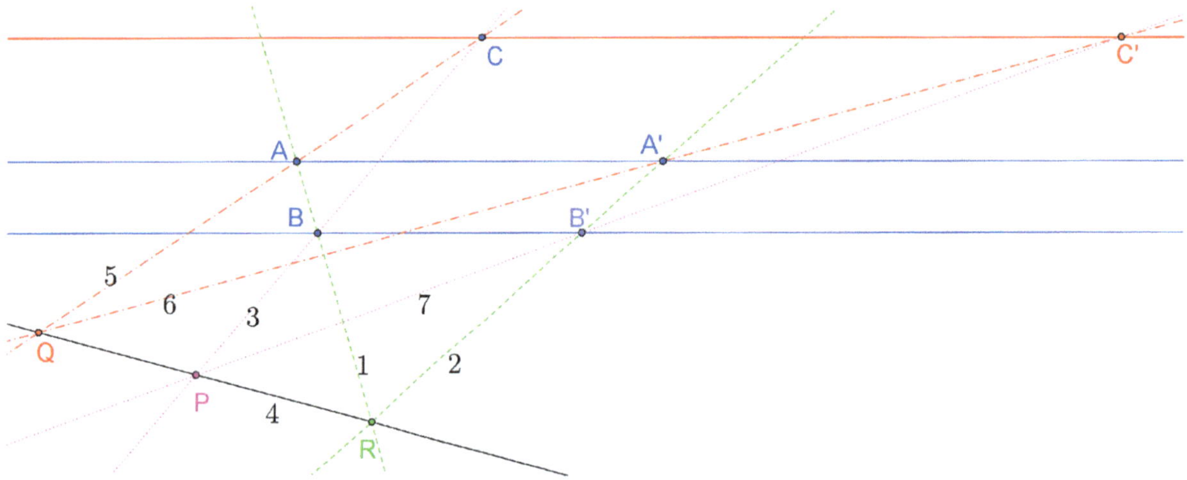

Figure I.4.6: Constructing the parallel to two given parallel lines.

The main outline for the procedure. In the figure on page 62, points and lines are constructed in the order

$$1\ 2\ R\ 3\ P\ 4\ 5\ Q\ 6\ 7\ C'$$

Get the intersection point $R = AB \cap A'B'$.

We choose any point P on the line BC.

Get the intersection point $Q = CA \cap PR$.

Get the intersection point $C' = A'Q \cap PB'$.

The line $CC' \parallel AA'$ is the parallel to be constructed.

$\sqcap$

Complete proof including considerations for the exceptional cases. In an affine plane of order $n \geq 4$ are given two different parallel lines $a = AA' \parallel b = BB'$ and the point C, not lying on either line.

In the special case that points A, B, C lie on a line, we replace A by a third point on the line a, different from A and A'. Thus we may assume that the triangle $\triangle ABC$ exists. On the line AB, we choose any point $R \neq A, B$ different from A and B. One see that $R \neq B'$ since R lies neither on the line A nor the line b. The line RB' exists and intersects the line b. By Proclus' lemma, the line RB' intersects the line a, too. We call the intersection point A', deleting the former point with this name. Since $B \neq B'$, we see that $A \neq A'$, too.

Let Q_0 be the intersection point of line CA with the parallel to line BC through point R. (It does not matter if Q_0 is improper.) On the line CA, there exists a (proper) point Q lying neither on line a nor line b and different from Q_0.

The points $Q \neq R$ are different,—otherwise the three points A, B, C would lie on a line, contradicting the choice of A, B, C made above. By the choice of point Q, the lines QR and BC intersect. We call the intersection point P.

Lemma I.4.6 (Claim). *I now claim that*

- *$P \neq B'$, and P does not lie on line b;*
- *$Q \neq A'$ and Q does not lie on line a;*

- *the lines PB' and QA' are not parallel.*

Reason for the lemma.

- $P = B'$ would imply that point B' would lie on the line BC and hence $B = B'$, which is impossible. P does not lie on line b,—otherwise $P = B'$;
- $Q = A'$ is impossible since Q does not lie on the line a.
- Assume towards a contradiction that lines $PB' \parallel QA'$ are parallel. We use the projective completion and let $C' = PB' \cap QA'$ be the improper intersection point in the completion. Now $C \neq C'$ since C is a proper point. The three lines $PB' \parallel QA' \parallel CC'$ are all parallel to each other.

  By the second part of the Theorem of Desargues I.4.2 in projective setting, the lines $PB' \parallel QA' \parallel CC' \parallel AA' \parallel BB'$ are all parallel to each other. Hence points P, B, B' lie on a line, which is line b. This is impossible as pointed out above. The points Q, A, A' lie on a line, which is line a. This is impossible as pointed out above.

□

By the lemma, the lines PB' and QA' are not parallel. Instead they have a proper intersection point C'.

By the second part of the Theorem of Desargues I.4.2 in projective setting, any two triangles with their sides lying in perspective, have their vertices lying in perspective.

The sides of the two triangles $\triangle ABC$ and $\triangle A'B'C'$ lie in perspective, as the line PRQ witnesses. Hence the vertices of these two triangles lie in perspective, too. Since the lines $AA' \parallel BB'$ are different and parallel, the center Z of perspective is an improper point. The conclusion that points C, C' and the improper point Z lie on a line yields $AA' \parallel BB' \parallel CC'$. Thus we have obtained the requested parallel. □

Theorem I.4.1 (The Little Theorem of Desargues). *Given is an affine plane, respectively its projective completion. We assume that*

- *the vertices of two triangles lie in perspective;*
- *the center of projection and intersections points on two pairs of corresponding sides lie on a line $\mathfrak{L}$.*

Then the intersection point of the third pair of corresponding sides lies on the line $\mathfrak{L}$, too. Thus the sides of the triangles lie in perspective.

Proposition I.4.4 (Converse of the Little Theorem of Desargues). *We assume that*

- *two triangles lie in the same plane and their sides are in perspective at the line $\mathfrak{L}$;*
- *the intersection point of two lines through corresponding vertices lies on the line $\mathfrak{L}$.*

These assumptions imply that the vertices of the triangles lie in perspective.

Remark. In an affine plane, we have to assert that the relevant intersections exist either as proper or improper points. So these theorems really refers to the projective completion of the affine plane.

Problem I.4.5. *Convince yourself that in any affine plane, the Little Theorem of Desargues implies its converse.*

Proposition I.4.5. *In any affine plane, the Little Theorem of Desargues implies its converse. The Theorem of Desargues implies the Little Theorem of Desargues.*

Definition I.4.3 (Configuration of scissors). A *scissor* is called the system of two distinct intersecting or parallel lines, with two points on each one of them, and the quadrilateral zig-zaging between these doublets.

A *configuration of scissors* consists of two such scissors lying between the same pair of lines, together with consistent bijections between the points, and the sides of the two scissors. An example is shown in the figure on page 64.

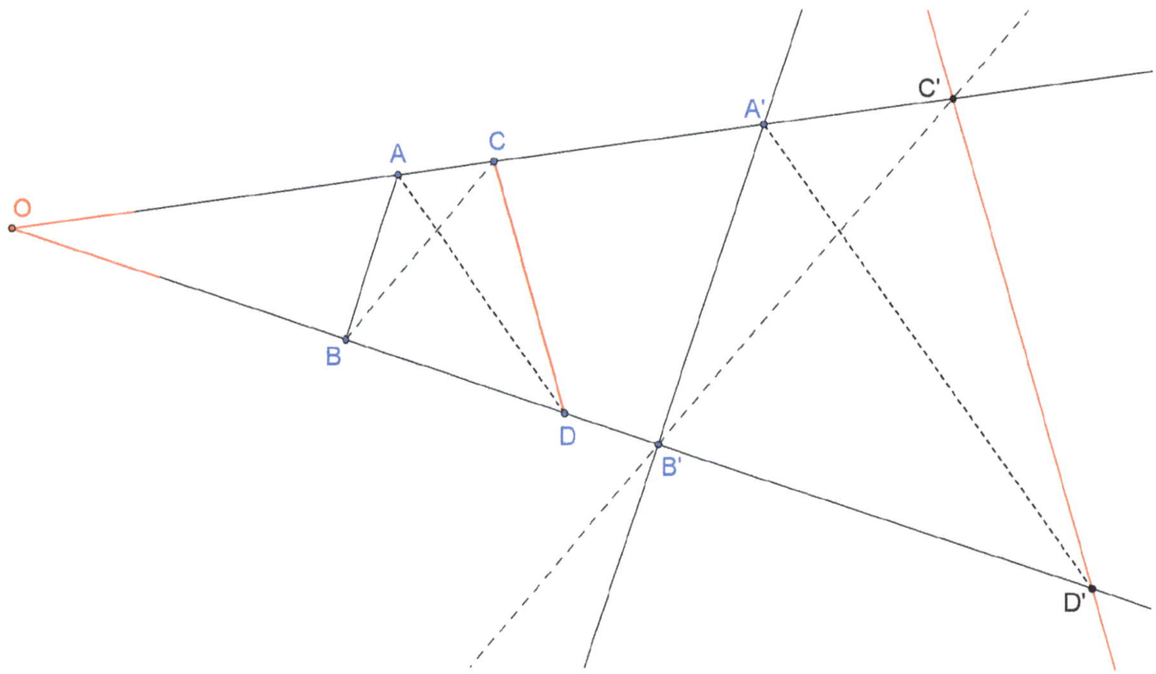

Figure I.4.7: The Theorem of scissors.

Theorem I.4.2 (Theorem of scissors). *If the two scissors from a configuration of scissors have three pairs of corresponding parallel sides, their fourth pair of corresponding sides are parallels, too. An illustration is given in the figure on page 64.*

Theorem I.4.3 (Little Theorem of scissors). *If a configuration of scissors lies between two parallel lines, and the two scissors have three pairs of corresponding parallel sides, their fourth pair of corresponding sides are parallels, too.*

Problem I.4.6. *Draw an illustration for the Little Theorem of scissors.*

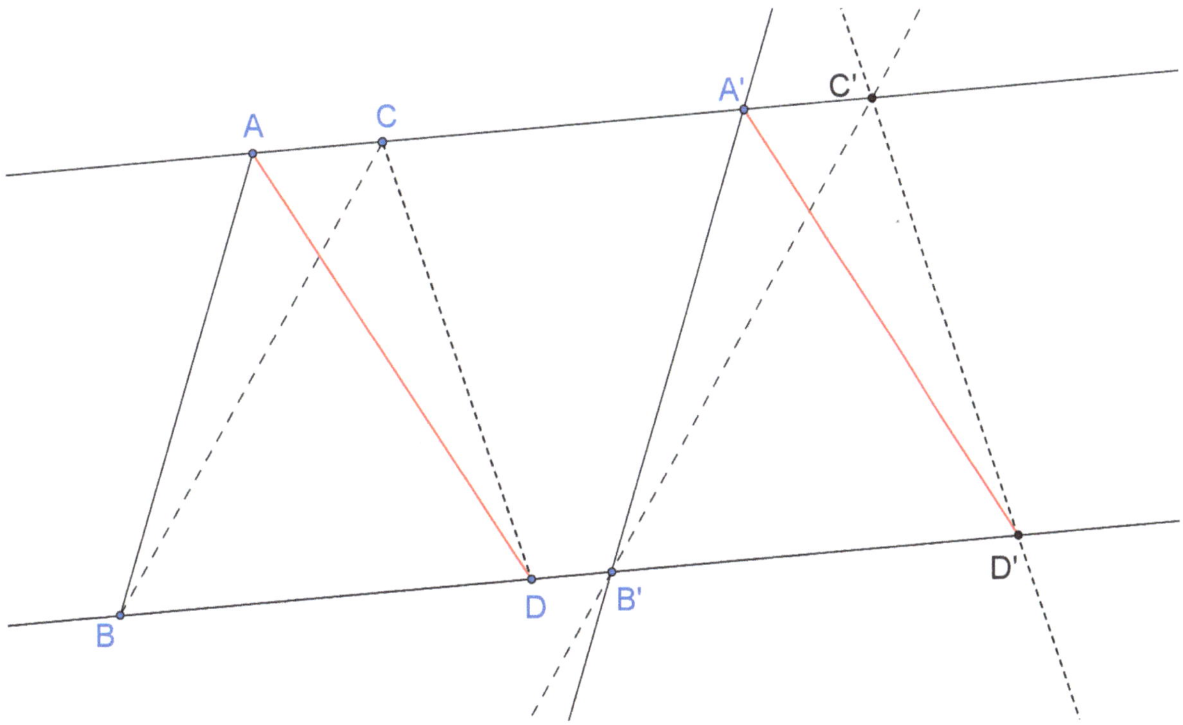

Figure I.4.8: The Little Theorem of scissors.

Answer. An illustration is given in the figure on page 65.

Theorem I.4.4 ("Scissors Theorem"). *For any affine plane,*

(a) *validity of the Desargues Theorem implies validity of the Theorem of scissors.*

(b) *validity of the Little Desargues Theorem implies validity of the Little Theorem of scissors.*

Problem I.4.7. *Give a proof of part* (a)*. You may assume the Desargues Theorem together with its converse. Into the figure on page 66, put the extra lines for the proof in green, and mark and name the extra points needed. Write a paragraph for the proof.*

Answer. Let X be the intersection point of lines AD and BC, and correspondingly let $X' = A'D' \cap B'C'$. The triangles $\triangle ABX$ and $\triangle A'B'X'$ have their sides, and hence, by the converse Desargues Theorem, their vertices in perspective. Thus the three lines AA', BB' and XX' intersect at one point. Hence the two triangles $\triangle XCD$ and $\triangle X'C'D'$ have their vertices in perspective. Hence, by the Desargues Theorem, the two triangles have their sides in perspective. Thus the lines CD and $C'D'$ are parallel, as claimed.

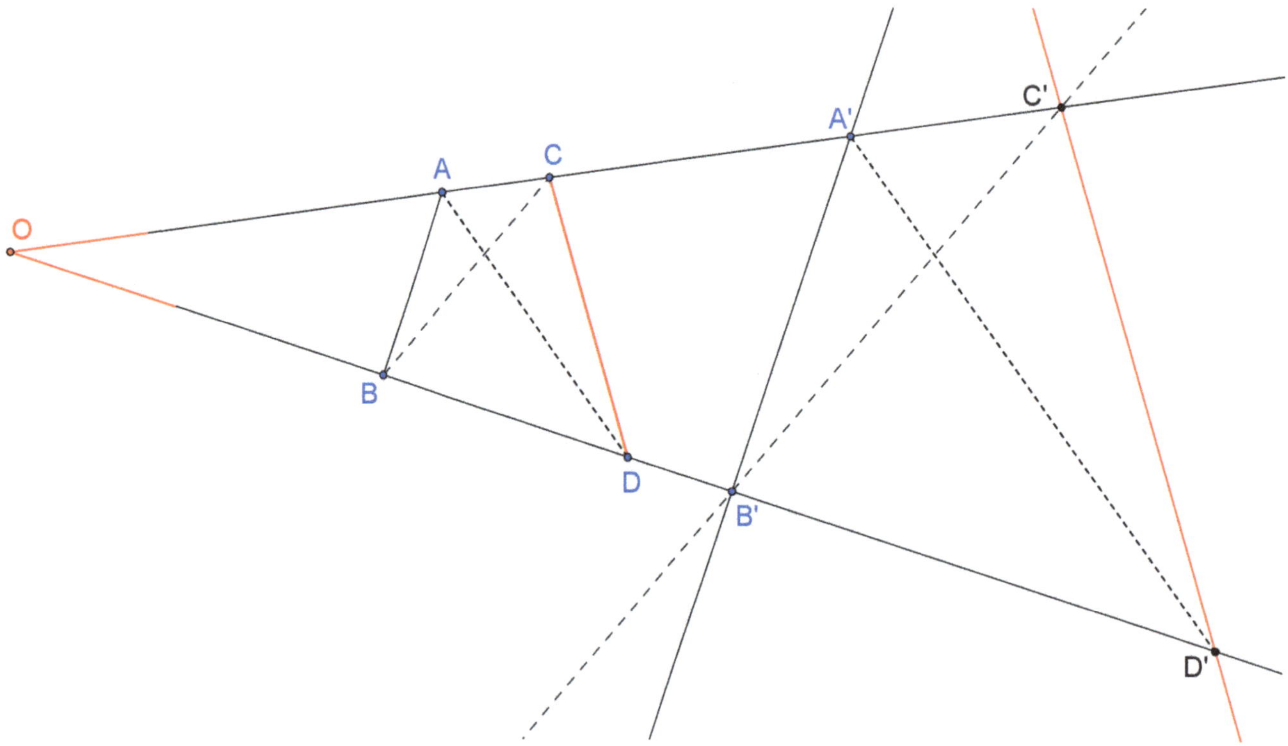

Figure I.4.9: Prove the Theorem of scissors.

Remark. In a more comprehensive treatment of the Theorems of Desargues and Pappus, and the corresponding Little Theorems, it is customary to use the setting of a projective plane. Neither the Theorem of Desargues, nor the Theorem of Pappus, nor the corresponding Little Theorems hold automatically in every projective plane. In the affine setting, assuming only the axioms of incidence (I.1) through (I.3) and the Euclidean parallel postulate (IV*), is not enough to deduct the Little Theorem of Desargues, or the Theorem of Desargues.

The first example of a non-Desarguesian plane is due to Frederic Moulton (1902). A detailed exposition is available in Hilbert's Foundations, and in the recent book by Stillwell [31]. In the Moulton plane, it is easy to get an instance, where even the Little Theorem of Desargues is violated.

Remark. The embedding into a three dimensional geometry is the most natural assumption leading to the Theorem of Desargues. For Hilbert and his followers this is a major motivation why to include three dimensions into the axiomatic setting.

Remark. It is an rather easy but important fact that the Theorem of Desargues holds for all incidence geometries obtained from coordinates with values in a field or skew field.

I.4.3 Tiling in perspective view

Given is a parallelogram, or even simply a square or rectangle. Clearly it is possible to tile the Euclidean plane with shifts of the parallelogram. How can one draw the perspective view of such a tiling?

From the given first quadrangle, we can find the horizon since two pairs of opposite sides intersect in two points on the horizon. To construct shifts of the first quadrangle, we use the fact that the shifted quadrangles have parallel diagonals. We need to use just one diagonal of the first quadrangle, and get in a unique way three adjacent ones.

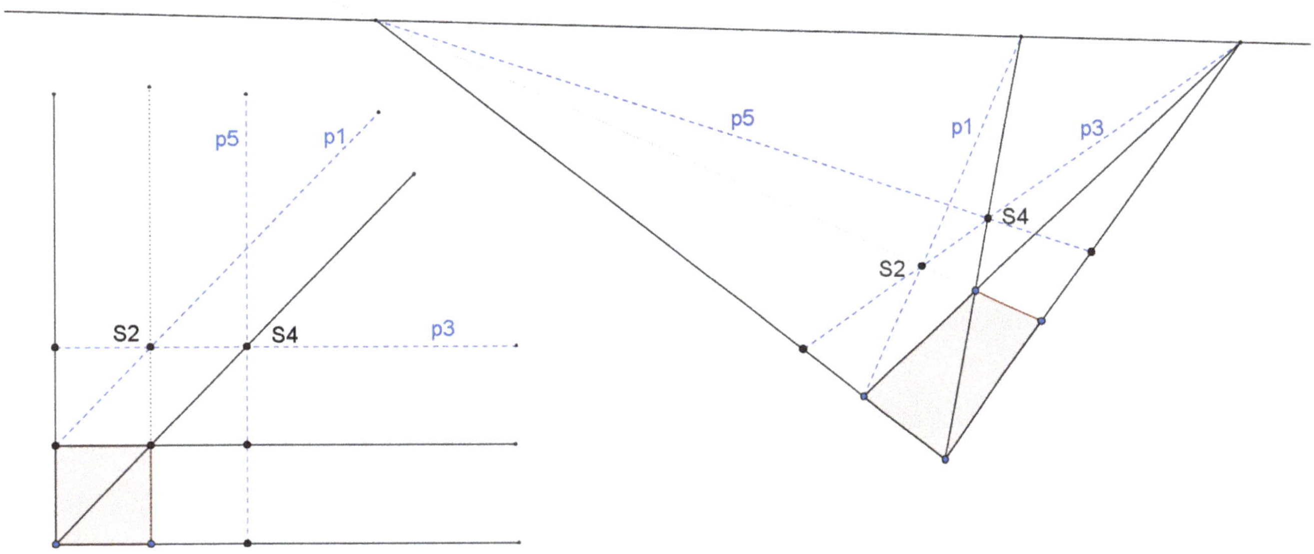

Figure I.4.10: How to construct a tiling.

The drawing on page 67 shows the process in perspective, as well as simply from the view of the Euclidean plane. In both figures is done the construction with the following steps:

- Construct parallel $p1$ to the diagonal,
- find intersection point $S2$,
- draw parallel $p3$ through $S2$ to the horizontal side,
- intersection point $S4$ with the diagonal,
- draw the parallel $p5$ through $S4$ to the vertical side.

The other intersection points which complete the four adjacent parallelograms are obviously given at that point.

A deeper mathematical question is the following. The process can obviously be done in any affine plane. Do there exist affine planes for which this process leads to contradictions. Especially, is the third diagonal automatically parallel. And how about the three other diagonals. Are they parallel? Are there indeed three points on the other long diagonal?

Proposition I.4.6. *Assume the Little Theorem of Desargues holds. Then the construction from above of the four translates of a quadrangle can be done such that the parallel sides and parallel diagonals of the translated quadrangles intersect on four points on the horizon. In other words, the obvious incidences asked for above do hold.*

The figures on page 68 show how to complete the proof in five steps. The first three steps depend on the Little Theorem of Desargues; the last two steps on the converse Little Theorem of Desargues.

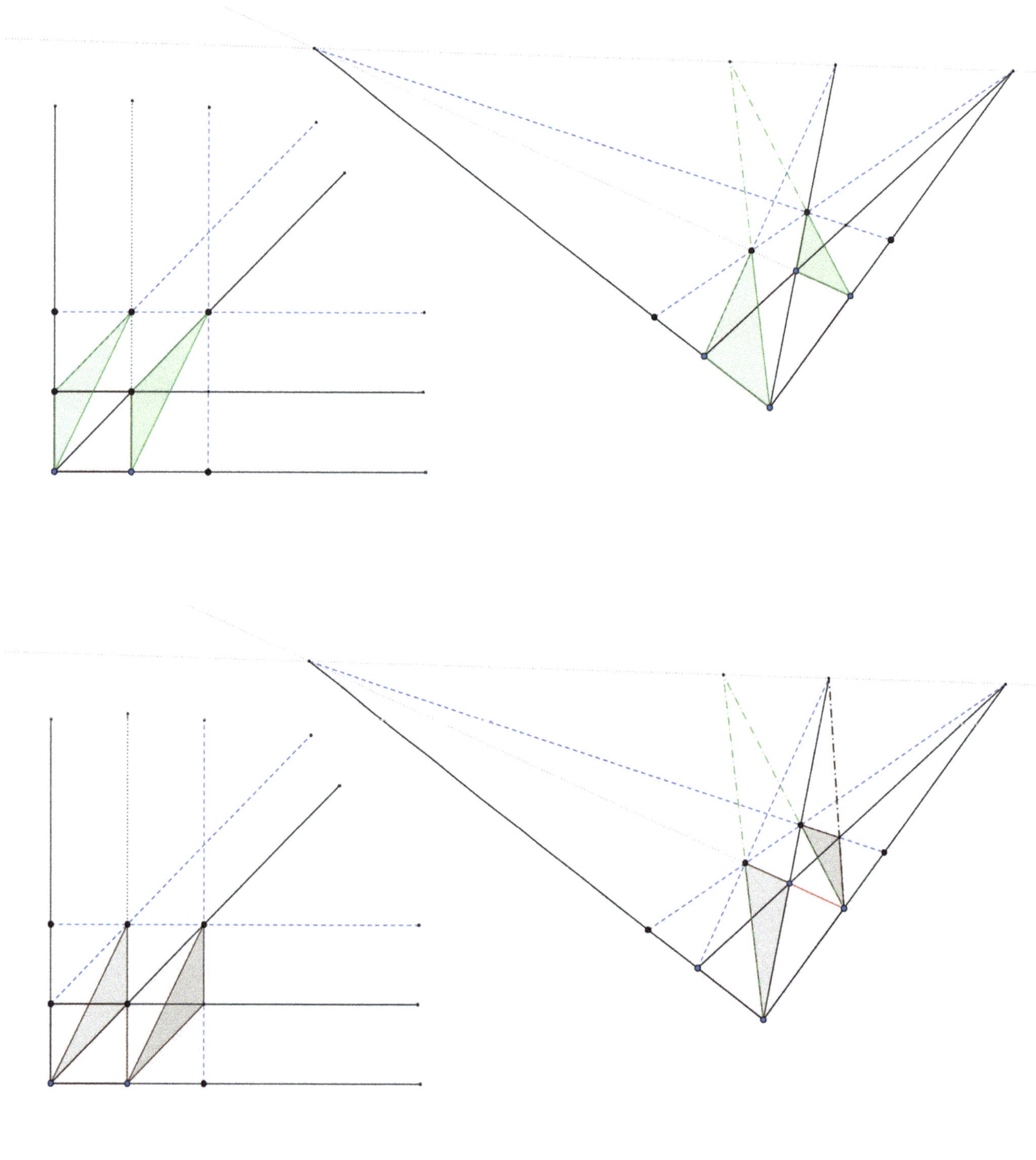

Figure I.4.11: How to prove the expected incidences;—step 1.

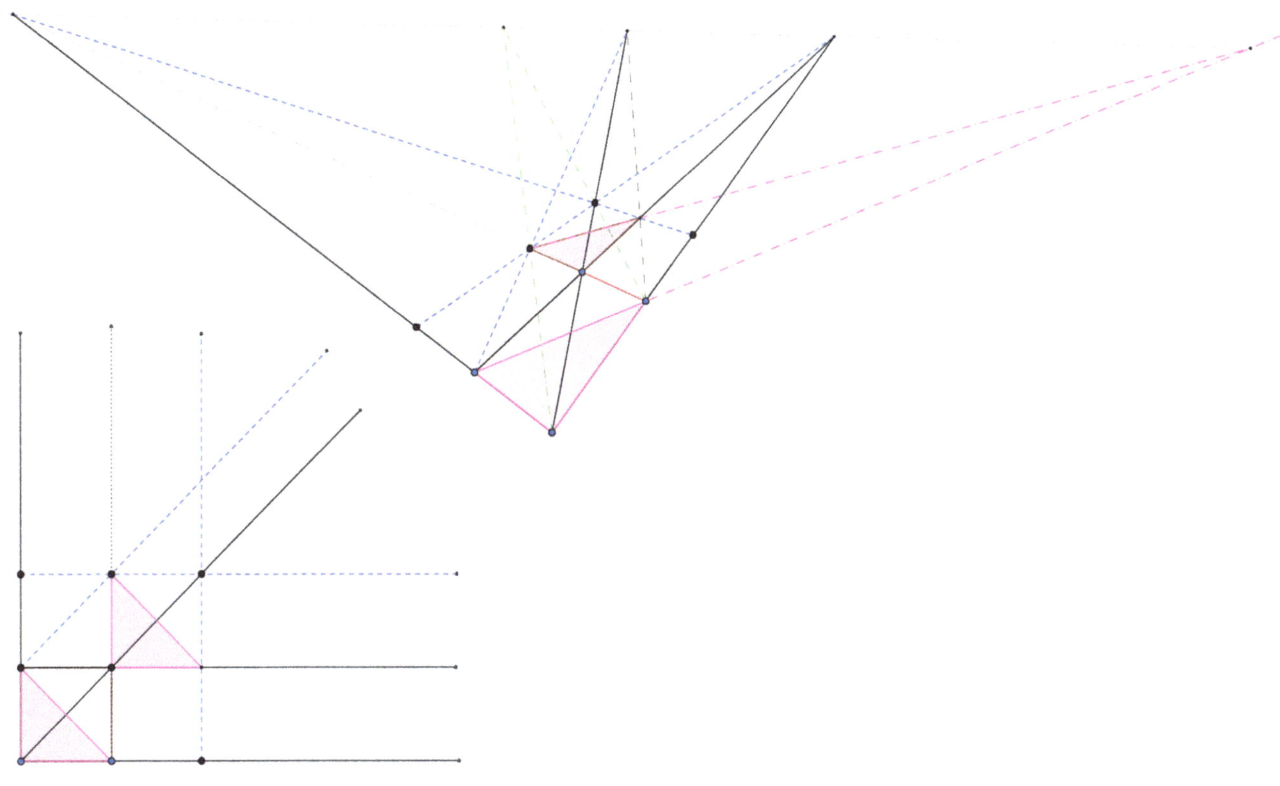

Figure I.4.12: How to prove the expected incidences;—step 2.

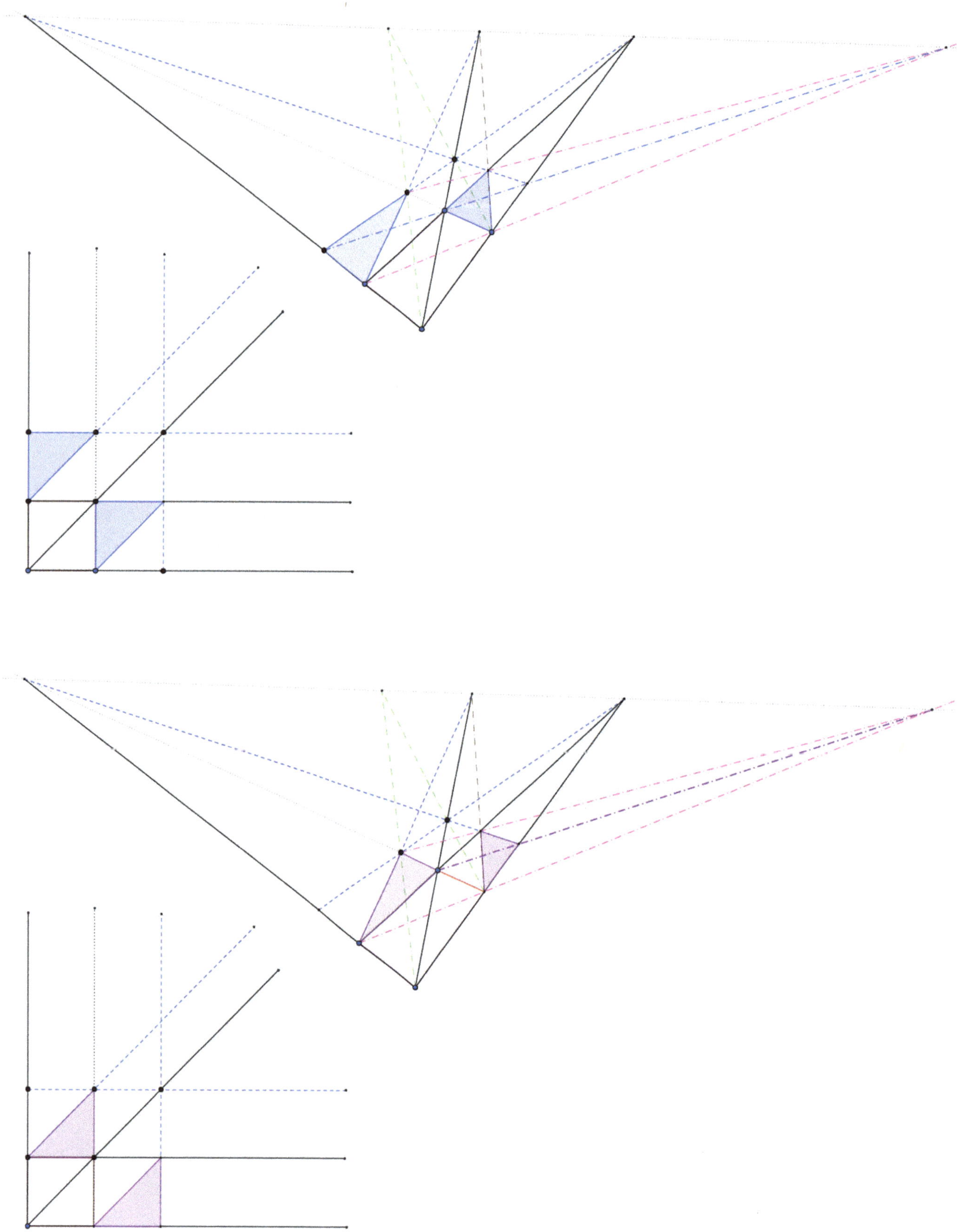

Figure I.4.13: How to prove the expected incidences;—step 3.

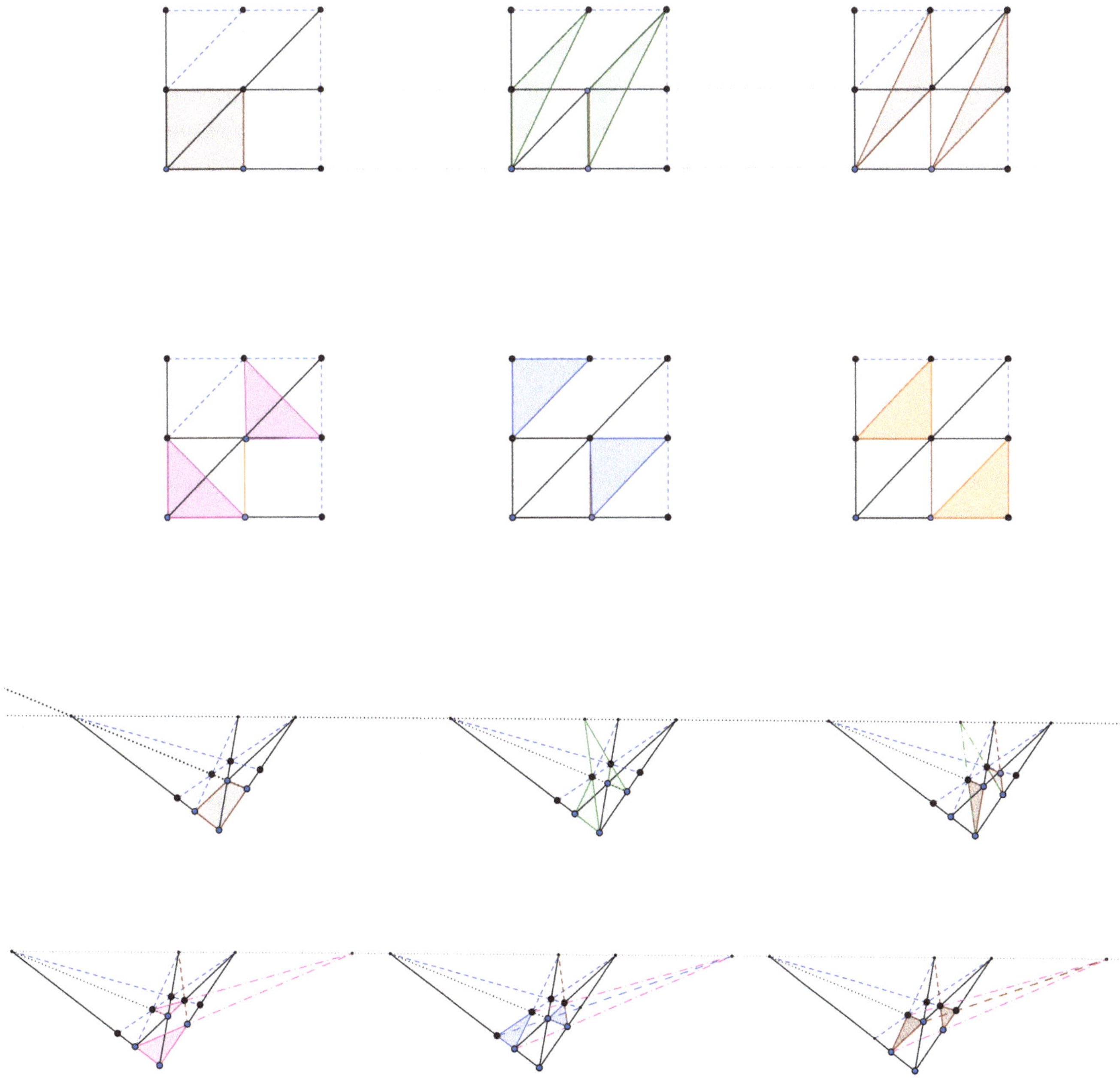

Figure I.4.14: How to prove the expected incidences;—all steps.

I.4.4 The Prime Power Conjecture about Non-Desarguesian Planes

From analytic geometry we are so much used to coordinates that one could easily overlook the main difficult and important questions arising now.

Problem I. From which properties can one recognize that a given projective plane is constructed as a plane over some field?

As Hilbert has shown, to answer this question, one needs two major theorems from projective geometry: the Theorems of Pappus and Desargues. We give only the main result and postpone a detailed proof to later sections. (see "Arithmetic of Segments—Hilbert's Road from Geometry to Algebra".)

Main Theorem 3. *A projective plane is isomorphic to a plane obtained from a skew field if and only if the Theorem of Desargues holds in this plane.*

Main Theorem 4. *A projective plane is isomorphic to a plane obtained from a field if and only if the Theorem of Pappus holds in this plane.*

Since the projective completion is possible for an arbitrary affine plane without the use of coordinates, the Main Theorem 4 implies the converse of the elementary Theorem I.3.2. Hence we can state:

Theorem I.4.5. *Any affine plane is isomorphic to a plane over a field if and only if the Theorem of Pappus holds in its projective completion.*

Problem II. Do there exist projective planes which are not coordinate planes over some field.

Indeed, <u>not all incidence geometries can be obtained from coordinates</u>. The first example of a non-Desarguesian plane is due to Frederic Moulton (1902), and is available in Hilbert's Foundations, and in the recent book by Stillwell [31].

Problem III. Do there exist finite projective planes which are not coordinate planes over some field.

The first example of a finite non-Desarguesian plane was published in 1907 by O. Veblen and J.H.M. Wedderburn [34]. Indeed, these authors have obtained four non-isomorphic projective planes of order 9. They have 81 points! In the section on latin squares we given an independent account of Veblen's arguments. [13]

Open Problem IV. How many non-isomorphic planes do exist of order any prime power?

Open Problem V, the Prime Power Conjecture. There do not exist projective planes of any order different from a prime power. This is a long-standing open conjecture.

[13]Not all of Veblen's results are reproduced.

I.4.5 Hilbert's investigation about the Theorem of Desargues

For Hilbert's detailed investigations about the Theorem of Desargues, it is enough to use the following formulation "in parallel setting". This is not any real restriction. By means of the projective completion, one can get back to the general Theorem of Desargues, spelled out in Proposition I.4.2 above. Recall definition I.3.13 of the affine plane: only the axioms of incidence (I.1),(I.2),(I.3) and the Euclidean Parallel Postulate I.3.3 are assumed. We begin with an overview of Hilbert's results, detailed proofs follow afterwards.

Proposition I.4.7. *If an affine plane is embedded into a* <u>*three-dimensional*</u> *incidence geometry, where the all axioms of incidence* (I.1) *through* (I.8) *and the strict parallel axiom* (IV*) *hold, then the Theorem of Desargues is valid in this affine plane.*

Proposition I.4.8. *In an affine plane, where the Theorem of Desargues in parallel setting Proposition I.4.1 is valid, too, one constructs the* Desarguan arithmetic of segments. *In this way, one does not always obtain a field, but at least a* <u>*skew field*</u>. *If additionally, the axioms of order* (II.1) *through* (II.4) *are valid, we get an ordered skew field.*

Proposition I.4.9. *From a given skew field* $\mathcal{D}$, *one constructs a coordinate plane. The steps have already been explained above in Theorem I.3.2 about the Cartesian plane over an arbitrary field. But now, the terms in the equation for a line*

$$ax + by + c = 0 \tag{I.3.2}$$

need to be kept in the order as written down.

For this coordinate plane, the axioms of incidence (I.1),(I.2),(I.3) *and the strict axiom of parallelism* (IV*), *and the Theorem of Desargues in parallel setting Proposition I.4.1 are valid. The skew field* $\mathcal{D}$ *to start with is isomorphic to the* Desarguan arithmetic of segments *one can construct, in a further step, from the coordinate plane.*

Moreover, if the skew field is ordered, the axioms of order (II.1) *through* (II.4) *are valid in the coordinate plane, too.*

Proposition I.4.10. *From a given skew field* $\mathcal{D}$, *one constructs even a three dimensional coordinate space. The steps are well-known from three dimensional analytic geometry. In the equations for lines and planes, the terms need to be kept in the order analogous to equation* (I.3.2).

In this geometry, all incidence axioms (I.1) *through* (I.8) *and the strict axiom of parallelism* (IV*) *are satisfied. Hence the Theorem of Desargues is valid, both the version in parallel setting Proposition I.4.1, as well as the general version given by Proposition I.4.2.*

Moreover, if the skew field is ordered, the axioms of order (II.1) *through* (II.4) *are valid, too.*

We see from these investigations:

Main Theorem 5. *Any projective plane is isomorphic to a plane obtained from a skew field if and only if the Theorem of Desargues holds in this plane.*

Remark. The Theorem of Desargues can only be proved using either all incidence axioms (I.1) through (I.8), <u>including</u> the <u>three-dimensional</u> ones. As a second road, the Theorem of Desargues can be proved in a Hilbert plane, using all congruence axioms, including the SAS axiom and the Euclidean parallel postulate. Indeed, by Theorem IV.19.1 from the section "Arithmetic of

Segments—Hilbert's Road from Geometry to Algebra", we know that any Pythagorean plane is isomorphic to the Cartesian plane $\mathbf{F}^2$ over its field $\mathbf{F}$ of segment lengths. By Proposition I.4.9, in a coordinate plane of a field or even skew field, the Theorem of Desargues in parallel setting Proposition I.4.1 is valid.

I.4.6 The Mouton plane

We proceed to the definition of the simplified Moulton plane. Over any ordered field $\mathbf{F}$, the Cartesian plane is constructed in the usual way. We know that the axioms of incidence (I.1)(I.2)(I.3a)(I.3b), and order (II.1) through (II.4) are valid,—in their usual interpretation. The complete explanation and justification are given by Definition I.3.15 and Theorem I.3.2. With real coordinates, the axioms of continuity (V.1)(V.2) are valid, too.

This ordered affine plane is now used as the underlying reality, and a new model is constructed by redefining the notion of "`line`". From the coordinate system, one obtains the x-axis, and a positive direction on it since the field is assumed to be ordered. We mark the left half-plane of line x as the upper half-plane $\{(x, y) : x, y \in \mathbf{F} \text{ and } y > 0\}$. The right half-plane is the lower half-plane $\{(x, y) : x, y \in \mathbf{F} \text{ and } y < 0\}$.

Let any line l which is not parallel nor equal to x, be given. Let x_+ be the ray from the intersection $L := x \cap l$ in the positive x-direction. Let k be the ray on the line l in the lower half-plane, and h be the ray on l in the upper half-plane, both with the same vertex L. For a line l with positive slope $m > 0$, let h' the ray with vertex L and slope $m/2$.

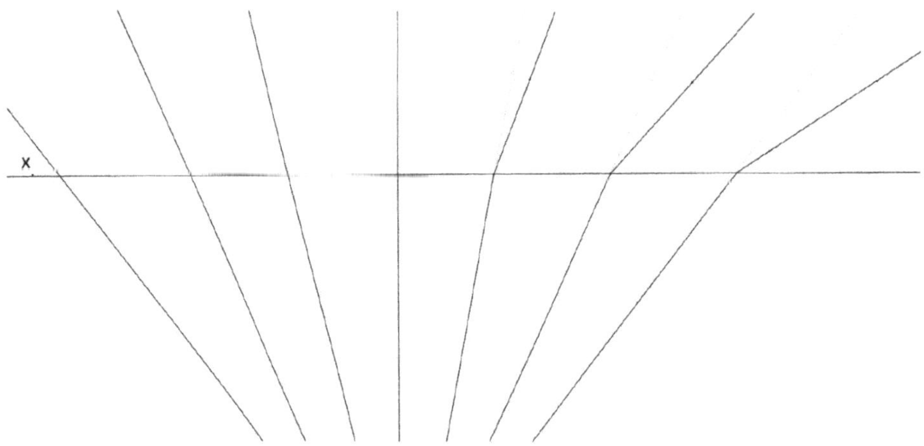

Figure I.4.15: Lines in the Moulton plane.

Definition I.4.4 (Simplified Moulton plane). A "`line`" of the new geometry is now each one of the following:

the x-axis;

any line parallel or perpendicular to the x-axis;

any line of negative slope;

for any line with positive slope the union of the two rays $k \cup h'$.

The "`order`" of the points on the new lines is defined in the obvious way.

Problem I.4.8. *Convince yourself that from the notions above, one has obtained a planar geometry for which the axioms of incidence* (I.1),(I.2),(I.3), *all axioms of order* (II), *and the Euclidean Parallel Postulate I.3.3 are all valid.*

Remark. This non-Desarguesian plane is due to Frederic Moulton (1902), and is available in Hilbert's Foundations. The simplified version given in Definition I.4.4 is used in the recent book by Stillwell [31].

Proposition I.4.11. *There exists an order affine plane where the following violations occur:*

Neither the Theorem of Desargues, not the Little Theorem of Desargues, nor the Theorem of Pappus are valid.

It is not possible to embed the planar geometry in a three-dimensional incidence geometry.

Proof. By the figures below on page 76, page 76, and page 81, one sees easily that neither the converse part of the Theorem of Desargues, nor the Little Theorem of Desargues, nor the Little Pappus Theorem are valid in the new affine plane. To get such easy counterexamples, it is convenient to use a case in which only very few "`lines`" of the new geometry are affected by their intersection with the x-axis.

The impossibility of an embedding into a three-dimensional incidence geometry follows via Proposition I.4.7. □

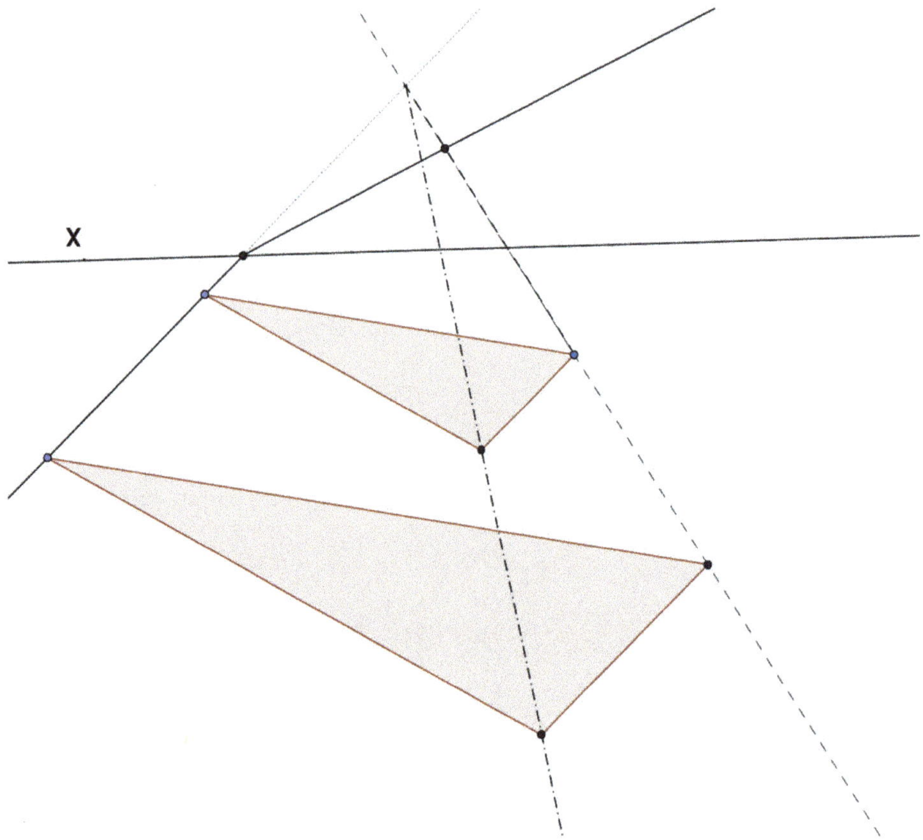

Figure I.4.16: In the Moulton plane, the Desargues Theorem is not valid.

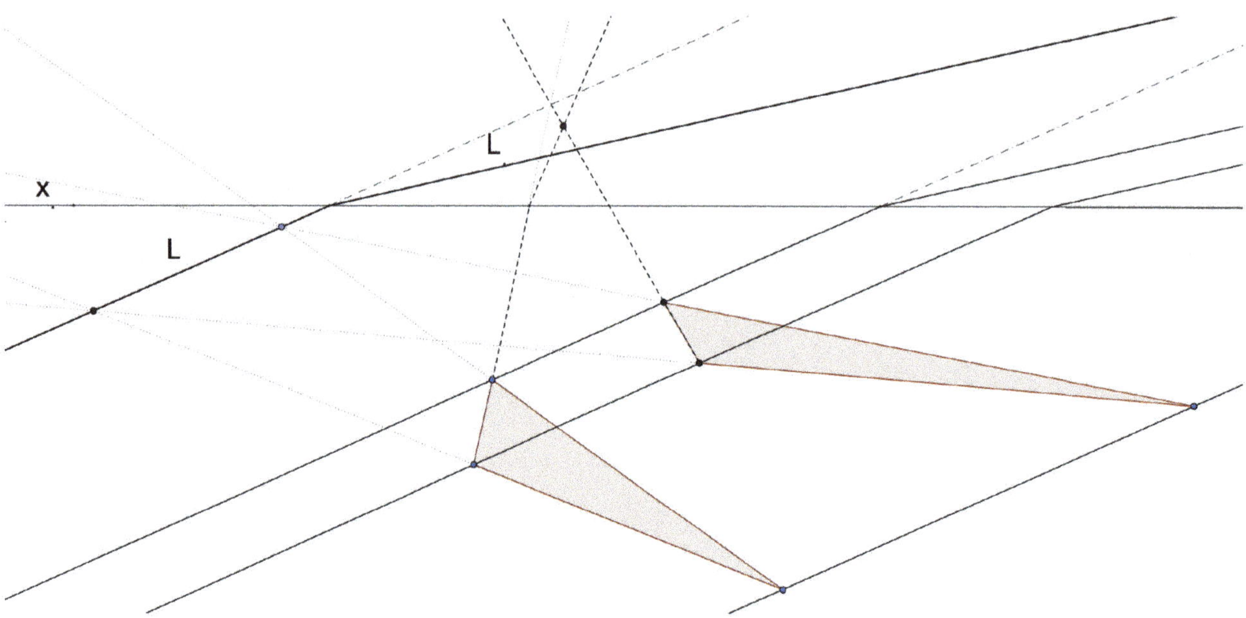

Figure I.4.17: In the Moulton plane, the Little Desargues Theorem is not valid.

I.4.7 Theorem of Pappus and related theorems

Definition I.4.5 (Pappus' configuration). We call the system of two distinct intersecting or parallel lines, with three points on each one of them, and the M-W hexagon zig-zaging between those triplets the *configuration of Pappus*.

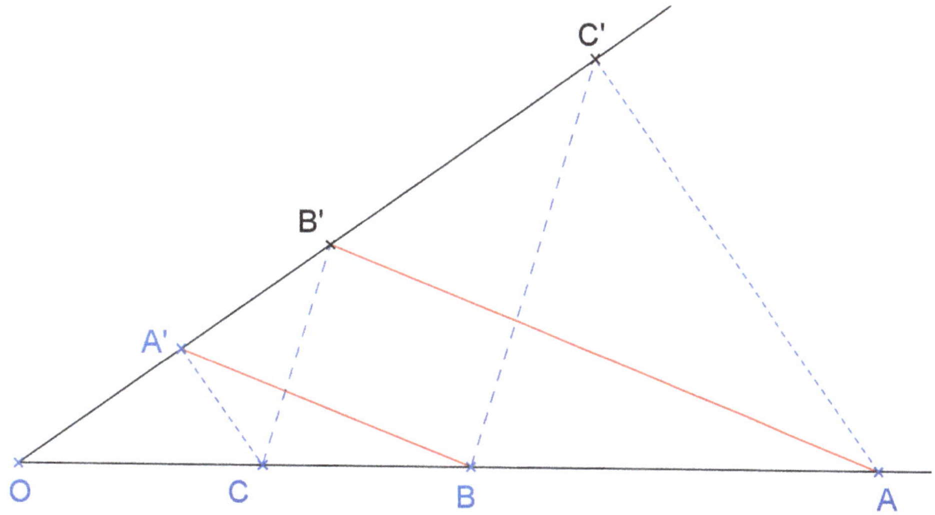

Figure I.4.18: For Pappus' configuration: If $BC' \parallel B'C$ and $AC' \parallel A'C$, then $AB' \parallel A'B$.

In the present figure, we have points A, B, C on the horizontal line, three points A', B', C' on the tilted line. In the hexagon $AB'CA'BC'$ there are three pairs of opposite sides: the three pairs of lines

$$\begin{array}{c} A'B \text{ and } AB' \\ B'C \text{ and } BC' \\ C'A \text{ and } CA' \end{array}$$

the parallelism of which has to be investigated. Thus we are lead to conjecture that the hexagon of Pappus' configuration containing two pairs of opposite parallel sides shall have a third pair of opposite parallel sides, too. A formulation with definitely specified variables is given here.

Theorem I.4.6 (Pappus Theorem in parallel setting). *Let A, B, C and A', B', C' be both three points on two intersecting lines, all different from the intersection point. If the lines BC' and $B'C$ are parallel, and the lines AC' and $A'C$ are parallel, then the lines AB' and $A'B$ and parallel, too.*

Theorem I.4.7 (Pappus' Theorem in projective setting). *Let A, B, C and A', B', C' be three points on two intersecting lines, respectively.The three intersection points $X = BC' \cap B'C$, $Y = AC' \cap A'C$ and $Z = AB' \cap A'B$ of three pairs of opposite sides of hexagon $AB'CA'BC'$ lie on a line.*

I.4.8 Theorem of Hessenberg

There are remarkably different routes to a proof of the Pappus' Theorem!

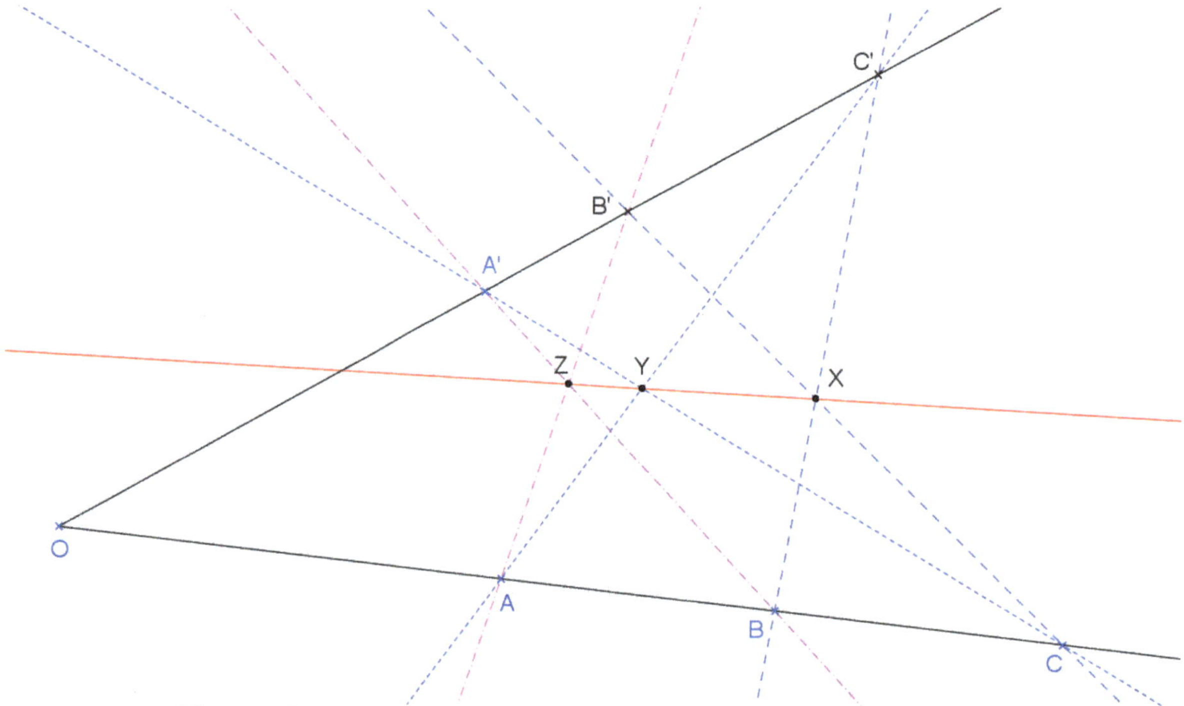

Figure I.4.19: The Theorem of Pappus in the projective setting.

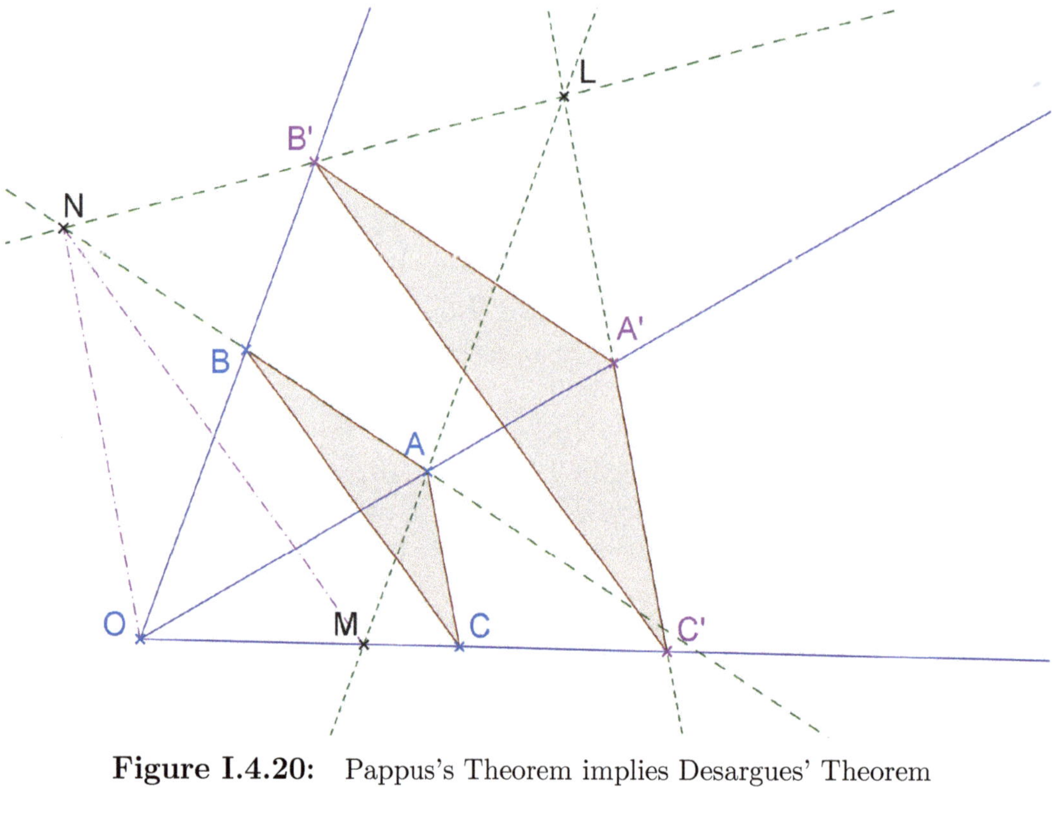

Figure I.4.20: Pappus's Theorem implies Desargues' Theorem

Theorem I.4.8 (Theorem of Hessenberg). *For an affine plane, validity of Pappus's Theorem implies Desargues' Theorem.*

Proof. We shall proof that the second part of Desargues' Theorem holds under the given assumptions. Furthermore, we shall assume that the two triangles $\triangle ABC$ and $\triangle A'B'C'$ are in perspective from point O, and that $AB \parallel A'B'$ and $AC \parallel A'C'$. We give the proof under the following

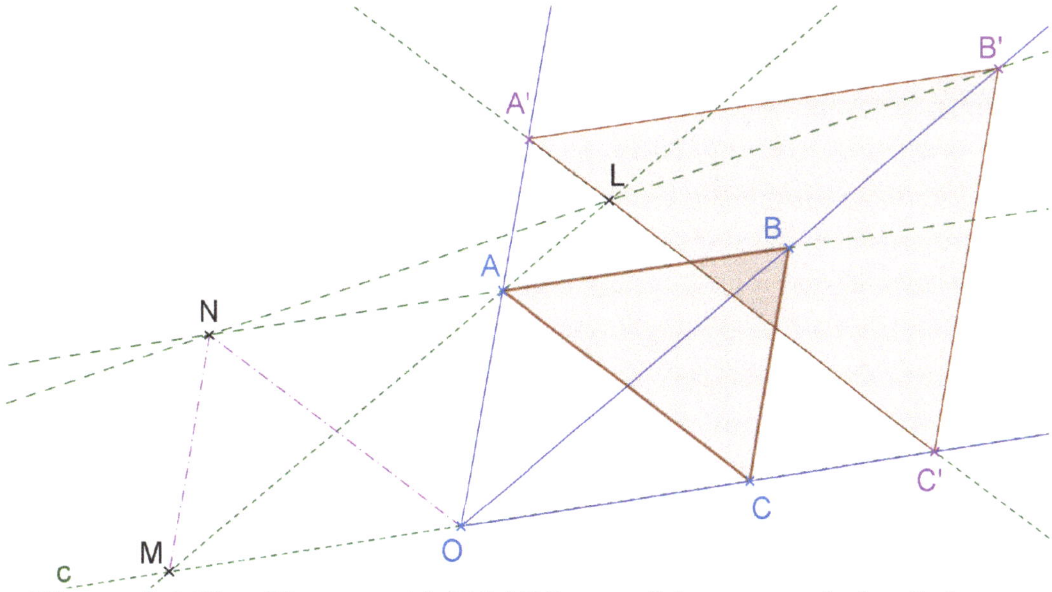

Figure I.4.21: The case with $\square OABC$ a parallelogram can be handled, too

simplifying assumption:

$$OB' \nparallel A'C' \tag{I.4.1}$$

Draw the parallel to line OB through point A. Let L be the intersection point of this parallel with line $A'C'$. Let M be the intersection point of the parallel with line OC. Let N be the intersection point of lines LB' and AB.

Question. Why do points L, M and N exist?

Answer.

We now use Pappus' Theorem for three different configurations:

Step 1: Use Pappus' Theorem in configuration $ONALA'B'$. Because $AB = NA \parallel A'B'$ and $AL \parallel B'O$, we conclude $ON \parallel LA'$.

Now $ON \parallel LA' = A'C'$ and $A'C' \parallel AC$ imply $ON \parallel AC$.

Step 2: Use Pappus' Theorem in configuration $ONMACB$. Because $ON \parallel AC$ and $MA \parallel BO$, we conclude $NM \parallel CB$.

Step 3: Use Pappus' Theorem in configuration $ONMLC'B'$. Because $ON \parallel LC' = LA'$ and $ML \parallel B'O$, we conclude $NM \parallel C'B'$.

Finally, $NM \parallel CB$ and $NM \parallel C'B'$ imply $CB \parallel C'B'$, as to be shown. □

I.4.9 Relations to the "Little Theorems"

Theorem I.4.9 (Little Pappus Theorem). *Let A, B, C and A', B', C' be both three points on two parallel lines. If the lines BC' and $B'C$ are parallel, and the lines AC' and $A'C$ are parallel, then the lines AB' and $A'B$ and parallel, too.*

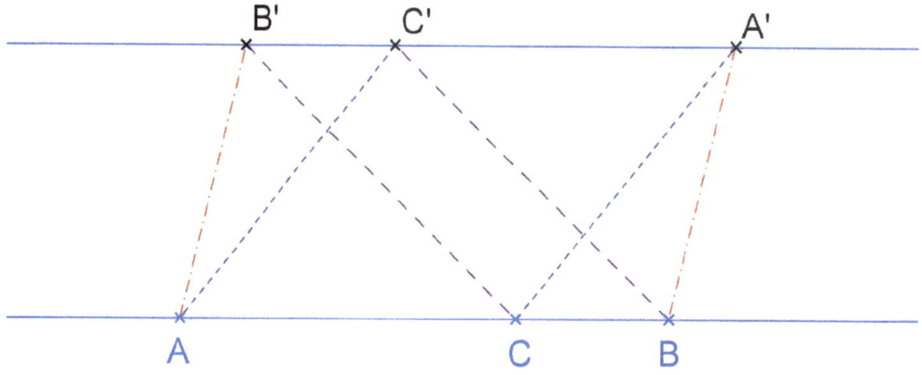

Figure I.4.22: The Little Pappus Theorem asserts for a hexagon $AC'BA'CB'$ that has its vertices alternating on two parallel lines: if $BC' \parallel B'C$ and $AC' \parallel A'C$, then $AB' \parallel A'B$.

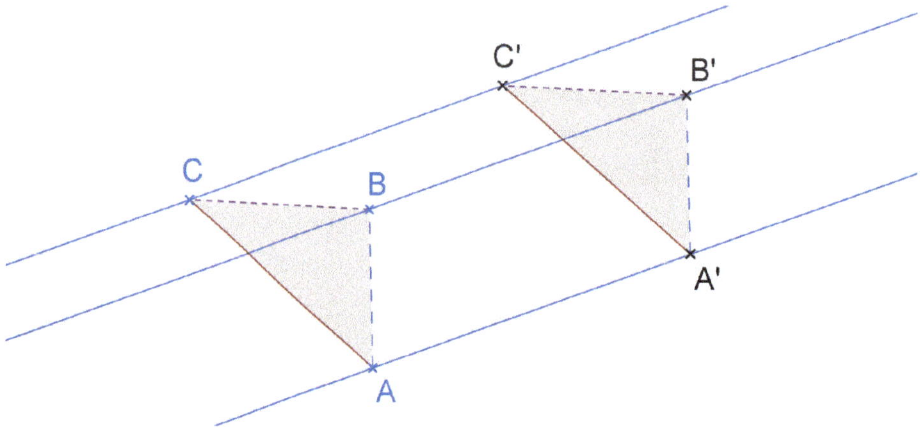

Figure I.4.23: The Little Desargues Theorem asserts for two triangles with vertices on three parallel lines: if $AB \parallel A'B'$ and $BC \parallel B'C'$, then $AC \parallel A'C'$.

Theorem I.4.10 (Little Desargues Theorem). *If corresponding vertices of two triangles lie on three parallel lines, and, furthermore two pairs of corresponding sides are parallel, then the third pair of sides are parallel, too.*

Theorem I.4.11 ("Little Hessenberg Theorem"). *For an affine plane, validity of the little Desargues Theorem implies validity of the little Pappus Theorem.*

Proof. Given are points A, B, C and A', B', C' on two parallel lines such that $BC' \parallel B'C$ and $AC' \parallel A'C$.

We draw the parallel to line AC' through point B', and the parallel to line BC' through point A'. These two lines intersect in a point D. The little Desargues Theorem is now applied to the two triangles $\triangle ACB'$ and $\triangle C'A'D$. Corresponding vertices are indeed joint by three parallel lines. Too, there are two pairs of parallel sides: $AC \parallel C'A'$ and $CB' \parallel A'D$. Hence the little Desargues Theorem assures that

$$AB' \parallel C'D$$

Secondly, one applies the little Desargues Theorem to triangles $\triangle CBA'$ and $\triangle B'C'D$. Again, corresponding vertices are indeed joint by three parallel lines. Two pairs of parallel sides are

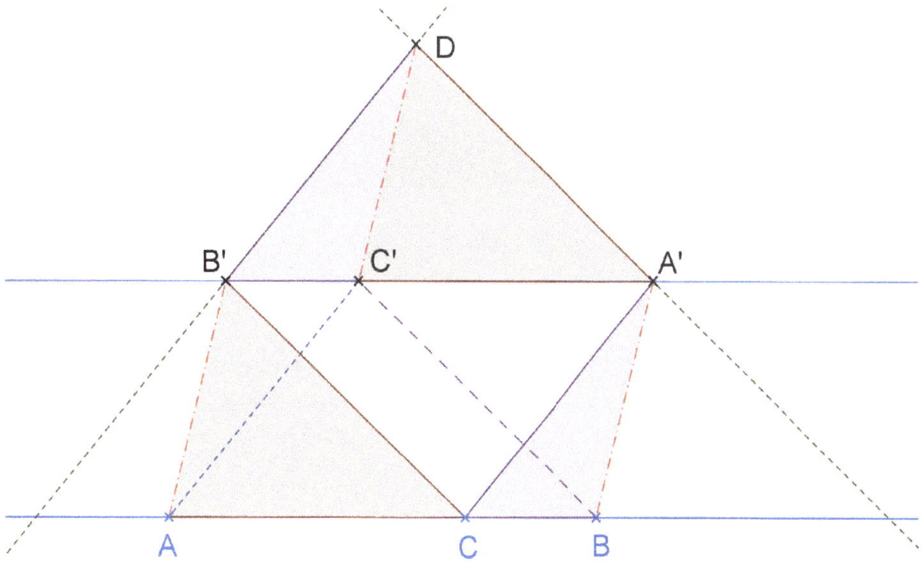

Figure I.4.24: The Little Desargues Theorem implies the Little Pappus Theorem.

$CB \parallel B'C'$ and $CA' \parallel B'D$. Hence the little Desargues Theorem assures that

$$BA' \parallel C'D$$

Both instances of the little Desargues theorem together imply $AB' \parallel BA'$, as to be shown. □

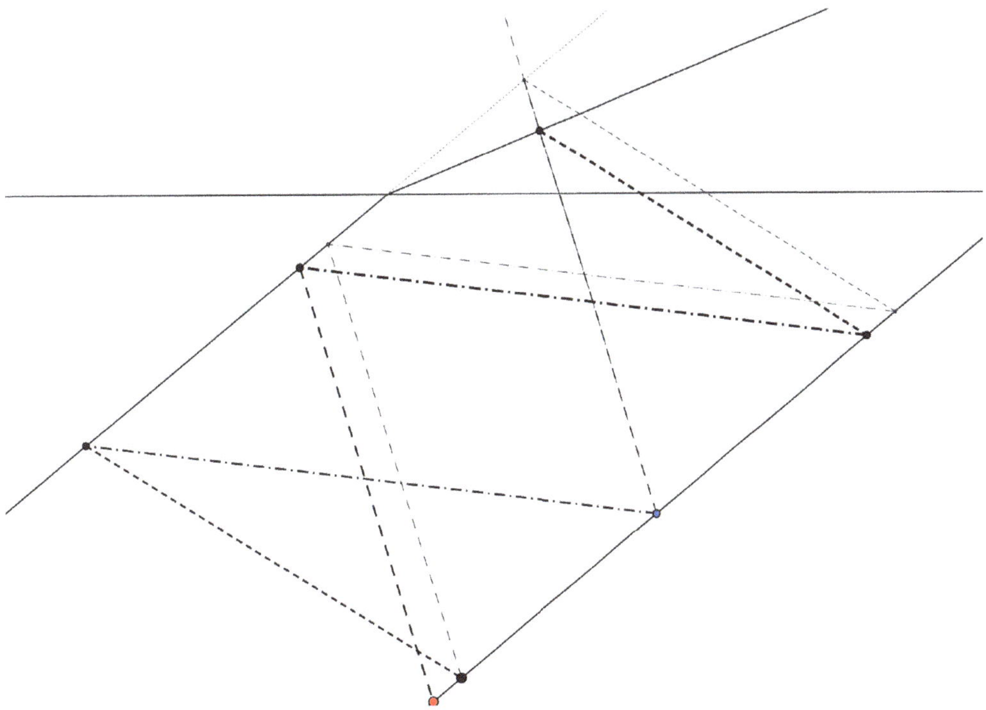

Figure I.4.25: In the Moulton plane, the Little Pappus Theorem is not valid.

Remark. For simplicity all four configurations— Pappus, Desargues and Little Pappus, Little Desargues— were given above in the *affine* version. The affine version deals with parallel lines. Hilbert's foundations [19] use the affine version.

As explained in definition I.3.11, improper elements can be adjoined to produce a projective plane from a given affine plane. The pencils of parallel lines are denoted as *improper points* for the different directions of these pencils. The line through all improper points is called the *improper line*.

All four configurations and theorems— Pappus, Desargues and Little Pappus, Little Desargues— have a corresponding version in the *projective* plane. In the projective version, the statement that any lines are parallel has to be replaced by the statement that they intersect on the improper line. [14]

Theorem I.4.12 (The logical relations between Pappus, Desargues and scissors). *All four configurations and theorems— Pappus, Desargues and Little Pappus, Little Desargues—are now taken in the* projective *plane. The logical relations between them are summed up in the following diagram:*

$$\begin{array}{ccccc} \textit{Pappus} & \xrightarrow{\textit{Hessenberg Th. I.4.8}} & \textit{Desargues} & \xrightarrow{\textit{Scissors Theorem I.4.4}} & \textit{scissors} \\ \downarrow & & \downarrow & & \downarrow \\ \textit{Little Pappus} & \xleftarrow{\textit{Theorem I.4.11}} & \textit{Little Desargues} & \xrightarrow{\textit{Scissors Theorem I.4.4}} & \textit{Little scissors} \end{array}$$

[14] Stillwell's book [31] states only the projective versions.

I.5 Finite Affine and Projective Incidence Planes and Latin Squares

I.5.1 Latin squares

Definition I.5.1. A $n \times n$ square matrix where each row and column is a permutation of the numbers $1, 2, \ldots, n$ is called a *Latin square.*

Definition I.5.2. Two Latin squares a_{ij} and b_{ij} are called *orthogonal* if the pairs (a_{ij}, b_{ij}) are different at all n^2 positions (i, j).

Equivalently, two Latin squares are orthogonal, if the mapping

$$(i, j) \in \{1, 2, \ldots, n\} \times \{1, 2, \ldots, n\} \mapsto (a_{ij}, b_{ij}) \in \{1, 2, \ldots, n\} \times \{1, 2, \ldots, n\} \tag{I.5.1}$$

is a bijection. The search for orthogonal Latin squares goes back to the following problem posed by Leonhard Euler in 1782:

> A very curious question, which has exercised for some time the ingenuity of many people, has involved me in the following studies, which seem to open a new field of analysis, in particular in the study of combinatorics. The question revolves around arranging 36 officers to be drawn from 6 different ranks, and at the same time from 6 different regiments, so that they are also arranged in a square so that in each line—both horizontal and vertical—there are 6 officers of different ranks and and regiments.

Euler's problem is just equivalent to finding two orthogonal Latin squares of side 6—one specifying the rank, the second specifying the regiment for each officer.

It is easy to find two orthogonal 3×3 Latin squares:

$$\begin{matrix} 1 & 2 & 3 \\ 2 & 3 & 1 \\ 3 & 1 & 2 \end{matrix} \quad \text{and} \quad \begin{matrix} 1 & 3 & 2 \\ 2 & 1 & 3 \\ 3 & 2 & 1 \end{matrix} \tag{I.5.2}$$

Problem I.5.1. *Solve Euler's problem for a 4×4 square: Take 16 cards with numbers $1, 2, 3, 4$ and suits $\diamondsuit, \heartsuit, \spadesuit, \clubsuit$ and put them into a 4×4 square such that each row and column contains exactly one number and one suit.*

Answer. Here is the solution I got in five minutes:

$$\begin{matrix} 1 & 2 & 3 & 4 \\ 2 & 1 & 4 & 3 \\ 3 & 4 & 1 & 2 \\ 4 & 3 & 2 & 1 \end{matrix} \quad \text{and} \quad \begin{matrix} 1 & 3 & 4 & 2 \\ 2 & 4 & 3 & 1 \\ 3 & 1 & 2 & 4 \\ 4 & 2 & 1 & 3 \end{matrix} \quad \text{The cards are:} \quad \begin{matrix} 1\diamondsuit & 2\spadesuit & 3\clubsuit & 4\heartsuit \\ 2\heartsuit & 1\clubsuit & 4\spadesuit & 3\diamondsuit \\ 3\spadesuit & 4\diamondsuit & 1\heartsuit & 2\clubsuit \\ 4\clubsuit & 3\heartsuit & 2\diamondsuit & 1\spadesuit \end{matrix} \tag{I.5.3}$$

Euler correctly proved that two orthogonal squares exist for every odd side $n \geq 3$ and for every side divisible by four. Furthermore, he wrote:

> I have examined a very great number of tables ... and I do not hesitate to conclude that one cannot produce an orthogonal pair of order 6, and that the same impossibility extends to $n = 6, 10, 14, \ldots$ and in general to all orders which are unevenly even.

Thus, Euler conjectured that no solution exist any $n \equiv 2 \mod 4$.

Problem I.5.2 (An attempt of solve Euler's officer problem). *Put 36 officers in a 6×6 square with places (i, j), were $i, j = 1, \ldots, 6$. We (and the officers) calculate modulo six. Regiment and rank numbers are given by*

$$\text{reg}_{ij} = \begin{cases} i + j & \textit{if } i \textit{ is even} \\ 2i + j - 1 & \textit{if } i \textit{ is odd} \end{cases}$$

$$\text{rank}_{ij} = \begin{cases} 2i + j - 2 & \textit{if } i \textit{ is even} \\ i + j - 2 & \textit{if } i \textit{ is odd} \end{cases}$$

Find out which combinations of regiment- and rank-numbers come up more than once. Which ones are missing. It is simplest to make up a 6×6 matrix indexed by the regiment- and rank-numbers and tally.

Answer. The matrices of the regiment- and rank-numbers are

$$\text{reg}_{ij} = \begin{array}{|cccccc|} \hline 2 & 3 & 4 & 5 & 6 & 1 \\ 3 & 4 & 5 & 6 & 1 & 2 \\ 6 & 1 & 2 & 3 & 4 & 5 \\ 5 & 6 & 1 & 2 & 3 & 4 \\ 4 & 5 & 6 & 1 & 2 & 3 \\ 1 & 2 & 3 & 4 & 5 & 6 \\ \hline \end{array} \quad \text{and} \quad \text{rank}_{ij} = \begin{array}{|cccccc|} \hline 6 & 1 & 2 & 3 & 4 & 5 \\ 3 & 4 & 5 & 6 & 1 & 2 \\ 2 & 3 & 4 & 5 & 6 & 1 \\ 1 & 2 & 3 & 4 & 5 & 6 \\ 4 & 5 & 6 & 1 & 2 & 3 \\ 5 & 6 & 1 & 2 & 3 & 4 \\ \hline \end{array}$$

To find out how often each combination of regiment and rank number comes up, and which do not come up at all, one has to tally these combinations. The result is:

$$\text{number of occurrences of } (\text{reg}, \text{rank}) = \begin{array}{|cccccc|} \hline 2 & 0 & 2 & 0 & 2 & 0 \\ 0 & 2 & 0 & 2 & 0 & 2 \\ 2 & 0 & 2 & 0 & 2 & 0 \\ 0 & 2 & 0 & 2 & 0 & 2 \\ 2 & 0 & 2 & 0 & 2 & 0 \\ 0 & 2 & 0 & 2 & 0 & 2 \\ \hline \end{array}$$

It is easy to check directly that $\text{reg}_{ij} + \text{rank}_{ij}$ is always even. From the tally, we see that all the combinations with $\text{reg}_{ij} + \text{rank}_{ij} = 2$ modulo 6 come up exactly twice.

About 177 years later, it was discovered that Euler's conjecture is right for order $n = 6$, but wrong for all larger orders.

Main Theorem 6 (Existence of orthogonal Latin squares). *Nonexistence of orthogonal Latin squares is a rare exception:*

G.Tarry, 1900 *There do not exist two orthogonal Latin squares of side 6.*

R.C.Bose, E.T.Parker, S.S. Shrikhande, 1959 *Two orthogonal Latin squares exist for all sides $n \neq 2$ and $n \neq 6$.*

Here are two orthogonal Latin squares of side ten:

$$\begin{array}{cccccccccc} 0&4&1&7&2&9&8&3&6&5\\ 8&1&5&2&7&3&9&4&0&6\\ 9&8&2&6&3&7&4&5&1&0\\ 5&9&8&3&0&4&7&6&2&1\\ 7&6&9&8&4&1&5&0&3&2\\ 6&7&0&9&8&5&2&1&4&3\\ 3&0&7&1&9&8&6&2&5&4\\ 1&2&3&4&5&6&0&7&8&9\\ 2&3&4&5&6&0&1&8&9&7\\ 4&5&6&0&1&2&3&9&7&8 \end{array} \quad \text{and} \quad \begin{array}{cccccccccc} 0&7&8&6&9&3&5&4&1&2\\ 6&1&7&8&0&9&4&5&2&3\\ 5&0&2&7&8&1&9&6&3&4\\ 9&6&1&3&7&8&2&0&4&5\\ 3&9&0&2&4&7&8&1&5&6\\ 8&4&9&1&3&5&7&2&6&0\\ 7&8&5&9&2&4&6&3&0&1\\ 4&5&6&0&1&2&3&7&8&9\\ 1&2&3&4&5&6&0&9&7&8\\ 2&3&4&5&6&0&1&8&9&7 \end{array} \tag{I.5.4}$$

The exact maximum number $N(n)$ of mutually orthogonal squares of side n is still not known in all cases. Especially, it is not known whether there are three mutually orthogonal squares of side 10. In 1988, after having completed the computer proof that $N(10) \leq 8$, Lam from the Concordia group writes:

> I suspect that finding three mutually orthogonal Latin squares of order 10 is even more work than what we have done.

Besides the obvious $N(3) = 2$, this would be the only case with $N(n) = 2$, possibly. [15] The following basic fact gives the connection with finite incidence geometry.

Proposition I.5.1 (The basic correspondence of affine planes and Latin squares). *From an affine plane of order n can be constructed $n-1$ pairwise orthogonal Latin squares of side n. Conversely, these $n-1$ orthogonal $n \times n$ Latin squares define an affine plane of order n.*

Reason. Take any two intersecting lines l and m. The n^2 intersection points (i,j) of these two lines and their parallels build a square grid of size $\{1,2,\dots,n\} \times \{1,2,\dots,n\}$.

Take now any pencil of n parallel lines, not parallel to neither l nor m— and number them in any way by $\{1,2,\dots,n\}$. A Latin square is defined by placing at grid point (i,j) the number of the line in the pencil passing through the grid point.

Two Latin squares a_{ij} and b_{ij} derived from two different pencils a and b cannot have the equal pairs (a_{ij}, b_{ij}) at two positions (i,j) and (i',j'). Indeed $a_{ij} = a_{i'j'}$ and $b_{ij} = b_{i'j'}$ would imply that the line through points (i,j) and (i',j') is both in the pencil a and the pencil b.

Conversely, these $n-1$ orthogonal Latin squares define $n-1$ pencils of parallel lines in $n-1$ different directions— all different from the horizontal and vertical lines of the grid. Together with the latter lines, one gets $n+1$ pencils of n parallel lines in $n+1$ different directions. These are all n^2+n lines of an affine plane with n^2 points. The order of the affine plane is n—equal to the side of the given Latin squares. □

I.5.2 Latin squares from finite fields

As explained in proposition I.5.1, there is a one-to-one correspondence between and finite affine plane of order n and a maximal set of $n-1$ orthogonal Latin squares of side n.

[15] This is my information from [6] and [27], p.783.

We now construct the Latin squares corresponding to the Cartesian plane over the finite Galois field $GF(p^r)$. The indices i and j of the rows and columns, and the matrix elements a_{ij} of the Latin squares are now elements of $GF(p^r)$.

Corollary 4 (Latin squares from finite fields). *Let the side $n = p^r$ be a prime power. For this side, there exist $n-1$ mutually orthogonal Latin squares. With addition and multiplication from the finite Galois field $GF(n)$, and slopes m running over all $p^r - 1$ non zero values $m \in GF(p^r) \setminus \{0\}$, their elements are*

$$a_{ij} = i + mj \tag{I.5.5}$$

for all $i, j \in GF(p^r)$.

Problem I.5.3. *Explain once more what are the points and lines of the affine plane given by these Latin squares. Confirm the basic counting from proposition I.3.4.*

Answer. The *points* are ordered pairs (i, j) of elements $i, j \in GF(n)$. The *lines* are the sets of points, specified as follows:
The horizontal lines are $\{(i, j) : j \in GF(n)\}$ for all $i \in GF(n)$
The vertical lines are $\{(i, j) : i \in GF(n)\}$ for all $j \in GF(n)$
For each slope $m \in GF(n) \setminus \{0\}$, a Latin square defines a set of parallel lines

$$\{(i, j) : i + mj = i_0 \ , \ i, j \in GF(n)\}$$

for all intercepts $i_0 \in GF(n)$.

(a) Each line has n points.

(b) The lines can be partitioned into $n+1$ classes— horizontal, vertical, and $n-1$ sloped pencils, each containing n parallel lines.

(c) In every point intersect $n+1$ different lines.

(d) There are $n^2 = p^{2r}$ points and $n(n+1)$ lines.

Problem I.5.4. *Find four mutually orthogonal Latin squares of side* 5.

Answer. Since 5 is a prime, the relevant Galois field is $GF(5) = \mathbf{Z}_5$. Thus the arithmetic is done modulo 5. With $m = 1, -1$, equation (I.5.5) yields the tables

0	1	2	3	4
1	2	3	4	0
2	3	4	0	1
3	4	0	1	2
4	0	1	2	3

and

0	4	3	2	1
1	0	4	3	2
2	1	0	4	3
3	2	1	0	4
4	3	2	1	0

With $m = 2, -2$, equation (I.5.5) yields the tables

0	2	4	1	3
1	3	0	2	4
2	4	1	3	0
3	0	2	4	1
4	1	3	0	2

and

0	3	1	4	2
1	4	2	0	3
2	0	3	1	4
3	1	4	2	0
4	2	0	3	1

Remark (Construction of the Galois field $GF(9)$). One needs to get a generator of the multiplicative group $GF(9) \setminus \{0\}$, which is known to be cyclic. The field $GF(9)$ is a 2-dimensional vector space over $\mathbf{Z}_3$. Hence the generator α is a zero of a quadratic irreducible polynomial $P(x) \in \mathbf{Z}_3[x]$. All elements of $GF(9)$ are zeros of the polynomial $x^9 - x$. The factoring

$$x^9 - x = x(x^4+1)(x^2+1)(x+1)(x-1)$$

implies that $P(x)$ has to be a divisor of x^4+1— otherwise the order of α would not be 8, as required for the generator. Since in $\mathbf{Z}_3[x]$ one can factor

$$x^4+1 = x^4 - 3x^2 + 1 = (x^2 - x - 1)(x^2 + x - 1)$$

there are two choices for the irreducible factor of which α is a zero. I choose $P(x) = x^2 - x - 1$, and require $\alpha^2 = \alpha + 1$. Because there exists only one unique finite field for an given order, there are always some arbitrary choices for the corresponding irreducible polynomial $P(x)$. Next we express the powers α^i in the basis $1, \alpha$:

α	α
α^2	$1+\alpha$
α^3	$1+2\alpha$
α^4	2
α^5	2α
α^6	$2+2\alpha$
α^7	$2+\alpha$
α^8	1

Problem I.5.5. *Take addition and multiplication as in* $\mathbf{Z}_2$. *The Galois field* $GF(4)$ *is* $\{0, 1, \alpha, 1+\alpha\}$, *where* α *is a generator of the multiplicative group. It satisfies* $\alpha^2 + \alpha + 1 = 0$. *Find three mutually orthogonal Latin squares of side* 4.

Answer. The equation (I.5.5), with $m = 1, \alpha, 1+\alpha$ yields the three tables below. Because of their unusual names, I have repeated the indices i and j in an extra column left and row on top. To get Latin squares in ordinary fashion, one can substitute $\alpha := 2$ and $1 + \alpha := 3$, and add one to all entries. In that way, we arrives once more at the solution given for problem I.5.1.

$m=1$	0	1	α	$1+\alpha$
0	0	1	α	$1+\alpha$
1	1	0	$1+\alpha$	α
α	α	$1+\alpha$	0	1
$1+\alpha$	$1+\alpha$	α	1	0

$m=\alpha$	0	1	α	$1+\alpha$
0	0	α	$1+\alpha$	1
1	1	$1+\alpha$	α	0
α	α	0	1	$1+\alpha$
$1+\alpha$	$1+\alpha$	1	0	α

$m=1+\alpha$	0	1	α	$1+\alpha$
0	0	$1+\alpha$	1	α
1	1	α	0	$1+\alpha$
α	α	1	$1+\alpha$	0
$1+\alpha$	$1+\alpha$	0	α	1

Problem I.5.6. *Find eight mutually orthogonal Latin squares of side* 9. *Put as indices of the rows and columns the field elements* $0, \alpha^1, \alpha^2, \ldots \alpha^8$. *Calculate with slopes* $m = \alpha^\mu$. *Get the symmetric square for* $\mu = 8$, *and explain how the other ones for* $\mu = 1, \ldots 7$ *are obtained from it.*

Answer. The elements of the Latin squares become

$$\begin{aligned} a_{00} &= 0 & & \\ a_{i0} &= \alpha^i & \text{for } & i = 1, \ldots, 8 \\ a_{0j} &= \alpha^{\mu+j} & \text{for } & j = 1, \ldots, 8 \\ a_{ij} &= \alpha^i + \alpha^{\mu+j} & \text{for } & i, j = 1, \ldots, 8 \end{aligned}$$

For $\mu = 8$ one gets a symmetric square

0	α	α^2	α^3	α^4	α^5	α^6	α^7	α^8
α	α^5	α^3	α^8	α^7	0	α^4	α^6	α^2
α^2	α^3	α^6	α^4	α	α^8	0	α^5	α^7
α^3	α^8	α^4	α^7	α^5	α^2	α	0	α^6
α^4	α^7	α	α^5	α^8	α^6	α^3	α^2	0
α^5	0	α^8	α^2	α^6	α	α^7	α^4	α^3
α^6	α^4	0	α	α^3	α^7	α^2	α^8	α^5
α^7	α^6	α^5	0	α^2	α^4	α^8	α^3	α
α^8	α^2	α^7	α^6	0	α^3	α^5	α	α^4

The other seven square are obtained by cyclic permutation of the last seven columns, leaving the first column fixed.

Problem I.5.7. *Follow the description from problem* (I.5.6) *and really write out the eight Desarguesian Latin squares of side* 9.

Answer. For $\mu = 8$ one gets a symmetric square. And for $\mu = 1$ one gets

0	1	2	3	4	5	6	7	8
1	5	3	8	7	0	4	6	2
2	3	6	4	1	8	0	5	7
3	8	4	7	5	2	1	0	6
4	7	1	5	8	6	3	2	0
5	0	8	2	6	1	7	4	3
6	4	0	1	3	7	2	8	5
7	6	5	0	2	4	8	3	1
8	2	7	6	0	3	5	1	4

0	2	3	4	5	6	7	8	1
1	3	8	7	0	4	6	2	5
2	6	4	1	8	0	5	7	3
3	4	7	5	2	1	0	6	8
4	1	5	8	6	3	2	0	7
5	8	2	6	1	7	4	3	0
6	0	1	3	7	2	8	5	4
7	5	0	2	4	8	3	1	6
8	7	6	0	3	5	1	4	2

For $\mu = 2$ and $\mu = 3$ one gets

0	3	4	5	6	7	8	1	2
1	8	7	0	4	6	2	5	3
2	4	1	8	0	5	7	3	6
3	7	5	2	1	0	6	8	4
4	5	8	6	3	2	0	7	1
5	2	6	1	7	4	3	0	8
6	1	3	7	2	8	5	4	0
7	0	2	4	8	3	1	6	5
8	6	0	3	5	1	4	2	7

0	4	5	6	7	8	1	2	3
1	7	0	4	6	2	5	3	8
2	1	8	0	5	7	3	6	4
3	5	2	1	0	6	8	4	7
4	8	6	3	2	0	7	1	5
5	6	1	7	4	3	0	8	2
6	3	7	2	8	5	4	0	1
7	2	4	8	3	1	6	5	0
8	0	3	5	1	4	2	7	6

For $\mu = 4$ and $\mu = 5$ one gets

0	5	6	7	8	1	2	3	4
1	0	4	6	2	5	3	8	7
2	8	0	5	7	3	6	4	1
3	2	1	0	6	8	4	7	5
4	6	3	2	0	7	1	5	8
5	1	7	4	3	0	8	2	6
6	7	2	8	5	4	0	1	3
7	4	8	3	1	6	5	0	2
8	3	5	1	4	2	7	6	0

0	6	7	8	1	2	3	4	5
1	4	6	2	5	3	8	7	0
2	0	5	7	3	6	4	1	8
3	1	0	6	8	4	7	5	2
4	3	2	0	7	1	5	8	6
5	7	4	3	0	8	2	6	1
6	2	8	5	4	0	1	3	7
7	8	3	1	6	5	0	2	4
8	5	1	4	2	7	6	0	3

For $\mu = 6$ and $\mu = 7$ one gets

0	7	8	1	2	3	4	5	6
1	6	2	5	3	8	7	0	4
2	5	7	3	6	4	1	8	0
3	0	6	8	4	7	5	2	1
4	2	0	7	1	5	8	6	3
5	4	3	0	8	2	6	1	7
6	8	5	4	0	1	3	7	2
7	3	1	6	5	0	2	4	8
8	1	4	2	7	6	0	3	5

0	8	1	2	3	4	5	6	7
1	2	5	3	8	7	0	4	6
2	7	3	6	4	1	8	0	5
3	6	8	4	7	5	2	1	0
4	0	7	1	5	8	6	3	2
5	3	0	8	2	6	1	7	4
6	5	4	0	1	3	7	2	8
7	1	6	5	0	2	4	8	3
8	4	2	7	6	0	3	5	1

I.5.3 Finite Non-Desarguesian planes

The first example of a finite non-Desarguesian plane was published in 1907 by O. Veblen and J.H.M. Wedderburn [34]. Indeed, these authors have obtained four non-isomorphic projective planes of order 9. In the following pages, I give an independent access to part of Veblen's arguments.

Remark (Conjugations in the Galois field $GF(9)$). A *conjugation* is defined as an automorphism of a field, which is its own inverse, but not the identity. The complex conjugation in the complex number field is the most common example. The Galois field $GF(9)$ has indeed a conjugation. The multiplicative group $GF(9) \setminus \{0\}$ is known to be cyclic. As above, we use a generator α. The mapping defined by

$$\overline{\alpha^i} := \alpha^{si} \quad \text{for all } i = 1, \dots, 8 \quad \text{and } \overline{0} := 0 \tag{I.5.6}$$

is an automorphism of the multiplicative group, for $s = 3, 5, 7$, since these numbers are relatively prime to the order 8 of the multiplicative group. Since one calculates modulo 3 in the field $GF(9)$

$$(\alpha^i + \alpha^j)^3 = \alpha^{3i} + \alpha^{3j} \quad \text{for all } i, j = 1, \dots, 8$$

follows directly from the binomial theorem. Hence the mapping (I.5.6) is a field automorphism. It is indeed the well-known Frobenius automorphism.

Now one can check that $\overline{1 + \alpha} = -(1 + \alpha)$, which makes $1 + \alpha$ take the role of the imaginary unit for the present conjugation.

Problem I.5.8. *Check that neither the choice $s = 5$ nor $s = 7$ leads to a field automorphism.*

Remark (The perfect squares in the Galois field $GF(p^r)$). Given is any $a \in GF(p^r)\setminus\{0\}$. A necessary condition for solvability of the equation

$$x^2 = a \tag{I.5.7}$$

follows from the fact that $p^r - 1$ is the order of the multiplicative group $GF(p^r) \setminus \{0\}$:

$$a^{\frac{p^r-1}{2}} = 1 \tag{I.5.8}$$

is necessary for equation (I.5.7) to have a solution. Since the equation (I.5.8) holds for exactly half of the members of the multiplicative group $GF(p^r) \setminus \{0\}$, and, too, the equation (I.5.7) is solvable for right-hands sides a in exactly half of the members of this group, one concludes that a is a perfect square if and only if relation (I.5.8) holds.

Remark (A second algebraic structure in the Galois field $GF(9)$). We define the addition as before, and the new *pseudo-multiplication* by setting

$$a\Diamond b := \begin{cases} a \cdot b, & \text{if } a \text{ is a perfect square} \\ a \cdot \bar{b}, & \text{if } a \text{ is not a perfect square} \end{cases} \tag{I.5.9}$$

With that new pseudo-multiplication, $GF(p^r) \setminus \{0\}$ is a <u>non-commutative</u> group. But only the right-handed distributive law

$$a\Diamond(b+c) = a\Diamond b + a\Diamond c \tag{I.5.10}$$

holds, whereas the left-handed distributive law is violated.

Question. Find a formula for the inverse of the pseudo-multiplication.

Question. Check that the pseudo-multiplication $\Diamond$ makes the set $GF(p^r) \setminus \{0\}$ a <u>non-commutative</u> group.

Question. Give the reason for the right-handed distributive law.

Question. Give an example where the left-handed distributive law does not hold.

Answer. Since for example $\alpha^2 + \alpha^4 = \alpha$, the sum of two perfect squares need not be a perfect square. Especially

$$(\alpha^2 + \alpha^4)\Diamond\alpha \neq \alpha^2\Diamond\alpha + \alpha^4\Diamond\alpha$$

Problem I.5.9. *As above, we use pairs of the field elements $0, \alpha^1, \alpha^2, \dots \alpha^8$ as coordinates, and as indices for the matrix elements of the Latin squares, too. Find eight Latin squares of side 9, which are essentially different from* (I.5.2). *Construct the sloped lines by means of the pseudo-multiplication.*

Answer. The elements of the of Latin squares become

$$\begin{aligned} a_{00} &= 0 \\ a_{i0} &= \alpha^i && \text{for } i = 1, \dots, 8 \\ a_{0j} &= \alpha^\mu\Diamond\alpha^j && \text{for } j = 1, \dots, 8 \\ a_{ij} &= \alpha^i + \alpha^\mu\Diamond\alpha^j && \text{for } i, j = 1, \dots, 8 \end{aligned} \tag{left-slope}$$

For even μ, one gets the same Latin squares as in the Desarguesian case given above in (I.5.2). The odd μ, the new j-th column is the old $(3j \bmod 8)$-th column. Thus columns 1 and 3, columns 2 and 6, as well as columns 5 and 7 are exchanged.

Problem I.5.10. *The following basic facts were checked in problem* (I.5.3) *for the Desarguesian case.*

(a) *The lines are the sets of $n = p^r$ points each.*

(b) *There are $n+1$ pencils of n parallel lines.*

(c) *In each point, $n+1$ lines intersect.*

(d) *Together, there are $n^2 = p^{2r}$ points and $n^2 + n$ lines.*

Convince yourself that these facts are still valid for non-Desarguesian Latin squares.

Answer.

Are the items (a) through (d) strong enough to get an affine plane? Clearly not, indeed, we need check axioms (A.1), (A.2) and (A.3).

Problem I.5.11. *Check axiom* (A.1)*—existence and uniqueness of the line between any two points (i, j) and (i', j').*

Answer. The case that either $i = i'$ or $j = j'$ leads to a unique horizontal or vertical line. We now assume that $i \neq i'$ and $j \neq j'$. If these two points lie on a line, it has to have some slope $m \neq 0$ such that

$$i + m\Diamond j = i' + m\Diamond j'$$

Because of the right-handed distributive law (I.5.10), we conclude that

$$i - i' = m\Diamond(j - j')$$

Because the pseudo-multiplication $\Diamond$ makes the set $GF(n) \setminus \{0\}$ a multiplicative group, and both $i - i' \neq 0$, $j - j' \neq 0$, one can solve this equation uniquely for the slope m. Now it is easy to get the unique line connecting the two given points (i, j) and (i', j').

Axiom (A.3) is indeed even easier to check.

Problem I.5.12. *Check axiom (A.3).*

Answer. The three points with coordinates $(0, 0), (0, n), (n, 0)$ do not lie on a line.

This is a good step forward. Still, we are left with the problem to check axiom (A.2)—existence and uniqueness of the parallel.

Axiom (A.1) clearly implies that two different lines can intersect in at most one point. The existence of a parallel follows to any given line through any given point is stated in item (c).

The hard part is to confirm uniqueness of the parallel. As Veblen has observed, here simple counting helps better than calculation.

Problem I.5.13. *Show that any finite system of points and lines, which satisfies items (a), (d) and axioms* (A.1) *and* (A.3), *satisfies axiom* (A.2), *too.*

Answer. Axiom (A.1) clearly implies that two different lines can intersect in at most one point. The existence of a parallel to any given line through any given point is stated in item (b).

The hard part is to confirm uniqueness of the parallel. Towards a contradiction, we assume that there exists a line l and a point P not on line l, and two parallels p and q to line l pass through point P. A contradiction arises because so many lines would lead to more points than there exist. Here are the details:

By item (a), there are n points $L_1, \dots, L_n$ on line l. On each of the n different lines $L_1P, \dots, L_nP$ there exist $n-2$ points different from $L_1, \dots, L_n, P$. All these points are different, yielding $n(n-1)+1$ points. Too, there are $n-1$ points on the parallel p, which are different from P and the points above. Altogether, we have found $n(n-1)+1+n-1 = n^2$ points. But by item (d), these are all the points that exist. Hence point P is the only point on the second parallel q, contradicting item (a).

Here is an alternative to Veblen's argument to confirm axiom (A.2).

Problem I.5.14. *Why does a direct calculation fail to yield uniqueness of parallels?*

Answer. One needs to show that two lines of different slopes $m \neq n$ always intersect. The case involving a horizontal or vertical line is easy. Let these two lines be given by equations

$$i + m\diamondsuit j = i_0 \quad \text{and} \quad i + n\diamondsuit j = i_1$$

One concludes that

$$m\diamondsuit j - n\diamondsuit j = i_0 - i_1 \tag{I.5.11}$$

But then one is stuck. Because there is no left-hand distributive law, one cannot solve for j. One cannot tell whether the two lines are parallel or do intersect.

Problem I.5.15. *How can one get around the problem, deal with equation* (I.5.11), *and confirm axiom (A.2) nevertheless?*

Answer. Again, we assume that $m \neq n$. One needs to show that the mapping

$$j \in GF(n) \mapsto m\diamondsuit j - n\diamondsuit j \in GF(n) \tag{I.5.12}$$

is a bijection. Because the field is finite, it is enough to show that this mapping is injective. Hence we assume

$$m\diamondsuit j - n\diamondsuit j = m\diamondsuit j' - n\diamondsuit j'$$

and check whether j and j' are equal. One gets immediately

$$m\diamondsuit j - m\diamondsuit j' = n\diamondsuit j - m\diamondsuit j'$$

But now the right-hand distributive law can be used, again. Hence

$$m\diamondsuit(j - j') = n\diamondsuit(j - j')$$

If $j \neq j'$, the difference $j - j'$ would have a multiplicative inverse, and $m = n$ would follow, contradicting the assumption $m \neq n$. Hence one concludes that $j = j'$, which confirms that the mapping defined by (I.5.12) is injective. As a map between finite sets of the same size, it is a bijection, too. Hence equation (I.5.11) can be uniquely solved for j. One has proved this fact even without being able to write a formula for the solution. Thus two lines with different slopes always intersect, confirming axiom (A.2)

Problem I.5.16. *Follow the description from problem* (I.5.9) *and really write out the eight non-Desarguesian Latin squares of side* 9.

Answer. For even μ, one gets the same Latin squares as in the Desarguesian case given above in (I.5.2). The odd μ, the new j-th column is the old $3j \mod 8$-th column. Thus columns 1 and 3, columns 2 and 6, as well as columns 5 and 7 are exchanged.

Veblen [34] even constructs two further non-Desarguesian affine planes of order 9. Finally, let's make the point! Veblen's ingenious example now begs the question:

Open Problem III. For which orders do there exist affine planes at all?

It is a long-standing conjecture that affine planes exist only if the order is a prime power. The best result toward a solution of this conjecture is the Bruck-Ryser Theorem:

Main Theorem 7 (Nonexistence of projective planes (R.H. Bruck and H.J.Ryser), 1949). *No projective plane with order M does exist, if the number number satisfies the following three conditions:*

(a) *M is not a prime power.*

(b) *$M \equiv 1 \mod 4$ or $M \equiv 2 \mod 4$.*

(c) *$M \neq a^2 + b^2$ for all integers a and b.*

Problem I.5.17. *Use the Bruck-Ryser Theorem to show that no projective planes of order $M \equiv 6 \mod 8$ can exist.*

Answer. One needs to check the three conditions:

(a) Indeed M is not a prime power, since $M = 2m$ and $m \equiv 3 \mod 4$ is odd.

(b) $M \equiv 2 \mod 4$ does hold.

(c) Assume towards a contradiction that $M = A^2 + B^2$. Both A and B would be odd, and $m = \frac{M}{2} = a^2 + b^2$ would hold with $a = \frac{A+B}{2}$ and $b = \frac{A-B}{2}$. Since $m \equiv 3 \mod 4$ is odd, one number among a and b is even, and the other one is odd. But this implies $a^2 + b^2 \equiv 1 \mod 4$ contradicting $m \equiv 3 \mod 4$.

The smallest orders for which the existence of a projective plane remains open at this point are

$$10, 12, 15, 18, 20, \ldots$$

Only the smallest number in this list has been settled:

Main Theorem 8 (Lam, Thiel and Swiercz, 1989). *There exist no projective plane of order* 10.

The computer proof of this theorem was started by Lam and McKay at Concordia in 1979. The final run on a Cray supercomputer was running from September 1986 until November 1988. Lam estimates that the proof involved *about two orders of magnitude more computing power* than the proof of the famous Four Color Theorem by Wolfgang Haken and Kenneth Appel in 1976. The Concordia group investigated different ways of attacking the problem. Lam writes

> The final program we wrote is based on a lot of experimentation. If we had just written a program without doing the estimation, we'd probably have a program that was too long to be run anywhere.

I.5.4 A note on projective spaces

The n-dimensional projective space over a field $\mathbf{F}$ is defined quite similar to a projective plane (see Definition I.3.20). The "points" of the projective space $\mathbf{PF}^n$ are the sets of equivalent $n+1$-tuples $(x_1\lambda, x_2\lambda, \dots x_{n+1}\lambda)$, where $(x_1, x_2, \dots x_{n+1}) \in \mathbf{F}^{n+1}$ and λ runs over the nonzero elements of $\mathbf{F}$. In this manner, one uses homogeneous coordinates and an extra $n+1$-th dimension. As a shorthand one writes

$$\mathbf{PF}^n = \frac{\mathbf{F}^{n+1} \setminus \{0\}}{\mathbf{F} \setminus \{0\}}$$

The k-dimensional subspaces of $\mathbf{PF}^n$ are given in homogeneous coordinates by the $k+1$-dimensional subspaces through the origin in $\mathbf{F}^{n+1}$. This interpretation works for $k = 0, 1, \dots n$.

The n-dimensional projective space over some finite Galois field $GF(q)$ is denoted by $PG(n, q)$.

Problem I.5.18. *Count the points of the projective space $PG(n, q)$.*

Answer. The n-dimensional projective space over the Galois field $GF(q)$ has $\frac{q^{n+1}-1}{q-1}$ points since the Galois field has q points.

In general, the number k-dimensional subspaces of $PG(n, q)$ is given by the product

$$\binom{n+1}{k+1}_q = \prod_{i=0}^{k} \frac{q^{n+1-i}-1}{q^{i+1}-1} = \frac{(q^{n+1}-1)(q^n-1)\cdots(q^{n+1-k}-1)}{(q-1)(q^2-1)\cdots(q^{k+1}-1)}$$

This is a Gaussian binomial coefficient, a q-analogue of a binomial coefficient. In the limit $q \to 1$, one gets back to the ordinary binomial coefficient.

Main Theorem 9 (Veblen-Young Theorem). *Every finite projective space of geometric dimension $n \geq 3$ is isomorphic to a n-dimensional projective space over some finite Galois field $GF(q)$; thus isomorphic to one to the spaces $PG(n, q)$.*

Remark. Note that this result holds only for dimension at least 3. In sharp contrast, the projective planes are much harder to classify, as not all of them are isomorphic with one of the $PG(2, q)$. The Desarguesian planes are those which are isomorphic with a $PG(2, q)$. These are exactly those planes satisfying Desargues's theorem. But there exist non-Desargesian planes—and even the prime power conjecture about their order is still open.

The smallest 3-dimensional projective space is built over the field $GF(2) = \mathbf{Z}_2$ and is denoted by $PG(3, 2)$. It has 15 points, 35 lines, and 15 planes. Each of the 15 planes contains 7 points and 7 lines. As geometries, these planes are isomorphic to the Fano plane. Every point of $PG(3, 2)$ is contained in 7 lines and every line contains three points. In addition, two distinct points are contained in exactly one line and two planes intersect in exactly one line. In 1892, Gino Fano was the first to consider such a finite geometry – a three dimensional geometry containing 15 points, 35 lines, and 15 planes, with each plane containing 7 points and 7 lines.

Problem I.5.19 (Kirkman's Schoolgirl Problem 1850). *Kirkman's schoolgirl problem is a problem in combinatorics proposed by* Rev. Thomas Penyngton Kirkman *in 1850 as Query VI in* The Lady's and Gentleman's Diary *(pg.48). The problem states:*

> *Fifteen young ladies in a school walk out three abreast for seven days in succession: it is required to arrange them daily so that no two shall walk twice abreast.*

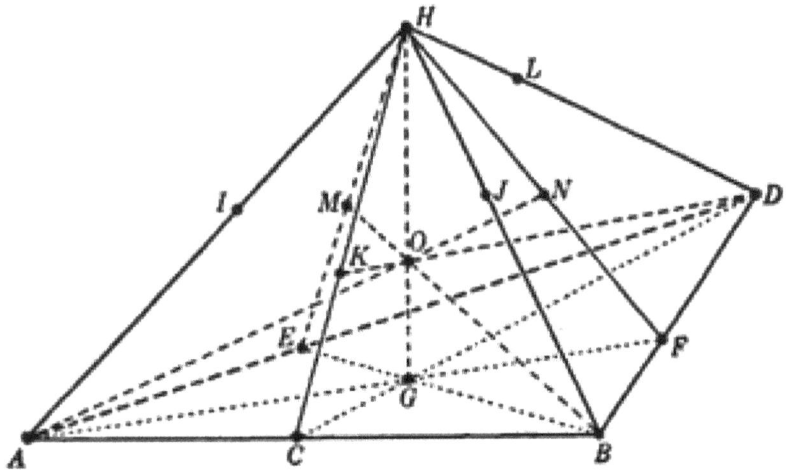

Figure I.5.1: $PG(3,2)$ but not all the lines are drawn.

Answer. There are 7 days of the week, and 3 girls in each of 5 group. There are $35 = 7 \times 5$ different combinations for three girls to walk together.

Two of the seven non-isomorphic solutions of Kirkman's schoolgirl problem provide a visual representation of the Fano 3-space $PG(3,2)$. Some diagrams for this problem can be found at:

Beutelspacher, Albrecht; Rosenbaum, Ute (1998), Projective geometry: from foundations to applications, Cambridge University Press,

ISBN 978-0-521-48277-6; 978-0-521-48364-3, MR1629468

Each color represents the day of the week (seven colors, blue, green, yellow, purple, red, black, and orange). The definition of a Fano space says that each line is on three points. The figure represents this showing that there are 3 points for every line. This is the basis for the answer to the schoolgirl problem. This figure is then rotated 7 times. There are 5 different lines for each day, multiplied by 7 (days) and the result is 35. Then, there are 15 points, and there are also 7 starting lines on each point. This then gives a representation of the Fano 3-space, PG(3,2).

http://mathworld.wolfram.com/KirkmansSchoolgirlProblem.html

gives the solution

Sun	ABC	DEF	GHI	JKL	MNO
Mon	ADH	BEK	CIO	FLN	GJM
Tue	AEM	BHN	CGK	DIL	FJO
Wed	AFI	BLO	CHJ	DKM	EGN
Thu	AGL	BDJ	CFM	EHO	IKN
Fri	AJN	BIM	CEL	DOG	FHK
Sat	AKO	BFG	CDN	EIJ	HLM

http://en.wikipedia.org/wiki/Kirkman%27s_schoolgirl_problem#Notes

http://en.wikipedia.org/wiki/Thomas_Kirkman

http://en.wikipedia.org/wiki/Finite_geometry

Remark. The problem can be generalized to n girls walking in triplets for $(n-1)/2$ days, with the requirement, again, that no pair of girls walk in the same row twice. The number n has to be an odd multiple of 3: thus $n \equiv 3 \mod 6$. It is this generalization of the problem that Kirkman discussed first, while the famous special case $n = 15$ was only proposed later. A complete solution to the general case was given by D. K. Ray-Chaudhuri and R. M. Wilson [15] in 1968, but had already been settled by Lu Jiaxi [16] in 1965.

Part II

Neutral Geometry

II.6 The Axioms of Order and Their Consequences

Definition II.6.1 (Segment). Let A and B be two distinct points. The *segment* AB is the set consisting of the points A and B and all points lying between A and B. The points A and B are called the *endpoints* of the segment, the points between A and B are called the *interior points*, and the remaining points on the line AB are called the *exterior* points of the segment.

Definition II.6.2 (Triangle). We define a *triangle* to be the union of the three segments AB, BC and CA. The three points A, B and C are assumed not to lie on a line. These three points are the *vertices*, and the segments BC, AC, and AB are the *sides* of the triangle.

For a segment, it is assumed that the two endpoints A and B are *different.* For a triangle $\triangle ABC$, it is assumed that the three vertices *do not lie on a line.*

Remark. For the correct statement of the Theorems about congruence, similarity of two triangles, as well as for the Theorem of Desargues, and for the purpose of fixing an orientation, we assign a definite order to the vertices and sides of the relevant triangles.

II.6.1 Order of points on a line

Proposition II.6.1 (Hilbert's Proposition 3). *For any two distinct points A and C, there exists at least one point B on the line AC lying between A and C.*

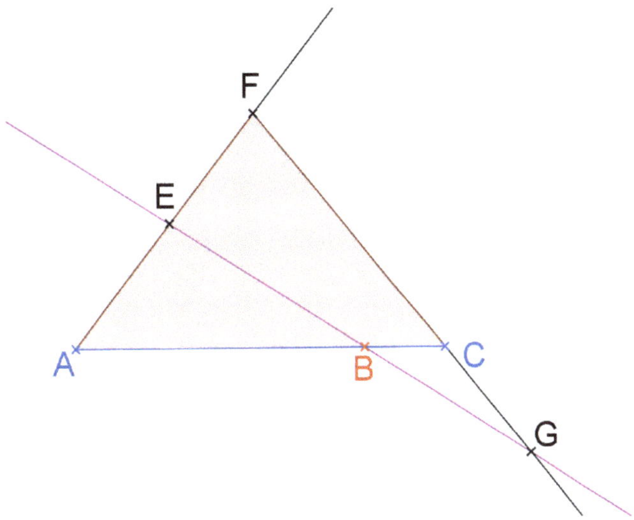

Figure II.6.1: How to get a point inside the given segment AC.

Proof. By the axiom of incidence (I.3), there exists a point E not lying on the line AC, and by axiom of order (II.2), there exists a point F such that E is a point inside the segment AF. By the same axiom, there exists a point G such that C is a point inside segment FG. Now we use Pasch's axiom (II.4) for the triangle $\triangle ACF$ and line EG.

Question. Why is $\triangle ACF$ a triangle?

Answer. Points A and C are two different points by assumption. The third vertex F does not lie on line AC— pother point E would lie on that line, too, contrary to the construction.

Question. Why are E and G two different points?

Answer. Assume towards a contradiction $E = G$. In that case, lines AF and CF would intersect in that point, hence $F = E = G$, contradicting the definition of point F.

For the triangle $\triangle ACF$ and line EG, Pasch's axiom yields that the line intersects a second side of the triangle besides side AF. But line EG does not intersect segment FC, because the intersection point of the lines EG and FC is G, which lies outside the segment FC. Hence line EG intersects the third side of triangle $\triangle ACF$, which is segment AC. The intersection point B is a point between A and C, existence of which was to shown. □

Definition II.6.3 (Ray). Given two distinct points A and B, the *ray* $\overrightarrow{AB}$ is the set consisting of the points A and B, the points inside the segment AB, and all points P on the line AB such that the given point B lies between A and P. The point A is called the *vertex* of the ray.

Problem II.6.1. *Explain why the ray $\overrightarrow{AB}$ contains more points than the segment AB.*

Answer. The axiom of order (II.2) implies that the ray $\overrightarrow{AB}$ contains a point C such that $A * B * C$. The axiom of order (II.1) implies that A, B, C are three distinct points lying on line AB. The axiom of order (II.3) implies that point C does not satisfy $A * C * B$ and hence does not lie in the segment AB.

Proposition II.6.2 (Hilbert's Proposition 4, also called "Three-point Theorem"). *Among any three different points A, B and C lying on a line, there exists exactly one lying between the two other points.*

Remark. In the first edition of the foundations of geometry, proposition II.6.2 had been postulated as an axiom. The proof given below was published in 1902 by E. H. Moore [9].

Proof. The axiom or order (II.3) states that at most one of the three points lies between the two others. We need to prove that actually one of the three points does lie between the two others. Assume that neither A not C lies between the two other points. The construction shows that B does lie between A and C. One chooses a point D not lying on line AC. Next we choose a point G on the ray $\overrightarrow{BD}$ such that D lies between B and G. Being used repeatedly, we have to mention the follow consequence of axioms (II.3) and (II.4):

A consequence of Pasch's axiom. If a line cuts one side of a triangle, and the extension of a second side, then the line cuts the third side.

Now we use Pasch's axiom for the triangle $\triangle BCG$ and the line AD. Because this line intersects side BG, but not side BC, the line intersects the third side CG, say in point E.

A similar application of Pasch's axiom, now for triangle $\triangle ABG$ and line CD yields existence of a point F between A and G.

Now use Pasch's axiom a third time—for triangle $\triangle AEG$ and the line CF. Because this line intersects side AG, but not side GE, the line intersects the third side, which is the <u>segment</u> AE. But this intersection point has to be the intersection of line CF with <u>line</u> AE, which his point D. Hence point D lies between A and E.

Finally use Pasch's axiom a fourth time—for triangle $\triangle AEC$ and the line GB. Because this line intersects side AE, but not side CE, the line intersects the third side AC. But this intersection point has to be the intersection of lines GB and AC, which his point B. Hence point B lies between A and C. □

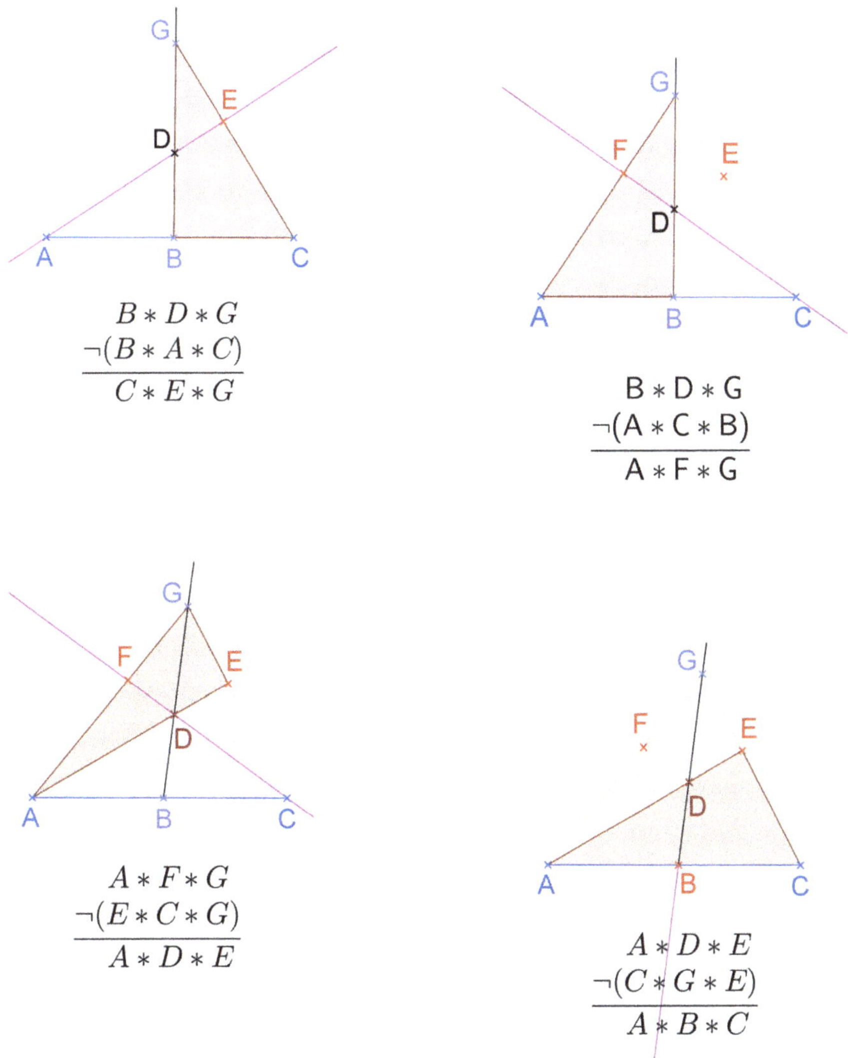

Figure II.6.2: Assume that neither A nor C lie between the two other of the three points A, B and C. The construction shows that B does lie between A and C.

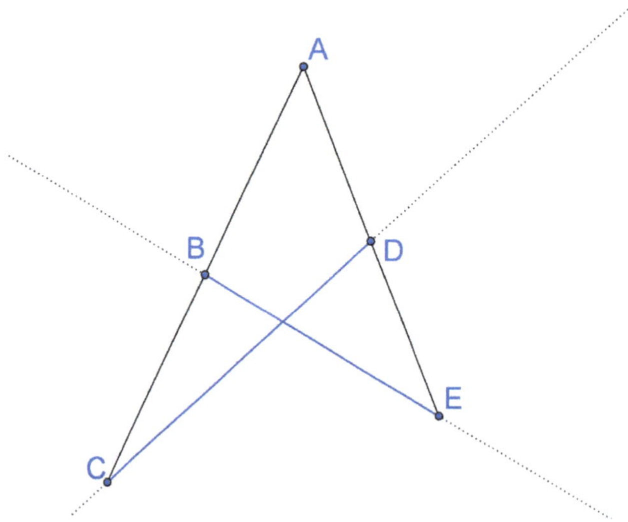

Figure II.6.3: Two intersecting segments.

Problem II.6.2. *We assume $A * B * C$ and $A * D * E$, and points A, C, E do not lie on a line. Prove that the segments BE and CD intersect each other. Provide a drawing.*

(i) *Why are the two lines BE and CD different?*

(ii) *Why does the segment BE intersect the line CD?*

(iii) *Why does the segment CD intersect the line BE?*

(iv) *Why are these two intersection points equal? Why do the segments BE and CD intersect each other?*

Answer. **(i)** The two lines CD and BE are distinct, otherwise all points A, B, C, D, E would lie on a line.

(ii) We use Pasch's axiom for triangle $\triangle ABE$ and line CD. This line intersects side AE, but not side AB. Hence line CD intersects the side BE, say at point X.

(iii) We use Pasch's axiom for triangle $\triangle ACD$ and line BE. This line intersects side AC, but not side AD. Hence line BE intersects the side CD, say at point Y.

(iv) By item (i), the lines CD and BE are different. Hence they have at most one intersection point. By items (ii) and (iii), they indeed intersect, and their unique intersection point is $X = Y$. Hence this point X is the intersection of the segments BE and CD.

II.6.2 Bernays' Lemma

One can say that Pasch's axiom tells: a line entering a triangle has to leave it, too. It can be proved that a line cannot intersect all three sides of a triangle.

Proposition II.6.3 (Bernays' Lemma). *A line cannot cut all three sides of a triangle.*

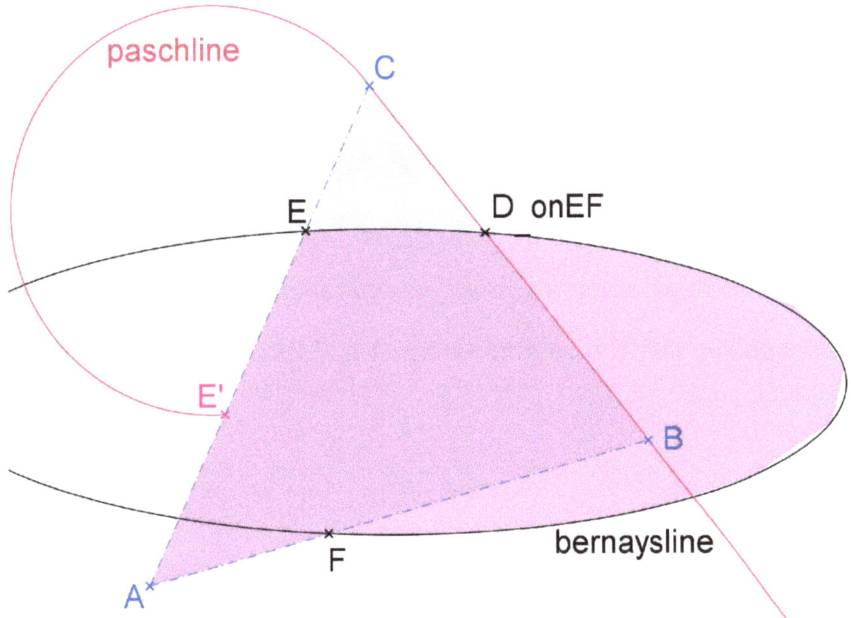

Figure II.6.4: Bernays' Lemma would not hold— and line A, E', E, C would be equal to line E', C, D, B—all points collapse onto one line.

Proof. [16] We assume that line l intersects all three sides of triangle $\triangle ABC$ and derive a contradiction. If the sides BC, CA and AB would intersect line l in the points D, E and F, these three points would be all different.

By Proposition 4, one of the three points lies between the two others. Without loss of generality, we can assume that point D lies between E and F. We now apply Pasch's axiom to the triangle $\triangle AEF$ [17] and the line BC. This line cut the side EF in point D. Hence it intersects a second side, say

side AE in point E' or

side AF in a point F'.

In both cases we get a contradiction.

Here are the details for the first case: $E' \neq C$, because E' lies inside the segment AE, but C lies on the ray $\overrightarrow{AE}$ outside that segment. The four points A, E', E, C lie on one line. Too, the four points E', C, D, B lie on one line. But these two lines have the two points E' and C in common. Hence they are identical. Thus all three points A, B, C lie on one line, contradicting the assumption that $\triangle ABC$ is a triangle.

In the second case, one gets a similar contradiction: $F' \neq B$, because F' lies inside the segment AF, but B lies on the ray $\overrightarrow{AF}$ outside that segment. The four points A, F', F, B lie on one line. Too, the four points F', B, D, C lie on one line. But these two lines have the two points F' and B in common. Hence they are identical. Thus all three points A, B, C lie on one line, contradicting the assumption that $\triangle ABC$ is a triangle. Hence a line cannot intersect all three sides of a triangle. □

[16] Hilbert credits this proof to his collaborator Paul Bernays.

[17] The figure does not show this triangle correctly

II.6.3 Plane separation

Definition II.6.4. Given is plane α and a line a lying in this plane. We say that the points A and A' lie on *the same side* of line a—or in the same half plane— in case that segment AA' does not intersect the line a. The points A and B lie *on different sides* of line a in case that the segment AB does intersect the line a. The two open regions one gets in this way are called the *half-planes* bounded by the line a. [18]

The next theorem shows that any line lying in plane separates this plane into *exactly two* half-planes.

Proposition II.6.4 (Hilbert's Proposition 8, the "Plane separation Theorem"). *Each line a, lying in a plane α, separates the points of this plane which do not lie on the line, into two nonempty regions $\mathcal{R}$ and $\mathcal{S}$ called* half planes *which have the following properties:*

(i) *every point lying in the plane α lies either on the line a, or in the half plane $\mathcal{R}$, or in the half-plane $\mathcal{S}$, but not in any two of these three sets;*

(ii) *for any two points A in the half-plane $\mathcal{R}$ and B in the half-plane $\mathcal{S}$, the segment AB intersects line a;*

(iii) *for any two points A and A' in the half-plane $\mathcal{R}$, the segment AA' does not intersect the line a;*

(iv) *for any two points B and B' in the half plane $\mathcal{S}$, the segment BB' does not intersect the line a.*

The following proof of the plane separation theorem uses only the axioms of order and Bernay's Lemma.

A simple proof of the plane separation theorem. By axiom (I.3a), there exist two points $P \neq Q$ on the line a. By Proposition I.3.9, there exist three points in the plane α, not lying on a line. [19]. Thus there exists a third point R in the plane α, not lying on the line a. The three points P, Q and R span the plane α.

We can now separate the points of the plane α into those lying on the line a and, additionally, the following two sets:

$$\mathcal{R} := \{A \in \alpha \,:\, A \notin a, \text{ the segment } RA \text{ does not intersect the line } a\}$$
$$\mathcal{S} := \{B \in \alpha \,:\, B \notin a, \text{ the segment } RB \text{ intersects the line } a\}$$

By axiom (II.2), there exist a point S such that $R * P * S$. In other words, the point S lies on the extension of the ray $\overrightarrow{RP}$ beyond point P.

Similarly, we get point T such that $R * Q * T$. Since the points P, Q and R do not lie on a line, neither the three points R, S and T lie on a line.

Since $R \in \mathcal{R}$, and $S, T \in \mathcal{S}$, both sets are nonempty. From the definition, it is clear that each point of the plane α lies in exactly one of the three sets $\mathcal{R}$, a or $\mathcal{S}$. Thus item (i) has been checked.

In the special case that $A = R$ or $A' = R$, the statements (ii) and (iii) follow from the definition. Moreover, we use repeatedly the following statement:

[18] More exactly: half-planes of the plane α bounded by the line a.

[19] For purely two-dimensional geometry, we get this statement directly from axiom (I.3b)

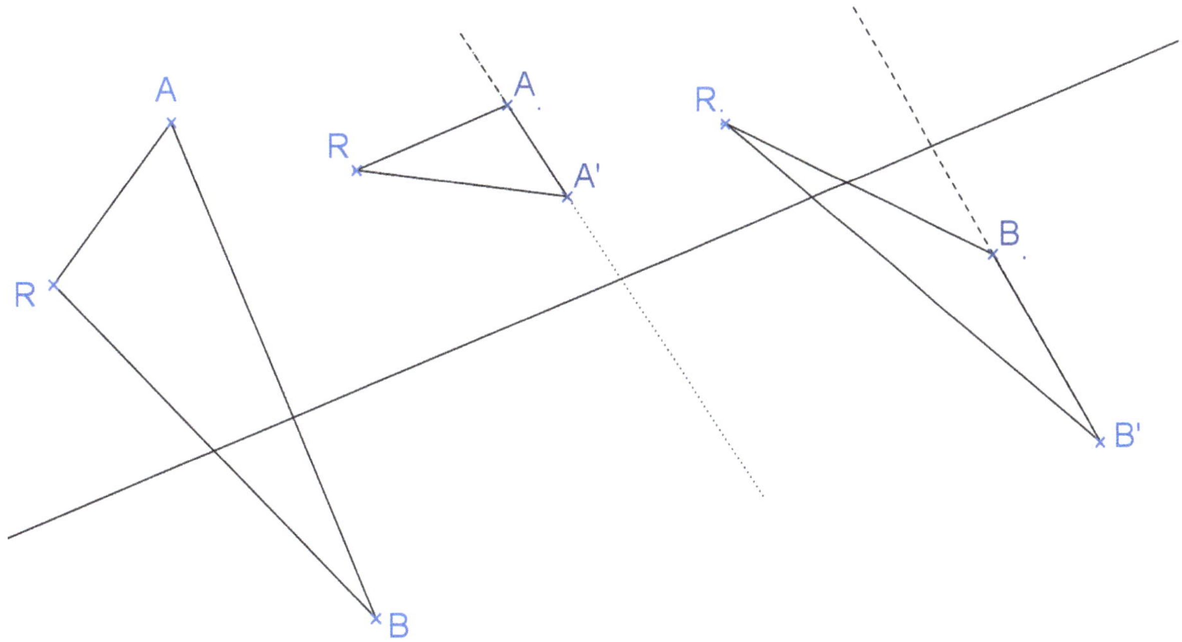

Figure II.6.5: Generic cases in the proof of the plane separation theorem.

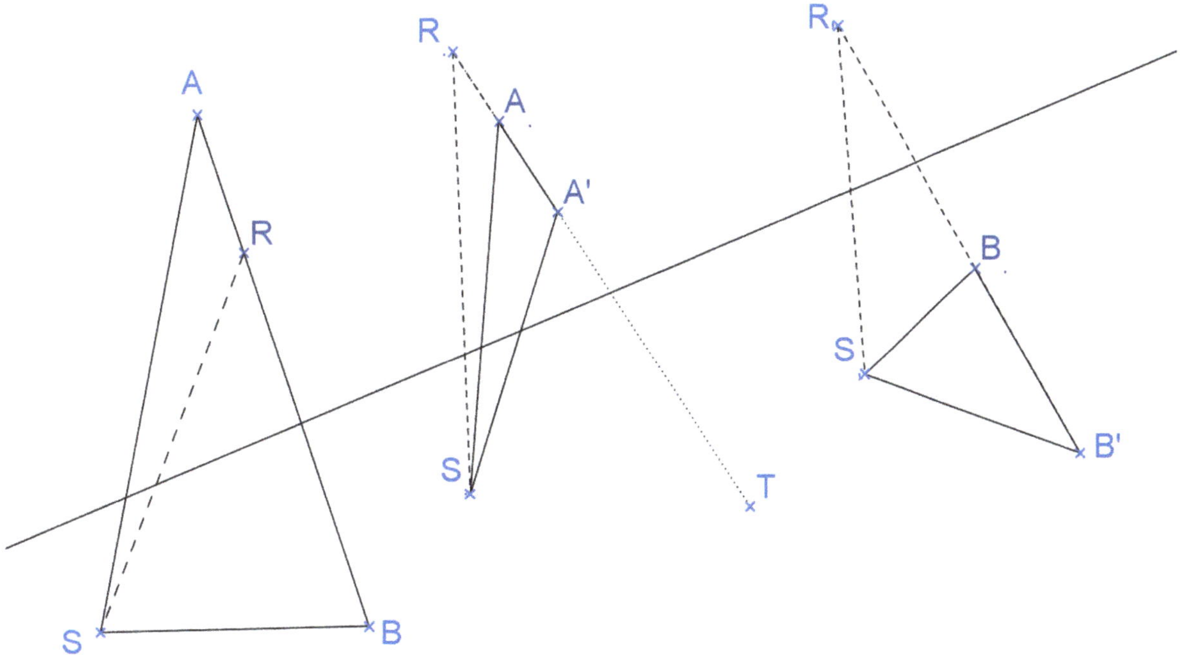

Figure II.6.6: Special cases in the proof of the plane separation theorem.

Lemma II.6.1. *Assume that three points $A \in \mathcal{R}$ and $B \in \mathcal{S}$ and R as above lie on a line r, but point S does not lie on this line. Then line a intersects segment AS, but does not intersect segment BS.*

Reason for the Lemma. In the triangle $\triangle ASR$, the line a intersects the side SR by the construction of point S, but does not intersect side AR. By Pasch's axiom the line a intersects the third side

AS, as to be shown.

In the triangle $\triangle BSR$, the line a intersects both sides SR and BR. By Bernay's Lemma, the line a cannot intersect all three sides of a triangle. Hence the line a does not intersect the third side BS, as to be shown. □

We now assume that $A \neq R$ and $A' \neq R$, and check statements (ii),(iii) and (iv).

(ii) Given are a point A in the region $\mathcal{R}$ and a point B in the region $\mathcal{S}$. We distinguish two cases:

- The three points R, A and B do not lie on a line. We apply Pasch's axiom to the triangle $\triangle RAB$ and line a. The assumptions $A \in \mathcal{R}$ and $B \in \mathcal{S}$ mean that line a does not intersect the side AR, but does intersect the side BR of the triangle. By Pasch's axiom the line a intersects a second side which can only be side AB, as to be shown.
- We consider now the special case that points R, A and B lie on a line. Either the three points S, A and B or the three points T, A and B do not lie on a line. Since the roles of S and T can be switched, I need only to consider the first possibility. We use Pasch's axiom for triangle $\triangle ABS$. Since line a intersects the side AS but not the side SB, the line a intersects the side AB, as to be shown.

(iii) Given are two points A and A' in the region $\mathcal{R}$. We distinguish two cases:

- The three points R, A and A' do not lie on a line. We apply Pasch's axiom to the triangle $\triangle RAA'$ and line a. The assumptions $A \in \mathcal{R}$ and $A' \in \mathcal{R}$ mean that line a does not intersect neither the side AR nor the side $A'R$. By Pasch's axiom the line a cannot intersect the third side AA', as to be shown.
- We consider now the special case that points R, A and A' lie on a line. Either the point S or point T does not lie on this line. Since the roles of S and T can be switched, I need only to consider the first possibility.

  We use Bernay's Lemma for triangle $\triangle AA'S$. Since line a intersects both sides AS and $A'S$, the line a does not intersect the side AA', as to be shown.

(iv) Given are two points B and B' in the region $\mathcal{S}$. We distinguish two cases:

- The three points R, B and B' do not lie on a line. We apply Pasch's axiom to the triangle $\triangle RBB'$ and line a. The assumptions $B \in \mathcal{S}$ and $B' \in \mathcal{S}$ mean that line a does not intersect both sides BR and $B'R$. By Bernay's Lemma, the line a cannot intersect the third side BB', as to be shown.
- We consider now the special case that points R, B and B' lie on a line. Either the point S or point T does not lie on this line. Since the roles of S and T can be switched, I need only to consider the first possibility.

  We use Pasch's axiom for triangle $\triangle BB'S$. Since line a does not intersect neither side BS nor side $B'S$, the line a does not intersect the side BB', as to be shown.

□

Question. Why is a plane mentioned in the plane separation theorem?

- These are just the points of any given plane that are separated into the points on the line and in the two half planes.
- Too, to keep all rigor, one needs the plane just to tell *which points* are separated by the theorem.
- The theorem is not true for a line in the three dimensional space.

Problem II.6.3 (Almost no problem). *Given is a plane α, any line a lying in this plane, and any point A on the line a. For any ray r with vertex A, either one of the following possibilities occurs:*

- *the entire ray lies on the line a;*
- *except for its vertex, the ray lies in one of the two half-planes of line a in the plane α;*
- *only the vertex of the ray lies in the plane α. Except for its vertex, the ray lies in one of the two half-spaces of plane α*

Remark. In the second case, we simply say "the ray r lies in the respective half-plane";—in the third case "the ray r lies in the respective half-space".

Answer. Let B a second point on the ray and assume $r = \overrightarrow{AB}$. By definition, the further points of the ray are the points P such that $A * P * B$, as well as the points Q such that $A * B * Q$. The following cases are possible:

- In the case that point B lies on the line a, we see by axiom (II.1) that the points P and B, respectively the points B and Q, lie on the line a, too.
- In the case that point B does not lie on the line a, but in the plane α, we see by the plane separation Theorem II.6.4 that point B lies in either one of the two half-planes. By the same Theorem II.6.4 the points P and B, respectively the points B and Q lie in the same half-plane as B. Thus the entire ray,—except its vertex A,—lies in this half-plane, too.
- In the case that point B does not lie in the plane α, we see use Proposition I.3.6, part (c). By this proposition, the plane α and the line AB have exactly the point A in common. Hence only the vertex of the ray lies in the plane α.

  We see by the space separation Theorem II.6.11 that point B lies in either one of the two half-spaces of plane α. By the same Theorem the points P and B, respectively the points B and Q lie in the same half-space as B. Thus the entire ray,—except its vertex A,—lies in this half-space, too.

Problem II.6.4. *In an ordered incidence plane α, any line a is given. Prove that both half planes of the line a contain at least three points which do not lie on a line.*

Proof. We begin as in the proof of the plane-separation theorem. By axiom (I.3a), there exist two points $P \neq Q$ on the line a. By axiom (I.3b), there exists a third point R in the plane α, not lying on the line l. In the case that this point lies in the given half-plane $\mathcal{H}$, we put $A := R$. Otherwise, we proceed as in the proof of the plane-separation theorem. By axiom (II.2), there exist a point S

such that $R*P*S$. In that case, we put $A := S$. In any case, we have obtained one point A in the given half-plane.

Now we twice use proposition II.6.1. There exists a point B lying between A and P. Since the line AB intersects the line a outside the segment AB, the plane separation theorem implies that points A and B lie in the same half-plane $\mathcal{H}$ of line a.

Too, there exists a point C lying between A and Q. All three points A, B, C lie in the same half-plane $\mathcal{H}$. Moreover points A, B, C cannot lie on a line,—otherwise all points A, B, C, P, Q would lie on a line. Hence we have obtained three points A, B, C as required. □

II.6.4 Four-point and n-point Theorems

Definition II.6.5. Given any n distinct points lying on a line. We say that the list $[A_1, A_2, \ldots, A_n]$ or $A_1 * A_2 * \cdots * A_n$ *represents the order* of the points $A_1, A_2, \ldots, A_n$ iff the order relations $A_i * A_j * A_k$ hold for all $1 \leq i < j < k \leq n$.

Proposition II.6.5 (Four-point Theorem). *Any four distinct points on a line can be put in exactly two ways into a list that represents their order. These two lists are just reversed to each other, as for example the lists $A * B * C * D$ and $D * C * B * A$ are.*

Reason for uniqueness. Assume we have the list $A * B * C * D$ showing the order, as well as any second list. Since points B and C both lie between two other points among A, B, C, D, they cannot appear in the second list as neither the first nor last item. Neither can the second list be $A*C*B*D$, nor $D * B * C * A$. Hence the reversed list $D * C * B * A$ is the only possible second list. □

Problem II.6.5. *Prove the Four-point Theorem II.6.5 directly from the Three-point Theorem II.6.2 and the Plane separation Theorem II.6.4.*

Answer. Given are four different points on a line l. Choose any three among them. By the Three-point Theorem, one of them lies between two others, and is now renamed C. We draw a line c through C different from l and use plane separation. In one half-plane of c lie two of the remaining points, in the other half-plane only the last remaining point, which we call D. The two points remaining points in the same half plane are named A and B. The three-point theorem holds for points A, B, C. Point C does not lie between A and B, as one sees from the plane separation theorem. We name B the point lying between C and the other one, finally named A. We have thus obtained the order relations

$$A * B * C, \quad A * C * D, \quad B * C * D$$

To confirm the fourth order relation, we draw a third line b through B different from l, and use plane separation once more. Point C and D lie in the same half-plane of b whereas A lies in the opposite half-plane from C. Hence A and D lie in opposite half-planes, confirming the order relation $A * B * D$.

Definition II.6.6. Let A, B and C be three points on the given line with B lying between A and C. The two rays $\overrightarrow{BA}$ and $\overrightarrow{BC}$ are called the *opposite rays* if their common vertex B lies between A and C. We say that two points in the same ray lie on *the same side* of the vertex. Two points in the opposite rays lie *on different sides* of the vertex.

Proposition II.6.6 ("Line separation Theorem"). *Given is a line and a vertex lying on it. Each point of the line except the vertex is contained in exactly one of the opposite rays originating from the vertex, and lying on the given line.*

Proof. Given is a line and a point B on this line. By axiom (I.3a) there exists a second point on the line, which we name A By axiom (II.2), there exists a third point C on the line such that $A * B * C$.

We want to check that an arbitrary point $P \neq B$ is contained in exactly one of the rays $\overrightarrow{BA}$ and $\overrightarrow{BC}$. If $P = A$ or $P = C$, the assertion follows from the three-point Theorem.

Otherwise, $P \neq A$, $P \neq B$ and $P \neq C$, and the assertion follows from the four-point Theorem, and its Corollary. Indeed, the four points $\{A, B, C, P\}$ can be put into an order representing list, with A preceding B. In that list, B precedes C, since $A * B * C$. For the order representing lists $[P, A, B, C]$ or $[A, P, B, C]$, the point P is contained in the ray $\overrightarrow{BA}$. For the order representing lists $[A, B, P, C]$ or $[A, B, C, P]$, the point P is contained in the ray $\overrightarrow{BC}$. □

Question. There are three pairs of order relations among three points from A, B, C, D on a line that imply all four alphabetic order relations. Assume the order representing list is $A * B * C * D$. Which are these three pairs?

Answer. If either

(1) $A * B * C$ and $B * C * D$ or
(2) $A * B * C$ and $A * C * D$ or
(3) $A * B * D$ and $B * C * D$

then all four alphabetic order relations follow.

Remark. The Line separation Theorem does not follow neither simply from the axioms (II.1)(II.2) and (II.3), nor from the Three-point Theorem. For the beginner, that may not be obvious. Indeed, the very first edition of Hilbert's foundations makes the four-point theorem an axiom.

The following counterexample has been constructed by Hartshorne. The `"points"` are the integers modulo 5. Hence there exist just five points. There is one `"line"` through all five points. The `"order relation"` is defined by requiring

(II.6.1) $$a * b * c \quad \text{if and only if} \quad 2b \equiv a + c \pmod 5$$

Question. Convince yourself that

axioms (II.1) and (II.2) are valid for this new ordering;

of any three distinct points $0, a, b$ exactly one lies between the two others. There are six cases to be checked:

$$(0, a, b) = (0, 1, 4), (0, 2, 3), (0, 1, 2), (0, 2, 4), (0, 1, 3), (0, 3, 4)$$

In each case, find the middle point and write down the valid order relation.

Show that of any three distinct points, exactly one lies between the other two.

Answer. Axiom (II.1) is valid for this new ordering. Indeed, $a*b*c$ implies $c*b*a$ since $a+c = c+a$.

Axiom (II.2) is valid. Given two points a and c, let $b :\equiv 2c - a \pmod 5$. Since $a + b \equiv 2c \pmod 5$, we get a point b (beyond the segment) such that point c lies between a and b.

Of any three distinct points $0, a, b$ exactly one lies between the two others. Indeed, for the six possible cases, the order turns out to be:

$$1 * 0 * 4\ ,\ 2 * 0 * 3\ ,\ 0 * 1 * 2\ ,\ 0 * 2 * 4\ ,\ 0 * 3 * 1\ ,\ 0 * 4 * 3$$

Given any three distinct points x, y, z, let $a := y - x$ and $b := z - x$. We order the three points $0, a, b$ and finally get for the three points x, y, z the corresponding order relation.

Problem II.6.6. *To see that the line separation Theorem does not follow neither simply from the axioms* (II.1)(II.2) *and* (II.3), *nor from the Three-point Theorem, we proceed as follows. Convince yourself that for the order defined above*

$$\begin{aligned} a * x * c &\quad \text{if and only if} \quad x \equiv 3(a + c) \pmod 5 \\ a * c * y &\quad \text{if and only if} \quad y \equiv 2c - a \pmod 5 \end{aligned}$$

For better reading I denote the five points now by P_0, P_1, P_2, P_3, P_4.

- *Of which three points consists the segment* P_0P_2.
- *Of which four points consists the ray* $\overrightarrow{P_0P_2}$.
- *Of which points consists the segment* P_0P_3.
- *Of which points consists the ray* $\overrightarrow{P_0P_3}$.
- *Why are* $\overrightarrow{P_0P_2}$ *and* $\overrightarrow{P_0P_3}$ *two opposite rays.*
- *Which points are common to the opposite rays* $\overrightarrow{P_0P_2}$ *and* $\overrightarrow{P_0P_3}$.
- *Why is the line separation Theorem not valid.*

The above order provides a counterexample to the Four-point Theorem, too:

$$\begin{array}{llllll} 1 & *2 & & *3 & & \text{and} \\ & 2 & *0 & *3 & & \text{are both valid; nevertheless} \\ 1 & & & *3 & *0 & \end{array}$$

But <u>not $1 * 0 * 3$</u> as naturally expected, in case that the Four-point Theorem would be valid.

Problem II.6.7. *In the affine plane* $\mathbf{Z}_5 \times \mathbf{Z}_5$ *we introduce the following* `"order relation"`. *For any three points with coordinates* $A = (a_1, a_2)$, $B = (b_1, b_2)$ *and* $C = (c_1, c_2)$, *we require*

(II.6.2) $\qquad A * B * C$ if and only if $2b_1 \equiv a_1 + c_1 \pmod 5$ and $2b_2 \equiv a_2 + c_2 \pmod 5$

Convince yourself that the axioms (II.1)(II.2) *and* (II.3) *are valid, but Pasch's axiom* (II.4) *is not valid.*

Answer. In the affine plane $\mathbf{Z}_5 \times \mathbf{Z}_5$, we may denote the points by (i, k) with $i = 0, 1, 2, 3, 4$ and $k = 0, 1, 2, 3, 4$. We leave the first part of the problem to the reader.

To show that Pasch's axiom is not valid, take thc triangle with vertices $A = (0, 0), B = (4, 0)$ and $C = (2, 2)$. The only interior points of the sides are the midpoints $M_c = (2, 0)$, $M_a = (3, 1)$ and $M_b = (1, 1)$. From Problem I.3.19 and the figure on page 35, we check which are the five points on each line. There exist 6 lines through point M_c:

- the horizontal one is AB, the vertical one is M_cC, which both go through a vertex and are excluded by the assumptions of Pasch's axiom;
- the lines M_cM_a and M_cM_b for which the assertion of Pasch's axiom hold;
- finally the lines M_cP and M_cQ where $P = (3, 2)$ and $Q = (1, 2)$. These two line meet only one side of the triangle $\triangle ABC$. Hence Pasch's axiom is not valid.

□

Recall definition II.6.5 for a list $[A_1, A_2, \ldots, A_n]$ or $A_1 * A_2 * \cdots * A_n$ that *represents the order* of n points on a line.

Lemma II.6.2. *There are exactly two lists which both represent the order of n points $A_1, A_2, \ldots, A_n$ lying on a line. The two lists are reversed of each other.*

Reason for uniqueness. In the case of 3 points, uniqueness follows directly from axioms (II.1) and (II.3). Here is the induction step "$n \mapsto n + 1$":

Assume the uniqueness lemma holds for orderings of up to n points on a line. Let two ordering, say $[A_1 * A_2 * \cdots * A_{n+1}] = \mathcal{O}$, and a second list $\mathcal{L}$, of $n + 1$ points on the line a be given. The points $A_2 \ldots A_n$ all lie between A_1 and A_{n+1}. Hence they may not occur as first nor last item of the second list $\mathcal{L}$. Thus list $\mathcal{L}$ has either A_1 or A_{n+1} as its first item. In the second case, we reverse the list, and get a second list $\mathcal{L}'$ starting with A_1 and ending with A_{n+1}, too. We delete the last item A_{n+1} from both lists to be compared, and get two lists of n points which represent the same ordering, both beginnig with point A_1. By the induction assumption, these two (truncated) lists are equal. Hence the two given lists $\mathcal{O}$ and $\mathcal{L}$ are either equal or reversed to each other. □

Proposition II.6.7 (The n-point Theorem). *Any n points on a line can be put in exactly two ways into a list that represents their order. These two lists are just reversed to each other.*

Problem II.6.8. *Prove proposition II.6.7 by induction from the three-point theorem II.6.2 and the plane separation theorem II.6.4.*

Induction start for $n = 3$: The proposition is just the three-point theorem, proposition II.6.2. □

Induction step "$n \mapsto n + 1$": Assume that proposition II.6.7 holds for ordering up to n points on a line. On the line a, any $n + 1$ points $A_1, A_2, \ldots, A_n$ and P are given. By the induction assumption, the points $A_1, A_2, \ldots, A_n$ may have been ordered. Thus we may assume the order relations

$$A_1 * A_2 * \cdots * A_n \tag{II.6.3}$$

to hold. We draw a line $p \neq a$ through the point P and use plane separation. By possibly reversing the order list (II.6.3), we may assume that point A_1 lies in the half plane $\mathcal{H}$ which contains more or at least equally many points as the opposite half plane. Let A_1 and A_j be the points lying in the half plane $\mathcal{H}$ for which the integer $j \in \{2 \ldots n\}$ is maximal. A special case occurs if all points $A_1 \ldots A_n$ lie in the half plane $\mathcal{H}$. By possibly reversing the order list, we may assume additionally the order relation $A_1 * A_n * P$ to hold. After these agreements, we are left with two cases (i) and (ii), for which we now finish the argument.

(i) The points A_k for $k \in \{j+1 \dots n\}$ lie in the opposite half plane. We draw a further line $a_j \neq a$ through point A_j. Using plane separation we see that

$$\begin{array}{lll} A_i{*}A_j{*} & \quad A_k & \text{and} \\ & A_j{*}P{*}A_k & \text{imply} \\ A_i{*}A_j{*}P & & \end{array}$$

for all $i \in \{1 \dots j-1\}$. Once more, we use plane separation with the line p through the point P to see that

$$\begin{array}{lll} A_i{*}A_j{*}P & & \text{and} \\ & A_j{*}P{*}A_k & \text{imply} \\ A_i{*} & \quad P{*}A_k & \end{array}$$

for all $i \in \{1 \dots j-1\}$ and all $k \in \{j+1 \dots n\}$.

Finally, we use plane separation with the line p through the point P to see that

$$\begin{array}{lll} A_j{*} & A_k{*}A_l & \text{and} \\ A_j{*}P{*}A_k & & \text{imply} \\ & P{*}A_k{*}A_l & \end{array}$$

for all $k \in \{j+1 \dots n\}$ and $l \in \{k+1 \dots n\}$.

(ii) A special case occurs if all points $A_1 \dots A_n$ lie in one half plane. Moreover $A_1 * A_n * P$. We draw a further line $a_n \neq a$ through point A_n and use plane separation to see that

$$\begin{array}{lll} A_1{*} & A_n{*}\Gamma & \text{and} \\ A_1{*}A_j{*}A_n & & \text{imply} \\ A_1{*}A_j{*} & \quad P & \end{array}$$

for all $j \in \{2 \dots n\}$. Next we use plane separation with the line $a_j \neq a$ through the point A_j to see that

$$\begin{array}{lll} A_1{*} & A_j{*}P & \text{and} \\ A_1{*}A_i{*}A_j & & \text{imply} \\ & A_i{*}A_j{*}P & \end{array}$$

for all $i \in \{1 \dots j-1\}$ and $j \in \{2 \dots n\}$.

In both cases, all instances of three point orders contained in the order

$$A_1 * A_2 * \cdots * A_j * P * A_{j+1} * \cdots * A_n$$

have been confirmed. Thus the inductive proof of proposition II.6.7 is completed. □

Proposition II.6.8. *For any different points $A \neq B$, and for all natural numbers n, there exist at least n points between A and B.*

Proof. We construct a sequence of points $A_1, A_2, \dots,$ all of which are different and lie between the two given points A and B. The existence of the point A_1 between A and B was proved in Proposition II.6.1.

Assume that n points $A_1, A_2, \dots, A_n$ between A and B have been constructed, and there order is represented by the list

$$[A, A_1, A_2, \dots, A_n, B]$$

We choose any point A_{n+1} between A_n and B. As shown in Proposition II.6.7, the order of the $n+1$ points is now represented by the list

$$[A, A_1, A_2, \dots, A_n, A_{n+1}, B]$$

Hence one has obtained $n+1$ different points between A and B. □

Corollary 5. *For any* finite *incidence plane, it is impossible to define an order relation for which all the axioms of order* (II.1),(II.2),(II.3),(II.4) *hold.*

Remark. Hilbert's Proposition 7 simply states: "there exist infinitely many points lying between any two different points." Such an assertion has not been exactly proved. We have only shown that "for any natural number n, there exist more than n different points on the given line."

Since the axiomatic set theory of Zermelo Fraenkel has become common knowledge, it is clear that existence of any infinite set *needs an extra axiom.* Such an axiom of infinity is part of Zermelo Fraenkel set theory, but not of Peano arithmetic. Indeed, no proof by induction yields existence of any infinite set.

A similar case occurs with respect to the common-place statement "there exist infinitely many primes." Any of the known proofs only verify the assertion "For any natural number n, there exist more than n prime numbers."

Remark. Of course, one wants to know more about the structure of a line than the assertion made by proposition II.6.8. To this end, we need to assume Peano arithmetic is true; and moreover use the important results obtained at several places below in these notes.

Thus one is lead to the following results. We fix a unit segment CD. There exist segments having the length given by any positive rational number $p/q \in \mathbf{Q}$, under either one of the following assumptions:

- for any affine plane coordinate plane over an ordered field;

- for any affine ordered Pappian plane;

- for any Hilbert plane where both axioms of continuity hold,—the Archimedean axiom (V.1) as well as some completeness axiom are both required;

- for any Pythagorean plane. In the latter case, a segment having length p/q is constructible for any positive rational number p/q.

Problem II.6.9. *State in detail which theorems are needed to confirm the assertions of this remark.*

II.6.5 Angles

Definition II.6.7 (Angle). An *angle* is the union of two rays with common vertex, and not lying on one line. The common vertex of these rays is called the *vertex* of the angle. The rays are called the *sides* of the angle.

For an angle, we may use the two ray notation $\angle(h,k)$;— the two rays h and k are supposed to have common vertex and no lie on a line.

As an alternative, we use the three letter notation. Here the *middle letter* denotes the vertex. Thus the angle $\angle BAC$ has the vertex A, and the sides $\overrightarrow{AB}$ and $\overrightarrow{AC}$.

Remark. In the section on congruence, we introduce by Definition II.7.14 the notion of *oriented angle.* In the section about similar triangles, we introduce by definition V.23.3 the *group of unwound angles.*

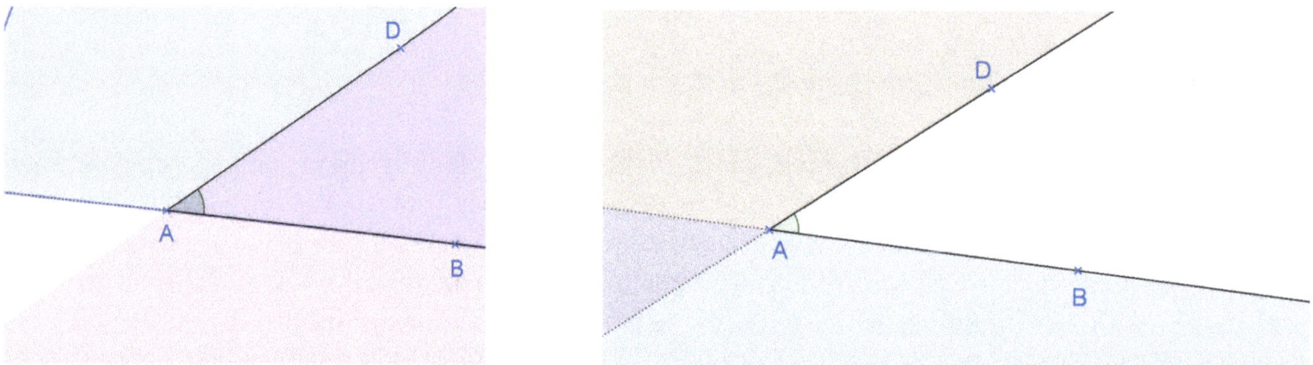

Figure II.6.7: Interior and exterior of an angle are intersection and union of two half planes.

Definition II.6.8 (Interior and exterior domain of an angle). The *interior domain* of an angle lying in a plane $\mathcal{A}$ is the intersection of two corresponding half planes—bordered by the sides of the angle, and containing points on the other side of the angle, respectively. The exterior of an angle is the union of two opposite half planes—-bordered by the sides of the angle, and not containing the points neither in the interior nor on the legs of the angle. Half planes, interior and exterior of an angle all do not include the lines or rays on their boundary.

Thus the interior of $\angle BAD$ is the intersection of the half plane of AB in which D lies, and the half plane of AD in which B lies. The exterior of $\angle BAD$ is the union of the half plane of AB opposite to D, and the half plane of AD opposite to B.

Remark. Note the "exterior of an angle" is defined differently from "exterior angle of a triangle". The latter is any supplementary angle to an interior angle of the triangle.

Problem II.6.10. *Given an angle $\angle(h,k)$ lying in a plane $\mathcal{A}$.*

Let $\mathcal{K}$ be the intersection of the half plane of h in which k does not lie with the half plane of k in which h does not lie. Explain what the set $\mathcal{K}$ is like and provide a drawing. Give a short exact description for this set.

Answer. From the drawing on page 115, one sees that the set $\mathcal{K}$ is the interior of an angle. Indeed, it is the interior of the vertical angle $\angle(h',k')$ corresponding to the given angle $\angle(h,k)$.

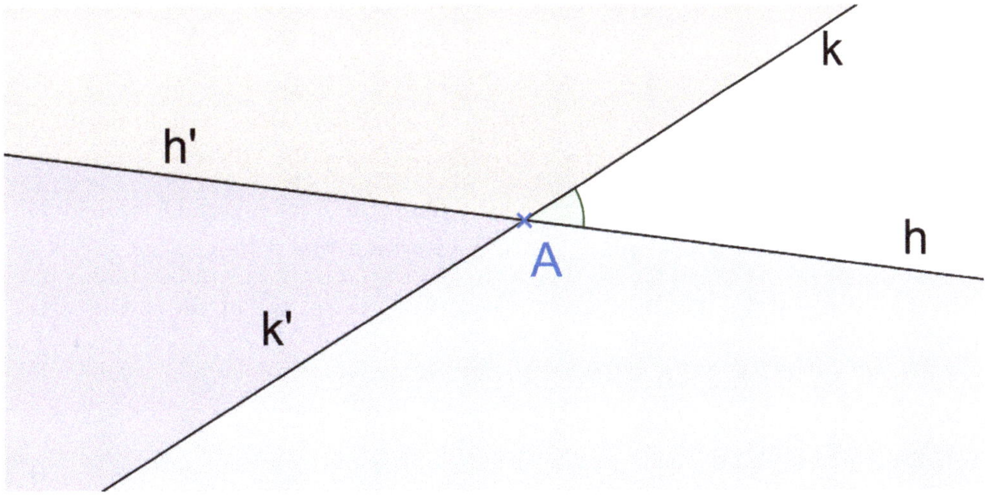

Figure II.6.8: Interior of the vertical angle.

Problem II.6.11. *Given an angle $\angle(h,k)$ lying in a plane $\mathcal{A}$.*

(a) *Let points $H \in h$ and $K \in k$ be given, both different from the vertex of the angle. Explain why all interior points of segment HK lie in the interior of the angle $\angle(h,k)$.*

(b) *Let $A = h \cap k$ be the vertex of the angle $\angle(h,k)$ and D be any point in the interior of the angle. Explain why all points of the ray $\overrightarrow{AD}$ except A lie in the interior of the angle $\angle(h,k)$.*

Answer. **(a)** Take any point P interior to segment HK, thus $H * P * K$. Point P lies in the same half plane of h in which point K, and hence the ray k lies. Similarly, point P lies in the same half plane of k in which point H, and hence the ray h lies. Hence point P lies in the interior of the angle $\angle(h,k)$. Indeed, all interior points of segment HK lie in the interior of the angle $\angle(h,k)$.

(b) Take any point $P \neq A$ on the ray $\overrightarrow{AD}$. Hence either $P = D$, or $A * P * D$, or $A * D * P$.

In each case, point P lies in the same half plane of h in which point D lies. But point D lies in the interior of the angle. Hence both point P and point D lie in the same half plane of h in which the ray k lies. Similarly, point P lies in the same half plane of k in which point D and hence the ray h lies. Hence point P lies in the interior of the angle $\angle(h,k)$. Indeed, all points of the ray $\overrightarrow{AD}$ except the vertex A lie in the interior of the angle $\angle(h,k)$.

A second way to answer. **(a)** Given are points $H \in h$ and $K \in k$, both different from the vertex of the angle. By the previous problem II.6.3, the entire ray $\overrightarrow{HK}$ lies in the same half-plane of h as point K. This is the half-plane of h in which ray k lies, too. We call it $\mathcal{K}$.

The entire ray $\overrightarrow{KH}$ lies in the same half-plane of k as point H. This is the half-plane of k in which ray h lies, too. We call it $\mathcal{H}$.

The interior of the angle $\angle(h,k)$ is the intersection of half-planes $\mathcal{K} \cap \mathcal{H}$. The intersection $\overrightarrow{HK} \cap \overrightarrow{KH}$ is the interior of the segment HK. Since

$$\overrightarrow{HK} \cap \overrightarrow{KH} \subseteq \mathcal{K} \cap \mathcal{H}$$

we see that the interior of the segment HK lies in the interior of the angle $\angle(h,k)$.

(b) Let $A = h \cap k$ be the vertex of the angle $\angle(h, k)$, and D be any point in the interior of the angle. By the previous problem II.6.3, the entire ray $\overrightarrow{AD} \setminus \{A\}$ lies in the same half-plane of h as point D. Since point D is in the interior of the angle, this is half-plane $\mathcal{K}$ of h in which ray k lies.

The entire ray $\overrightarrow{AD} \setminus \{A\}$ lies in the same half-plane of k as point D. Since point D is in the interior of the angle, this is half-plane $\mathcal{H}$ of k in which ray h lies.

Hence the ray $\overrightarrow{AD} \setminus \{A\}$ lies in the intersection $\mathcal{K} \cap \mathcal{H}$, which is the interior of the angle $\angle(h, k)$.

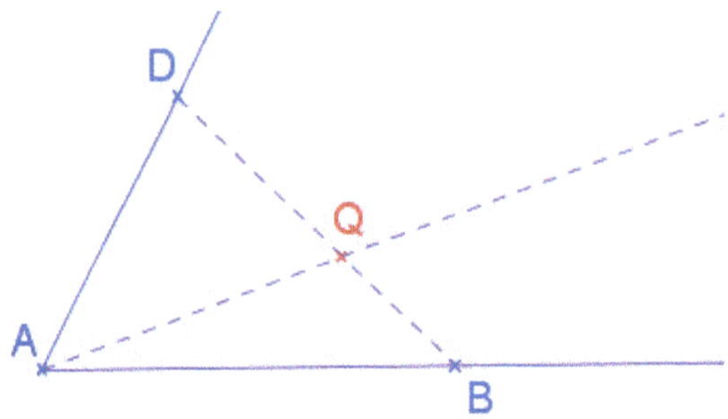

Figure II.6.9: A ray interior of an angle intersects a segment from side to side.

Proposition II.6.9 (The Crossbar Theorem). *A segment with endpoints on the two sides of an angle and a ray emanating from its vertex into the interior of the angle intersect.*

Proof. Let ray $r = \overrightarrow{AC}$ lie in the interior of the given angle $\angle BAD$, and let B and D be arbitrary points on the two sides of this angle. We have to show that the ray $\overrightarrow{AC}$ intersects the segment BD.

Let F be any point on the ray opposite to $\overrightarrow{AB}$. We apply Pasch's axiom to triangle $\triangle FBD$ and line $l = AC$. The line intersects the side FB of the triangle in point A, and does not pass through neither one of the vertices F, B, D. We check that side FD does not intersect line AC.

Indeed, the points inside segment FD and the points inside ray $\overrightarrow{AC}$ lie on different sides of line AD. But the points inside segment FD and inside the opposite ray lie on different sides of line AB. (Points A, D, F are exceptions, but neither can segment FD and line AC intersect in any of these points.)

Hence—by Pasch's axiom—the third side BD intersects line AC, say at point Q. Segment BD, and hence the intersection point Q are in the interior of $\angle BAD$. Since only the ray $\overrightarrow{AC}$, but not its opposite ray, lies in the interior of $\angle BAD$, the intersection point Q lies on the ray $\overrightarrow{AC}$. Here is a drawing, to show how Pasch's axiom is applied. □

Warning. A point in the interior of an angle may not lie on any segment going from side to side of the angle.

To spell it out precisely: Given an angle $\angle BAD$ and a point Q in its interior, there *does not always* exist any two points G and H on the rays $\overrightarrow{AB}$ and $\overrightarrow{AD}$ such that Q lies on the segment GH.

Indeed, in hyperbolic geometry, the union of all segments from side to side of an angle span only the interior of an asymptotic triangle—and this is a proper subset of the interior of the angle.

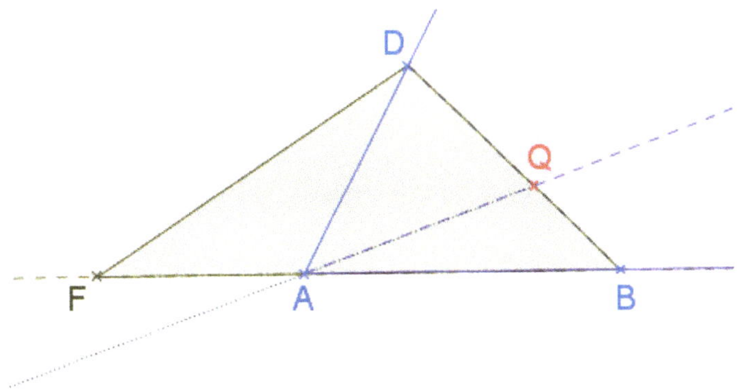

Figure II.6.10: The Crossbar Theorem is proved using Pasch's axiom.

Lemma II.6.3. *Given are two opposite rays r and s with common vertex O, one half-plane $\mathcal{H}$ of line $r \cup s$, and two distinct rays a and b from vertex O lying in this half-plane $\mathcal{H}$. There the two following cases to occur:*

(i) *either the ray a lies in the interior of angle $\angle(r, b)$ and the ray b lies in the interior of angle $\angle(s, a)$;*

(ii) *or the ray b lies in the interior of angle $\angle(r, a)$ and the ray a lies in the interior of the angle $\angle(s, b)$.*

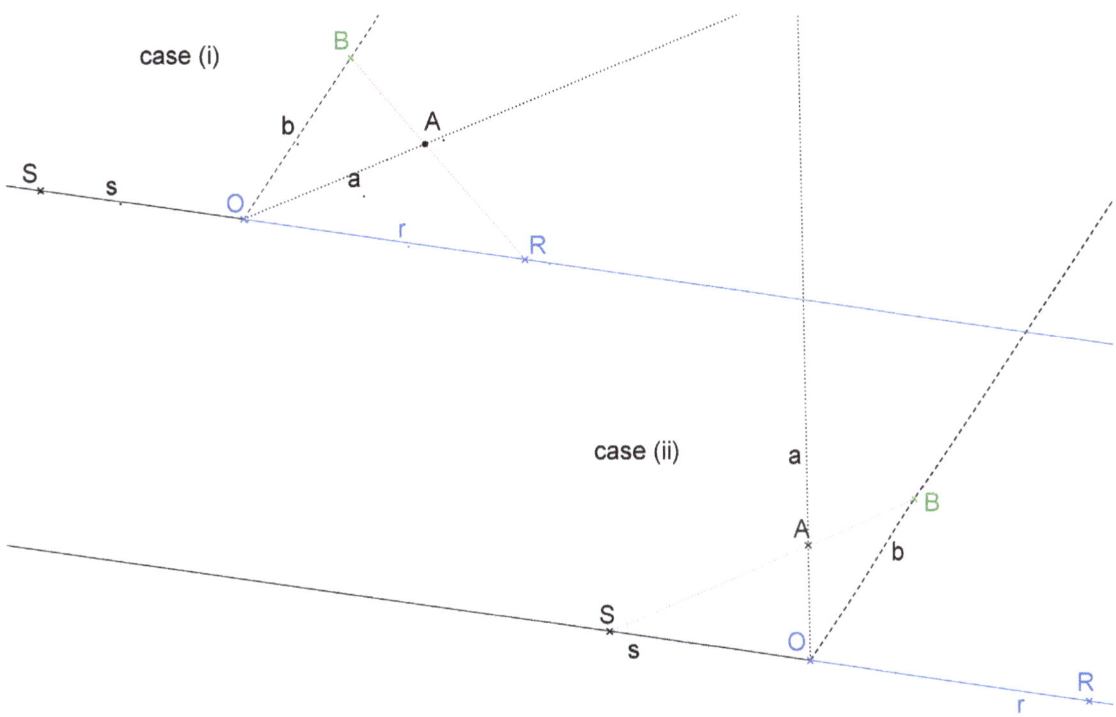

Figure II.6.11: About any two angles.

Proof. The entire figure lies in the half-plane $\mathcal{H}$ of line $r \cup s$. We may choose points R and S such that $r = \overrightarrow{OR}$ and $s = \overrightarrow{OS}$. We choose any point B on the ray b such that $b = \overrightarrow{OB}$. We use the plane separation theorem with the line of ray b. Asking about the ray a, either one of the two following cases occur. Both are shown in the figure on page 117.

(i) The rays r and a lie in the same half plane of ray b. Hence the ray a lies in the interior of angle $\angle(r,b)$. The Crossbar Theorem implies that the ray a intersects the segment BR, say in point A. From the order relation $B*A*R$, we see that points B and R lie on different sides of ray a. Hence points B and S lie on the same side of ray a. Hence ray $b=\overrightarrow{OB}$ lies in the interior of angle $\angle(s,a)$.

(ii) The rays r and a lie in opposite half planes of ray b. Hence the ray a lies in the interior of angle $\angle(s,b)$. The Crossbar Theorem implies that the ray a intersects the segment BS, say in point A. From the order relation $B*A*S$, we see that points B and S lie on different sides of ray a. Hence points B and R lie on the same side of ray a. Hence ray b lies in the interior of angle $\angle(s,a)$.

□

One may at this point already introduce the notions of *supplementary angles* and *vertical angles.*

Definition II.6.9 (Supplementary Angles). Two angles are called *supplementary angles* or a *linear pair*, if and only if they have a common vertex, they both have one side on a common ray, and the two other sides are the opposite rays on a line.

Definition II.6.10 (Vertical Angles). Two angles are called *vertical angles*, iff they have a common vertex, and their sides are two pairs of opposite rays on two lines.

Proposition II.6.10 (About n points in a plane). *Among any $n\ge 3$ points lying in an ordered incidence plane but not on a line, there exist three points P,Q,R such that all n points lie in the interior or on the boundary of the angle $\angle PQR$.*

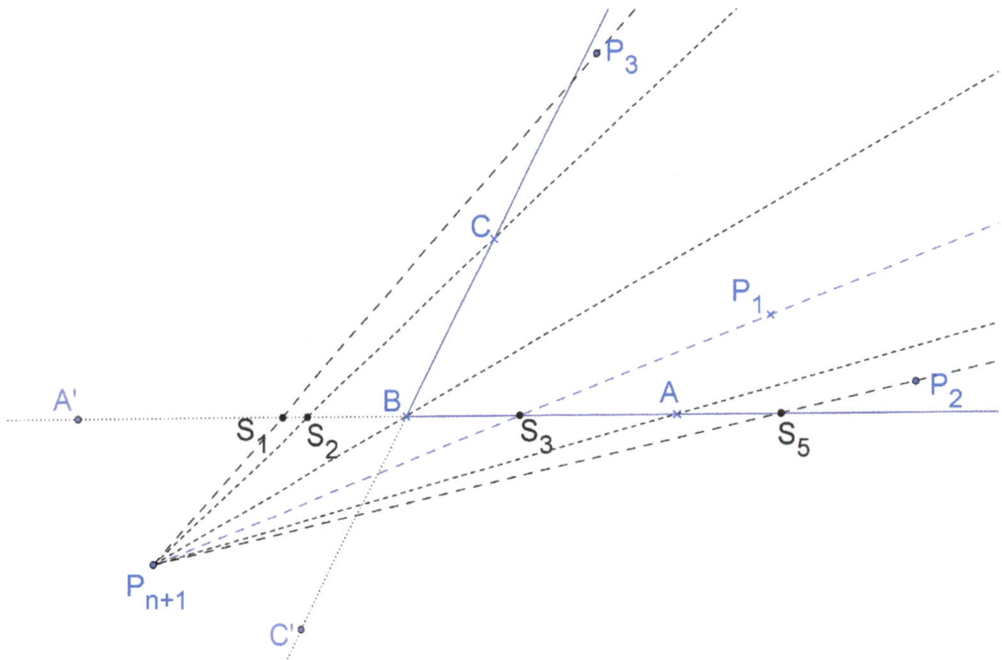

Figure II.6.12: Any n points lie in the interior or on the boundary of an angle.

Induction start for $n = 3$: The three given points P, Q, R do not lie on a line, hence the angle $\angle PQR$ exists. □

Induction step "$n \mapsto n + 1$": Assume that proposition holds for any n points.
Given are now these points, and an extra point P_{n+1}. We distinguish these cases:

(i) The points $P_1 \dots P_n$ lie on a line l. By the n-point proposition, they can be put into an ordered list $P_1 * P_2 * \cdots * P_n$. By assumption point P_{n+1} does not lie on the line l. We put $P := P_1$, the first item in the list; $R := P_n$, the last item in the list; and choose $Q := P_{n+1}$ as vertex of the angle. Thus all $n + 1$ points lie inside or on the boundary of the angle $\angle PQR$.

(ii) The points $P_1 \dots P_n$ do not lie on a line l. By the induction assumption, there are points A, B, C among them such that points $P_1 \dots P_n$ lie in the interior or on the boundary of the angle $\angle ABC$.

- In case that the extra point P_{n+1} lies in the interior or on the boundary of the angle $\angle ABC$, we are ready.
- In case that the extra point P_{n+1} lies in the interior of a supplementary angle of the angle $\angle ABC$, all points $P_1 \dots P_{n+1}$ lie in the interior or on the boundary of either angle $\angle ABP_{n+1}$ or angle $\angle P_{n+1}BC$.
- We choose any points $A * B * A'$ and $C * B * C'$. Assume that the extra point $Q := P_{n+1}$ lies in the interior of the vertical angle $\angle A'BC'$ or on the ray $\overrightarrow{BA'}$.

  This case in shown in the figure on page 118. We draw the segments P_iP_{n+1} for $i = 1 \dots n$, obviously including the segments AP_{n+1}, BP_{n+1} and CP_{n+1}. Since the points $P_1 \dots P_n$ either lie on the line AB, or the opposite side of AB as point P_{n+1}, all these segments intersect the line AB .

  By the n-point theorem we may order these intersection points into a list $S_1 * \cdots * S_n$. Let P be a point among $P_1 \dots P_n$ on the *first* ray $\overrightarrow{P_{n+1}S_1}$, and R be a point among $P_1 \dots P_n$ on the last ray $\overrightarrow{P_{n+1}S_n}$.

  Now all points $P_1 \dots P_{n+1}$ lie in the interior or on the boundary of the angle $\angle PQR$.

□

II.6.6 Space separation

Proposition II.6.11 (Hilbert's Proposition 10, "Space separation Theorem"). *Each plane α separates the remaining points of the space into two nonempty regions $\mathcal{R}$ and $\mathcal{S}$ with the following properties:*

- *Every point not lying in the plane α lies either in region $\mathcal{R}$, or in region $\mathcal{S}$, but not in both regions.*
- *for any points A in region $\mathcal{R}$ and B in region $\mathcal{S}$, segment AB intersects plane α;*
- *for two points A and A' in the same region, segment AA' does not intersect α.*

Definition II.6.11. In case that segment AA' does not intersect the plane α, we say that the points A and A' lie on *the same side* of plane α—or in the same half space. The points A and B lie *on different sides* of plane α in case that the segment AB does intersect this plane.

The two open regions one gets in this way are called the *half spaces* bounded by the plane α.

Proof. Take any point A not lying in the plane α. Since by axiom I.8, there exist at least four points which do not lie in a plane, such a point A does exist. We can now separate the points not lying in the plane α into the following two sets:

$$\mathcal{R} := \{B : B \notin \alpha, \text{ the segment } AB \text{ does not intersect the plane } \alpha\}$$
$$\mathcal{S} := \{C : C \notin \alpha, \text{ the segment } AC \text{ intersects the plane } \alpha\}$$

Each point lies in exactly one of the three sets $\mathcal{R}$, α or $\mathcal{S}$. Since $A \in \mathcal{R}$ this set is nonempty.

Question. Explain why the set $\mathcal{S}$ is nonempty.

Answer. Take a point $P \in \alpha$. By axioms (I.1) and (II.2), there exists a point Q on the line AP such that $A * P * Q$. Since the segment AQ intersects the plane α, the point Q lies in the set $\mathcal{S}$.

We check the following facts:

(a) Take any two points B and B' in the domain $\mathcal{R}$. Then the segment BB' does not intersect α.

(b) Take any two points C and C' in the domain $\mathcal{S}$. Then the segment CC' does not intersect α.

(c) Take points B in domain $\mathcal{R}$ and C in domain $\mathcal{S}$. Then the segment BC intersects the plane α.

Consider assertion (a). The three points A, B and B' lie either a line l or span a plane β. If this line or plane does not intersect the plane α, we are ready.

Take the case that points A, B and B' lie on a line l intersecting plane α in a point D. Assumption (a) implies that A and B lie in the same ray of line l from vertex D. Similarly, points A and B' lie in the same ray from D. By line separation (Proposition II.6.6), we conclude that points B and B', too, lie in the same ray from D. Hence segment BB' does not intersect the plane α.

Take the case that points A, B and B' do not lie on a line. By the axioms I.4 and I.5, they determine a unique plane β. Take the case that the planes β and α intersect. By Hilbert's Proposition 1 (Proposition I.3.6), the intersection $\alpha \cap \beta$ is a line k. The assumption (a) implies that points A and B lie on the same side of k in the plane β. Similarly A and B' lie in the same half-plane. We use plane separation (Proposition II.6.4) for the plane β, and conclude that points B and B' lie in the same half-plane. Hence segment BB' does not intersect the plane α.

Consider assertion (c). Take points $B \in \mathcal{R}$ and $C \in \mathcal{S}$. The segment AC intersects the plane α in a point D. If $A = B$, we are ready.

The points A, B and C lie either on a line l or span a plane β. Take the case that points A, B and C lie on a line. This line l intersects plane α in point D. Assumption (c) implies that A and B lie in the same ray of line l from vertex D. But points A and C lie on opposite rays from D. By line separation (Proposition II.6.6), we conclude that points B and C, too, lie on opposite rays from D. Hence segment BC intersects the plane α in point D.

Take the case that points A, B and C do not lie on a line. By the axioms I.4 and I.5, they determine a unique plane β. The planes β and α intersect in point D. By Hilbert's Proposition 1 (Proposition I.3.6), the intersection $\alpha \cap \beta$ is a line k. The assumption (c) implies that points A and B lie on the same side of k in the plane β, but A and C lie in opposite half-planes. We use

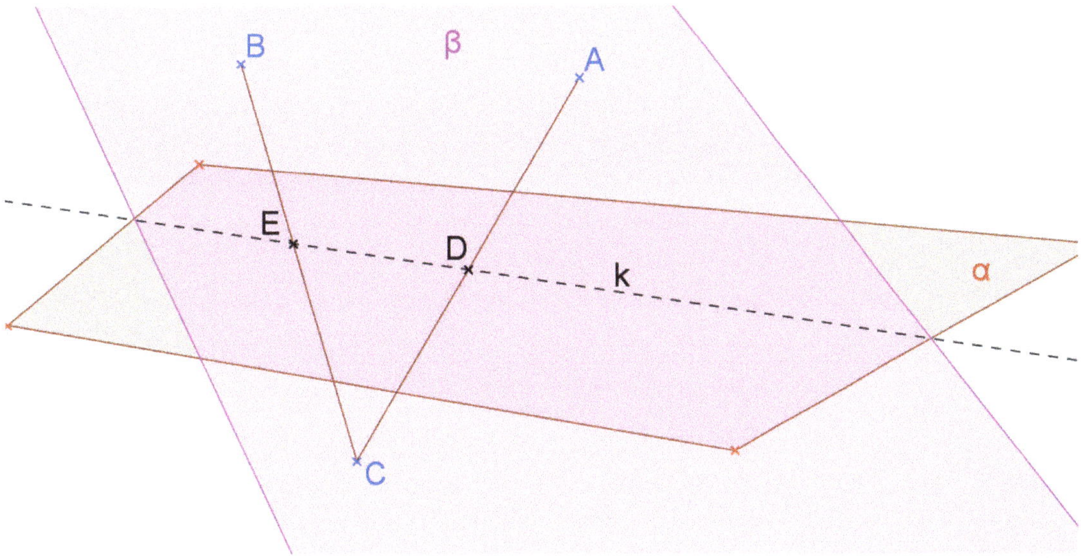

Figure II.6.13: The segment between two points on different sides of a plane intersects the plane.

plane separation (Proposition II.6.4) for the plane β, and conclude that points B and C lie opposite half-planes. Hence segment BC intersects the line k, and hence the plane α in a point E.

We leave it to the reader to check case (b). □

II.6.7 Interior and exterior of a triangle

Definition II.6.12 (Interior and exterior of a triangle). The *interior domain* of a triangle is the intersection of the three half-planes of the sides in which the respectively opposite vertices lie. By $\mathrm{Int}_{\triangle}ABC$ we denote the set of interior points of triangle $\triangle ABC$.

The *exterior* domain of a triangle consists of the points of the plane which do not lie neither in the interior domain, nor on the sides, nor at the vertices of the triangle.

Problem II.6.12. *Show that the interior domain of a triangle is nonempty.*

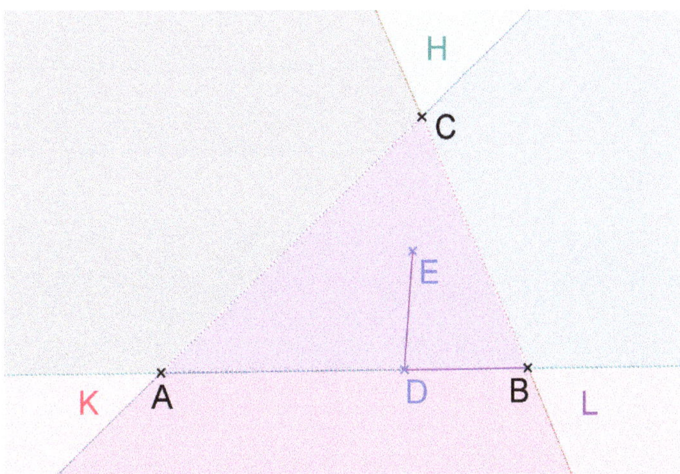

Figure II.6.14: The interior of a triangle is the intersection of three half planes.

Reason why the interior domain of a triangle is nonempty. Take a triangle $\triangle ABC$. Let $\mathcal{H}$ be the half plane of line AB in which point C lies, $\mathcal{K}$ be the half plane of line BC in which point A lies, and $\mathcal{L}$ be the half plane of line CA in which point B lies. By definition, the interior domain of the triangle $\triangle ABC$ is the intersection $\mathcal{H} \cap \mathcal{K} \cap \mathcal{L}$.

By definition, the interior domain of the triangle $\triangle ABC$ is the intersection $\mathcal{H} \cap \mathcal{K} \cap \mathcal{L}$. By Proposition II.6.1 (Hilbert's Proposition 3), there exists a there exists a point D lying between A and B. By the same reasoning, there exists a point E lying between C and D.

I claim that $E \in \mathcal{H} \cap \mathcal{K} \cap \mathcal{L}$, which is the interior of triangle $\triangle ABC$. Indeed, the segment EC lies entire in $\mathcal{H}$ since $D * E * C$ and point E lies in the same half plane $\mathcal{H}$ of line AB as point C. Because of $A * D * B$ and $D * E * C$, segments DB and DE lie entirely in $\mathcal{L}$. Hence points B, D and E lie in the same half plane $\mathcal{L}$ of line CA.

Similarly, since segments AD and DE lie in the half plane $\mathcal{K}$, we see that points A, D and E lie in the same half plane $\mathcal{K}$ of line BC. □

Problem II.6.13. *By choosing an appropriate point on a side of the triangle, and an appropriate point on the segment from this point to the opposite vertex, one can obtain any point in the interior of a triangle. Give a reason.*

Proof. We use a reasoning reversing the steps of problem II.6.12. Let E be any point in the interior of triangle $\triangle ABC$. The ray $r = \overrightarrow{CE}$ lies in the interior of the angle $\angle ACB$. By the Crossbar Theorem, the ray r intersects the side AB. Let the intersection point be D. Since the given point E lies in the interior of the triangle, points E and C lie on the same side of AB, and hence $C*E*D$.

Thus point E has been obtained as described in the problem: by choosing point D on a side AB of the triangle, and then point E on the segment DC to the opposite vertex. □

The exterior domain of a triangle consists of the points of the plane which do not lie neither in the interior domain, nor on the sides, nor at the vertices of the triangle. It is equal to the union $\mathcal{H}' \cup \mathcal{K}' \cup \mathcal{L}'$ of the three opposite half planes.

Problem II.6.14. *Show that the half-planes of the three sides of a triangle, which are opposite to the third vertex, respectively, do not intersect.*

Answer. The three half planes $\mathcal{H}', \mathcal{K}', \mathcal{L}'$ opposite to $\mathcal{H}, \mathcal{K}, \mathcal{L}$ have empty triple intersection. Indeed, the intersection $\mathcal{H}' \cap \mathcal{K}'$ is the interior of angle $\angle C'BA'$. This is the vertical angle of $\angle ABC$— I have chosen any points $A * B * A'$ and $C * B * C'$.

The rays $\overrightarrow{BA'}$ and $\overrightarrow{BC'}$ lie entirely inside the half plane $\mathcal{L}$. Hence the interior of the angle $\angle A'BC'$ which is the intersection $\mathcal{H}' \cap \mathcal{K}'$ is a subset of $\mathcal{L}$ and does not intersect $\mathcal{L}'$.

The purpose of the following exercises is to check the Jordan curve Theorem in the case that the closed curve is a triangle.

Problem II.6.15. *Given is a triangle and a line through an interior point of the triangle. Show that the line either intersects two sides of the triangle, or it intersects only one side and goes through the opposite vertex.*

Answer. Let the line l go through the point P in the interior of triangle $\triangle ABC$. We distinguish two cases:

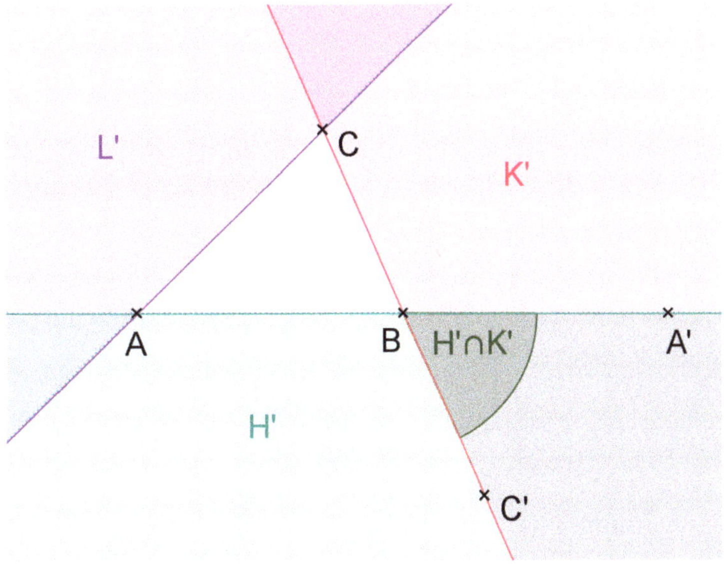

Figure II.6.15: The three half planes $\mathcal{H}', \mathcal{K}', \mathcal{L}'$ opposite to the interior ones have empty intersection.

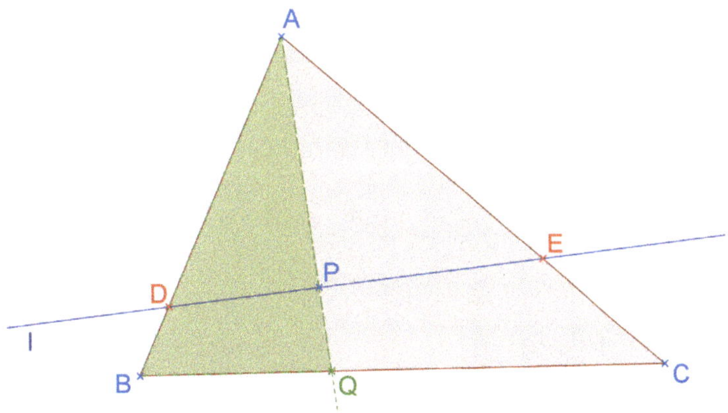

Figure II.6.16: A line through an interior point of a triangle intersects either two sides, or goes through a vertex and intersects the opposite side.

(a) **The line goes through a vertex of the triangle.** Say the line goes through vertex A. Because point P lies in the interior of angle $\angle BAC$, the Crossbar theorem shows that the ray $\overrightarrow{AP}$ intersects the opposite side BC at some point, say Q.

(b) **The line does not go through any vertex of the triangle.** We draw the ray $\overrightarrow{AP}$. Because point P lies in the interior of angle $\angle BAC$, the Crossbar theorem shows that the ray $\overrightarrow{AP}$ intersects the opposite side BC at some point, say Q.

We now apply Pasch's axiom to the triangle $\triangle ABQ$ and the line l. This line intersects side AQ at point P. Hence it intersects a second side, either $BQ \subset BC$ or AB, say at point D. In both cases, the line l intersects one side of the given triangle $\triangle ABC$. By Pasch's axiom it intersects a second side, say at point E. By Bernays' lemma, the line l intersects exactly two sides of the given triangle.

Problem II.6.16 ("non-Pasch"). *Give a reason why a line which intersects the extensions of two different sides of a triangle lies entirely in the exterior domain of the triangle.*

Answer. Since the given line l is different from the lines on all three sides, it can only intersect a given side of the triangle or its extension, but not both. Assume line l intersects both the extensions of sides AB and BC.

Could line l still intersect the third side CA? No, in this case, by Pasch's axiom, it would intersect a second side, either AB or BC, too, and hence not its extension.

Problem II.6.17. *Given is a triangle.*

(a) *Prove that any segment PR from an interior point P to an exterior point R on the extension of a side of the triangle intersects a second side of the triangle.*

(b) *Prove that any segment PQ from an interior point P to an exterior point Q intersects either a vertex or a side of the triangle.*

Answer. **(a)** Given is a segment PR from an interior point P of triangle $\triangle ABC$ to an exterior point R on the extension of a side. We may assume that $R * A * B$, the other cases being similar.

Points R and P lie on different sides of line AC. Hence segment RP intersects line AC, say in point S.

Why does point S lie on segment AC? By the previous problem, the line PQ cannot intersect the extensions of two different sides of the triangle. Hence point S lies on the side AC.

(b) Given is a segment PQ from an interior point P to an exterior point Q of the triangle. The two points P and Q lie either in different half planes of one side of the triangle, say AB, or point Q lies on the extension of a side AB.

In the second case we have shown in part (a) that segment PQ intersects a side of the triangle. In the first case, segment PQ intersects line AB, by plane separation. Let the intersection point be R, hence $P * R * Q$. There are two cases: If point R lies on the side AB, we are ready. If point R lies on the extension of side AB, we go back to part (a) and conclude that segment RP intersects a second side of the triangle.

Remark. In hyperbolic geometry, it is possible that a line passes through the extension of one side, and is parallel (even asymptotically parallel!) to the two other sides of a triangle.

Problem II.6.18. *Given are any three points on a line, where B lies between A and C. Assume point D does not lies on this line. Show that the interior of triangle $\triangle ACD$ is the union of the interiors of the triangles $\triangle ABD$ and $\triangle BCD$ and the interior of the segment BD.*

II.6.8 Convexity

Definition II.6.13 (Convex set). A set of points is called *convex*, if and only if with any two points P and Q, the segment PQ is contained in the set.

Lemma II.6.4. *Any segment is a convex set.*

Reason. This statement is a consequence of the four- and five-point theorems. Given is any segment AB, and two points P and Q from this segment. Let R be any point from the segment PQ. We need to consider two cases, the remaining ones are similar.

(i) $P = A$ is an endpoint, and Q an interior point of the segment.

Thus the order relations $A * Q * B$ and $P = A * R * Q$ hold. The four-point theorem implies the ordered list $A * R * Q * B$. Hence we conclude $A * R * B$, telling the point R lies in the segment AB, as to be shown.

(ii) P and Q are both interior points of the segment.

Thus the order relations $A * P * B$, $A * Q * B$, and $P * R * Q$ hold. The five points can appear in either of two ways in an ordered list: one obtains either $A * P * R * Q * B$ or $A * Q * R * P * B$. In both cases we conclude $A * R * B$, telling the point R lies in the segment AB, as to be shown.

□

Lemma II.6.5. *Any ray is a convex set.*

Lemma II.6.6. *Any half-plane is a convex set.*

Reason. Given is any plane α, a line l lying in this plane, and a half-plane $\mathcal{H} \subset \alpha$ bounded by this line. Let P and Q be any two points from the half-plane $\mathcal{H}$. By the plane separation theorem, the segment PQ does not intersect the line l.

Let R be any point from the segment PQ. The plane separation theorem implies that point R does not lie on the line l.

By the previous lemma II.6.4, any segment is a convex set, and hence $PR \subseteq PQ$. Hence the line l does not intersect the (shorter) segment PR, neither. By the plane separation theorem, we conclude that points P and R lie in the same half-plane. Thus we have checked that $R \in \mathcal{H}$, as claimed. □

Lemma II.6.7. *The intersection of any convex sets is convex. Hence the interior of a triangle is a convex set.*

Definition II.6.14. A polygon or polygonal curve is called *strictly convex*, iff for any segment $AB, BC, CD, \ldots, KL$ the remaining vertices lie in one half plane of the respective segment.

A polygon or polygonal curve is called *convex*, iff for any segment $AB, BC, CD, \ldots, KL$ the remaining vertices lie in one half plane or on the line of the respective segment.

It is easy to see that any closed and strictly convex polygon is simple.

Too, any closed and convex polygon is simple, provided its vertices which happen to lie on one line appear in their natural order, such that the corresponding segments intersect only at their endpoints.

Lemma II.6.8. *Assume no three of the four points A, B, C, D lie on a line. For the quadrilateral $\square ABCD$ the following statements are equivalent:*

(a) *The quadrilateral $\square ABCD$ is strictly convex.*

(b) *Any three of the following statements are true:*

  (i) *Vertices C and D lie in one half plane of side AB;*

(ii) *vertices* D *and* A *lie in one half plane of side* BC*;*

(iii) *vertices* A *and* B *lie in one half plane of side* CD*;*

(iv) *vertices* B *and* C *lie in one half plane of side* DA*.*

(c) *The diagonals* AC *and* BD *intersect at an interior point* Q*.*

(d) *The triangles* $\triangle ABC$ *and* $\triangle ADC$ *lie on different sides of line* AC*; and the triangles* $\triangle BCD$ *and* $\triangle BAD$ *lie on different sides of line* BD*.*

Proof that (i)(ii)(iv) *imply* (c). Items (i) and (ii) imply that point D lies in the interior of the angle $\angle ABC$. By the Crossbar Theorem, segment AC intersects the ray $\overrightarrow{BD}$ at some point Q.

Items (i) and (iv) imply that point C lies in the interior of the angle $\angle BAD$. By the Crossbar Theorem, segment BD intersects the ray $\overrightarrow{AC}$ at some point R.

Since no three of the four points A, B, C, D lie on a line, the lines AC and BD are distinct, and hence have a unique intersection point. Hence $Q = R$, and this point is the intersection of the diagonals AC and BD. □

Proof that (c) *implies* (a). By the assumptions $A * Q * C$ and $B * Q * D$, the three points C, D and Q lie in one half plane of line AB, confirming (i). We get three similar valid reasonings by the cyclic permutation $A \mapsto B \mapsto C \mapsto D \mapsto A$. Thus we confirm items (ii),(iii) and (iv). The four statements (i) through (iv) mean that the quadrilateral is strictly convex. □

It can now be left to the reader to complete the entire proof.

II.6.9 Topology of the ordered incidence plane

Problem II.6.19. *In an ordered incidence plane* α*, any point* P *is given. Prove there exists three rays* a, b, c *with vertex* P*, lying in the plane* α *such that*

- a *and* b *lie in different half planes of* c*;*
- b *and* c *lie in different half planes of* a*;*
- c *and* a *lie in different half planes of* b*.*

I say these three rays make a "star shape". Provide a drawing to illustrate your proof.[20]

Answer. A drawing is given on page 127. By axiom (I.3b), there exist three points in the plane α, not lying on a line. As explained in Lemma I.3.2, we deduct that there exist two points A and B, different from P, for which the lines PA and PB are different. Hence the rays $a = \overrightarrow{PA}$ and $b = \overrightarrow{PB}$ are different and lie on different lines.

By axiom (II.2), there exists a point A' such that $A * P * A'$. Similarly, there exists a point B' such that $B * P * B'$. Now we use Hilbert's proposition 3. By this proposition, there exists a point C lying between A' and B'. Define the ray $c = \overrightarrow{PC}$.

We now check that these three rays make a "star shape". Use plane separation by line of c. Points A' and B' lie in different half-planes of c; points A and A' lie in different half-planes of c;

[20] I find it rather impossible to understand the reasoning without a drawing,—hence without a drawing the problem cannot be counted.

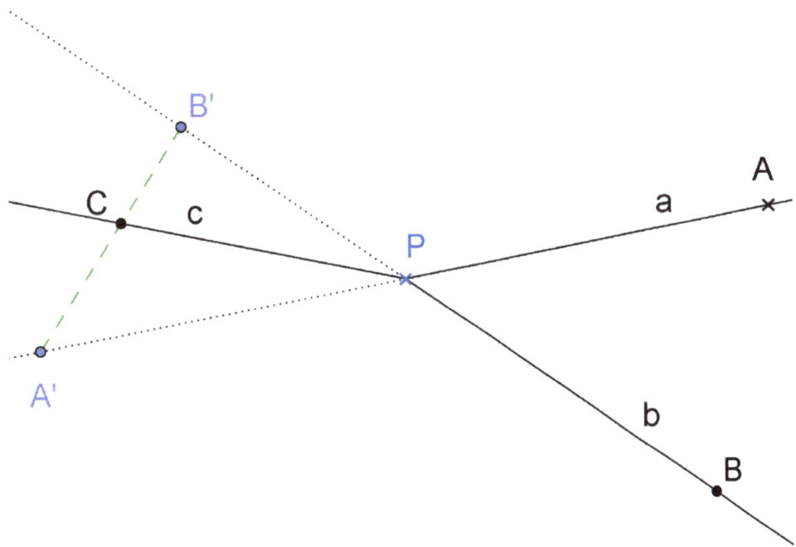

Figure II.6.17: How to get three rays from given vertex P in a star-shape.

points B and B' lie in different half-planes of c. Hence points A and B lie in different half-planes of c, and hence rays a and b lie in different half-planes of c.

Use plane separation by line of a. Points B and B' lie in different half-planes of a; points B' and C lie in the same half-plane of a. Hence points B and C lie in different half-planes of a, and hence rays b and c lie in different half-planes of a.

Use plane separation by line of b. Points A and A' lie in different half-planes of b; points A' and C lie in the same half-plane of b. Hence points A and C lie in different half-planes of b, and hence rays a and c lie in different half-planes of b.

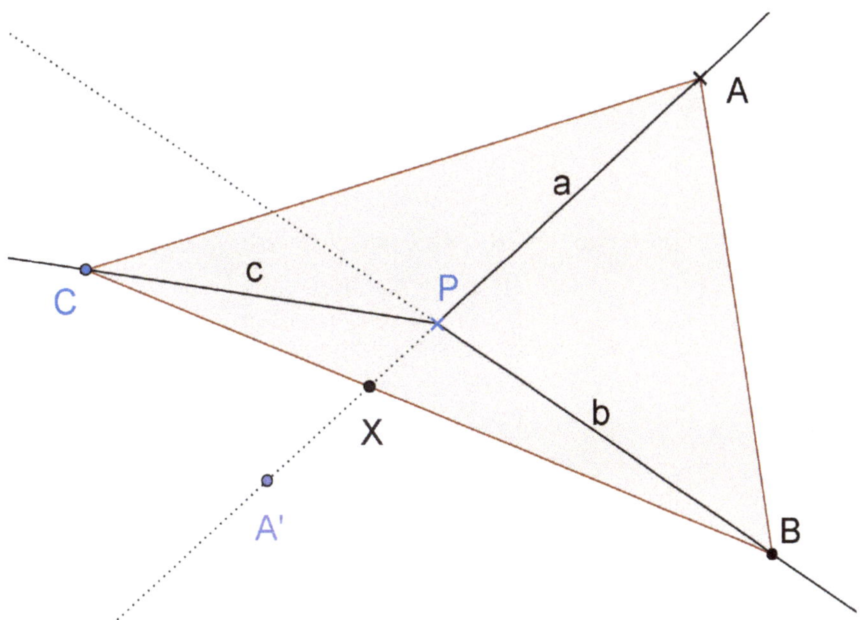

Figure II.6.18: The three ray star makes a triangle with common vertex in the interior, and conversely.

Lemma II.6.9. *Assume the three rays $a := \overrightarrow{PA}$, $b := \overrightarrow{PB}$ and $c := \overrightarrow{PC}$ are a star shape. Then any point X not lying on the three rays lies in the interior of exactly one of the angles $\angle APB$, $\angle BPC$, or $\angle CPA$.*

Proof. The reader may check that plane separation leave no other possibilities. □

Lemma II.6.10. *Given are a point P and three rays a, b, c with vertex P lying in an ordered incidence plane α. We assume these rays to make a star shape, as explained in the previous problem II.6.19. Let A, B and C be any three points lying on the three rays a, b, and c, respectively, and different from vertex P. Then the vertex P lies in the interior of triangle $\triangle ABC$.*

Proof. By axiom (II.2) there exists a point A' such that $A * P * A'$. The ray $a' := \overrightarrow{PA'}$ is the ray opposite to ray $a := \overrightarrow{PA}$.

Because of the star shape, rays $\overrightarrow{PA}$ and $\overrightarrow{PB}$ lie on different sides of line PC. By construction, the opposite rays $\overrightarrow{PA}$ and $\overrightarrow{PA'}$ lie on different sides of line PC. Hence plane separation implies that the rays $\overrightarrow{PA'}$ and $\overrightarrow{PB}$ lie on the same side of line PC.

The next argument is similar, just exchanging B and C. Because of the star shape, rays $\overrightarrow{PA}$ and $\overrightarrow{PC}$ lie on different sides of line PB. By construction, the opposite rays $\overrightarrow{PA}$ and $\overrightarrow{PA'}$ lie on different sides of line PB. Hence plane separation implies that the rays $\overrightarrow{PA'}$ and $\overrightarrow{PC}$ lie on the same side of line PB.

Both arguments together show that the ray $\overrightarrow{PA'}$ lies in the interior of angle $\angle BPC$. Because of the star shape the rays b and c do not lie on a line and this angle exists.

By the Crossbar Theorem, the segment BC intersects the ray $\overrightarrow{PA'}$. We call the intersection point X and see that $B * X * C$. Moreover $X * P * A$ since X lies on the ray $\overrightarrow{PA'}$. Thus points A and P lie on the same side of line BC. By switching A, B, C cyclically, we see that

Points A and P lie on the same side of line BC.

Points B and P lie on the same side of line CA.

Points C and P lie on the same side of line AB.

All three statements together imply that point P lies in the interior of triangle $\triangle ABC$. □

Lemma II.6.11. *Given is a triangle $\triangle ABC$ and any interior point P of the triangle. Then the three rays $a := \overrightarrow{PA}$, $b := \overrightarrow{PB}$ and $c := \overrightarrow{PC}$ are a star shape.*

Proof. No two of the three rays a, b, c lie on a line,—otherwise point P would lie on a side of the triangle. The point P and hence the ray a lie in the interior of angle $\angle BAC$. The Crossbar Theorem yields that the (opposite) ray $\overrightarrow{AP}$ intersects the side BC, say at point X. Hence the rays b and c lie on opposite sides of the line PA. By switching a, b, c cyclically, we see that

Rays a and b lie on opposite sides of line PC.

Rays b and c lie on opposite sides of line PA.

Rays c and a lie on opposite sides of line PB.

All three statements together imply that the three rays $a := \overrightarrow{PA}$, $b := \overrightarrow{PB}$ and $c := \overrightarrow{PC}$ are a star shape. □

Lemma II.6.12. *Given is a triangle and a ray r emanating from an interior point P of the triangle. The ray r either intersects exactly intersects exactly one of the vertices or one side of the triangle. Let this intersection point be called X. The interior points of segment PX lie in the interior domain of the given triangle.*

Proof. Let the ray r have vertex P in the interior of triangle $\triangle ABC$.

Note that the ray cannot intersect a vertex and still another side of the triangle,—otherwise point P would lie on a side of the triangle. In the case that the ray goes through a vertex of the triangle, we are ready.

Now assume that the ray r does not go through any vertex of the triangle. Take any point $R \neq P$ on the ray. By Lemma II.6.9 the point R, and hence the ray r lies in the interior of exactly one of the angles $\angle APB$, $\angle BPC$, or $\angle CPA$. We may assume the ray r lies in the interior of angle $\angle APB$. Hence the Crossbar theorem shows that the ray intersects the side BC at some point, say X. I leave it to the reader to check that any point between P and X lies in the interior of triangle $\triangle ABC$. □

Lemma II.6.13. *Given are three points A, B, C in the interior of a triangle $\triangle EFG$. Then the sides and vertices from triangle $\triangle ABC$, as well as the interior of $\triangle ABC$ lie in the interior of triangle $\triangle EFG$.*

Proof. The interior of any triangle is the intersection of three half-planes which are known to be convex. As already seen by Lemma II.6.7, this intersection is convex, too. Hence all three segments AB, BC and CA lie in the interior of triangle $\triangle EFG$.

In the special case that points A, B, C lie on a line we are ready. Now we assume that triangle $\triangle ABC$ exists. We take any point P in the interior of triangle $\triangle ABC$, and any line l through this point. By the result of problem II.6.15, the line either intersects two sides of the triangle, or it intersects only one side and goes through the opposite vertex. Let X_1 and X_2 be the two intersection points of the line l with the triangle. All three segments AB, BC and CA and hence the points X_i lie in the interior of triangle $\triangle EFG$. Since the interior of a triangle is convex, the entire segment X_1X_2 lies in the interior of triangle $\triangle EFG$. □

By $\mathrm{Int}_\triangle ABC$ we denote the set of interior points of triangle $\triangle ABC$.

Definition II.6.15 (The topology of an ordered incidence plane). Any subset O of an ordered incidence plane α is called *open*, if and only if for every point $P \in \alpha$, there exists a triangle $\triangle ABC$ such that P lies in the interior of this triangle and $\mathrm{Int}_\triangle ABC \subseteq O$.

Problem II.6.20. *Check for the topology from definition II.6.15 the postulates for the open sets are valid:*

(O1) *$\emptyset$ and X are open sets.*

(O2) *If U, V are open, their intersection $U \cap V$ is open.*

(O3) *If U_α is a family of open sets, with indexes $\alpha \in I$, their union $\bigcup_{\alpha \in I} U_\alpha$ is open.*

Answer. **(O1)** $\emptyset$ open is vacuously true. The entire plane α is open. Indeed given any point $P \in \alpha$, by problem II.6.19 there exist three rays a, b, c with vertex P making a star shape. As in Lemma II.6.10, we choose any three points A, B and C lying on the three rays a, b, and c, respectively, and different from vertex P. By Lemma II.6.10 the vertex P lies in the interior of triangle $\triangle ABC$. On the other hand, $\mathrm{Int}_\triangle ABC \subseteq \alpha$ holds.

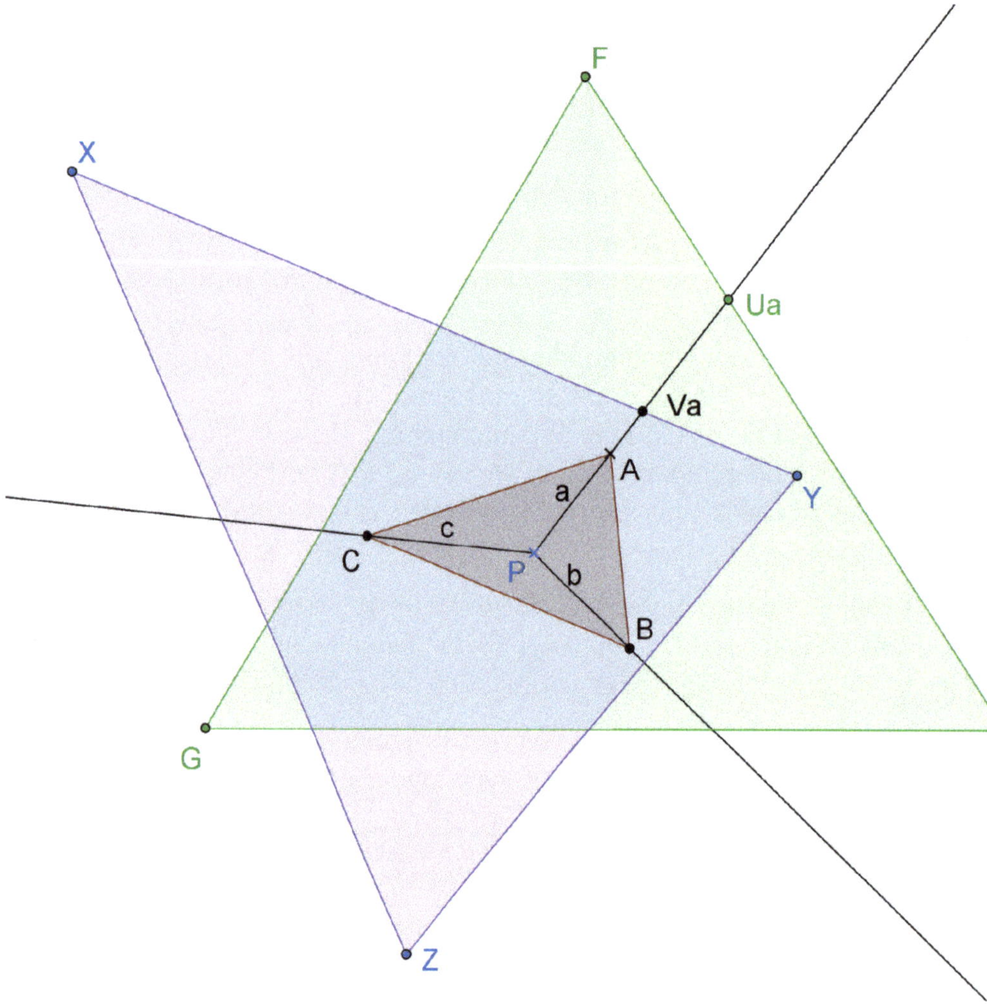

Figure II.6.19: The nonempty intersection of interiors of two triangles contains a triangle.

(O2) Take any open sets U and V. To check whether their intersection is open, take any point $P \in U \cap V$. There exist triangles $\triangle EFG$ and $\triangle XYZ$ such that $\mathrm{Int}_{\triangle} EFG \subseteq U$ and $\mathrm{Int}_{\triangle} XYZ \subseteq V$, and point P lies in the interior of both triangles.

By problem II.6.19 there exist three rays a, b, c with vertex P making a star shape. By Lemma II.6.12, the ray a intersects the (union of the sides from) triangle $\triangle EFG$ at one point U_a, and the triangle $\triangle XYZ$ at one point V_a.

By the Three-point Theorem II.6.2, either $P * U_a * V_a$ or $P * V_a * U_a$ or $U_a = V_a$, exclusively. Take the first case. By Proposition II.6.1, there exists a point A such that $P * A * U_a$. Now the Four-point Theorem II.6.5 implies the both order relations $P * A * U_a$ and $P * A * V_a$ to hold. Moreover, by Lemma II.6.12 the point A lies in the interior of both triangles $\triangle EFG$ and triangle $\triangle XYZ$.

Similarly, there exists a point B such that both $P * B * U_b$ and $P * B * V_b$ hold, and B lies in the interior of both triangles $\triangle EFG$ and triangle $\triangle XYZ$.

Finally, there exists a a point C such that both $P * C * U_c$ and $P * C * V_c$ hold, and C lies in the interior of both triangles $\triangle EFG$ and triangle $\triangle XYZ$.

By Lemma II.6.13, the sides and vertices from triangle $\triangle ABC$, as well as the interior of

$\triangle ABC$ lie in the interior of triangle $\triangle EFG$, as well as of triangle $\triangle XYZ$.

By Lemma II.6.10, the vertex P lies in the interior of triangle $\triangle ABC$. We already know that $\triangle EFG \subseteq U$ and $\triangle XYZ \subseteq V$. Altogether, we have obtained

$$P \in \mathrm{Int}_\triangle ABC \subseteq \mathrm{Int}_\triangle EFG \cap \mathrm{Int}_\triangle XYZ \subseteq U \cap V$$

We see the intersection $U \cap V$ contain a triangle $\triangle ABC$ around each one of its points, and hence is open, as to be checked.

(O3) Let U_α be a nonempty family of open sets, with indexes $\alpha \in I$. To check whether their union $U := \bigcup_{\alpha \in I} U_\alpha$ is open, take any point $x \in U$. There exist $\beta \in I$ and a triangle such that $P \in \mathrm{Int}_\triangle ABC \subseteq U_\beta$. Hence

$$P \in \mathrm{Int}_\triangle ABC \subseteq U_\beta \subseteq \bigcup_{\alpha \in I} U_\alpha$$

We see the union U contains a triangle around each one of its points, and hence is open, as to be checked.

II.6.10 Left and right, orientation

Definition II.6.16 (Coterminal rays). Two rays are called *coterminal* if one of them can be obtained from the other one by extension or deletion of a segment.

Problem II.6.21. *It is easy to see that being coterminal defines an equivalence relation among rays. Explain the details.*

Problem II.6.22. *The opposite rays of two coterminal rays are coterminal, too. Explain the details.*

Definition II.6.17 (Oriented line). An *oriented line* is a line AB together with the direction specified by the ray $\overrightarrow{AB}$, where A and B are any points lying on the line.

In other words, an oriented line is a line together with an equivalence class of coterminal rays lying on it.

Problem II.6.23. *Convince yourself that there are exactly two equivalence classes of coterminal rays on a given line. Hence there are two possibilities to assign an orientation to a line.*

Theorem II.6.1 (Orientation Theorem). *In any given plane, one can assign a left and a right half plane to every oriented line in a consistent way such that*

- *reversing the orientation of a line switches its left and right half plane;*
- *any point in the interior of an angle lies in the left half plane of one of its sides, and the right half plane of the other side of the angle.*

Hence there exist exactly two possibilities to assign a left and a right half plane to all oriented lines in a plane.

For a single first ray $\overrightarrow{OA}$, we define—rather arbitrarily or really determined by physics— one of the half-planes of line OA to be the left half-plane, and the other one the right half-plane. Typically, we take as ray $\overrightarrow{OA}$ the positive x-axis, and agree the half-plane $\{(x,y) : y > 0\}$ to be lying left to the positive x-axis. Once the left and right half planes have been agreed on for one ray, the left and a right half-plane are uniquely determined for any ray $\overrightarrow{AB}$, too.

Justification. We begin by an agreement what are to be the left and right half-plane for one fixed ray $\overrightarrow{OA}$. After that agreement, for any second ray $\overrightarrow{OB}$ with the same vertex, the left and right half-planes are determined. Indeed, if point B is left of ray $\overrightarrow{OA}$, then point A lies in the right half plane of ray $\overrightarrow{OB}$. One checks that for the opposite rays, the half-planes get exchanged.

Problem II.6.24. *Convince yourself that the statement of Theorem II.6.1 holds for any two rays $\overrightarrow{OB}$ and $\overrightarrow{OC}$ from vertex O. Begin with the case that B and C are two points in opposite half planes of $\overrightarrow{OA}$.*

Answer. It is enough to confirm the statement:

$$\text{If}\quad B \in \text{left}\,[\overrightarrow{OA}] \text{ and } C \in \text{right}\,[\overrightarrow{OA}] \text{ and } B \in \text{left}\,[\overrightarrow{OC}]$$
$$\text{then}\quad C \in \text{right}\,[\overrightarrow{OB}]$$

Point B is neither in the interior of angle $\angle AOC$ nor in the interior of the vertical angle. Hence point B lies in the interior of one of the supplementary angles of $\angle AOC$, and points A and C lie on the same side of line OB. Furthermore A is right of $\overrightarrow{OB}$. Hence C is right of $\overrightarrow{OB}$.

Problem II.6.25. *Convince yourself that the statement of Theorem II.6.1 holds for any two rays $\overrightarrow{OB}$ and $\overrightarrow{OC}$ from vertex O for which points B and C lie in the same half plane of $\overrightarrow{OA}$.*

Answer. I need to consider the cases that B and C are two points in the same half plane of $\overrightarrow{OA}$. It is enough to confirm the statement:

$$\text{If}\quad B, C \in \text{left}\,[\overrightarrow{OA}] \text{ and } B \in \text{left}\,[\overrightarrow{OC}]$$
$$\text{then}\quad C \in \text{right}\,[\overrightarrow{OB}]$$

Point B is neither in the interior of angle $\angle AOC$ nor in the interior of the vertical angle. Hence point B lies in the interior of one of the supplementary angles of $\angle AOC$, and points A and C lie on the same side of line OB. Because of the position of B, points A and C are on different sides of line OB. Furthermore A is right of $\overrightarrow{OB}$. Hence C is right of $\overrightarrow{OB}$.

In the next step, take any oriented line $\overrightarrow{AB}$, on which point O does not lie. Now left- and right half-plane have already determined by the requirement of the orientation theorem. Indeed

- if point B happens to lie in the right half plane of ray $\overrightarrow{OA}$, then B lies in the left half plane of the opposite ray $\overrightarrow{AO}$ and hence point O lies in the right half plane of ray $\overrightarrow{AB}$;
- in the opposite case, it turns out that point O lies in the left half plane of ray $\overrightarrow{AB}$ provided that point B lies in the left half plane of ray $\overrightarrow{OA}$.

We still need to verify that no contradiction can arise.

Problem II.6.26. *Convince yourself that the above definition gives the left and right planes in the same way, if the ray $\overrightarrow{AB}$ is replaced by any coterminal ray $\overrightarrow{A'B'}$.*

Is the requirement of the orientation theorem true for any given angle $\angle CAB$? Does a point in the interior of the angle lie in the left half plane of on of its sides and the right half plane of the other side of the angle?

Problem II.6.27. *Assume that point O lies in the interior of angle $\angle CAB$.*

Convince yourself that point C lies in the left half plane of ray $\overrightarrow{AB}$ if and only if point B lies in the right half plane of ray $\overrightarrow{AC}$. Hence the requirement of the orientation theorem that any point in the interior of an angle lies in the left half plane of on of its sides and the right half plane of the other side of the angle is satisfied.

Solution. By the Crossbar Theorem, the segment BC and the ray $\overrightarrow{AO}$ intersect. Hence

$$B \in \text{left}\,[\overrightarrow{OA}] \quad \Leftrightarrow \quad C \in \text{right}\,[\overrightarrow{OA}]$$

By the definition above, the point O lies in the left half plane of ray $\overrightarrow{AB}$ if and only if point B lies in the left half plane of ray $\overrightarrow{OA}$. Too, the agreement states that the point O lies in the right half plane of ray $\overrightarrow{AC}$ if and only if point C lies in the right half plane of ray $\overrightarrow{OA}$. In the end, we get the equivalences:

$$\begin{aligned} C \in \text{left}\,[\overrightarrow{AB}] \quad &\Leftrightarrow \quad O \in \text{left}\,[\overrightarrow{AB}] \quad \Leftrightarrow \quad B \in \text{left}\,[\overrightarrow{OA}] \\ \Leftrightarrow \quad C \in \text{right}\,[\overrightarrow{OA}] \quad &\Leftrightarrow \quad O \in \text{right}\,[\overrightarrow{AC}] \quad \Leftrightarrow \quad B \in \text{right}\,[\overrightarrow{AC}] \end{aligned}$$

□

The remaining cases correspond to point O lying in the vertical or one of the supplementary angles of the given angle $\angle CAB$. By applying the reasoning above to the requested vertical or one of the supplementary angle, we can verify the requirement of the orientation theorem for each of these cases.

We have confirmed there exist exactly two ways to assign a left and a right half plane to every oriented line, and have checked that the assignment is consistent. Indeed, if any point C lies in the left half-plane of any ray $\overrightarrow{AB}$, then point B lies in the right half-plane of ray $\overrightarrow{AC}$. □

Theorem II.6.2. *The interior domain of a triangle $\triangle ABC$ is either the intersection of the left half-planes of the three rays $\overrightarrow{AB}, \overrightarrow{BC}, \overrightarrow{CA}$, or the intersection of the three right half-planes of these rays. In either case, the intersection of the three opposite half-planes is empty.*

Definition II.6.18 (Orientation of a triangle). If the interior domain of a triangle $\triangle ABC$ is the intersection of the left half-planes of the three rays $\overrightarrow{AB}, \overrightarrow{BC}, \overrightarrow{CA}$, we say that the triangle is oriented *counterclockwise* or *positive*.

If the interior domain of a triangle $\triangle ABC$ is the intersection of the right half-planes of the three rays $\overrightarrow{AB}, \overrightarrow{BC}, \overrightarrow{CA}$, we say the triangle is oriented *clockwise* or *negative*.

Proposition II.6.12. *Assume the points $A \neq B$ and $P \neq Q$ lie on the same line and the points C and D do not lie on this line.*

The orientation of the triangles $\triangle ABC$ and $\triangle ABD$ are equal if C and D lie in the same half-plane of AB, and opposite if C and D lie in opposite half-planes.

The orientation of the triangles $\triangle ABC$ and $\triangle PQC$ are equal if the rays $\overrightarrow{AB}$ and $\overrightarrow{PQ}$ point in the same direction, and opposite if the rays point in opposite directions.

Indication of reason. If $A*B*C$, and point P does not lie on line ABC, then the triangles $\triangle PAB$, $\triangle PAC$ and $\triangle PBC$ all have the same orientation. □

Problem II.6.28. *Given is a triangle $\triangle ABC$ and three rays a with vertex A, ray b with vertex B, and ray c with vertex C, pointing into the interior of the triangle.*

(a) *Given a reason why any two of these rays, say rays a and b, intersect in the interior of the triangle. Define the intersection points*

$$Z := a \cap b\,,\ X := b \cap c\,,\ Y := c \cap a$$

(b) *If two of the points X, Y, Z are equal, then all three are equal. Explain why. Provide a drawing for this case.*

(c) *If the points X, Y, Z are distinct, they do not lie on a line. Explain why.*

Answer. By the Crossbar Theorem, the ray a intersects the opposite side BC, say in a point P. Similarly the ray b intersects the opposite side CA, say in a point Q.

It is convenient to introduce a notation for proofs using the plane separation theorem repeatedly. Point C and Q lie on the same side of line a, but points C and B lie on different sides of line a. We summarize these two facts by writing

$$B\,|_a\, C, Q$$

By the plane separation theorem, we conclude that points B and Q lie on on different sides of line a. Hence the segment QB intersects the line a. The intersection point R even lies on the ray a. This is confirmed in the same way as in the proof of the Crossbar Theorem. Hence point R lies the interior of angle $\angle BAC$. Switching the roles of rays a and b, we see that point R lies in the interior of angle $\angle ABC$, too. Hence point R lies in the interior of triangle $\triangle ABC$.

If two of the points X, Y, Z are equal, this point is the intersection of all three rays a, b, c. Hence $X = Y = Z$ confirming item (b). To check item (c), assume towards a contradiction that $X \neq Y$ and the three points X, Y, Z lie on a line. On this line would lie all three rays a, b, c and all three vertices A, B, C of triangle $\triangle ABC$, contradicting the definition of a triangle.

We continue the previous problem and now assume that the triangle $\triangle XYZ$ exists. Two cases can occur. Either (i) $A * Z * Y$ or (ii) $A * Y * Z$.

Problem II.6.29. *Begin with the case* (i). *The order of the points C, X, Y and the points B, X, Z is determined. Explain how this happens.*

Answer. By the assumption (i) $A * Z * Y$ and point Z lies on line b. Hence points A and Y lie on different sides of line b. Too, the points A and C lie on different sides of line b. As a shorthand we write

$$A\,|_b\, Y, C$$

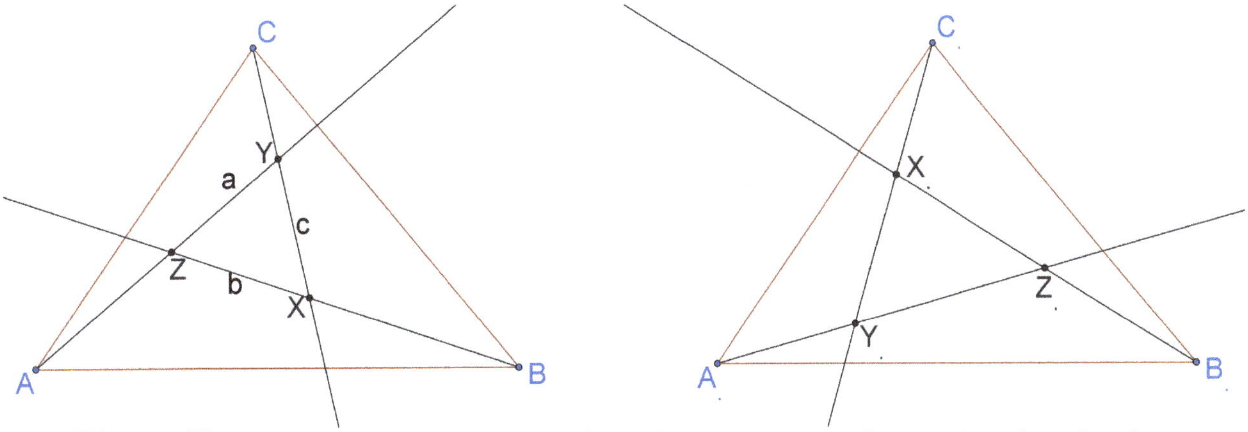

Figure II.6.20: Three rays pointing from the vertices into the interior of a triangle.

We use the plane separation theorem with line b and conclude that the points Y and C lie on the same side of line b. Now we determine the order of the points C, X, Y. Since point X lies on line b, and points Y and C lie on the same side of this line, we conclude that either $C * Y * X$ or $Y * C * X$. The last possibility is excluded since the points X, Y lie on the ray c emanating from point C. We conclude that $C * Y * X$.

To get the order of points B, X, Z, we use the half planes of line a. Points C and X are in different half planes since $C * Y * X$ and $Y \in a$. Too, points B and C are in different half planes. Indeed

$$C \,|_a\, X, B$$

Hence points X and B are in the same half plane of line a. Since point Z is on this line, one concludes $B * X * Z$.

Problem II.6.30. *Take now the case* (ii). *Determine the order of the point B, X, Z and the points C, X, Y.*

Answer. You can confirm

$$A \,|_c\, Z, B \quad \text{and} \quad B \,|_a\, X, C$$

By the assumption (ii) $A * Y * Z$. We use the plane separation theorem with line c. Since point $Y \in c$, points A and Z are on different sides of line c. Too, the points A and B lie on different sides of line c. Hence we conclude

$$A \,|_c\, Z, B$$

Hence points B, Z in the same half plane on c. Thus we determine the order of the points B, X, Z. The points X, Z lie on the ray b emanating from point B, and point X lies on line c. We conclude that $X * Z * B$.

To get the order of points C, X, Y, we use the half planes of line a. The points B and C lie on different sides of line a. Moreover point Z lies on line a and points X and B are in different half planes of line a. Hence

$$B \,|_a\, X, C$$

and points X and C are in the same half plane of line a. Since point Y is on this line, one concludes $C * X * Y$.

Problem II.6.31. *We assume that triangle $\triangle ABC$ is oriented counterclockwise. Convince yourself that each one of the two cases* (i)(ii) *considered above, the triangle $\triangle XYZ$ is oriented counterclockwise, too*

Answer. **(i)** $A * Z * Y$ and $A \in \text{left}\,[\overrightarrow{BC}]$ imply $A \in \text{left}\,[\overrightarrow{XY}]$, hence $Z \in \text{left}\,[\overrightarrow{XY}]$. Hence the triangle $\triangle XYZ$ is oriented counterclockwise.

(ii) $A * Y * Z$ and $A \in \text{right}\,[\overrightarrow{CB}]$ imply $A \in \text{right}\,[\overrightarrow{XY}]$, hence $Z \in \text{left}\,[\overrightarrow{XY}]$. Hence the triangle $\triangle XYZ$ is oriented counterclockwise.

II.6.11 The restricted Jordan Curve Theorem

Definition II.6.19 (Open and closed polygon). A list of segments $AB, BC, CD, \ldots, KL$ is called a *polygon* or polygonal curve. The points inside the segments and their endpoints are called the points of polygon. If the endpoint L is equal to the beginning point $A = L$, the polygon is called *closed*; if $A \neq L$ the polygon is *open*. The points $A, B, C, \ldots, K$ are called the *vertices*, and the segments $AB, BC, CD, \ldots, KL$ are called the *sides* of the polygon.

Definition II.6.20. A polygon or polygonal curve is called *simple*, if the vertices—with possible exception of the first and last—are all different, and any two sides do not intersect except at their common endpoint when they follow each other in the list $AB, BC, CD, \ldots, KL$.

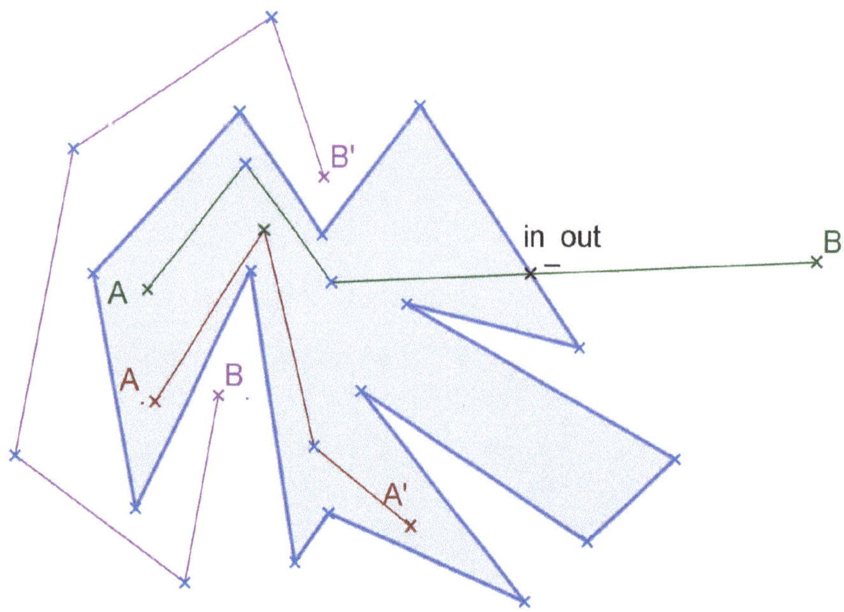

Figure II.6.21: The restricted Jordan Curve Theorem.

Proposition II.6.13 (Restricted Jordan Curve Theorem). *A simple closed polygonal curve $\mathcal{C}$ partitions the points of the plane not lying on the polygon into two regions, called the* interior *and* exterior *of the polygon, which have the following properties:*

- *If A is a point of the interior, and B a point of the exterior region, then each polygonal curve con9o090onecting A to B has at least one point in common with the curve $\mathcal{C}$.*
- *If A and A' are two points of the interior region, then there exists a polygonal curve connecting A to A' which has no point in common with the curve $\mathcal{C}$.*
- *Similarly, if B and B' are two points of the exterior region, there exists a polygonal curve connecting B to B' which has no point in common with the curve $\mathcal{C}$.*
- *There exists a line which does not cut the interior region.*
- *Every line cuts the exterior region.*

Sketch of a proof for the Euclidean plane. [21] The set $\mathbf{R}^2 \setminus \mathcal{C}$ consists of finitely many components. As one follows the polygon $\mathcal{C}$, the points near to the right, are all in the same component as some reference point R. Similarly, the points near to the left are all in the same component as some reference point L. Furthermore, one can choose points R and L near to the same side of the polygon $\mathcal{C}$ and to each other. Any point X in $\mathbf{R}^2 \setminus \mathcal{C}$ can be connected by a polygonal curve inside this set $\mathbf{R}^2 \setminus \mathcal{C}$ to one of these these two points R and L. Hence $\mathbf{R}^2 \setminus \mathcal{C}$ consists of at most two connected components.

To prove that the points on the left and the right lie in different components, consider rays. We call a ray *exceptional*, if it contains a vertex of the curve $\mathcal{C}$. For a given vertex P, there exist at most finitely many directions of exceptional rays.

Each ray that is not exceptional, has only finitely many intersection points with the curve $\mathcal{C}$. As one turns the ray around a fixed vertex P, the number of intersection points changes only at the exceptional directions. Nearby, before and after passing through an exceptional direction, the parity of the number of intersection points is the same. Hence we can assign a *even parity* or *odd parity* to the vertex P, depending of the number of intersection points being even or odd.

Next we show that the parity is constant inside each component of the set $\mathbf{R}^2 \setminus \mathcal{C}$. Let X and Y be two points in the same component. There exists a polygonal curve $\mathcal{P}$ connecting X and Y inside this component. We can assume that the rays along the sides of curve $\mathcal{P}$ are not exceptional. This can be achieved by an arbitrarily small adjustment of the curve $\mathcal{P}$.

After this adjustment, the two endpoints of each segment of $\mathcal{P}$ have the same parity, since the intersection set of the ray with $\mathcal{C}$ changes continuously as one moves the vertex of the ray along one side of $\mathcal{P}$, keeping the ray parallel. Hence any two points in the same component of $\mathbf{R}^2 \setminus \mathcal{C}$ have the same parity.

The endpoints of a small segment crossing the polygon $\mathcal{C}$ have opposite parities. Points far away from the polygon have even parity, because there exists a ray not intersecting the curve $\mathcal{C}$. Hence the unbounded outside component of $\mathbf{R}^2 \setminus \mathcal{C}$ has even parity, and the inside bounded component has odd parity. □

Remark. A complete proof in the present context is given by G. Feigl [11]. The really interesting point is that none of the further axioms of congruence, continuity or parallels are needed. Especially, the Jordan Curve Theorem holds for the hyperbolic plane, too.

On the other hand, the Restricted Jordan Curve Theorem II.6.13 is not valid for the projective plane, and hence neither in double elliptic geometry. The reason is that the improper points and line of the projective plane change the topological structure.

[21] This proof follows H. Tverberg [32].

Remark. The more general *Jordan curve theorem* says that any closed simple continuous curve separates the Euclidean plane into an interior and exterior domain. It is credited to Camille Jordan (1836-1922), but Jordan's original argument was in fact inadequate. The first correct proof is credited to Oswald Veblen (1880-1960).

II.7 Congruence of Segments, Angles and Triangles

This section deals with David Hilbert's axiomatization of neutral geometry, following Hilbert's foundations of geometry with slight modifications.

II.7.1 Congruence of segments

Proposition II.7.1 (Congruence is an equivalence relation). *Congruence is an equivalence relation on the class of line segments.*

Question. Which three properties do we need to check for a congruence relation?

Answer. For an arbitrary relation to be a congruence relation, we need to check

(a) **reflexivity:** Each segment is congruent to itself, written $AB \cong AB$.

(b) **symmetry:** If $AB \cong A'B'$, then $A'B' \cong AB$.

(c) **transitivity:** If $AB \cong CD$, and $CD \cong EF$, then $AB \cong EF$.

Proof of reflexivity, for a bottle of wine from Hilbert personally. [22] We need to show that each segment is congruent to itself. Take any given segment AB. We transfer the segment to a ray starting from any point C. By Hilbert's axiom (III.1) , there exists a point D on that ray such that $AB \cong CD$. Now use Hilbert's axiom (III.2) :

"If $A'B' \cong AB$ and $A''B'' \cong AB$, then $A'B' \cong A''B''$."

Hence, for a case with other notation, $AB \cong CD$ and $AB \cong CD$ imply $AB \cong AB$. □

Proof of symmetry. Assume that $AB \cong CD$. Because of reflexivity $CD \cong CD$. Once more, we use Hilbert's axiom (III.2) :

"If $A'B' \cong AB$ and $A''B'' \cong AB$, then $A'B' \cong A''B''$."

Hence $CD \cong CD$ and $AB \cong CD$ imply $CD \cong AB$. □

Proof of transitivity. Assume $AB \cong CD$ and $CD \cong EF$. Because of symmetry $EF \cong CD$. Now we use Hilbert's axiom (III.2) :

"If $A'B' \cong AB$ and $A''B'' \cong AB$, then $A'B' \cong A''B''$."

For a case with other notation, this means that $AB \cong CD$ and $EF \cong CD$ imply $AB \cong EF$, as to be shown. □

[22] which one can take gracefully as a thank-you—for translating from page 15 of the millenium edition of "Grundlagen der Geometrie"–etc.

Proposition II.7.2 (Existence and uniqueness of segment transfer). *Given a segment AB and given a ray r originating at point A', there exists a unique point B' on the ray r such that $AB \cong A'B'$.*

Question. Which part of this statement is among Hilbert's axioms? Which axiom is used?

Answer. The existence of the segment $A'B'$ is postulated in Hilbert's axiom of congruence (III.1).

Question. How does the uniqueness of segment transfer follow? Which axioms are needed for that part?

Answer. The uniqueness of segment transfer follows from the uniqueness of angle transfer, stated in (III.4), and the SAS Axiom (III.5).

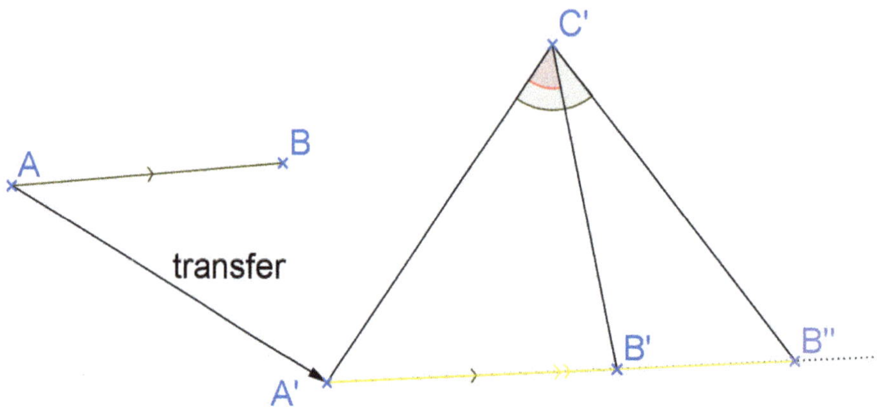

Figure II.7.1: Uniqueness of segment transfer

Proof of uniqueness. Assume the segment AB can be transferred to the ray r from A' in two ways, such that both $AB \cong A'B'$ and $AB \cong A'B''$. We choose a point C' not on the line $A'B'$. One obtains the congruences

$$A'B' \cong A'B'' \,,\; A'C' \cong A'C' \,,\; \angle B'A'C' \cong \angle B''A'C'$$

We are using axiom (III.2), the fact that a segment is congruent to itself, and an angle is congruent to itself by axiom (III.4c). By the axiom (III.5), this implies $\angle A'C'B' \cong \angle A'C'B''$. By the uniqueness of angle transfer, as stated in axiom (III.4b), this implies that the rays $\overrightarrow{C'B'} = \overrightarrow{C'B''}$ are equal. Hence $B' = B''$ is the unique intersection point of the two different rays $r = \overrightarrow{A'B'}$ and $\overrightarrow{C'B'}$. We have shown that $AB \cong A'B'$ and $AB \cong A'B''$ imply $B' = B''$. Thus segment transfer is unique. □

Proposition II.7.3 (Subtraction of segments). *Given are three points on a line such that $A * B * C$, and two points B' and C' on a ray $\overrightarrow{A'B'}$ emanating from A'. Suppose that*

$$AB \cong A'B' \,,\; AC \cong A'C'$$

then $BC \cong B'C'$ and $A' * B' * C'$ follow.

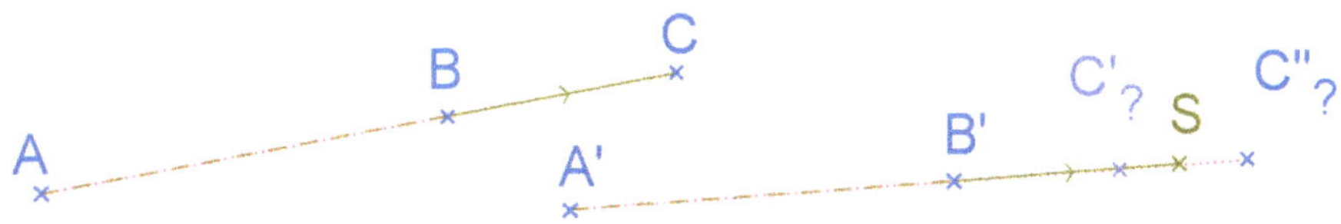

Figure II.7.2: Segment subtraction

Proof. On the ray originating from B', opposite to $\overrightarrow{B'A'}$, we transfer segment BC. We get the seventh point S such that $BC \cong B'S$, and $A' * B' * S$. By Hilbert's axiom (III.3) on segment addition, $AB \cong A'B'$ and $BC \cong B'S$ imply now $AC \cong A'S$.

On the other hand, it is assumed that $AC \cong A'C'$. Both points S and C' lie on the same ray $\overrightarrow{A'B'}$. As explained in Proposition II.7.2, the transfer of segment AC is unique. Hence we get $C' = S$. Finally we substitute equals. Thus $BC \cong B'S$ and $A' * B' * S$ imply $BC \cong B'C'$ and $A' * B' * C'$, as to be shown. □

Definition II.7.1 (Segment comparison). For any two given segments AB and CD we say that AB is less than CD, iff there exists a point E between C and D such that $AB \cong CE$ and $C * E * D$. In this case, also say that CD is greater than AB. We write $CD > AB$ or $AB < CD$. equivalently.

Proposition II.7.4 (Segment comparison holds for congruence classes). *Assuming $AB \cong A'B'$ and $CD \cong C'D'$, we get: $AB < CD$ if and only if $A'B' < C'D'$.*

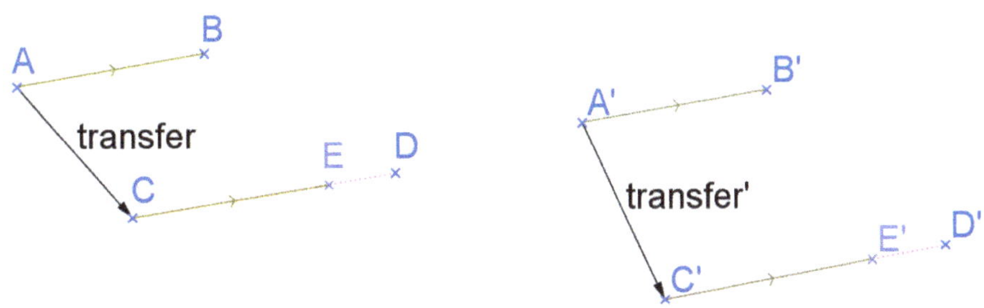

Figure II.7.3: Segment comparison for congruence classes

Proof. Transfer segment AB onto ray $\overrightarrow{CD}$, and get $AB \cong CE$. The assumption $AB < CD$ implies that $C * E * D$. Too, we transfer segment $A'B'$ onto ray $\overrightarrow{C'D'}$, and get $A'B' \cong C'E'$.

We now use segment subtraction for points C, E, D and C', E', D'. Since $CD \cong C'D'$ by assumption, and $CE \cong AB \cong A'B' \cong C'E'$—by assumption and construction and transitivity—, and $C * E * D$; segment subtraction yields $ED \cong E'D'$ and $C' * E' * D'$. Thus E' lies between C' and D', from which we conclude $A'B' < E'D'$. □

Proposition II.7.5 (Transitivity of segment comparison). *If $AB < CD$ and $CD < EF$, then $AB < EF$.*

Proof. I can assume that all three segments lie on the same ray $\overrightarrow{AB}$, and $A = C = E$. (Some transferring of segments will produce segments congruent with the given ones which satisfy these requirements.) Now, having done this, we get $AB < AD$ and $AD < AF$. By definition this means $A * B * D$ and $A * D * F$.

Question. Express these order relations in words.

Answer. Point B lies between A and D, and point D lies between A and F.

By Theorem 5 from Hilbert's foundations, any four points on a line can be notated in a way that all four alphabetic order relations hold. (I shall that statement the "Four point Theorem"). Now the four points A, B, D, F satisfy the order relations $A * B * D$ and $A * D * F$, which implies they are already notated in alphabetic order. Hence the two other order relations follow: $A * B * F$ and $B * D * F$. But $A * B * F$ means by definition that $AB < AF$, as to be shown. □

Figure II.7.4: Transitivity of segment comparison

Corollary 6. *If $AB < AD$ and $AD < AF$, then $BD < BF$.*

Proposition II.7.6 (Any two segments are comparable). *For any two segments AB and CD, one and only one of the three cases* (i)(ii)(iii) *occurs:*

Either (i) $AB < CD$ *or* (ii) $AB \cong CD$ *or* (iii) $CD < AB$.

Proof. Transfer segment AB onto the ray $\overrightarrow{CD}$. We get $AB \cong CB'$. Now the two segments CB' and CD start at the same vertex and lie on the same ray.

By Theorem 4 from Hilbert's foundations, for any given three points on a line, exactly one lies between the other two. Hence the three points C, B', D can satisfy

either (i) $C * B' * D$. or (ii) $B' = D$. or (iii) $C * D * B'$. These cases correspond to the three cases as claimed. Indeed, in case (i), $C * B' * D$ implies by definition $AB < CD$. Indeed, in case (ii), $B' = D$ implies $AB \cong CB' = CD$ by construction. In case (iii), we get $CD < CB'$. Since $AB \cong CB'$, Proposition II.7.4 implies $CD < AB$, as to be shown. □

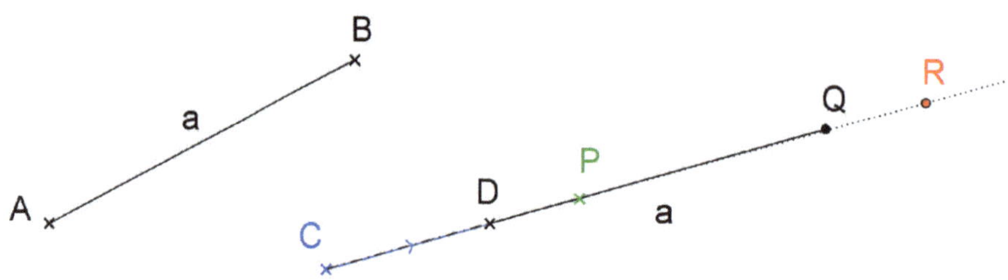

Figure II.7.5: To extend a line—.

Lemma II.7.1 ("simple fact"). *Given a segment AB and a ray $\overrightarrow{CD}$, there exists a point R on this ray such that the segment DR is longer than the segment AB.*

Proof. By the axiom of order (II.2), there exists a point P such that $C * D * P$. By the axiom of congruence (III.1), there exists a point Q on the line CD, lying on the same side of D as point P, such that $AB \cong DQ$. By the axiom of order (II.2), there exists a point R such that $P * Q * R$.

In the drawing on page 142, the given ray $\overrightarrow{CD}$ is extended to obtain a point P outside the segment CD. Next we transfer the given segment AB onto the extension ray $\overrightarrow{DP}$. Finally, the ray $\overrightarrow{PQ}$ can be extended even more to get a point R outside the segment PQ.

The construction has produced points C, D, Q, R on a line satisfying the first two order relations below. We invoke the Four point Theorem, Proposition II.6.5.

$$\begin{array}{ll} C{*}D{*}Q & \text{and} \\ \quad D{*}Q{*}R & \text{imply} \\ C{*}D{*}\quad R & \end{array}$$

The third order relation confirms that point R lies on the ray $\overrightarrow{CD}$. The order relation $D{*}Q{*}R$ means that the segment DR is longer than the segment DQ. By construction, the segments $AB \cong DQ$ are congruent. We see that

$$DR > DQ \cong AB$$

By Proposition II.7.4, segment comparison holds for congruence classes. We conclude $DR > AB$, as required. □

Remark. What does Euclid's Second Postulate *"To extend a line"* really mean?

A bid of thought shows that this postulate means more than just Hilbert's axiom of order (II.2). Even extenbding as many times as one wants needs not lead very far. Remember Achilles and the tortoise! But clearly we often need to extend a given line beyond a given segment. This process involves a combination of the axiom of order (II.2) with the axiom of congruence (III.1) for the segment transfer.

Another situation occurs when one knows, for any other reason, some point E to exist beyond which one has to extend the line. Even in that situation, we need to invoke the Four-point Theorem (Hilbert's Proposition 5) for the possibility of further extension beyond E. So we end up with different possible interpretations of Euclid's postulate!

Definition II.7.2 (Sum of segments). The sum of any two segments AB and EF is defined to be the segment AC were C is the point such that $A * B * C$ and $EF \cong BC$.

In other words, the sum segment $AB + EF$ is obtained by extending the ray $\overrightarrow{AB}$ to the point C by a segment BC congruent to the second summand EF.

Proposition II.7.7 (Congruence classes of segments are an ordered Abelian group). *The sum of segments is defined on equivalence classes of congruent segments because of axiom (III.3) and the existence and uniqueness of segment transfer, as shown in Proposition II.7.2. These equivalence classes have the following properties:*

Commutativity $a + b = b + a$.

Associativity $(a + b) + c = a + (b + c)$

Comparison *Any two segments a and c satisfy either $a < c$, or $a = c$ or $a > c$.*

Difference *$a < c$ if and only if there exists b such that $a + b = c$.*

Comparison of Sums *If $a < b$ and $c < d$, then $a + c < b + d$.*

Put together with their negatives $-a$ for each congruence class a, and finally the neutral element 0, these equivalence classes are an ordered Abelian group.

Proof. By Hilbert's axiom of congruence (III.3), the sum is defined on equivalence classes of congruent segments. Now we check the items stated:

Commutativity: Let segment AB represent the equivalence class a. By axiom (III.1), we can choose point C such that $A*B*C$, and segment BC represents the equivalence class b. Since $BC \cong CB$, and $AB \cong BA$, these latter two segments represent the classes b and a. By the definition of segment addition $CA = CB + BA$, and the segment CA represents $b + a$. Since $AC \cong CA$, we conclude $a + b = b + a$.

Associativity: We construct $(a + b) + c$. To this end, choose a segment AB of congruence class a, in short $AB \in a$. Again, choose point C such that $A * B * C$, and segment $BC \in b$. Furthermore, choose point D such that $A * C * D$, and segment $CD \in c$. We now know that

$$AD \in (a + b) + c$$

On the other hand, we begin by constructing at first $b+c$. By the four-point theorem, $A*B*C$ and $A * C * D$ imply $B * C * D$. Hence $BD \in b + c$. The construction of $a + (b + c)$ is finished by finding point H such that $A * B * H$, and segment $BH \in b + c$. We now know that

$$AH \in a + (b + c)$$

The uniqueness of segment transfer implies $H = D$. Finally, we see that $AD = AH$, and hence $(a + b) + c = a + (b + c)$.

Comparison: Let any segments a and c be given. These equivalence classes can be represented by segments $AB \in a$ and $AC \in c$ on the same ray $\overrightarrow{AB} = \overrightarrow{AC}$. The three point theorem implies that either $B = C$, or $A * B * C$, or $A * C * B$. Hence either $a = c$, or $a < c$, or $c < a$. In the last case $a > c$ as shown above.

Difference: Assume $a < c$. These equivalence classes can be represented by segments $AB \in a$ and $AC \in c$ such that $A * B * C$. Now BC represents the class b such that $a + b = c$. The converse is as obvious.

Comparison of Sums The proof is left to the reader.

□

Proposition II.7.8 (Comparison of supplements). *Let a segment PQ and two points A, B in it be given. Assume that $PA < PB$. Then $BQ < AQ$.*

Proof. The reader has to try himself. □

II.7.2 Some elementary triangle congruences

Definition II.7.3 (Triangle, Euler's notation). For a *triangle* $\triangle ABC$, it is assumed that the three vertices do not lie on a line. I follow *Euler's conventional notation* for vertices, sides and angles: in triangle $\triangle ABC$, let A, B and C be the vertices, let the segments $a = BC$, $b = AC$, and $c = AB$ be the sides and the angles $\alpha := \angle BAC$, $\beta := \angle ABC$, and $\gamma := \angle ACB$ be the angles.

For a segment AB, it is assumed that the two endpoints A and B are <u>different</u>. For a triangle $\triangle ABC$, it is assumed that the three vertices <u>do not lie on a line</u>.

Definition II.7.4 (Congruence of triangles). Two triangles are called *congruent* iff all three pairs of corresponding sides, and all three pairs of corresponding angles are congruent. Thus the congruence of triangles $\triangle ABC \cong \triangle A'B'C'$ means the six congruences

$$\begin{gathered} AB \cong A'B' \,,\; BC \cong B'C' \,,\; CA \cong C'A' \;\text{ and} \\ \angle ABC \cong \angle A'B'C' \,,\; \angle BCA \cong \angle B'C'A' \,,\; \angle CAB \cong \angle C'A'B' \end{gathered} \tag{II.7.1}$$

Note that one has to put the vertices, sides and angles of the two triangles into a definite order.

Definition II.7.5 (Isosceles triangle, equilateral triangle, right triangle). A triangle with two congruent sides is called *isosceles.* A triangle for which all three sides are congruent, is called *equilateral.* A triangle with at least one right angle is called a right triangle.

Problem II.7.1. *For which geometries does exist an isosceles right triangle?*

Answer. An isosceles right triangle exists in any Hilbert plane, especially in Euclidean and hyperbolic geometry. In spherical geometry, too, exists an isosceles right triangle.

Proposition II.7.9 (Isosceles Triangle Proposition). [Euclid I.5, and Hilbert's theorem 11] *In a triangle with two congruent sides, the opposite angles are congruent. An isosceles triangle has congruent base angles.*

Question. Formulate the theorem with specific quantities from a triangle $\triangle ABC$ like drawn in the figure on page 146.

Answer. If $b \cong c$, then $\beta \cong \gamma$.

Proof. Assume that the sides $AB \cong AC$ of triangle $\triangle ABC$ are congruent. We need to show that the base angles $\beta = \angle ABC$ and $\gamma = \angle BCA$ are congruent. Define a second triangle $\triangle A'B'C'$ by setting [23]

$$A' := A \,,\; B' := C \,,\; C' := B$$

To apply SAS congruence, we match corresponding pieces:

(1) $\angle BAC \cong \angle CAB$ since by axiom (III.4d), an angle is congruent to the angle with the sides reversed. As the points are defined, we know $\angle CAB = \angle B'A'C'$. Hence $\angle BAC \cong \angle B'A'C'$.

(2) $AB \cong A'B'$.

[23] It does not matter that the second triangle is just "on top" of the first one.

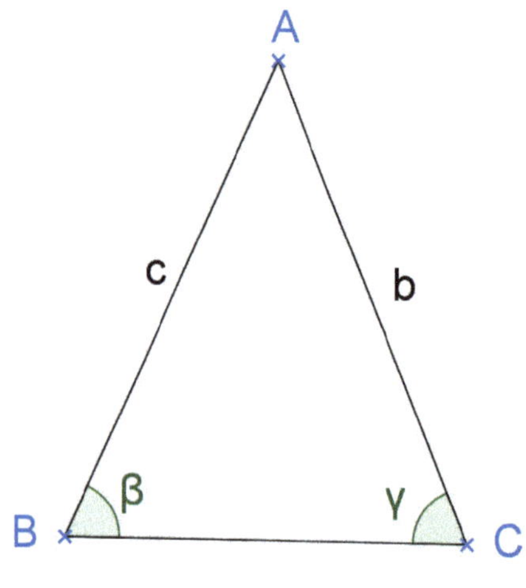

Figure II.7.6: An isosceles triangle

Question. Explain why this holds.

Answer. $AB \cong AC$ because we have assumed the triangle to be isosceles, and $AC = A'B'$ by construction. Hence $AB \cong AC$.

(3) Similarly, we show $AC \cong A'C'$: Indeed, $AC \cong AB$ because we have assumed the triangle to be isosceles, and congruence is a symmetric relation, and $AB = A'C'$ by construction. Hence $AC \cong A'C'$.

Finally, we use axiom (III.5). Items (1)(2)(3) imply $\angle ABC \cong \angle A'B'C' = \angle ACB$. But this is just the claimed congruence of base angles. □

Remark. This proof is far simpler than Euclid's proof for the congruence of the base angles of an isosceles triangle. It is attributed to Pappus of Alexandria, A.D. 300. Hilbert only gives the hint that Axiom (III.4d) and (III.5) are used for the proof.

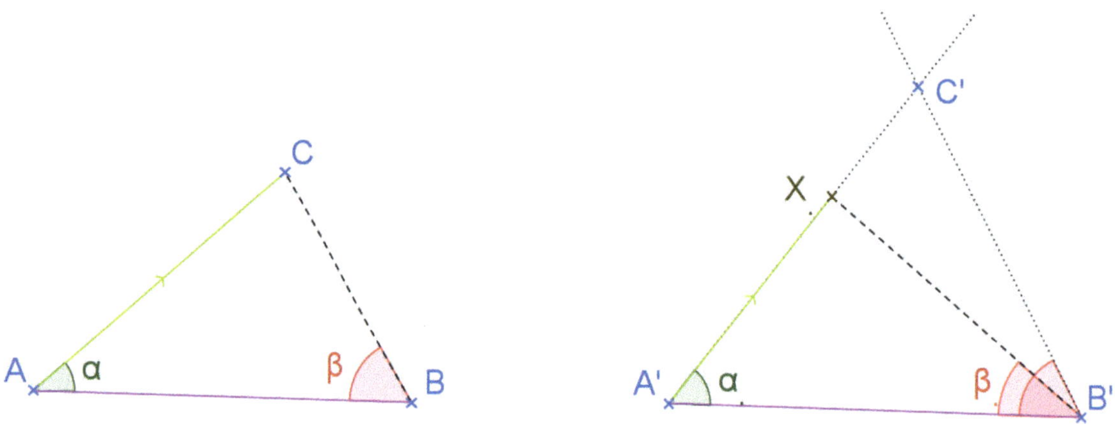

Figure II.7.7: How to simply get ASA congruence

Proposition II.7.10 (ASA Congruence). [Theorem 13 in Hilbert] *Two triangles with a pair of congruent sides, and pairwise congruent adjacent angles are congruent.*

Independent proof of ASA congruence. Given are the triangles $\triangle ABC$ and $\triangle A'B'C'$, with congruent sides $AB \cong A'B'$ and two pairs of congruent adjacent angles $\angle BAC \cong \angle B'A'C'$ and $\angle ABC \cong \angle A'B'C'$. The segment AC is transferred onto the ray $\overrightarrow{A'C'}$. On this ray, one gets a point X such that

$$AC \cong A'X \tag{II.7.2}$$

Now axiom (III.5) is applied to the triangles $\triangle ABC$ and $\triangle A'B'X$. In the figure on page 146, the three pairs of matching pieces used to prove the congruence are stressed in matching colors. From axiom (III.5) one concludes

$$\angle ABC \cong \angle A'B'X \tag{II.7.3}$$

We now have obtained both $\angle ABC \cong \angle A'B'X$, and $\angle ABC \cong \angle A'B'C'$ as assumed, too. Hence uniqueness of angle transfer, postulated by axiom (III.4b), implies the rays to be equal:

$$\overrightarrow{B'X} = \overrightarrow{B'C'}$$

Thus the point X lies both on this ray, and by construction on the ray $\overrightarrow{A'X} = \overrightarrow{A'C'}$, too. By axiom (I.2), these two rays have a unique intersection point, since they do not lie on the same line, and hence $X = C'$. By substitution of equals, equations (II.7.2) and (II.7.3) imply

$$AC \cong A'C' \quad \text{and} \quad \angle ABC \cong \angle A'B'C' \tag{II.7.4}$$

Finally we show $BC \cong B'C'$ by a similar argument with vertices A and B exchanged. Hence the two triangles are congruent, as to be shown. □

Proposition II.7.11 (SAS Congruence). [Theorem 12 in Hilbert] *Given are two triangles. We assume that two sides and the angle* between *these sides are congruent to the corresponding pieces of the second triangle. Then the two triangles are congruent.*

Short proof of SAS-congruence Proposition II.7.11. Given are two triangles $\triangle ABC$ and $\triangle A'B'C'$. We assume that the angles at A and A' as well as two pairs of adjacent sides are matched:

$$AB \cong A'B' \,,\; AC \cong A'C' \,,\; \angle BAC \cong \angle B'A'C'$$

By axiom (III.5) we conclude

$$\angle ABC \cong \angle A'B'C'$$

Now the ASA-congruence stated in Proposition II.7.10 above yields the triangle congruence $\triangle ABC \cong \triangle A'B'C'$ to be shown. □

Question. We see from Proposition II.7.11 that Hilbert's SAS-axiom (III.5) implies the full SAS congruence theorem. Why is Hilbert's Axiom (III.5) weaker than the SAS congruence theorem?

Answer. In Hilbert's Axiom, only congruence of a further pair of angles is postulated. In the SAS congruence theorem, all pieces of the two triangles are stated to be pairwise congruent.

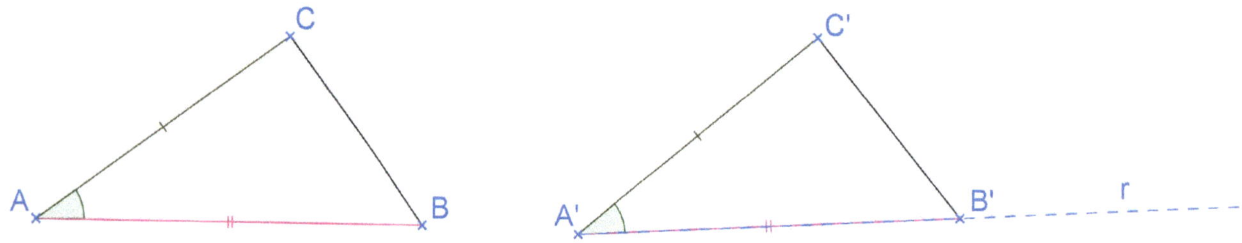

Figure II.7.8: Transfer of a triangle.

Proposition II.7.12 (Transfer of a triangle). *Any given triangle $\triangle ABC$ can be transferred into any given half-plane $\mathcal{H}'$ of any given ray r, such that one obtains a congruent triangle $\triangle A'B'C'$ lying in the prescribed half-plane, and the given ray is $r = \overrightarrow{A'B'}$, emanates from vertex A' and lies on the side $A'B'$.*

Problem II.7.2. *Provide a drawing for Proposition II.7.12 and prove the Proposition, starting from the axioms and ASA-congruence. State clearly which axioms you use, and where one has to use the ASA-congruence.*

Proof. Using axiom (III.1), we transfer the segment AB onto the ray r to obtain the congruent segment $AB \cong A'B'$. Next we use axiom (III.4), and transfer the angle $\angle BAC$ into the given half-plane $\mathcal{H}'$ to obtain the congruent angle $\angle BAC \cong \angle(r,k)$. Thus the newly produced angle has the given ray r as one side. Onto the other side k of the newly produced angle, we transfer the segment AC and obtain the congruent segment $AC \cong A'C'$. Since by construction $r = \overrightarrow{A'B'}$ and $k = \overrightarrow{A'C'}$, we see that $\angle BAC \cong \angle B'A'C'$.

From axiom (III.5)—which we now use even literally!—we conclude the congruence $\angle ABC \cong \angle A'B'C'$. We are now in the position to use the ASA-congruence, stated in Proposition II.7.10 for the triangles $\triangle ABC$ and $\triangle A'B'C'$. Hence we conclude that the triangle congruence $\triangle ABC \cong \triangle A'B'C'$, as to be shown. □

Proposition II.7.13 (Extended ASA-Congruence). *Given is a triangle, a segment congruent to one of its sides, and a half-plane $\mathcal{H}'$ bounded by this segment. The two angles at the vertices of this side are transferred to the endpoints of the segment, and reproduced in the same half plane. Then the newly constructed rays intersect, and one gets a second congruent triangle.*

Short proof of the extended ASA-congruence Proposition II.7.13. Given are the triangle $\triangle ABC$ and the segment $A'B' \cong AB$. We define the ray $r = \overrightarrow{A'B'}$. By Proposition II.7.12 about the transfer of a triangle, we transfer the triangle $\triangle ABC$ into the given half-plane $\mathcal{H}'$. We obtain a congruent triangle $\triangle A'B'C'$ with vertex C' lying in the prescribed half-plane. □

Question. In which point does the extended ASA congruence theorem extend the usual ASA congruence theorem?

Answer. The extended ASA theorem differs from the usual ASA-congruence theorem, because existence of a second triangle is not assumed—besides the first triangle, only a second segment is given. It is proved that the two newly produced rays do intersect.

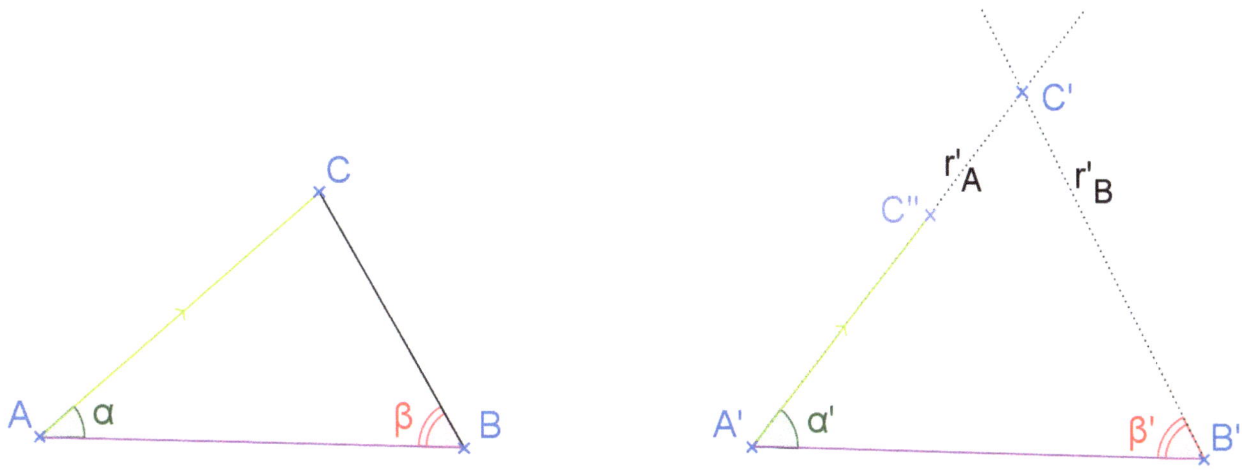

Figure II.7.9: Extended ASA congruence

Proposition II.7.14 (Preliminary Converse Isosceles Triangle Proposition). *If the two base angles of a triangle are congruent to each other, the triangle is isosceles.*

Problem II.7.3.

- *Formulate the proposition II.7.14 above with specific quantities from a triangle $\triangle ABC$.*
- *Provide a drawing.*
- *Prove the theorem as stated.*

Proof. I use the formulation: "If $\beta \cong \gamma$ and $\gamma \cong \beta$, then $b \cong c$." I may still use the drawing from page 146. We use ASA-congruence to prove the proposition. Assume that the angles $\beta = \angle ABC$ and $\gamma = \angle ACB$ are congruent in $\triangle ABC$. We need to show that the two sides $b = AC$ and $c = AB$ are congruent. Define a second triangle $\triangle A'B'C'$ by setting once more

$$A' := A\,,\ B' := C\,,\ C' := B$$

To apply ASA congruence, we match corresponding pieces:

(1) $BC \cong B'C'$**:** Indeed $BC = C'B' = B'C'$, because a segment is congruent to itself and the order of the endpoints of a segment is arbitrary. Hence $BC \cong B'C'$.

(2) $\beta \cong \beta'$**:** Indeed, $\beta \cong \gamma$ by assumption, and $\gamma = \angle ACB = \angle A'B'C' = \beta'$ by construction. Hence $\beta \cong \beta'$.

(3) Similarly, one shows that $\gamma \cong \gamma'$: $\gamma \cong \beta$ by assumption, and $\beta = \angle ABC = \angle A'C'B' = \gamma'$ by construction. Hence $\gamma \cong \gamma'$.

Via ASA congruence, items (1)(2)(3) imply that $\triangle ABC \cong \triangle A'B'C'$, and hence especially $AB \cong A'B' = AC$ as to be shown. □

Remark. Later on in Proposition II.7.31, we shall avoid the awkward assumption: "If the two base angles of a triangle are congruent to each other". Hilbert proves the converse isosceles triangle theorem as his Theorem 24, after having the exterior angle theorem at his disposal.

II.7.3 Congruence of angles

Recall definition II.6.9 of *supplementary angles*, and definition II.6.10 of *vertical angles*.

Remark. After having proved the results below, we may also correctly say that a pair of supplementary angles adds up to two right angles, or 180°. But this statement is not an acceptable definition for supplementary angles. Indeed one has produced a bad example for *circular reasoning.* The notion of right angle can only be defined,—by definition II.7.8 below,—to be an angle congruent to its supplementary angle. Such a definition requires that one already knows what supplementary angles are.

Remark. Nevertheless, it has become common usage to talk about angles adding to 180° as supplementary angles in a relaxed sense. In order to distinguish the relaxed notion from the strict definition II.6.9, the term "linear pair" has been suggested for the primary basic notion. That practice is only followed by some authors, but neither by Hilbert nor by Hartshorn.

Proposition II.7.15 (Congruence of Supplementary Angles). [Theorem 14 of Hilbert] *If an angle $\angle ABC$ is congruent to another angle $\angle A'B'C'$, then its supplementary angle $\angle CBD$ is congruent to the supplementary angle $\angle C'B'D'$ of the second angle.*

Proof. The three steps of the proof each identify a new pair of congruent triangles.
Step 1: One can choose the points A', C' and D' on the given rays from B such that

$$AB \cong A'B' \,,\; CB \cong C'B' \,,\; DB \cong D'B'$$

Because of the assumption $\angle ABC \cong \angle A'B'C'$, SAS congruence (Theorem 12 of Hilbert, see Proposition II.7.11 above) now implies that $\triangle ABC \cong \triangle A'B'C'$. In the drawing, the three pairs of matching pieces used to prove the congruence are stressed in matching colors. Congruence of the

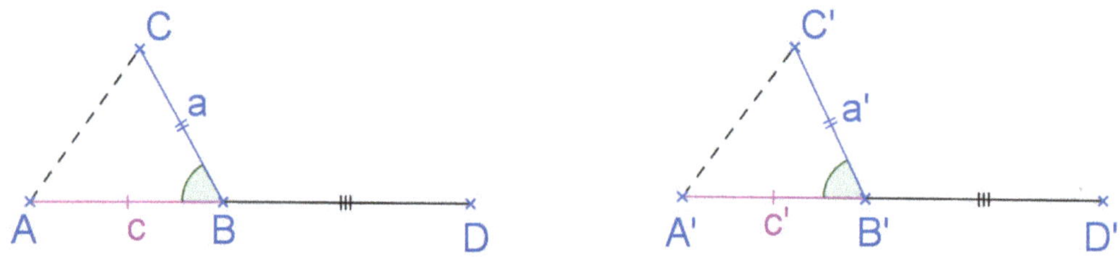

Figure II.7.10: Congruence of supplementary angles, the first pair of congruent triangles

two triangles implies

$$(1) \qquad AC \cong A'C' \text{ and } \quad \angle BAC \cong \angle B'A'C'$$

Step 2: By axiom (III.3), adding congruent segments yields congruent segments. Hence the segments AD and $A'D'$ are congruent. Now SAS congruence (Theorem 12 of Hilbert, see Proposition II.7.11 above) implies that the (greater) triangles, too, are congruent in the two figures below. In the drawing, the three pairs of matching pieces used to prove the congruence are stressed in matching colors. The congruence $\triangle CAD \cong \triangle C'A'D'$ implies

$$(2) \qquad CD \cong C'D' \text{ and } \quad \angle ADC \cong \angle A'D'C'$$

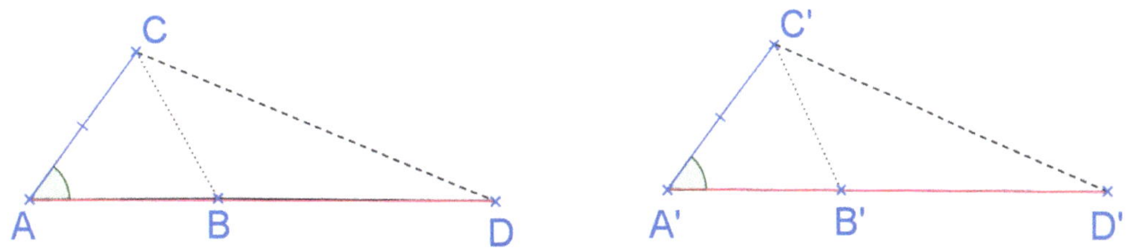

Figure II.7.11: Congruence of supplementary angles, the second pair of congruent triangles

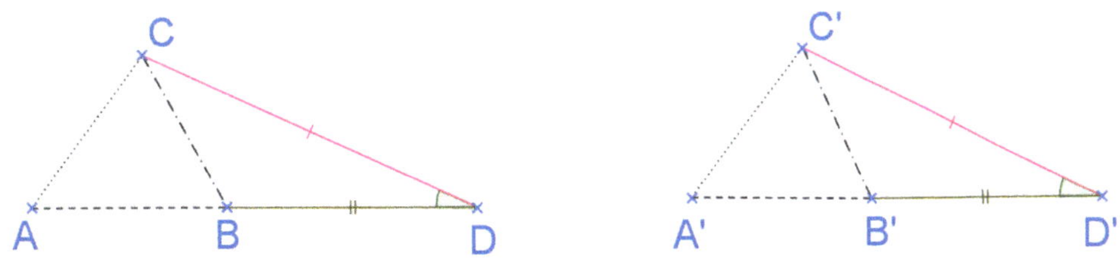

Figure II.7.12: Congruence of supplementary angles, the third pair of congruent triangles

Step 3: At last we consider the two triangles $\triangle BCD$ and $\triangle B'C'D'$ on the right side. In the drawing, the three pairs of matching pieces used to prove the congruence are stressed in matching colors. Again by using SAS congruence (Theorem 12 of Hilbert, see Proposition II.7.11 above), we see the two triangles are congruent. Finally $\angle CBD \cong \angle C'B'D'$, as to be shown. □

Corollary 7. *Adjacent angles congruent to supplementary angles are supplementary, too.*

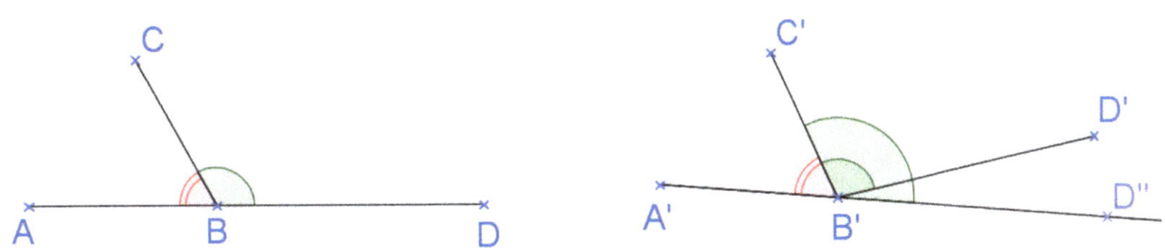

Figure II.7.13: Supplementary angles yield points on a line

Proof. Given are supplementary angles $\angle ABC$ and $\angle DBC$, a further congruent angle $\angle ABC \cong \angle A'B'C'$, and a point D' with A' and D' lying on different sides of line $B'C'$.

We shown that the angles $\angle A'B'C'$ and $\angle D'B'C'$ are supplementary if and only if $\angle CBD \cong \angle C'B'D'$. Above, we have already shown one direction: If the angles $\angle A'B'C'$ and $\angle D'B'C'$ are supplementary, then $\angle CBD \cong \angle C'B'D'$.

Now we show the converse. Assume that $\angle CBD \cong \angle C'B'D'$. We have to check whether point B' lies between A' and D'. Choose any point D'' on the ray opposite to $\overrightarrow{B'A'}$. Congruence of supplementary angles (Hilbert's Theorem 14, see Proposition II.7.15 above) implies $\angle CBD \cong \angle C'B'D''$. On the other hand, $\angle CBD \cong \angle C'B'D'$ is assumed. Angle $\angle CBD$ is transferred uniquely along ray $\overrightarrow{B'C'}$ into the half plane opposite to A'. Indeed, by axiom III.4, angle transfer produces a unique new ray. Hence $\overrightarrow{B'D''} = \overrightarrow{B'D'}$. and the four points A', B', D', D'' lie on one line, with point B' between A' and D'. Hence angles $\angle A'B'C'$ and $\angle D'B'C'$ are supplementary, too. □

Proposition II.7.16 (Congruence of Vertical Angles). [Euclid I.15] *Vertical angles are congruent.*

Proof. This is an easy consequence of Hilbert's Theorem 14 about supplementary angles (see Proposition II.7.15 above). Take any two vertical angles $\angle ABC$ and $\angle A'BC'$. We assume that vertex B lies between points A and A' on one line, as well as between the two points C and C' on a second line. As shown in the figures, angle $\angle ABC'$ has angle (i) $\angle ABC$ as supplementary angle. Secondly,

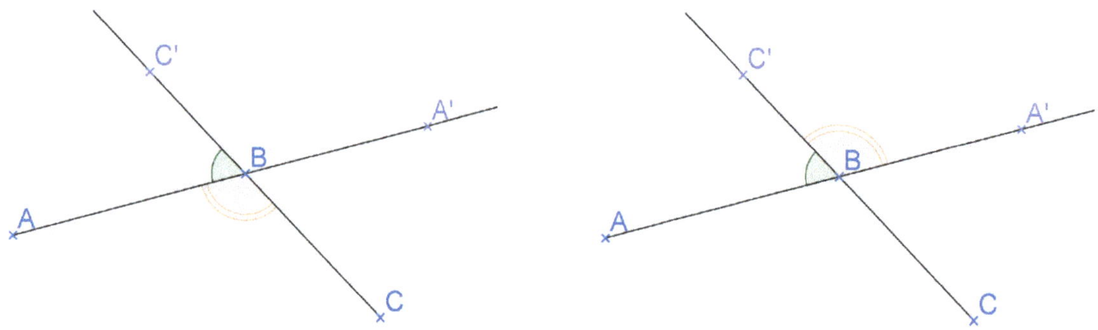

Figure II.7.14: Two pairs of supplementary angles yield vertical angles

angle $\angle ABC'$ has (ii) $\angle A'BC'$ as supplementary angle, too.

An angle is congruent to itself, as stated in Axiom (III.4). Hence especially $\angle ABC' \cong \angle ABC'$. By Theorem 14 of Hilbert (see Proposition II.7.15 above), angles supplementary to congruent angles are congruent, too. Hence we conclude congruence of the two vertical angles: $\angle ABC \cong \angle A'BC'$, as to be shown. □

Proposition II.7.17 (The main case of angle-addition). *Given is an angle $\angle ABC$ and a ray $\overrightarrow{BG}$ in its interior, as well as a second angle $\angle A'B'C'$. We assume that points A' and C' lie in different half planes of ray $\overrightarrow{B'G'}$. Furthermore, we assume that $\angle CBG \cong \angle C'B'G'$ and $\angle ABG \cong \angle A'B'G'$.*
Then the ray $\overrightarrow{B'G'}$ lies in the interior of angle $\angle A'B'C'$, and the two angles $\angle ABC \cong \angle A'B'C'$ are congruent.

Proof. By the Crossbar Theorem, a segment going from one side of an angle to the other, and a ray in the interior of that angle always intersect. Hence there exists a point H such that $\overrightarrow{BG} = \overrightarrow{BH}$ and $A \star H \star C$. For simplicity, we choose $G = H$ from the beginning. Too, we may assume that A', C' and G' are chosen such that $BA \cong B'A'$, $BC \cong B'C'$ and $BG \cong B'G'$. The proof uses three pairs of congruent triangles, in step (1)(2)(4), respectively. Step (3) is needed to show $A' * G' * C'$ and the ray $\overrightarrow{B'G'}$ lies in the interior of angle $\angle A'B'C'$.

Step **(1)**: The SAS-congruence axiom implies $\triangle BGC \cong \triangle B'G'C'$ because of $\angle GBC \cong \angle G'B'C'$, $GB \cong G'B'$ and $BC \cong B'C'$ which hold by assumption and the remarks above. In the drawing, the three pairs of matching pieces used to prove the congruence are stressed in matching colors. From the congruence of these two triangles, we conclude that

$$\angle BCG \cong \angle B'C'G' \tag{1}$$

$$\angle BGC \cong \angle B'G'C' \tag{1b}$$

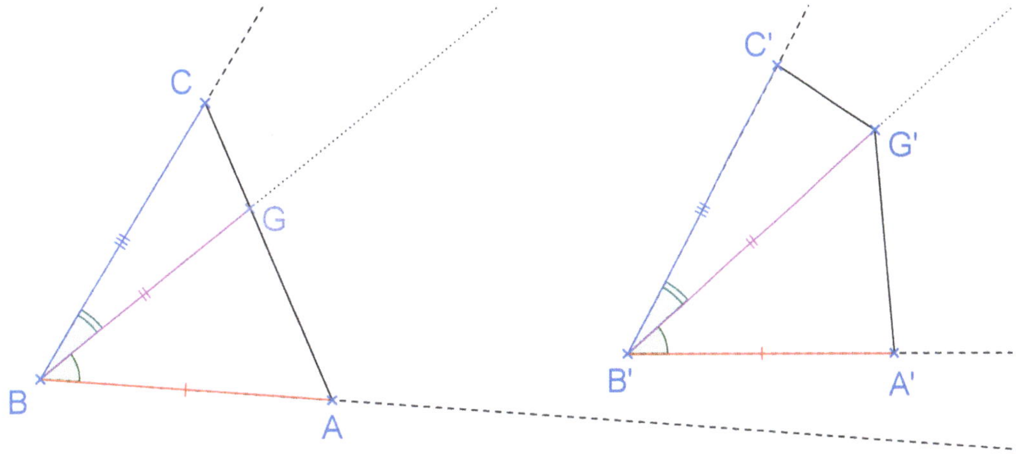

Figure II.7.15: Angle addition

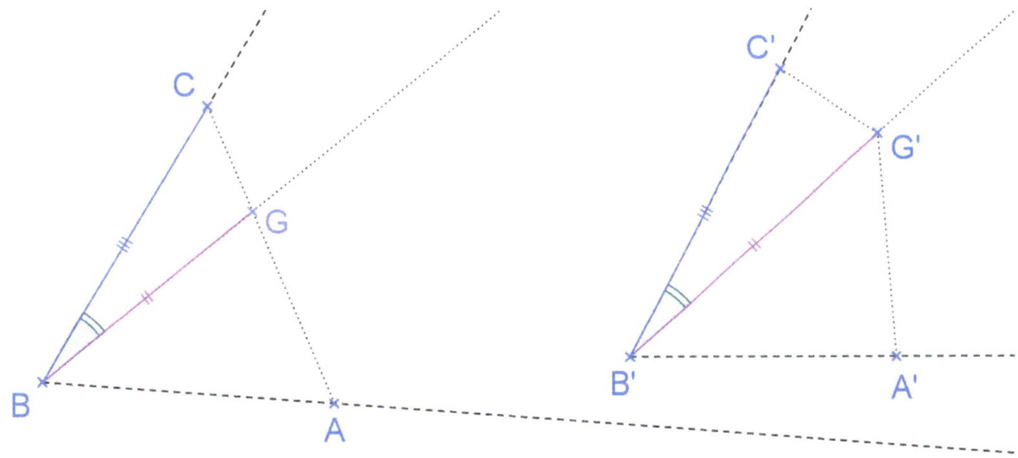

Figure II.7.16: the first pair of congruent triangles

Step **(2)**: As a second step, the SAS-congruence axiom implies $\triangle BGA \cong \triangle B'G'A'$ because of $\angle GBA \cong \angle G'B'A'$, $GB \cong G'B'$ and $BA \cong B'A'$ which hold by assumption and the remarks above. Again, in the drawing, the three pairs of matching pieces used to prove the congruence are stressed in matching colors.

Step **(3)**: It is assumed that points A' and C' lie in different half planes of ray $\overrightarrow{B'G'}$. Hence the angles $\angle B'G'A'$ and $\angle B'G'C'$ are *adjacent angles*. They are congruent to the *supplementary angles* $\angle BGA$ and $\angle BGC$. The congruence $\angle BGA \cong \angle B'G'A'$ follows from step (2), and $\angle BGC \cong \angle B'G'C'$ follows from step (1).

As derived in our Corollary 7, adjacent angles congruent to supplementary angles are supplementary, too. Hence the two angles $\angle B'G'A'$ and $\angle B'G'C'$ are supplementary. Hence point G' lies between points A' and C', the three points A', G' and C' lie on a straight line, and the ray $\overrightarrow{B'G'}$ lies in the interior of angle $\angle A'B'C'$.

Step **(4)**: By segment addition (Hilbert's axiom (III.3)), $AG \cong A'G'$ and $GC \cong G'C'$, and $A * G * C$, $A' * G' * C'$ imply

$$AC \cong A'C' \tag{2}$$

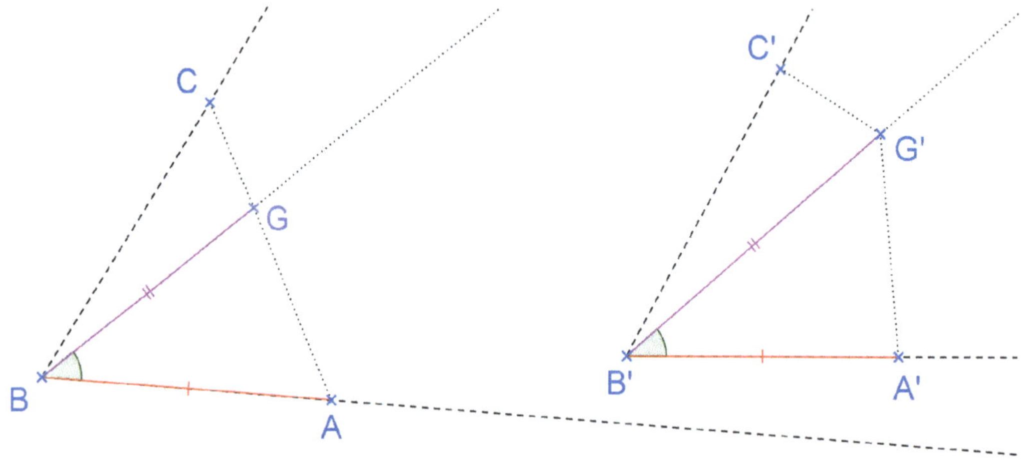

Figure II.7.17: the second pair of congruent triangles

To set up a third pair of congruent triangles, we use (1) (2) and the assumption

$$BC \cong B'C' \tag{3}$$

The SAS axiom shows that $\triangle ABC \cong \triangle A'B'C'$ because of (1)(2)(3). I have stressed these three pairs of matching pieces in matching colors. Finally, $\angle ABC \cong \angle A'B'C'$ follows from $\triangle ABC \cong$

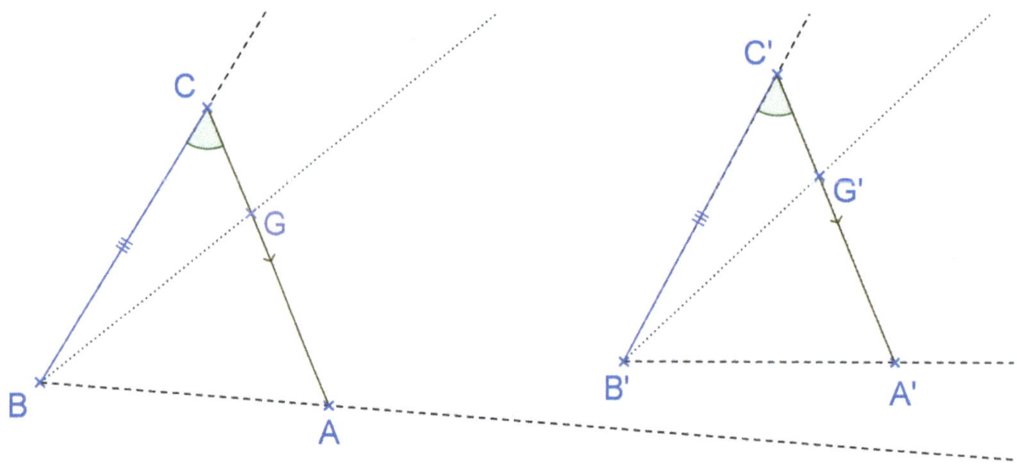

Figure II.7.18: the third pair of congruent triangles

$\triangle A'B'C'$, as to be shown. □

Proposition II.7.18 (The main case of angle subtraction). *Given is an angle $\angle ABC$ and a ray $\overrightarrow{BG}$ in its interior, as well as a second angle $\angle A'B'C'$ and a ray $\overrightarrow{B'G'}$ in its interior. If $\angle CBG \cong \angle C'B'G'$ and $\angle ABC \cong \angle A'B'C'$ then $\angle ABG \cong \angle A'B'G'$.*

Proposition II.7.19 (Angle-Addition and Subtraction). [Theorem 15 in Hilbert] *Given are three rays h, k, l with the common vertex B lying in a plane a, and three rays h', k', l' with the common vertex B' lying in a plane a'. There are the cases of angle subtraction or angle addition.*

angle addition: *in this case the rays h and k lie in the different half-planes of l, and the rays h' and k' lie in different half-planes of l'.*

angle subtraction: *in this case the rays h and k lie in the same half-plane of l, and the rays h' and k' lie in the same half-plane of l';*

Assume either angle addition or angle subtraction occurs and

$$\angle(h,l) \cong \angle(h',l') \quad \text{and} \quad \angle(l,k) \cong \angle(l',k'),$$

Then, either

$$\angle(h,k) \cong \angle(h',k')$$

or both h and k are opposite rays, and h' and k', too, are opposite rays.

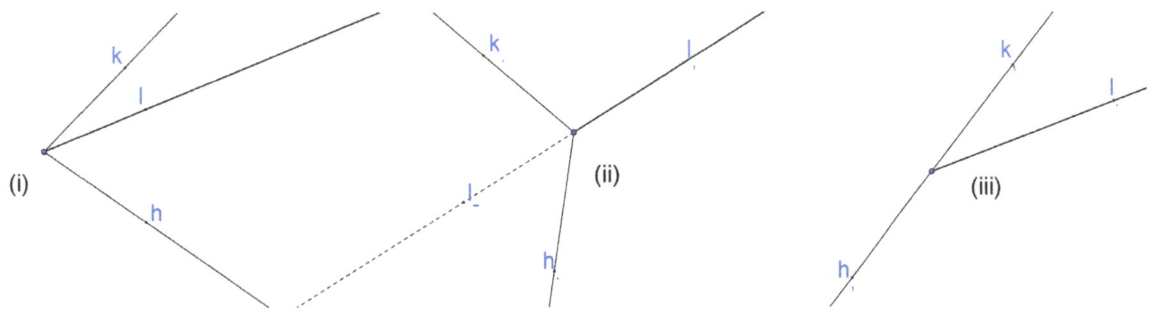

Figure II.7.19: Three subcases for angle addition.

Proof. We explain only the case of angle addition. As shown in the figure on page 155, we led to three different subcases.

(i) the two rays k and l lie in one half-plane of ray h;

(ii) the two rays k and l lie in different half-planes of ray h;

(iii) the rays h and k are opposite rays.

In the special case (iii), the angles $\angle(h,l)$ and $\angle(l,k)$ are adjacent supplements. Hence by Corollary 7, the congruent angles $\angle(h',l')$ and $\angle(l',k')$ are supplementary, too. In other words we get opposite rays h' and k'.

We consider the two main cases. In the first case (i), we use Proposition II.7.17 about the main case of angle-addition. As see in this Proposition, $A' * G' * C'$ and the ray $l' = \overrightarrow{B'G'}$ lies in the interior of angle $\angle A'B'C'$. Hence the two rays $l' = \overrightarrow{B'G'}$ and $k' = \overrightarrow{B'C'}$ lie in one half-plane of ray h'. The claim $\angle(h,k) \cong \angle(h',k')$ follows by angle addition as in Proposition II.7.17.

In the second case (ii), we use the opposition rays l_- of l and l'_- from l'. Now Proposition II.7.15 about the congruence of supplementary angles implies

$$\angle(h,l_-) \cong \angle(h',l'_-) \quad \text{and} \quad \angle(l_-,k) \cong \angle(l'_-,k')$$

Since we are back to case (i), we can conclude the claim $\angle(h,k) \cong \angle(h',k')$.

The case of angle subtraction does not lead to subcases, and can be left to the reader. □

Proposition II.7.20 (Transfer of three rays). *Given are three different rays h, k, l with the common vertex B lying in a plane a, and two rays h', k' with the common vertex B' lying in a plane a'. We assume that the three rays h, k and l lie on three different lines. Finally, we assume the angles $\angle(h,k) \cong \angle(h',k')$ to be congruent.*

Then there exists a unique *ray l' lying in the plane a', emanating from the vertex of $\angle(h',k')$ such that*

$$\angle(h,l) \cong \angle(h',l') \quad \text{and} \quad \angle(l,k) \cong \angle(l',k') \tag{II.7.5}$$

This ray l' lies in the interior of angle $\angle(h',k')$ if and only if the ray l lies in the interior of angle $\angle(h,k)$.

Proof. We begin with the case that the ray l lies in the interior of angle $\angle(h,k)$. The common vertex of the rays h, k, l is the point B. We choose any points A on the ray h and C on the ray k. By the Crossbar Theorem, there exists the intersection point G of *segment AC* with the *ray* of l.

We transfer the triangle $\triangle ABC$ onto the ray $\overrightarrow{B'A'} = k'$, into the half plane containing ray k'. By the assumption $\angle(h,k) \cong \angle(h',k')$ and the uniqueness of angle transfer, the point C' lies on the ray k'.

We transfer the triangle $\triangle ABG$ onto the ray $\overrightarrow{B'A'} = k'$, into the half plane containing ray k' to obtain the congruent triangle such that $\triangle ABG \cong \triangle A'B'G'$. Thus we obtain the new ray $l' := \overrightarrow{B'G'}$. The uniqueness of angle transfer implies $\overrightarrow{A'G} = \overrightarrow{A'C'}$. Moreover, $A * G * C$ implies $A' * G' * C'$.

We see that the new ray l' lies in the interior of angle $\angle(h',k')$, and the claimed congruences $\angle(h,l) \cong \angle(h',l')$ and $\angle(l,k) \cong \angle(l',k')$ indeed hold.

It is left to the reader to check the uniqueness of the ray l'.

In the case that the ray l does not lies in the interior of angle $\angle(h,k)$, it lies in the interior of either the vertical angle of one of the two supplementary angles of angle $\angle(h,k)$. Here we have used the assumption that the three rays h, k, and l lie on three different lines. By the argument above, we obtain the ray l', and we see that it lies in in the interior of either the vertical angle of one of the two supplementary angles of angle $\angle(h',k')$. The congruences (II.7.5) can now be obtained by means of Proposition II.7.15 about the congruence of supplementary angles. □

II.7.4 SSS congruence

The "Theorem about the symmetric kite" is needed as a preparation to get SSS congruence.

Definition II.7.6 (The Symmetric Kite). A *symmetric kite* or simply *kite* is a quadrilateral $\square XZ_1YZ_2$, where Z_1 and Z_2 are two points on different sides of diagonal XY, and $XZ_1 \cong XZ_2$ and $YZ_1 \cong YZ_2$.

Proposition II.7.21 (The Symmetric Kite). [Theorem 17 of Hilbert] *Let Z_1 and Z_2 be two points on different sides of line XY, and assume that $XZ_1 \cong XZ_2$ and $YZ_1 \cong YZ_2$. Under these assumptions*

(i) *the angles $\angle XZ_1Y \cong \angle XZ_2Y$ are congruent;*

(ii) *the angles $\angle XYZ_1 \cong \angle XYZ_2$ are congruent;*

(iii) *the triangles $\triangle XYZ_1 \cong \triangle XYZ_2$ are congruent.*

Proof. The congruence of the base angles of isosceles $\triangle XZ_1Z_2$ yields $\angle XZ_1Z_2 \cong \angle XZ_2Z_1$. Similarly, one gets $\angle YZ_1Z_2 \cong \angle YZ_2Z_1$. By plane separation, the segment Z_1Z_2 intersects the <u>line</u> XY. Let F be the point of intersection. According to the three-point theorem, we distinguish the cases

(a) $X * F * Y$

(b) $X * Y * F$

(c) $Y = F$

(d) $F * X * Y$

(e) $X = F$

By angle addition in case (a) or substraction in cases (b) and (d), respectively, we conclude that

$$(*) \qquad \angle XZ_1Z_2 \cong \angle XZ_2Z_1 \text{ and } \angle YZ_1Z_2 \cong \angle YZ_2Z_1 \quad \text{imply } \angle XZ_1Y \cong \angle XZ_2Y$$

Angle addition, as proved in Proposition II.7.17 is needed in case ray $\overrightarrow{Z_1Z_2}$ lies both inside the angle $\angle XZ_1Y$ and the angle $\angle XZ_2Y$. Angle subtraction from Proposition II.7.18 is used for the case that ray $\overrightarrow{Z_1Z_2}$ lies both outside the $\angle XZ_1Y$ and the $\angle XZ_2Y$.

In the special cases (c) and (e) that either point X or Y lies on the line Z_1Z_2, one gets the same conclusion even easier. Here are drawings for the three cases (a),(b) and (c). One now applies SAS

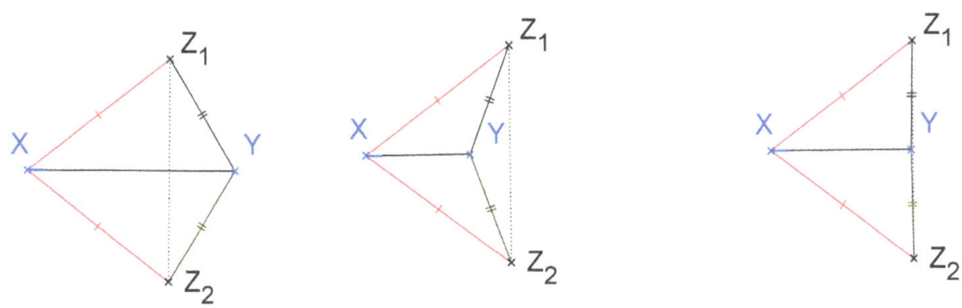

Figure II.7.20: The symmetric kite

congruence to the triangles $\triangle XZ_1Y$ and $\triangle XZ_2Y$. Indeed, the angles at Z_1 and Z_2 and the adjacent sides are pairwise congruent. Hence the assertion $\angle XYZ_1 \cong \angle XYZ_2$ follows from axiom (III.5). Moreover, SAS congruence theorem II.7.11 yields the triangle congruence $\triangle XYZ_1 \cong \triangle XYZ_2$. □

Before getting the general SSS-congruence, I consider one further special case.

Lemma II.7.2 (Lemma for SSS Congruence). [Euclid I.7] *Assume that the two triangles $\triangle ABC$ and $\triangle AB'C$ have a common side AC, and all three corresponding sides are congruent, and the two vertices B and B' lie on the same side of line AC. Then the two triangles are identical.*

Proof. We transfer the angle $\angle BAC$ onto the ray $\overrightarrow{AC}$, on the side of line AC <u>opposite</u> to B and B'. On the newly produced ray, we transfer segment AB, starting at vertex A. Thus we get point B'', and segment $AB'' \cong AB$. From SAS-congruence (Theorem 12 of Hilbert, see Proposition II.7.11

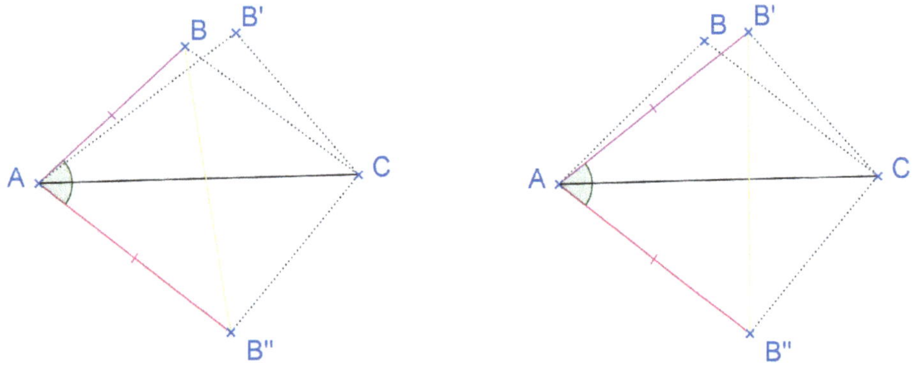

Figure II.7.21: Which kite is symmetric?

above), one concludes that $BC \cong B''C$. As stressed in the first drawing, one has constructed a symmetric kite with the four points

$$X := A, Y := C, Z_1 := B, Z_2 := B''$$

Hence by Theorem 17 (see Proposition II.7.21 above), $\angle B''AC \cong \angle BAC$

But wait! Another choice of four points to get a kite is

$$X := A, Y := C, Z_1 := B', Z_2 := B''$$

—replacing B by B'. Because of the assumptions, and the construction of the first kite, $AB' \cong AB \cong AB''$ and $B'C \cong BC \cong B''C$. We see that the second kite, stressed in the second drawing, satisfies the assumptions of Hilbert's Theorem 17 (see Proposition II.7.21 above), too. Now we conclude from Hilbert's Theorem 17 (see Proposition II.7.21 above) that $\angle B''AC \cong \angle B'AC$. Finally, the uniqueness of angle transfer implies that $\overrightarrow{AB} = \overrightarrow{AB'}$. Since $AB \cong AB'$, and segment transfer was shown to produce a unique point, we have confirmed that $B = B'$. Thus the two triangles $\triangle ABC$ and $\triangle AB'C$ are identical, as to be shown. □

Remark. We do not need to assume that angle congruence is an equivalence relation. But therefore we have to consider two kites, by choosing two of the three triangles, as shown in two drawings. Then we can use that transferring an angle gives a unique ray.

From Hilbert's Theorem 17 (see Proposition II.7.21 above), and the Lemma II.7.2 above, we see that SSS-congruence holds for any two triangles with a common side. Now we can easily get the general case of

Proposition II.7.22 (SSS Congruence). [Theorem 18 in Hilbert's Foundations] *Two triangles with three pairs of congruent sides are congruent.*

End of the proof of SSS-congruence. Assume the two triangles $\triangle ABC$ and $\triangle A'B'C'$ have corresponding sides which are congruent. We transfer the angle $\angle BAC$ onto the ray $\overrightarrow{A'C'}$, at vertex A', to the same side of $A'C'$ as B'. On the newly produced ray, we transfer segment AB, starting at vertex A'. Thus we get point B_0, such that $A'B_0 \cong AB$. Because of SAS-congruence (Theorem 12 of Hilbert, see Proposition II.7.11 above), we get

$$(*) \qquad \triangle ABC \cong \triangle A'B_0C'$$

Hence especially, $AB \cong A'B_0 \cong A'B'$ and $BC \cong B_0C' \cong B'C'$. We can now apply the Lemma II.7.2 to the two triangles $\triangle A'B'C'$ and $\triangle A'B_0C'$. Hence $B' = B_0$, and the assertion follows from (*). □

Proposition II.7.23 (The diagonals of the rhombus bisect each other perpendicularly). *Given are four* different *points A, B, C, D lying in one plane such that the segments $AB \cong BC \cong CD \cong DA$ are congruent. Then the segments AC and BD bisect each other perpendicularly at their common midpoint.*

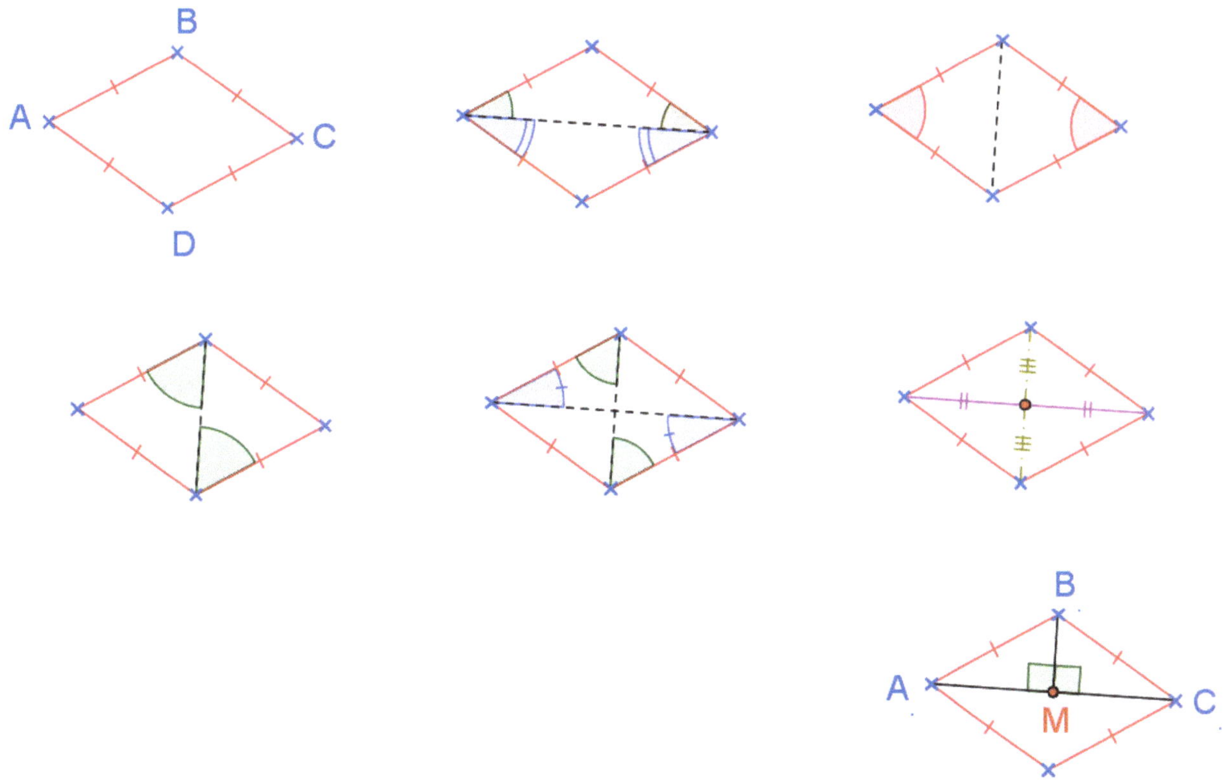

Figure II.7.22: The diagonals of the rhombus bisect each other.

Answer. Points A and C lie on different sides of line BD. Otherwise the Lemma for SSS-congruence would imply $A = C$, contrary to the assumption that four different points A, B, C, D are given. Similarly, we see that points B and D lie on different sides of the other diagonal AC. Hence the two segments AC and BD intersect. Let point M be their intersection point.

We draw the horizontal diagonal AC. The upper- and lower triangles have two pairs of congruent bases angles (see second figure). Angle addition implies $\angle ABD \cong \angle CBD$. By SAS congruence, the left- and right triangles $\triangle ABD \cong \triangle CBD$ are congruent (see third figure).

Furthermore, the latter two triangles are both isosceles. Hence $\triangle ABD \cong \triangle CDB$ holds, too, and $\angle ABD \cong \angle CDB$ (see first figure in the second row).

Similarly, we prove $\angle CAB \cong \angle ACD$. (Why?) By ASA congruence, we get $\triangle MAB \cong \triangle MCD$ (see second figure in the second row). Hence the diagonals AC and BD bisect each other (see last figure in the second row).

Since the isosceles triangle $\triangle ABC$ has congruent base angles, SAS congruence yields $\triangle MAB \cong \triangle MCB$. Hence the angles $\angle AMB \cong \angle CMB$ are congruent supplements, and hence they are right angles. Hence the diagonals AC and BD are perpendicular to each other.

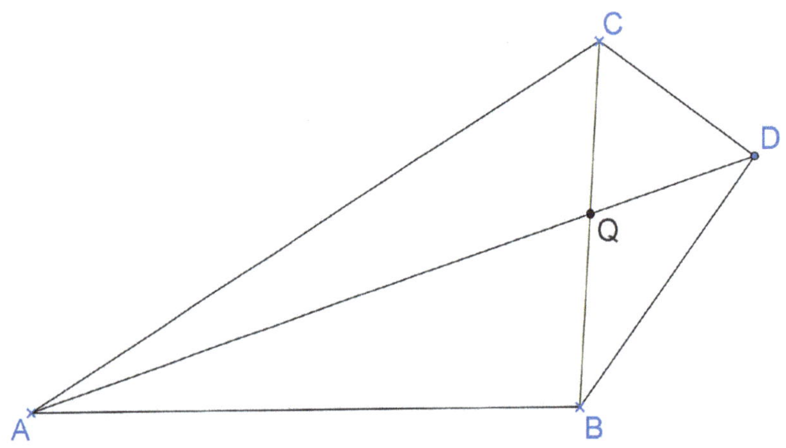

Figure II.7.23: Completing Euclid's proof of (I.7).

Problem II.7.4. *Provide the missing parts to complete Euclid'd proof of* (I.7), *in the case he considers. As shown in the drawing on page 160, we assume that the ray* $\overrightarrow{AD}$ *is in the interior of angle* $\angle CAB$, *and point* D *is outside the triangle* $\triangle ABC$. *We need to confirm the following items:*

(i) *the ray* $\overrightarrow{DA}$ *lies in the interior of angle* $\angle CDB$;

(ii) *the ray* $\overrightarrow{CB}$ *lies in the interior of angle* $\angle ACD$.

Use the axioms of order, the Crossbar Theorem, and so on, to provide a detailed justification.

Answer. By the Crossbar Theorem, the ray $\overrightarrow{AD}$ intersects the segment BC, say in point Q. Hence $B * Q * C$ and $A * Q * D$. Since $B * Q * C$ and point Q lies on the ray $\overrightarrow{DA}$, this ray lies in the interior of angle $\angle CDB$. Since $A * Q * D$ and point Q lies on the ray $\overrightarrow{CB}$, this ray lies in the interior of angle $\angle ACD$.

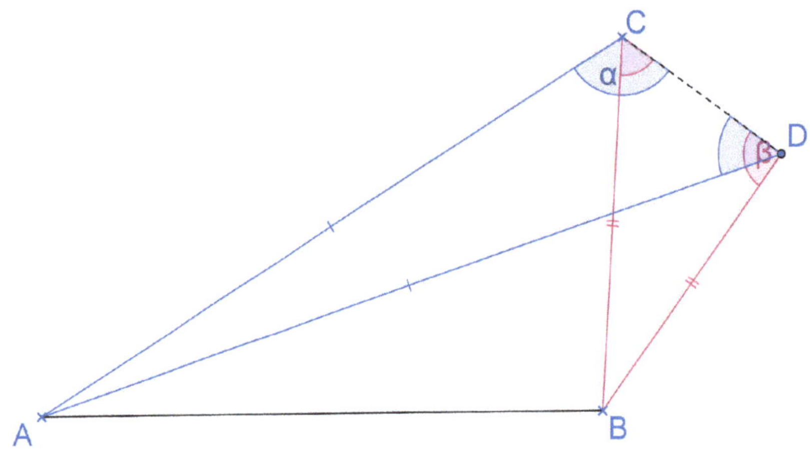

Figure II.7.24: Completing Euclid's proof of (I.7).

Problem II.7.5. *We now finish Euclid'd proof of* (I.7)*, in the case he considers. The triangles* $\triangle ABC$ *and* $\triangle ABD$ *are assumed to lie on the same side of line* AB*. Moreover, as shown in the drawing on page 160, we assume that the ray* $\overrightarrow{AD}$ *is in the interior of angle* $\angle CAB$*, and point* D *is outside the triangle* $\triangle ABC$*.*

Show that the congruences $AC \cong AD$ *and* $BC \cong BD$ *cannot both hold. Show that these congruences would yield two isosceles triangles. Get from the previous problem II.7.4 their base angles would satisfy both* $\alpha < \beta$ *and* $\beta < \alpha$*,—which is impossible. Nevertheless, provide a drawing to explain this impossible case.*

Answer. The assumed congruences yields two isosceles triangles: $\triangle ACD$ with congruent base angles α, and $\triangle BCD$ with congruent base angles β. Angle comparison at vertex D yields $\alpha < \beta$ since by item (i), the ray $\overrightarrow{DA}$ lies in the interior of angle $\angle CDB \cong \beta$.

Angle comparison at vertex C yields $\beta < \alpha$ since by item (ii), the ray $\overrightarrow{CB}$ lies in the interior of angle $\angle ACD \cong \alpha$.

II.7.5 Right, acute and obtuse angles

Proposition II.7.24 (Theorem 19 in Hilbert's Foundations). *If two angles* $\angle(h', k')$ *and* $\angle(h'', k'')$ *are congruent to a third angle* $\angle(h, k)$*, then the two angles* $\angle(h', k')$ *and* $\angle(h'', k'')$ *are congruent, too.*

Proof for a bottle of wine from Hilbert to A. Rosenthal. [24] Let the vertices of the angles be O, O', O''. Choose points A, A', A'' on one side of the three angles, respectively, such that $O'A' \cong OA$ and $O''A'' \cong OA$. Similarly choose points B, B', B'' on the three remaining sides, respectively, such that $O'B' \cong OB$ and $O''B'' \cong OB$. Now the assumption of SAS congruence (Theorem 12 of Hilbert, see Proposition II.7.11 above) are met for both $\triangle A'O'B'$ and $\triangle AOB$, as well as $\triangle A''O''B''$ and $\triangle AOB$. Hence

$$A'B' \cong AB\,,\ A''B'' \cong AB$$

We use axiom (III.2) :

```
"two segments congruent to a third one are congruent to each other".
```

Hence $\triangle A'B'O'$ and $\triangle A''B''O''$ have three pairs of congruent sides. By the SSS-congruence Theorem 18, we conclude that $\angle(h', k') \cong \angle(h'', k'')$, as to be shown. □

Proposition II.7.25 (Congruence is an equivalence relation). *Congruence is an equivalence relation on the class of angles.*

Question. Which three properties do we need to check for a congruence relation?

Answer. For an arbitrary relation to be a congruence relation, we need to check

- **(a) reflexivity:** Each angle is congruent to itself, written $\alpha \cong \alpha$.
- **(b) symmetry:** If $\alpha \cong \beta$, then $\beta \cong \alpha$.
- **(c) transitivity:** If $\alpha \cong \beta$, and $\beta \cong \gamma$, then $\alpha \cong \gamma$.

[24] Hilbert gives credit for this proof to A. Rosenthal (Math. Ann. Band 71)

Question. How can we claim reflexivity?

Answer. Reflexivity is given by the last part of axiom (III.4)

Proof of symmetry, for a bottle of wine from A. Rosenthal to Hilbert. [25] By Proposition II.7.24 above, $\angle(h',k') \cong \angle(h,k)$ and $\angle(h'',k'') \cong \angle(h,k)$ imply $\angle(h',k') \cong \angle(h'',k'')$. Hence, with just other notation, we see that $\beta \cong \beta$ and $\alpha \cong \beta$ imply $\beta \cong \alpha$. □

Proof of transitivity. Assume $\alpha \cong \beta$ and $\beta \cong \gamma$. Because of symmetry $\gamma \cong \beta$. Now we use Proposition II.7.24 above. Hence, just with other notation, $\alpha \cong \beta$ and $\gamma \cong \beta$ imply $\alpha \cong \gamma$. □

Definition II.7.7 (Angle comparison). Given are two angles $\angle BAC$ and $\angle B'A'C'$. We say that $\angle BAC$ is <u>less</u> than $\angle B'A'C'$, iff there exists a ray $\overrightarrow{A'G}$ in the interior of $\angle B'A'C'$ such that $\angle BAC \cong \angle B'A'G$. In this case, we also say that $\angle B'A'C'$ is greater than $\angle BAC$. We write $\angle B'A'C' > \angle BAC$ and $\angle BAC < \angle B'A'C'$, equivalently.

Given is a ray r with vertex O and a half-plane $\mathcal{H}$ of this ray. We may compare any given angles α and β, by transferring them onto the ray r, with their second side in the half plane $\mathcal{H}$. We obtain the congruent angles $\angle(r,a) \cong \alpha$ and $\angle(r,b) \cong \beta$. Because of the postulate (III.4), the relative position of the rays a and b can be used for comparison of the angles α and β.

Lemma II.7.3. *In the above situation, no matter how the ray r and the half plane $\mathcal{H}$ are chosen, the following equivalence holds:*
$\alpha < \beta$ holds if and only if the ray a lies in the interior of the angle $\angle(r,b)$.

As a consequence, one obtains the following propositions.

Proposition II.7.26 (Angle comparison holds for congruence classes). *Assume that $\alpha \cong \alpha'$ and $\beta \cong \beta'$. $\alpha < \beta$ if and only if $\alpha' < \beta'$.*

Proof. The reader should do it on her own. □

Proposition II.7.27 (Transitivity of angle comparison). *If $\alpha < \beta$ and $\beta < \gamma$, then $\alpha < \gamma$.*

Proof. After having done some transfer of angles, I can assume that all three angles α, β, γ have the common side $\overrightarrow{AB}$, and lie on the same side of AB. Choose any point E on the second side of $\gamma = \angle BAE$.

Because $\beta < \gamma$, the second side of β is in the interior of the largest angle γ. Hence, by the Crossbar Theorem the segment BC intersects that second side of β, say at point D, and $\beta = \angle BAD$ as well as $B * D * E$.

Because $\alpha < \beta$, the second side of α is in the interior of β. Hence, by the Crossbar Theorem the segment BD intersects that second side of α, say at point E, and $\alpha = \angle BAE$ as well as $B * C * D$. Any four points on a line can be ordered in a way that all four alphabetic order relations hold. (see Theorem 5 in Hilbert, which I call the four-point Theorem II.6.5). Now the four points B, C, D, E satisfy the order relations $B * C * D$ and $B * D * E$. Therefore they are already put in alphabetic order. Hence $B * C * E$. This shows by definition that $\alpha = \angle BAC < \angle BAE = \gamma$. □

Let s be the ray opposite to ray r. As already explained in lemma II.6.3, plane separation by the ray b leads to the following cases to be distinguished:

[25] Hilbert gave credit and got back a good bottle

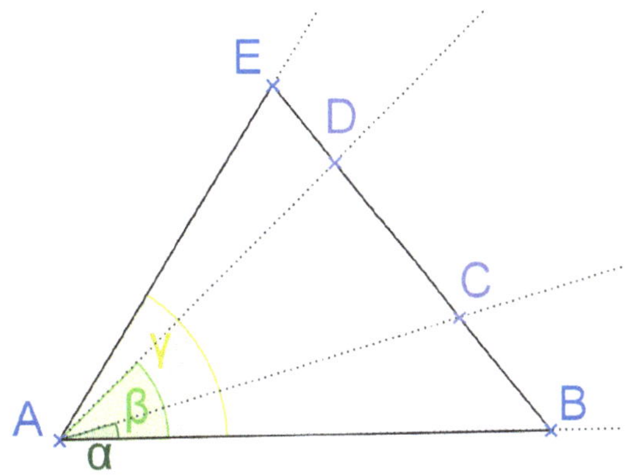

Figure II.7.25: Transitivity of comparison of angles

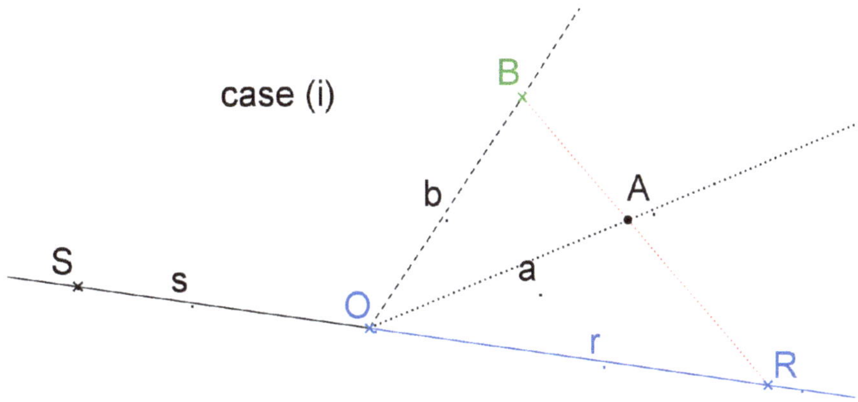

Figure II.7.26: The case (i) where $\alpha < \beta$.

(i) either the ray a lies in the interior of angle $\angle(r, b)$ and the ray b lies in the interior of angle $\angle(s, a)$, as shown in the figure on page 163;

(ii) or the ray b lies in the interior of angle $\angle(r, a)$ and the ray a lies in the interior of the angle $\angle(s, b)$, as shown in the figure on page 164;

(iii) or the two rays a and b are equal.

We have obtained the following result:

Proposition II.7.28 (All angles are comparable). *For any two angles α and β, one and only one of the three cases* (i)(ii)(iii) *occurs:*

Either (i) $\alpha < \beta$ *or* (ii) $\beta < \alpha$ *or* (iii) $\alpha \cong \beta$.

With r and s opposite rays, the above figures contain the supplementary pairs

$$\alpha \cong \angle(r, a) \quad \text{and} \quad S(\alpha) \cong \angle(s, a)$$
$$\beta \cong \angle(r, b) \quad \text{and} \quad S(\beta) \cong \angle(s, b)$$

Hence lemma II.6.3 yields the result

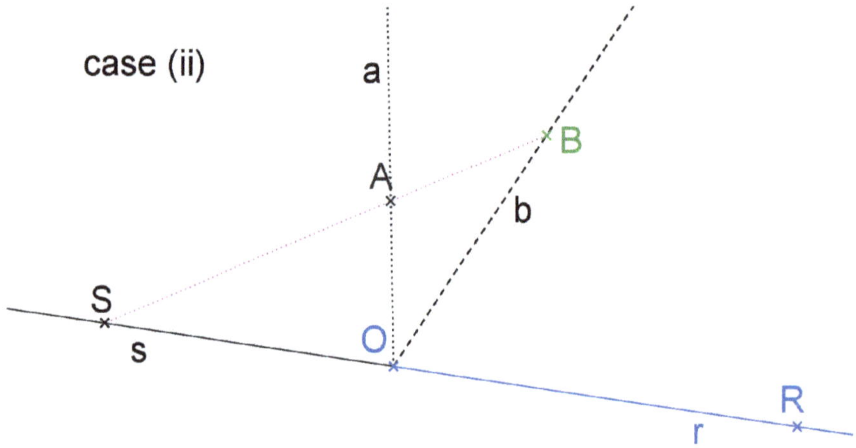

Figure II.7.27: The case (ii) where $\beta < \alpha$.

Proposition II.7.29 (Comparison of supplements). *If $\alpha < \beta$, then their supplements $S(\alpha)$ and $S(\beta)$ satisfy $S(\alpha) > S(\beta)$.*

Definition II.7.8 (right angle). A *right* angle is an angle congruent to its supplementary angle.

Proposition II.7.30 (All right angles are congruent (Proclus 410-485 A.D.)). *All right angles are congruent*

Question. How is a right angle defined?

Answer. A right angle is an angle congruent to its supplementary angle.

Proof is easy with the preparation above. Let α be a right angle, and $\beta = S(\alpha) \cong \alpha$ be its congruent supplement. Similarly, we consider a second pair of right angles α' and $\beta' = S(\alpha') \cong \alpha'$ being its congruent supplement.

The question is whether both right angles α and α' are congruent to each other. By Proposition II.7.28 all angles are comparable. Hence the two angles α and α' satisfy just one of the following relations.

Either (i) $\alpha < \alpha'$ or (ii) $\alpha' < \alpha$ or (iii) $\alpha \cong \alpha'$.

We need to rule out cases (i) and (ii) by deriving a contradiction. I only need to explain case (i), because (ii) is similar. We now assume $\alpha < \alpha'$ towards a contradiction. By the comparison of supplements from Proposition II.7.29, we get $S(\alpha) > S(\alpha')$. Because a right angle is congruent to its supplement we get $\alpha \cong S(\alpha) > S(\alpha') \cong \alpha'$. As explained in Proposition II.7.26, angle comparison holds for congruence classes. Hence we get $\alpha > \alpha'$.

Thus, from the assumption $\alpha < \alpha'$, we have derive $\alpha > \alpha'$. Transitivity holds by proposition II.7.27 and implies $\alpha < \alpha$. But axiom (III.4c) states that an angle is congruent to itself, and hence $\alpha \cong \alpha$. By proposition II.7.28, all angles are comparable. It cannot happen that both $\alpha < \alpha$ and $\alpha \cong \alpha$ would hold.

Thus case (i) leads to a contradiction. Similarly, case (ii) leads to a contradiction. The only remaining possibility is case (iii): Any two given right angles α and α' are congruent. □

Definition II.7.9 (acute, right and obtuse angles). A *right* angle is an angle congruent to its supplementary angle. An *acute* angle is an angle less than a right angle. An *obtuse* angle is an angle greater than a right angle.

Remark. If an angle α is acute, its supplement $S(\alpha)$ is obtuse. Furthermore, the angle is less than its supplement: $\alpha < S(\alpha)$.

If an angle β is obtuse, its supplement $S(\beta)$ is acute. Furthermore, the angle is larger than its supplement: $\beta > S(\beta)$.

Reason. Assume $\alpha < R$, where R denotes a right angle. From the comparison of supplements done in Proposition II.7.29, we conclude $S(\alpha) > S(R)$. But by the definition of a right angle $R \cong S(R)$. Hence $S(\alpha) > S(R) \cong R$, and because comparison is for congruence classes, we get $S(\alpha) > R$. Hence $\alpha < R < S(\alpha)$, and by transitivity (Problem 10.2), we conclude $\alpha < S(\alpha)$.

Similarly, we explain that $\beta > R$ implies $S(\beta) < R$ and $S(\beta) < \beta$. □

Proposition II.7.31 (Converse Isosceles Triangle Proposition). [Euclid I.6, Theorem 24 of Hilbert] *A triangle with two congruent angles is isosceles.*

Question. Formulate the theorem with specific quantities from a triangle $\triangle ABC$.

Answer. If $\alpha \cong \beta$, then $a \cong b$.

Proof. Given is a triangle $\triangle ABC$ with $\alpha \cong \beta$. By Hilbert's Theorem 19 (see Proposition II.7.24 above) and proposition II.7.25, congruence of angles is an equivalence relation. Hence $\alpha \cong \beta$ implies $\beta \cong \alpha$. In the preliminary version given as Proposition II.7.14, we have shown that $\alpha \cong \beta$ and $\beta \cong \alpha$ together imply $a \cong b$. Hence the $\triangle ABC$ is isosceles, as to be shown. □

Proposition II.7.32 (The Hypothenuse Leg Theorem). *Two right triangles for which the two hypothenuse, and one pair of legs are congruent, are congruent.*

Question. Of which one of SAS, SSA, SAA, ASA, SSS congruence contains the hypothenuse-leg theorem as a special case? Is there a corresponding unrestricted congruence theorem?

Answer. The hypothenuse-leg theorem is a special case of SSA congruence. There is no unrestricted SSA congruence theorem.

First proof of the hypothenuse-leg theorem II.7.32 . Given are two right triangles $\triangle ABC$ and $\triangle A'B'C'$. As usual, we put the right angles at vertices C and C'. We assume congruence of the hypothenuses $AB \cong A'B'$ and of one pair of legs $AC \cong A'C'$.

We build a kite out of two copies of triangle $\triangle ABC$ and two copies of triangle $\triangle A'B'C'$. To this end, one transfers angle $\angle C'A'B'$ onto the ray $\overrightarrow{AC}$ into the opposite half plane and gets by transfer of triangle (see Proposition II.7.12) the congruence $\triangle C'A'B' \cong \triangle CAB''$. Especially

$$B'C' \cong B''C \tag{II.7.6}$$

Since all right angles are congruent (see Proclus's Proposition II.7.30), the two supplementary right angles at vertex C imply that points B, C and B'' lie on a line.

Next we transfer segment CA onto the ray opposite to $\overrightarrow{CA}$ and get point D such that $CA \cong CD$. We produce the two triangles to the right, as shown in the second row of the figure on page 166. Using the right angle, SAS congruences imply $\triangle ABC \cong \triangle DBC$ and $\triangle AB''C \cong \triangle DB''C$.

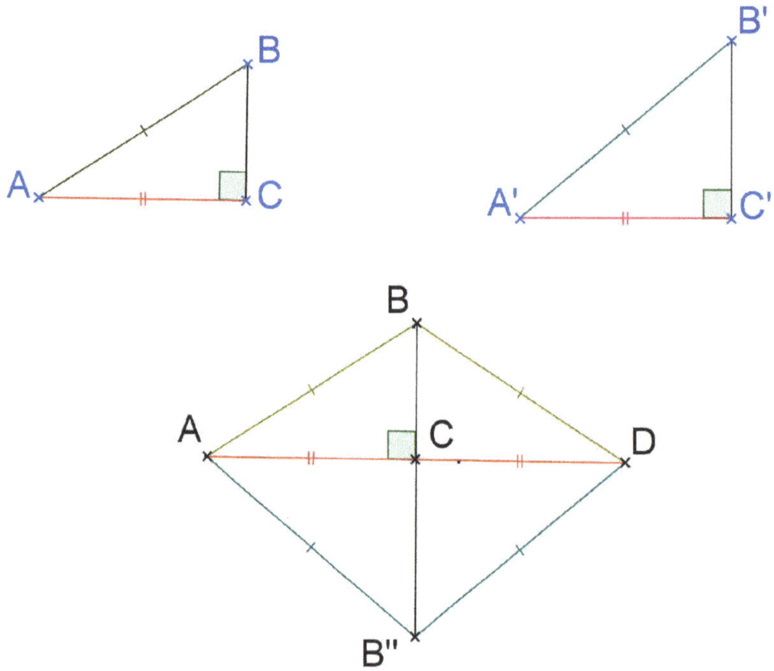

Figure II.7.28: First proof of the hypothenuse leg theorem

Since $AB \cong A'B' \cong AB''$ and $DB \cong AB \cong AB'' \cong DB''$, one has constructed a symmetric kite with the four points

$$X := A,\ Y := D,\ Z_1 := B,\ Z_2 := B''$$

We can apply Hilbert's kite-theorem II.7.21 and conclude

$$\angle DAB = \angle XYZ_1 \cong \angle XYZ_2 = \angle DAB''$$

The congruence of the same angles is $\angle CAB = \angle CAB''$. Hence SAS congruence implies now $\triangle CAB \cong \triangle CAB''$ and

$$BC \cong B''C \tag{II.7.7}$$

Together, the formulas (II.7.6) and (II.7.7) imply $B'C' \cong BC$. Since all right angles are congruent, a final SAS congruence using the right angles implies congruence $\triangle ABC \cong \triangle A'B'C'$ of the originally given triangles, as to be shown. □

II.7.6 Constructions with Hilbert tools

Definition II.7.10 (Hilbert tools). By incidence axiom (I.1) and (I.2), we can draw a unique line between any two given points. By axiom (III.1) and Proposition II.7.2 from the very beginning, we transfer a given segment uniquely to a given ray. Finally by axiom III.4, we transfer a given angle uniquely on a given ray into the specified half plane. Constructions done using only these means are called constructions by *Hilbert tools.*

Now we show that a right angle actually exists, and do some further basic constructions with the Hilbert tools. The only means of construction are those granted by Hilbert's axioms.

Problem II.7.6 (Drop a Perpendicular). *Given is a line OA and a point B not on this line. We have to drop the perpendicular from point B onto line OA.*

Construction II.7.1 (Drop a Perpendicular). *Draw ray $\overrightarrow{OB}$. Transfer angle $\angle AOB$, into the half plane opposite to B, with ray $\overrightarrow{OA}$ as one side. On the newly produced ray, transfer segment OB to produce a new segment $OC \cong OB$. The line BC is the perpendicular, dropped from point B onto line OA.*

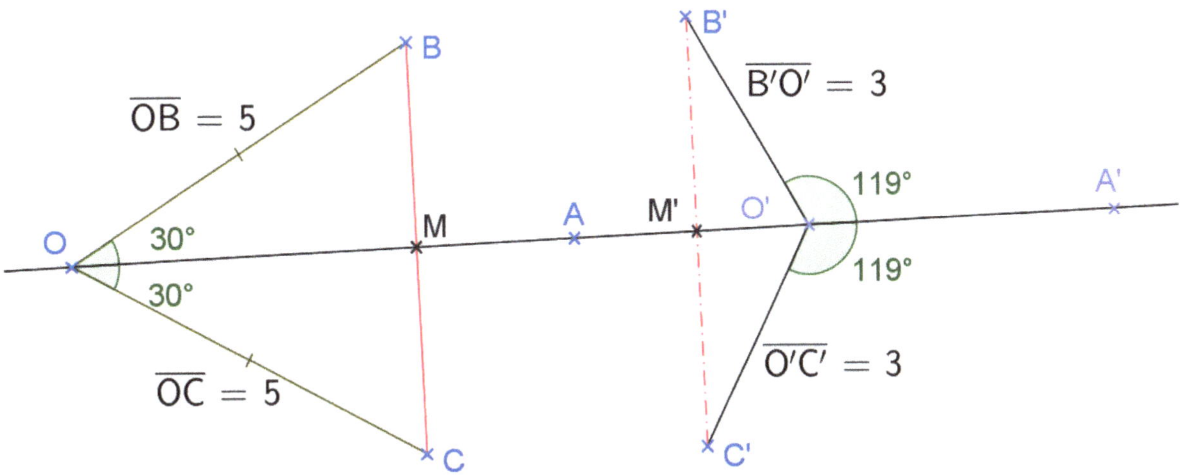

Figure II.7.29: Drop the perpendicular, the two cases

Proof of validity.

Question. Why do the line OA and the segment BC intersect.

Answer. Line OA and the segment BC intersect because points B and C lie on different sides of line OA.

I call the intersection point M.

Question. It can happen that $O = M$. How does one proceed in this special case.

Answer. In the special case that $O = M$, we know that $A \neq M$ and

(**) $$\angle AMB \cong \angle AMC$$

But these is a pair of congruent supplementary angles, because M lies between B and C. By definition, an angle congruent to its supplementary angle is a right angle. Hence $\angle AMB$ is a right angle.

In the generic situation $O \neq M$, we distinguish two cases:

(i) points M and A lie on the same ray with vertex O;

(ii) point O lies between M and A.

In both cases we show the two triangles $\triangle OMB$ and $\triangle OMC$ are congruent. Indeed, the angles $\angle MOB \cong \angle MOC$ at vertex O are congruent, both in case (i) and (ii).

Question. In case (i), one gets this congruence directly from the construction since the rays $\overrightarrow{OA} = \overrightarrow{OM}$ are equal. How does one proceed in case (ii).

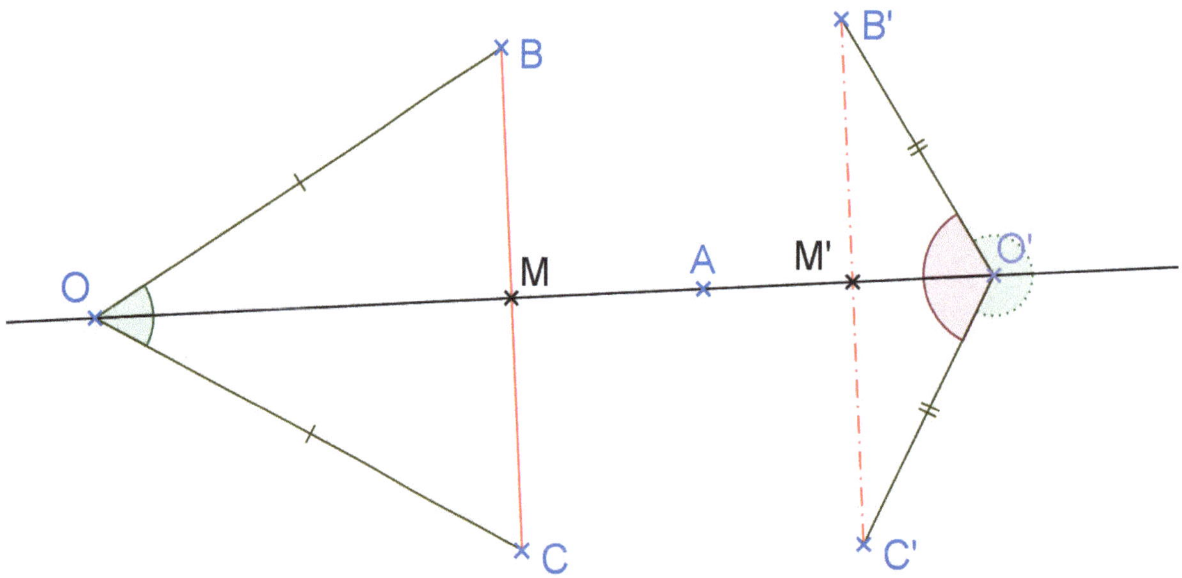

Figure II.7.30: Congruences needed in the two cases.

Answer. In case (ii), the rays $\overrightarrow{OA}$ and $\overrightarrow{OM}$ are opposite. Thus $\angle MOB$ and $\angle AOB$, as well as $\angle MOC$ and $\angle AOC$ are supplementary angles. Again $\angle AOB \cong \angle AOC$ because of the angle transfer done in the construction. Now Theorem 14 of Hilbert (see Proposition II.7.15 above) tells that supplements of congruent angles are congruent. Hence we get $\angle MOB \cong \angle MOC$ once again.

Question. Finish the argument showing the triangle congruence $\triangle MOB \cong \triangle MOC$.

Answer. The adjacent sides $OB \cong OC$ are congruent by construction, too. Finally the common sides OM is congruent to itself. Hence the congruence follows by SAS congruence. The drawing stresses the pieces matched to prove the congruence.

Because the two triangles are congruent, the corresponding angles at vertex M are congruent, too. Hence

$$(*) \qquad \angle OMB \cong \angle OMC$$

which is a pair of congruent supplementary angles.

Question. How is a right angle defined. How do we see that $\angle OMB$ is a right angle.

Answer. A right angle is, by definition, an angle congruent to its supplementary angle . Hence $\angle OMB$ is a right angle because of formula (*).

□

Corollary 8. *A right angle exists and is unique. Hence it deserve a special name.*

Proof. By axiom (I.3b) there exist three points A, B, C not lying on a line. We use the construction II.7.1 to drop the perpendicular from point C onto the line AB. At the foot point of the perpendicular, we obtain a right angle. It has been proved in lemma II.7.30 that all right angles are congruent (Proclus 410-485 A.D.) □

Remark. Traditionally, we put a right angle equal to 90°. But note that such a convention is only allowable after one has proved that a right angle exists and is unique.

Proposition II.7.33 (Existence of an Isosceles Triangle). *For any given segment AB, on a given side of line AB, there exists an isosceles triangle.*

Remark. Of course, the isosceles triangle one has obtained, is not unique.

Construction II.7.2 (Construction of an isosceles triangle). *Given is a segment AB. Choose any point P not on line AB, in the half plane specified. Compare the two angles $\angle BAP$ and $\angle ABP$. In the drawing, $\angle BAP$ is the smaller angle. Transfer the smaller angle, with ray $\overrightarrow{BA}$ as one side, and the newly produced ray r_B in the same half plane as P. Ray r_B and segment AP do intersect. The intersection point C lies in the half planen as required, and $\triangle ABC$ is isosceles.*

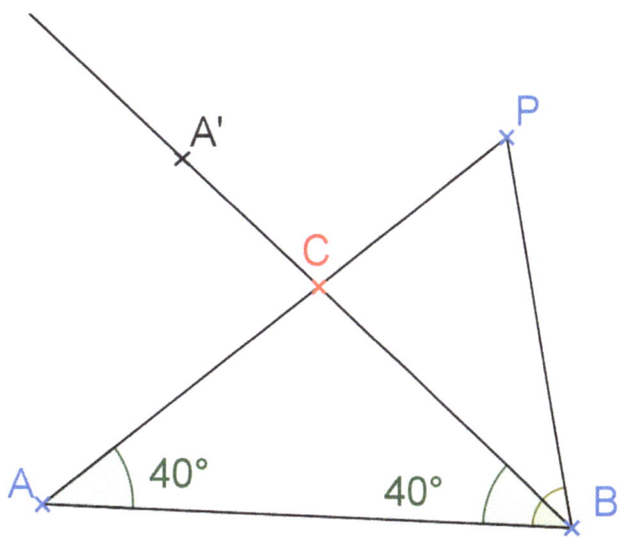

Figure II.7.31: Construction of an isosceles triangle

Proof of validity. By Proposition II.7.28, all angles are comparable. Comparison of the two angles $\angle BAP$ and $\angle ABP$ leads to one of the three possibilities:

Either (i) The two angles are congruent

or (ii) $\alpha = \angle BAP < \angle ABP = \beta$

or (iii) $\alpha = \angle BAP > \angle ABP = \beta$.

It is enough to consider cases (i) and (ii). In case (i) let $C = P$. In case (ii) we have transferred the smaller angle $\angle BAP$, onto ray $\overrightarrow{BA}$, and the newly produced ray r_B in the same half planen as P. By the Crossbar Theorem, a segment going from one side of an angle to the other, and a ray in the interior of an angle always intersect. By the Crossbar Theorem, ray r_B and segment AP intersect. The intersection point C lies between A and P, hence both points C and P lie in the same half plane of line AB. Hence point C lies in the half planen as required.

By construction, the base angles of triangle $\triangle ABC$ are congruent. Hence the triangle is isosceles by Euclid I.6 (Converse Isosceles Triangles). □

Remark. One may suggest to make it part of the construction <u>how</u> to compare the two angles $\angle BAP$ and $\angle ABP$. One possibility is to transfer <u>both</u> angles to the other vertex, B or A, respectively. Only the newly produced ray from the smaller angle does intersect the opposite side of $\triangle ABP$.

Problem II.7.7 (Erect a Perpendicular). *Given is a line l and a point R on this line. We have to erect the perpendicular at point R onto line l.*

Construction II.7.3. *Choose any point $A \neq R$ on line l. Transfer segment AR onto the ray opposite to $\overrightarrow{RA}$ to get a segment $RB \cong AR$. As explained in Construction II.7.2, construct any isosceles triangle $\triangle ABC$. The line CR is the perpendicular to the given line l through point R.*

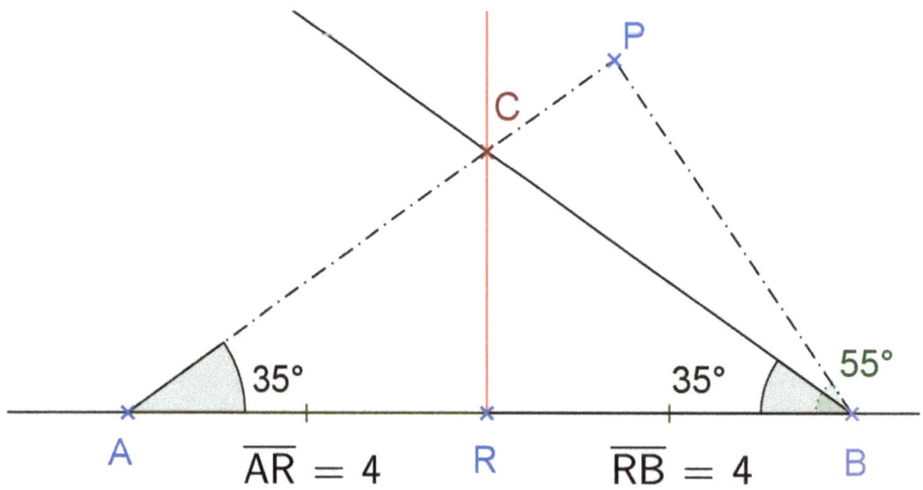

Figure II.7.32: Erect a perpendicular

Reason for validity. The two triangles $\triangle RAC$ and $\triangle RBC$ are congruent by SAS congruence. Indeed, $\angle RAC \cong \angle RBC$ by the construction of the isosceles triangle. Furthermore, we have a pair of congruent adjacent sides: $AR \cong BR$ by the construction above, and $AC \cong BC$ because $\triangle ABC$ is isosceles.

Now the triangle congruence $\triangle RAC \cong \triangle RBC$ implies that $\angle ARC \cong \angle BRC$. But these two angles are supplementary angles. Because congruent supplementary angles are right angles, we have confirmed that $\angle ARC$ is a right angle, as to be shown. □

Definition II.7.11 (The perpendicular bisector). The line through the midpoint of a segment and perpendicular to the segment is called the *perpendicular bisector.*

Construction II.7.4 (Construction I of the perpendicular bisector). *Construct two different isosceles triangles over segment AB. The line connecting the third vertices of the two isosceles triangles is the perpendicular bisector.*

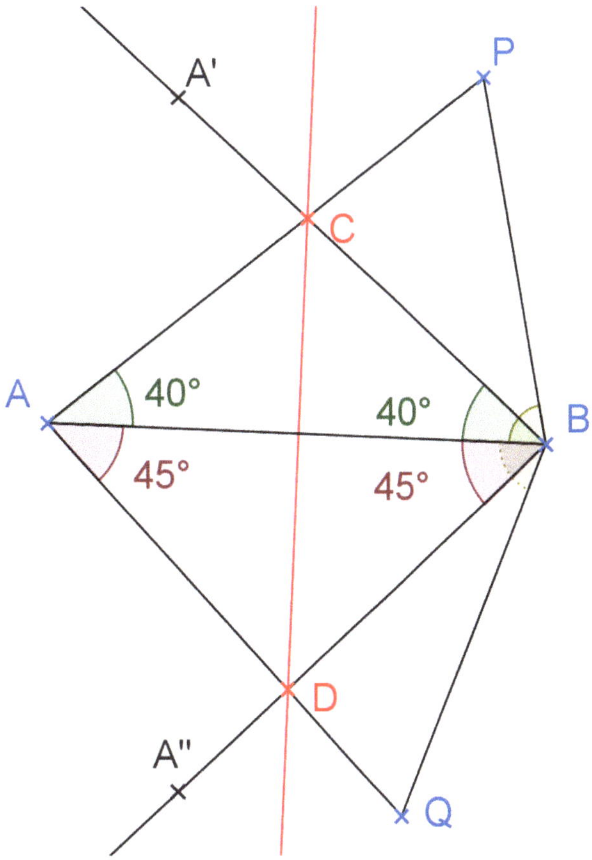

Figure II.7.33: The perpendicular bisector via the kite

Proof of validity. Let $\triangle ABC$ be the first isosceles triangle, and $\triangle ABD$ be the second one.

Question. Why do the points A and B lie on different sides of CD? We need just to use the Lemma II.7.2 to SSS-congruence!

Answer. If points A and B would lie on the same side of line CD, the congruence $\triangle ACD \cong \triangle BCD$ would imply $A = B$, contradicting to the endpoints of a segment being different.

Because A and B lie on different sides of line CD, the segment AB intersects the line CD, say at point M. Now we have the symmetric kite $\square ACBD$ with two congruent triangles (left and right in the figure). By the kite Theorem (Proposition II.7.21 above), we conclude the congruence $\triangle ACD \cong \triangle BCD$.

Of the three points M, C, D exactly one lies between the two others. We can assume that C does not lie between D and M, this assumption can be achieved by possibly interchanging the names C and D.

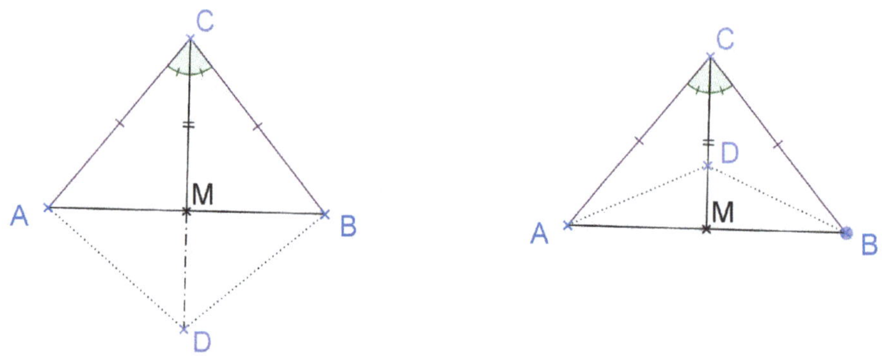

Figure II.7.34: Get the perpendicular bisector via: (a) a convex kite, (b) a nonconvex kite

Question. Give a detailed reason for the triangle congruence $\triangle ACM \cong \triangle BCM$.

Answer. Indeed, the triangles $\triangle ACM$ and $\triangle BCM$ have congruent angles at the common vertex C, and two pairs of congruent adjacent sides $AC \cong BC$ and $MC \cong MC$. Now we get by SAS congruence that $\triangle ACM \cong \triangle BCM$. Matching pieces are stressed in the drawing.

Question. Use the triangle congruence $\triangle ACM \cong \triangle BCM$ to confirm that line CD is the perpendicular bisector of the given segment AB.

Answer. From the last triangle congruence we get $\angle AMC \cong \angle BMC$. Because M lies between A and B, these are two congruent supplementary angles. Hence they are right angles. Too, the triangle congruence implies $AM \cong BM$. Hence M is the midpoint of segment AB. We have shown that the segment AB and the line CD intersect perpendicularly at the midpoint of segment AB. Hence CD is the perpendicular bisector.

□

Remark. There are the possibilities of an (a) convex kite, or (b) a non convex kite.

(a): If the two isosceles $\triangle ABC$ and $\triangle ABD$ lie on different sides of AB, points C and D are on different sides of AB, and it is clear that the segment CD intersects the line AB.

(b): Too, it is possible to use two different isosceles triangles on the same side of AB. To get the second isosceles triangle $\triangle ABD$, one transfers as base angles any two congruent angles which are less than the base angles of the first triangle $\triangle ABC$. The figure $\square ACBD$ is still a symmetric—but non convex—kite. As proved in Hilbert's kite Proposition II.7.21, the two triangles left and right in the figure are still congruent: $\triangle ACD \cong \triangle BCD$. The line CD still intersects the segment AB, because points A and B are on different sides of CD, but the two segments AB and CD do not intersect each other.

Corollary 9. *The perpendicular bisector of any segment is unique. A point lies on the perpendicular bisector of segment AB if and only if its distances AP and BP to the endpoints of the segment are congruent.*

Proof. Since the midpoint of a segment is unique, and all right angles are congruent by Proclus' proposition II.7.30, the uniqueness of angle transfer by axiom (III.4) implies uniqueness of the perpendicular bisector.

Given any point P for which $AP \cong BP$. If $P = M$ turns out to be the midpoint of segment AB, the distances $AP \cong BP$ are congruent. Assume that $M \neq P$. In that case, points A, B and M do not lie on a line since the midpoint is unique. One can use the isosceles triangle $\triangle ABP$ for the construction of the *unique* perpendicular bisector p. Hence point P lies on the bisector.

Conversely given any point $P \neq M$ on the perpendicular bisector p. We may use SAS congruence to show the triangle congruence $\triangle AMP \cong \triangle BMP$. Indeed, they both have a right angles at vertex M, and the adjacent sides $MP \cong MP$ and $MA \cong MB$ are pairwise congruent. Hence $AP\ congBP$, as claimed. □

Problem II.7.8. *Given any plane α and two points A and B in it. Give an independent proof showing that any point of the plane with congruent distances to these two points lies on the perpendicular bisector of segment AB.*

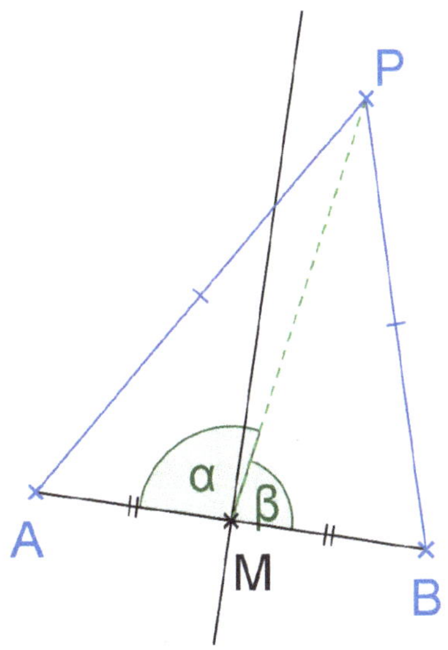

Figure II.7.35: A point with congruent distances to both endpoints lies on the perpendicular bisector.

Proof. Assume point P lies in the plane α and has congruent distances $AP \cong BP$ to the endpoints of the segment. Let M be the midpoint of segment AB. If $P = M$, we are ready since M is a point on the perpendicular bisector. Hence we may assume $P \neq M$. The congruence

$$\triangle AMP \cong \triangle BMP$$

follows from SSS congruence . Indeed, $PM \cong PM$ since a segment is congruent to itself, $AP \cong BP$ by assumption, and $AM \cong BM$ holds since M is the midpoint of AB. Hence $\angle AMP \cong \angle BMP$ are congruent supplementary angles, and hence right angles. Since the right angle is unique by Proposition II.7.30, and angle transfer produces a unique ray by axiom (III.4), the point P lies on the perpendicular bisector, as to be shown. □

Construction II.7.5 (Construction II of the perpendicular bisector— *the way one really wants it*). *Construct any isosceles triangle over the given segment. Drop the perpendicular from its third vertex.*

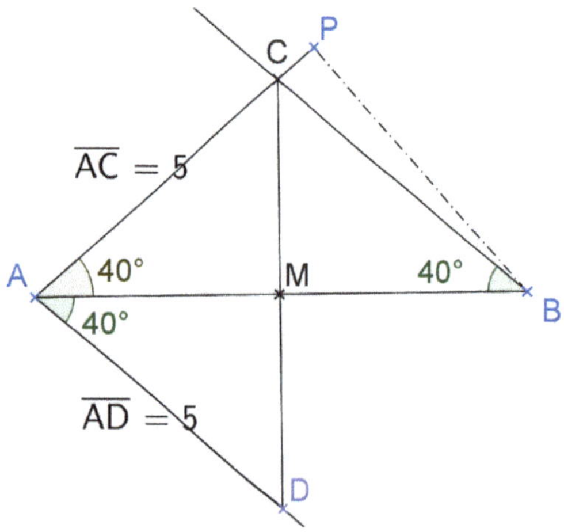

Figure II.7.36: Construction of the perpendicular bisector—the natural way

Problem II.7.9. *Give a short proof of validity for construction II of the perpendicular bisector. Use the construction done above and the hypothenuse leg Theorem II.7.32*

Short proof of validity of construction II. Let $\triangle ABC$ be the isosceles triangle constructed at the first step. Next we drop the perpendicular from point C onto line AB. As explained in construction II.7.1, we get the reflected point D and the perpendicular CD. The foot point M is the intersection of segment CD and line AB.

Question. Use the hypothenuse leg Theorem to confirm the triangle congruence $\triangle AMC \cong \triangle BMC$.

Answer. We know from construction II.7.1 above that the angle $\angle AMC$ is a right angle, and hence $\angle AMC \cong \angle BMC$. We know from construction II.7.2 of an isosceles triangle that $AC \cong BC$. Too, the segment $MC \cong MC$ is congruent to itself. Hence the hypothenuse leg Theorem II.7.32 yields the triangle congruence $\triangle AMC \cong \triangle BMC$.

This triangle congruence implies that the segments $AM \cong BM$ are congruent. This is impossible for a point on the line AB outside the segment AB. Hence point M lies between A and B, and is the midpoint of segment AB. Thus we have confirmed that CD is the perpendicular bisector of segment AB, as to be shown. □

Definition II.7.12 (The angular bisector). The ray in the interior of an angle which bisects the angle into two congruent angles is called the *angular bisector*.

Construction II.7.6 (Construction of the angular bisector). *One transfers two congruent segments AB and AC onto the two sides of the angle, both starting from the vertex A of the angle. The perpendicular, dropped from the vertex A onto the segment BC, is the angular bisector.*

Question. Reformulate the description of this construction, including the details for all steps.

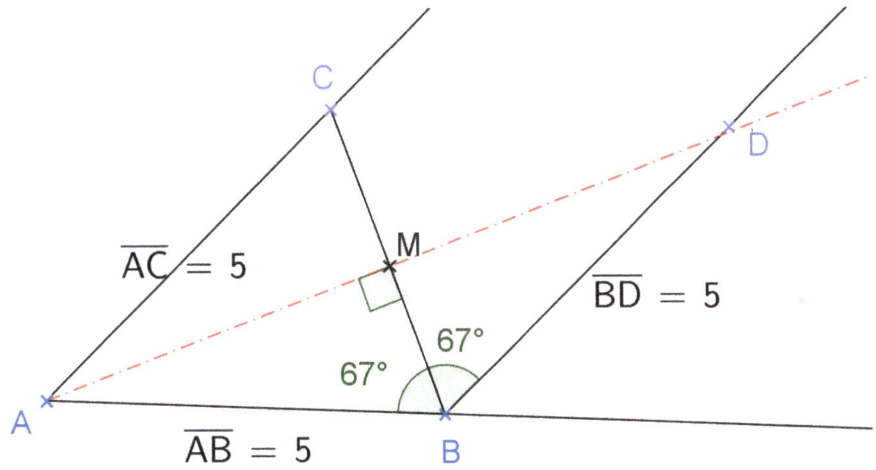

Figure II.7.37: The angular bisector

Answer. We transfer congruent segments $AB \cong AC$ onto the sides of the given angle. Draw the line BC, and transfer the base angle $\angle ABC$ to the ray $\overrightarrow{BC}$, on the side of line BC opposite to vertex A. On the new ray, we transfer segment AB to get the new segment $BD \cong BA$. The ray $\overrightarrow{AD}$ is the bisector of the given angle $\angle BAC$.

Short proof of validity. By assumption, the three points A, B, C do not lie on a line. Let M be the foot point of the perpendicular dropped from point A onto line BC.

Question. Confirm the triangle congruence $\triangle AMB \cong \triangle AMC$.

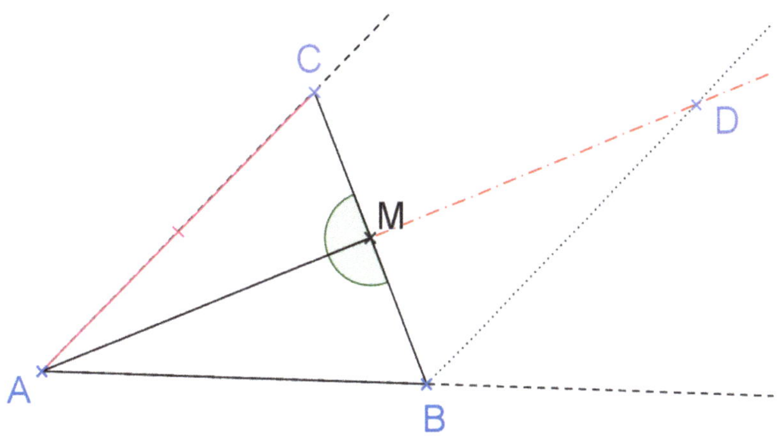

Figure II.7.38: The essential pair of congruent triangles

Answer. Indeed $\angle AMB \cong \angle AMC \cong R$, because a right angle is congruent to its supplement. It does not matter whether M lies on the ray $\overrightarrow{BC}$ or the opposite ray. Too, we have by construction the pair of congruent sides $AB \cong AC$. Finally $AM \cong AM$. The hypothenuse leg Theorem II.7.32 yields the triangle congruence $\triangle AMB \cong \triangle AMC$.

Question. Explain why the ray $\overrightarrow{AM}$ lies inside the given $\angle BAC$ and bisects this angle.

Answer. From the triangle congruence $\triangle AMB \cong \triangle AMC$, we get $\angle MAB \cong \angle MAC$, and $MB \cong MC$. Since point M lies on the line BC, the last congruence shows that M lies between B and C, too. Hence ray $\overrightarrow{AM}$ lies inside the given angle $\angle BAC$ and bisects this angle.

□

Proposition II.7.34 (Existence of the angular bisector). *For any angle $\angle BAC$, there exists a ray $\overrightarrow{AD}$ inside the given angle such that $\angle DAB \cong \angle DAC$.*

II.7.7 Remarks about angles

We have already defined supplementary angles (see Definition II.6.9.) Moreover, a right angle is defined to be an angle congruent to its supplement (see Definition II.7.8.) Because we have a valid construction II.7.1 to drop a perpendicular from a point onto a line, we know a right angle really exists. Moreover, we have proved that all right angles are congruent (see Proposition II.7.30.)

Hence it is legitimate to introduce a special notation. I use either the capital letter R, or the conventional 90°. Such a convention can neither replace the geometric Definition II.7.8, nor the proofs of existence and uniqueness!

Definition II.7.13 (The sum of two angles). Let an angle $\angle ABC$ and a ray $\overrightarrow{BG}$ in its interior. Angle $\angle ABC$ is called the *sum* of the angles $\angle ABG$ and $\angle GBC$. One writes $\angle ABC = \angle ABG + \angle GBC$.

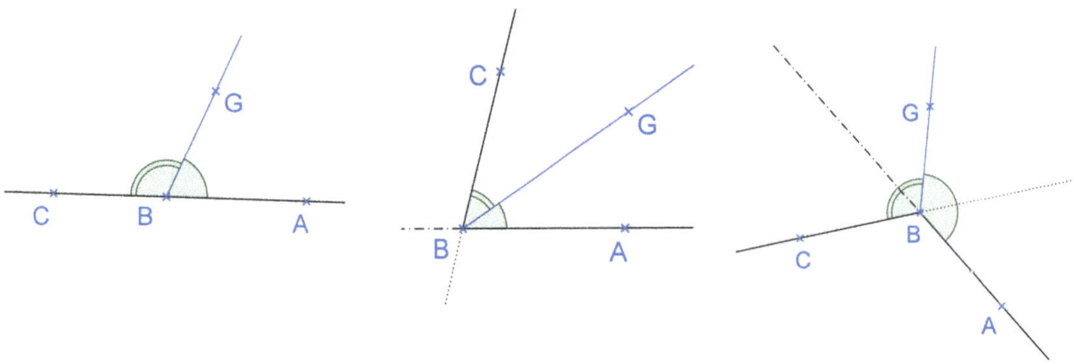

Figure II.7.39: Three cases for adding angles: they (i) are supplementary (ii) can be added (iii) or cannot be added

Lemma II.7.4. *Given are two angles $\angle ABG$ and $\angle GBC$ with same vertex B lying on different sides of a common ray $\overrightarrow{BG}$. Exactly one of three possibilities occur about the angles $\angle ABG$ and $\angle GBC$:*

(i) They are supplementary. *The three points A, B and C lie on a line.*

(ii) They can be added. *Their sum is $\angle ABG + \angle GBC = \angle ABC$. Points A and G lie on the same side of line BC. Points C and G lie on the same side of line AB.*

(iii) They cannot be added. *Points A and G lie on different sides of line BC. Points C and G lie on different sides of line AB. (The sum would be an over-obtuse angle.)*

Proof. Suppose that neither case (i) nor (ii) occurs. Under that assumption, either

(a) points A and G lie on different sides of line BC—or

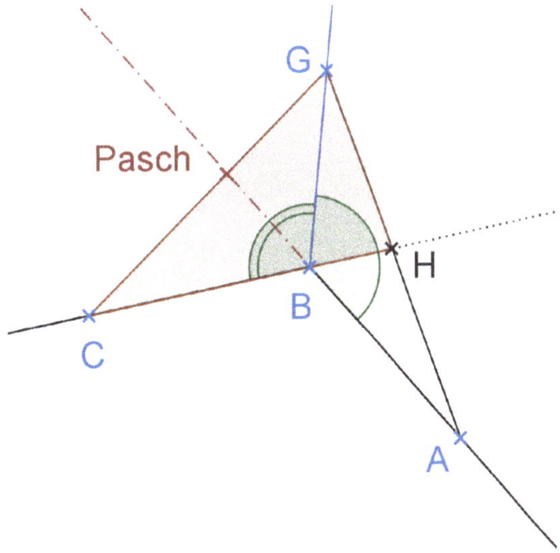

Figure II.7.40: If A and G lie on different sides of BC, then C and G lie on different sides of AB.

(b) points C and G lie on different sides of line AB.

Suppose case (a) occurs. Segment AG intersects line BC, say in point H. We use Pasch's axiom for triangle $\triangle CHG$ and line AB. This line does not intersect side GH, but intersects side CH. Indeed, point B lies between C and H, since C and A, and hence C and H are on different sides of line BG.

Now Pasch's axiom—for triangle $\triangle CHG$ and line AB—implies that this line intersects side CG, say at point *Pasch*. Thus points C and G lie on different sides of line AB. We have shown case (iii) to occur.

Suppose case (b) occurs. The same argument—with A and C exchanged— shows that case (iii) occurs once more. □

Remark. Because of these theorems, angle addition and subtraction are defined for *congruence classes* of angles.

Proposition II.7.35 (About sums of angles). *The sum of angles is defined on equivalence classes of congruent angles. It satisfies the following properties:*

Commutativity *If $\alpha + \beta$ exists, then $\beta + \alpha$ exists and $\beta + \alpha \cong \alpha + \beta$.*

Associativity $(\alpha + \beta) + \gamma \cong \alpha + (\beta + \gamma)$ *and existence of one side implies existence of the other one.*

Difference $\alpha < \gamma$ *if and only if there exists β such that $\alpha + \beta \cong \gamma$.*

Comparison of Sums 1 *If $\alpha \cong \beta$ and $\gamma < \delta$ and $\beta+\delta$ exists, then $\alpha+\gamma$ exists and $\alpha+\gamma < \beta+\gamma$.*

Comparison of Sums 2 *If $\alpha < \beta$ and $\gamma < \delta$ and $\beta+\delta$ exists, then $\alpha+\gamma$ exists and $\alpha+\gamma < \beta+\gamma$.*

Proof. By Hilbert's Theorem 15 and Proposition II.7.25, the sum of angles is defined on equivalence classes of congruent angles. Now we check the items stated:

Commutativity: Let $\alpha \cong \angle ABG$ and $\beta \cong \angle GBC$ where points G and C lie both on the same side of line AB. With that setup, $\alpha + \beta \cong \angle ABC$, which is assumed to exist. Since the order of the two rays of an angle is defined to be chosen arbitrarily,

$$\angle ABC = \angle CBA \cong \angle CBG + \angle GBA \cong \beta + \alpha$$

hence the latter angle exists and $\beta + \alpha \cong \alpha + \beta$.

Associativity: The proof is left to the reader.

Difference: Let $\alpha \cong \angle ABG$ and $\gamma \cong \angle ABC$ where points G and C lie both on the same side of line AB. With that setup, $\alpha < \gamma$ iff the ray $\overrightarrow{BG}$ lies inside the angle $\angle ABC$ iff $\alpha + \beta \cong \gamma$ with $\beta \cong \angle GBC$.

Comparison of Sums 1: Assuming that $\alpha \cong \beta$ and $\gamma < \delta$ and $\beta + \delta$ exists, we know there exists ε such that $\gamma + \varepsilon \cong \delta$. Hence

$$(\alpha + \gamma) + \varepsilon \cong \alpha + (\gamma + \varepsilon) \cong \beta + \delta$$

and $\alpha + \gamma < \beta + \gamma$ where the former is shown to exist.

Comparison of Sums 2: Assuming $\alpha < \beta$ and $\gamma < \delta$, we know there exist angles η and ε such that $\alpha + \eta \cong \beta$ and $\gamma + \varepsilon \cong \delta$. Hence

$$(\alpha + \gamma) + \eta + \varepsilon \cong (\alpha + \eta) + (\gamma + \varepsilon) \cong \beta + \delta$$

and $\alpha + \gamma < \beta + \gamma$ where the former is shown to exist.

□

Problem II.7.10. *Explain why the angular bisectors of two supplementary angles are perpendicular to each other.*

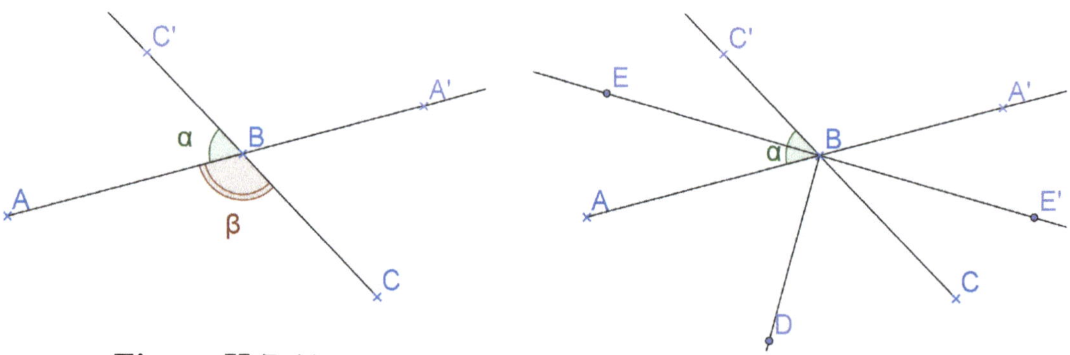

Figure II.7.41: Bisecting a pair of supplementary angles yields perpendicular rays.

Answer. Take any pair of supplementary angles, for instance $\alpha = \angle ABC'$ and $\beta = \angle ABC$ in the first figure on page 178.

We bisect these two angles, and draw the vertical angle $\angle A'BC$. Opposite rays $\overrightarrow{BE}$ and $\overrightarrow{BE'}$ are the bisectors of the vertical angles $\alpha = \angle ABC'$ and angle $\angle A'BC$. Let $\overrightarrow{BD}$ be the bisector of the supplement $\beta = \angle ABC$.

Both angles $\angle EBD$ and its supplement $\angle DBE'$ are the sum of angles $\alpha/2$ and $\beta/2$. Hence they are congruent. Thus they are a pair of congruent supplementary, and hence right angles, as to be shown.

Proposition II.7.36 (Supplements of acute and obtuse angles). *Let R denote a right angle. For any angle γ and its supplement $S(\gamma)$ exactly one of the following three cases occurs: Either* (1) *or* (2) *or* (3).

(1) $\gamma < R$, $S(\gamma) > R$ *and* $\gamma < S(\gamma)$

(2) $\gamma \cong R$, $S(\gamma) \cong R$ *and* $\gamma \cong S(\gamma)$

(3) $\gamma > R$, $S(\gamma) < R$ *and* $\gamma > S(\gamma)$.

Proof. All angles are comparable (see Proposition II.7.28). For angle γ and the right angle R, exactly one of the three cases holds:

Either (i) $\gamma < R$ or (ii) $\gamma \cong R$ or (iii) $\gamma > R$.

From SAS congruence (Theorem 12 of Hilbert, see Proposition II.7.11 above), we know that $\alpha \cong \beta$ implies $S(\alpha) \cong S(\beta)$. From Proposition II.7.29 about comparison of supplements, we know that $\alpha < \beta$ implies $S(\alpha) > S(\beta)$. Hence with the help of Propositions II.7.25 and II.7.26 we get:

in case (i), one concludes $S(\gamma) > S(R) \cong R > \gamma$, and hence (a) $\gamma < S(\gamma)$,

in case (ii), one concludes $S(\gamma) \cong S(R) \cong R \cong \gamma$, and hence (b) $\gamma \cong S(\gamma)$,

in case (iii), one concludes $S(\gamma) < S(R) \cong R < \gamma$, and hence (c) $\gamma > S(\gamma)$.

By Proposition II.7.28 above, all angles are comparable. Hence the two angles γ and its supplement $S(\gamma)$ satisfy either (a) or (b) or (c). This observation allows one to get the converse of the conclusions shown above. For example, (a) excludes both (ii) and (iii), and hence implies (i). Thus we get

(a) implies (i), (b) implies (ii), (c) implies (iii); and finally
(a) if and only if (i), (b) if and only if (ii), (c) if and only if (iii).

This leads to the mutually exclusive cases (1) (2) (3), as originally stated. □

Corollary 10 (All right angles are congruent).

Proof. The proposition II.7.36 about supplements of acute and obtuse angles yields an easy proof that all right angles are congruent:

Let R be the right angle as has been constructed in Construction II.7.1. Now suppose $\gamma = R'$ is another right angle. By definition of a right angle, this means that $\gamma \cong S(\gamma)$.
Hence case (b) above occurs, which implies (ii): $\gamma \cong R$. You see that <u>existence</u> of a right angle make proving its <u>uniqueness</u> a bit easier. □

Proposition II.7.37 (Uniqueness of angular bisector). *The angular bisector is unique.*

Proof. Suppose both rays $\overrightarrow{BG}$ and $\overrightarrow{BG'}$ bisect angle $\angle BAC$. The angles $\alpha := \angle ABG$ and $\alpha' := \angle ABG'$ are comparable, because all angles are comparable by Proposition II.7.28.

Suppose towards a contradiction that $\alpha < \alpha'$. As specified in Proposition II.7.35, one can add inequalities of angles. Hence we conclude that

$$(?) \qquad \angle ABC = \angle ABG + \angle GBC \cong \alpha + \alpha < \alpha' + \alpha' \cong \angle ABG' + \angle G'BC \cong \angle ABC$$

which is impossible. Similarly, the case $\alpha > \alpha'$ can be ruled out. Hence $\alpha \cong \alpha'$, and hence by uniqueness of angle transfer $\overrightarrow{BG} = \overrightarrow{BG'}$, as to be shown. □

II.7.8 Orientated angles

Definition II.7.14 (Orientated angle). An *orientated angle* is an ordered pair (h, k) of two rays h and k with a common vertex. It is allowed that h and k are equal or opposite rays. For emphasis, I denote the orientated angle by $\curvearrowleft (h, k)$. Let an *orientated angle* between two lines p and q be an ordered pair (p, q).

Definition II.7.15 (Degenerate angle). An orientated angle $\curvearrowleft (h, k)$ is called *degenerate* if h and k are either equal or opposite rays. These degenerate angles are the zero and the straight angle.

An orientated angle $\curvearrowleft (h, k)$ is called *non-degenerate* if the two rays h and k do not lie on the same line.

We agree to allow degenerate angles only for orientated and unwound angles, but not for common angles without orientation. As shown by the orientation theorem II.6.1, in every ordered incidence plane, it is possible to define and fix an orientation. In every ordered incidence plane, it is possible to define an orientation. The orientation is fixed, as soon as one has agreed which are the left and right half plane for one fixed ray.

Definition II.7.16 (Two orientations for angles). Two orientated non-degenerate angles $\curvearrowleft (h, k)$ and $\curvearrowleft (h', k')$ have the *same orientation* if and only if either

- ray k lies in the left half-plane of ray h, and ray k' lies in the left half-plane of ray h' (they are both counter clockwise), or
- ray k lies in the right half-plane of ray h, and ray k' lies in the right half-plane of ray h' (they are both clockwise).

Two orientated non-degenerate angles $\curvearrowleft (h, k)$ and $\curvearrowleft (h', k')$ have the *opposite orientation* if and only if either

- ray k lies in the left half-plane of ray h, and ray k' lies in the right half-plane of ray h', or
- ray k lies in the right half-plane of ray h, and ray k' lies in the left half-plane of ray h'.

Definition II.7.17 (Congruence of orientated angles). Two orientated angles $\curvearrowleft (h, k)$ and $\curvearrowleft (h', k')$ are called *congruent* in either one of the following cases

- $h = k$ and $h' = k'$, or
- h and k as well as h' and k' are opposite rays, or
- they are congruent in the usual sense $\angle(h, k) \cong \angle(h', k')$ and the orientated angles $\curvearrowleft (h, k)$ and $\curvearrowleft (h', k')$ have the same orientation.

Problem II.7.11. *We have seen in Proposition II.7.25 that the congruence of ordinary angles is an equivalence relation. Convince yourself that the congruence of orientated angles in an equivalence relation, too.*

We denote the congruence class of right angles by R. The positively orientated angles take values between 0 and $2R$. The straight angle $\curvearrowleft (k, k')$ where k and k' are opposite rays takes by convention the value $2R$. The negatively orientated angles take values between $-2R$ and 0. We have seen in Proposition II.7.26 that the comparison of ordinary angles is well defined for congruence classes.

Problem II.7.12. *Convince yourself of the following facts:*

- *The congruence of orientated angles is an equivalence relation.*
- *All zero angles* $\curvearrowleft (h, h)$ *are congruent.*
- *All straight angles* $\curvearrowleft (h, k)$ *with two opposite rays* h *and* k *are congruent.*
- *The orientated angles* $\curvearrowleft (h, k)$ *and* $\curvearrowleft (k, h)$ *are congruent if and only if the two rays* h *and* k *are either equal or opposite.*
- *There exist right angles in two different orientations.*
- *Only orientated right angles with the same orientation are congruent.*

The sum of two orientated angles is defined simpler than the sum of common angle from definition II.7.13.

Definition II.7.18 (The sum of orientated angles). Any two orientated angles $\curvearrowleft (h, k)$ and $\curvearrowleft (k, l)$ have the sum

$$\curvearrowleft (h, k) + \curvearrowleft (k, l) = \curvearrowleft (h, l)$$

The operation of addition is compatible with congruence.

Proposition II.7.38. *The addition of congruence classes of orientated angles is well-defined. With this addition, the congruence classes of orientated angles are an Abelian group. This is not a free group since* $4R = 0$*. Hence it is neither an ordered group.*

Proposition II.7.39. *The group of orientated angles is isomorphic to the group of rotations around a fixed point.*

II.7.9 The exterior angle theorem

Definition II.7.19 (Exterior angle of a triangle). The *exterior angle of a triangle* is the supplement of an interior angle.

Proposition II.7.40 (The Exterior Angle Theorem). [Euclid I.16. Theorem 22 in Hilbert] *The exterior angle of a triangle is greater than both nonadjacent interior angles.*

Proof. For the given $\triangle ABC$, we can choose point D on the ray opposite to $\overrightarrow{AB}$, such that $AD \cong CB$. We compare the two nonadjacent interior angles $\gamma = \angle ACB$ and $\beta = \angle ABC$ to the exterior angle $\delta = \angle CAD$. As a first step, we show

Lemma II.7.5. *The exterior angle* $\delta = \angle CAD$ *is not congruent to the interior angle* $\angle ACB = \gamma$.

Proof of Lemma'II.7.5. Assume towards a contradiction that $\delta \cong \gamma$.

The supplementary angle of δ is $\alpha = \angle CAB$. The supplementary angle of γ is $\angle ACE$, with a point E on the ray opposite to $\overrightarrow{CB}$. Supplementary angles of congruent angles are congruent (Theorem 14 in Hilbert, see Proposition II.7.15 above). Hence

$$\angle CAB \cong \angle ACE \tag{?1}$$

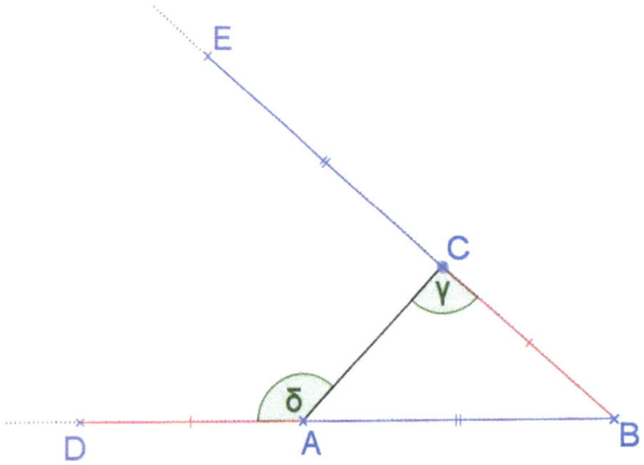

Figure II.7.42: The impossible situation of a congruent exterior angle

On the other hand, we use SAS-congruence for the two triangles $\triangle ABC$ and $\triangle A'B'C'$ with

$$A' := C\,,\; B' := D\,,\; C' := A$$

Those two triangles would be congruent by SAS congruence. Indeed the angles at C and C' are congruent by our assumption $\gamma = \angle ACB \cong \angle CAD = \delta$. Too, the adjacent sides are pairwise congruent because of $CB \cong AD = C'B'$ by construction, and $CA \cong AC = C'A'$ by axiom (III.1b). [26] From Axiom (III.5), we conclude that the corresponding angles at A and $A' = C$ are congruent:

$$\angle CAB \cong \angle C'A'B' = \angle ACD \tag{?2}$$

We have shown that angle $\angle CAB$ is congruent to both (?1) $\angle ACE$ and (?2) $\angle ACD$.

Now we use the uniqueness of angle transfer. Transferring angle $\angle CAB$ with one side $\overrightarrow{CA}$, into the half plane not containing B, yields as second side of the angle once the ray $\overrightarrow{CE}$ and the second time $\overrightarrow{CD}$. Hence these two rays are the same. Hence points E and D both lie on the line $l_a := BC$. In other words, the four points B, C, D, E all lie on one line. Because B and D lie on <u>opposite</u> rays with vertex A, these two points are <u>different</u>: $B \neq D$. [27]

Because a line is uniquely specified by two of its points, this implies $l_a = BC = BD$. We have just seen that point C lies on this line. By construction, point A lies on the line BD, too. Hence all the points A, B, C, D, E lie on the same line. This contradicts the definition of a triangle. By definition of a triangle, its three vertices do not lie on one line. This contradiction confirms the original claim $\delta \not\cong \gamma$. □

Remark. Given are <u>any</u> three different points A, B, C not lying on a line. We choose point D on the ray opposite to $\overrightarrow{AB}$, such that $AD \cong CB$. We choose the point E on the ray opposite to $\overrightarrow{CB}$ for which $AB \cong CE$.

The assumption that the exterior angle $\delta = \angle CAD$ is congruent to the interior angle $\angle ACB = \gamma$ has the following consequences:

- one gets congruent triangles $\triangle ABC \cong \triangle CDA$;
- $D = E$

[26] I would prefer to add the item (III.1b) $AB \cong BA$ to Hilbert's axioms.

[27] Note that just this simple remark is not true in elliptic geometry!

- the points B, A, D lie on a line;
- the points B, C, D lie on a line.

Remark. No contradiction arises from the assumption $\delta \cong \gamma$ in the cases that either (a) or (b) hold:

(a) $D \neq B$, but the line DB is not unique;

(b) $D = B$.

Remark (**Cutting a slice of orange in a strange way**). In spherical geometry, the case (a) is possible. The figure constructed does exist. We do not get a contradiction, the conclusion is just that the two points B and D are antipodes, but still $D \neq B$. The two congruent triangles

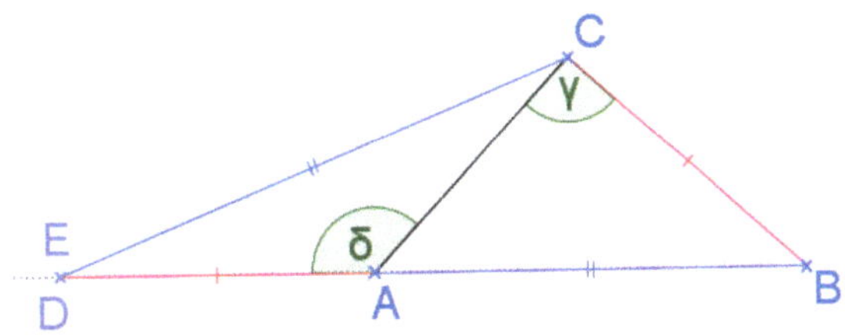

Figure II.7.43: Two ways from B to D

$\triangle ABC \cong \triangle A'B'C' = \triangle CDA$ are obtained from each other by a rotation of 180° around the midpoint M of segment AC. The vertices A and C lie on two different lines BAD and BCD through these two antipodes. Too, the sum of the segments is congruent going both ways from B to D. Hence $BC \cong DA$ and $CD \cong AB$, and hence the sums $BC + CD \cong DA + AB$ are congruent. Both turn out to be 180°. The union of the triangles $\triangle ABC \cong \triangle A'B'C' = \triangle CDA$ has the shape as the peel of a slice of orange, which is cut along a line through M. This line need not be perpendicular to neither one of the great circles BAD nor BCD.

Remark. In double elliptic geometry, the case (b) is possible. A model of double elliptic geometry can be obtained by identifying antipodes on the sphere.

The next Lemma rules out the possibility $\delta < \gamma$.

Lemma II.7.6. *The exterior angle $\delta = \angle CAD$ is not smaller that the interior angle $\angle ACB = \gamma$.*

Proof of Lemma II.7.6. Suppose towards a contradiction that $\delta < \gamma$. We transfer the exterior angle

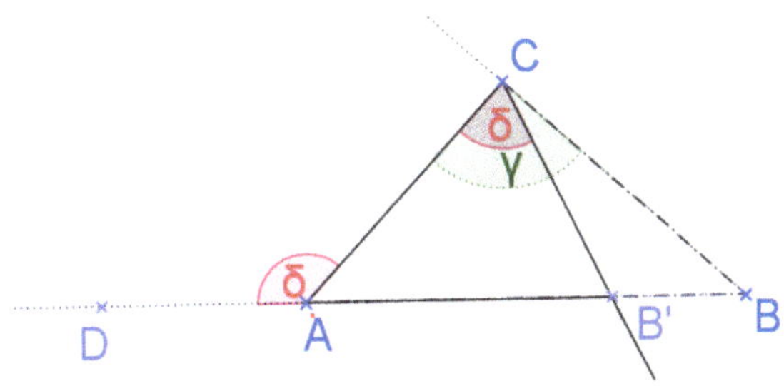

Figure II.7.44: The impossible situation of a smaller exterior angle

$\delta = \angle CAD$ with one side $\overrightarrow{CA}$, into the half planen containing B. The second side of the angle we get is a ray $\overrightarrow{CB'}$ inside the angle $\angle ACB$. By the Crossbar Theorem, it meets the segment AB in a point B'.

We can now apply the first Lemma II.7.5 to $\triangle AB'C$. This smaller triangle would have exterior angle δ congruent to the interior angle $\gamma' = \angle ACB'$. This is impossible by Lemma II.7.5. □

Thus we have both ruled out the possibility that $\delta = \gamma$, or that $\delta < \gamma$. By Proposition II.7.28, all angles are comparable. Hence there remains only the possibility that $\delta > \gamma$.

Lemma II.7.7. *The exterior angle* $\delta = \angle CAD$ *is greater than the interior angle* $\beta = \angle ABC$.

Proof of Lemma II.7.7. We compare angle $\beta = \angle ABC$ to the exterior angle $\delta = \angle CAD$. Choose

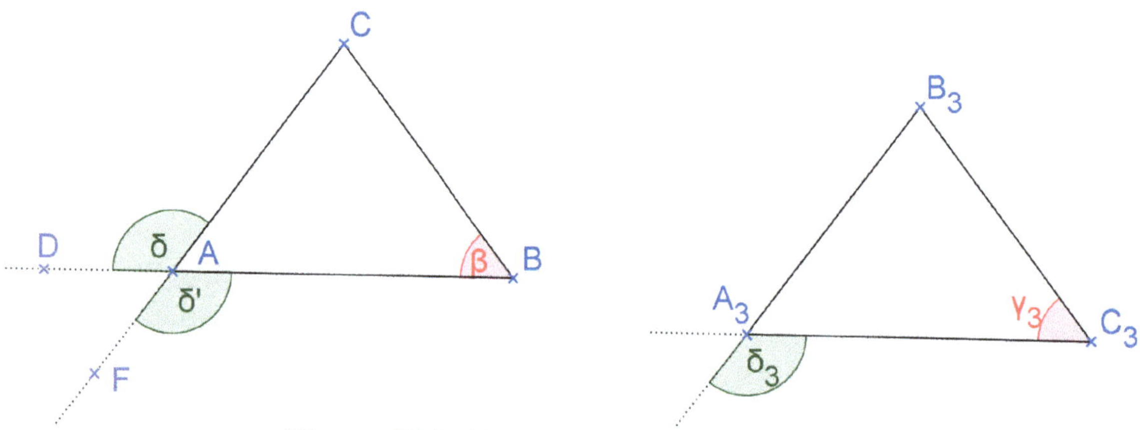

Figure II.7.45: Comparing the other nonadjacent angle

any point F on the ray opposite to $\overrightarrow{AC}$. The vertical angles $\delta = \angle CAD \cong \angle FAB = \delta'$ are congruent by Euclid I.15. Now we are back to the case already covered, because the interior angle $\beta = \angle ABC$ and exterior angle $\angle FAB$ lie on opposite sides of triangle side AB, which is also part of one side of each of these angles. By Lemma II.7.5 and II.7.6, we conclude $\angle FAB > \angle ABC$ and hence $\delta > \beta$. □

Remark. Here is an other way to prove Lemma II.7.7: Define $\triangle A_3B_3C_3$ by setting $A_3 := A$, $B_3 := C$, $C_3 := B$. Now we can apply LemmaII.7.5 and II.7.6 to this new triangle and get $\delta' = \delta_3 > \gamma_3 = \beta$. By congruence of the vertical angles $\delta = \angle CAD \cong \angle FAB = \delta'$ we get $\delta > \beta$.

Thus the proof of the exterior angle theorem is finished. □

Below I give a second proof of the exterior angle Theorem, following Euclid's ideas. This proof is presupposing a valid construction to bisect a segment. To avoid a circular reasoning, there are two possibilities:

- Either one is following Hilbert's line of reasoning. He proves at first the exterior angle Theorem by the arguments on Lemmas II.7.5, II.7.6, II.7.7 above, and afterwards gives the construction II.7.7 below for the bisection of a segment. The justification of his method of bisection presupposes the exterior angle theorem.

- One may at first give valid constructions for an isosceles triangle, and the perpendicular bisector of a segment. This line of reasoning is based on the converse isosceles triangle proposition II.7.31, already proved above, in a way different from Hilbert's proof. Moreover, one needs the construction II.7.2 of an isosceles triangle as well as the construction II.7.5 of the perpendicular bisector. After the validity of these steps is provided, one may uses Euclid's idea explained below to prove the exterior angle theorem.

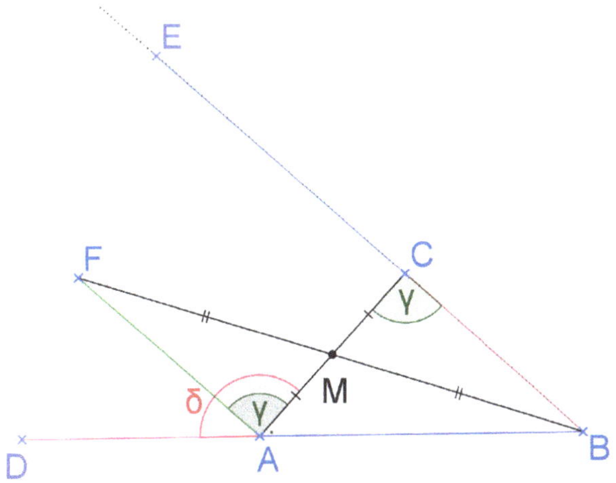

Figure II.7.46: A proof of the exterior angle theorem presupposing bisection of a segment.

Proof of the exterior angle Theorem following Euclid's ideas. In the given triangle $\triangle ABC$, we bisect the side AC and let M be the midpoint. We extend the ray $\overrightarrow{BM}$ and construct on the extension a point F such that $MB \cong MF$. As one easily checks via SAS congruence, one obtains congruent triangles

$$\triangle MCB \cong \triangle MAF \tag{II.7.8}$$

Indeed, these triangles have congruent vertical angles at vertex M, and by construction two pairs of congruent sides $MC \cong MA$ and $MB \cong MF$. From the triangle congruence (II.7.8) we obtain congruent angles $\gamma := \angle ACB \cong \angle CAF$.

We choose any point D on the ray opposite to $\overrightarrow{AB}$. Since by construction $A * M * C$ and $B * M * F$, all three points C, M and F lie on the same side of line AB. Hence the ray $\overrightarrow{AF}$ lies in the interior of the angle $\delta := \angle CAD$, which is an exterior angle of triangle $\triangle ABC$. One obtains $\gamma \cong \angle CAF < \delta \cong \angle CAD$. As to be shown, the exterior angle δ is indeed greater than the nonadjacent interior angle γ. □

Proposition II.7.41. [Euclid I.17] *The sum of any two interior angles of a triangle is less than two right angles.*

II.7.10 Congruence of z-angles

Definition II.7.20 (Alternate interior angles or z-angles). Let a transversal t intersect two lines a and b at the points A and B. A pair of *alternate interior angles* or simply *z-angles* are two angles

$$\alpha := \angle(\overrightarrow{AB}, r_A) \quad \text{and} \quad \beta := \angle(\overrightarrow{BA}, r_B)$$

They have vertices A and B. One of their respective sides are the rays $r_A \subseteq a$ and $r_B \subseteq b$ lying on the lines a respectively b, and lying in opposite half-planes of the transversal t. Their other sides are the rays $\overrightarrow{AB}$ and $\overrightarrow{BA}$, which both contain the segment AB.

Proposition II.7.42 (Congruent z-angles imply parallels). [Euclid I.27] *If two lines form congruent z-angles with a transversal, they are parallel.*

Proof. This is an immediate consequence of the exterior angle theorem. We argue by contradiction. Suppose the two lines would intersect in point C. One of the z-angles is an exterior angle of triangle $\triangle ABC$, the other one is a nonadjacent interior angle. Lemma II.7.5 does imply that the two z-angles are not congruent— contradicting the assumption from above. Hence the two lines a and b cannot intersect. □

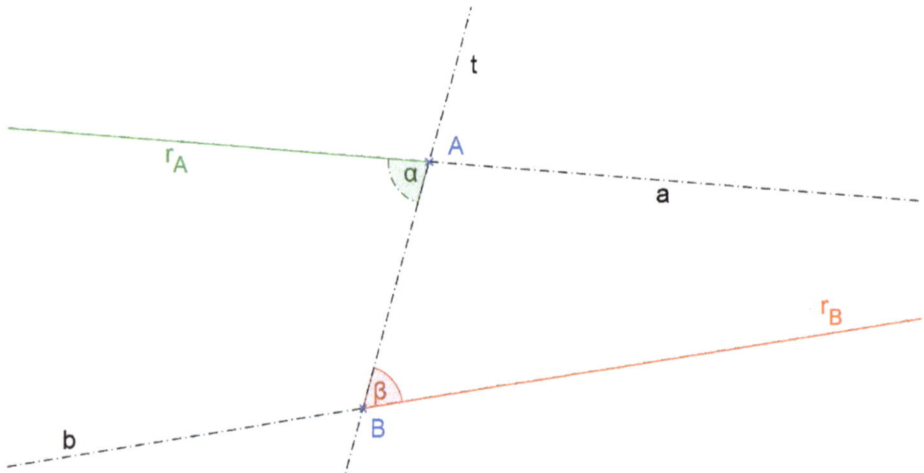

Figure II.7.47: A pair of z-angles.

Remark. Referring to the pair of z-angles shown in the figure on page 186 the Proposition II.7.42 tells that

$$\alpha \cong \beta \Rightarrow a \parallel b$$

This is equivalent to its contrapositive

$$a \nparallel b \Rightarrow \alpha \ncong \beta$$

which relates even more directly to the exterior angle theorem.

As a consequence of Euclid I.27, the existence of a parallel can be proved in neutral geometry. One parallel to l through point P is conveniently constructed as "double perpendicular".

Proposition II.7.43 (Existence of a parallel). *For every line l and for every point P lying not on l, there exists at least one parallel m to l through point P.*

Proof. Given is line l and a point P not on l. One drops the perpendicular from point P onto line l and denotes the foot point by F. Next, one erects at point P the perpendicular to line PF. Thus, one gets a line, which we call m. Because the two lines l and m form congruent z-angles with the transversal $\overleftrightarrow{PF}$, Euclid I.27 or I.28 imply that l and m are parallel. This leaves open the question whether or not m is the unique parallel to line l through point P. □

Remark. Similar to the exterior angle theorem, these are facts of *neutral geometry.* They are is valid both in Euclidean and hyperbolic geometry. Note that Euclid I.2 through I.28 are theorems that hold for every Hilbert plane. Indeed Euclid I.29 is the first theorem in Euclid's elements that uses the Euclidean parallel postulate.

Warning. The converse of Euclid I.27 is does not hold in hyperbolic geometry. Indeed, in hyperbolic geometry, parallels can form non-congruent z-angles with any transversal.

Recall that the *Euclidean Parallel Postulate* I.3.3 postulates both existence and uniqueness of the parallel to a given line through a given point. On the other hand, *Hilbert's Parallel Postulate for plane geometry* I.3.4 postulates only the <u>uniqueness</u> of the parallel to a given line through a given point. Because of Proposition II.7.43, Euclid's and Hilbert's parallel postulate turn out to be equivalent for any Hilbert plane. In other word, we can state the definition equivalent to definition I.2.2:

Definition II.7.21 (Pythagorean plane). A *Pythagorean plane* is a Hilbert plane for which the Euclidean parallel postulate holds.

Proposition II.7.44 (In a Pythagorean plane, parallels imply congruent z-angles). [Euclid I.29] *In a Pythagorean plane, any two parallel lines form congruent z-angles with any transversal. Referring to the pair of z-angles shown in the figure on page 186*

$$a \parallel b \Rightarrow \alpha \cong \beta$$

Proof. We transfer the angle $\alpha := \angle(\overrightarrow{AB}, r_A)$ onto the ray $\overrightarrow{BA}$ into the same half plane as the ray r_B from angle $\beta := \angle(\overrightarrow{BA}, r_B)$. We obtain a new ray s such that $\alpha \cong \angle(\overrightarrow{BA}, s)$.

Now the line a and the line on ray s form congruent z-angles with the transversal AB. According to proposition II.7.42 they are parallel. On the other hand, the lines a and b are parallel by assumption. Finally, the line a has a *unique* parallel through point B since the plane is <u>assumed</u> to be <u>Pythagorean.</u>

Hence we conclude that the line b and the line on ray s are equal. Since the rays r_B and s lie both on the line b, have vertex B and lie in the same half-plane of AB, they are equal. We conclude that as to be shown

$$\alpha \cong \angle(\overrightarrow{BA}, s) = \angle(\overrightarrow{BA}, r_B) \cong \beta$$

□

Proposition II.7.45 (A Hilbert plane where parallel lines are a means to transport of an angle, is Pythagorean). *Assume that in the Hilbert plane to be considered, any two parallel lines form congruent z-angles with any transversal. Then for every line l and for every point P lying not on l, there exists <u>a unique</u> one parallel m to l through point P. In other words, the plane is Pythagorean.*

Remark. By the force of proposition II.7.44 and II.7.45, we get the following insight:

> <u>Only in a Pythagorean plane</u> can parallel lines be used as a means for the transport of an angle.

Corollary 11. *In any Hilbert plane are equivalent*

(i) *Hilbert's Parallel Postulate I.3.4;*

(ii) *the Euclidean Parallel Postulate] I.3.3.*

II.7.11 Consequences of the exterior angle theorem

Proposition II.7.46 (**Immediate consequences of the exterior angle theorem**).

(i) *Every triangle can have at most one right or obtuse angle.*

(ii) *The base angles of an isosceles triangle are acute.*

(iii) *The foot point of a perpendicular is unique.*

(iv) *Given a line l and a point O not on l. At most two points of l have the same distance from O.*

(v) *A circle and a line can intersect in at most two points.*

Proof. **(i):** Suppose angle α of $\triangle ABC$ is right or obtuse: $\alpha \geq R$. By Proposition II.7.36, its supplement is right or acute: $S(\alpha) \leq R$. But the supplement is an exterior angle: $\delta = S(\alpha)$. By the exterior angle theorem, the two other nonadjacent interior angles of the triangle are less than that exterior angle. Hence they are acute: $\beta, \gamma < \delta = S(\alpha) \leq R$, and hence $\beta, \gamma < R$.

(ii): The two base angles being congruent, by (i), they cannot be both right or obtuse.

(iii): Given a line l and a point O not on l. Suppose there are two perpendiculars, with foot points F and G. The $\triangle OFG$ would have two right angles, contradicting (i).

(iv): Given a line l and a point O not on l. Suppose *towards a contradiction* there are three points

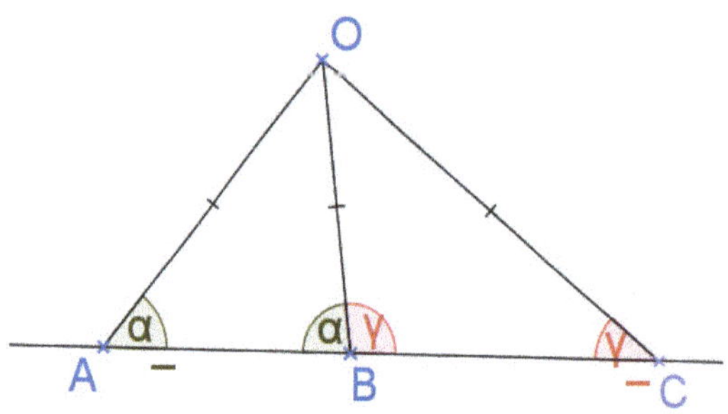

Figure II.7.48: No three points have the same distance from a line

A, B, C on the line l such that $OA \cong OB \cong OC$. Of any three points, one lies between the two others (Theorem 4 of Hilbert). We may suppose by renaming that $A * B * C$.

The base angle α of triangle $\triangle OBC$ is an exterior angle of the other triangle $\triangle OAB$. Hence the exterior angle theorem yields $\gamma > \alpha$. But the role of the two triangles can be switched: The base angle γ of triangle $\triangle OAB$ is an exterior angle of the other triangle $\triangle OBC$. Hence the exterior angle theorem yields $\alpha < \gamma$. Both inequalities together are impossible. This contradiction implies there can exist at most two points on a line which have congruent distances from point O.

(v) For the case that the center O of the circle does not lie on the line, the statement is confirmed by item (iv) above.

For the case that the center O of the circle lies on the line, the statement follows from line separation and the uniqueness of segment transfer stated in Proposition II.7.2.

□

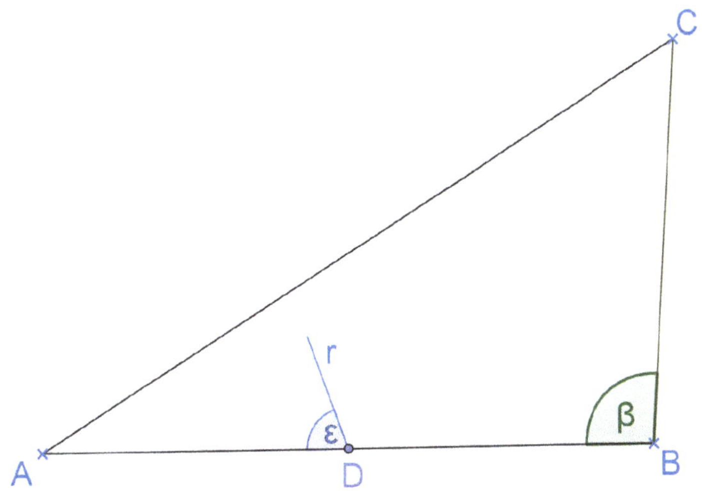

Figure II.7.49: Assumed is $\varepsilon \leq \beta$. Why does the ray r intersect the side AC.

Problem II.7.13. *As shown in the figure on page 189, given is a triangle* $\triangle ABC$, *a point* D *on the side* AB, *and a ray* r *emanating from point* D *into the interior of the triangle. We assume that this ray forms the angle* $\varepsilon \leq \beta$ *with the ray* $\overrightarrow{DA}$.

Use the exterior angle theorem and the crossbar theorem to show, in neutral geometry, that the ray intersects the side AC.

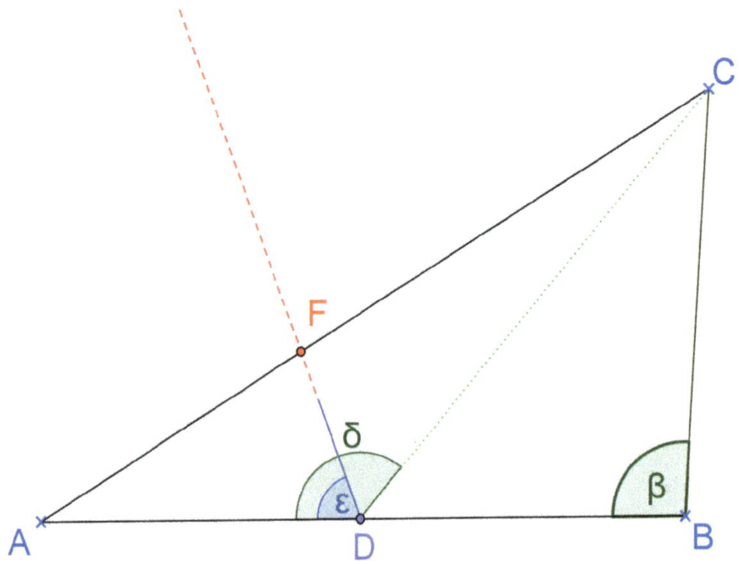

Figure II.7.50: Since $\varepsilon \leq \beta < \delta$ the ray r intersects the side AC.

Answer. We draw the segment DC and use the exterior angle theorem for triangle $\triangle DBC$. In this triangle the angle $\delta = \angle(\overrightarrow{DA}, r)$ is an exterior angle and $\beta = \angle ABC$ is a nonadjacent interior angle. One concludes $\beta < \delta$. Hence the assumptions imply $\varepsilon \leq \beta < \delta$, which means that the ray r lies in the interior of the angle δ. We use the crossbar theorem for the triangle $\triangle ADC$, and conclude that the ray r intersects the opposite side AC, as to be shown.

Problem II.7.14. *Given is a strictly convex quadrilateral with two pairs of congruent opposite sides. As shown in Lemma II.6.8, each diagonal separates the quadrilateral into two triangles in opposite half planenns of their common side. Prove in* neutral geometry *that the two pairs of opposite sides are parallel. Provide a drawing.*

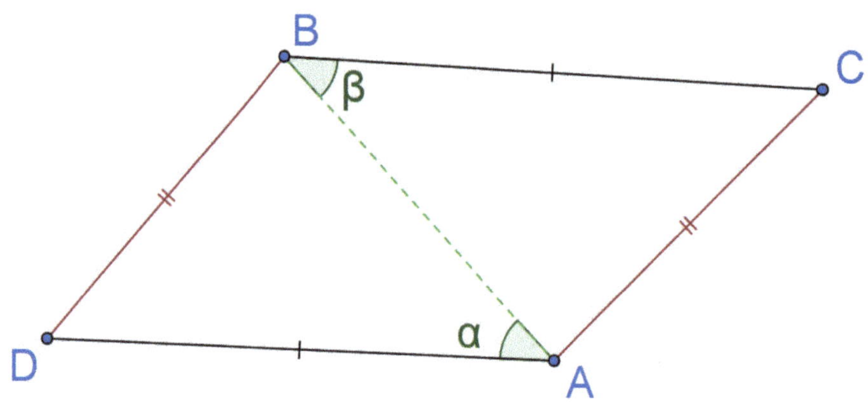

Figure II.7.51: A parallelogram in neutral geometry.

Answer. Given is the quadrilateral $\square ACBD$ with two pairs of congruent opposite sides $AC \cong BD$ and $BC \cong AD$. We draw the diagonal AB and get the congruent triangles $\triangle ABC \cong \triangle BAD$, as one confirms by SSS congruence. Hence the angles $\beta = \angle ABC$ and $\alpha = \angle BAD$ are congruent. We see that the diagonal AB transverses the lines of the opposite sides DA and BC with congruent z-angles. By Euclid I.27, we conclude that the opposite sides are parallel.

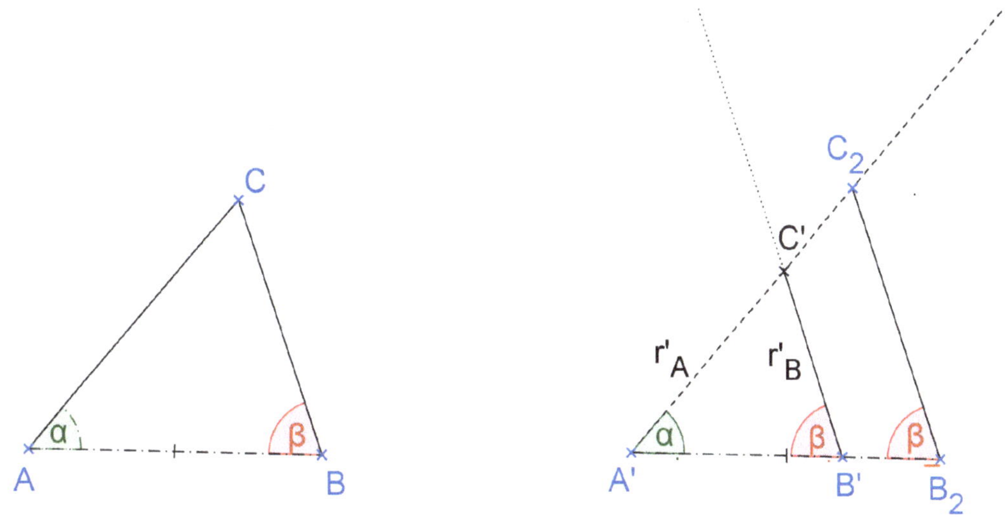

Figure II.7.52: Addendum to the extended ASA congruence.

Proposition II.7.47 (Addendum to the extended ASA-Congruence). *Given is a triangle $\triangle ABC$, a* shorter *segment $A'B' < AB$ and a half-plane $\mathcal{H}'$ bounded by this segment. The two angles at the vertices A and B are transferred to the endpoints of the segment $A'B'$, and reproduced in the same half plane $\mathcal{H}'$. Then the newly obtained rays r'_A and r'_B intersect, say at point C'. One obtains a triangle $\triangle A'B'C'$, lying in the half planen $\mathcal{H}'$, and having at vertices A' and B' the angles $\angle B'A'C' \cong \angle BAC$ and $\angle A'B'C' \cong \angle ABC$.*

Problem II.7.15. *Prove proposition II.7.47, using extended ASA-congruence, Euclid I.27 (see proposition II.7.42), and Pasch's axiom.*

Proof. One transfers segment AB to the ray $\overrightarrow{A'B'}$ and gets a segment $A'B_2 \cong AB$ with $A' * B' * B_2$.

Question. Which assumption has been used to get $A' * B' * B_2$?

Answer. Since $A'B' < AB$, we get $A' * B' * B_2$.

The two angles at the vertices A and B have been transferred to the endpoints of the segment $A'B'$, and reproduced in the same half planen $\mathcal{H}'$. We transfer the segment AC onto the newly obtained ray r'_A and obtain a congruent segment $A'C_2 \cong AC$. Thus we get a triangle $\triangle A'B_2C_2$ and check by SAS congruence that $\triangle A'B_2C_2 \cong \triangle ABC$.

Now we apply Pasch's axiom to triangle $\triangle A'B_2C_2$ and line l' on the ray r'_B. This line intersects the triangle side $A'B_2$ at point B', by construction. Line l' does not intersect the segment B_2C_2.

Question. Why are the lines l' and B_2C_2 parallel?

Answer. The lines $l_2 = B_2C_2$ and l' form congruent angles β with the base $A'B'$. By Euclid I.27, the two lines are parallel.

Hence Pasch's axiom implies that line l' intersects the third side $A'C_2$ of $\triangle A'B_2C_2$.

We call the intersection point C'. Thus we have produced a $\triangle A'B'C''$, the angles of which at vertices A' and B' are congruent to the corresponding angles of $\triangle ABC$. $\square$

Remark. In Euclidean geometry, the two rays r'_A and r'_B always intersect, no matter whether $A'B' < AB$, $A'B' \cong AB$, or $A'B' > AB$. In all three cases, the $\triangle A'B'C'$ is similar to $\triangle ABC$.

Remark. Here is what happens in hyperbolic geometry: In hyperbolic geometry, similar triangles are always congruent, this implies that $\triangle A'B'C'$ and $\triangle ABC$ are not similar if either $A'B' < AB$ or $A'B' > AB$!

In the case just considered, $A'B' < AB$ and $\gamma' > \gamma$. In the opposite case that $A'B' > AB$, the second triangle $\triangle A'B'C'$ has either a different angle $\gamma' < \gamma$, or does not exist at all. The last possibilities occurs because the two rays r'_A and r'_B do not intersect at all. This happens always, once the segment $A'B'$ is long enough.

Proposition II.7.48 (Comparison of sides implies comparison of angles). [Euclid I.18, Theorem 23 of Hilbert] *In any triangle, across the longer side lies the greater angle.*

Proof. In $\triangle ABC$, we assume for sides AB and BC that $c = AB > BC = a$. The issue is to compare the angles $\alpha = \angle CAB$ and $\gamma = \angle ACB$ across these two sides.

We transfer the shorter side BC at the common vertex B onto the longer side. Thus one gets a segment $BD \cong BC$, with point D between B and A. Because the $\triangle BCD$ is isosceles, it has two congruent base angles

$$\delta = \angle CDB \cong \angle DCB$$

Because $B * D * A$, we get by angle comparison at vertex C

$$\delta = \angle DCB < \gamma = \angle ACB$$

Now we use the exterior angle theorem for $\triangle ACD$. Hence

$$\alpha = \angle CAB < \delta = \angle CDB$$

By transitivity, these three equations together imply that $\alpha < \gamma$. Hence the angle α across the smaller side CB is smaller than the angle γ lying across the greater side AB. In short, we have shown that $c > a \Rightarrow \gamma > \alpha$. $\square$

Proposition II.7.49 (Comparison of angles implies comparison of sides). [Euclid I.19] *In any triangle, across the greater angle lies the longer side.*

Proof. In $\triangle ABC$, we assume for two angles that $\alpha = \angle CAB < \gamma = \angle ACB$. The issue is to compare the two sides BC and AB lying across the angles and show $a = BC < AB = c$. As shown in Proposition II.7.6, any two segments are comparable.

Question. Please repeat the reasoning for convenience.

Answer. We transfer segment BC along the ray $\overrightarrow{BA}$, and get a segment $BD \cong BC$. By Theorem 4 in Hilbert, of the three points A, B, D on a line AB, exactly one lies between the two others. This leads to the following three cases: Either

(i) $A * D * B$ and $a < c$. or (ii) $A = D$ and $a = c$. or (i) $D * A * B$ and $a > c$.

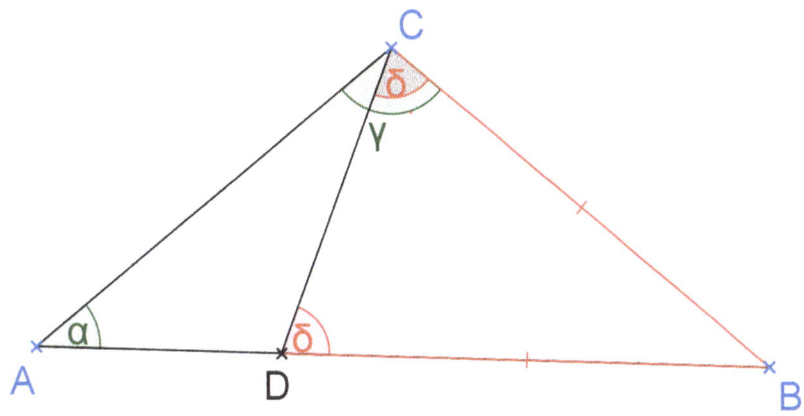

Figure II.7.53: Across the longer side lies the greater angle

We can now rule out cases (ii) and (iii).

In case (ii), the $\triangle ABC$ is isosceles, and Euclid I.5 would imply $\alpha = \gamma$, contrary to the hypothesis about these two angles.

In case (iii), Euclid I.18 or Theorem 23 of Hilbert, would imply $\alpha > \gamma$, contrary to the hypothesis.

Hence only case (i) is left, and $a < c$, as to be shown. In short, we have shown that $\gamma > \alpha \Rightarrow c > a$. □

Question. Explain how Euclid I.18 and Euclid I.19 are logically related.

Answer. Euclid I.18 in shorthand: $c > a \Rightarrow \gamma > \alpha$.

Euclid I.19 in shorthand: $\gamma > \alpha \Rightarrow c > a$.

Euclid I.19 is the converse of Euclid I.18.

Question. Does Euclid I.19 follow from Euclid I.18 by pure logic? Why not?

Answer. No, the converse does not follow purely by logic.

Question. Which fact does the proof above work nevertheless?

Answer. Because any two segments are comparable, we get the converse, nevertheless.

Corollary 12. *A triangle with two congruent angles is isosceles.*

Proof. Assume $\alpha \cong \gamma$ for the given triangle. Since $c > a \Rightarrow \gamma > \alpha$ and $c < a \Rightarrow \gamma < \alpha$, and any two segments are comparable, only $a \cong c$ is possible. □

Proposition II.7.50. *The hypothenuse is the longest side of a right triangle.*

Proof. By Proposition II.7.46(i), a triangle can have at most one right or obtuse angle. Hence a right triangle $\triangle ABC$ with the right angle at vertex C has acute angles at the vertices A and B. As shown in Proposition II.7.49 (Euclid I.19), comparison of angles of a triangle implies comparison of its sides. Hence the two legs across of the acute angles A or B are shorter than the hypothenuse across the right angle. □

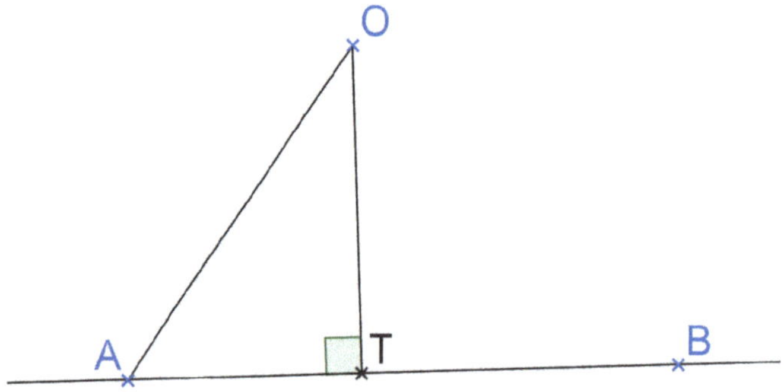

Figure II.7.54: The foot point has the shortest distance

Proposition II.7.51 (The foot point has the shortest distance). *Given is a line and a point O not on the line. The foot point F of a perpendicular from O to the line is the unique among all points of the line, which has the shortest distance.*

Proof. Call the given line l and let O be the given point not on l. Let T be the foot point of the perpendicular dropped from O onto the line l. Take any different point $A \neq T$ on the line l. The $\triangle ATO$ has a right angle at vertex T, which is its largest angle by item (i) from proposition II.7.46. The hypothenuse AO lies across to T, hence by Euclid I.19 it is the longest side. $\square$

Remark. Because all other point on l except the foot point have strictly greater distance from O, we can once more conclude that the foot point of the perpendicular is unique.

Proposition II.7.52 (The Triangle Inequality). [Euclid I.20] *In any triangle, the sum of two sides is greater than the third side.*

Proof. In $\triangle ABC$, we compare the sum $AB+BC$ to the third side AC. According to definition II.7.2, the sum $AB+BC$ is obtained by transfer of the segment BC onto the ray opposite to $\overrightarrow{BA}$. Thus we get segment $BD \cong BC$ and the sum $AB + BC = AD$. Too, the construction yields an additional triangle $\triangle DBC$, which turns out to be *isosceles*. Its two congruent base angles are $\angle CDB$ and $\angle DCB$. In the figure on page 195, both are notated by δ. Because point B lies between A and D, angle comparison at vertex C yields

$$\delta \cong \angle DCB < \angle DCA = \eta$$

Euclid I.19 tells that in a triangle, across a larger angle lies a longer side. We use Euclid I.19 for the larger triangle $\triangle ACD$. Hence the side AD lying across the greater angle η is longer than the side AC, which lies across the smaller angle δ. We have obtained $AC < AD$. By construction, the latter segment is $AD = AB + BD = AB + BC$. This confirms that the sum of two sides $AB + BC$ of the given triangle is larger than the third side AC. $\square$

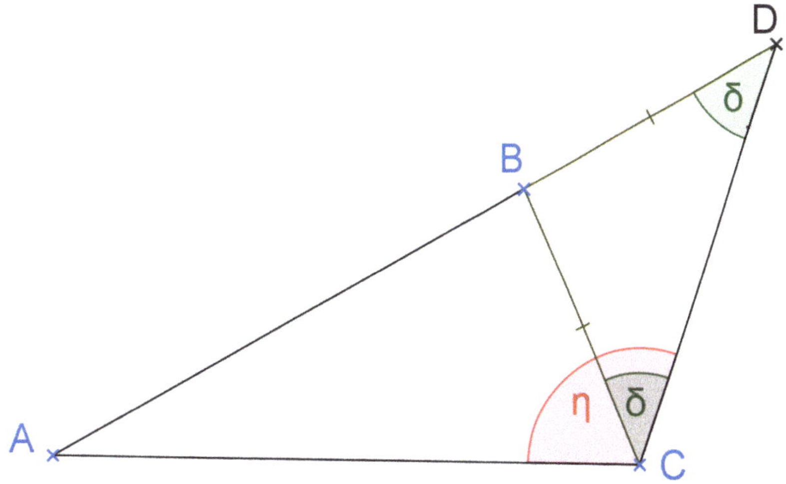

Figure II.7.55: The triangle inequality

Corollary 13. *For any three points A, B, C hold the inequalities*

$$AC \leq AB + BC \quad \text{and} \quad |AC| \leq |AB| + |BC|$$

for the segments, and their lengths, respectively. Moreover, $AC \cong AB + BC$ respectively $|AC| = |AB| + |BC|$ holds if and only if point B lies between A and C or is equal to one of them.

Proposition II.7.53. **(i)** *The sum of the segments of any polygon is equal to or longer than the distance of its first and last vertex.*

(ii) *The sum of the segments $A_0A_1, A_1A_2, A_2A_3, \ldots, A_{n-1}A_n$ is equal to the segment from its first vertex A_0 to its last vertex A_n if and only if all vertices lie between them and occur in the order $A_0 * A_1 * A_2 * A_3 * \cdots * A_n$.*

Proof of item (i). Recall definition II.6.19 of a *polygon*. We proceed by induction on the number of its segments $A_0A_1, A_1A_2, A_2A_3, \ldots, A_{n-1}A_n$. The assertion holds for a polygon A_0A_1, A_1A_2 with two segments since $A_0A_2 \leq A_0A_1 + A_1A_2$ according to the triangle inequality, and its corollary 13.

For the induction step, we proceed from n to $n+1$ segments. We know that

$$A_0A_{n+1} \leq A_0A_n + A_nA_{n+1}$$

by the triangle inequality and

$$A_0A_n \leq A_0A_1 + A_1A_2 + \cdots + A_{n-1}A_n$$

by the induction assumption. The associativity of the segment addition has already been stated as Proposition II.7.7. Hence

$$\begin{aligned} A_0A_{n+1} &\leq A_0A_n + A_nA_{n+1} \leq (A_0A_1 + A_1A_2 + \cdots + A_{n-1}A_n) + A_nA_{n+1} \\ &\cong A_0A_1 + A_1A_2 + \cdots + A_{n-1}A_n + A_nA_{n+1} \end{aligned}$$

as to be shown. □

Proof of item (ii). We proceed by induction on the number of the polygon's segments. For simplicity, I may assume that all these points are different.

The assertion holds for a polygon A_0A_1, A_1A_2 with two segments since

$$A_0A_2 = A_0A_1 + A_1A_2 \Leftrightarrow A_0 * A_1 * A_2 \text{ or } A_1 = A_0 \text{ or } A_1 = A_2$$

according to corollary 13.

For the induction step, we proceed from n to $n+1$ segments. Assume the equality

$$A_0A_1 + A_1A_2 + \cdots + A_nA_{n+1} = A_0A_{n+1} \tag{II.7.9}$$

to hold. In that case

$$A_0A_{n+1} \le A_0A_n + A_nA_{n+1} \le (A_0A_1 + A_1A_2 + \cdots + A_{n-1}A_n) + A_nA_{n+1} = A_0A_{n+1} \tag{II.7.10}$$

and hence equality occurs everywhere and

$$A_0A_n = A_0A_1 + A_1A_2 + \cdots + A_{n-1}A_n$$

Now the induction assumption implies that the first n vertices lie in the order

$$A_0 * A_1 * A_2 * A_3 * \cdots * A_n$$

Moreover the statement (II.7.10) implies

$$A_0A_{n+1} = A_0A_n + A_nA_{n+1}$$

and hence according to corollary 13, we get the order relation

$$A_0 * A_n * A_{n+1}$$

Via the n-point Theorem II.6.7 we get the order relation

$$A_0 * A_1 * A_2 * A_3 * \cdots * A_n * A_{n+1} \tag{II.7.11}$$

for all $n+1$ segment as claimed. The converse tells that statement II.7.11 implies equality (II.7.9), and is even easier to check. □

Proposition II.7.54 (SAA-Congruence Theorem). [Theorem 25 in Hilbert] *Assume two triangles have a pair of congruent sides, one pair of congruent angles across these sides, and a second pair of congruent angles adjacent to these sides. Then the two triangles are congruent.*

Proof. Let the triangles be $\triangle ABC$ and $\triangle A'B'C'$, and assume that $AB \cong A'B'$, $\angle BAC \cong \angle B'A'C'$ and $\angle ACB \cong \angle A'C'B'$. We choose a point C'' on the ray $\overrightarrow{A'C'}$ such that $AC \cong A'C''$. By axiom (III.5)

$$\gamma = \angle ACB \cong \angle A'C''B'$$

Indeed, the SAS congruence even implies

$$\triangle ABC \cong \triangle A'B'C'' \tag{*}$$

On the other hand, by assumption

$$\gamma = \angle ACB \cong \angle A'C'B'$$

If $C' \neq C''$, an exterior angle of $\triangle C'C''B'$ would be congruent to a nonadjacent interior angle, which is impossible by the exterior angle theorem. Hence we can conclude that $C' = C''$. Because of (*), this implies $\triangle ABC \cong \triangle A'B'C'$, as to be shown. □

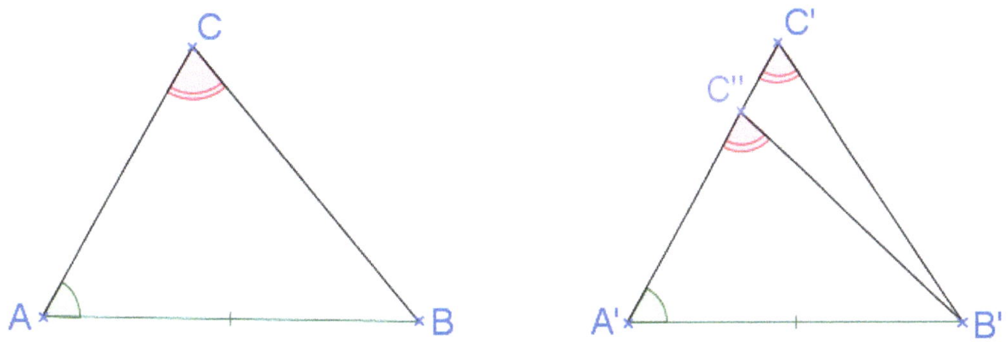

Figure II.7.56: SAA congruence

Question. How does this theorem differ from the ASA congruence (Hilbert's Theorem 13, and Proposition II.7.10 above)?
Answer. Of the two pairs of angles that are given (or compared), one pair lies across the pair of given sides.

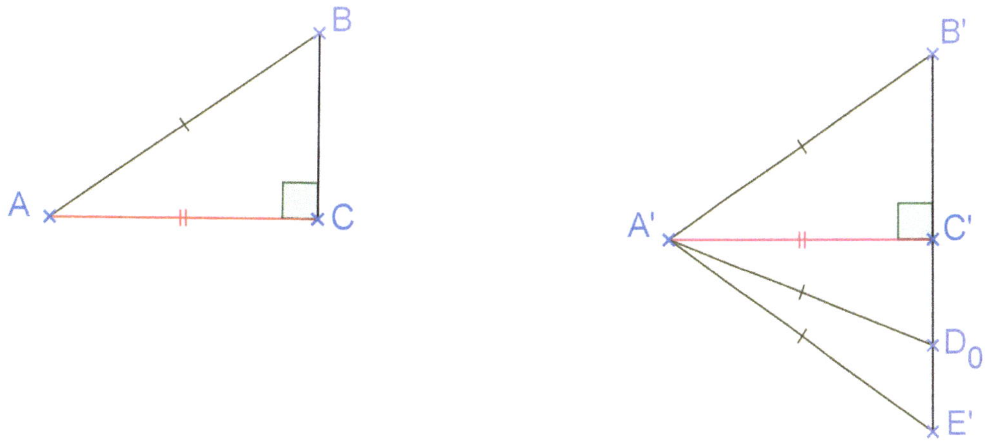

Figure II.7.57: The hypothenuse leg theorem

Second proof of the hypothenuse-leg theorem II.7.32. Take two right $\triangle ABC$ and $\triangle A'B'C'$ with $b = b', c = c'$ and $\gamma = \gamma' = R$. One transfers segment CB and segment $C'B'$, both on the ray opposite to $\overrightarrow{C'B'}$, and get new segments $C'D_0 \cong CB$ as well as $C'E' \cong C'B'$. From the construction and SAS congruence, we conclude

$$\triangle ABC \cong \triangle A'D_0C \ , \quad \triangle A'B'C' \cong \triangle A'E'C' \tag{1}$$

Hence especially

$$A'B' \cong A'D_0 \cong A'E' \tag{2}$$

By Proposition II.7.46(iv), at most two points can have the same distance from a line. Hence not all three points B', D_0, E' can be different. The only possibility left is $D_0 = B'$, because they both lie on the opposite side of $A'C'$ than B'.

Question. For completeness, explain once more. Assume that $D_0 \neq E'$ towards a contradiction. Take the case that $B' * D_0 * E'$ shown in the drawing. (The other cases can be dealt with similarly.)

Independent answer. There are two isosceles $\triangle B'A'D_0$ and $\triangle DA'E'$.

Question. What would happen with their base angles at vertex D_0?

Answer. The two base angles would be supplementary.

Question. Why is this impossible?

Answer. This is impossible, because the exterior angle theorem would imply that each one of them is larger than the other one.

This contradiction leave only the possibility that $D_0 = E'$.

Now the required congruence $\triangle ABC \cong \triangle A'B'C'$ follows from $D_0 = E'$ and (1). □

Proposition II.7.55 (The Hinge Theorem). [Euclid I.24] *Increasing the angle between two constant sides increases the opposite side of a triangle.*

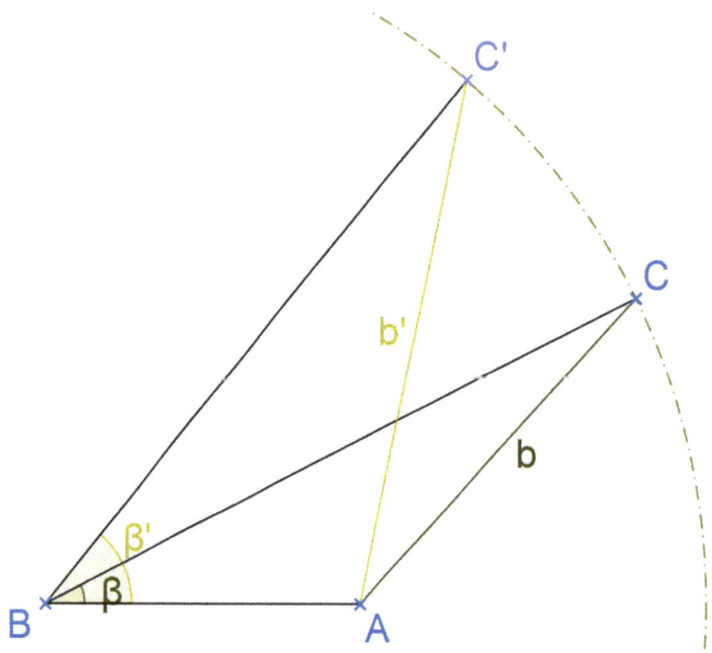

Figure II.7.58: The Hinge Theorem

Corollary 14 states the equivalence we get by taking from [Euclid I.24] and [Euclid I.25] together:

Corollary 14. *Given are two triangles with two pairs of congruent sides. In the first triangle, the angle between them is smaller, congruent, or greater than in the second one if and only if the opposite side is smaller, congruent, or greater in the first triangle.*

Proof. Given are $\triangle ABC$ and $\triangle A'B'C'$ with $a = a'$ and $c = c'$. Assuming $\beta < \beta'$, we have to check whether $b < b'$. One can assume that $A = A'$, $B = B'$, and put the two points C and C' lie on the same side of line AB. By the hypothesis $\beta < \beta'$, the two points A and C' lie on different sides of line BC. Hence this line intersects the segment AC'.

Indeed, as stated by the Crossbar theorem, the ray $\overrightarrow{BC}$ intersects the segment AC'. Let D be the intersection point. We now distinguish three cases:

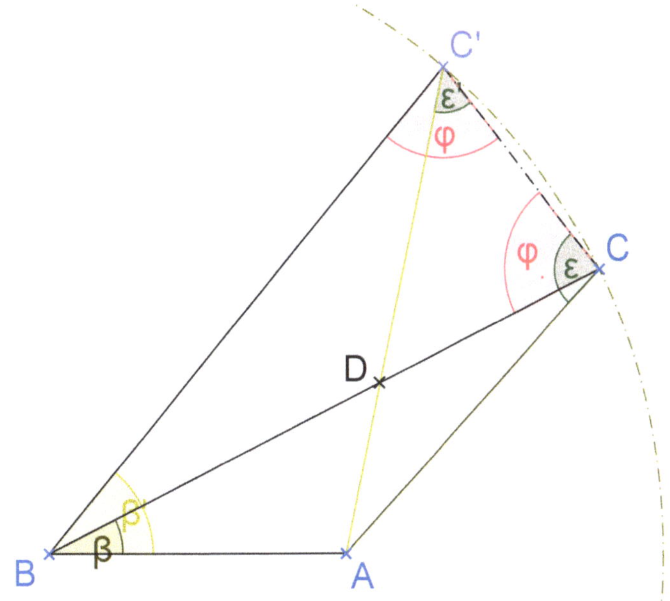

Figure II.7.59: The Hinge Theorem: turning a long side

(a) $B * D * C$. This case occurs if the triangle side BC to be rotated is long enough.

(b) $C = D$.

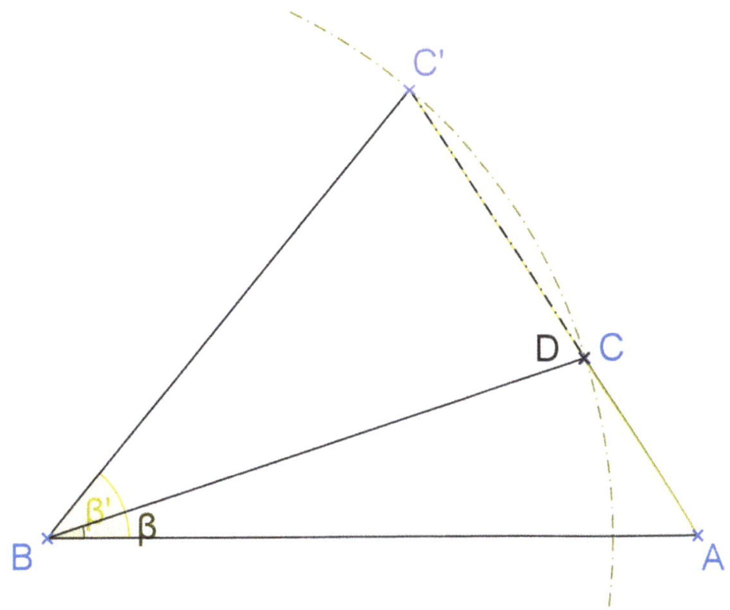

Figure II.7.60: The Hinge Theorem: the borderline case

(c) $B * C * D$. In this case the triangle side BC to be rotated is rather short.

In the border line case **(b)**, we see directly that $\beta < \beta'$ implies $A * C * C'$ and hence $AC < AC'$, as to be shown. In the two other cases, we apply Euclid I.19 to triangle $\triangle ACC'$. Thus it is enough to show that this triangle has a larger angle $\varepsilon = \angle ACC'$ at vertex C than the angle $\varepsilon' = \angle AC'C$ at vertex C'. The $\triangle BCC'$ is isosceles. Hence, by Euclid I.5, it has two congruent base angles, which we denote by φ.

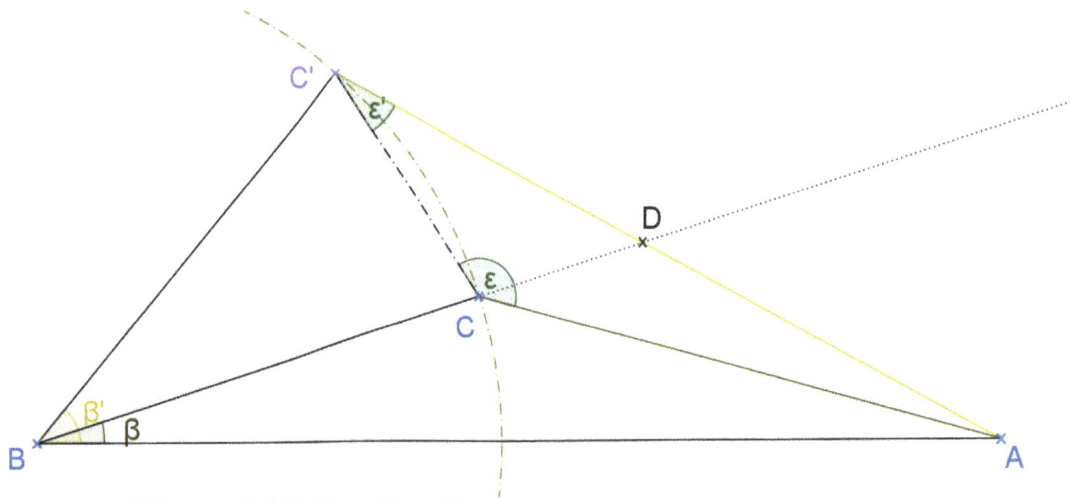

Figure II.7.61: The Hinge Theorem: turning a short side

In case **(a)**, we proceed as follows: Because B and C lie on opposite sides of AC', comparison of angles at vertex C' yields

(1) $$\varepsilon' < \varphi$$

And because A and C' lie on opposite sides of line BC, comparison of angles at vertex C yields

(2) $$\varphi < \varepsilon$$

Because of transitivity, (1)(2) together imply $\varepsilon' < \varepsilon$ and hence $AC' > AC$.
In case **(c)**, we show that ε is an obtuse, and ε' is an acute angle. Because A and C' lie on opposite sides of line BCD, we get

(3) $$\angle C'CD < \angle C'CA = \varepsilon$$

but $\angle C'CD$ is an exterior angle of the isosceles $\triangle BCC'$. Because the base angles of an isosceles triangle are acute, its supplement $\angle C'CD$ is obtuse. Hence by (3), ε is obtuse, too. Since the triangle $\triangle ACC'$ can have at most one right or obtuse angle, the angle ε' is acute. Now $\varepsilon' < R < \varepsilon$ implies again $\varepsilon' < \varepsilon$, and hence Euclid I.19 yields $AC' > AC$. □

Proposition II.7.56 (The Midpoint of a Segment). [Theorem 26 in Hilbert] *Any segment has a midpoint.*

Construction II.7.7 (Construction of the Midpoint of a Segment with Hilbert Tools). *Given is a segment AB. Choose any point C not on the line AB. Transfer angle $\angle CAB$ into the half-plane of AB opposite to point C, with the second endpoint B as vertex and ray $\overrightarrow{BA}$ as one side. Next we transfer congruent segments $AC \cong BD$ onto the newly produced legs of these two angles. Finally, line $\overleftrightarrow{AB}$ and segment CD intersect at the midpoint M.*

(a) Here is a drawing.

(b) Explain why lines AB and CD intersect.

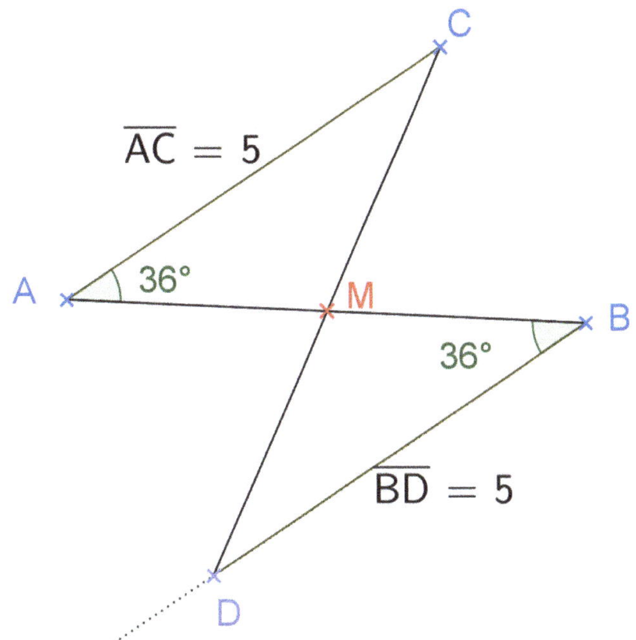

Figure II.7.62: Hilbert's construction of the midpoint

Answer. By construction, points C and D lie on different sides of line AB. Hence, by the plane separation theorem, line AB and segment CD intersect.

Let M be the intersection point. From the plane separation theorem, too, it follows that the intersection point M lies between C and D. But it turns out to be harder to see why M lies between A and B!

(c) Show that $M = A$ is impossible.

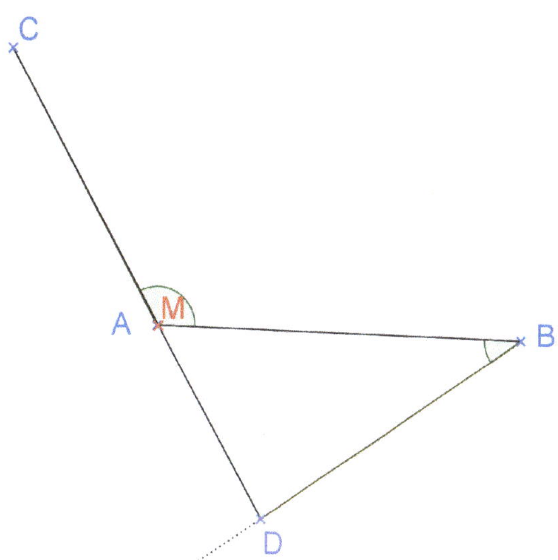

Figure II.7.63: $M = A$ is impossible

Answer. In that case line $l = AC$ would go through point D. The ray $\overrightarrow{AC}$ would be an extension of side AD of the triangle $\triangle ABD$. This triangle would have the interior angle

$\angle ABD$ congruent to the exterior angle $\angle CAB$, contradicting the exterior angle theorem. (Remember: an exterior angle is always greater than a nonadjacent interior angle.)

(d) Show that $M * A * B$ is impossible.

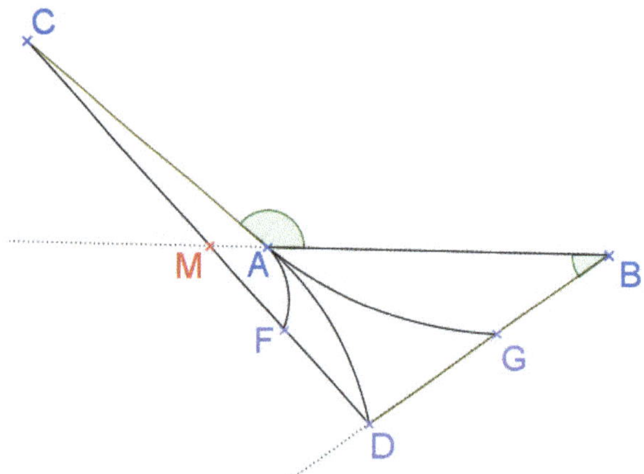

Figure II.7.64: Point A lying between M and B is impossible

Answer 1. We use Pasch's axiom for $\triangle MBD$ and line $l = CA$. Which conclusion do you get?

Answer. The line CA enters $\triangle MBD$ on the side MB. By Pasch's axiom, this line either (i) intersects side DB, or (ii) goes through point D, or (iii) intersects side MD.

In all three cases, we derive a contradiction:

Case (i): suppose line l intersects segment DB, say at point G. The $\triangle ABG$ would have the interior $\angle ABG$ congruent to the exterior $\angle CAB$. This contradicts the exterior angle theorem, which tells an exterior angle is always greater than a nonadjacent interior angle.

Case (ii): suppose line l goes through point D. The $\triangle ABD$ would have the interior $\angle ABD$ congruent to the exterior $\angle CAB$. This contradicts the exterior angle theorem, which tells an exterior angle is always greater than a nonadjacent interior angle.

Case (iii): suppose line l intersects segment MD, say at point F. Points $C \neq F$ are different, because they lie on different sides of AB. The lines AC and MD intersect both in point C and in point F,. Because this are two different points, they determine a line uniquely. Hence all five points C, M, D, A, F lie on one line. Hence we are back to the case $M = A$, ruled out earlier.

Answer following Hilbert. We apply the exterior angle theorem twice. Here is a sketchy drawing for that impossibility. [28] In triangle $\triangle ABD$, the interior angle at vertex B is $\beta = \angle DBM$, which is smaller than the exterior angle at vertex $\varepsilon = \angle BMC$. In triangle $\triangle AMC$, the angle ε from above is interior angle at vertex M, and hence smaller than the exterior angle $\alpha = \angle BAC$ at vertex A.

Now transitivity yields $\beta < \varepsilon < \alpha$. On the other hand, the angles $\angle DBA = \beta$ and $\alpha = \angle CAB$ are congruent by construction. This contradiction rules out the case $M * A * B$.

[28] The drawing, too, occurs in the millenium edition of "Grundlagen der Geometrie", page 26.

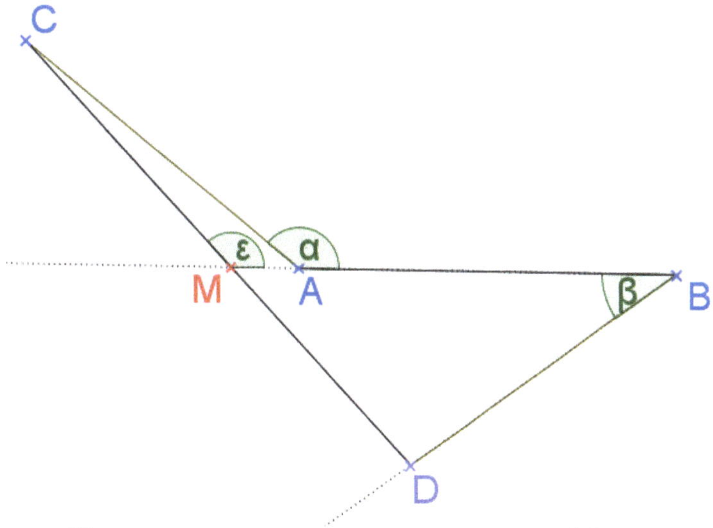

Figure II.7.65: Point A lying between M and B is impossible

(e) Now we know that M lies between A and B, we finally can prove that M is the midpoint.

Question. Which congruence theorem is used for which triangles?

Answer. One uses SAA congruence for $\triangle AMC$ and $\triangle BMD$. Indeed, the angles at A and

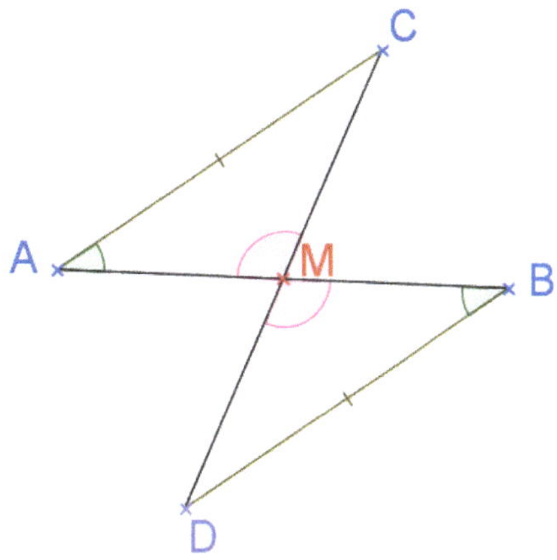

Figure II.7.66: Apply the SAA congruence

B are congruent by construction, and the angles at vertex M are congruent vertical angles. (Both statements are only true <u>because</u> M lies between A and B!) The sides AC and BD opposite to those angles are congruent by construction. Hence the two triangles are congruent, and especially $AM \cong MB$

Problem II.7.16. *Given is a convex quadrilateral with two pairs of congruent opposite sides. Give a short proof that the diagonals bisect each other.*

You may use the result of problem II.7.14 and assume Hilbert's construction of the midpoint of a segment is valid. Provide a drawing.

Answer. Since the quadrilateral is convex, the opposite vertices C and D lie on different sides of the diagonal AB.

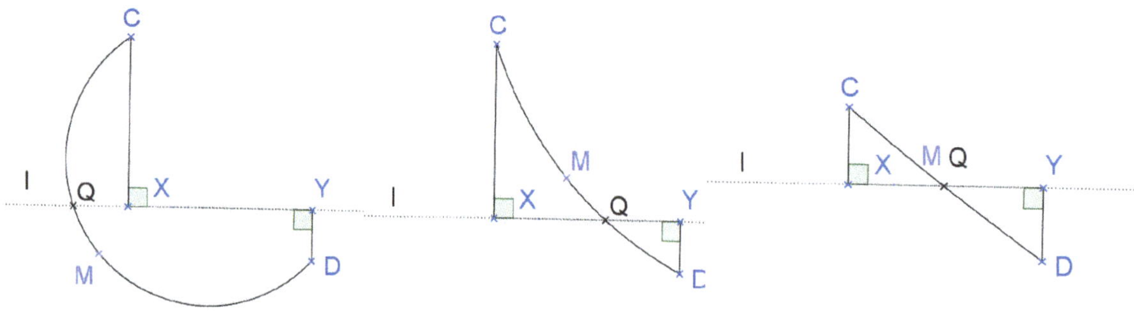

Figure II.7.67: The generic situation for Proposition II.7.57, for which we prove: Two segments CX and DY on different sides of XY are congruent if and only if midpoint M of segment CD lies on line l. The third figure shows the case with both conditions true.

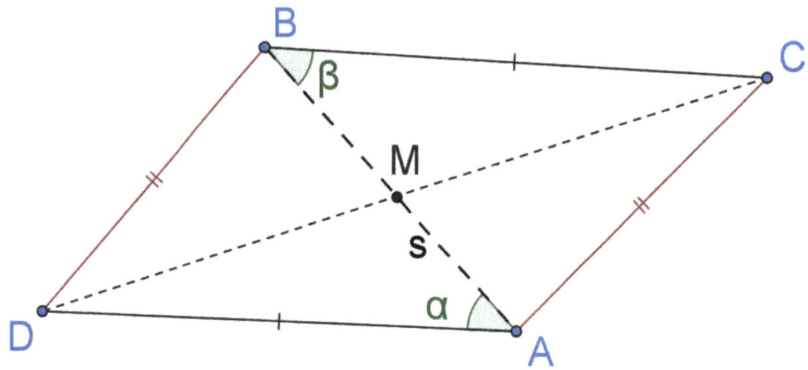

Figure II.7.68: The diagonals of a parallelogram bisect each other.

We have proved in problem II.7.14 that the angles $\beta = \angle ABC$ and $\alpha = \angle BAD$ are congruent. Hence the figure repeats exactly Hilbert's construction of the midpoint of segment AB. Hence the diagonal segments intersect in the midpoint M of diagonal AB. Similarly, we prove that M is the midpoint of the other diagonal CD, too.

Proposition II.7.57. *The midpoint of a segment lies on a given line l if and only if the two endpoints of a segment have the same distance to the line, and lie on the opposite sides of it.*

Proof. Assume that the two endpoints C and D of the given segment lie on opposite sides of l. Furthermore, assume $CX \cong DY$ are the congruent segments to the foot points X and Y.

Let Q be the intersection point of line l and segment CD, which exists by plane separation. In the special case $X = Y$, we are ready immediately. Otherwise, we need to see why Q lies between X and Y! This is done is the same way as in Hilbert's construction II.7.11 of the midpoint, which is now applied to the segment $AB = XY$.

Finally, one obtains the triangle congruence

$$\triangle CQX \cong \triangle DQY \qquad \text{(oneflier)}$$

via SAA congruence, using the right angles at X and Y, vertical angles at vertex M, and the congruent segments $CX \cong DY$. Hence $CQ \cong DQ$, confirming that $Q = M$ is the midpoint of segment CD.

Conversely, assume that the midpoint M of segment CD lies on the line l. In the special case $X = Y$ we are ready immediately. Otherwise, we need to confirm, once more, why M lies between X and Y! Again one needs to use the exterior angle theorem. Finally, one obtains the triangle congruence

(oneflier) $$\triangle CMX \cong \triangle DMY$$

via SAA congruence, using the right angles at X and Y, vertical angles at vertex M, and the congruent segments $CX \cong DY$, as to be shown. □

II.7.12 SSA congruence

Next we study the possibilities and difficulties with SSA congruence. Thus the matching pieces of the two given triangles are two pairs of sides, and one pair of angles opposite to one of these sides. I follow Euler's conventional notation: in triangle $\triangle ABC$, let $a = BC$, $b = AC$, and $c = AB$ be the sides and $\alpha := \angle BAC$, $\beta := \angle ABC$, and $\gamma := \angle ACB$ be the angles.

Problem II.7.17. ***Investigate an example*** *Use straightedge and compass to construct in Euclidian geometry all triangles (up to congruence) with* $\gamma = \angle ACB = 30°$, *side* $AC = 10$ units *and side* AB *given as below. How many non-congruent solutions (none, one, two?) do you get in each case? How many of them are acute, right or obtuse triangles? Measure and report the angle* $\beta = \angle ABC$ *for all your solutions. Make clear by the drawings what happens in all cases, especially how many acute, right, and obtuse triangles you get as solutions.*

Hint: It is convenient to construct all triangles with one common side AC. Check carefully to find the obtuse angles! (a) $AB = 4$ units.

Answer (a). One has to begin the construction by drawing a segment AC of length 10, and a ray r with vertex C, forming an angle of 30° with $\overleftarrow{CA}$. The point B has to lie on the ray r, as well as on a circle of radius AB around A. In case (a), the circle does not intersect this ray, hence there is no solution.

(b) $AB = 5$ units.

Answer. In case (b), the circle just touches the ray r, hence there is one solution, with a right angle at B.

(c) $AB = 5.5$ units.

Answer. In case (c), the circle intersect the ray r in two points, hence there are two non congruent solutions. At vertices B and B', I measure the angles about $\beta = 67°$ and $\beta' = 113°$. One solution is an acute, the other an obtuse triangle.

(d) $AB = 6$ units.

Answer. Again in case (d), the circle intersects the ray r in two points, and there are two non congruent solutions. At vertices B and B', I measure the angles about $\beta = 57°$ and $\beta' = 123°$. Both solutions are obtuse triangles. Indeed, the first solution has the obtuse angle $\alpha = 180° - \beta - \gamma = 93°$.

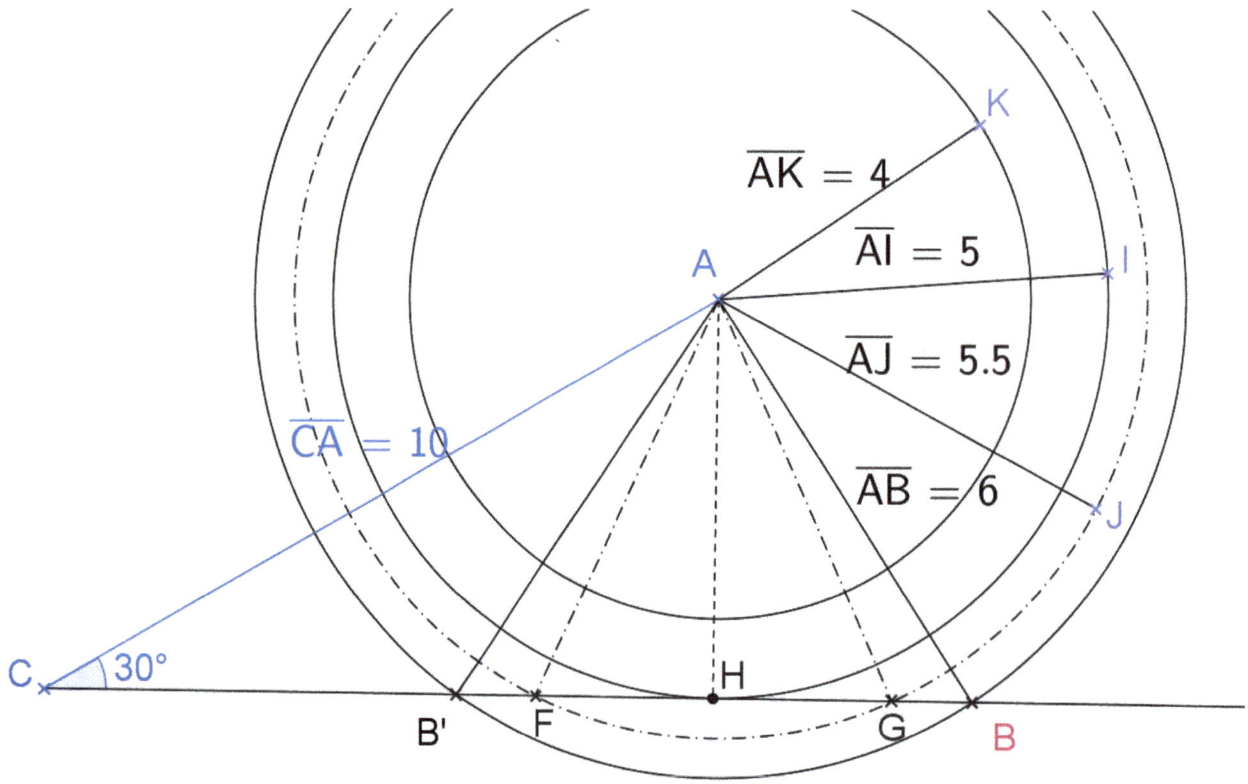

Figure II.7.69: A Euclidean example for SSA triangle construction

Proposition II.7.58 (SSA Matching Proposition). *Given are two triangles. Assume that two sides of the first triangle are pairwise congruent to two sides of the second triangle, and that the angles across to one of these pairs are congruent.*

Then either the two triangles are congruent, or the angles across the other pair of congruent sides add up to two right angles.

Proof. It is assumed that two sides and the angle *across* one of these sides of $\triangle ABC$ can be matched to congruent pieces of $\triangle A'B'C'$:

(SSA) $$c = AB \cong A'B' = c', b = AC \cong A'C' = b', \gamma = \angle ACB \cong \angle A'C'B' = \gamma'$$

We have to show that either (a) or (b) holds:

(a) $\triangle ABC \cong \triangle A'B'C'$.

(b) The two angles across the second matched side $\beta = \angle ABC$ and $\beta' = \angle A'B'C'$ are (congruent to) supplementary angles. One of them is acute and the other one is obtuse.

We begin by reducing the problem to the special case that $A = A'$, $C = C'$ and that B and $B' = D$ lie on the same side of AC. We use the SAS congruence to construct a triangle $\triangle ADC$, such that $\triangle A'B'C' \cong \triangle ADC$, with points B and D on the same side of AC. The drawing below shows that procedure.

Question. Explain how you get the triangle $\triangle ADC$ congruent to $\triangle A'B'C'$.

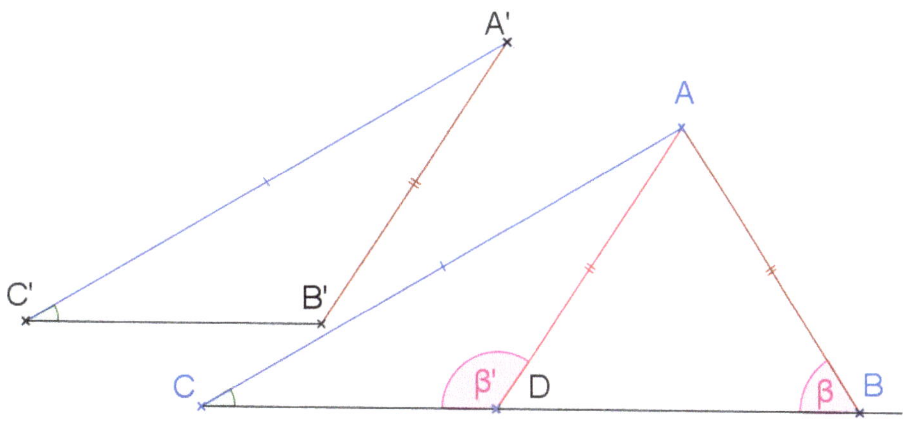

Figure II.7.70: Matching two triangles with SSA

Answer. I just transfer segment $C'B'$ onto the the ray $\overrightarrow{CB}$ and get $C'B' \cong CD$. The congruence $\triangle A'B'C' \cong \triangle ADC$, follows from SAS, given in Theorem 12 of Hilbert (see Proposition II.7.11 above).

Assume that congruence (a) does not hold. We have to prove that case (b) occurs. We know that $\alpha \neq \alpha'$, since otherwise the SAS-congruence theorem implies $\triangle ABC \cong \triangle A'B'C'$, which again would be case (a) just ruled out. Without loss of generality, we can assume $\alpha' < \alpha$. Since $\angle DAC = \alpha' < \alpha = \angle BAC$, the point D lies between B and C, as shown in the drawing.

Since $AD \cong A'B' \cong AB$ by assumption, triangle $\triangle ABD$ is isosceles, with baseline BD. By Euclid I.5, the two base angles of an isosceles triangle are congruent. One of them is the angle $\beta = \angle ABD$. By Proposition II.7.46(ii), these base angles are always acute.

The angle $\beta' = \angle A'B'C' \cong \angle ADC$ is the supplement to the second base angle $\angle ADB \cong \angle ABD = \beta$. Hence β and β' are congruent to supplementary angles. The first one is acute, the second one is obtuse, as claimed in (b). $\square$

Corollary 15 (SSAA Congruence). *Two triangles which have two pairs of congruent sides, and two pairs of congruent angles across to the latter, are congruent.*

Proposition II.7.59 (Restricted SSA Congruence, case of a unique solution). *As in the matching Proposition above, we assume that two sides and the angle across one of these sides of $\triangle ABC$ can be matched to congruent pieces of $\triangle A'B'C'$:*

(SSA) $$AB \cong A'B'\,,\ AC \cong A'C' \quad \text{and} \quad \angle ACB \cong \angle A'C'B'$$

Under the additional assumption that either (i) *or* (ii) *hold,*

(i) *The given angle γ lies across a side longer or equal to the other given side: $c = AB \geq AC = b$.*

(ii) *The given angle $\gamma = \angle ACB$ is right or obtuse.*

congruence holds among all triangles matched by (SSA). *In all these cases there do not exist any non congruent triangles with congruent sides b, c and angle γ. Actually assumption* (ii) *implies* (i).

Proof of Proposition II.7.59. Assume that (ii) holds. Indeed, by Proposition II.7.46(i), a triangle can have at most one angle which is right or obtuse. Hence assumption (ii) implies $\gamma \geq \beta$. By Euclid I.19, across the greater angle of any triangle lies the longer side. Hence $\gamma \geq \beta$ implies $c \geq b$, which is assumption (i).

We now assume that $c = AB \geq AC = b$ holds, as stated by (i). By Euclid I.18, across the longer side of a triangle lies the greater angle. Hence $c \geq b$ implies $\gamma \geq \beta$. A triangle cannot have two angles which are right or obtuse. Hence β and β' are both acute or right. Now congruence follows because case (b) in the matching Proposition is ruled out. □

Corollary 16 (SSA Congruence for isosceles and right triangles). (i') *Two isosceles triangles with congruent legs and congruent base angles are congruent.*

(ii') *Any two right triangles with congruent hypothenuses and one pair of congruent legs are congruent.*

Proof. Assumption (i') is a special case of (i), and (ii') a special case of (ii). □

Proposition II.7.60 (Restricted SSA Congruence Theorem, case with non uniqueness). *Again we assume that two sides and the angle across one of these sides of $\triangle ABC$ can be matched to congruent pieces of $\triangle A'B'C'$:*

(SSA) $$AB \cong A'B'\,,\ AC \cong A'C' \quad \textit{and} \quad \angle ACB \cong \angle A'C'B'$$

Complementary to Proposition II.7.59, we assume that neither (i) *nor* (ii) *nor $\beta \cong R$ holds. In other words, we assume that the given angle $\gamma = \angle ACB$ is* <u>*acute*</u> *and lies across the* <u>*shorter*</u> *given side: $c = AB < AC = b$, and angle β is not a right angle. Under these additional assumptions* **(a)** *There exist two non congruent triangles matched by* (SSA). **(b)** *Nevertheless, equivalent are*

(1) *The two given triangles $\triangle ABC$ and $\triangle A'B'C'$ are congruent.*

(2) *(SSAA) The two triangles can be matched in two sides and the two angles across both given sides.*

(3) *For both triangles, the angles β and β'' across the longer given side b are both acute or both right or both obtuse.*

(4) *Both triangles are acute, or both are right, or both triangles are obtuse with the two obtuse angles at* <u>*corresponding vertices*</u>.

In this case just the two given triangles $\triangle ABC$ and $\triangle A'B'C'$—not all other triangles with those given angle and sides γ, c, b—are congruent.

Proof for part (a). We drop the perpendicular from vertex A onto the opposite side CB. The foot point $F \neq B$, because otherwise β would be a right angle. Transfer segment FB on the ray opposite to $\overrightarrow{FB}$ to produce segment $FB'' \cong FB$. The two right triangles $\triangle ABF$ and $\triangle AB''F$ are congruent

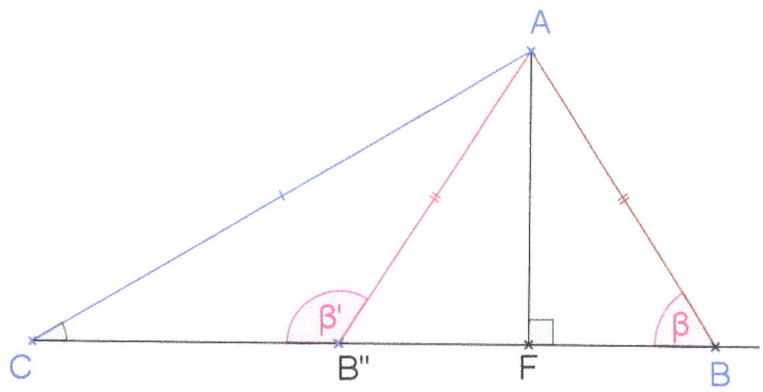

Figure II.7.71: Establish two solutions for SSA

by the Hypothenuse-Leg Theorem 15. Hence $AB \cong AB''$. Thus we got two non congruent triangles $\triangle ABC$ and $\triangle AB''C$ which nevertheless satisfy

$$\text{(SSA")} \qquad AB \cong AB''\,,\ AC \cong AC \quad \text{and} \quad \angle ACB \cong \angle ACB''$$

For these non congruent triangles, the two angles β and β'' are supplementary. One of them is acute, the other one is obtuse. □

Proof for part (b). Clearly (1) implies (2) implies (3) implies (4). If (4) holds, the matching Proposition excludes non congruent solutions.

Indeed, if both triangles are acute, both angles β and β' are acute. Congruence follows by the SSA matching Proposition. If both triangles are right, both angles β and β' are acute or right. In both cases, congruence follows by the SSA matching Proposition.

If both triangles are obtuse, and obtuse angle occurs at corresponding vertices, then either both angles β and β' are obtuse, or both angles α and α' are obtuse. In the second case both angles both angles β and β' are acute. In both cases, congruence follows by the SSA matching Proposition. □

Remark. Earlier on, we have carefully studied triangles with

$$c = AB = 5\,,\ 5.5\,.\,6 \quad b = AC = 10 \quad \text{and} \quad \gamma = \angle ACB = 30^\circ$$

in Euclidean geometry. In the first case $c = 5$, one gets a unique right triangle as solution. In the second case $c = 5.5$ and $c = 6$, there are two non-congruent solutions, one of which is an acute triangle, the other one is an obtuse triangle with $\beta' > R$.

In the third case $c = 6$, there are two non-congruent solutions again. But both are obtuse triangles! One gets one triangle with $\alpha > R$ and $\beta < R$, as well as a second one with $\alpha' < R$ and $\beta' > R$.

II.7.13 Reflection

Definition II.7.22 (Reflection across a line). Given is a line l. The reflection across line l is the mapping $r : P \mapsto P'$ which maps an arbitrary point P to a point P' meeting the following requirements:

- If P lies on the symmetry axis l, we put $P' := P$.

- If P does not lie on the symmetry axis, the point P' is specified by requiring
  (i) P and P' lie on different sides of l.
  (ii) the lines l and PP' are perpendicular, intersecting at F.
  (iii) $PF \cong P'F$.

Construction II.7.8 (of reflection point). *Choose any two points A and B on the line l. Draw ray $\overrightarrow{AP}$. Transfer angle $\angle BAP$, into the half plane opposite to P, with ray $\overrightarrow{AP}$ as one side. On the newly produced ray, transfer segment AP to produce a new segment $AP' \cong AP$. The point P' is the reflection of the given point P across the line AB.*

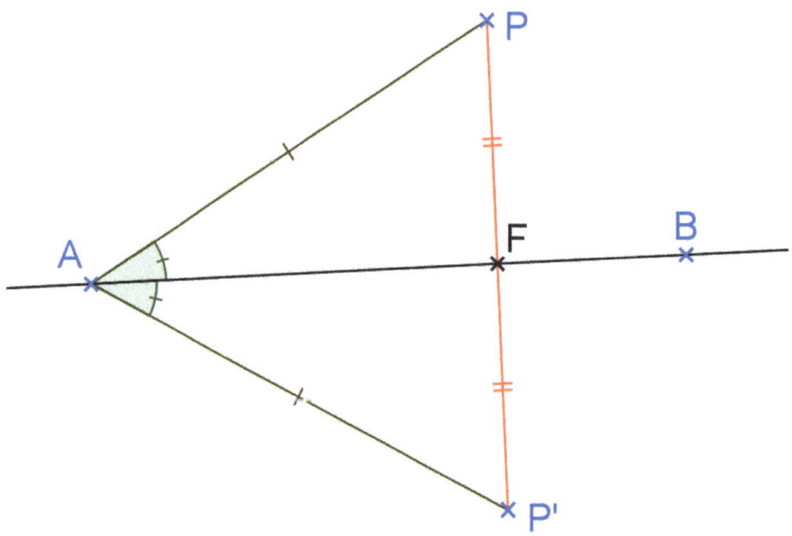

Figure II.7.72: Reflect a point

Proof of validity. The line AB and the segment PP' intersect, because P and P' lie on different sides of the line. I call the intersection point F.

We still have to distinguish several cases depending on the order of the three points A, B and F :

- as drawn in the figure on page 210, the point F lies on the ray $\overrightarrow{AB}$;
- point F lies on the ray opposite to $\overrightarrow{AB}$;
- the special case that point F is equal to point A

or B, In the first two cases (generic cases), we show by the SAS-congruence that the two triangles $\triangle FAP$ and $\triangle FAP'$ are congruent.

- Indeed, the angles at vertex A are congruent—either directly by construction, or in the second case by construction and because Proposition II.7.15 tells that supplements of congruent angles are congruent.
- The common side $FA \cong FA$ is congruent to itself,
- and the other adjacent side $AP \cong AP'$ by construction.

From the congruence of the two triangles we conclude $\angle AFP \cong \angle AFP'$ which is a pair of congruent supplementary, and hence right angles. Moreover $FP \cong FP'$. Hence all three requirements (i)(ii) and (iii) in the definition of the reflected point are satisfied.

In the special case that point F turns out to be equal to point A, the result is proved even simpler. ☐

Proposition II.7.61. *Given are two points P, Q on the same side of l and their images P', Q'. We show that*

(a) $PQ \cong P'Q'$

(b) *If the lines $m := \overleftrightarrow{PQ}$ and l intersect at M, then P', Q' and M lie on a second line m'.*

Question. For this exercise, you can if you like, do drawing and proof, in neutral geometry, on your own.

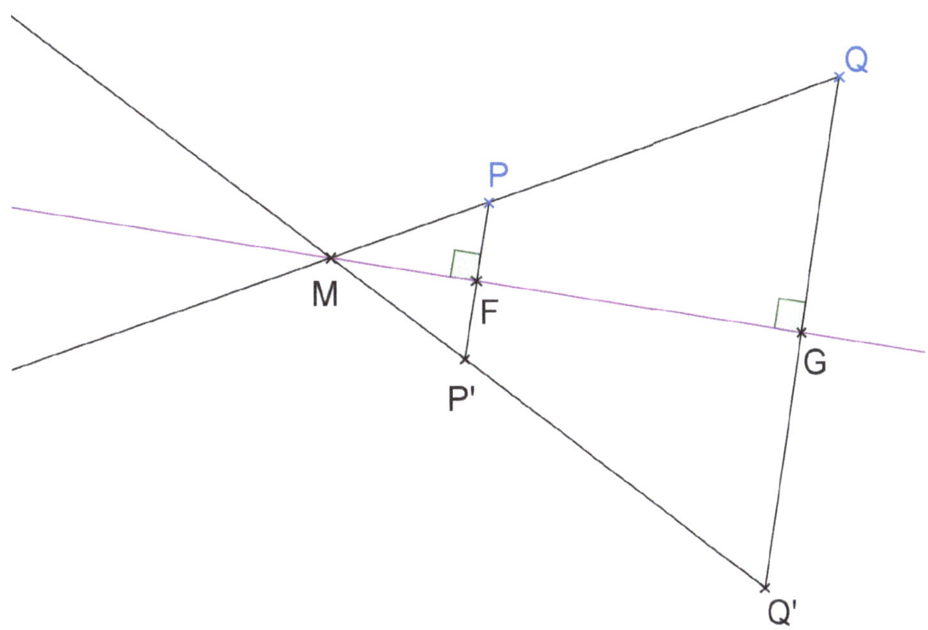

Figure II.7.73: Reflection by a line l

Proof of (a). Let F and G be the foot points of the perpendiculars dropped from P and Q onto the symmetry axis l. We use SAS-congruence for the two triangles $\triangle QGF$ and $\triangle Q'GF$.

Question. Indeed both triangles have a right angle at G—. Explain why.

Answer. $\angle QGF$ is right by definition of a reflection. Since a right angle is congruent to its supplementary angle, the supplementary angle $\angle Q'GF$ is a right angle, too.

—a common side GF and congruent sides $GQ \cong GQ'$, by definition of a reflection. Hence SAS congruence implies

$$\triangle QGF \cong \triangle Q'GF \tag{1}$$

Because of congruence (1), we get

$$\alpha = \angle QFG \cong \angle Q'FG\,, \tag{2}$$

$$QF \cong Q'F \tag{3}$$

Since $\angle GFP \cong \angle GFP'$ is a right angle, angle subtraction yields

$$\beta = \angle QFP \cong \angle Q'FP' \tag{4}$$

By construction

$$PF \cong P'F \tag{5}$$

Using (3)(4) and (5), SAS-congruence implies that

$$\triangle QFP \cong \triangle Q'FP' \tag{6}$$

Hence especially (a) holds. □

Proof of (b). We use SAS-congruence for the two triangles $\triangle QGM$ and $\triangle Q'GM$. Indeed both triangles have a right angle at G: $\angle QGM$ is right by definition of a reflection. Since a right angle is congruent to its supplementary angle, the supplementary angle $\angle Q'GM$ is a right angle, too. Furthermore, they have a common side GM and congruent sides $GQ \cong GQ'$, by definition of a reflection. Hence SAS congruence implies

$$\triangle QGM \cong \triangle Q'GM$$

For the angle x between the line l of reflection and the given line m, one gets

$$x = \angle QMG \cong \angle Q'MG \tag{8}$$

Now we argue similarly for the two triangles $\triangle PFM$ and $\triangle P'FM$. Thus we get

$$y = \angle PMF \cong \angle P'MF = \angle P'MG \tag{9}$$

Now by assumption, the three points P, Q and M lie on a line m.

$$x = \angle QMG = \angle PMF = y \tag{10}$$

From (8)(9),(10) and transitivity of angle congruence, we get

$$\angle Q'MG = \angle P'MG \tag{11}$$

Now by Hilbert's axiom, the angle transfer produces a <u>unique</u> ray. Hence $\overrightarrow{MQ'} = \overrightarrow{MP'}$. Thus the uniqueness of angle transfer implies that the three points M, P' and Q' lie on one line. □

An alternative for the proof of (b) . (6) above implies

$$\gamma = \angle QPF \cong \angle Q'P'F \tag{7}$$

We apply the extended ASA congruence to triangle $\triangle PFM$ and segment FP'. The angles to be transferred to the endpoints of that segment are a right angle at F and γ at P'. Transferring produces as second sides of these angles part of the line l and the ray $\overrightarrow{P'Q'}$. Hence those rays intersect, say at point M'. By ASA-congruence

$$\triangle MFP \cong \triangle M'FP'$$

Hence $FM \cong FM'$. The points Q and Q' and hence M and M' lie on the same side of PP'. Hence uniqueness of segment transfer implies $M = M'$ as to be shown. □

Theorem II.7.1. *The mapping of reflection preserves incidence and maps segments and angles to congruent ones. Hence it is an isometry.*

II.7.14 Applied problems

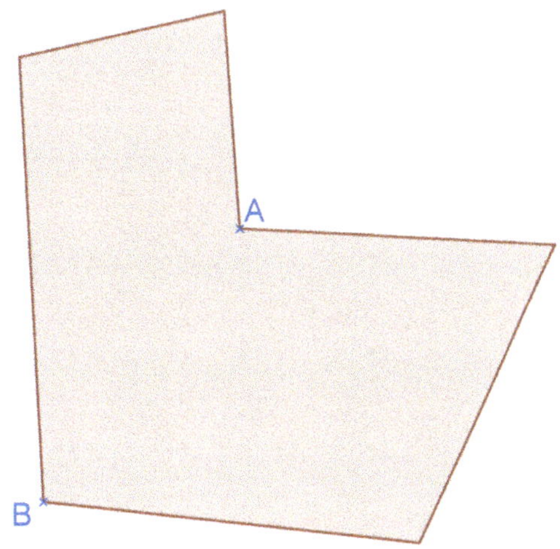

Figure II.7.74: Where on the boundary can guys A and B meet the quickest?

Problem II.7.18. *Two guys who are equally good runners stand at points A and B on the border of an L-shaped place shown in the figure on page 213. They are allowed to go across the place, but they want to meet at any point on the boundary. Which point do they need to choose to meet as quick as possible. Give a construction and explain your reasoning.*

Answer. The runners pass congruent segments AM and BM until they meet at some point M. Hence they meet where the perpendicular bisector of AB intersects the boundary. There are two such points, of which they choose the nearer one.

Problem II.7.19. *Repeat the problem II.7.18 for the situation drawn in the two figures on page 214.*

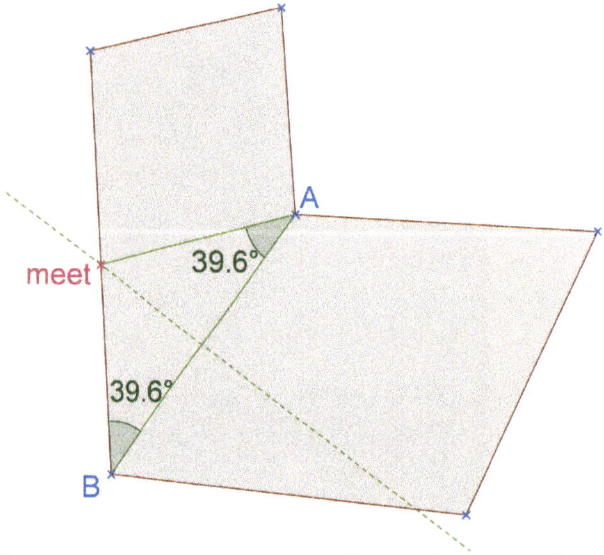

Figure II.7.75: They meet where the perpendicular bisector of AB intersects the boundary.

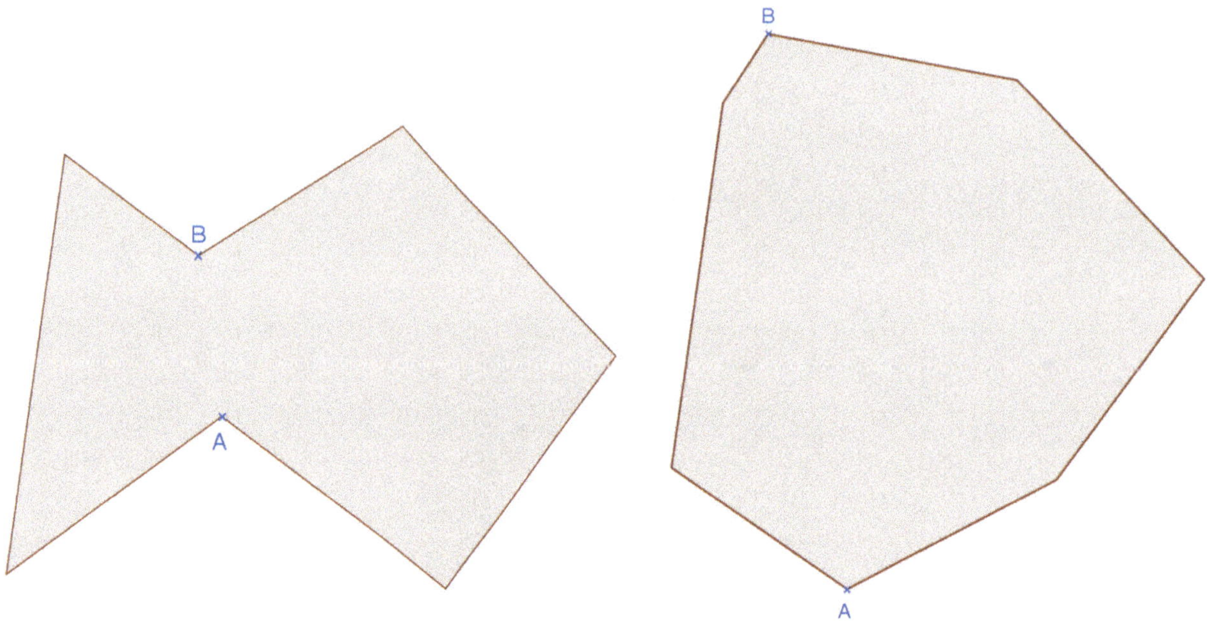

Figure II.7.76: Where on the boundary can guys A and B meet the quickest?

Answer. Until they meet, the runners pass congruent distances. They need not pass any distance longer than their mutual distance AB. Hence their meeting point lies inside the two equilateral triangles erected on both sides of segment AB. They can meet either in point A, or point B, or a point where the perpendicular bisector of AB intersects the boundary.

For the example drawn in the left part figure on page 215, these are the points N and F. But these points do not lies inside the two equilateral triangles erected on both sides of segment AB. Hence it is more advantageous for the runners to simply meet at point A or point B.

For the example drawn in the right part figure on page 215, it is most advantageous for the runners to simply meet at the point of the perpendicular bisector denoted nearest.

Problem II.7.20. *Given is a line l which is a plain mirror, a bundled light source at point P, and*

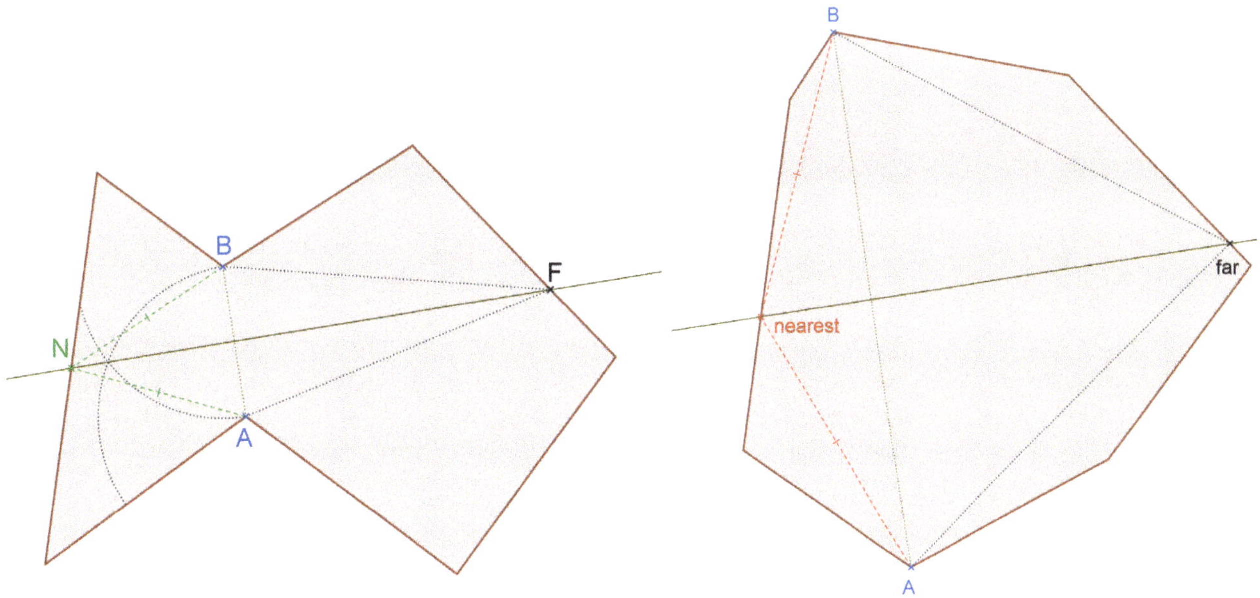

Figure II.7.77: Where on the boundary can guys A and B meet the quickest?

a point Q on the same side of the mirror, to which one wants the light ray to be reflected.

Give a construction of the light ray from P to Q, reflected by the mirror. Provide a drawing and an explanation.

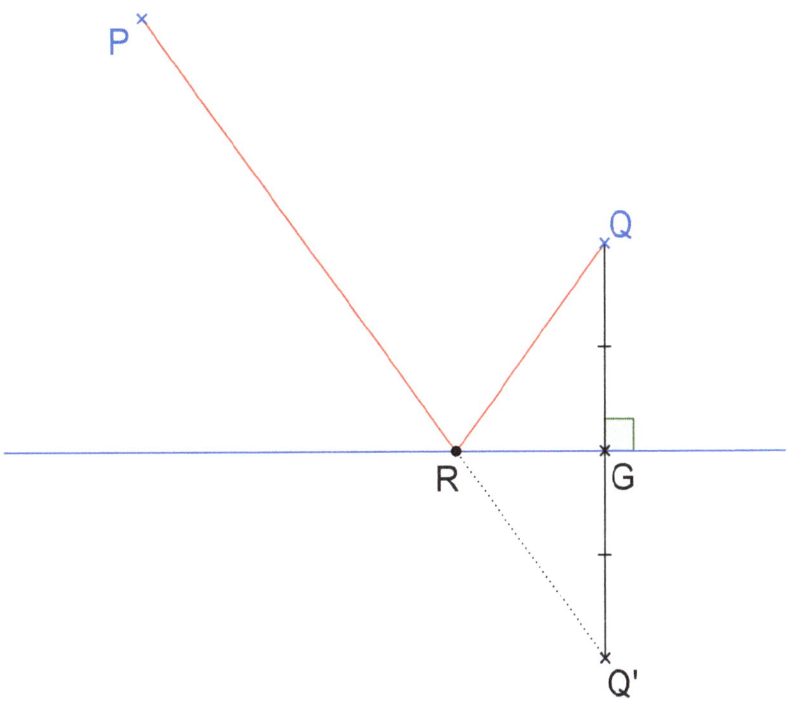

Figure II.7.78: Reflect a light ray from P to a given point Q.

Answer. We construct the reflected point Q' of the point Q to which the light ray is to be directed.

The light has to be directed along the segment PQ'. The intersection of this segment with the (line of the) mirror gives the point R where the light ray is reflected. The reflected light goes from from point R to the given target point Q.

Proof of validity. Let G be the foot-point of the perpendicular dropped from point Q onto the mirror. The angles $\angle QRG \cong \angle GRQ'$ are congruent. The angle of the incident light ray with the mirror is a vertical angle of $\angle GRQ'$, and the angle of the reflected light ray with the mirror is $\angle QRG$. Hence they are congruent, as required for the reflection of a light ray. □

Problem II.7.21. *In many technical and physical applications—for example for sensitive measurement of electrical current—one uses a small mirror attached to a twisting thin wire. A light beam shines onto the mirror and its reflected beam is depicted on a scale, which can be rather far away. Additionally it is assumed that the incoming light ray, and the perpendicular lines to the mirror in both positions lie in a plane. Thus we deal with the two-dimensional problem.*

Give the reason why turning the mirror by an angle θ results in turning the reflected beam by angle 2θ. Provide a drawing.

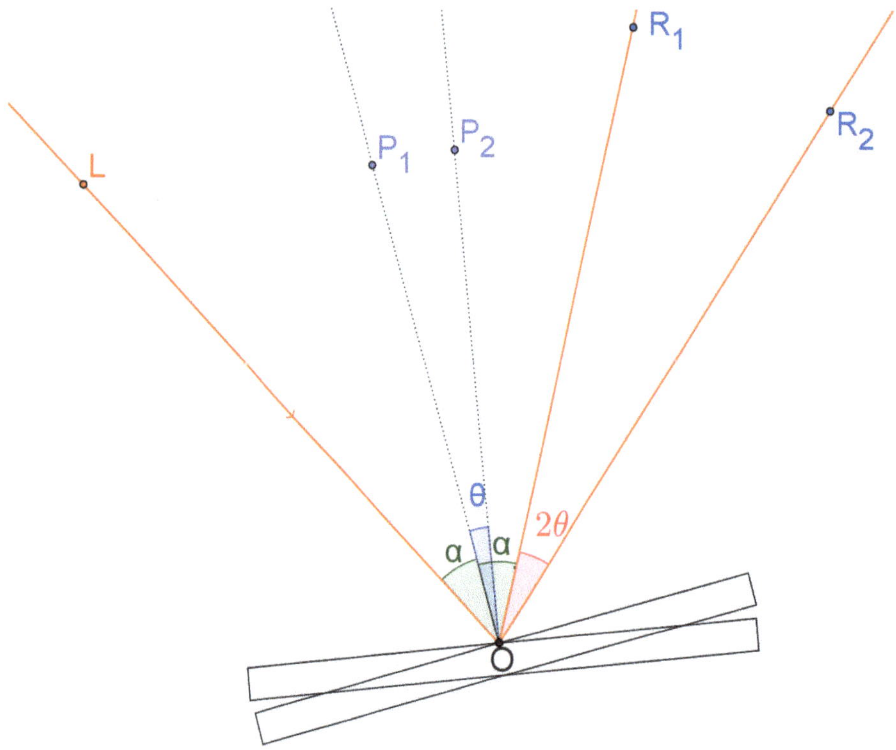

Figure II.7.79: Turning the mirror by angle θ turns the reflected ray by the angle 2θ.

Answer. The problem is easiest under the assumption that the light ray hints the mirror exact at its point of rotation (respectively in three dimensions on its axis of rotation). By the law of reflection of light, the angle between the perpendicular to the mirror and the reflected ray is congruent to the angle between the incoming light ray and the perpendicular. Hence $\alpha = \angle LOP_1 \cong \angle P_1OR_1$ in the figure on page 216. The angle between the incoming light ray and its reflected ray is the sum $\angle LOR_1 = 2\alpha$ of these two angles.

As the mirror is turning by the angle θ, its perpendicular is turning by the congruent angle $\theta = \angle P_1OP_2$ between the perpendiculars; shown in blue. The angle between the incoming light ray and the perpendicular to the mirror in its second position is $\angle LOP_2 = \alpha + \theta$, with plus sign for the situation as drawn. The angle between the perpendicular and the reflected ray increases by the same angle θ.

After the turn of the mirror has occurred, the angle between the incoming light ray and its reflected ray is $\angle LOR_2 = 2(\alpha + \theta)$. The reflected light ray has been turned by the difference angle $\angle R_1OR_2 = \angle LOR_2 - \angle LOR_1 = 2\theta$.

II.7.15 Independence of the SAS-axiom

Already in a commentary to Euclid, written about 1557 by Peletier, the author writes about the side-angle-side (SAS) congruence for triangles, Euclid I.4:

> The truth of this proposition belongs among the common notions, because to superimpose one figure to another is mechanics, not mathematics.

The *"principle of superposition"* used by Euclid for the justification of side-angle-side (SAS) congruence of triangles, has been sharply criticised by Bertrand Russell (1872-1970). In his acerbic manner, he speaks of superposition—in no uncertain terms—as a *tissue of nonsense.* [29]

Indeed, although the principle of superposition may seem reasonable enough in dealing with material triangles of small laboratory size, its legitimacy is questionable for working with conceptional entities. The assumption that moving a triangle is possible without alteration of its internal structure is just based on experience, not logic. We talk today of a *physical thought experiment*, instead of the historic term "mechanics" (still used by Gauss in such philosophical considerations).

For the mathematician, the above discussion cannot be the endpoint. As already made clear in the preamble, the goals of Hilbert's foundations of geometry are not only

- to set up a *complete* and *as simple as possible* system of axioms;
- to deduct from them the most important geometric theorems;
- but finally, to clarify the meaning and importance of the different axioms and their consequences.

Indeed, Hilbert proves the independence of the SAS axiom (III.5), and thus establishes the necessity of this axiom beyond any doubt.

Theorem II.7.2 (Hilbert). *There exists a spatial geometry for which the axioms* I, II, IV,V *as well as* (III.1),(III.2),(III.3),(III.4) *are all satisfied, but the SAS axiom* (III.5) *is violated. Moreover, Euclid I.5 is violated.*

But the following Propositions are still valid in this model:

Proposition II.7.15 about the congruence of supplementary angles;

Proposition II.7.19 about angle-addition and subtraction;

Proposition II.7.20 about the transfer of three rays;

[29] In his autobiography, Bertrand Russell penned this remarkable recollection: "At the age of eleven, I began Euclid, with my brother as tutor. This was one of the great events of my life, as dazzling as first love."

Proposition II.7.30 Euclid's postulate that all right angles are congruent;

Proposition II.7.25 telling that congruence is an equivalence relation on the class of angles.

We see that it is impossible to deduct the SAS axiom (III.5) *from the remaining axioms* I, II, (III.1),(III.2),(III.3),(III.4), IV,V, *nor from the Propositions enumerated above.*

Proof. We define the `"points"`, `"lines"`, `"planes"` of our model to be the points, lines, planes of the common three-dimensional analytic geometry with coordinates from any real-closed field. [30] It is well-known that the axioms of incidence, order, parallelism and continuity are valid. The delicate point are of course the axioms of congruence.

The transfer of angles is defined as in analytic geometry. Indeed, we take as usual Defintion II.7.17 for the congruence of orientated angles and Definition IV.20.11 for congruence of angles. Thus, the transfer of angle is well defined and axiom (III.4) is valid.

On the other hand, the length of the segment between the points $A = (a_1, a_2, a_3)$ and $B = (b_1, b_2, b_3)$, is measured by a non-Pythagorean distance

$$dn(A,B) = \sqrt{(a_1 - b_1 + a_2 - b_2)^2 + (a_2 - b_2)^2 + (a_3 - b_3)^2}$$

We define the congruence of any two segments AB and $A'B'$ by postulating

$$AB \cong A'B' \Leftrightarrow dn^2(A,B) = dn^2(A',B')$$

Distance and congruence is thus defined differently from a Pythagorean plane. Nevertheless, one easily checks that congruence of segment is an equivalence relation and hence Hilbert's axioms (III.1) and (III.2) hold.

Moreover, one can still add two adjacent segments on a line and check that Hilbert's axiom (III.3) is valid. Indeed, it is easy to use and modify the procedure from Problem IV.20.3 from the section "Coordinate Planes—Descartes' Road from Algebra to Geometry", subsection about congruence in a Pythagorean plane. This—a bid tedious but not difficult exercise—is left to the reader.

As the "coup de grâce", we now construct a counterexample showing that the axiom (III.5) is violated. Take the points

$$O = (0,0,0)\,,\ A = (1,0,0)\,,\ B = (-1,0,0)\,,\ C = (0, \frac{\sqrt{2}}{2}, 0)$$

In the new distance

$$dn(O,A) = dn(O,B) = dn(O,C) = 1$$

We get two right triangles $\triangle AOC$ and $\triangle COB$ for which hold the congruences

$$\angle AOC \cong \angle COB\,, \quad OA \cong OC\,, \quad OC \cong OB$$

But, contrary to the conclusion from axiom (III.5), the angles $\angle OAC$ and $\angle OCB$ are not congruent.

Moreover, the SAS congruence of the two triangle is violated for the sides opposite to the right angles since

$$dn^2(A,C) = 2 - \sqrt{2} \quad \text{and} \quad dn^2(B,C) = 2 + \sqrt{2}$$

are different. Finally the two triangles $\triangle AOC$ and $\triangle COB$ are isosceles, but do not have congruent base angles. Thus Euclid I.5 is not valid. □

[30] Hilbert did simply use real coordinates.

Remark. A counterexample for two-dimensional geometry has been constructed by Hilbert as follows. In the three-dimensional Euclidean space, we take the planes

$$\alpha = \{(x, y, z) : z = 0\} \quad \text{and} \quad \beta = \{(x, y, z) : x = z\}$$

and project the plane β onto α by the parallel projection

$$(x, y, x) \in \beta \mapsto (x, y, 0) \in \alpha$$

In the plane α, we define as new "`distance`" by measuring the Euclidean distance of the pre-image in the plane β. But we keep the Euclidean angle as "`angle`" in the plane α. We get in the plane α a new geometry, for which all axioms except (III.5) hold.

II.7.16 The Moulton plane

Theorem II.7.3 (Hilbert's Proposition 54). *There exists a planar geometry for which the axioms of incidence* (I.1),(I.2),(I.3)*, all axioms of order* (II.1),(II.2),(II.3),(II.4)*, the axioms of congruence* (III.1),(III.2),(III.3),(III.4) *but not the SAS-axiom* (III.5)*, and the Euclidean Parallel Postulate I.3.3 are all valid. But the following violations occur:*

The SAS axiom (III.5) *is violated.*

Moreover, neither the Theorem of Desargues, not the Little Theorem of Desargues, nor the Theorem of Pappus are valid.

It is not possible to embed the planar geometry in a three-dimensional incidence geometry.

Remark. If one wishes, even the axioms of continuity (V) can be assured to be valid. We see once more that it is impossible to deduct the SAS axiom (III.5) from the remaining axioms I, II, (III.1),(III.2),(III.3),(III.4), IV,V.

Remark. This non-Desarguesian plane is due to Frederic Moulton (1902), and is available in Hilbert's Foundations. The simplified version given in Definition I.4.4 is mentioned in the recent book by Stillwell [31].

Proof. We begin with any Pythagorean plane. Thus we have a Hilbert plane, for which the axioms of incidence (I.1)(I.2)(I.3a)(I.3b), order (II.1) through (II.4), and congruence (III.1) through (III.5) hold. Moreover, the axiom of parallelism (IV.1) is assumed to hold. By Proposition II.7.43, the existence of a parallel is shown. Thus the Euclidean Parallel Postulate I.3.3 is valid.

This Pythagorean plane is now used as the underlying reality, and a new model is constructed by redefining the notions "`lines`", "`congruence of segments`", "`congruence of angles`". The part of the construction without congruence has already been explained in Definition I.4.4, but is repeated for convenience. We choose one line x, a positive direction on it, and mark the left half-plane as the upper half-plane, the right half-plane as the lower half-plane. [31] Let any line l which is not parallel nor equal to x, be given. Let x_+ be the ray from the intersection $L := x \cap l$ in the positive x-direction. Let k be the ray on the line l in the lower half-plane, and h be the ray on l in the upper half-plane. For a line l with positive slope, we choose any point P on the ray h and

[31]In case of a construction from a real coordinate plane, this line is the x-axis. With real coordinates, one is including the validity of axioms of continuity (V.1)(V.2).

drop the perpendicular onto the x-axis. Let F be the foot point, M be the midpoint of the segment FP, and $h' = \overrightarrow{LM}$. In other words, this is the ray with the same vertex L as h but half its slope.

A "`line`" of the new geometry is now each one of the following:

the x-axis;

any line parallel or perpendicular to the x-axis;

any line with obtuse angle $\angle(h, x_+)$;

in case the angle $\angle(h, x_+)$ is acute, the "`line`" is the union of the two rays $k \cup h'$.

The "`order`" of the points on the new lines is defined in the obvious way. The "`distance`" of any two points is defined to be the Euclidean distance, but as measured along the new lines. Two segments are `congruent`, if the `distances` between their endpoints are equal.

Problem II.7.22. *Convince yourself that from the notions above, one has obtained a planar geometry for which the axioms of incidence* (I.1),(I.2),(I.3), *all axioms of order* (II), *the axioms of congruence* (III.1),(III.2),(III.3), *and the Euclidean Parallel Postulate I.3.3 are all valid.*

The angles which do not have their vertex on the x-axis are measured by constructing an angle with vertex on the x-axis and having its sides parallel to the given angle. To measure an angle $\angle(m', l')$ with vertex on the x-axis, we replace any of its side with positive slope by a ray from the same vertex with double its slope. Thus one goes make to the original unbroken lines.

Problem II.7.23. *Convince yourself that from the notions above, all parts of axiom of congruence* (III.4) *are valid.*

By the figures below on page 76, page 76, and page 81, one sees easily that neither the converse part of the Theorem of Desargues, nor the Little Theorem of Desargues, nor the Little Pappus Theorem are valid in the new incidence plane. To get such easy counterexamples, it is convenient to use a case in which only very few "`lines`" of the new geometry are affected by their intersection with the x-axis. □

II.7.17 Restriction of SAS congruence

The SAS axiom restricted to triangles of the same orientation is discussed by Hilbert in an appendix to the Foundations of Geometry.

Note that the triangles $\triangle ABC$ and $\triangle A'B'C'$ have the *same orientation* if either both point C lies in the left half-plane of ray $\overrightarrow{AB}$ and point C' lies in the left half-plane of ray $\overrightarrow{A'B'}$; or both point C lies in the right half-plane of ray $\overrightarrow{AB}$ and point C' lies in the right half-plane of ray $\overrightarrow{A'B'}$. Hilbert postulates SAS congruence restricted to triangles with the same orientation. and two additional axioms:

III.5* Assume that the two triangles $\triangle ABC$ and $\triangle A'B'C'$ have the same orientation and the congruences

$$AB \cong A'B' \, , \; AC \cong A'C' \, , \;\; \angle BAC \cong \angle B'A'C'$$

hold, then the congruence $\angle ABC \cong \angle A'B'C'$ is also satisfied.

III.6 If the angles $\angle(h', k')$ and $\angle(h'', k'')$ are both congruent to angle $\angle(h, k)$, then the angles $\angle(h', k')$ and $\angle(h'', k'')$ are congruent to each other.

III.7 If the rays c and d have the same vertex as the angle $\angle(h, k)$ and lie in the interior of that angle, then the angles $\angle(h, k)$ and $\angle(c, d)$ are not congruent.

Hilbert then announces the following result without a proof:

Proposition II.7.62 (The importance of Euclid I.5). *We assume only the restricted SAS axiom* (III.5*), *axioms* (III.6) *and* (III.7), *and finally we assume Euclid* (I.5) *to be valid: the base angles of any isosceles triangle are congruent. These assumptions imply the unrestricted SAS axiom* (III.5)*,—for any triangles independent of their orientation.*

Problem II.7.24 (A hard problem, only for fans). *Provide some drawings and prove this proposition.*

Definition II.7.23 (Skew-Hilbert plane). A *skew-Hilbert plane* is any model for two-dimensional geometry where Hilbert's axioms of incidence (I.1)(I.2)(I.3a)(I.3b), order (II.1) through (II.4), and congruence (III.1) through (III.4) , the restricted SAS-axiom (III.5*), and the axioms (III.6) and (III.7) hold.

Lemma II.7.8 (Transfer of a triangle with the same orientation). *Any given triangle $\triangle ABC$ can be transferred into either the left or the right half-plane of any given ray r, to obtain a congruent triangle with one vertex at the vertex of this ray, one side lying on the given ray, and having the* same orientation *as the original triangle.*

Lemma II.7.9 (Preparations in a skew-Hilbert plane). *The following Propositions are valid in any skew-Hilbert plane*

- *The angle congruence is an equivalence relation. Too, Proposition* II.7.26 *holds: angle comparison holds for congruence classes.*
- *Proposition* II.7.11 *about SAS congruence holds for two triangles with the same orientation.*
- *As stated in Lemma* II.7.8*, the transfer of a triangle is still possible under the restriction that the orientation is preserved.*
- *Proposition* II.7.19 *about angle-addition and subtraction (Theorem 15 in Hilbert) remains valid.*
- *The Exterior Angle Theorem (Euclid* (I.16)*. Theorem 22 in Hilbert, see Proposition* II.7.40) *is valid.*
- *The SAA-Congruence Theorem, Proposition* II.7.54 *(Theorem 25 in Hilbert) is valid for triangles of the same orientation.*
- *The midpoint of a segment can be obtained by the method used in Theorem 26 of Hilbert (see Proposition* II.7.56*). Any segment has a midpoint.*

Lemma II.7.10. *The following Propositions are valid in a skew-Hilbert plane where additionally Euclid* (I.5) *is assumed.*

- *From the Theorem* II.7.21 *about the symmetric kite, part* (i) *holds.* [32]
- *Euclid*(I.18) *(Theorem 23 of Hilbert, see Proposition* II.7.48 *above) is valid: in any triangle, across the longer side lies the greater angle.*
- *Euclid* (I.19) *(Proposition* II.7.49 *above) is valid: in any triangle, across the greater angle lies the longer side.*
- *We get Euclid* (I.6), *the converse of Euclid* (I.5).

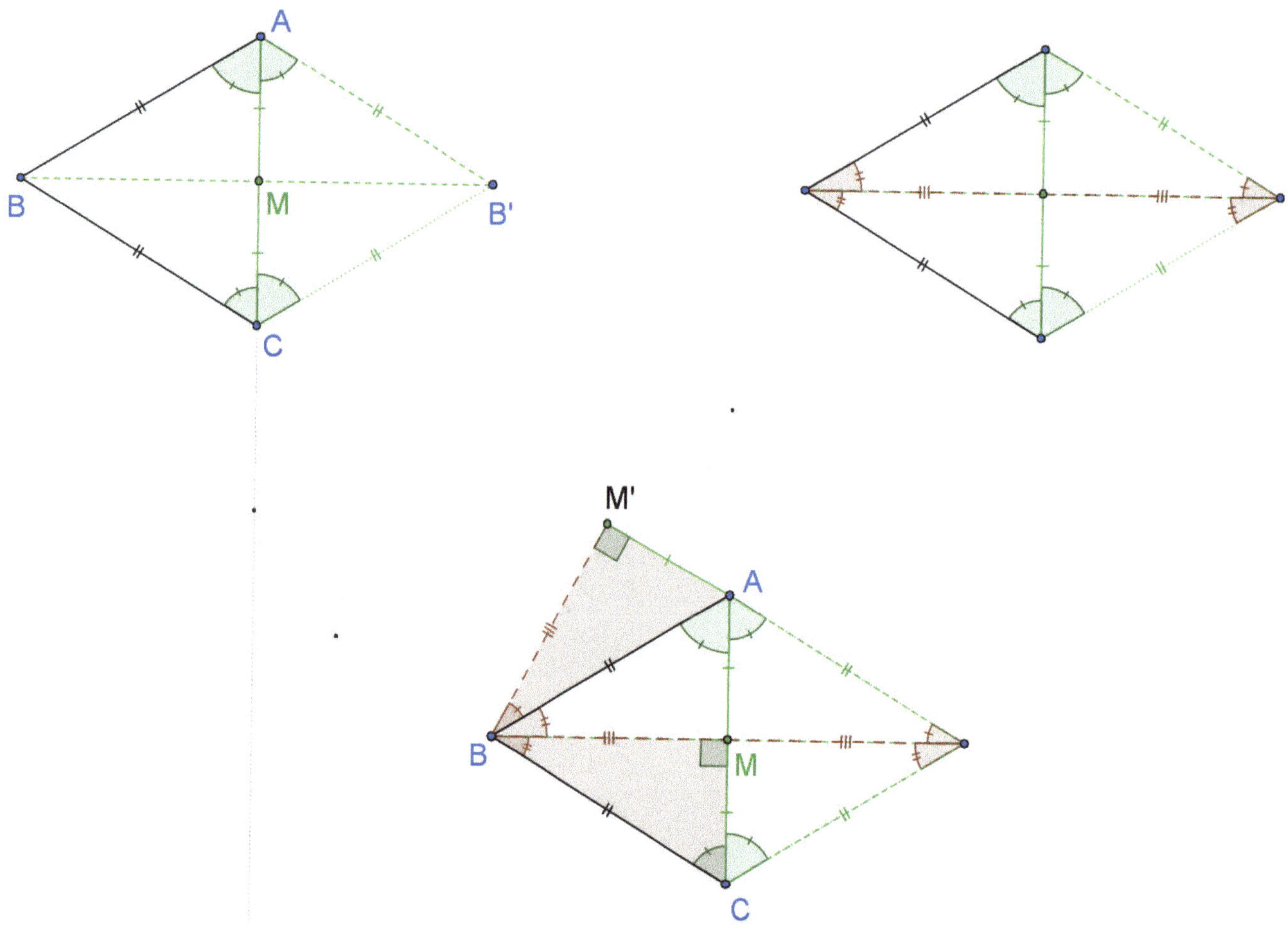

Figure II.7.80: To get a rhombus without use of reflection.

Lemma II.7.11. *The diagonals of the rhombus bisect each other, they bisect the angles at the vertices of the rhombus, and they are perpendicularly to each other. For any isosceles triangle $\triangle ABC$, we can construct a rhombus $\square ABCD$ which inherits three of its vertices from the triangle.*

Proof of the Lemma. The midpoint M of side AC can be obtained with Hilbert's construction. From SAS axiom (III.5*) we derive the triangle congruences

$$\triangle AMB \cong \triangle CMB'' \quad \text{and} \quad \triangle CMB \cong \triangle AMB''$$

[32]Part (ii) cannot be recovered at this point but only later.

Hence we have obtained a rhombus $\square ABCB''$. Too, get four congruent angles at vertices A and C and see that the diagonal AC bisects the angles at these two vertices. Beginning with the isosceles triangle $\triangle BAB''$ we repeat a similar argument. Hence the diagonals of a rhombus bisect each other, and they bisect the angles at the vertices of the rhombus.

It remains to check that the diagonals are perpendicular to each other. To this end, we transfer the hypothenuse BC of triangle $\triangle BMC$ on the segment BA and obtain the congruent triangle $\triangle BM'A \cong \triangle MBC$ with the same orientation. One needs only axioms (III.1),(III.4) and the restricted SAS axiom (III.5*). From the first item (i) of the Theorem II.7.21 about the symmetric kite, we conclude $\angle BMA \cong \angle BM'A$. From the triangle congruence we conclude we conclude $\angle BMC \cong \angle BM'A$. Hence $\angle BMC \cong \angle BMA$ are congruent supplementary, and hence right angles. □

Lemma II.7.12. *All points P of the perpendicular bisector of two points B and C have congruent distances $BP \cong CP$ from these two points.*

Proof. We may suppose that point P does not lie on the line BC. We take the smaller one of the two angles $\angle CBP$ and $\angle BCP$ and transfer it to the vertex and onto the side of the other one. Suppose to be definite that $\angle BCP \leq \angle CBP$. We transfer $\angle BCP$ onto the ray $\overrightarrow{BC}$. By the Crossbar Theorem, the newly produced ray intersects the segment BP, say at point Q. The triangle $\triangle CQB$ has congruent base angles, and hence from Euclid (I.6) we conclude $BQ \cong CQ$.

It remains to show that $Q = P$. Let M be the midpoint of segment BC. From the Proposition about the rhombus (Lemma II.7.11) we know that line QM is the perpendicular bisector of segment BC. Both P and Q are the intersection point of the two different lines QM and BP, hence $P = Q$. Finally we have confirmed $BP \cong CP$. □

Lemma II.7.13. *In a skew-Hilbert plane Euclid* (I.5) *has been assumed. Theorem* II.7.21 *about the symmetric kite, parts* (i) *and* (ii) *both hold. Indeed, for a kite with congruent sides $XZ_1 \cong XZ_2$ and $YZ_1 \cong YZ_2$, and Z_1 and Z_2 on different sides of XY, the triangles $\triangle XZ_1Y \cong \triangle XZ_2Y$ are congruent.*

End of the proof of the Theorem II.7.21 *about the symmetric kite.* Let Z_1 and Z_2 be two points on different sides of line XY, and assume that $XZ_1 \cong XZ_2$ and $YZ_1 \cong YZ_2$.

As explained above, from the isosceles triangle $\triangle Z_1XZ_2$, we construct a rhombus Z_1XZ_2X'. Its diagonals XX' and Z_1Z_2 intersect perpendicularly in the midpoint M of segment Z_1Z_2.

We begin a similar procedure with the isosceles triangle $\triangle Z_1YZ_2$ and obtain a second rhombus $Z_1Y'Z_2Y$. Its diagonals YY' and Z_1Z_2 intersect perpendicularly in the midpoint M of segment Z_1Z_2.

Hence all four points X, X', Y, Y' lie on the perpendicular to segment Z_1Z_2, erected at its midpoint M,—which we recognize as the perpendicular bisector of Z_1Z_2. Since we have shown that the diagonals XX' and YY' bisect the angles at X and Y, we get the remaining angle congruences for triangles $\triangle XZ_1Y \cong \triangle XZ_2Y$. □

Lemma II.7.14. *Assume that the triangles $\triangle APB$ and $\triangle APC$ lie on different sides of AP, have congruent sides $AB \cong AC$ and congruent angles $\angle PAB \cong \angle CAP$ between these sides and their common side. Then the two triangles are congruent.*

Proof. From the Proposition about the rhombus (Lemma II.7.11) we know that line AP is the perpendicular bisector of segment BC. From Lemma II.7.12, we know that $BP \cong CP$. From the

Theorem II.7.21 about the symmetric kite we conclude that the triangles $\triangle APB \cong \triangle APC$ are congruent. $\square$

End of the proof of the Proposition about importance of Euclid (I.5). Combine
Lemma II.7.14 with the Lemma II.7.8 about the transfer of a triangle. Assume the two triangles $\triangle ABC$ and $\triangle A'B'C'$ have different orientations, have corresponding congruent sides $AB \cong A'B'$ and $AC \cong A'C'$, and congruent angles $\angle BAC \cong \angle B'A'C'$ between them.

We transfer the angle $\angle BAC$ onto the ray $\overrightarrow{A'C'}$, at vertex A', to the same side of $A'C'$ as B'. On the newly produced ray, we transfer segment AB, starting at vertex A'. Thus we get point B_0, such that $A'B_0 \cong AB$. Because of Lemma II.7.8 we get

$$\triangle ABC \cong \triangle A'B_0C'$$

By Lemma II.7.14 we conclude that

$$\triangle A'B'C' \cong \triangle A'B_0C'$$

Hence $\triangle ABC \cong \triangle A'B'C'$ as to be shown. $\square$

II.8 Measurement and Continuity

Hilbert postulates two axioms of continuity: the axiom of Archimedes, and the axiom of completeness.

II.8.1 The Archimedean axiom

Given are any segments AB and CD. We now use the second one as a measure unit. The Archimedean axiom states existence of natural number n and as finite sequence of points $A = A_0, A_1, A_2, \ldots, A_n$ on the ray $\overrightarrow{AB}$ such that

(II.8.1) $$CD \cong A_{k-1}A_k \quad \text{for} \quad k = 1, 2, \ldots n \quad \text{and} \quad A * A_{k-1} * A_k \quad \text{for} \quad k = 2, 3 \ldots n$$

and the point B either lies between A_{n-1} and A_n, or $B = A_{n-1}$. In Hilbert's words,

> "n segments congruent to CD constructed contiguously from A, along a ray from A through B, will pass beyond B."

As a shorthand notation, we shall write

(II.8.2) $$n \cdot CD > AB$$

More colloquially, one can say that no segment is too *long* as to be measured in terms of a given unit segment. As a first consequence of this fact, we note that no segment is so *short* that it cannot

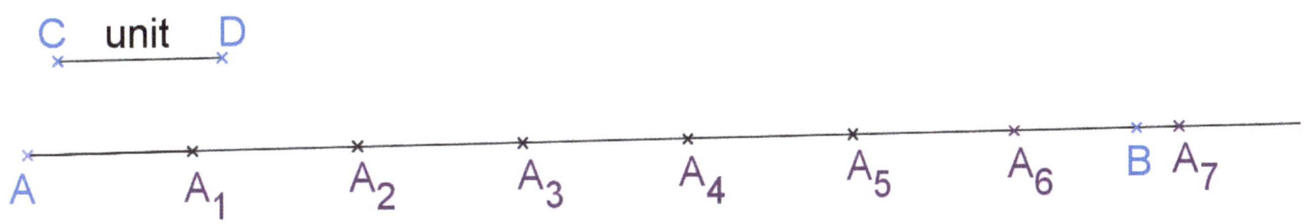

Figure II.8.1: The Archimedean axiom. For the case drawn in the figure, it turns out that $n = 7$.

be measured in terms of a given unit segment.

(Repeated bisection produces arbitrarily short segments). *Given any segments PQ and EF, there exists a natural number s such that s successive bisections produce a segment $P_sQ_s = \frac{PQ}{2^s}$ shorter that segment EF, written as a formula*

$$\frac{PQ}{2^s} < EF$$

Especially, not both points E and F can lie inside segment P_sQ_s or on its endpoints.

Proof. We apply the Archimedean axiom to the segments $AB := EF$ and $CD := PQ$. Hence there exists a natural number n such that

(II.8.3) $$n \cdot EF > PQ$$

Let s be an integer such that $2^s \geq n$. We get

$$2^s \cdot EF > PQ$$

Bisection of all segments involved leads inductively to the inequalities

$$2^{s-1} \cdot EF > \frac{PQ}{2}, \quad 2^{s-2} \cdot EF > \frac{PQ}{4}, \quad 2^{s-3} \cdot EF > \frac{PQ}{8}, \ldots, EF > \frac{PQ}{2^s}$$

as to be shown. □

The Archimedean axiom allows the measurement of segments and angles using real numbers. These real numbers occur during the measurement process in the form of binary fractions. Since Hilbert, this axiom is also known as the axiom of measurement. To start the measuring process, one assigns the length one to some arbitrary—but convenient—segment. We shall call it the *unit segment.*

Main Theorem 10 (Measurement of segments). *Given is a* unit segment OI. *There exists a unique way of assigning a* length *to any segment. We denote the length of segment AB by $|AB|$.* [33] *We construct the length as an (finite or infinite) binary fraction. Too, this length is called the* distance *of points A and B. The length has the following properties:*

(1) Positivity *If $A \neq B$, then $|AB|$ is a positive real number, and $|OI| = 1$.*

(2) Congruence *$|AB| = |CD|$ if and only if $AB \cong CD$.*

(3) Ordering *$|AB| < |CD|$ if and only if $AB < CD$.*

(4) Additivity *The distances are additive for any three points on a line. With point B lying between A and C, we get $|AC| = |AB| + |BC|$.*

Proof. We begin by assuming existence of the measurement function, and really construct it. In a second step, we confirm that the construction has produced the result as claimed. Let M be the midpoint of the unit segment OI. Since $OM \cong MI$, the additivity of segment lengths implies $|OM| + |MI| = 1$, and hence $|OM| = |MI| = \frac{1}{2}$. Successive bisections now produce segments of lengths $\frac{1}{4}, \frac{1}{8}, \ldots$. A length is assigned to any given segment AB by the process indicated in the

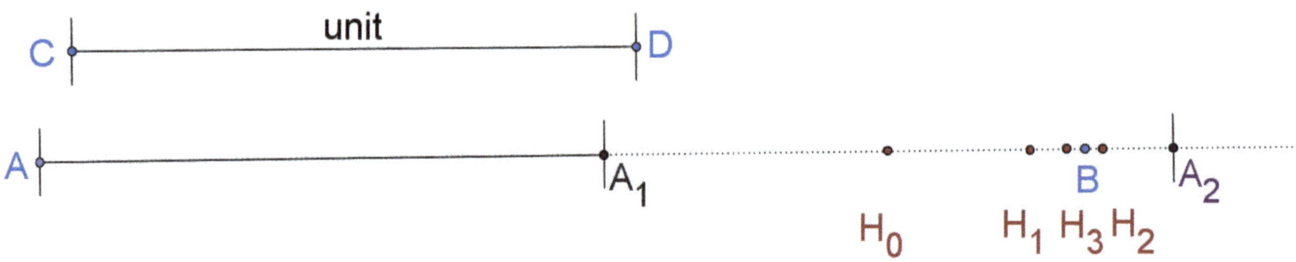

Figure II.8.2: This measurement yields $|AB| = 1.1101\ldots$ as a binary fraction.

figure on page 226. As in the Archimedean axiom, one begins by constructing a finite sequence of points $A_0 = A, A_1, A_2, \ldots, A_n$ on the ray $\overrightarrow{AB}$ such that

$$OI \cong A_{k-1}A_k \quad \text{and} \quad \overrightarrow{AA_k} = \overrightarrow{AB} \quad \text{for } k = 1, 2, \ldots n \tag{II.8.4}$$

[33] In projective geometry and other contexts, one uses a signed length, which I shall denote by $\overline{AB}$.

This part of the process stops, and the integer part of the length $|AB|$ is determined, as soon as point B either lies between A_{n-1} and A_n, or $B = A_{n-1}$. In the latter case, one gets the length $|AB| = n-1$ exactly, and the measure process is finished. In general, the former case occurs. One concludes that $n-1 < |AB| < n$, and needs to construct the digits d_s of an finite or infinite binary fraction

$$n-1\,.\,d_1 d_2 d_3 \ldots$$

in order to achieve a measurement with more and more precision. [34] To begin the process, we set the lower bound $L_0 := A_{n-1}$, and the upper bound $U_0 := A_n$. The measurement uses sequences for the lower bound, digit, and upper bound, which are constructed inductively by repeated bisections.

Inductive measurement step to determine d_s. For $s = 1, 2, 3, \ldots$, let H_{s-1} be midpoint of segment $L_{s-1}U_{s-1}$.

If point B lies between L_{s-1} and H_{s-1}, one puts

new digit $d_s := 0$

lower bound $L_s := L_{s-1}$

upper bound $U_s := H_{s-1}$

If point B lies between H_{s-1} and U_{s-1}, or $B = H_s$, one puts

new digit $d_s := 1$

lower bound $L_s := H_{s-1}$

upper bound $U_s := U_{s-1}$

Hence the digit d_s depends on whether point B lies left or right of H_{s-1}. In both cases, the approximation to $|AB|$ obtained by the s-th step is

$$|AL_s| = n - 1 + \frac{d_1}{2} + \frac{d_2}{4} + \frac{d_3}{8} + \cdots + \frac{d_s}{2^s}$$

It can happen that $B = H_{s-1}$, in which case this approximation is exact, and the measure process is finished. In general, the measurement process does not stop, but produces an infinite fraction. By Cantor's principal of boxed intervals, the infinite binary fraction <u>is</u> indeed a real number.

Question. Explain once more, how the first digit d_1 is determined.

Answer. One puts the lower bound $L_0 := A_{n-1}$ and the upper bound bound $U_0 := A_n$, with A_{n-1} and A_n obtained via the Archimedean axiom. Let H_0 be the midpoint of segment L_0U_0.

If point B lies between L_0 and H_0, one puts

new digit $d_1 := 0$

current approximation $|AL_1| = n - 1$

[34]Of course, any real life measure has to stop such an infinite process, and hence cannot confirm an exact real number as result, but only a finite number digits, limited by the technical possibilities.

lower bound $L_1 := L_0$

upper bound $U_1 := H_0$

If point B lies between H_0 and U_0, or $B = H_0$, one puts

new digit $d_1 := 1$

current approximation $|AL_1| = n - 1 + \frac{1}{2}$

lower bound $L_1 := H_0$

upper bound $U_1 := U_0$

Finally, one wants to be convinced that the congruence, ordering, and additivity properties stated as (2),(3) and (4) are really satisfied. We start by confirming

(2a) If $AB \cong CD$, then $|AB| = |CD|$.

(3a) If $AB < CD$ then $|AB| < |CD|$. If $AB > CD$ then $|AB| > |CD|$.

It is rather straightforward to deduct (2a) from Hilbert's axiom III.3. Let any two segments $AB < AC$ be given. Repeated bisection produces arbitrarily short segments. Hence there exists a number s such that

(II.8.5)
$$L_sU_s \cong \frac{OI}{2^s} < BC$$

Because we go through the measurement process explained above for segment AB, we know that $B = L_s$ or $L_s * B * U_s$. By the segment comparison above, not both points B and C can lie in the interval L_sU_s, but indeed $L_s * U_s * C$. Hence the binary fractions for points B and C differ in the sth place, and hence have less than s common places, furthermore $|AB| \neq |AC|$.

Assume that t is the first place where the measurement fractions of points B and C are different:

(II.8.6)
$$b_1 = c_1, b_2 = c_2, \ldots, b_{t-1} = c_{t-1}, \quad \text{but} \quad b_t \neq c_t$$

It can happen by accident that t from equation (II.8.6) is much smaller than any number s for which equation (II.8.5) holds. The four-point theorem yields the natural order $L_{t-1} * B * C * U_{t-1}$, and its generalization to five or more points leads to the natural order $L_{t-1} * B * H_{t-1} * C * U_{t-1}$, or $C = H_{t-1}$. Hence $|AB| < |AC|$ follows by using the t-th approximations. The second part of (3a) is quite similar to check.

Since all segments are comparable, items (2a) and (3a) implies the converse statements, and hence (2) and (3) are confirmed.

The additivity (4) can at first be checked by induction to hold for all segments of integer lengths. Next, one can prove additivity for segments of which lengths are finite binary fractions, using induction on the number of digits occurring. I only explain the further case that the length of segment AB is given by a finite binary fraction, but segment BC by an infinite fraction. For any arbitrary lower and upper bound finite fractions $BL < BC < BU$, we get

$$BL < BC < BU$$
$$|BL| < |BC| < |BU|$$
$$|AB| + |BL| < |AB| + |BC| < |AB| + |BU|$$
$$|AL| < |AB| + |BC| < |AU|$$

and on the other hand

$$AB + BL < AB + BC < AB + BU$$
$$AL < AB + BC < AU$$
$$|AL| < |AB + BC| < |AU|$$

Both statements

$$|AL| < |AB| + |BC| < |AU|$$
$$|AL| < |AB + BC| < |AU|$$

hold for all lower and upper bounds L and U. Hence the Archimedean axiom for the real numbers implies $|AB| + |BC| = |AB + BC|$. ☐

Corollary 17. *Given any three points A, B and C, the equality*

$$|AB| + |BC| = |AC|$$

holds if and only if the three points lie on a line and either two of them are equal or B lies between A and C.

The measurement of angles is done quite similar to the measurement of segments. Contrary to the situation for segments, there exists a well defined and convenient unit of measurement, which is the *right angle.* The traditional measurement in degrees is obtained by assigning the value 90° to the right angle.

Main Theorem 11 (Measurement of angles). *There exists a unique way of assigning a* degree measurement *to any angle. We denote the value of angle $\angle ABC$ by $(\angle ABC)^\circ$.* [35]

For additivity, we consider a ray $\overrightarrow{BG}$ in the interior angle $\angle ABC$. Following definition II.7.13, the angle $\angle ABC$ is the sum of the angles $\angle ABG$ and $\angle GBC$, written as $\angle ABC = \angle ABG + \angle GBC$.

The measurement has the following properties:

- **(1a) Positivity** *For any angle $\angle ABC$, the degree measurement is a positive number and $0^\circ < (\angle ABC)^\circ < 180^\circ$.*
- **(1b) Right angle** *The measurement of a right angle is 90°.*
- **(2) Congruence** *$(\angle ABC)^\circ = (\angle A'B'C')^\circ$ if and only if $\angle ABC \cong \angle A'B'C'$.*
- **(3) Ordering** *$(\angle ABC)^\circ < (\angle A'B'C')^\circ$ if and only if $\angle ABC < \angle A'B'C'$.*
- **(4) Additivity** *If $\angle ABC = \angle ABG + \angle GBC$, then $(\angle ABC)^\circ = (\angle ABG)^\circ + (\angle GBC)^\circ$.*

Question. Why does bisecting any angle produce an acute angle?

Answer. It is shown in the figure on page 230, how to bisect an angle. Let A be the vertex of the angle. One transfers two congruent segments AB and AC onto the two sides of the angle, both starting from the vertex A. The perpendicular, dropped from the vertex A onto the segment BC, is the angular bisector.

Let F be the foot point of the perpendicular. As a consequence of the exterior angle theorem, we have shown that the two further angles in a right triangle are acute. Hence the angle $\angle BAF$ is acute, but this is just the bisected angle.

[35] In calculus and differential geometry, one uses often measurement using the arc length on the unit circle. In that case, the right angle is assigned the value $\frac{\pi}{2}$.

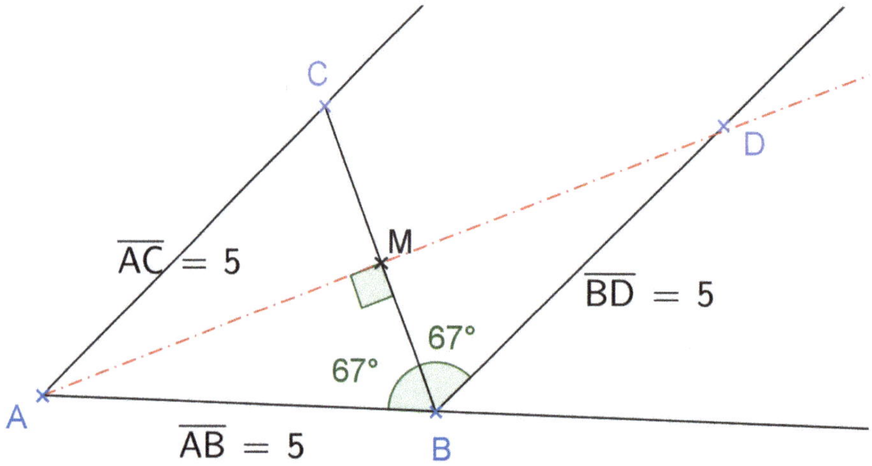

Figure II.8.3: The angular bisector

Theorem II.8.1 (The Archimedean property for angles, Hilbert's Proposition 34). *Assume that the Archimedean Axiom holds. For any two angles α and ε there exists a natural number r such that*

$$\frac{\alpha}{2^r} < \varepsilon \tag{II.8.7}$$

Proof of Proposition II.8.1. The angle $\frac{\alpha}{2}$ is acute. If $\frac{\alpha}{2} \leq \varepsilon$, assertion (IV.20.3) holds with $r = 1$, or $r = 2$ in case of equality. We need a construction for the remaining case $\frac{\alpha}{2} > \varepsilon$. Choose any point C on one side of the angle $\frac{\alpha}{2}$ and drop the perpendicular on the other side of that angle. Let B be the foot point of that perpendicular, and let A be the vertex of $\frac{\alpha}{2}$. Next we transfer angle ε at vertex A, with one side AB, and the other side inside the angle $\frac{\alpha}{2} = \angle CAB$. By the Crossbar Theorem, there exists a point D where this other side intersects segment BC.

By the Archimedean axiom, there exists a natural number n such that

$$n \cdot BD > BC \tag{II.8.8}$$

Now, one transfers the angle ε, repeating n times, always with vertex A, using common sides, and turning away from segment AB. Let $C_0 = B, C_1 = D$. Let the new sides of transferred angle ε intersect $\overrightarrow{BC}$ at points $C_1 = D, C_2, C_3, \ldots$. We distinguish two cases (II.8.9) or (II.8.10):

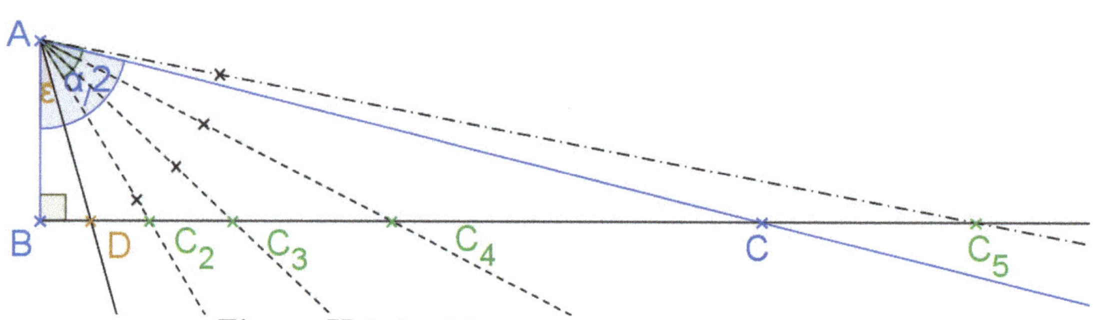

Figure II.8.4: Many consecutive small angles surpass any angle.

(II.8.9) It can happen that one of the new sides do no longer intersect ray $\overrightarrow{BC}$. Let m be the smallest number of angle transfers for which this happens. Since the second leg $\overrightarrow{AC}$ of angle $\frac{\alpha}{2}$ does intersect ray $\overrightarrow{BC}$, we conclude that

$$m \cdot \varepsilon > \frac{\alpha}{2} \tag{II.8.9}$$

Otherwise the Crossbar Theorem would lead to a contradiction.

(II.8.10) The other possible case is that the new sides produced by transferring angle ε all n times intersect ray $\overrightarrow{BC}$, say at points $C_1 = D, C_2, C_3, \ldots, C_n$.

Let $E = C_n$ be the point where the new side of nth angle ε intersects $\overrightarrow{BC}$. The segments $C_0C_1, C_1C_2, C_2C_3, \ldots, C_{n-1}C_n$ cut on line BC by these angles $\varepsilon = \angle C_kAC_{k-1}$ satisfy

$$C_{k-1}C_k < C_kC_{k+1}$$

for $k = 1, 2, 3, \ldots, n-1$, as follows from Proposition 33 given below. Hence a simple induction argument yields $C_{k-1}C_k \geq BD$ and $BC_k > k \cdot DB$ for $k = 1, 2, 3, \ldots, n-1$. Hence (II.8.8) implies $BC_n > n \cdot BD > BC$. Because $\angle BAC_n = n \cdot \varepsilon$, we get

$$n \cdot \varepsilon > \frac{\alpha}{2} \tag{II.8.10}$$

Let r be an integer such that $2^{r-1} \geq m$ or $2^{r-1} \geq n$, in case (a) or (b), respectively. Now (II.8.9) or (II.8.10) imply

$$2^{r-1} \cdot \varepsilon > \frac{\alpha}{2}$$

Bisection of those two angles leads inductively to the inequalities

$$2^{r-2} \cdot \varepsilon > \frac{\alpha}{4}, \; 2^{r-3} \cdot \varepsilon > \frac{\alpha}{8}, \ldots, \varepsilon > \frac{\alpha}{2^r}$$

This finishes the proof of Proposition 34. □

Here is still the missing proposition, already used above.

(**Hilbert's Proposition 33**). *Let triangle $\triangle OPZ$ have a right angle at vertex P. Let X and Y be two points on segment PZ such that $\angle XOY = \angle YOZ$. Then $XY < YZ$.*

Proof of Proposition 33. One transfers segment OX onto the ray $\overrightarrow{OZ}$ at vertex O. One gets the segment $OX' \cong OX$. Because segment OZ is the side opposite to the obtuse angle in $\triangle OXZ$, and the obtuse angle is the largest angle of any triangle, we get $OZ > OX$ from Euclid I.19. Hence the point X' lies between O and Z.

From the exterior angle theorem (Euclid I.16) for $\triangle OYZ$ one gets $\angle OZY < \angle OYX$. From the exterior angle theorem for $\triangle OX'Y$ one gets $\angle OYX' < \angle YX'Z$. Hence

$$\angle X'ZY = \angle OZY < \angle OYX \cong \angle OYX' < \angle YX'Z \tag{II.8.11}$$

In figure II.8.1, we see that $\alpha < \beta < \gamma$. By Euclid I.19, the side opposite to a larger angle is larger. We use this theorem for $\triangle X'YZ$. Hence (II.8.11) implies $X'Y < YZ$ and hence $XY \cong X'Y < YZ$ as to be shown. □

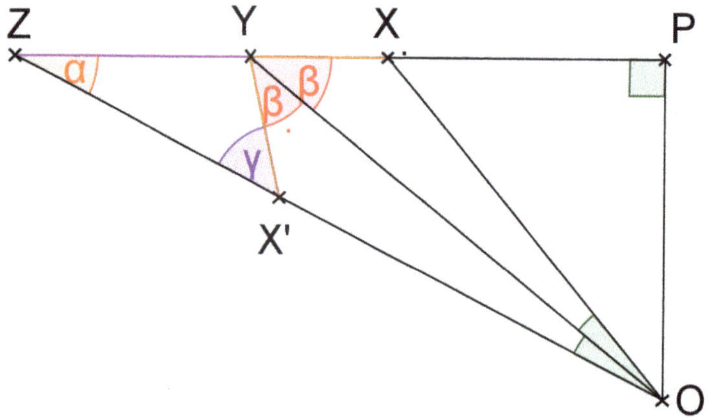

Figure II.8.5: Congruent angles cut longer and longer segments from a line.

Sketch of the proof for the measurement of angles. Let the given angle be $\angle BAC$. We erect the perpendicular to the ray $\overrightarrow{AB}$ at vertex A. Depending on whether the perpendicular is inside the supplementary right angle, or inside the given angle, or the perpendicular coincide with the second side $\overrightarrow{AC}$, the given angle is acute, obtuse or right. The measurement process is now most easily explained by taking the right angle as unit. Starting with that one of the two complementary right angles, inside of which the ray $\overrightarrow{AC}$ lies, one successively bisects angles—always keeping the ray $\overrightarrow{AC}$ inside of them. The left bounds of the bisected angles correspond to an infinite fraction

$$0 \,.\, d_1 d_2 d_3 \ldots$$

for an acute, or

$$1 \,.\, d_1 d_2 d_3 \ldots$$

for an obtuse angle. To obtain a measurement in the traditional degrees, these binary fractions have to be multiplied by 90 [36] The further details are so similar to the case of segment measurement that we do not need to repeat. □

II.8.2 Axioms related to completeness

The clear-cut understanding of this material was achieved only in the late nineteenth century. There are several axioms for completeness, with very similar implications, having slight but deep differences. It is hard to say what is the most natural one of these axioms. Even Hilbert has suggested different axioms of continuity in different editions of his foundations of geometry.

Cantor's axiom

Following Baldus and Löbell [7], p.43, I state my favorite version of Cantor's axiom.

Cantor's Axiom. There exists at least one segment A_1B_1 with the following property: Given any sequence A_iB_i of boxed subsegments for $i = 2, 3, \ldots$ such that

(II.8.12) $$A_{i-1} * A_i * B_i * B_{i-1} \quad \text{for all} \quad i = 2, 3, \ldots$$

[36] In binary notation $90 = 64 + 16 + 8 + 2 = \overline{1011010}$

there exists a point X^* such that

$$A_i * X^* * B_i \quad \text{for all} \quad i = 1, 2, 3, \dots \tag{II.8.13}$$

For any every Hilbert plane, the following similar, but a bid more general statement is an immediate consequence:

Cantor's principle of boxed intervals For every sequence of boxed segments A_iB_i such that either

$$A_{i-1}*A_i*B_i*B_{i-1} \quad \text{or } A_{i-1} = A_i\,,\ A_{i-1}*B_i*B_{i-1} \quad \text{or } A_{i-1}*A_i*B_{i-1}\,,\ B_{i-1} = B_i, \tag{II.8.14}$$

for all $i = 2, 3, \dots$, there exists a point X^* such that

$$\text{either } A_i * X^* * B_i \quad \text{or } X^* = A_i \quad \text{or } X^* = B_i \quad \text{for all} \quad i = 1, 2, 3, \dots \tag{II.8.15}$$

Problem II.8.1. *Convince yourself, and explain that Cantor's axiom implies Cantor's principal of boxed intervals, just assuming the axioms of incidence, order and congruence.*

Main Theorem 12 (Completeness of measured segments and angles). *Assume the axioms of incidence, order and congruence, as well as the Archimedean and Cantor's axiom.*

Given is a unit segment *OI of length* $|OI| = 1$. *For every real number* $r > 0$, *there exists a segment segment AB of length* $|AB| = r$.

For every real number $0 < d < 180$, *there exists an angle, which has the measurement* $\angle ABC = d$, *in traditional degrees.*

Remark. As far as the continuity axioms are concerned, one needs to assume the Archimedean axiom (V.1) and Cantor's principle of boxed intervals (V.2). I have stated these axioms of continuity II.14.1 among a simplified axiomatization of geometry II.14.1.

Indication of reason. The real number r can be given as an finite or infinite binary fraction

$$r = n - 1\,.\,d_1d_2d_3\dots$$

The boxed intervals are constructed as explained in the measuring process. Let L_0U_0 be a segment of unit length such that $|AL_0| = n - 1$ and $|AU_0| = n$. We define a sequence of boxed intervals L_sU_s by setting

$$L_s := L_{s-1}\,,\ U_s := H_{s-1} \quad \text{if } d_s = 0$$

$$L_s := H_{s-1}\,,\ U_s := U_{s-1} \quad \text{if } d_s = 1$$

for $s = 1, 2, 3, \dots$. Here H_{s-1} is midpoint of segment $L_{s-1}U_{s-1}$. In both cases, the left lower approximation to $|AB|$ obtained by the s-th step is

$$|AL_s| = n - 1 + \frac{d_1}{2} + \frac{d_2}{4} + \frac{d_3}{8} + \dots + \frac{d_s}{2^s}$$

In the special case of a finite fraction, one gets $X^* = H_{s-1}$, in which case this approximation is exact, and the measure process is finished.

For an infinite binary fraction, we need Cantor's principal of boxed intervals: There exists a point X^* lying is in the *infinite intersection* of all intervals L_sU_s. The interval L_0X^* has length $|L_0X^*| = r$ equal to the originally given number r. □

Dedekind's axiom

This is now most popular axiom for continuity. Let us start with a definition.

Definition II.8.1 (Dedekind Cut). A *Dedekind cut* of the line l is a pair of sets $\{\Sigma_1, \Sigma_2\}$ such that the set of points on a line l is the disjoint union of the two nonempty sets Σ_1 and Σ_2, which have the following property:

(P) If P_1 and P_3 are any two points in Σ_i, and the third point P_2 lies between P_1 and P_3, then P_2 is in the same set Σ_i, for $i = 1, 2$.

Dedekind's Axiom. Assume that $\{\Sigma_1, \Sigma_2\}$ is a Dedekind cut of the line l. Then there exists a *cut point* K^* on the line l such that $\Sigma_1 \cup \{K^*\}$ and $\Sigma_2 \cup \{K^*\}$ are the two opposite rays on the line l with vertex K^*.

In his recent book [1], Greenberg writes on p.260 about this axiom:

> We bring in our *deus ex machina*, as classical Greek theater called it—a god comes down from heaven to save the day.

This axiom was proposed by J.W.R. Dedekind in 1871. Here is what Dedekind says about continuity in "Stetigkeit und irrationale Zahlen" (1872):

> I find the essence of continuity in the following principle: "If all the points of a line fall into two classes in such a way that each point of the first class lies to the left of each point of the second class, then there exists one and only one point that gives rise to this division of all the points into two classes, cutting of the line into two pieces."
>
> As mentioned before, I believe I am not wrong if I assume that everyone will immediately admit truth of this assertion; most of my readers will be very disappointed to realize that by this triviality the mystery of continuity will be revealed. I am very glad if everyone finds the above principle so clear and so much in agreement with our own conception of a line; for I am not in a position to give any kind of proof of its correctness; nor is anyone else.
>
> The assumption of this property of a line is nothing else than an axiom by which we first recognize continuity of the line, through which we think continuity into the line.
>
> If space has any real existence at all, it does not necessary need to be continuous; countless properties would remain the same if it were discontinuous. And if we knew for certain that space was discontinuous, still nothing could hinder us, if we so desired, from making it continuous in our thought by filling up its gaps; this filling up would consist in the creation of new point-individuals, and would have to be carried out in accord with the above principle.

Indeed, Dedekind's axiom is a very strong axiom. It essentially introduces the real numbers into our geometry, which is not in the spirit of Euclid, but useful and even necessary from the engineering point of view.

Dedekind's axiom is by no means necessary to do interesting mathematics— on the contrary—it spoils many fine points of algebra and set theory. Many more subtle distinctions and questions about constructibility are obscured by Dedekind's axiom.

Too, the strength of Dedekind's axiom becomes apparent because of its many consequences, some of which are now explained.

Theorem II.8.2. *Dedekind's axiom implies the Archimedean axiom.*

Theorem II.8.3. *Dedekind's axiom implies Cantor's axiom.*

Main Theorem 13 (Dedekind's axiom implies completeness). *The elements of geometry— the points, lines and planes— which satisfy the axioms of incidence, order, congruence, and Dedekind's axiom have no extension to any larger system, for which all these axioms still hold.*

To facilitate the proofs of theorem II.8.2 as well as theorem II.8.3, we introduce still another suggested axiom:

Weierstrass' Axiom. Let A_i for $i = 2, 3, \ldots$ be sequence of a points , and B a point, lying all on one line, such that

$$A_{i-1} * A_i * B \quad \text{for} \quad i = 2, 3, \ldots \tag{II.8.16}$$

Then there exists a point K^* such that

(i) either $K^* = B$ or

$$A_i * K^* * B \quad \text{for} \quad i = 1, 2, 3, \ldots \tag{II.8.17}$$

(ii) Furthermore, every other point X such that

$$A_i * X * B \quad \text{for} \quad i = 1, 2, 3, \ldots \tag{II.8.18}$$

satisfies

$$A_i * K^* * X * B \quad \text{for} \quad i = 1, 2, 3, \ldots \tag{II.8.19}$$

Remark. In terms of the real number line, one has the following: The A_i are an increasing sequence bounded by B. Hence this sequence has the limit K^*.

Proposition II.8.1. *Dedekind's axiom implies Weierstrass' axiom.*

Proof that Dedekind's axiom implies Weierstrass's axiom. We define a Dedekind cut as follows: Let Σ_1 consist of all points P such that

$$\text{either} \quad P * A_1 * B \quad \text{or} \quad P = A_1 \quad \text{or there exists } i \geq 1 \text{ such that} \quad A_1 * P * A_i$$

As follows logically from Hilbert's four-point theorem (II.6.5), the complement Σ_2 on the line A_1B consists of all points such that

$$A_1 * A_i * P \quad \text{for all } i = 2, 3 \ldots$$

The cut point K^* of the Dedekind cut $\{\Sigma_1, \Sigma_2\}$ needs to lie in Σ_2. It is either $K^* = B$ or satisfies $A_i * K^* * B$ for $i = 1, 2, 3, \ldots$. This confirms item (i) of the Weierstrass axiom. Furthermore, if one assumes that any point X satisfies

$$A_i * X * B \quad \text{for} \quad i = 1, 2, 3, \ldots \tag{II.8.20}$$

then $X \in \Sigma_2$. Hence $K^* * X * B$ because K^* is the vertex of the ray producing Σ_2. Now relation (II.8.20) and the four-point theorem imply

$$A_i * K^* * X * B \quad \text{for} \quad i = 1, 2, 3, \ldots \tag{II.8.19}$$

as to be shown, in order to confirm item (ii) of Weierstrass' axiom. □

Proposition II.8.2. *Weierstrass' axiom implies Cantor's axiom.*

Proof that Weierstrass's axiom implies Cantor's axiom. Within the assumptions to setup Cantor's axiom, Hilbert's n-point theorem (II.6.7) implies inductively

$$A_1 * A_i * B_j * B_1 \quad \text{for all } i, j \geq 2$$

The first item (i) of Weierstrass' axiom implies

$$A_i * K^* * B_1 \quad \text{for} \quad i = 1, 2, 3, \ldots$$

Fix some index j. Since $A_i * B_j * B_1$ for all $i = 1.2. \ldots$, the second item (ii) from Weierstrass' axiom with $X := B_j$ implies

$$A_i * K^* * B_j * B_1 \quad \text{for} \quad i, j = 1, 2, 3, \ldots$$

as to be shown. □

Proposition II.8.3. *Weierstrass' axiom implies the Archimedean axiom.*

Proof. As in the Archimedean axiom, two segments CD and AB are given. We use CD as unit of measurement and make the following

Definition II.8.2. We say that a point P on the ray $\overrightarrow{AB}$ *can be reached with the unit of measurement* CD if and only if—as in the Archimedean axiom—, there exists a natural number n for which construction of the finite sequence of points $A_0 = A, A_1, A_2, \ldots, A_n$ such that

$$CD \cong A_{k-1}A_k \quad \text{and} \quad \overrightarrow{AA_k} = \overrightarrow{AB} \quad \text{for } k = 1, 2, \ldots n \tag{II.8.4}$$

leads up to a point A_n such that point P either lies between A_{n-1} and A_n, or $P = A_{n-1}$.

Let $A_1, A_2, A_2, \ldots$ be the sequence of points constructed by the Archimedean measurement (II.8.4). If the Archimedean axiom would not hold, then there would exist a point F which cannot be reached with the unit measurement CD. We can now apply Weierstrass' Axiom for the sequence A_i and point F. Hence there exists a point K^* such that

$$A_i * K^* * F \quad \text{for} \quad i = 1, 2, 3, \ldots \tag{II.8.21}$$

(K^* is the limit of the increasing sequence $i \mapsto K_i$. This limit exists since the sequence is bounded by F.) Let K_- and K_+ be the points on both sides of K^* such that $K^*K_- \cong K^*K_+ \cong CD$, and (1*): $K_- * K^* * F$.

It is clear that the point K_+ cannot be reached by measurement since

$$A_{i-1} * A_i * K_+ \quad \text{for} \quad i = 1, 2, 3, \ldots \tag{II.8.22}$$

Hence K_- cannot be reached by measurement, neither. Indeed

$$A_{i-1} * A_i * K_- * F \quad \text{for} \quad i = 1, 2, 3, \ldots \tag{II.8.23}$$

We get now a contradiction to item (ii) of Weierstrass' axiom, K^* is the point most to the left that cannot be reached by measurement. Every (other) point X such that

$$A_i * X * F \quad \text{for} \quad i = 1, 2, 3, \ldots \tag{II.8.20}$$

satisfies

(II.8.19) $$A_i * K^* * X * F \quad \text{for} \quad i = 1, 2, 3, \dots$$

With $X := K_-$ we get $A_i * K^* * K_- * F$. This contradicts (1*): $K_- * K^* * F$ from above. This contradiction implies that the Archimedean axiom does hold. □

In order to defend my preference of Cantor's axiom—and trying to take some of the deus ex machina image away from Dedekind—I prove now:

Main Theorem 14. *Assuming both the Archimedean axiom, as well as Cantor's axiom, Dedekind's axiom follows.*

Proof. Given is a line AB with a Dedekind cut $\{\Sigma_1, \Sigma_2\}$ of it. Let OI be any measurement unit. We may assume $A \in \Sigma_1$ and $B \in \Sigma_2$. As explained in the theorem of measurement, one measures the distance from $A \in \Sigma_1$ to the cut point. Thus one can construct a sequence of approximations to the cut point: $L_s \in \Sigma_1$ and $U_s \in \Sigma_2$ such that

$$|AL_s| = n - 1 + \frac{d_1}{2} + \frac{d_2}{4} + \frac{d_3}{8} + \cdots + \frac{d_s}{2^s}$$

and $|L_s U_s| = 2^{-s}$ for $s = 0, 1, 2, \dots$. Cantor's principle of boxed intervals yields a point X such that $L_s * X * U_s$, which is the cut point. □

Hilbert's axiom of completeness

The following axiom of completeness is suggested in the millenium edition of Hilbert's foundations:

V.2 (Axiom of linear completeness) An extension of a set of points on a line, with its order and congruence relations existing among the original elements as well as the fundamental properties of line order and congruence that follow from Axioms I-III and from V.1, is impossible.

Remark. The insight that it is enough to require the extension of a set of points on a line goes back to Paul Bernays.

(Theorem of completeness, Hilbert's Proposition 32). *The elements of geometry—the points, lines and planes— have no extension to any larger system, under the assumptions that the axioms of incidence, order, congruence, and the Archimedean axiom still hold for the extension.*

The parallel axiom may be assumed or not, it does not interact with completeness at all. The axiom of completeness is not a consequence of the Archimedean axiom. But, on the contrary, the Archimedean axiom *needs to be assumed* for completeness to be meaningful to hold.

Main Theorem 15. *Together, the Archimedean axiom (V.1), and the axiom of linear completeness (V.2) imply Cantor's axiom and Dedekind's axiom.*

There exist infinitely many models for the axioms I through IV, and V.1. But only one model satisfies the axiom of completeness, too—this the Cartesian geometry.

Historic context. In the very earliest edition, Hilbert proposed the conclusion of the Theorem of completeness (II.8.2) as an axiom. Only the German edition of the foundation of 1903 contains the axiom of linear completeness. Even earlier, the axiom of linear completeness had already appeared in May 1900 in the French translation of the foundation. It appeared for the very first time on October 12th 1899, in Hilbert's lecture "Über den Zahlbegriff".

Hilbert's axiom of completeness has given rise to many positive as well as negative comments by important mathematicians.

"An axiom about axioms with a complicated logical structure" (Schmidt)

"An unhappy axiom" (Freudenthal)

"The axioms of continuity are introduced by Hilbert, to show that they are really unnecessary." (Freudenthal)

Hilbert's completeness axiom is obviously not a geometric statement, and not a statement formalizable in the language used previously—so what does it accomplish? (M.J. Greenberg, 2010)

"The foundations of geometry contain more than insight in the nature of axiomatic." (Freudenthal)

"The most original creation in Hilbert's axiomatic" (Baldus)

"Hilbert has made the philosophy of mathematics take a long step in advance." (H. Poincaré)

II.9 Legendre's Theorems

Recall that a Hilbert plane is a geometry, where the axioms of incidence, order, and congruence are assumed, as stated in Hilbert's Foundations of Geometry. Neither the axioms of continuity (Archimedean axiom and the axiom of completeness), nor the parallel axiom need to hold for an arbitrary Hilbert plane. The proposition numbering is taken from Hilbert's Foundations of Geometry. For simplicity, I am using the letter R to denote a right angle.

II.9.1 The First Legendre Theorem

Proposition II.9.1 (The First Legendre Theorem, Hilbert's Proposition 35). *In a Hilbert plane, for which the Archimedean Axiom holds, every triangle has angle sum less or equal two right angles.*

Plan of proof. The proof relies on three ideas:

(a) The exterior angle theorem Euclid I.16 given by Proposition II.7.40, from which one concludes Euclid I.17 (Proposition II.7.41): the sum of any two angles of a triangle is less than two right angles.

(b) The construction from Euclid's proof II.7.9 of the exterior angle theorem. From a given triangle, this construction yields a second triangle, with the same angle sum as the first one, as explained in Lemma II.9.1. Additionally, one of its angles is less or equal half of an angle of the original triangle.

(c) The Archimedean property for angles given in Hilbert's Proposition II.8.1.

□

Lemma II.9.1. *For any given $\triangle ABC$, there exists a $\triangle A'B'C'$ such that*

$$\alpha' + \beta' + \gamma' = \alpha + \beta + \gamma \quad \text{and} \quad \alpha' \leq \frac{\alpha}{2} \tag{II.9.1}$$

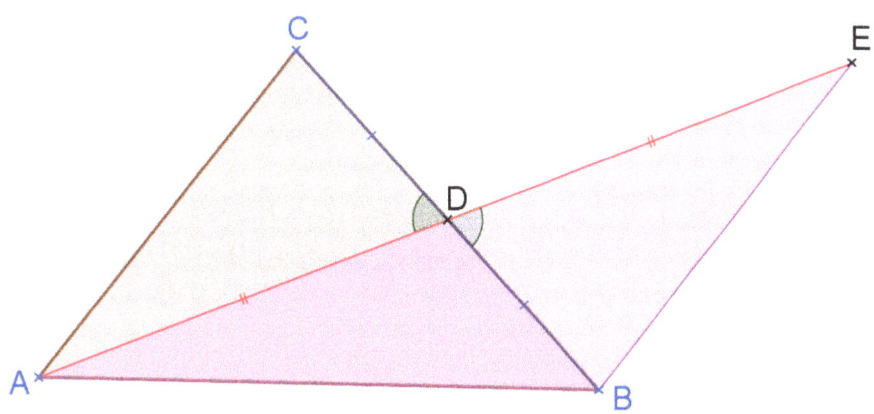

Figure II.9.1: Two triangles with same angle sum—and area.

Proof of the Lemma. Let D be the midpoint of side BC. Extend the ray $\overrightarrow{AD}$ and transfer segment AD to get point E such that $AD \cong DE$. We need to consider two cases:

(i) If $\angle EAB \leq \angle CAE$, the new $\triangle A'B'C'$ has vertices $A' = A, B' = B$ and $C' = E$.

(ii) If $\angle EAB > \angle CAE$, the new $\triangle A'B'C'$ has vertices $A' = A, B' = C$ and $C' = E$.

We explain the details for case (i). By SAS congruence, $\triangle ADC \cong \triangle EDB$, because the vertical angles at vertex D are congruent, and the adjacent sides are pairwise congruent by construction. The congruence of the two triangles yields two pairs of congruent angles

$$\gamma = \angle ACD \cong \angle DBE \qquad \text{and} \qquad \gamma' = \angle DEB \cong \angle DAC$$

From angle addition at vertices A and B, respectively, and a final addition of formulas, one concludes

$$\begin{aligned} \alpha &= \alpha' + \gamma' \\ \beta + \gamma &= \beta' \\ \alpha + \beta + \gamma &= \alpha' + \beta' + \gamma' \end{aligned}$$

A similar result is concluded in the second case (ii) via the congruence $\triangle ADB \cong \triangle EDC$.

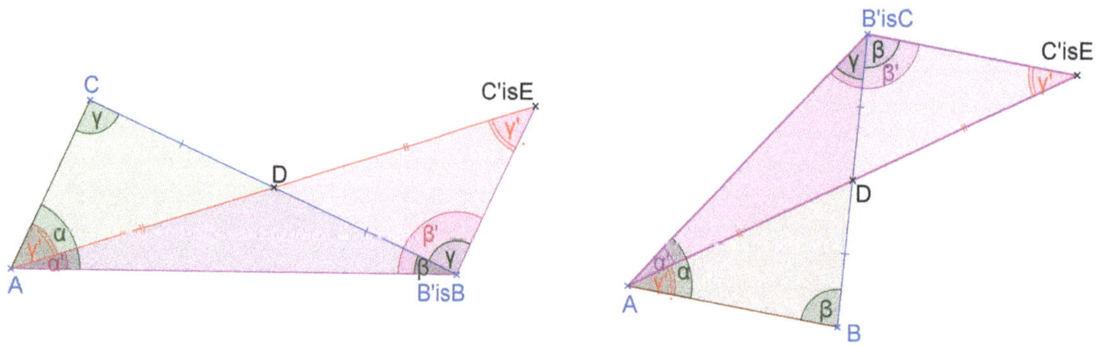

Figure II.9.2: Two triangles with same angle sum, cases (i) and (ii).

$\square$

Remark. Equivalent conditions all leading to case (i) are

$$\beta \leq \gamma \Leftrightarrow AC \leq AB \Leftrightarrow EB \leq AB \Leftrightarrow \alpha' \leq \gamma'$$

Reason. The first equivalence is proved by Euclid I.18 and Euclid I.19 (Propositions II.7.48 and II.7.49) for triangle $\triangle ABC$. Moreover $AC \cong EB$ because of the triangle congruence $\triangle ADC \cong \triangle EDB$. The third equivalence is proved by Euclid I.18 and Euclid I.19 (Propositions II.7.48 and II.7.49) for triangle $\triangle ABE$. $\square$

Corollary 18. *The two triangles $\triangle ABC$ and $\triangle A'B'C'$ have the same area.*

Reason. To produce the new triangle $\triangle A'B'C'$ from the old triangle $\triangle ABC$, one needs to take away triangle $\triangle ADC$ and add the congruent triangle $\triangle EDB$. $\square$

End of the proof of the First Legendre Theorem. Let $\triangle ABC$ be any triangle. We use the first Lemma repeatedly to get a sequence of triangles

$$\triangle A'B'C' = \triangle A_1B_1C_1\,,\ \triangle A_2B_2C_2\,,\ \triangle A_3B_3C_3\,,\dots \triangle A_nB_nC_n\,,\dots$$

such that $\alpha+\beta+\gamma = \alpha_n+\beta_n+\gamma_n$ and $\alpha_n \le \frac{\alpha}{2^n}$ for all natural numbers $n \ge 0$. The exterior angle theorem implies $\beta_n+\gamma_n < 2R$ and hence

$$\alpha+\beta+\gamma = \alpha_n+\beta_n+\gamma_n < \frac{\alpha}{2^n} + 2R \tag{II.9.2}$$

for all $n \ge 0$. Now a limiting process with $n \to \infty$ implies the result.

An accurate version of this part of the argument uses the Archimedean property for angles. We argue by contradiction, assuming that the angle sum would be $\alpha+\beta+\gamma > 2R$. Let

$$\varepsilon := \alpha+\beta+\gamma - 2R \tag{II.9.3}$$

which, because of the assumption $\alpha+\beta+\gamma > 2R$, would be an angle $\varepsilon > 0$. By the Archimedean property for angles Proposition II.8.1, there would exist a natural number r such that

$$\frac{\alpha}{2^r} < \varepsilon \tag{II.9.4}$$

Now (II.9.2), (II.9.4) and (II.9.3) together would imply

$$\alpha+\beta+\gamma = \alpha_r+\beta_r+\gamma_r < \frac{\alpha}{2^r} + 2R \le \varepsilon + 2R = \alpha+\beta+\gamma$$

which is impossible. Because any two angles are comparable, $\alpha+\beta+\gamma \le 2R$ is the only possibility left, as was to be shown. □

II.9.2 The Second Legendre Theorem

Proposition II.9.2 (The Second Legendre Theorem, Hilbert's Proposition 39). *Given is any Hilbert plane. If one triangle has angle sum $2R$, then every triangle has angle sum $2R$.*

It turns out to be easier, to prove at first a Proposition about quadrilaterals. We define some special quadrilaterals.

Definition II.9.1. A *Saccheri quadrilateral* $\square ABCD$ has right angles at two adjacent vertices A and B, and the opposite sides $AD \cong BC$ are congruent. A *Lambert quadrilateral* has three right angles. A *rectangle* has four right angles.

Proposition II.9.3 (Saccheri and Lambert quadrilaterals, Hilbert's Proposition 36). *Assume that $\square ABCD$ is a Saccheri quadrilateral with right angles at vertices A and B—and congruent opposite sides $AD \cong BC$. Let M be the midpoint of segment AB. The perpendicular bisector m of segment AB intersects the opposite side CD at right angles, say at point N. One gets two congruent Lambert quadrilaterals $\square AMND$ and $\square BMNC$.*

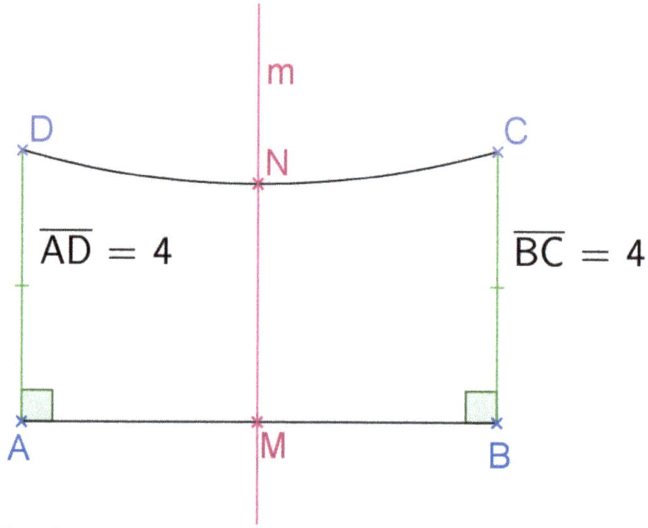

Figure II.9.3: A Saccheri quadrilateral is bisected into two Lambert quadrilaterals.

Figure II.9.4: The steps to get symmetry of a Saccheri quadrilateral.

Proof. To show that m intersects the opposite side CD, one can use plane separation with line m. Indeed, points A and B lie on different sides of m by construction. By Euclid I.27, the three lines AD and m and BC are parallel. Hence points A and D lie on the same side of m. By the same reasoning, B and C lie on the same side of line m. Hence C and D lie on different sides of m. Thus segment DC intersects m. By SAS congruence, $\triangle MAD \cong \triangle MBC$, since the right angles at vertices A and B and the adjacent sides match pairwise. Next we see that the triangles $\triangle MDN \cong \triangle MCN$, again by SAS congruence, since the angles at the common vertex M and the adjacent sides match. Hence we conclude that $\angle ADM \cong \angle BCM$ and $\angle MDN \cong \angle MCN$. Now angle addition yields

$$\angle ADC \cong \angle BCD$$

Hence the quadrilateral $ABCD$ has two congruent angles at vertices C and D. At vertex N, the angles $\angle MND \cong \angle MNC$ are congruent supplementary angles. Hence they are both right angles. This finishes the proof that $\square AMND$ and $\square BMNC$ are congruent Lambert quadrilaterals. $\square$

Definition II.9.2. The *reflection by line l* is defined as follows: From a given point P the perpendicular is dropped onto l, and extended by a congruent segment $FP' \cong PF$ beyond the foot point F. Then P' is called the reflected point of P.

Corollary 19. *By the perpendicular bisector of its base line, a Saccheri quadrilateral is bisected into two Lambert quadrilaterals, which are reflection symmetric to each other.*

Proposition II.9.4 (Hilbert's Proposition 37). *Assume* $\square ABCD$ *is a rectangle. Drop the perpendicular from a point* E *of line* $\overleftrightarrow{CD}$ *onto the opposite side* AB*. Let* F *be the foot point. Then the quadrilaterals* $\square ADEF$ *and* $\square BCEF$ *are both rectangles.*

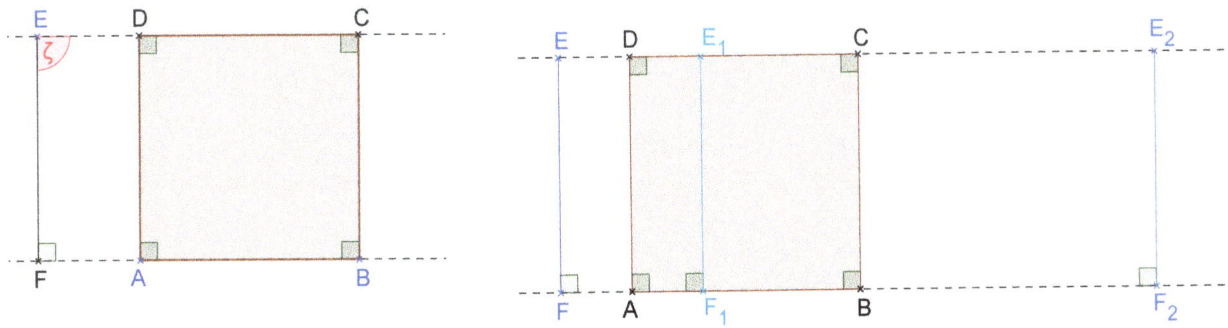

Figure II.9.5: Getting rectangles of arbitrary width— Is ζ a right angle?

Proof. One reflects the segment EF, both by line $\overleftrightarrow{AD}$ and $\overleftrightarrow{BC}$. Let E_iF_i for $i = 1, 2$ be the mirror images. By Proposition II.9.3, both segments are congruent to EF. Because $\square ABCD$ was assumed to be a rectangle, both points E_i lie on line CD and both F_i lie on line AB. The assumptions of Proposition II.9.3 hold for all three quadrilaterals $\square EFF_1E_1, \square EFF_2E_2$ and $\square E_1F_1F_2E_2$.

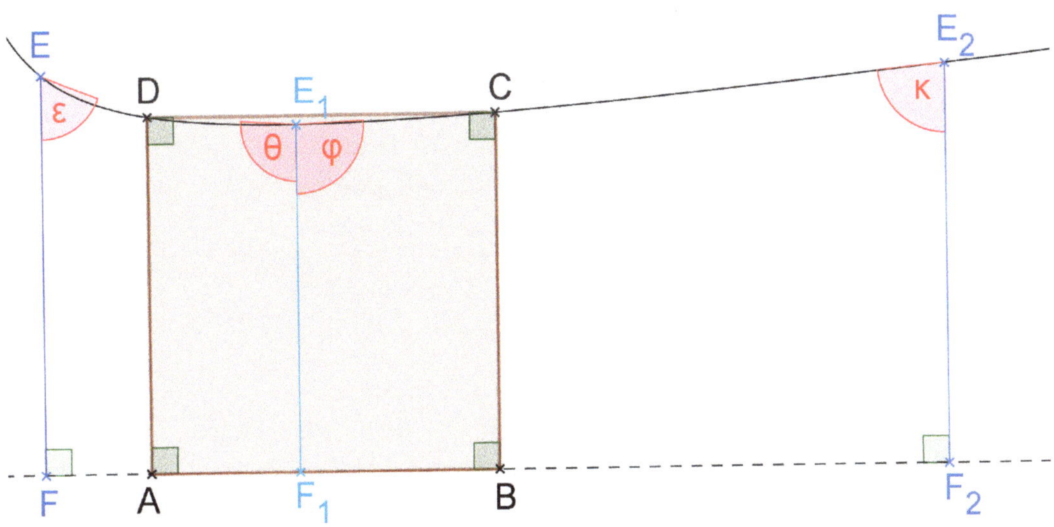

Figure II.9.6: ε is congruent to both θ and φ, hence there are two supplementary right angles θ and φ!

Hence all three are Saccheri quadrilaterals. We get four congruent angles with vertices E_1, E_2 and E—denoted in figure II.9.2 by $\varepsilon, \theta, \varphi$ and κ.

One of these three points E_1, E_2 and E lies between the other two. At that vertex, one gets a pair of supplementary angles, which are congruent. Hence they are right angles, and the other angles just mentioned are right angles, too. $\square$

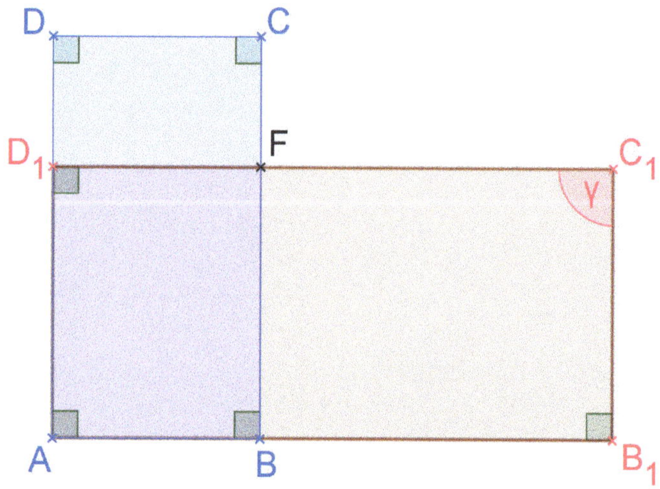

Figure II.9.7: If one rectangle exists, why are all Lambert quadrilaterals rectangles?

Proposition II.9.5 (Hilbert's Proposition 38). *If there exists a rectangle $\square ABCD$, then every Lambert quadrilateral is a rectangle.*

Proof. Assume that the Lambert quadrilateral $A_1B_1C_1D_1$ has its three right angles at vertices A_1, B_1 and D_1. One may assume that $A = A_1$, and the three points A, B, B_1 as well as A, D, D_1 lie on a line. We have drawn the case that $A * B * B_1$ and $A * D_1 * D$. The easy modification of the proof to the other possible orders of points A, B, B_1 and A, D_1, D is left to the reader. As in Proposition 36, we show that segment D_1C_1 and line BC do intersect, say at point F. Now Proposition 37 implies that these two lines are perpendicular to each other. Hence $\square ABFD_1$ is a rectangle. Applying Proposition 37 a second time, we conclude that the quadrilateral $\square AB_1C_1D_1$ is a rectangle, too. □

Corollary 20. *If there exists a rectangle $\square ABCD$, then every Saccheri quadrilateral is a rectangle. There exist rectangles of arbitrarily prescribed height and width.*

Reason. Because of Proposition II.9.3, a Saccheri quadrilateral is subdivided into two Lambert quadrilaterals by its midline. By Proposition II.9.5, those are rectangles. Hence the Saccheri quadrilateral is a rectangle, too. The height and width of a Saccheri quadrilateral can be prescribed arbitrarily, and they are all rectangles. Hence there exists rectangles of arbitrarily prescribed height and width. □

Note of caution. In hyperbolic geometry, the height and width of a Lambert quadrilateral cannot be prescribed arbitrarily. Indeed if one chooses the width for a Lambert quadrilateral, the height has an upper bound.

Proposition II.9.6 (Hilbert's Proposition 39). *To every $\triangle ABC$ with angle sum 2ω, there corresponds a Saccheri quadrilateral with two top angles ω.*

Proof. The Saccheri quadrilateral is $\square ABGF$, with right angles at F and G and congruent opposite sides $AF \cong BG$, and congruent angles ω at A and B.

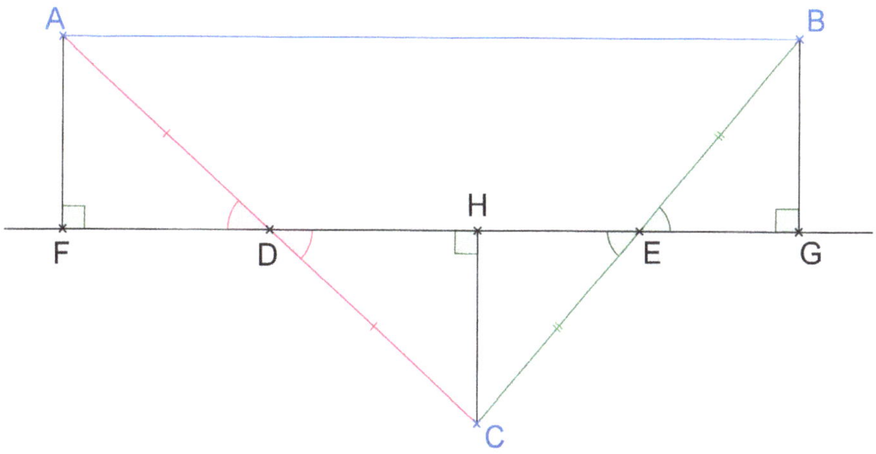

Figure II.9.8: To every triangle corresponds a Saccheri quadrilateral.

The drawing indicates, how the Saccheri quadrilateral is obtained. One connects the midpoint D of segment AC and midpoint E of segment BC, by line l. Then one drops the perpendiculars from all three vertices A, B, C onto line l. Let F, G and H be the foot points of the perpendiculars, respectively.

By the SAA congruence theorem, $\triangle AFD \cong \triangle CHD$, because of the right angles at vertices F and H, the congruent vertical angles at vertex D, and because segments $AD \cong DC$ are congruent by construction. By similar reasoning we get $\triangle BGE \cong \triangle CHE$. From these triangle congruences

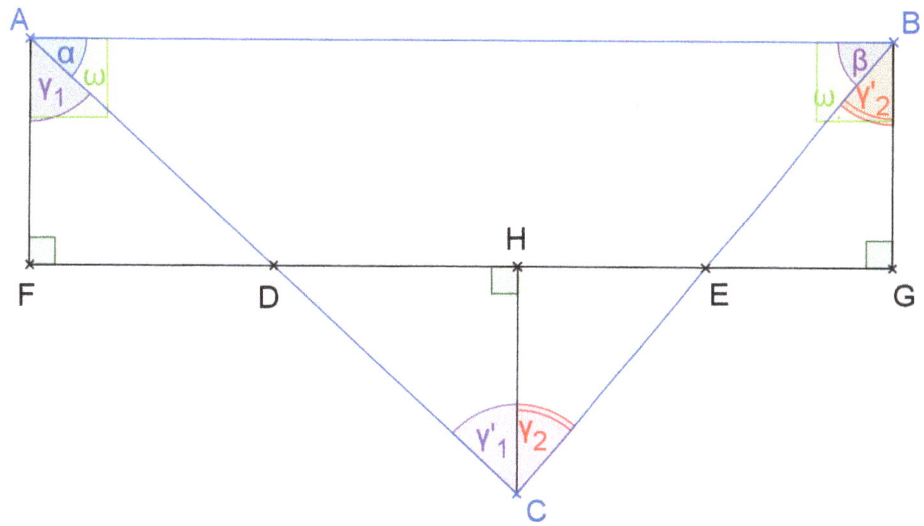

Figure II.9.9: How to get two pairs of congruent triangles, and how to get the top angles.

we conclude that $HC \cong FA$, and $HC \cong GB$. Hence $FA \cong GB$, which implies that $\square ABGF$ is a Saccheri quadrilateral.

Its top angles at vertices A and B were shown to be congruent in Proposition 36. We denote them by ω. From the triangle congruences, too, we get $\gamma_1 = \angle DCH \cong \angle DAF$ and $\gamma_2 = \angle ECH \cong \angle EBG$. The sum of the angles of $\triangle ABC$ is

$$\alpha + \beta + \gamma = \alpha + \gamma_1 + \beta + \gamma_2 = \angle FAB + \angle GBA = 2\omega$$

as to be shown. □

Corollary 21. *A triangle has angle sum $2R$ if and only if the corresponding Saccheri quadrilateral is a rectangle.*

Proof. This is an immediate special case of Proposition II.9.6. □

End of the proof of the Second Legendre Theorem. Assume triangle $\triangle ABC$ has angle sum $2R$.

Question. What can you say about the corresponding Saccheri quadrilateral?

Answer. The Saccheri quadrilateral corresponding to $\triangle ABC$ is a rectangle by the Corollary.

Hence, by Proposition 38, every Lambert quadrilateral is a rectangle. Now let $\triangle XYZ$ be any triangle.

Question. What can you say about the corresponding Saccheri quadrilateral?

Answer. As explained in Proposition II.9.3, the Saccheri quadrilateral corresponding to $\triangle XYZ$ is bisected into two congruent Lambert quadrilaterals. By Proposition II.9.5, these are both rectangles, since a rectangle already exists. Hence the Saccheri quadrilateral corresponding to $\triangle XYZ$ is a rectangle.

We have just shown that the Saccheri quadrilateral corresponding to the arbitrarily chosen $\triangle XYZ$ is a rectangle, too. Hence by Proposition II.9.6, the angle sum of $\triangle XYZ$ is $2R$. □

II.9.3 The alternative of two geometries

Finally, one wants to link the angle sum of triangles to uniqueness of parallels.

Proposition II.9.7. *If a triangle $\triangle ABC$ has angle sum $\alpha + \beta + \gamma < 2R$, then there exist two parallels to line AB through point C.*

The (equivalent) contrapositive: if there is a unique parallel to line AB through point C, then the triangle $\triangle ABC$ has angle sum $2R$.

Corollary 22. *In any Pythagorean plane, each triangle has angle sum $2R$. In other words, each Pythagorean plane is semi-Euclidean.*

Proof. We form two congruent z-angles $\alpha = \angle BAC$ by transferring that angle with one side ray $\overrightarrow{CA}$ as transversal. The second side m_1 of the new angle α has to lie the opposite side of $\overrightarrow{CA}$ as point B. Similarly one transfers angle $\beta = \angle ABC$ with one side ray $\overrightarrow{CB}$, and gets as second side the ray m_2.

The angle formed by the rays m_1 and m_2 is $\alpha + \beta + \gamma < 2R$. Hence m_1 and m_2 do not lie on the same line. On the other hand, the rays m_1 and m_2 are both (parts of) parallels to line l through point P. This follows from Euclid's I.27 (Alternate interior angles imply parallels). We have thus constructed two different parallels to line $l = AB$ through point C. □

Proposition II.9.8 (The Third Legendre Theorem). *Given is a Hilbert plane, for which the Archimedean Axiom is assumed to hold. If the angle sum of every triangle is $2R$, then the Euclidean Parallel Postulate holds.*

Idea of the proof. Given is line l and a point P not on l. We need to check the uniqueness of the parallel to line l through point P.

Inductively, there is constructed a sequence of isosceles triangles $\triangle PF_{n-1}F_n$. To start, let $F_0 = F$, and let F_1 be any of the two points on line l such that $F_0F_1 \cong PF$. Next let F_2 be the point on ray $\overrightarrow{FF_1}$ such that $F_1F_2 \cong PF_1$. Inductively, assume that $F_1, F_2, \ldots, F_{n-1}$ have been constructed, and let F_n is the point on ray $\overrightarrow{FF_1}$ such that $F_{n-1}F_n \cong PF_n{-}1$ and $F * F_{n-1} * F_n$.

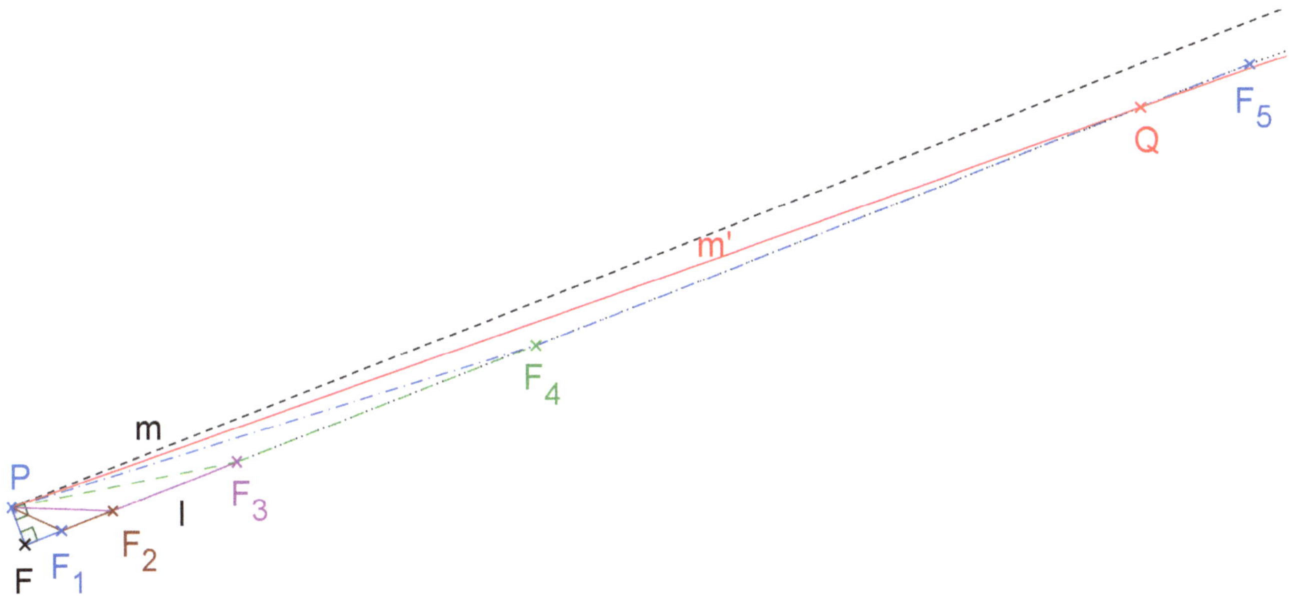

Figure II.9.10: The angle sum $2R$ implies uniqueness of the parallel.

The angles $\phi_n = \angle F_{n-1}PF_n$ can all be calculated by means of the following

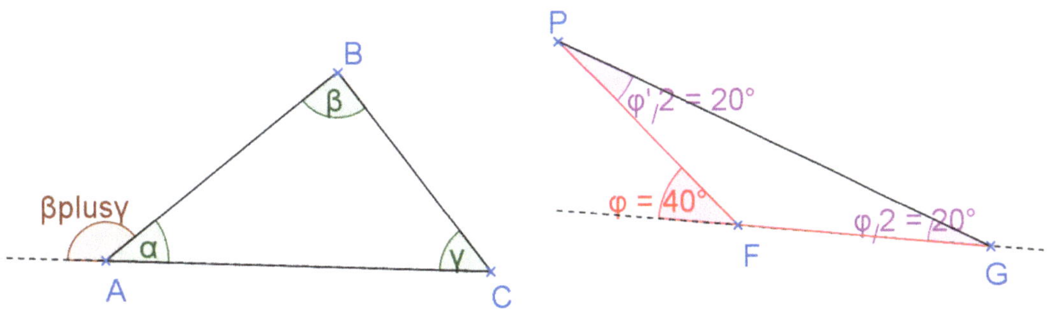

Figure II.9.11: Halfing an angle with an isosceles triangle.

Proposition II.9.9. *Assume that the angle sum for every triangle is $2R$. Then the base angle of an isosceles triangle is half of the exterior angle at the top.*

Reason for proposition II.9.9. Let δ be the exterior angle at the top vertex A of triangle $\triangle ABC$. Thus δ is the supplement of the interior angle α at that vertex. It was assumed that the angle sum of any triangle is $2R$. Hence

$$\delta = 2R - \alpha = \alpha + \beta + \gamma - \alpha = \beta + \gamma$$

which is the sum of the two nonadjacent interior angles. For an isosceles triangle with top $F = A$, the two base angles β and γ are congruent by Euclid I.5. Hence the exterior angle at the top is $\delta = 2\beta$ double the base angle. Hence the base angle $\beta = \frac{\delta}{2}$ is half of the exterior angle δ. □

End of the proof of the third Legendre Theorem. $\triangle F_0PF_1$ has a right angle on top. (It is a right isosceles triangle, for which all three angles can be calculated in Euclidean geometry.) Indeed the Proposition yields

$$\phi_0 = \angle F_0PF_1 = \frac{R}{2} \tag{1.1}$$

The other base angle of $\triangle F_0PF_1$ is the exterior angle on the top of $\triangle F_1PF_2$. Since this triangle is isosceles, too, the Lemma implies that its base angle is

$$\phi_1 = \angle F_1PF_2 = \frac{\phi_0}{2} = \frac{R}{4} \tag{1.2}$$

Inductively, we get that

$$\phi_n = \angle F_{n-1}PF_n = \frac{R}{2^n} \tag{1.n}$$

for all $n = 0, 1, 2, \ldots$. Here is the induction step:

Assume that (1.n-1) has been shown. The exterior angle on the top of $\triangle F_{n-1}PF_n$ is also a base angle of $\triangle F_{n-2}PF_{n-1}$, which is $\phi_{n-1} = \frac{R}{2^{n-1}}$ by the induction assumption. Since $\triangle F_{n-1}PF_n$ is isosceles, too, the Proposition implies that its base angle is half of that angle. Hence

$$\phi_n = \frac{\phi_{n-1}}{2} = \frac{R}{2^{n-1} \cdot 2} = \frac{R}{2^n}$$

which confirms (1.n). By angle addition, formulas (1.n) imply

$$\angle FPF_n = \phi_1 + \phi_2 + \cdots + \phi_n = R\left(\frac{1}{2} + \frac{1}{4} + \cdots + \frac{1}{2^n}\right) = R\left(1 - \frac{1}{2^n}\right) \tag{2}$$

One parallel m to l through point P is conveniently constructed as "double perpendicular", as explained above. We now give an argument to show that m is the *unique* parallel to l through P. We assume towards a contradiction that $m' \neq m$ is a different parallel to l through P. Let $\overrightarrow{m'}$ be one of the two rays on m' starting at point P that forms an acute angle with the perpendicular $\overrightarrow{PF}$. Let $\overrightarrow{m}$ be the ray on m starting at point P that lies on the same side of PF as $\overrightarrow{m'}$.

Legendre's triangle construction needs to be done on that side of PF where the two rays $\overrightarrow{m}$ and $\overrightarrow{m'}$ lie. Let $\varepsilon > 0$ be the angle between $\overrightarrow{m}$ and $\overrightarrow{m'}$. By the Archimedean axiom for angles, there exists a natural number r such that

$$\frac{R}{2^r} < \varepsilon$$

Hence the (complementary) angle between the rays $\overrightarrow{PF}$ and $\overrightarrow{m'}$ satisfies

$$R - \varepsilon < R - \frac{R}{2^r} = \angle FPF_r$$

This shows that the ray $\overrightarrow{m'}$ lies in the interior of $\angle FPF_r$. Hence, by the Crossbar Theorem II.6.9, the ray $\overrightarrow{m'}$ intersects the segment FF_r, say at point Q. Thus the line m' is not a parallel to line l, which is a contradiction. Hence there exists only one parallel to line l through point P. □

For completeness, we restate

Proposition II.9.10 (The Crossbar Theorem). *A segment with endpoints on the two sides of an angle and a ray emanating from its vertex into the interior of the angle intersect.*

Taking the three Legendre Theorems together one gets the

Main Theorem 16 (Alternative of Two Geometries). *In a Hilbert plane, for which the Archimedean Axiom is assumed, there occurs one of the two possibilities:*

(a) *All triangles have angle sum two right angles. All Lambert and all Saccheri quadrilaterals are rectangles. The Euclidean parallel postulate holds. One has arrived at the Euclidean geometry.*

(b) *All triangles have angle sum less than two right angles. All Lambert quadrilaterals have an acute angle. All Saccheri quadrilaterals have two congruent acute top angles. Rectangles do not exist. For every line l and point P not on l, there exist two or more parallels to line l through point P. In this case, one gets a semi-hyperbolic geometry.* [37]

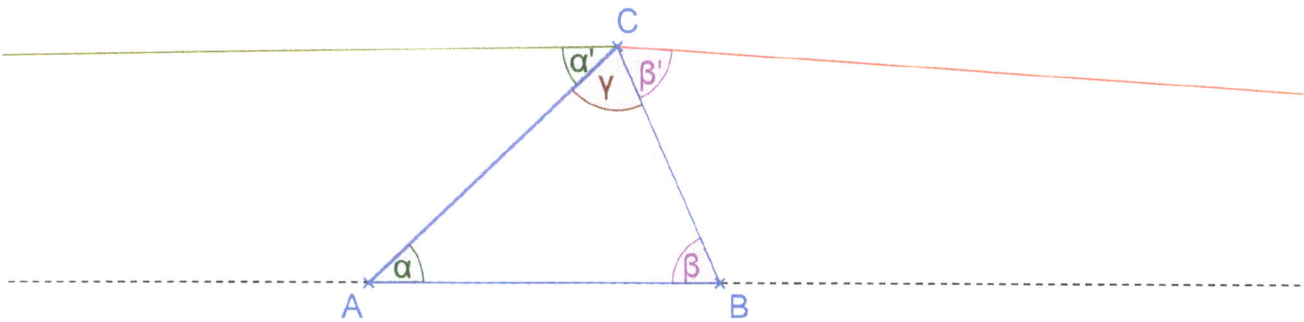

Figure II.9.12: For angle sum less $2R$, there exist two different parallels.

Reason. Pure logic tells that either (1) or (2) holds:

(1) All triangles have angle sum two right angles.

(2) There exists a $\triangle ABC$ with angle sum not equal than two right angles.

By the third Legendre Theorem, alternative (1) implies the Euclidean parallel postulate. By Proposition 39, we conclude that all Lambert and Saccheri quadrilaterals are rectangles. Thus case (a), the usual Euclidean geometry occurs.

Now we assume alternative (2), and derive all the conclusions stated in (b). By the first Legendre Theorem, the $\triangle ABC$ has angle sum less than two right angles. By the contrapositive of the second Legendre Theorem, we conclude that no triangle can have angle sum two right. Hence, again by the first Legendre Theorem, all triangles have angle sum less than two right angles. The statement about quadrilaterals follows from Proposition II.9.6. The Proposition below gives the well known construction, leading two different parallels. Thus all assertions of case (b) hold—we have arrived at the hyperbolic geometry. □

[37] The existence of the aymptotically parallel rays cannot be proved, therefore only *semi*-hyperbolic.

II.9.4 What is the natural geometry?

A legitimate question remains open at this point:

> *Is there a clear cut, suggestive or self-evident postulate that would better replace the Euclidean postulate?*

Legendre's contribution to this discussion is his investigation of the following postulate:

Definition II.9.3 (Legendre's Postulate). The exists an angle such that every point in its interior lies on a segment going from one side of the angle to the other one.

Proposition II.9.11 (The Fourth Legendre Theorem). *Given is a Hilbert plane, for which the Archimedean Axiom is assumed to hold. If Legendre's postulate holds, there exists a triangle with angle sum two right.*

By combining the result with Legendre's second Theorem, we conclude that every triangle has angle sum two right.

Corollary 23. *A Hilbert plane were the Archimedean Axiom and Legendre's postulate hold is semi-Euclidean.*

Proof. The proof relies on three ideas:

(a) The additivity of the defect of triangles.

(b) A construction doubling the defect of a triangle.

(c) The Archimedean property for angles given in Hilbert's Proposition 34.

Definition II.9.4 (Defect of a triangle). The *defect* of a triangle is the deviation of its angle sum from two right angles. We write
$\delta(ABC) = 2R - \alpha - \beta - \gamma$.

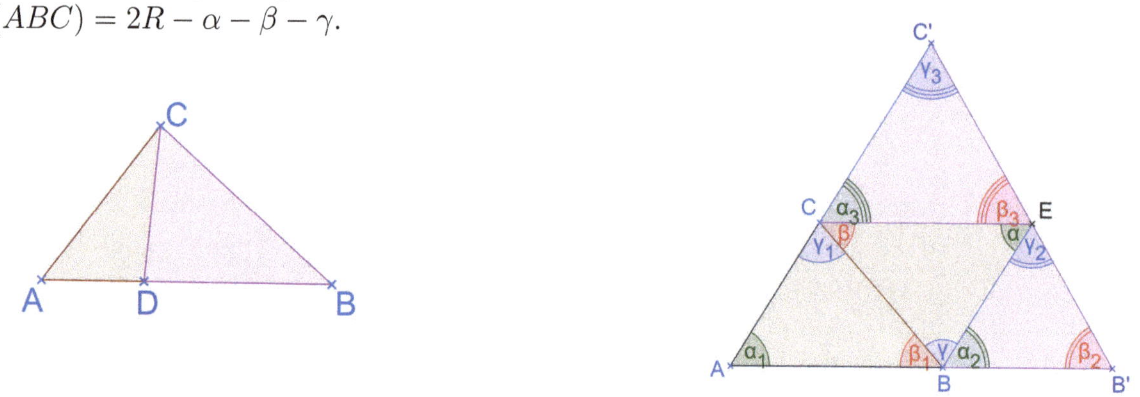

Figure II.9.13: The defect and the area of triangles are both additive.

Lemma II.9.2 (Additivity of the defect). *Let $\triangle ABC$ be any triangle and D be a point on segment AB. Then*

$$\delta(ABC) = \delta(ADC) + \delta(DBC)$$

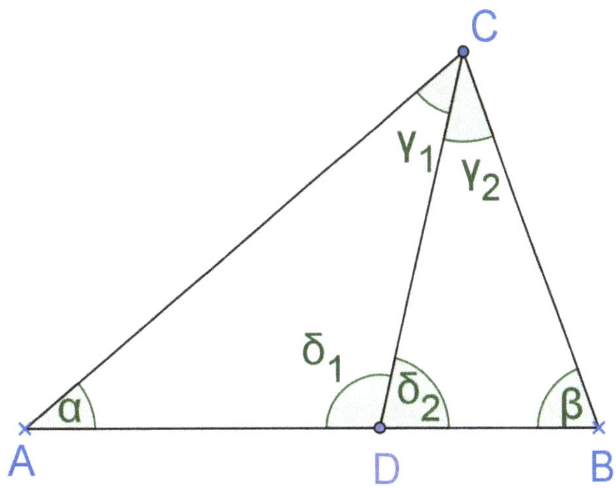

Figure II.9.14: Additivity of the defect.

Proof. With the notation of the figure on page 251, and the fact that δ_1 and δ_2 are supplementary angles.

$$\delta(ABC) = 2R - \alpha - \beta - \gamma = 2R - \alpha - \beta - \gamma_1 - \gamma_2$$
$$= (2R - \alpha - \gamma_1 - \delta_1) + (2R - \beta - \gamma_2 - \delta_2) = \delta(ADC) + \delta(DBC)$$

□

Suppose that a triangle is partitioned into four smaller triangles by three points lying on its sides. Then the defect of larger triangle is the sum of the defects of the four smaller triangles.

Lemma II.9.3 (Doubling the defect). *Given is a triangle $\triangle BAC$ where Legendre's postulate holds for the angle $\angle BAC$. Then there exists a triangle $\triangle B'AC'$, with the same vertex A and angle at A such that*

$$\delta(B'AC') \geq 2\,\delta(BAC) \quad \text{(II.9.5)}$$

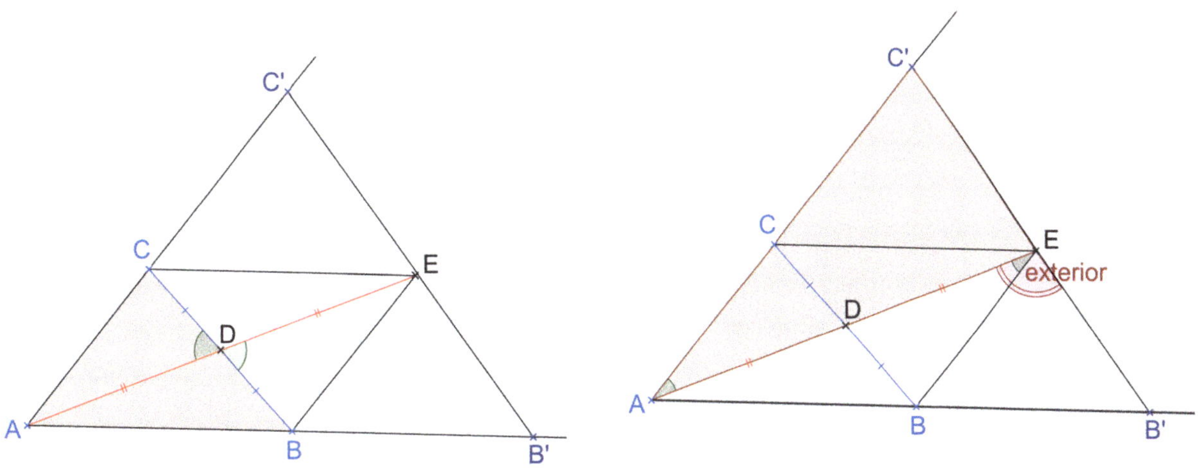

Figure II.9.15: Doubling the defect of a triangle.

Proof. One starts similarly to the procedure for the first Legendre Theorem. Let D be the midpoint of side BC, and extend the ray $\overrightarrow{AD}$ and transfer segment AD to get point E such that $AD \cong DE$. The congruences

$$\triangle ADC \cong \triangle EDB \quad \text{and} \quad \triangle ADB \cong \triangle EDC \tag{II.9.6}$$

are shown by SAS congruence. Next, the congruence

$$\triangle ABC \cong \triangle ECB \tag{II.9.7}$$

is shown by ASA congruence. By Legendre's postulate, there exists a point B' on ray $\overrightarrow{AB}$ and a point C' on ray $\overrightarrow{AC}$ such that point E lies on the segment $B'C'$.

We next claim point B lies between A and B'. This follows from the exterior angle theorem: Indeed, the triangle $\triangle AEC'$ has the interior angle $\angle EAC'$, which is less than a non adjacent exterior angle, thus

$$\angle EAC' < \angle AEB'$$

On the other hand, the triangle congruence (II.9.6) implies

$$\angle EAC' = \angle DAC \cong \angle BED = \angle AEB$$

Hence $\angle AEB < \angle AEB'$. Since B and B' lie on the same side of line AE, this implies that B lies between A and B'. Similarly, one shows that C lies between A and C'.

The larger triangle $\triangle B'AC'$ is partitioned into four smaller triangles by the segments between the three points B, E and C lying on its sides. The additivity and positivity of the defect, and finally the congruence (II.9.11) yield

$$\begin{aligned} \delta(AB'C') &= \delta(ABC) + \delta(ECB) + \delta(EBB') + \delta(ECC') \\ &\geq \delta(ABC) + \delta(ECB) = 2\,\delta(ABC) \end{aligned} \tag{II.9.8}$$

as to be shown. □

End of the proof of the Fourth Legendre Theorem. Let $\triangle ABC$ be such that Legendre's postulate holds for the angle $\angle BAC$. We use Lemma II.9.3 repeatedly to inductively get a sequence of triangles

$$\triangle AB'C' = \triangle AB_1C_1\,,\ \triangle AB_2C_2\,,\ \triangle AB_3C_3\,, \ldots \triangle AB_nC_n\,, \ldots$$

all with the same vertex A and angle $\angle BAC$. Since the defect has obviously the upper bound $2R$, one concludes

$$2R \geq \delta(AB_nC_n) \geq 2^n\,\delta(ABC)$$

and after dividing one gets

$$\delta(ABC) \leq \frac{2R}{2^n}$$

for all natural numbers n. Now the Archimedean property for angles implies $\delta(ABC) = 0$ and hence $\alpha + \beta + \gamma = 0$, as to be shown. □

Proof of the Corollary. If Legendre's postulate holds for one angle ε, one easily checks that it holds for any smaller angle, For the doubled angle 2ε, the postulate holds, too.

Given any other angle α, we use the Archimedean property. There exists a natural number r such that

$$\frac{\alpha}{2^r} < \varepsilon \tag{II.9.9}$$

Now we successively conclude that the Legendre postulate holds for the angles

$$\frac{\alpha}{2^r}, \ 2\frac{\alpha}{2^r}, \ldots, \ \frac{\alpha}{2}, \ \alpha$$

. Hence Legendre's postulate holds for every angle. By the proof above, we see that every triangle has angle sum two right. □

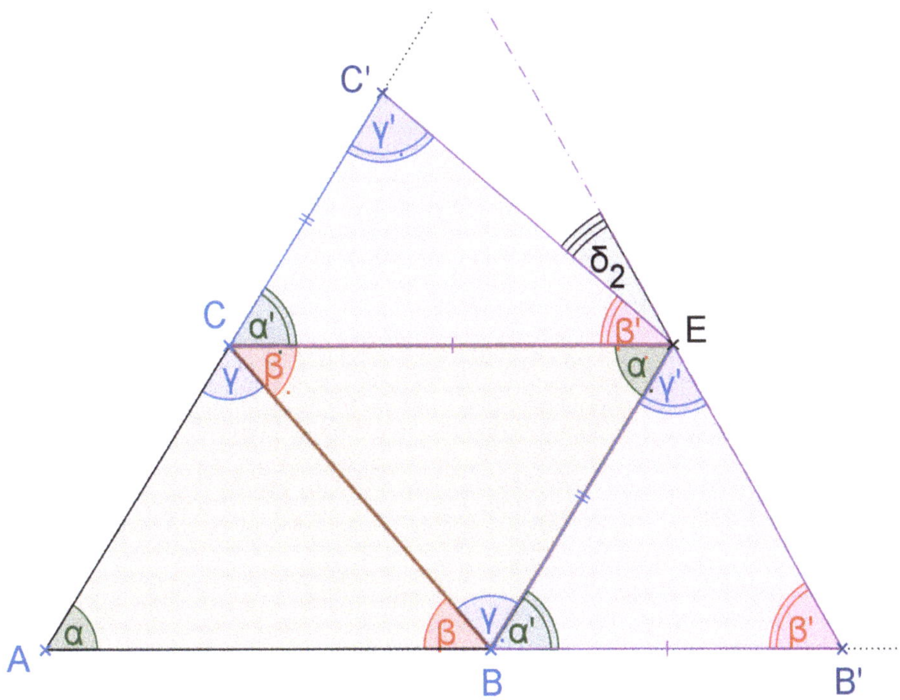

Figure II.9.16: Hyperbolic geometry produces an angle at vertex E.

Problem II.9.1. *Start a similar construction in hyperbolic geometry. Instead of relying on Legendre's postulate, it is natural to choose the points B' and C' on the two sides of angle $\angle BAC$ such that $BB' \cong CE$ and $CC' \cong BE$. Which congruences hold among the four triangles appearing in formula* (II.9.8)*? Show that the quadrilateral $\square AB'EC'$ is convex and that its total defect is* $2(2R - \angle B'EC')$.

Solution. As above,

$$\triangle ABC \cong \triangle ECB \tag{II.9.10}$$

is shown by ASA congruence and

$$\triangle BB'E \cong \triangle CEC' \tag{II.9.11}$$

is shown by SAS congruence. The total defect is

$$\delta(AB'EC') = \delta(ABC) + \delta(EBC) + \delta(EBB') + \delta(ECC') = 2\delta(ABC) + 2\delta(EBB')$$

Angle addition at vertex B yields $2R = \alpha' - \beta - \gamma$ and hence

$$\begin{aligned}\delta_2 := \frac{\delta(AB'EC')}{2} &= (2R - \alpha - \beta - \gamma) + (2R - \alpha' - \beta' - \gamma') \\ &= (2R - \alpha' - \beta - \gamma) + (2R - \alpha - \beta' - \gamma) \\ &= 2R - \alpha - \beta' - \gamma' = 2R - \angle B'EC'\end{aligned}$$

Half of the defect of quadrilateral $\square AB'EC'$ equals its exterior angle at vertex E. $\square$

/

II.10 Neutral Geometry of Circles and Continuity

Definition II.10.1 (Circle). A *circle* with center O through point R in a plane $\mathcal{A}$ is the set of all points P in the plane $\mathcal{A}$ such that $OR \cong OP$. A point P from the plane $\mathcal{A}$ lies *inside* the circle $\mathcal{C}$ if $OP < OR$. A point Q from the plane $\mathcal{A}$ lies *outside* the circle $\mathcal{C}$ if $OQ > OR$. The set of points inside a circle is called the *interior* or *disk* of the circle. The interior of circle $\mathcal{C}$ is denoted by $\operatorname{Int}(\mathcal{C})$.

II.10.1 Immediate consequences of neutral geometry

We begin by stating some properties of circles that follow simply from Hilbert's axioms. We make repeated use of the exterior angle theorem, its consequences, and the triangle inequality.

Proposition II.10.1.

(i) *A circle intersects a line in at most two points.*

(ii) *If a line contains points inside a circle, the foot point of the perpendicular dropped from the center onto the line lies inside the circle, or the line goes through the center.*

Question. How does the assumption that three points A, B and C lie both on a circle and a line lead to a contradiction. Give a detailed reasoning, based on Proposition II.7.46 item (iv).

Answer. Of any three points on a line, one lies between the two others (Theorem 4 in Hilbert's foundations). We can suppose that $A * B * C$. Let the isosceles $\triangle OAB$ and $\triangle OBC$ have base angles α and γ, respectively. Because these are the base angles of the isosceles $\triangle OAC$, too, they are congruent and acute. Congruent base angles $\alpha \cong \gamma$ occur as supplements at the middle vertex B. This would imply the base angles are right angles, which is impossible.

Proof of item (ii). One may assume that the center O does not lie on the line l, for otherwise the assertion is clear. We drop the perpendicular from O onto l. As assumed, let P be a point on the line l inside the circle. To show that the foot point F lies inside $\mathcal{C}$, consider the right triangle $\triangle OPF$. The right angle is the largest angle of any right triangle, as follows from the exterior angle theorem. Hence by Euclid I.19, the hypothenuse OP is the longest side. Hence $OF < OP < OR$ with R any point on circle $\mathcal{C}$. This shows that the foot point F lies inside circle $\mathcal{C}$. □

Proposition II.10.2. *All points between two distinct points inside or on a circle lie* <u>*inside*</u> *the circle.*

Proof. Let A and B be two points inside or on the circle. Let P be a point inside the segment AB. One of the two triangles $\triangle OAP$ or $\triangle OBP$ is obtuse or right, because of the supplementary angles at vertex P. Because a triangle can have only one obtuse angle, the side opposite to the obtuse angle is the longest side of a triangle. We conclude that either $OB > OP$ or $OA > OP$, which both mean that point P lies inside the circle.

The proof for the special case that segment AB goes through the center uses Hilbert's Proposition II.6.7. Any finite set of points on a line can be put into an ordered list such that their order relations hold according to the ordering of the list. Possible cases for such a list are $O * A * P * B$, $A * O * P * B$, $A * P * O * B$, $A * P * B * O$. In each of these cases, it is easy to see that point P lies inside the circle. □

Definition II.10.2 (Chord and diameter). The segment between two points of a circle is called a *chord* or *secant.* A chord through the center of the circle is called a *diameter.*

Corollary 24. *All points of a chord besides the endpoints lie inside the circle.*

Proposition II.10.3. *Let Q_1 and Q_3 be two points* <u>*outside*</u> *or on circle $\mathcal{C}$, and let F be the foot point of the perpendicular dropped from center O onto the line Q_1Q_3.*

- **(i)** *If the foot point lies on or outside the circle, then all other points of line Q_1Q_3 lie outside the circle.*
- **(ii)** *If the foot point F does not lie between the points Q_1 and Q_3, then all points inside the segment Q_1Q_3 lie outside the circle.*

Proof. To check item (i), consider the right triangle $\triangle OFQ$. The hypothenuse OQ is the longest side. Hence $OQ > OF$.

To check item (ii), let Q be a point in segment Q_1Q_3. The order of the four points is $F*Q_1*Q*Q_3$ or $F * Q_3 * Q * Q_1$. Take the first case. Angle $\angle OQ_1Q$ is obtuse, because it is supplementary to an acute angle of the right triangle $\triangle OFQ_1$. Because a triangle can have at most one obtuse angle, the side OQ opposite to the obtuse angle is the longest side. Hence $OQ > OQ_1$, and point Q lies outside the circle $\mathcal{C}$. □

Proposition II.10.4. *Suppose a circle and a line intersect in two points A and B. Recall that the segment between the two intersection points is called a chord.*

- **(iii)** *The foot point of the perpendicular dropped from the center to the line bisects the chord.*
- **(iv)** *The chord has at most the double length as the radius.*
- **(v)** *The points on the line AB outside the chord AB lie outside the circle, and the points inside the chord lie inside the circle.*

Proof. Item (iii) uses once more the figure on page 258. The two triangles $\triangle OFA$ and $\triangle OFB$ are congruent by the hypothenuse leg theorem. Hence $AF \cong BF$, confirming that the foot point F is the midpoint of segment AB. To confirm item (iv), note that the hypothenuse is the longest side of a right triangle. Hence $OA > AF \cong BF$, and $2OA > AF + FB \cong AB$. The equality occurs in the exceptional case that AB is a diameter. Item (v) is left to the reader. □

Definition II.10.3 (Short and long arc). Given is a circle with center O and a chord AB that is not a diameter. The *long arc* is the part of the circle lying in the *same* half plane of line AB as the center of the circle. The *short arc* is the part of the circle lying in the *opposite* half plane of line AB as the center of the circle.

Proposition II.10.5. *Given is a circle with center O and a chord AB that is not a diameter. A point S on the circle which lies in the interior of the central angle $\angle AOB$ lies on the short arc, and a point L on the circle which lies in the exterior of the central angle lies on the long arc of chord AB.*

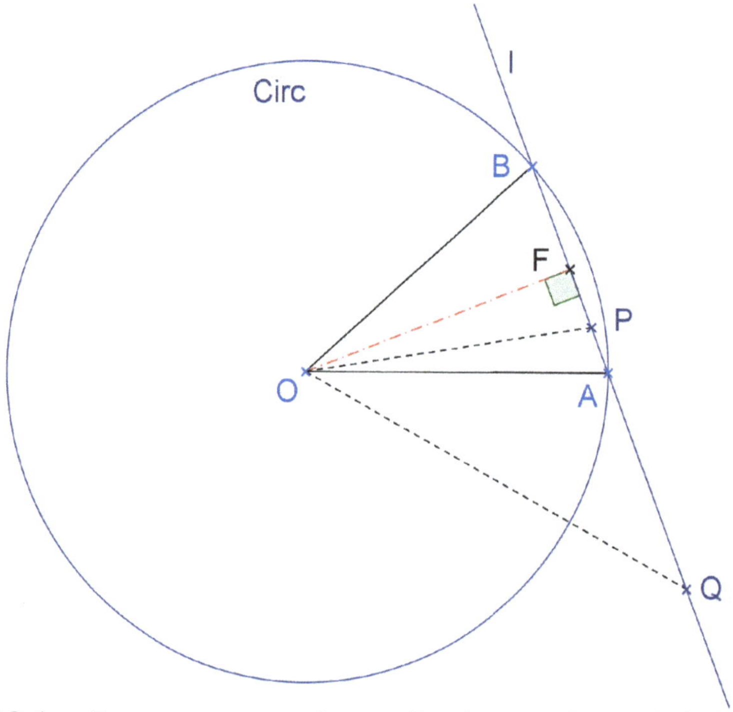

Figure II.10.1: Some statements about a line intersecting a circle in two points.

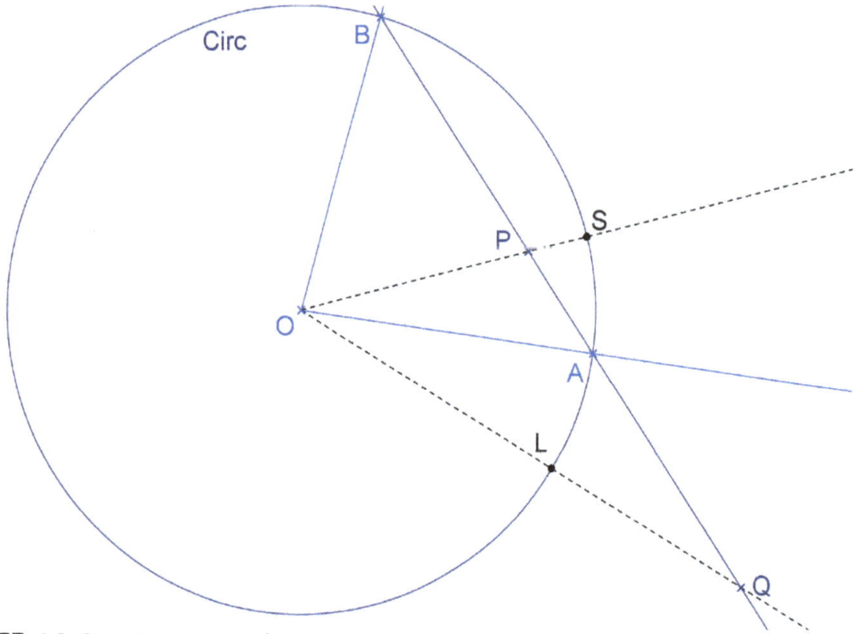

Figure II.10.2: The short/long arc lies in the interior/exterior of the central angle.

Proof. Take a point S lying at the intersection of the circle with the interior of angle $\angle AOB$. By the Crossbar Theorem, the ray $\overrightarrow{OS}$ intersects the segment AB. Let P be the intersection point. By the previous Proposition II.10.4 item (v), all points inside the segment AB lie inside the circle. Hence $OP < OA \cong OS$ and $O * P * S$. In other words, point S lie in the half-plane of line AB opposite to the center O, and hence on the short arc.

Take a point L lying at the intersection of the circle with the exterior of angle $\angle AOB$. Either the ray $\overrightarrow{OL}$ does not intersect the line AB, or there is an intersection point Q. In the first case, point L lie in the same half-plane of line AB as the center O, and hence on the long arc. In the second case, we use the previous Proposition II.10.4 item (v). We know that all points on the line AB outside the segment AB lie outside the circle. Hence $OP > OA \cong OL$ and $O * L * Q$. Once more, point L lie in the same half-plane of line AB as the center O, and hence on the long arc. □

Problem II.10.1 (A chord of given length). *Given is a circle $\mathcal{C}$ with center O, a point P inside the circle and a chord $A'B'$.*

(i) *Construct a chord congruent to $A'B'$ through point P.*

(ii) *Describe your construction.*

(iii) *Construct all solutions—none, one or two—and explain how you decide which case occurs. Draw examples for all three cases.*

Answer. **(i)** The construction of a chord $AB \cong A'B'$ through the point P is done in the figure on page 260.

(ii) Through the given point P, we draw a second circle $\mathcal{D}$ around O. Assume this circle intersects the chord $A'B'$, and let P' be an intersection point. The axial reflection which maps P to P' has the perpendicular bisector of PP' as axis of symmetry. For the construction of this axis, we draw two further circles around P and P', both through point O. They intersect at a second point Q, and the axis is the line $r = OQ$. The chord AB to be constructed is the mirror image of the given chord $A'B'$. To obtain AB, let Q be the intersection point of line $A'B'$ with the axis of symmetry r. The line QP intersects the given circle at points A and B.

(iii) The circle $\mathcal{D}$ can either intersect the chord $A'B'$ at two points, touch it at one point, or not intersect it at all. Depending on these cases, we get two, one or no solution of the construction problem.

Proposition II.10.6. *The perpendicular t to a radius OT, erected at point T, has only one point with a circle in common. The circle lies on one side of this line.*

Proof. By Proposition II.7.51, the foot point T of the perpendicular from O to the line t is the unique point on the line t, with shortest distance from O.

Hence all points $P \neq T$ on the tangent line t have distance $OP > OT$ from the center of the circle. Hence any radial segment OR on a line OP is shorter than segment OP. Because of $OR < OP$, these radial segments do not intersect the tangent t. By the plane separation theorem II.6.4, this fact implies that the center O and all points $R \neq T$ of the circle lie on the same side of tangent t. Too, we see that the tangent t has only the foot point T in common with the tangent line t. □

Definition II.10.4 (Tangent to a circle, an ad hoc definition). The perpendicular erected at a point of the circle to the radius is called the *tangent* of the circle at this point.

Proposition II.10.7. *A line which has a point in common with a circle, and is different from the tangent, intersects the circle in exactly two points.*

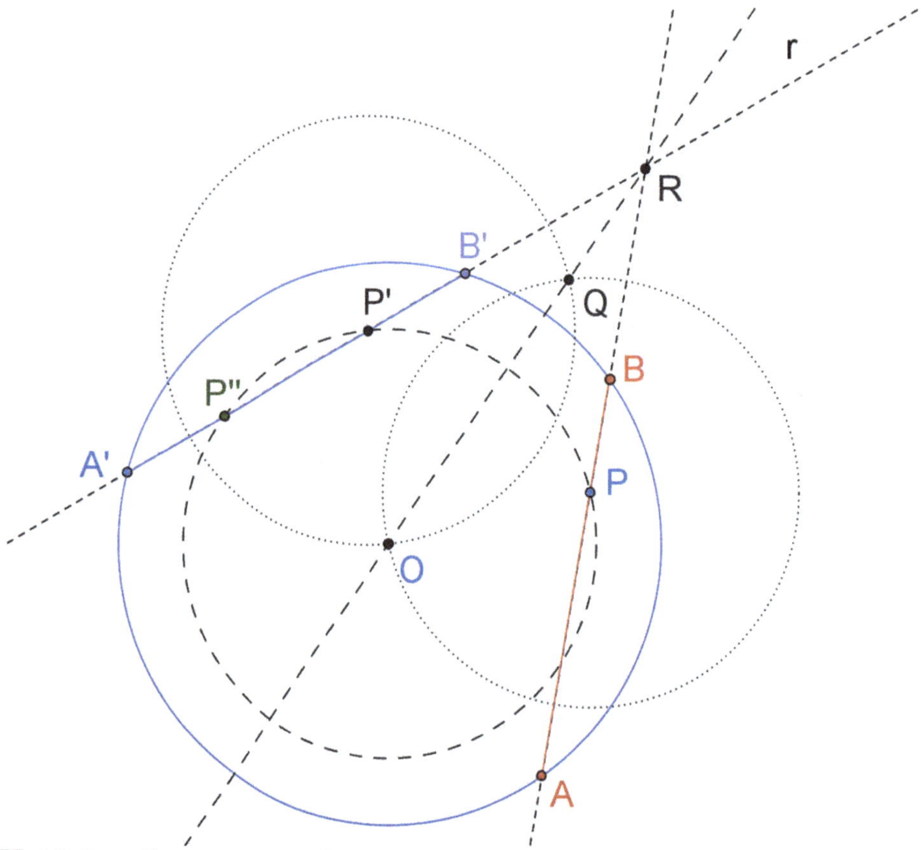

Figure II.10.3: Construction of a chord through the given point P of given length $A'B'$.

Proof. Take a circle $\mathcal{C}$ with center O, a point T on the circle, the perpendicular t erected at point T onto the radius OT, and a second line $l \neq t$. We drop the perpendicular from the center O onto line l, and let F be the foot point. Clearly $F \neq T$, since $l \neq t$. We transfer the segment TF to get a congruent segment $TF \cong FA$ with $T * F * A$. The two triangles $\triangle OFT \cong \triangle OFA$ are congruent by SAS congruence. Hence $OA \cong OT$, and point $A \neq T$ lies on the circle. Existence of a third intersection point has been ruled out above by item (v) from proposition II.7.46. □

Proposition II.10.8. *A line has only one point in common with a circle if and only if it is a tangent.*

Construction II.10.1 (A neutral construction of the tangent to a circle). *Let point A be the intersection of the segment OP with the given circle $\mathcal{C}$. At point A, we erect the perpendicular p onto segment OP. We need a second circle around O through point P. Let B be an intersection point of this second circle with the perpendicular p.*

Finally, we need an intersection point T of circle $\mathcal{C}$ with segment OB. This intersection point is the touching point of a tangent from point P to the given circle $\mathcal{C}$.

Validity of the Construction. The triangles $\triangle OAB \cong \triangle OTP$ are congruent. This can be seen by SAS congruence, because of the common angle $\angle AOB = \angle POT = \angle TOP$ and two pairs of congruent adjacent sides $OA \cong OT$ and $OB \cong OP$.

Hence we get congruent right angles $\angle OTP \cong \angle OAB = R$. Since point T lies on the circle $\mathcal{C}$, too, the segment OT is a radius of that circle. Since TP is perpendicular to the radius OT, it is a tangent of circle $\mathcal{C}$. □

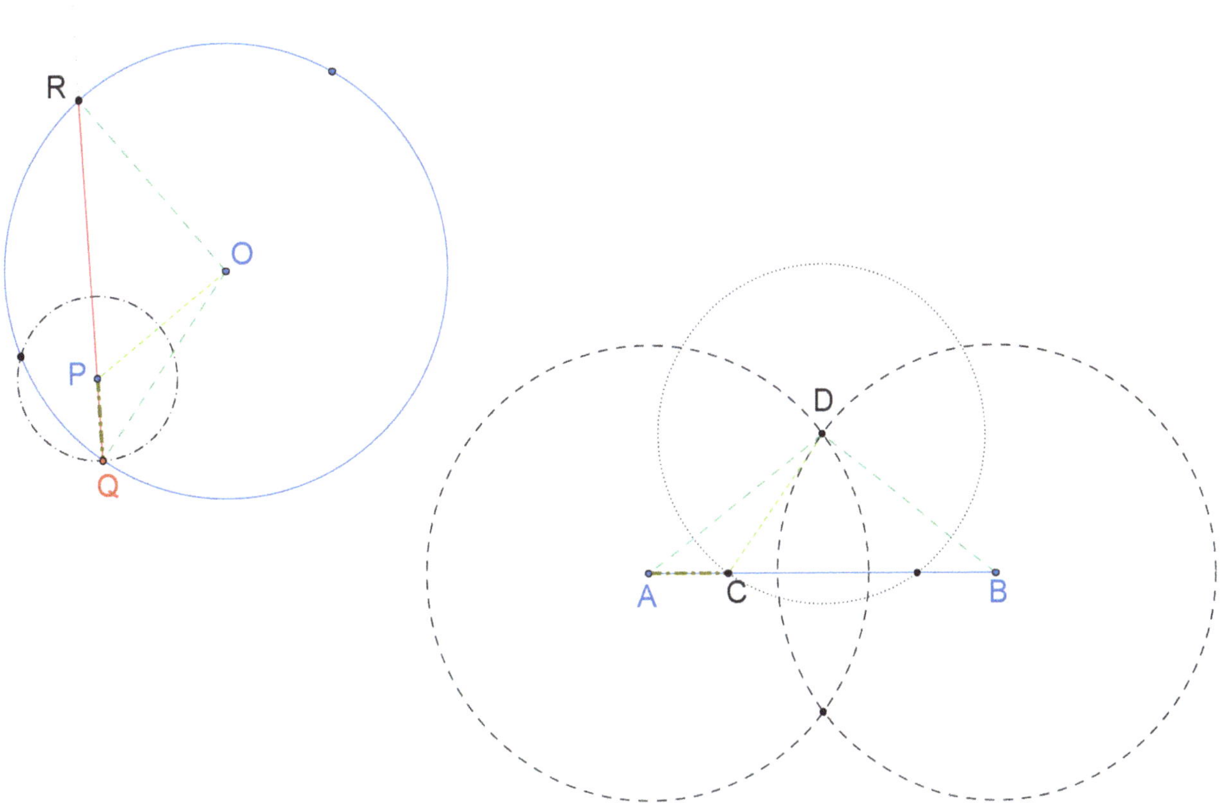

Figure II.10.4: Can you see why the construction works?

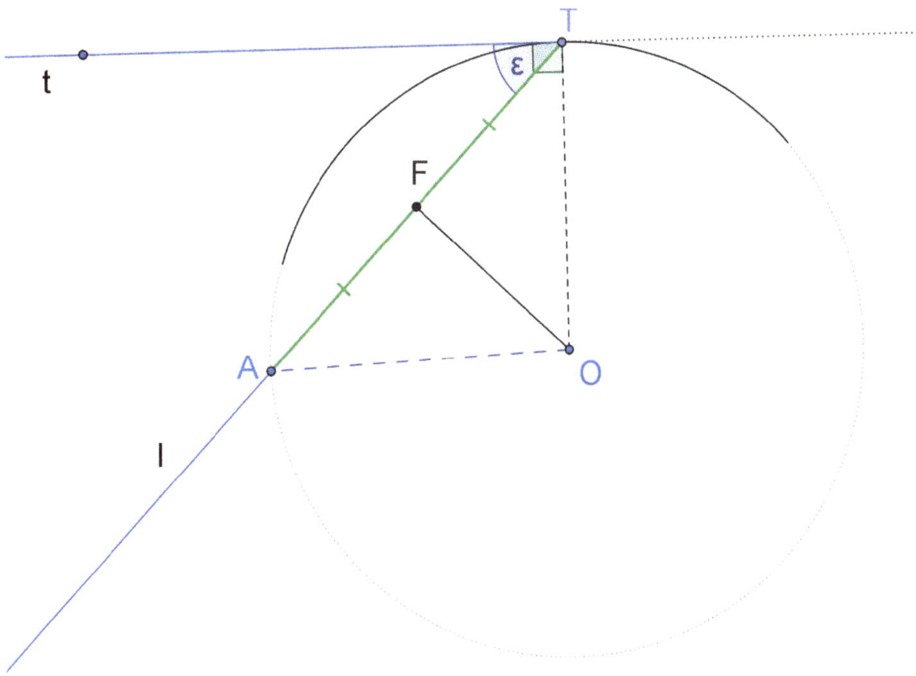

Figure II.10.5: How to get the second intersection point of a circle with a line.

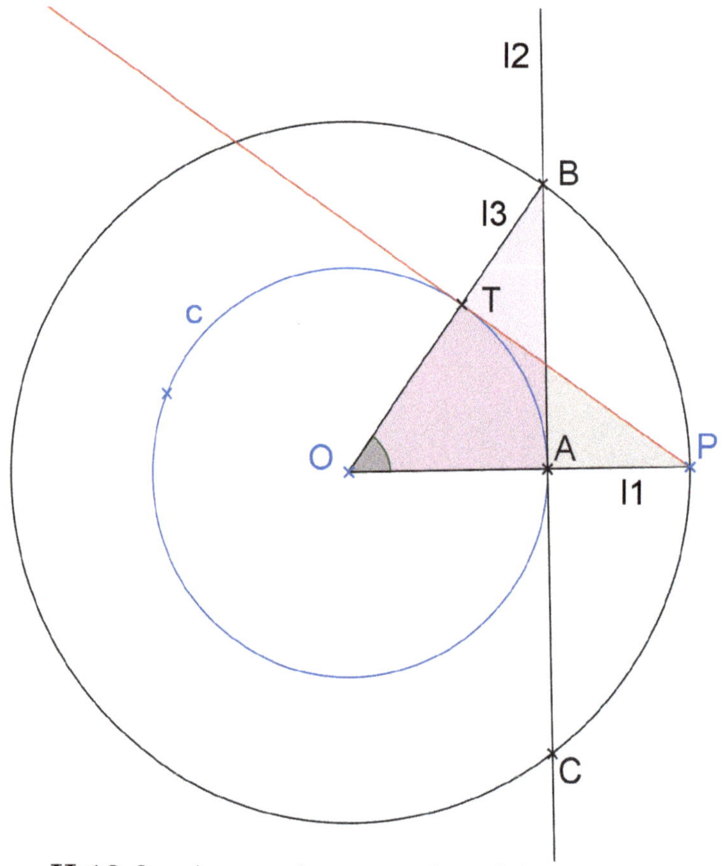

Figure II.10.6: A neutral construction of the tangents to a circle

Proposition II.10.9. *For any given Hilbert plane, we assume that an ad hoc construction of the tangent to a circle is possible and legitimate. Then the circle-line intersection property II.10.5 holds.*

Proof. We proceed as shown in the figure on page 263. Given are the circle *circ* and the line l that contains a point inside the circle and does not go through the center O. We drop the perpendicular p from center O onto the line l and get the foot point A.

Let P be the intersection point of the circle with the ray $\overrightarrow{OA} \perp l$ perpendicular to line l. Because of axiom (III.1), the point P can be obtained by transfer of any radial segment which has the radius of the circle as length. We draw a second circle c through the foot point A around the same center O. By Proposition II.10.1 item(ii), the foot point A lies inside the circle *circ*. Hence the order of points $O * A * P$ occurs, and the point P lies outside the (new smaller) circle c. By assumption, an ad hoc construction of the tangent from point P to the circle c is possible and legitimate. We drop the perpendicular from center O onto the tangent t and obtain the foot point T.

Let point B be the intersection of the radial ray $\overrightarrow{OT}$ with the given line l. The triangle congruence $\triangle OTP \cong \triangle OAB$ holds by ASA congruence. Indeed, the existence of the intersection point B is confirmed by the extended ASA-congruence, see Proposition II.7.13. Hence $OP \cong OB$, thus point B lies on the given circle *circ* and is an intersection point of circle *circ* with line l. □

II.10.2 The tangent is the limiting position of a secant

Imagine that a point P on the circle "moves" towards a fixed point T on the circle. We claim that the secant TP moves towards the tangent t.

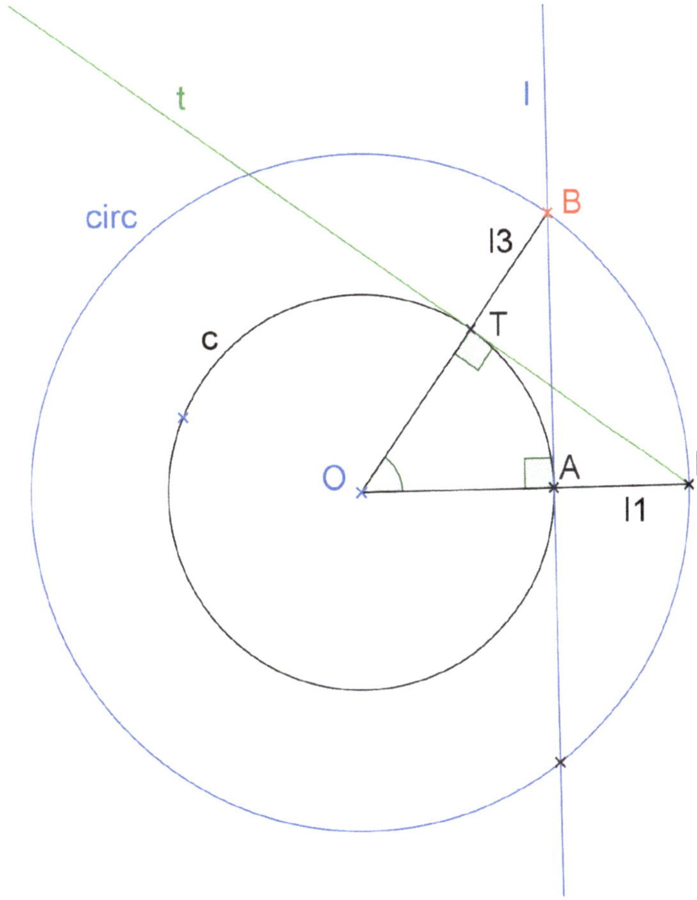

Figure II.10.7: If any ad hoc construction of the tangents to a circle is available, the circle-line intersection property holds.

A shorter, but more precise claim is: with the tangent at its endpoint, a secant which is short enough forms an angle smaller than any prescribed angle. We need the following theorem to justify the above claim.

Theorem II.10.1 (The tangent to a circle is the limiting position of a short secant). *Take any small angle $\varepsilon := \angle(\overrightarrow{TA}, t)$ between the secant TA and the tangent t. If for another point P on the circle, the secant $TP < TA$ is shorter, then the point P lies inside the angle $\angle(\overrightarrow{TA}, t)$:*

$$TP < TA \quad \textit{implies} \quad \angle(\overrightarrow{TP}, t) < \angle(\overrightarrow{TA}, t)$$

We see now that the claim from above holds. Given is any (small) angle ε by which we want to rotate the tangent. We draw the secant TA which forms this angle $\varepsilon := \angle(\overrightarrow{TA}, t)$ with the tangent t at touching point T. We obtain the second intersection point A, and put δ equal to the length of the chord TA. If any other secant TP is shorter than TA, then the point P lies inside the angle $\angle(\overrightarrow{TA}, t)$. Indeed, the tangent to a circle, as given by the perpendicular to the radius TO is the limiting position of a short secant.

Theorem II.10.2. *Given is a circle around O, a point P on the circle, and a line l through point P. The following three statements are equivalent. All three can be used as a definition of the tangent to a circle.*

The line l is perpendicular to the radius OP.

The point P is the only common point of the line and the circle.

The line is the limiting position of a secant PQ in the limit $Q \to P$.

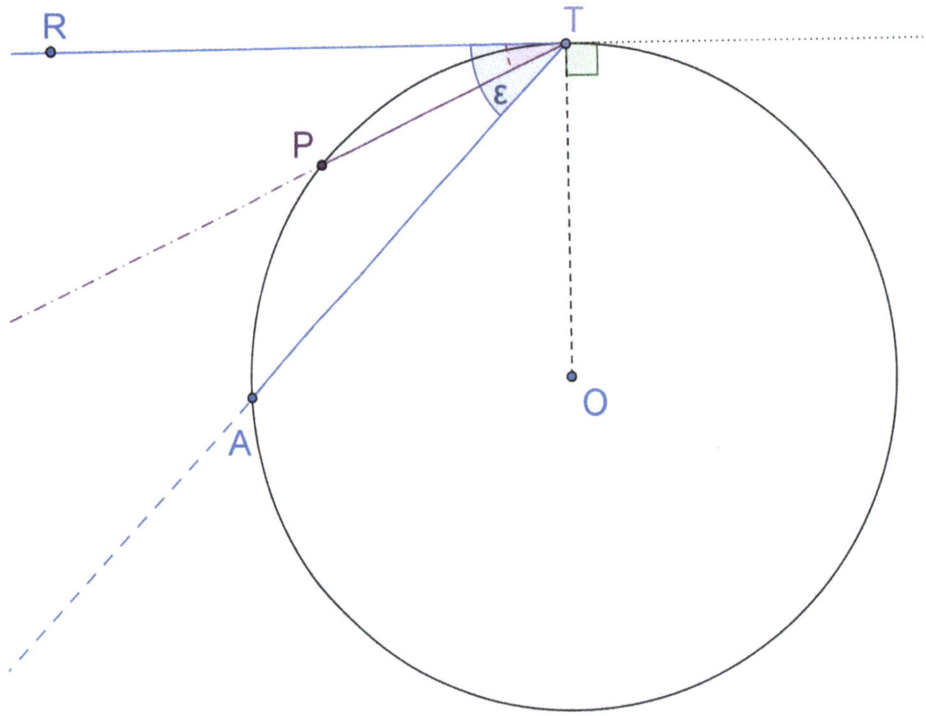

Figure II.10.8: The tangent is the limiting position of a short secant. Hence the shorter secant TP forms with the tangent the smaller angle $\angle RTP$.

Plan of proof for Theorem II.10.1. Given is a circle $\mathcal{C}$ with center O, a point T on the circle, and a ray $t = \overrightarrow{TR}$ perpendicular to the radius OT. Take any small angle ε.

Let e be a second ray with vertex T, lying on the same side on tangent t as the center O, and such that $\angle(t, e) = \varepsilon$.

Question. Why does the ray e intersect the circle in a second point A.

Answer. The rays e and t lie on different lines, since they are assumed to form an angle $\varepsilon = \angle(e, t)$. By Proposition II.10.7, the line on e which has a point in common with a circle, and is different from the tangent t, does intersect the circle in exactly two points.

We claim that any secant $TP < TA$ forms with the tangent at T an angle smaller than ε. It is enough to consider the case that points P and A lie on the same side of radius OT. In that case, we show that the secant TP lies inside the angle $\angle(e, t)$, and $\angle PTR < \angle ATR$. The argument is done in several steps:

(i) At first, we show that $TP < TA$ is equivalent to $\angle TOP < \angle TOA$. In other words, the length of a chord depends monotone on the central angle.

(ii) Secondly, we show that $\angle TOP < \angle TOA$ implies $\angle OTA < \angle OTP$.

(iii) Finally, $\angle OTA < \angle OTP$ implies the ray $\overrightarrow{TP}$ lies inside the angle $\angle RTA$, and hence $\angle RTP < \angle RTA$.

□

Detailed proof of Theorem II.10.1.

Question. By which proposition do we know that

$$\begin{aligned} &\angle TOP < \angle TOA \quad \text{is equivalent to } TP < TA \\ &\angle TOP > \angle TOA \quad \text{is equivalent to } TP > TA \\ &\angle TOP \cong \angle TOA \quad \text{is equivalent to } TP \cong TA \end{aligned} \tag{II.10.1}$$

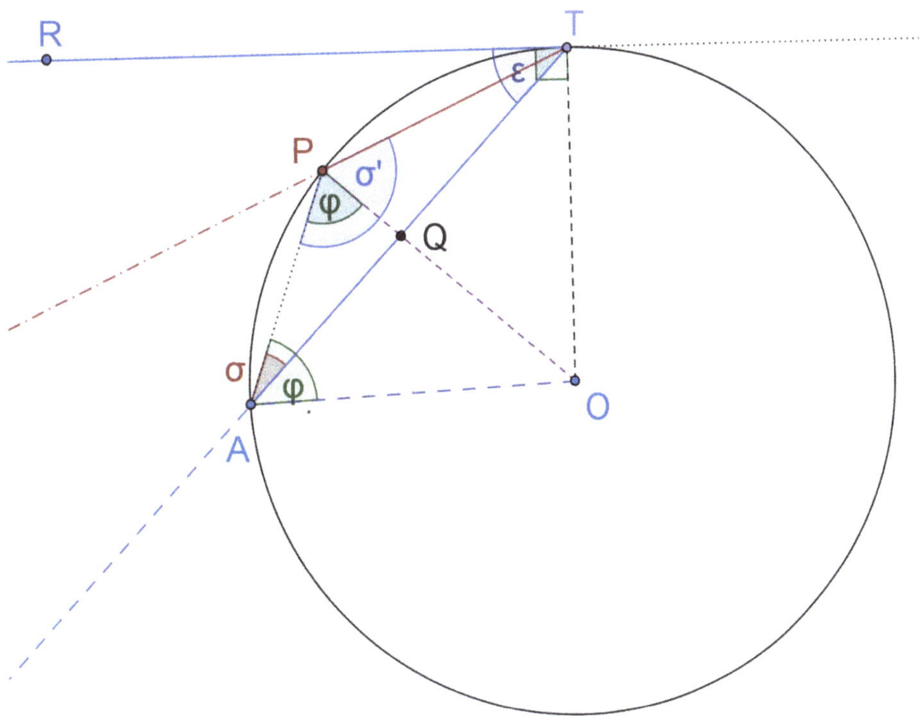

Figure II.10.9: We use the Hinge theorem.

Answer. The Hinge Theorem II.7.55 tells that increasing the angle between two constant sides increases the opposite side of a triangle [Euclid I.24].

Too, Corollary 14 states the equivalence: if and only if the angle between the two constant sides increases, is constant or decreases, then the opposite side does the same.

For completeness, we give the exact detailed reasoning in the present case.

Proof of the Hinge Theorem in the present case. Assume that the triangles $\triangle TOP$ and $\triangle TOA$ satisfy $\angle TOP < \angle TOA$.

Question. Show that the segments TA and OP intersect each other.

Answer. Since points P and A are assumed to lie on the same side of line TO, the ray $\overrightarrow{OP}$ lies inside the angle $\angle TOA$. Hence the Crossbar Theorem implies that this ray intersects the segment TA. Let Q be the intersection point. By Corollary 24, all points of a chord besides the endpoints lie inside the circle. Hence, we have checked that

$$O * Q * P \quad \text{and} \quad T * Q * A \tag{II.10.2}$$

Question. Use Euclid I.19 to give a reasoning why $TP < TA$.

Answer. To compare segments TP and TA, we now use Euclid I.19 for the triangle $\triangle TAP$. We need to compare the angles $\sigma = \angle TAP$ and $\sigma' = \angle TPA$ opposite to these two segments. Let ϕ be the base angle of the isosceles triangle $\triangle AOP$.

$$\begin{aligned} O * Q * P &\quad \text{implies } \sigma < \phi \\ T * Q * A &\quad \text{implies } \phi < \sigma' \end{aligned}$$

Hence $\sigma < \sigma'$ follows by transitivity, and $TP < TA$ by Euclid I.19.

Question. Why does now, conversely, $TP < TA$ implies $\angle TOP < \angle TOA$.

Answer. The argument above shows that $\angle TOP < \angle TOA$ implies $TP < TA$. Similarly, one gets that $\angle TOP > \angle TOA$ implies $TP > TA$. By SAS congruence, $\angle TOP \cong \angle TOA$ implies $TP \cong TA$.

All three statements together allow to get the corresponding converse statements—since all angles are comparable, and all segments, too.

□

Thus the Hinge Theorem yields item (i). Next, we check item (ii) that $\angle TOP < \angle TOA$ implies $\angle OTP < \angle OTA$. This is now easy, since the reasoning above shows that both $\angle TOP < \angle TOA$ and $TP < TA$. Hence point Q is the intersection point of segments segments TA and OP, as stated in equation (II.10.2). Obviously, $O * Q * P$ implies $\angle OTA = \angle OTQ < \angle OTP$.

Finally, we check item (iii). The angle $\angle OTP < \angle OTR = R$, since $\angle OTP$ is the base angle of the isosceles triangle $\triangle OTP$ and hence acute, but angle $\angle OTR$ is right by assumption. Since by assumption, points P and R lie on the same side of radius OT, the ray $\overrightarrow{TP}$ lies inside the angle $\angle RTA$. We see that $\angle RTP < \angle RTA$, as claimed. □

II.10.3 Mutual placement of two circles

We conclude this small survey with some simple facts about the mutual relations of two circles. The set of points inside a circle is called the *interior* or *disk* of the circle, The interior of circle $\mathcal{C}$ is denoted by $\operatorname{Int}(\mathcal{C})$.

Proposition II.10.10. *Let $\mathcal{C}$ be a circle with center C and radius c, and $\mathcal{D}$ be a circle with center D and radius d.*

(i) *The interiors of the two circles $\mathcal{C}$ and $\mathcal{D}$ are disjoint if and only if $|CD| \geq c + d$.*

(ii) *The interior of circle $\mathcal{C}$ is a subset of the interior of circle $\mathcal{D}$ if and only if $|CD| \leq d - c$.*

(iii) *The interior of circle $\mathcal{D}$ is a subset of the interior of circle $\mathcal{C}$ if and only if $|CD| \leq c - d$.*

(iv) *The interiors of the two circles intersect, but neither one is a subset of the other if and only if* $\max\{d-c, c-d\} < |CD| < c+d$.

In that case, each circle intersects both the interior and exterior of the other one. The interior of each circle intersects the exterior of the other circle.

With inequalities and equalities interpreted as segment comparison and congruence. This Proposition holds in any Hilbert plane, without assuming *neither the Archimedean axiom nor the circle-circle intersection property, nor the circle-line intersection property.*

Proof. It is left to the reader to check these statements for the case $C = D$ of *concentric* circles. We now assume $C \neq D$, and let C_1 and C_3 be the two points on the line CD lying on the circle $\mathcal{C}$. Similarly, let D_1 and D_3 be the two points on the line CD lying on the circle $\mathcal{D}$. These points exist, because they can be constructed by transferring any radial segment. We can assume the order relations $C_1 * C * D * D_3$, $C_1 * C * C_3$ and $D_1 * D * D_3$.

(i) Assume the interiors of the two circles $\mathcal{C}$ and $\mathcal{D}$ are disjoint. Then no point can lie both inside segment C_1C_3, and inside segment D_1D_3. Hence either $C_1 * C * C_3 * D_1 * D * D_3$, or $C_3 = D_1$. In the former case, $|CD| > c + d$. In the latter case $|CD| = c + d$.

Conversely, assume that the interiors of the two circles $\mathcal{C}$ and $\mathcal{D}$ intersect. Let P be a point inside both circle. The weak triangle inequality implies $|CD| \leq |CP| + |PD| < c + d$.

(ii) Assume that the interior of circle $\mathcal{C}$ is a subset of the interior of circle $\mathcal{D}$. In that case, every point lying inside segment C_1C_3 lies inside segment D_1D_3, too. Hence either

$$
\begin{aligned}
&C_1 = D_1 \text{ and } D_1 * D * D_3 * C_3, \quad \text{or}\\
&C_1 * D_1 * D * D_3 \text{ and } C_3 = D_3, \quad \text{or}\\
&C_1 * D_1 * D * D_3 * C_3.
\end{aligned}
$$

The first two cases imply $|CD| = d - c$, and the last case implies $|CD| < d - c$.

Conversely, if $|CD| < d - c$, then $C_1 * D_1 * D * D_3 * C_3$. Similarly $|CD| = d - c$ implies either $C_1 = D_1$ and $D_1 * D_3 * C_3$, or $C_3 = D_3$ and $C_1 * D_1 * D_3$.

Now let P be any point in the interior of circle $\mathcal{C}$. The weak triangle inequality implies $|PD| \leq |PC| + |CD| < c + d - c = d$, and hence point P lies inside the circle $\mathcal{D}$, too, and hence $\text{Int}(\mathcal{C}) \subseteq \text{Int}(\mathcal{D})$.

(iii) (iv) These are just logical consequences.

□

Proposition II.10.11. *Let $\mathcal{C}$ be a circle with center C and radius c, and $\mathcal{D}$ be a circle with center D and radius d.*

(a) *If the two circles intersect in a point not on the line between the two centers, then* $\max\{d - c, c - d\} < |CD| < c + d$.

(b) *The two circles intersect in exactly one point if and only if they are not concentric—$C \neq D$—, and equality holds in either*

(i) $|CD| = c + d$.

(ii) $|CD| = d - c$.

(iii) $|CD| = c - d$.

from the previous proposition. In that case, the intersection point and the two centers lie on a line.

Proof. **(a)** Assume the two circles intersect in point P. Item (a) follows from the strict triangle inequality for triangle $\triangle CDP$. Indeed

$$\begin{aligned} |CD| &< |CP| + |PD| = c + d \\ c - d = |CP| - |DP| &< |CD| + |DP| - |DP| = |CD| \\ d - c = |DP| - |CP| &< |DC| + |CP| - |CP| = |CD| \end{aligned}$$

(b) First we assume that the two circles intersect in exactly one point. This point needs to lie on the line CD, because otherwise, one would get a second intersection point P' symmetric to line CD. Hence the intersect point is either C_3 or D_1, and equality holds in either (i), (ii) or (iii).

Conversely, if equality holds in either (i), (ii) or (iii), and $C \neq D$, then the two circles intersect in either point C_3 or D_1, and no further point on the line CD. Because of item (a), they cannot intersect in any point P which is not on the line CD, neither.

□

Thus, we have shown that in the case

(iv) $\max\{d - c, c - d\} < |CD| < c + d$.

the interiors of the two circles intersect, and the interior of each circle intersects the exterior of the other circle.

Of course, this immediately suggests that the two circles intersect? But such a statement does not follow just from Hilbert's axioms!

II.10.4 Continuity principles for circles

Much more cannot be deducted about intersection of circles and lines, simply from Hilbert's axioms. Indeed, in theorem II.10.3 below, we give a precise statement of this sharply negative result.

One needs to introduce and use some continuity principle to obtain sharper, and more intuitive results about circles and lines. We discuss the line-circle and the circle-circle intersections.

Definition II.10.5 (Line-circle intersection property). A line that contains a point inside a circle does intersect the circle.

Proposition II.10.12. *Assume that the line circle intersection property holds.*

(i) *A line that contains a point inside a circle intersect it in exactly two points.*

(ii) *Let P be a point inside circle $\mathcal{C}$ and Q a point outside the circle. Then the segment PQ intersects the circle in exactly one point, and the line PQ intersects the circle in two points.*

Proof. In the special case that the center O of the circle lies on the line l, the result follows easily—even just from Hilbert's axioms. Indeed, the two intersection points A, B can be obtained by transferring the radial segment OR onto the two opposite rays on line l with vertex O.

Following Hilbert's Proposition II.6.7, any finite set of points on a line can be put into an ordered list such that their order relations hold according to the ordering of the list. Possibly by exchanging the names of A and B, such a list is either $A * O * P * B * Q$ or $A * P * O * B * Q$.

Let A be one intersection point of the given circle and line l. Reflecting point A by the line $OF \perp l$ yields a second intersection point B. In proposition II.10.1, it was shown that a line and a circle cannot intersect in three points. Hence the given line l intersects the circle in two points A and B.

We now check item (ii). Recall the order of the points A, P, B, Q on line l. Proposition II.10.1 implies that point P lies inside the segment AB, and point Q lies outside that segment. By Hilbert's four-point theorem, we conclude that either the list $A * P * B * Q$ or $Q * A * P * B$ yields the order of these points. Hence the segment PQ contains exactly one of the intersection points A and B of the given circle $\mathcal{C}$ and the line PQ. $\square$

Proposition II.10.13. *Assume that the line circle intersection property holds. If a circle has points on both sides of a line, then it intersects the line in two points.*

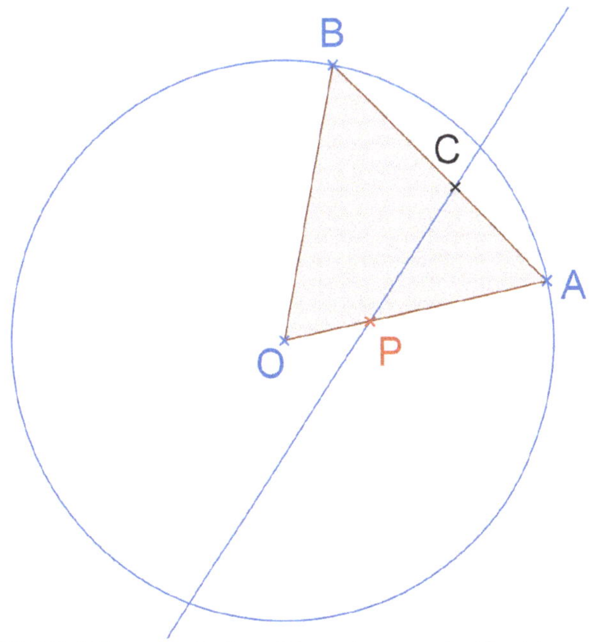

Figure II.10.10: A circle with points on both sides of a line.

Proof. Let A and B be points on different sides of line l, both lying on a circle around center O.

There are some special cases: In the special case that the center O lies on the line, the result follows immediately, by transferring radial segments. In the special case the three points A, O and B lie on a line, two of them, say A and O, lie on different sides. By the plane separation theorem, the line intersects the segment OA. The intersection point P of line l and segment OA lies inside the circle. Hence the line-circle intersection property implies that the line and the circle intersect.

Now we deal with the generic case: assume that center O does not lie on the line, and the three points A, O and B do not lie on a line. By the plane separation theorem, the segment AB intersects the line l. Hence, by Pasch's axiom, this line intersects a second side of the triangle OAB, say segment OA. Again, the intersection point P of line l and segment OA lies inside the circle. Hence the line-circle intersection property implies that the line and the circle intersect. □

Definition II.10.6 (Circle-circle intersection property). Given are two circles $\mathcal{C}$ and $\mathcal{D}$. Let P, Q be two points on circle $\mathcal{D}$. Assume that point P lies inside circle $\mathcal{C}$ and Q lies outside circle $\mathcal{C}$. Then the two circles $\mathcal{C}$ and $\mathcal{D}$ intersect each other.

Proposition II.10.14. *The two circles considered in the circle-circle intersection property have exactly two intersection points.*

Proof. Existence of the second intersection point follows using reflection across the line OO' connecting the two centers. The existence of a third intersection point was excluded in Lemma II.7.2 for the SSS congruence theorem. □

Corollary 25. *Let $\mathcal{C}$ be a circle with center C and radius c, and $\mathcal{D}$ be a circle with center D and radius d.*

The two circles intersect in exactly two points if and only if $\max\{d - c, c - d\} < |CD| < c + d$, *where the inequalities are to be interpreted as segment comparison.*

Proposition II.10.15. *The circle-circle intersection property implies the line-circle intersection property.*

Proof. Assume that the circle-circle intersection property holds. Let $l = AB$ be the line between the two given points A inside, and B outside the circle $\mathcal{C}$. The ray $\overrightarrow{OB}$ intersects the circle in a point Q inside the segment OB. A special situation occurs in case that the center O of the circle lies on the line l. In this case, the intersection point $Q = X$ of the ray $\overrightarrow{OB}$ with the circle is already the required intersection of line and circle. We now exclude that special case and assume that the center O does not lie on line l. The ray $\overrightarrow{OA}$ from the center intersects the circle in a point P outside the segment OA (since $O \neq A$).

Let O' and P', Q' be the reflection images of points O and P, Q across the axis $l = AB$. We construct the circle $\mathcal{C}'$ with center O' through point P'. Points P' and Q' both lie on the circle $\mathcal{C}'$—which is the reflection image of circle $\mathcal{C}$.

In order to apply the circle-circle intersection property, we check that point P' lies inside and point Q' lies outside the original circle $\mathcal{C}$. Indeed

$$|OP'| = |O'P| \leq |O'A| + |AP| = |OA| + |AP| = |OP|$$

We have used that the lengths of symmetric segments $OP' \cong O'P$ are equal (Proposition II.7.61), the weak triangle inequality for $\triangle O'AP$ (which may degenerate), the fact that the symmetric points O and O' have the same distance from point A on the axis, and the assumption that point A lies inside circle $\mathcal{C}$. We need to exclude the equality $|OP'| = |OP|$. But that assumption would imply $|O'P| = |OP|$. Hence point $P = P'$ and hence the center O would lie on the axis AB, too, which we did exclude. This contradiction confirms $|OP'| < |OP|$, and P' lies inside the circle $\mathcal{C}$.

With a bid modified argument we check that Q' lies outside circle $\mathcal{C}$.

$$|OQ| + |QB| = |OB| = |O'B| < |O'Q| + |QB|$$

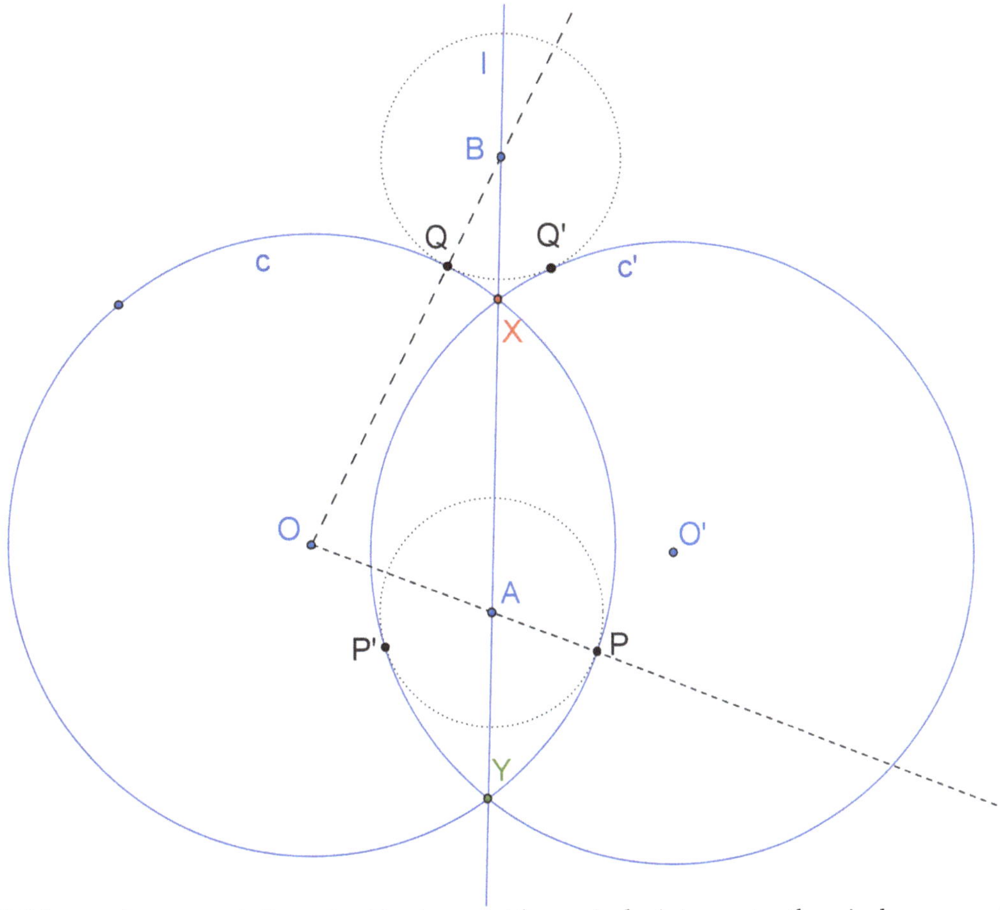

Figure II.10.11: A segment from inside to outside a circle intersects the circle—proved from the circle-circle intersection.

and hence $|OQ| < |O'Q| = |OQ'|$ which implies that Q' lies outside circle $\mathcal{C}$. In this case, the triangle inequality holds strictly, since points O', Q and B do not lie on a line. (We would get a contradiction: all five points O, O', Q, Q', B would lie on this line, which would be line l. Hence we would have the case excluded above.)

Because point P' lies inside, but point Q' lies outside the circle $\mathcal{C}$, and both points lie on the reflected circle $\mathcal{C}'$, the circle-circle intersection property implies that the two circles intersect in some point X. Since it has the same distance from both centers O and O', by proposition II.13.1 point X lies on the axis l. Thus we have obtained an intersection point of the given circle $\mathcal{C}$ and line AB.

A second intersection point Y is obtained by reflecting X across the perpendicular p to line AB through point O—which is in the generic case just line OO'. One of the two points X and Y lies in the same half plane of p as either point A or point B. It is left to the reader to check that this intersection lies between A and B. □

Proposition II.10.16. *The line-circle intersection property together with the* <u>*parallel axiom*</u> *imply the circle-circle intersection property.*

A proof using analytic geometry is given in the section about coordinate planes.

Problem II.10.2 (Construction with classical tools in neutral geometry). *Give a construction of an equilateral triangle, and an angle of* 60° *with compass and straightedge in neutral geometry.*

(a) *Do the construction and give a description.*

(b) *What can one say about the angles at the vertices of the triangle, in* neutral geometry*? Why are extra steps needed for the construction of the* 60° *angle?*

(c) *At which vertex can you get the angle of* 60° *even in neutral geometry, nevertheless?*

(d) *Justify your answer.*

(e) *Convince yourself once more that all reasoning has been done in neutral geometry. Additional to Hilbert's axioms, which intersection property do you need.*

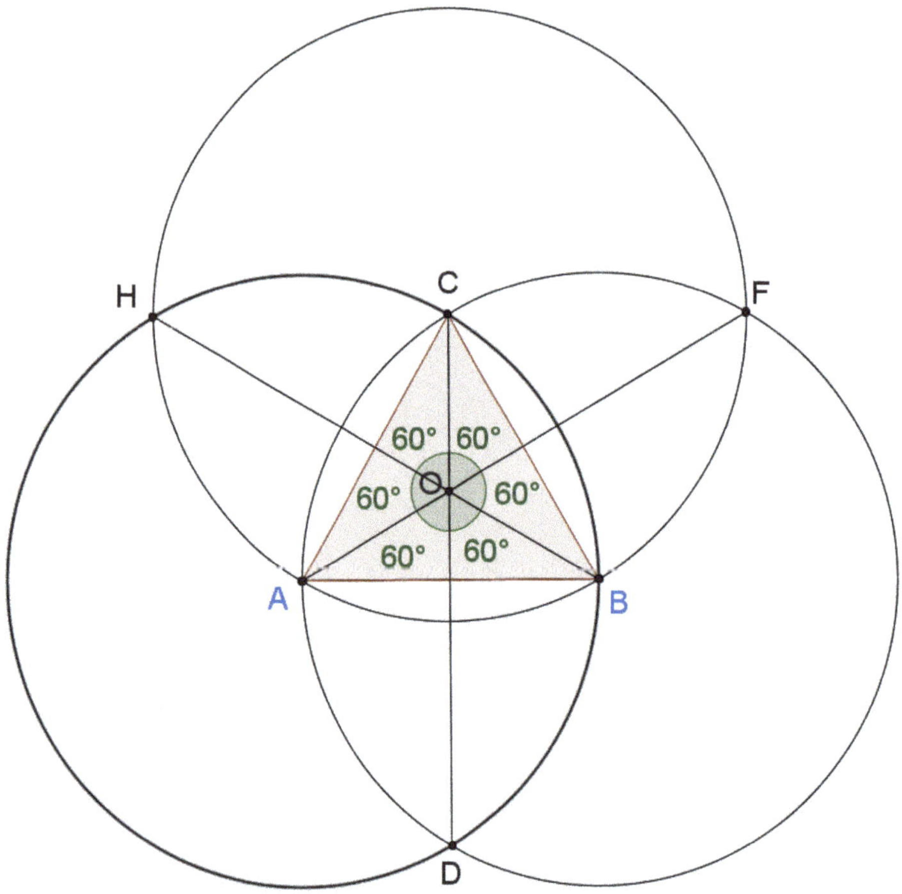

Figure II.10.12: The construction of a 60° angle at the center O is possible with straightedge compass in neutral geometry.

Answer. **(a) Description of the construction.** Draw any segment AB. The circles about A through B, and about B through A intersect in two points C and D. We get an equilateral triangle $\triangle ABC$. Now we draw a third circle about C through point A, which passes through point B, too. With the two circles drawn earlier, the third circle has additional intersection points H and F. Finally, we draw the segments AF, BH and CD. These are the perpendicular bisectors of the sides of triangle $\triangle ABC$. All three intersect in one point O inside triangle $\triangle ABC$. At vertex O, one gets six congruent angles which add up to 360°, hence they are all 60°.

(b) **The angles at the vertices may not be 60°.** Extra steps are needed, because, in neutral geometry, the angle sum of a triangle may be less than two right angles. All we can conclude about the angles at the vertices A, B, C, is their congruence. This follows because the angles opposite to congruent sides are congruent. Still they may all three be less than 60°.

(c) **The 60° angles appear at the center.** At vertex O, one gets six congruent angles which add to 360°, hence they are all 60°.

(d) **Justification.** The construction of the perpendicular side bisectors for triangle $\triangle ABC$ still works in neutral geometry because of proposition II.13.1. By plane separation, they intersect the corresponding sides of the triangle. Too, they lie inside the angles of the triangle. Hence the Crossbar Theorem implies that they intersect each other, as explained in theorem II.13.3 (about the in-circle).

The intersection point O is the center of the in-circle of triangle $\triangle ABC$, because it has equal distances from the three sides. The same point is the center of the circum-circle, because it has equal distances from the vertices.

The three side bisectors form six congruent angles at vertex O. Indeed, the ray $\overrightarrow{OD}$ bisects the angle $\angle AOB$. since SSS-congruence yields the triangle congruence $\triangle AOD \cong \triangle BOD$. The opposite ray $\overrightarrow{OC}$ bisects the angle $\angle HOF$. This can for example be obtained from the triangle congruence

$$\triangle HOC \cong \triangle HOA \cong \triangle BOF \cong \triangle OCF$$

(e) **The circle-circle intersection property is needed.** All these justifications above have been done in neutral geometry. Besides Hilbert's axioms, we only need the circle-circle intersection property II.10.6.

Corollary 26. *Assume Hilbert's axioms* (I.1)–(I.3), (II.1)–(II.4) *and* (III.1)–(III.5), *together with the circle-circle intersection property holds. Then for each given segment AB and half plane of line AB, there exists a unique equilateral triangle ABC. An angle of 60° is constructible.*

Problem II.10.3 (Construction with Hilbert tools in Euclidean geometry). *Give a construction of an equilateral triangle, and an angle of 60° with Hilbert tools in Euclidean geometry. Use the Theorem of Pythagoras.*

Answer. We construct an equilateral triangle with side length $|AB| = |BC| = |CA| = 2$. The Theorem of Pythagoras implies that the altitude of this triangle has length $\sqrt{3}$.

We begin by constructing the perpendicular bisector of segment AB. With its midpoint M as vertex, we transfer segment MA onto the bisector and get segment $MP \cong MA$.

Next we transfer segment AP onto the ray $\overrightarrow{MP}$ and get a segment $MQ \cong AP$. The Theorem of Pythagoras implies and congruence imply

$$|MQ| = |AP| = \sqrt{2}$$

In a further similar step, we transfer segment BQ onto the ray $\overrightarrow{MP}$ and get a segment $MC \cong BQ$. The Theorem of Pythagoras implies and congruence imply

$$|MC| = |BQ| = \sqrt{|BM|^2 + |MQ|^2} = \sqrt{3}$$

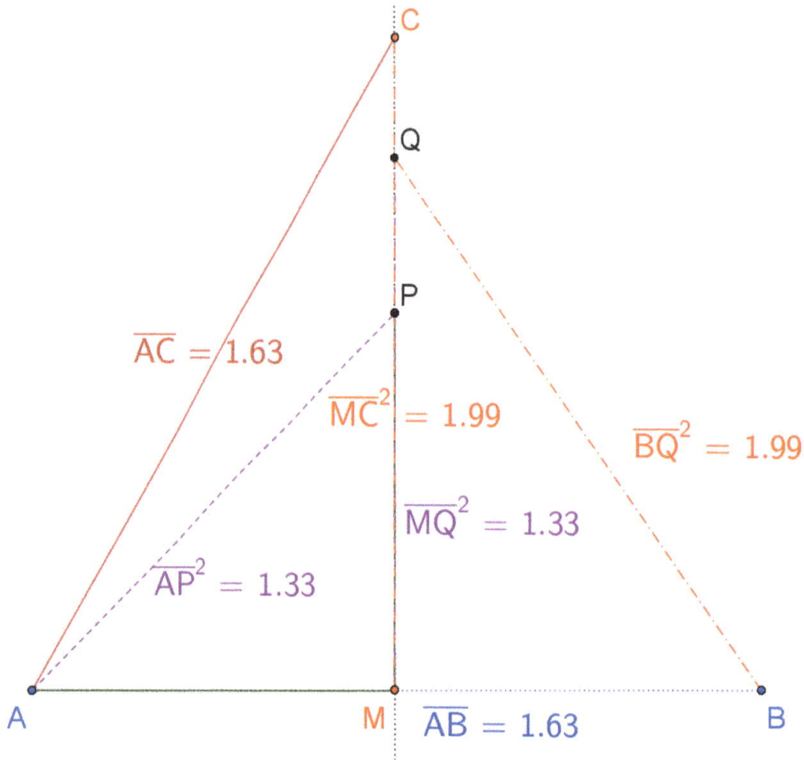

Figure II.10.13: This construction of an equilateral triangle with Hilbert tools works only in Euclidean geometry

Using, the Theorem of Pythagoras a third time, we check that

$$|AC| = \sqrt{|AM|^2 + |MC|^2} = 2$$

Hence the triangle $\triangle ABC$ is equilateral.

Corollary 27. *Assume Hilbert's axioms* (I.1)–(I.3), (II.1)–(II.4) *and* (III.1)–(III.5), *as well as the Euclidean parallel axiom. The circle-circle intersection property needs* not *be assumed to hold.*

Then for each given segment AB and half plane of line AB, there exists a unique equilateral triangle ABC. An angle of $60°$ *is constructible.*

Proof. Note that the Euclidean parallel axiom implies the Pythagorean theorem. Now the construction of the equilateral triangle ABC over a given segment AB can be done with Hilbert tools. The details are left to the reader. □

Problem II.10.4. *Is an angle of* $60°$ *constructible just with Hilbert tools is neutral geometry? Neither the parallel axiom, nor the circle-line intersection, nor the circle-circle intersection property is assumed.*

II.10.5 Continuity principles for circles are independent of Hilbert's axioms

We now give a strikingly negative result:

Theorem II.10.3. *Neither the circle-line intersection property, nor the circle-circle intersection property hold in every Hilbert plane. They cannot be deducted simply from the axioms of incidence, order and congruence.*

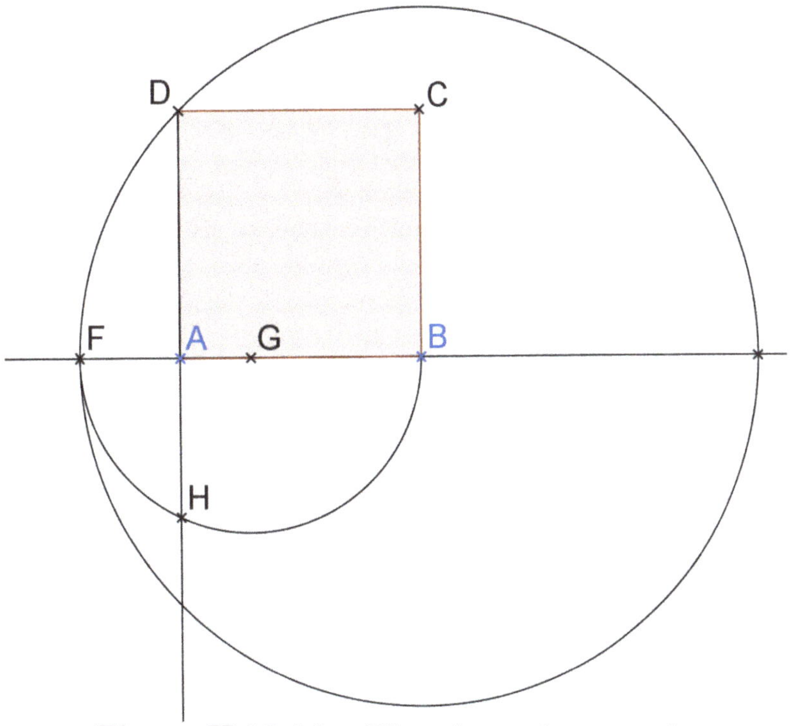

Figure II.10.14: What the teacher wanted.

Proposition II.10.17 (Counterexample). *The number $z = \sqrt{\sqrt{2} - 1}$ is not totally real. A segment of that length cannot be constructed by Hilbert tools, just starting from a given unit segment.*

In the figure on page 275, points A through G are constructible with Hilbert tools, but not point H is not constructible.

Proof. Hilbert's basic counterexample is given in more details in proposition II.10.17 in the section on Euclidean constructions with restricted means.

Indeed, because the number $z = \sqrt{\sqrt{2} - 1}$ is not totally real, a segment of that length cannot be constructed by Hilbert tools, just starting from a given unit segment, with $|AB| = 1$.

Let $\mathcal{H} \subset \mathbf{R}^2$ be the Hilbert plane consisting of all points in the Euclidean plane that can be constructed with Hilbert tools in finitely many steps, with only any two points A and B being given at the start.

In the figure on page 275, points A through G are in the Hilbert plane $\mathcal{H} \subset \mathbf{R}^2$, since they are constructible with Hilbert tools. But point H is not constructible, and hence $H \notin \mathcal{H}$.

Problem II.10.5. *Does the segment-circle intersection property hold in the Hilbert plane $\mathcal{H}$? Which (nonempty) segment and circle provide a counterexample?*

Answer. Let $\mathcal{C}$ be the circle around G through point B. The segment AD has one endpoint A inside, and the other endpoint D outside of circle $\mathcal{C}$. This segment is nonempty, too. But it does not intersect that circle, because point H is not in the Hilbert plane $\mathcal{H}$.

□

Proposition II.10.18. *There exist models of hyperbolic geometry where an equilateral triangles does not exist over every segment.*

Proof. See Hartshorne's book p. 373. The proof is omitted. It needs a bid of abstract algebra, and Poincaré's disk model with a virtual line of infinity. □

II.10.6 Derivation of continuity principles from Dedekind's axiom

What does Dedekind's axiom imply about intersection of circles and lines? For convenience of the reader, we prove at first the weaker statement of theorem II.10.4, and then the stronger theorem II.10.5.

Theorem II.10.4. *Dedekind's axiom implies the line-circle intersection property.*

Theorem II.10.5. *Dedekind's axiom implies the circle-circle intersection property.*

Proof of Theorem II.10.4: Dedekind's axiom implies the line-circle intersection property. Given is a circle $\mathcal{C}$, and a line l. We assume that point P lies on the line l and inside the circle $\mathcal{C}$. In the special case that the line goes through the center of the circle, existence of the intersection points of line and circle can be deducted directly via Hilbert's axioms.

Hence we may assume that the center O of the circle does not lie on the line. Let Q be a point on the line l such that $PQ \cong 2OR$.

Question. Why does the point Q lie outside the circle $\mathcal{C}$.

Answer. Indeed, the triangle inequality II.7.52 for the triangle $\triangle OPQ$ implies $PO + OQ > PQ$, and hence $OQ > PQ - OP > 2OR - OR \cong OR$.

We now construct the *Dedekind cut* $\{\Sigma_1, \Sigma_2\}$ of the line l. Let F be the foot point of the line dropped from center O onto the line l. Let the set Σ_1 consist of all points of the line l that lie either on the ray opposite to $\overrightarrow{FQ}$, or inside the circle $\mathcal{C}$. Let the set Σ_2 consist of all points on the ray $\overrightarrow{FQ}$ that lie outside the circle $\mathcal{C}$.

In the case that the ray $\overrightarrow{FQ}$ and the circle $\mathcal{C}$ intersect, we are ready anyway. Otherwise, we show that $\{\Sigma_1, \Sigma_2\}$ is a Dedekind cut.

Let P_2 be a point between two points $P_1, P_3 \in \Sigma_1$. We now check that point P_2 lies in Σ_1. If both P_1 and P_3 lie both on the ray opposite to $\overrightarrow{FQ}$, we are ready. So we need consider the case that $P_1 * F * P_3$. Hence either $P_1 * P_2 * F * P_3$ or $P_1 * F * P_2 * P_3$. In the first P_2 lies inside the circle, and hence in Σ_1. If P_2 lies on the ray opposite to $\overrightarrow{FQ}$, and $P_2 \in \Sigma_1$—we are ready.

From item (ii) of proposition II.10.2, we see that a point Q_2 between two points $Q_1, Q_3 \in \Sigma_2$ lies in the set Σ_2, too.

All assumptions occurring in Dedekind's axiom have been checked. Hence we conclude from Dedekind's axiom existence of a cut point K^*. By its definition, the cut point does not lie between two other points of Σ_1, and neither between two other points of Σ_2. On the other hand, a point inside a circle lies between two other points inside the circle, and any point outside the circle lies between two other points outside the circle, and on the same side of foot point F.

This contradiction rules out that K^* could lie inside or outside the circle. Hence the cut point K^* lies on the circle, and is an intersection point of the given circle and line.

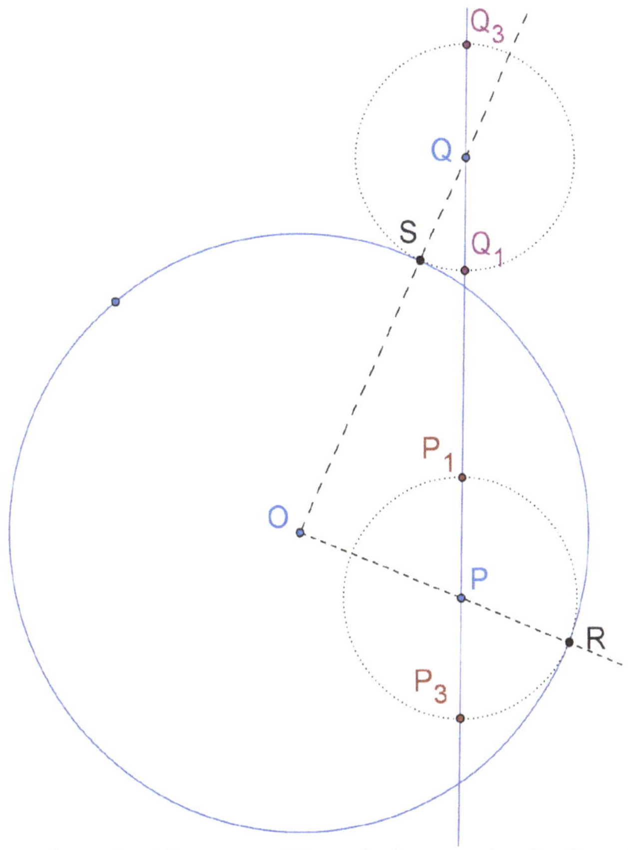

Figure II.10.15: A segment from inside to outside a circle contains both points inside and outside the circle.

Lemma II.10.1. *Let P be a point inside and Q a point outside the circle $\mathcal{C}$.*

Then any point inside the circle lies between two other points on the line PQ and inside the circle, similarly, every point outside the circle lies between two other points on the line PQ outside the circle.

Especially, the segment PQ contains both points inside and outside the circle.

Proof. Let R be the intersection of the circle with the central ray $\overrightarrow{OP}$. The point R can be constructed with Hilbert tools by transferal of any radial segment. Let P_1 and P_3 be the points on the line PQ, on both sides of point P, such that $PR \cong PP_1$ and $PR \cong PP_3$. We claim that these points lie inside the circle $\mathcal{C}$. Indeed, this follows from the triangle inequality since $OP_i < OP + PP_i \cong OP + PR = OR$, for $i = 1, 3$.

A similar argument is used to get the points outside the circle. Let S be the intersection of the circle with the central ray $\overrightarrow{OQ}$. Let Q_1 and Q_3 be the points on the line PQ, on both sides of point Q, such that $SQ \cong Q_1Q$ and $SQ \cong Q_3Q$. These points both lie outside the circle $\mathcal{C}$, since the triangle inequality yields $OS + SQ = OQ < OQ_i + Q_iQ$ and hence $OS < OQ_i$, for $i = 1, 3$. □

□

Proof of Theorem II.10.5: Dedekind's axiom implies the circle-circle intersection property. Given are two circles: circle $\mathcal{C}$, with center O, and circle $\mathcal{D}$ with center D—and two points P, Q on circle $\mathcal{D}$.

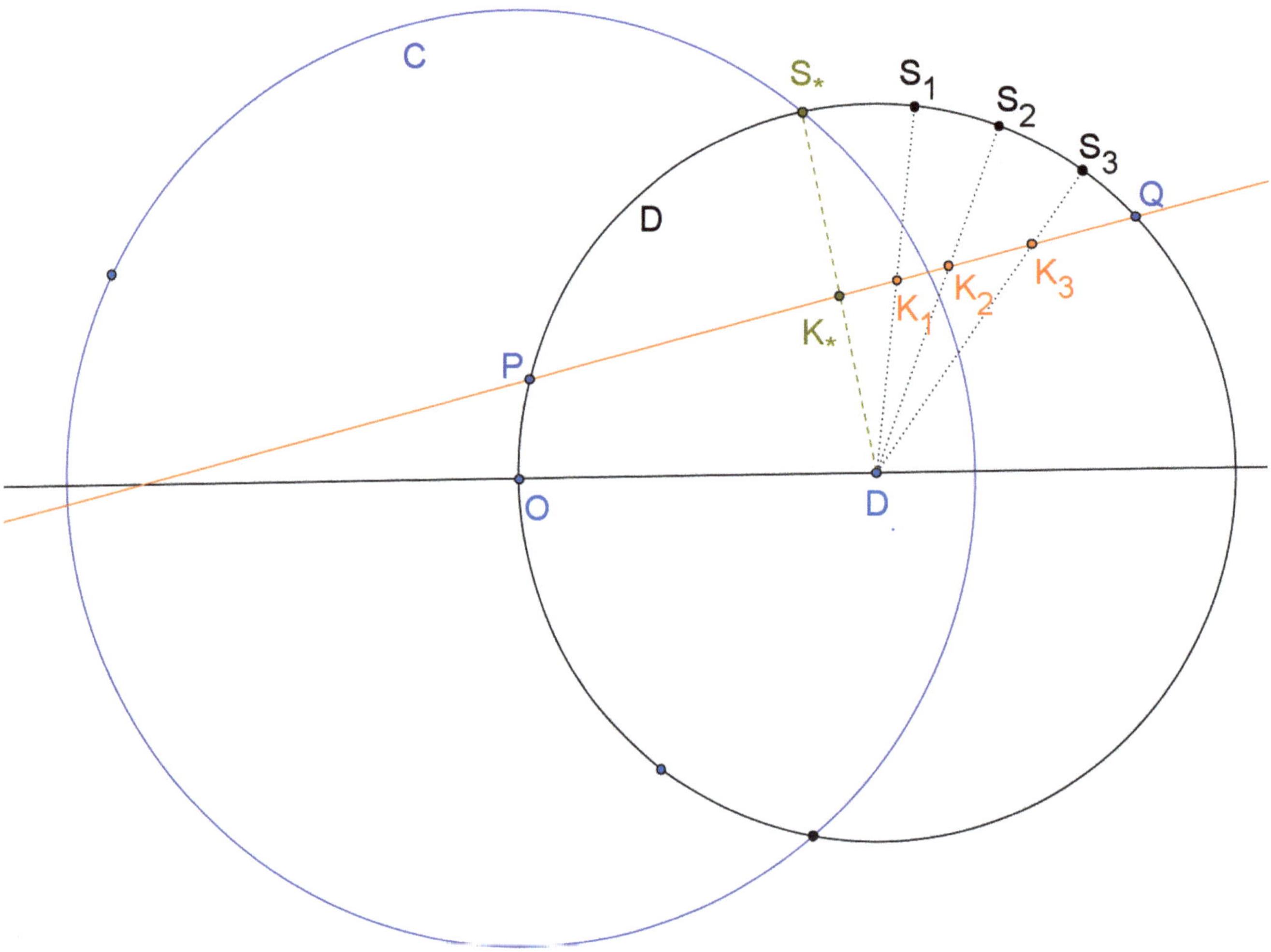

Figure II.10.16: Dedekind's axiom implies existence of the intersection point K^* of two circles.

Assume that point P lies inside circle $\mathcal{C}$ and Q lies outside circle $\mathcal{C}$. We need to show that the two circles do intersect.

To begin with, I reduce the proof to the convenient case that the two points P and Q lie on the same side of line OD, but not both on that line.

Indeed, I begin by eliminating the special case that both points P, Q lie on line CD. One transfers an arbitrary angle at vertex D and produces a third ray with vertex D, different from $\overrightarrow{DP}$ and $\overrightarrow{DQ}$. Transferring the radius of circle $\mathcal{D}$ onto this ray yields a third point R on this circle, which is not on line CD. Depending on whether R lies outside or inside circle $\mathcal{C}$, we now can use points P, R or Q, R for the construction to follow.

Next I eliminate the case that points P, Q lie on opposite sides of line CD. Let P' be the reflection of point P across line CD. Because P and P' lie on opposite sides of CD, plane separation implies that P' and Q lie on the same side of CD. Furthermore, because $CP \cong CP'$ and $DP \cong DP'$, the point P' lies on circle $\mathcal{D}$ and inside circle $\mathcal{C}$, same as point P. The points P' and Q are used for the construction to follow.

For any point $K \neq D$, let $S(K)$ be the unique intersection point of circle $\mathcal{D}$ with ray $\overrightarrow{DK}$. Let l be the line through points P and Q and let K be a generic point on line l. Take any three points

K_i for $i = 1, 2, 3$ and assume $P * K_1 * K_3 * K_2 * Q$. Since those points lie all on the same side of line CD, we get $\angle CDP < \angle CDK_1 < \angle CDK_2 < \angle CDK_3 < \angle CDQ$. Now we apply the Hinge theorem to the triangles $\triangle CDS(P)$, $\triangle CDS(K_i)$, with $i = 1, 2, 3$, and $\triangle CDS(Q)$. All these triangles have the common side CD, and congruent sides with vertex D. Hence the Hinge Theorem implies

$$\overline{CS(P)} < \overline{CS(K_1)} < \overline{CS(K_2)} < \overline{CS(K_3)} < \overline{CS(Q)}$$

In the special case that point P or Q lie on the line CD, the first or last inequality follows from the triangle inequality.

A Dedekind cut Σ_1, Σ_2 of line l is now defined in the following way:

$K \in \Sigma_1$ if K is a point of the segment PQ and $S(K)$ lies inside or on the circle $\mathcal{C}$, or if $K * P * Q$.

$K \in \Sigma_2$ if K is a point of the segment PQ and $S(K)$ lies outside of circle $\mathcal{C}$, or if $P * Q * K$.

First we check that this definition does produce a Dedekind cut. It is enough to explain one case. Let $r(\mathcal{C})$ be the radius of circle $\mathcal{C}$. Assume that $K_1, K_3 \in \Sigma_1$, and K_1, K_3 lie both in the segment PQ. By definition that means that $\overline{CS(K_1)} \leq r(\mathcal{C})$ and $\overline{CS(K_3)} \leq r(\mathcal{C})$. Hence formula (*) implies $\overline{CS(K_2)} \leq r(\mathcal{C})$ and hence $K_2 \in \Sigma_1$, as to be shown.

The remaining cases are even more easy to dispose. For example, assume that $K_1 * P * Q$ and only $K_3 \in \Sigma_1$ lies in the segment PQ. In that case, either $K_2 * P * Q$ or $K_2 = P$ and hence $K_2 \in \Sigma_1$ and we are finished. Or, $P * K_2 * K_3 * Q$, and we proceed as above, with $K_1 := P$.

Because Σ_1, Σ_2 is a Dedekind cut, by Dedekind's axiom there exists a cut point K^* on the line l. Now $\Sigma_1 \cup \{K^*\}$ and $\Sigma_2 \cup \{K^*\}$ are the two opposite rays on the line l with vertex K^*. Let $S^* := S(K^*)$. Recall that this is the point on circle $\mathcal{D}$ and the ray $\overrightarrow{DK^*}$.

The claim is that S^* lies on the circle $\mathcal{C}$, too. To confirm this claim, we need the following Lemma.

Lemma II.10.2. *Let S and S_{out} be points be a point inside and outside the circle $\mathcal{C}$, both lying on the arc of the second circle $\mathcal{D}$ between points P and Q.*

Every such point S_{out} lies between two other points S_1 and S_3 on the circle $\mathcal{D}$ and outside circle $\mathcal{C}$.

Similarly, point S lies between two other points S_4 and S_2 on the circle $\mathcal{D}$ and inside circle $\mathcal{C}$.

Moreover, the corresponding projections onto the line PQ have the order $K_4 * K * K_2 * K_1 * K_{out} * K_3$.

Proof. Let K_{out} be the intersection of the circle $\mathcal{C}$ with the central ray $\overrightarrow{CS_{out}}$—which can be constructed with Hilbert tools by transferal of any radial segment. The tangent line to circle $\mathcal{D}$ at point S_{out} is the perpendicular to the radius DS_{out} and can easily be constructed.

Let T_1 and T_3 be the points on the tangent line, on both sides of point S_{out}, such that $S_{out}T_1 < S_{out}K_{out}$ and $S_{out}T_3 < S_{out}K_{out}$. We claim that T_1 and T_3 lie outside the circle $\mathcal{C}$. Indeed, this follows from the weak triangle inequality since

$$CT_i \geq CS_{out} - S_{out}T_i \cong CK_{out} + K_{out}S_{out} - S_{out}T_i > CK_{out} \tag{II.10.3}$$

for $i = 1, 3$.

Let S_1 and S_3 be the points where the central rays $\overrightarrow{DT_i}$ intersect the circle $\mathcal{D}$. We claim that these points S_i lie outside the circle $\mathcal{C}$, too. Indeed, the triangle $\triangle CS_{out}S_i$ is isosceles, and hence

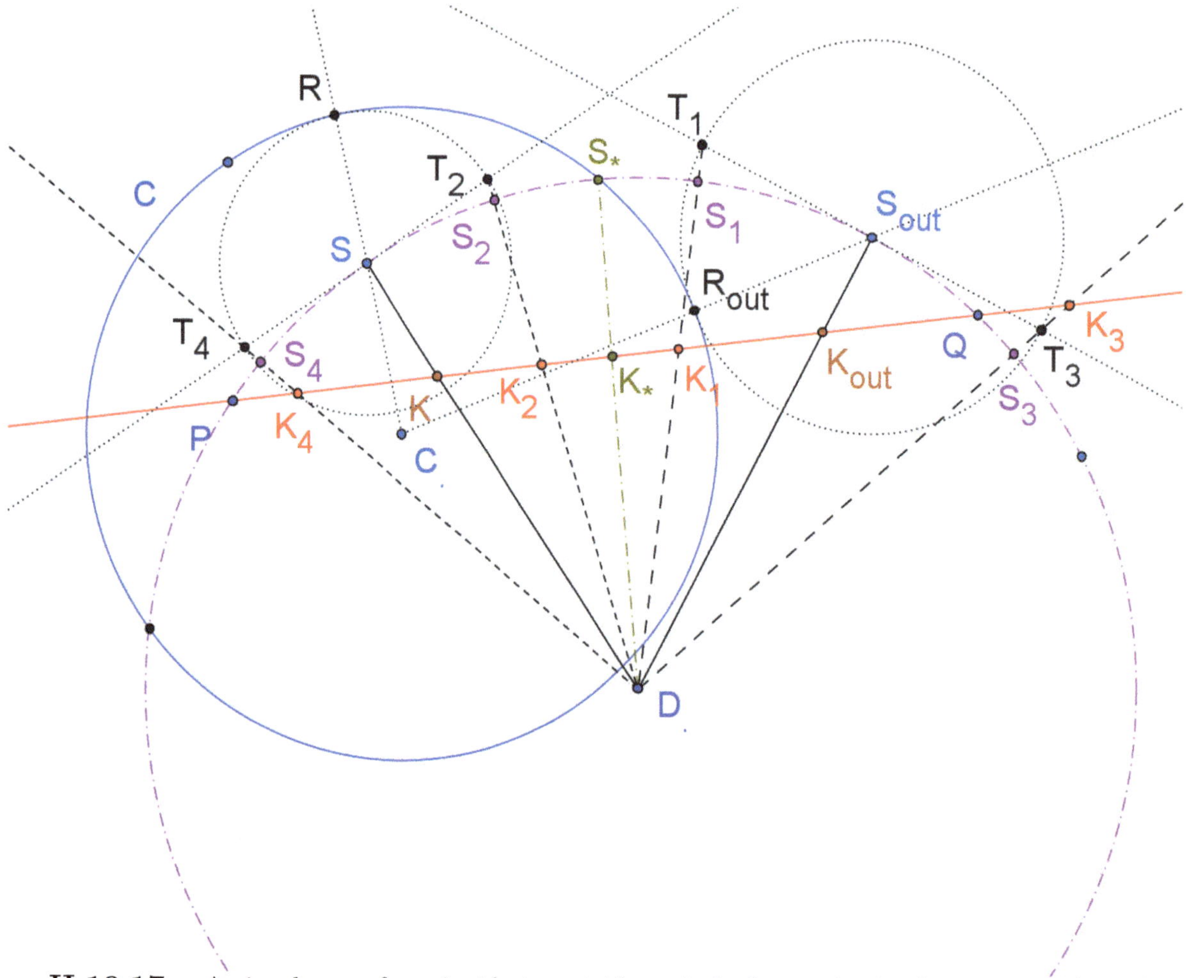

Figure II.10.17: A circular arc from inside to outside a circle $\mathcal{C}$ contains both points inside and outside that circle.

has acute base angles. Hence the supplementary angles $\angle S_i S_{out} T_i$ are obtuse. Because the side across an obtuse angle is the longest side in any triangle, we conclude $S_{out}S_i < S_{out}T_i$. Hence these points lie inside the circle drawn around S_{out} through point K_{out}. By the same reasoning as used in equation (II.10.3), they lie outside the circle $\mathcal{C}$.

Finally, the intersection points K_i of the central rays $\overrightarrow{DT_i}$ with the line PQ lie on different sides of point K.

A similar argument is used to get the points S_4 and S_2 on the circle $\mathcal{D}$ and inside circle $\mathcal{C}$, lying on different sides of line DS. □

Finally, we confirm that point S^* lies on the circle $\mathcal{C}$, too.

The cut point satisfies $P * K^* * Q$. Now, assume towards a contradiction that point S^* would lie inside circle $\mathcal{C}$. Then Lemma (II.10.2) would yield existence of points $P, K_2 \in \Sigma_1$ such that $P * K^* * K_2 * Q$. Then P and K_2 would be two points in Σ_1 lying on opposite sides of the cut point K^*. This is impossible. Similarly, using Lemma (II.10.2), we confirm that it is impossible that S^* lies outside circle $\mathcal{C}$.

The only remaining possibility is that S^* lies on circle $\mathcal{C}$, and hence it is a intersection point of the two circles $\mathcal{C}$ and $\mathcal{D}$. A second intersection point of the two circles is constructed by reflection across the line CD. □

II.11 Towards a Natural Axiomatization of Geometry

The title begs the question: what is meant in mathematics by the attribute "natural"? By itself, this clearly is a philosophical question beyond mathematics. First of all, mathematical postulates are and will in the future surely be chosen,—even more often,—based on criteria from history, or rooted in the convictions held by the scientific community. After that being understood, which place can be left to the requirement of "natural"? Nevertheless, the author of these notes wants to express his conviction, that "natural" for a mathematical postulate is important, and means the following:

- A postulate which uses notions connected to the everyday life experience, or are well motivated by the applications, may be called natural.
- A postulate which connects notions used by the proofs, or which facilitates the mathematical developments, may be called natural.
- A postulate which allows surprising connections between different mathematical theories, or shows an underlying unity of mathematics, may be called natural.

The material of this section is selected with such considerations in mind.

II.11.1 The Uniformity Theorem

Main Theorem 17 (Uniformity Theorem). *Any Hilbert plane is of either one of the following three types:*

semi-Euclidean *The angle sum of every triangle is two right angle, and every Lambert or Saccheri quadrilateral is a rectangle.*

semi-hyperbolic *The angle sum of every triangle is less than two right angle, and every Lambert or Saccheri quadrilateral has one respectively two acute angles.*

semi-elliptic *The angle sum of every triangle is larger than two right angle, and every Lambert or Saccheri quadrilateral has one respectively two obtuse angles.*

The Uniformity Theorem motivates the following definition:

Definition II.11.1 (Three basic types of Hilbert planes). According to the three cases occurring in the Uniformity Theorem—

A *semi-Euclidean plane* is a Hilbert plane for which the angle sum of every triangle is two right angles.

A *semi-hyperbolic plane* is a Hilbert plane for which the angle sum of every triangle is less than two right angles.

A *semi-elliptic plane* is a Hilbert plane for which the angle sum of every triangle is larger than two right angles.

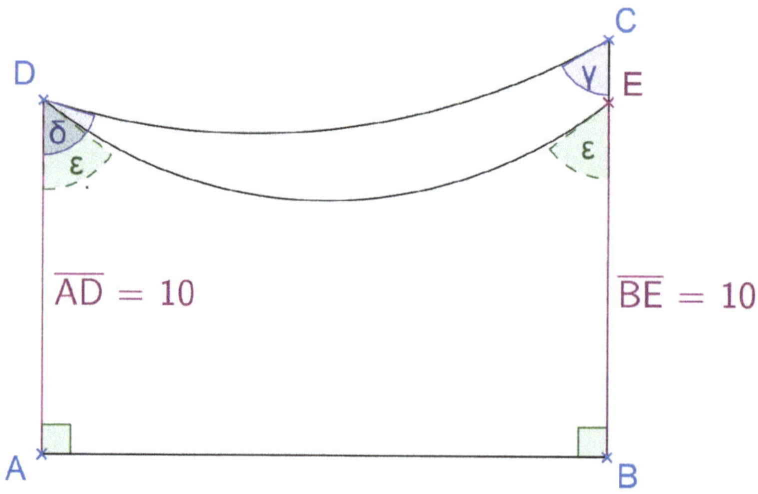

Figure II.11.1: Across the longer side BC is the larger angle δ.

Lemma II.11.1. *Given is any quadrilateral $\square ABCD$ with right angles at vertices A and B, vertices C and D lying on the same side of line AB. The angles γ at vertex C and δ at vertex D satisfy*

$$\begin{aligned} \gamma < \delta & \quad \textit{if and only if} \quad BC > AD \\ \gamma \cong \delta & \quad \textit{if and only if} \quad BC \cong AD \\ \gamma > \delta & \quad \textit{if and only if} \quad BC < AD \end{aligned}$$

Proof. As shown in the figure, we assume $BC > AD$. We have to check whether $\gamma < \delta$. By transferring segment AD, we produce the congruent segment $BE \cong AD$ such that $\square ABED$ is a Saccheri quadrilateral. Let its top angles be congruent to ε.

In the triangle $\triangle DEC$, the exterior angle theorem yields $\gamma < \varepsilon$. Since points A and C lie on different sides of line DE, angle comparison at vertex D implies $\varepsilon < \delta$. Hence $\gamma < \varepsilon < \delta$, and transitivity of angle comparison yields $\gamma < \delta$, as to be shown.

By a similar argument, we prove that $BC < AD$ implies $\gamma > \delta$. Finally $BC \cong AD$ implies $\gamma \cong \delta$, since the top angles of a Saccheri quadrilateral are congruent. □

Lemma II.11.2. *Given is a Saccheri quadrilateral $\square ABCD$, with right angles at vertices A and B. From any point P inside the top segment CD, the perpendicular is dropped onto the segment AB, with foot point Q. The following three cases can occur:*

$$\begin{aligned} \gamma < R & \quad \textit{and} \quad PQ < AD \\ \gamma = R & \quad \textit{and} \quad PQ \cong AD \\ \gamma > R & \quad \textit{and} \quad PQ > AD \end{aligned}$$

Proof. At vertex P there occur the supplementary angles $\alpha = \angle DPQ$ and $\beta = \angle CPQ$.

We begin by assuming $PQ < AD$, and look for a result about the top angle γ. Using the previous Lemma II.11.1 for the quadrilateral $\square AQPD$, we conclude $\delta < \alpha$. Using the Lemma II.11.1 once more for the quadrilateral $\square QBCP$, we conclude $\gamma < \beta$. Hence angle addition yields

$$2\gamma = \gamma + \delta < \beta + \alpha = 2R$$

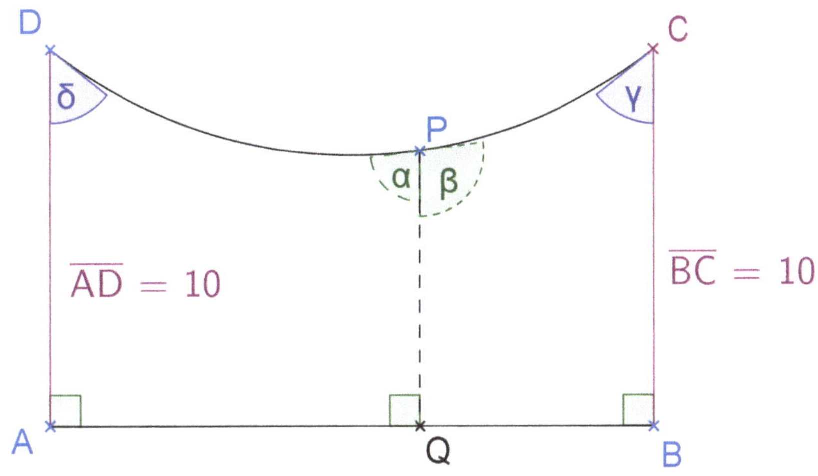

Figure II.11.2: For an acute top angle, the perpendicular dropped from a point P inside the top segment CD is shorter than the opposite sides.

and hence $\gamma < R$.

By a similar argument, the assumption $PQ > AD$ implies $\gamma > R$, and finally, indeed, the assumption $PQ \cong AD$ implies $\gamma = R$. ☐

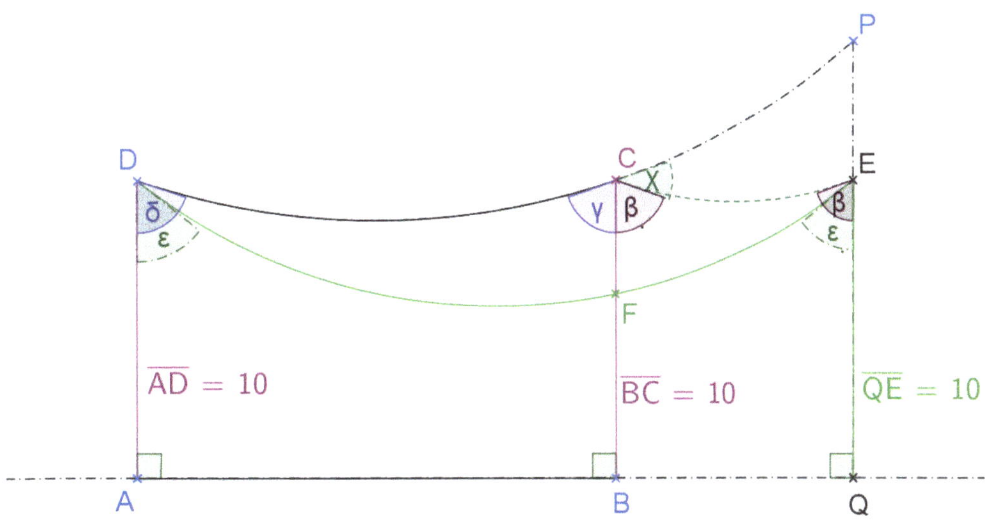

Figure II.11.3: For an acute top angle γ, the perpendicular PQ dropped from a point outside the top segment is longer than the opposite sides BC.

Lemma II.11.3. *Given is a Saccheri quadrilateral $\square ABCD$, with right angles at vertices A and B. From any point P on the ray $\overrightarrow{DC}$ outside of the top segment CD, the perpendicular is dropped onto the segment AB, with foot point Q. The following three cases can occur for the top angle:*

$$\gamma < R \quad \textit{and} \quad PQ > AD$$
$$\gamma = R \quad \textit{and} \quad PQ \cong AD$$
$$\gamma > R \quad \textit{and} \quad PQ < AD$$

Proof. By transferring segment AD, we produce the congruent segment $QE \cong AD$ such that $\square AQED$ is a Saccheri quadrilateral. Let its top angles be congruent to ε. We get the third Saccheri quadrilaterals $\square BQEC$, denote its top angles by β. The line BC intersects segment DE (Why?). Let F be the intersection point.

At first, we assume $QP > AD$, and check that $\gamma < R$. The three points A, B and Q on the base line lie on the same side of top line DE, since baseline and top line of the Saccheri quadrilateral $\square AQED$ have the middle line as their common perpendicular, and hence are parallel. Since $QP > AD \cong QE$, points Q and P lie on different sides of line DE. Hence points P and C lie on the (upper) side of DE, whereas A, B and Q lie on the lower side. The intersection point F of line DE with line BC lies between B and C.

Angle addition at vertex C yields $\gamma + \beta + \chi = 2R$. In the triangle $\triangle DEC$, the exterior angle $\chi > \delta - \varepsilon = \gamma - \varepsilon$. Finally, comparison of angles at vertex E shows that $\beta > \varepsilon$. Put together, we get

$$2R = \gamma + \beta + \chi > \gamma + \beta + \gamma - \varepsilon > 2\gamma$$

and hence $\gamma < R$, as to be shown.

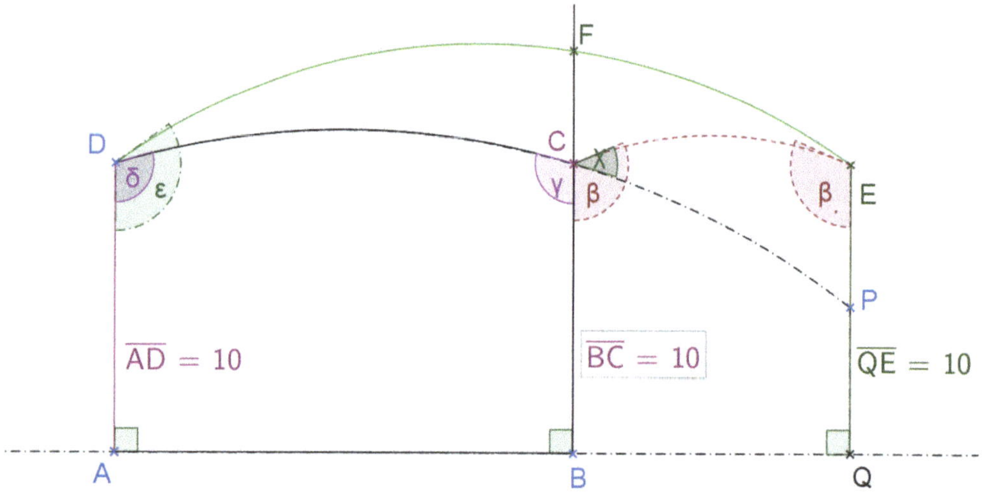

Figure II.11.4: For an obtuse top angle γ, the perpendicular PQ dropped from a point outside the top segment is shorter than the opposite sides BC and AD.

Under the assumption $PQ < AD$, several modifications occur, as can be seen in the figure on page 284. Indeed, because of $QP < AD \cong QE$, points Q and P lie on the same side of line CE. Hence all five points A, B, Q, E and D lie on the same (lower) side of top line DE. The intersection point F of line DE with line BC lies outside segment BC.

Angle addition at vertex C yields $\gamma + \beta - \chi = 2R$. In the triangle $\triangle DEC$, the exterior angle $\chi > -\delta + \varepsilon = -\gamma + \varepsilon$. Finally, comparison of angles at vertex E shows that $\beta < \varepsilon$. Put together, we get

$$2R = \gamma + \beta - \chi < \gamma + \beta + \gamma - \varepsilon < 2\gamma$$

and hence $\gamma > R$, as to be shown. □

Lemma II.11.4. *Given are two Saccheri quadrilaterals $\square ABCD$ and $\square A'B'C'D'$ with a common middle segment MN. Then there top angles γ and γ' are either both acute, both right, or both obtuse.*

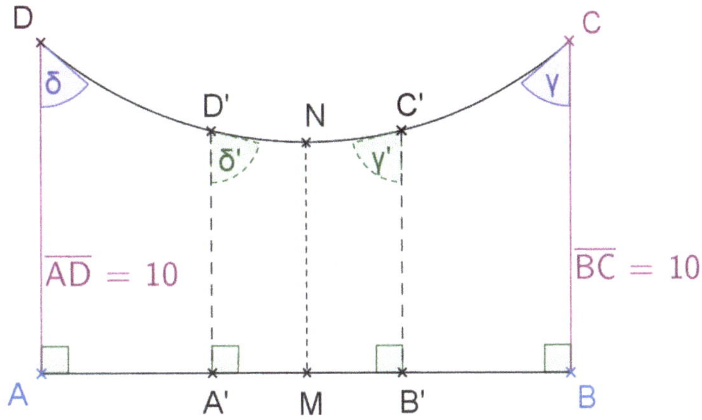

Figure II.11.5: Two Saccheri quadrilaterals with a common middle line MN have either both acute, both obtuse, or both right top angles.

Proof. Without loss of generality we may assume $A * A' * M * B' * B$ as order of the vertices on the base line. Using the previous Lemma II.11.2 for the Saccheri quadrilateral $\square ABCD$, we get the three equivalences

$$\begin{aligned} \gamma < R \quad &\text{if and only if} \quad A'D' < BC \\ \gamma \cong R \quad &\text{if and only if} \quad A'D' \cong BC \\ \gamma > R \quad &\text{if and only if} \quad A'D' > BC \end{aligned}$$

Using Lemma II.11.3 for the Saccheri quadrilateral $\square A'B'C'D'$, we get the three equivalences

$$\begin{aligned} \gamma' < R \quad &\text{if and only if} \quad A'D' < BC \\ \gamma' \cong R \quad &\text{if and only if} \quad A'D' \cong BC \\ \gamma' > R \quad &\text{if and only if} \quad A'D' > BC \end{aligned}$$

Put together, we see that angles γ and γ' are either both acute, both right, or both obtuse, as to be shown. $\square$

End of the proof of the Uniformity Theorem. Given a Saccheri quadrilateral $\square ABCD$ with middle line MN, and any second Saccheri quadrilateral. We transfer the middle line of the second Saccheri quadrilateral onto the ray $\overrightarrow{MB}$. By means of congruent triangles, it is straightforward to verify that we can produce a Saccheri quadrilateral $\square QPRS$ which is congruent to the second given one, and has the bottom line $SQ = MN$ and the middle line $MK = AB$.

The two lines CD and PR do intersect (Why?). With the intersection point L, we get a Lambert quadrilateral $\square MKLN$. Let $\lambda = \angle KLN$ be its top angle.

By Lemma II.11.4, we conclude that the top angles γ and λ of the quadrilaterals $\square ABCD$ and $\square MKLN$ are either both acute, right, or obtuse. Similarly, we conclude that the top angles φ and λ of the quadrilaterals $\square QPRS$ and $\square MKLN$ are either both acute, right, or obtuse. Hence the top angles γ and φ of the two given Saccheri quadrilaterals are either both acute, right, or obtuse. $\square$

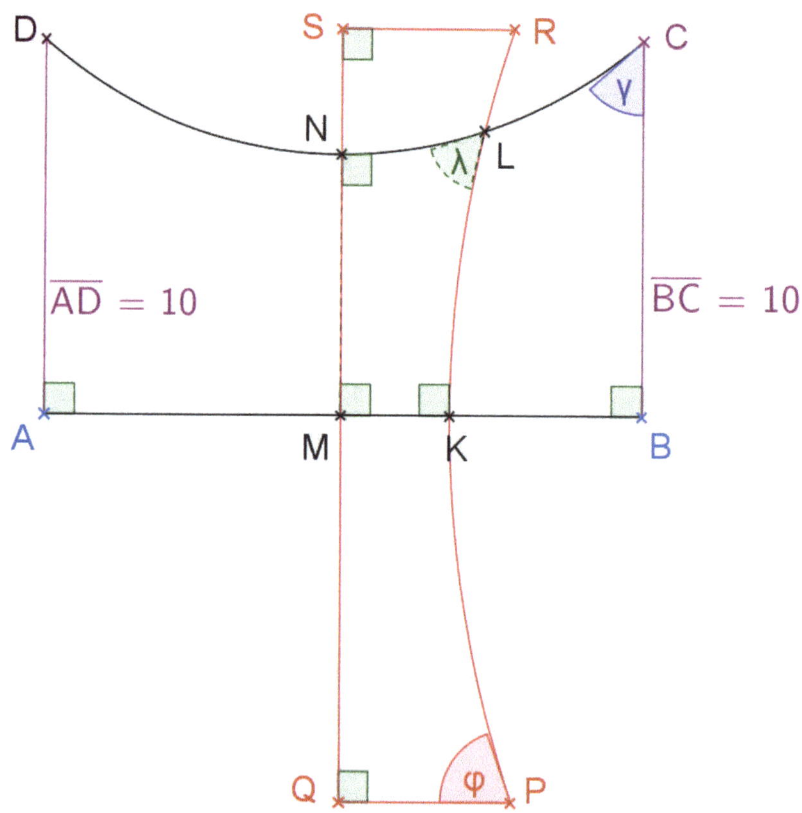

Figure II.11.6: Any two Saccheri quadrilaterals can be put into a position such that the middle line of the first one is the base line of the second one, and vice versa.

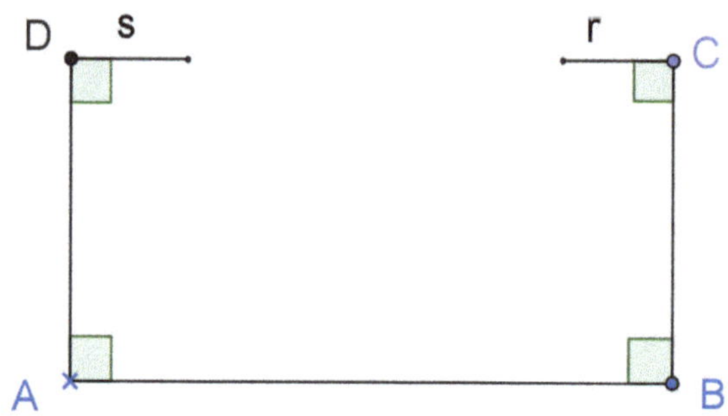

Figure II.11.7: Does the figure close to a rectangle?

II.11.2 Some strange polygons

Problem II.11.1. *In the figure on page 286 is shown how to start a construction of a rectangle, in neutral geometry. In Euclidean geometry, the two rays r and s have a common segment, and the rectangle closes. But in neutral geometry, there are other possibilities. Indeed, it is both possible that these two rays intersect, or they lie on parallel lines. I claim: if the two rays r and s intersect but do not lie on the same line, their intersection point lies on the perpendicular bisector of segment AB. Give a convincing argument to confirm this statement.*

Answer. Assume that these two rays do not lie on the same line and intersect at a point $P = r \cap s$. We reflect both rays across the perpendicular bisector p of segment AB. Because of the symmetry of the figure, one obtains the rays $r' = s$ and $s' = r$,— the two rays have simply been exchanged by the reflection. The intersection of the reflected rays is $P' = r' \cap s'$.

The points P and P' are mirror images of each other. Since the two rays lie on different lines, by axiom (I.2), their intersection point is unique. Hence $P = r \cap s = r' \cap s' = P'$ and thus the two rays intersect on the perpendicular bisector.

Problem II.11.2. *In the figure on page 286 is shown how to start a construction of a rectangle, in neutral geometry. Indicate in several figures all different possible cases.*

Answer. We draw the segment CD and obtain the Saccheri quadrilateral $\square ABCD$. By the Uniformity Theorem, this Saccheri quadrilateral has either right, acute, or obtuse top angles.

In the case of a right top angle, one obtains a rectangle $\square ABCD$. In this case, both rays r and s lie on the line CD. In the case of acute top angles, the ray r lies outside the angle $\angle BCD$. Ray r may or may not intersect the perpendicular bisector.

In the case of obtuse top angles, the ray r lies inside the angle $\angle BCD$, and ray s inside the angle $\angle ADC$. Together, one obtains either one of the cases indicated in the figure on page 287.

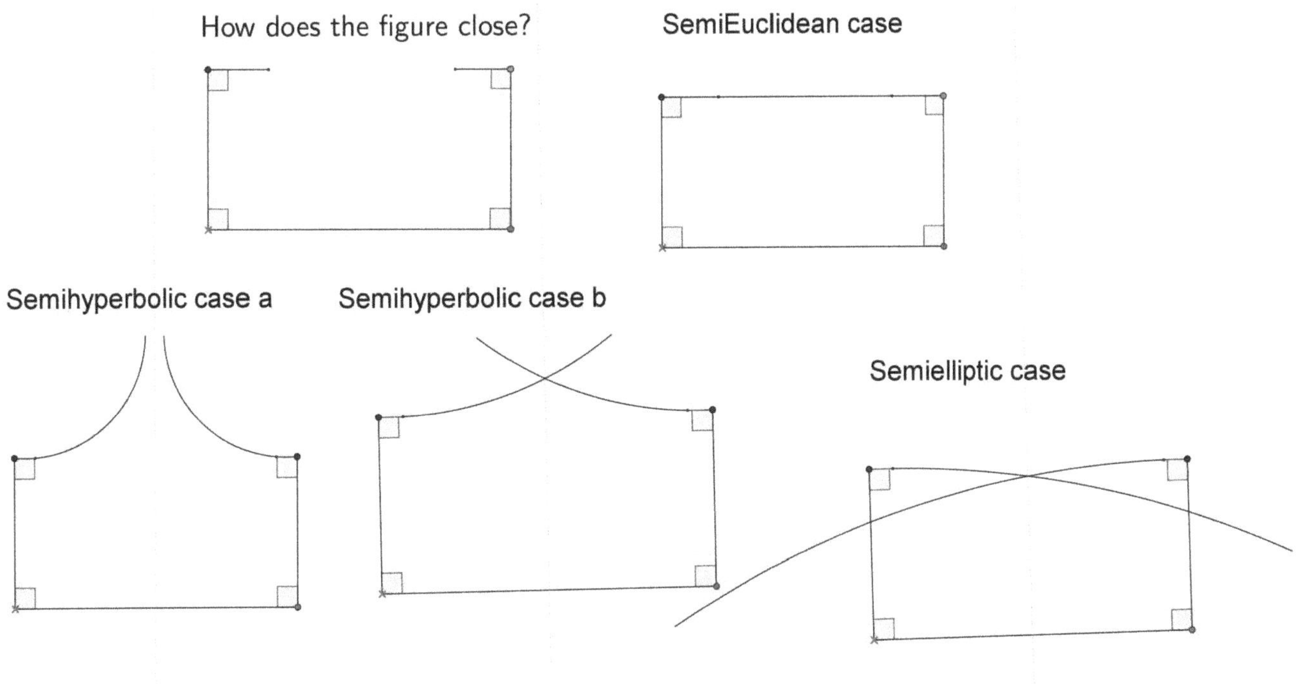

Figure II.11.8: Still not always a rectangle.

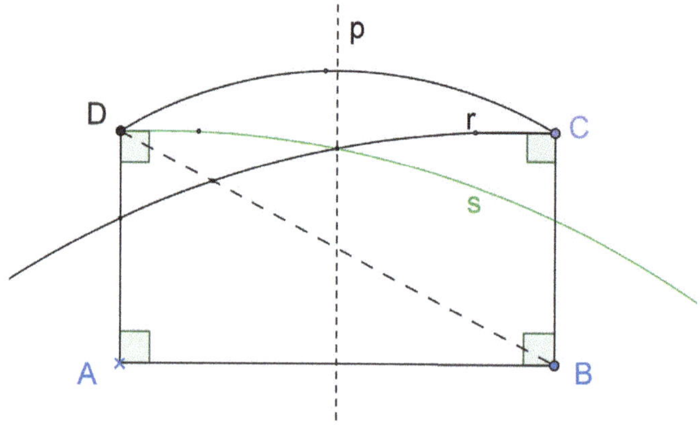

Figure II.11.9: Following the ray r.

Problem II.11.3. *While starting the construction of a rectangle for the case where Saccheri quadrilaterals have obtuse top angles, the "accident" shown in the figure on page 288 is happening. Complete the paragraph below. Indicate in the figure the three points X, Y, Z mentioned in the paragraph below.*

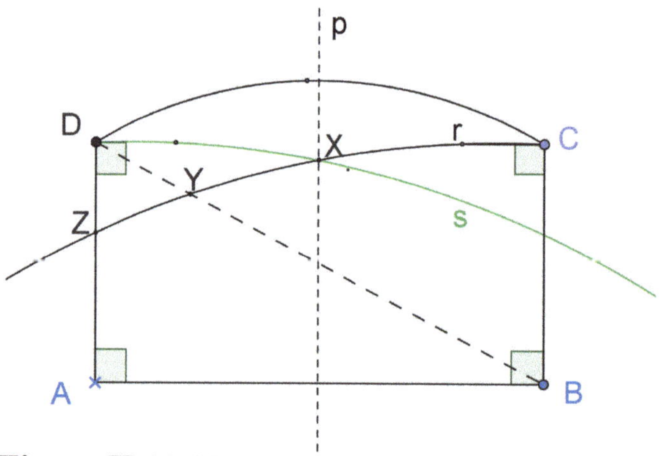

Figure II.11.10: The points X, Y, Z on the ray r.

Answer. We draw the segment CD and obtain the Saccheri quadrilateral $\square ABCD$. In the case of obtuse top angles, the ray r lies inside the obtuse angle $\angle BCD$, and ray s inside the angle $\angle ADC$. Indeed the rays r and s intersect the perpendicular bisector p of segment AB, as one may confirm as follows. By the Crossbar theorem , the ray r intersects side BD of triangle $\triangle BCD$, say at point Y. Hence by Pasch's axiom , ray r intersects two sides of triangle $\triangle ABD$. But ray r and segment AB cannot intersect, otherwise one would obtain a triangle with two right angles at vertices B and C, which is impossible. Hence ray r intersects segment AD, say at point Z. Since the points C and Z lie on different sides of the perpendicular bisector p, the ray r intersects the bisector p, say at point X. Since ray s is the mirror image of ray r, and the intersection point $r \cap s$ is unique, point $X = p \cap r = p \cap s = r \cap s$ is the intersection of the rays r and s, too.

Problem II.11.4. *Suppose a quadrilateral $\square ABCD$ has right angles at vertices A and B, and congruent sides $AB \cong CD$. Moreover, we may assume $d = AD \leq BC = c$.*

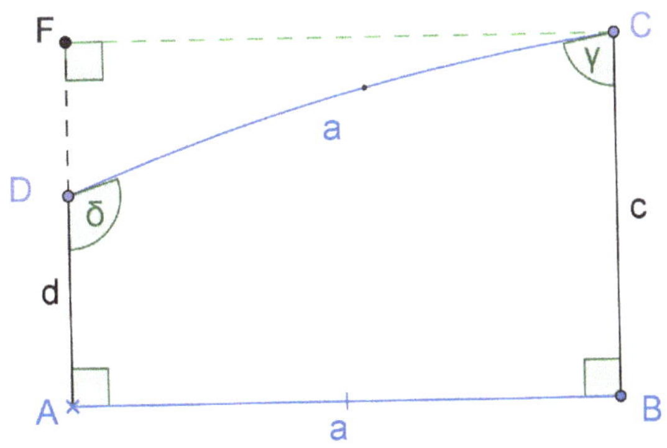

Figure II.11.11: What kind of quadrilateral does one get?

(i) *We drop the perpendicular from point C onto side AD, and let F be the foot-point. Give a reason why either (a) $F = D$ or (b) $CF < AB$.*

(ii) *Use Lemma II.11.1 to conclude that in case (a) the quadrilateral is a rectangle.*

(iii) *Use Lemma II.11.1 to conclude that in case (b) the angle $\angle FCB$ is obtuse.*

(iv) *Use the Uniformity Theorem to conclude that in case (a) the Hilbert plane is semiEuclidean, and in case (b) the Hilbert plane is semielliptic. Exclude the semihyperbolic case.*

(v) *We may assume $d = AD \le BC = c$. Use Lemma II.11.1 to conclude that in case (b) $A*D*F$.*

(vi) *Use Lemma II.11.1 to conclude that in case (b) $c = AD \le BC = d < AF$.*

Answer. **(i)** If $F = D$ we are ready, so we assume $F \neq D$. Hence the right triangle $\triangle FCD$ exists. We know that the hypothenuse is the longest side.
Hence $CF < DC \cong AB$ which is case (b).

(ii) In case (a), we get $\delta = 90°$ and $d = c$. Hence Lemma II.11.1 yields $\gamma = \delta$, and the quadrilateral $\square ABCD$ is a rectangle.

(iii) In case (b), Lemma II.11.1 yields $\angle FCB > \angle CFA$. Since $\angle CFA = 90°$, we conclude that $\angle FCB$ is obtuse.

(iv) The Uniformity Theorem yields that in case (a) the Hilbert plane is semi-Euclidean, and in case (b) the Hilbert plane is semi-elliptic. Since all possibilities have been exhausted, we have excluded the semi-hyperbolic case.

(v) We may assume $d = AD \le BC = c$. The angle $\angle CDF$ is an interior angle of a right triangle and hence acute. But the angle δ is obtuse in case (b). Hence the angles $\angle CDF$ and δ are distinct. This excludes the order $A * F * D$. Since points F and D are on the same side of line AB, only the order $A * D * F$ is possible.

(vi) We know by item (iii) that $\angle FCB$ is obtuse. Hence Lemma II.11.1 implies $BC < AF$, and indeed $c = AD \le BC = d < AF$.

II.11.3 Defect and AAA congruence

Proposition II.11.1 (Semi-hyperbolic AAA-Congruence Theorem). *In a semi-hyperbolic plane, two triangles having three pairs of congruent angles are congruent.*

Proof. The Uniformity Theorem 17 implies for a semi-hyperbolic geometry, any triangle $\triangle ABC$ has a positive defect

$$\delta(ABC) = 180° - \alpha - \beta - \gamma$$

Moreover we use Lemma II.9.2 about the additivity of the defect for non-overlapping triangles. By

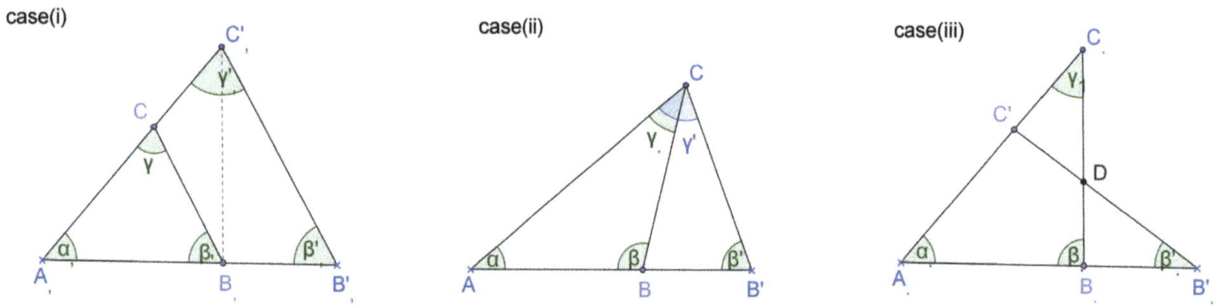

Figure II.11.12: About the semi-hyperbolic AAA congruence.

transfer of one of the triangles, we reduce the theorem to the case where the two triangles have a common vertex and the adjacent sides lie on the same two rays. Hence we get two triangles $\triangle ABC$ and $\triangle AB'C'$ such that,—possibly after some renaming,—either one of the cases (0), (i), (ii) or (iii) occurs:

(0) $B = B'$ and $C = C'$;

(i) $A * B * B'$ and $A * C * C'$;

(ii) $A * B * B'$ and $C = C'$;

(iii) $A * B * B'$ and $A * C' * C$.

- Case (i) is the most interesting one to be excluded. We use Lemma II.9.2 about the additivity and positivity of the defect to get

  $$\delta(AB'C') = \delta(ABC) + \delta(BCC') + \delta(BB'C') > \delta(ABC)$$

  and hence $\beta' + \gamma' > \beta + \gamma$, contradicting the assumption that the two triangles are equiangular

- The case (ii) is straightforward to exclude and is left out.

- In case (iii), as explained in Problem II.6.2, the segments BC and $B'C'$ intersect, say at point D. For the triangle $\triangle BDB'$, the exterior angle theorem tells $\beta > \beta'$. This contradicts the assumption $\beta = \beta'$ and excludes this case.

Hence only the case (0) is possible, showing that the triangles $\triangle ABC$ and $\triangle AB'C'$ are indeed congruent . □

Hence only the case (0) is possible, showing that the triangles $\triangle ABC$ and $\triangle AB'C'$ are indeed congruent . □

Remark. Alternatively, the cases (ii) and (iii) are excluded since the z-angle theorem implies that the sides $BC \parallel B'C'$ are parallel. As already seen by the addendum to the extended ASA-Congruence, Proposition II.7.47 only cases (0) and (i) are possible.

Remark. In any semi-elliptic plane, too, AAA-congruence holds. In this case, the Uniformity Theorem 17 implies that all triangles have a positive excess of the angle sum

$$\varepsilon(ABC) := \alpha + \beta + \gamma - 180^\circ$$

Again, the additivity lemma holds. Since the excess is strictly positive and additive, the AAA-congruence holds in the semi-elliptic geometry, too. We see that AAA-congruence holds in the semi-elliptic as well as the semi-hyperbolic geometry, but not in the semi-Euclidean geometry.

II.11.4 A hierarchy of planes

As we have learned in definition I.2.1:

Definition (I.2.1). A Hilbert plane is any model for two-dimensional geometry where Hilbert's axioms of incidence (I.1)(I.2)(I.3a)(I.3b), order (II.1) through (II.4), and congruence (III.1) through (III.5) hold.

Neither the axioms of continuity—Archimedean axiom and the axiom of completeness—nor the parallel axiom need to hold for an arbitrary Hilbert plane. The Uniformity Theorem brings confidence that these few axioms are a correct start towards a natural axiomatization of full-blown Euclidean geometry.

Which postulates do we want to be added to the Hilbert plane axioms? As the next step towards a natural axiomatization of Euclidean geometry, we investigate the Euclidean parallel property. This section gives an account of some recent achievements, mainly of M.J. Greenberg in the axiomatization of geometry. We begin with the axioms of a Hilbert plane, and successively postulate few additional axioms with the effort to arrive at a system as close as possible to Euclid's system. By further axioms—mainly of continuity—the system is narrowed to a system that has the real Cartesian plane as its unique model.

II.11.5 Wallis' axiom

Already in the section on Legendre's theorem, we have addressed the question to find a clear cut, suggestive or self-evident postulate that would better replace the Euclidean postulate.

John Wallis (1616-1703), in a treatise on Euclid published in 1693, was astute enough to propose a new postulate that he believed to be more plausible than Euclid's parallel postulate. He phrased it as follows:

> Finally (supposing the nature of ratio and of the science of similar figures already known), I take the following as a common notion: to every figure there exists a similar figure of arbitrary magnitude.

We want to make Wallis' idea precise in the context of axiomatic geometry, beginning with Hilbert's axiom. Euclid's theorems about similar triangles depends on the Archimedean axiom—via his definition of equality of ratios. Hence it is better to restrict attention to equiangular triangles instead of similar figures. Recall that equiangular triangles have three pairs of congruent angles. Both Greenberg and Hartshorne suggest a modification of Wallis' axiom along these lines. In a more common language, the first postulate says that equiangular triangles exist in any size, the second one that equiangular triangles exist in any arbitrarily large size.

(Wallis-Greenberg Postulate). *Given any triangle $\triangle ABC$ and segment DE there exists a triangle with DE as one side such that the triangles $\triangle ABC$ and $\triangle DEF$ are equiangular.*

(Wallis-Hartshorne Postulate). *Given any triangle $\triangle ABC$ and segment DE there exists a triangle $\triangle A'B'C'$ have side $A'B' \geq DE$ such that the triangles $\triangle ABC$ and $\triangle A'B'C'$ are equiangular.*

In this paragraph, we show the following theorem:

Theorem II.11.1. *The following three postulates are equivalent in any Hilbert plane:*

(a) *the Euclidean parallel postulate*

(b) *the Wallis-Greenberg postulate*

(c) *the Wallis-Hartshorne postulate*

Proposition II.11.2 (Equiangular triangles exist in all given sizes). *The Euclidean parallel postulate implies that equiangular triangles exist in any given size, and hence the Wallis-Greenberg postulate.*

In particular, for any triangle $\triangle ABC$ and segment $A'B'$ there exists a triangle with $A'B'$ as one side, and lying in a given half plane of this side, such that the triangles $\triangle ABC$ and $\triangle A'B'C'$ are equiangular.

Proof. Given is triangle $\triangle ABC$ and segment $A'B'$, as shown in the figure on page 293. We transfer segment AB onto the ray $\overrightarrow{A'B'}$ and get the congruent segments $AB \cong A'E'$.

As in the extended ASA-theorem, we construct the congruent triangles $\triangle ABC \cong \triangle A'E'F'$. Indeed, as already explained in the proof of the extended ASA-theorem, the ray r'_A forming the angle $\angle BAC$ with ray $\overrightarrow{A'B'}$ and the ray r'_E forming the angle $\angle ABC$ with ray $\overrightarrow{E'A'}$, both constructed in the same half-plane of line $A'B'$, intersect at point F'.

Additionally, we construct the ray r'_B forming the same angle $\angle ABC$ with ray $\overrightarrow{B'A'}$, again in the same half-plane of line $B'A'$. The rays r'_B and r'_E are parallel by [Euclid I.27], see II.7.42. The ray r'_A intersects one of these parallel rays. Hence by Proclus' Lemma I.3.3, it intersects the line of the second ray r'_B, too. Indeed, otherwise, the line of r'_B would have the two different parallels r'_A and r'_E both through point F', which is impossible.

Let now C' be the intersection point of rays r'_A and r'_B. The triangles $\triangle ABC$ and $\triangle A'B'C'$ have two pairs of congruent angles at vertices A and A', as well as vertices B and B', by construction.

We have seen in Proposition II.9.7 above that uniqueness of parallels implies that every triangle has angle sum 180°—every Pythagorean plane is semi-Euclidean.

Hence the triangles $\triangle ABC$ and $\triangle A'B'C'$ have a third pair of congruent angles at vertices C and C' and are equiangular, as to be shown. □

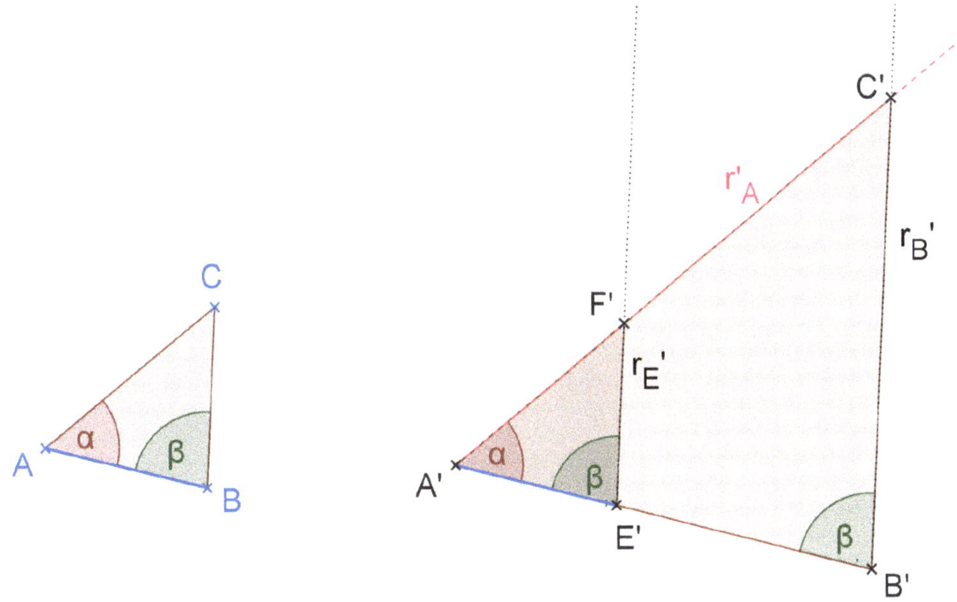

Figure II.11.13: Uniqueness of parallels implies Wallis' postulate.

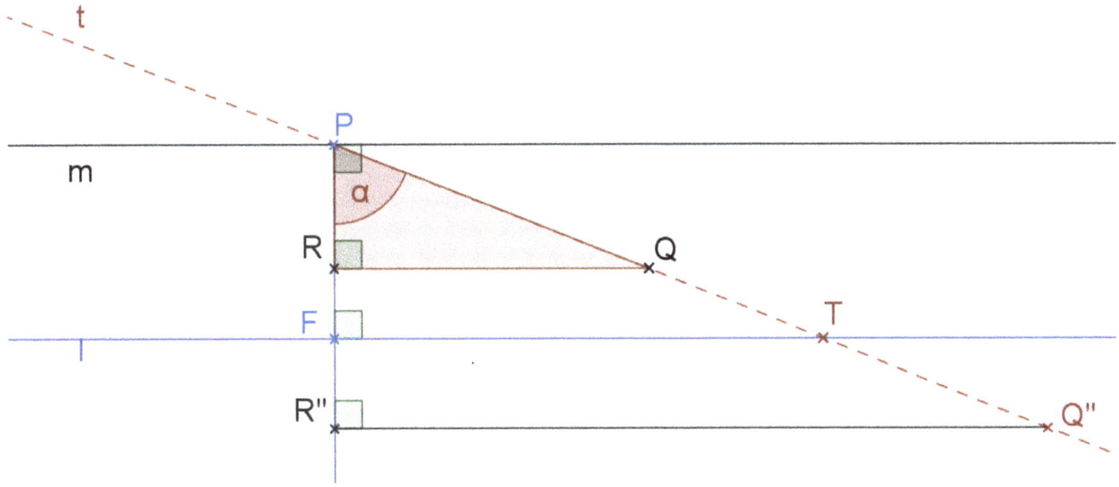

Figure II.11.14: Wallis' postulate implies uniqueness of parallels.

Proposition II.11.3. *The Wallis-Hartshorne postulate implies the Euclidean parallel postulate.*

Proof. Given is line l and a point P not on l. One parallel can be obtained as the "double perpendicular" as already explained in Proposition II.7.43. One drops the perpendicular p from point P onto line l and denotes the foot point by F. Next, one erects at point P the perpendicular to line PF. Thus, one gets a line m parallel to the given line l.

We now suppose towards a contradiction that there exists a second line t through point P which is parallel to line l, too. Let $\overrightarrow{t}$ be the ray on line t with vertex P and lying between the two parallels l and m. We choose any point $Q \neq P$ on this ray, and drop the perpendicular onto the line PF. The foot point R lies in the segment PF, since lines t and l do not intersect.

We now apply the Wallis-Hartshorn axiom to the triangle $\triangle PQR$ and the segment PF. Hence there exists an equiangular triangle $\triangle P'Q'R'$ with side $P'R' \geq PF$.

We transfer this segment and then the triangle onto the ray $\overrightarrow{PF}$ and get a triangle $\triangle PQ''R''$, which is congruent to $\triangle PQ'R'$ and hence equiangular to triangle $\triangle PQR$. We can put points Q and Q'' on the same side of line PF. Because of congruence of the angles $\alpha = \angle R''PQ'' \cong \angle RPQ$, the uniqueness of angle transfer imply that these rays are equal: $\overrightarrow{PQ} = \overrightarrow{PQ''}$. Hence point Q'' lies on the line $t = PQ$.

Since $PR'' \geq PF$, we conclude that either $R'' = F$ or R'' and P lie on different sides of F. We consider the second case. The foot points R and R'' lies on different sides of F and hence the line l. The lines RQ, l and $R''Q''$ are parallel, being all three perpendicular to PF. Hence points R and Q lie on one side of line l and points R'' and Q'' lie on the other side. Hence the segment QQ'' intersects the line l, say in point T. In the first case, we conclude that point $Q'' =: T$ lies on line l.

Hence, in both cases, the lines l and t do intersect. We have confirmed no second parallel t to line l through point P can exist. □

II.11.6 Proclus' Theorem

We shall repeatedly use

Lemma II.11.5 (Proclus lemma). *If one of two parallel lines is intersected by a third line, the other one is intersected, too.*

What is the missing link leading from the semi-Euclidean plane to a plane with uniqueness of parallels? Based on Proclus commentaries to Euclid, Greenberg has suggested the following angle unboundedness axiom. This axiom goes back to Aristole's book I of the treatise *De Caelo* ("On the heavens").

Definition II.11.2 (Aristole's Angle Unboundedness Axiom). For any acute angle θ and any segment PQ, there exists a point X on one side of the angle such that the perpendicular XY dropped onto the other side of the angle is longer than the given segment: $XY > PQ$.

Main Theorem 18 (Proclus' Theorem). *In any Hilbert plane, the Euclidean parallel postulate is equivalent to the following two requirements:*

- *Every triangle has angle sum two right angles.*
- *For every acute angle and every given segment, there exists a right triangle with the given angle and the leg across from it longer than the given segment.*

Corollary 28. *A Hilbert plane is Pythagorean if and only if it is semi-Euclidean and Aristole's axiom holds.*

We have seen in Proposition II.9.7 above that uniqueness of parallels implies that every triangle has angle sum $2R$. Hence every Pythagorean plane is semi-Euclidean. Indeed, uniqueness of parallels implies Aristole's axiom, too.

Proposition II.11.4. *Hilbert's parallel postulate implies Aristole's axiom.*

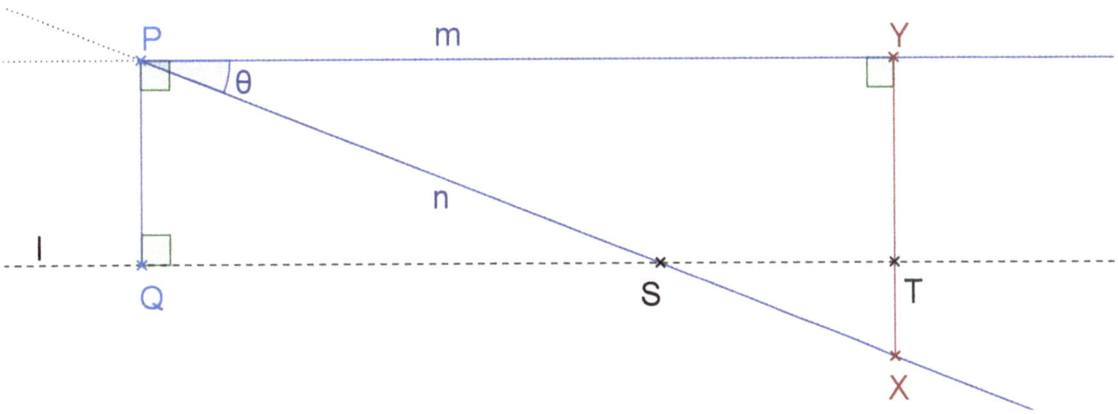

Figure II.11.15: Uniqueness of parallels implies unbounded opening of an angle.

Proof. Let the acute angle $\theta = \angle(m, n)$ and a segment be given. We transfer the segment onto the perpendicular erected at the vertex of the angle onto the side m and get the segment PQ. Next we erect the perpendicular l onto line PQ at Q. The lines l and m are parallel since they are both perpendicular to line PQ.

Because uniqueness of parallels has been assumed, the lines n and l are not parallel. Let S be their intersection point. Take any point X on the line n such that S is between P and X. We drop the perpendicular from X onto line m and let Y be the foot point. Since X and Y are on different sides of line l, the segment XY intersects the line l. Call T the intersection point.

The Lambert quadrilateral $\square QTYP$ is indeed a rectangle, as shown by the Uniformity Theorem. Its opposite sides are congruent by Lemma II.11.1. Hence we get

$$PQ \cong YT < YX$$

as claimed by Aristole's angle unboundedness axiom. $\square$

By Proposition II.11.4 and Proposition II.9.7, we know that Aristole's Axiom and the angle sum $2R$ for triangles are both necessary for uniqueness of parallels to occur. To complete the proof of Proclus' Theorem, we now show that these two assumptions are sufficient.

Proposition II.11.5. *A semi-Euclidean Hilbert plane, for which the Aristole's Axiom holds is Pythagorean.*

Proof. Given is a line l and a point P not on this line. As explained in Proposition II.7.43, we use the standard "double perpendicular" m to get a parallel to line l through point P. Let Q be the foot point of the perpendicular dropped from point P onto line l, and let m be the perpendicular erected at point P onto the first perpendicular PQ.

We need to show that m is the *unique parallel* through point P to line l. We take any other distinct line $n \neq m$ through point P and check whether it is parallel to line l. For the case that lines n and m are perpendicular, we get $n = PQ$, and hence $Q = n \cap l$. So lines n and l are not parallel.

We exclude this case and assume that lines n and m are not perpendicular. There exists rays n_m and r_n on lines m and n such that the angle $\theta = \angle(r_m, r_n)$ is acute, has vertex P, and secondly, the interior of the angle θ lies in the same half plane of line m as line l.

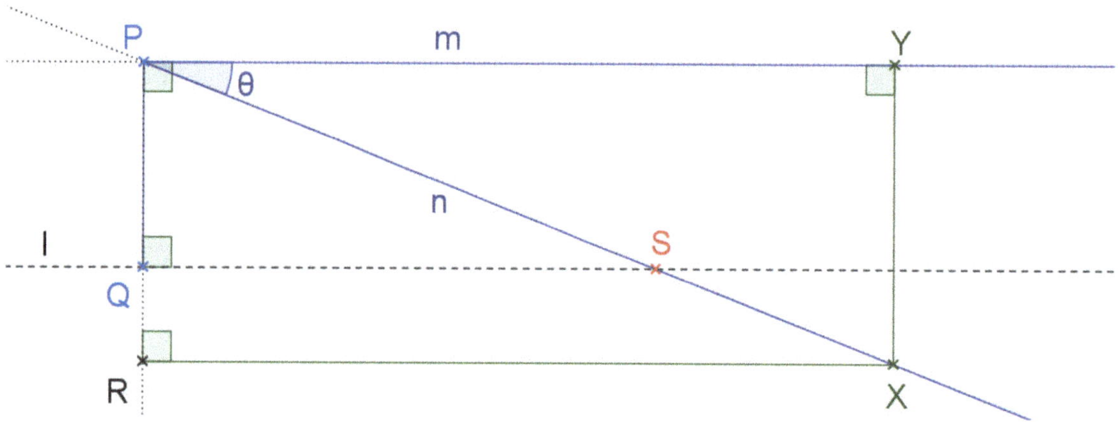

Figure II.11.16: The second line $n \neq m$, different to the double perpendicular m, drawn through point P is along the side of a triangle $\triangle PXR$, and hence by Pasch's axiom intersects line l.

We now explicitly prove existence of an intersection point of ray r_n and line l. Let $XY > PQ$ be the segment between the sides of angle $\theta = \angle XAY$, as postulated by Aristole's Axiom. We can choose point X to lie on the ray r_n, and drop the perpendicular from point X onto the line PQ, obtaining the foot point R. We have obtained a Lambert quadrilateral $\square RXYP$. In a semi-Euclidean plane, this is a rectangle. Its opposite sides are congruent by Lemma II.11.1. Hence we conclude

$$PR \cong XY > PQ$$

and points P and R lie on different sides of point Q.

We now use Pasch's axiom for line l and triangle $\triangle PXR$. Line l intersects the side PR in point Q and is parallel to side RX, since the lines l and RX are both perpendicular to PQ. Hence by Pasch's axiom, the line l intersects the third side PX of the triangle, say in point S.

Thus we have explicitly checked existence of an intersection point lines $n = PX$ with line l, as to be shown. □

II.11.7 More about Aristole's axiom

Proposition II.11.6. *In a semi-elliptic Hilbert plane, Aristole's Axiom does not hold. Especially, it does not hold for the angle between any two parallels to the same line, nor for any angle given by the excess of the angle sum of a triangle over two right angles.*

Proof. Given is a line l and a point P not on this line. Again, let Q be the foot point of the perpendicular dropped from point P onto line l, and let m be the perpendicular erected at point P onto the first perpendicular PQ. As explained in Proposition II.7.43, "double perpendicular" m is a parallel to line l through point P.

Let n be a second parallel to line l through point P. Take any point X on the line n such that X and Q lie on the same side of line m. We drop the perpendicular from point X onto the line PQ and let Z be its foot point. All three line m, ZX and l are perpendicular to PQ, and hence parallel. Hence points X and Z both lie between the parallels l and m. Hence point Z lies between P and Q, and hence.

$$PQ > PZ$$

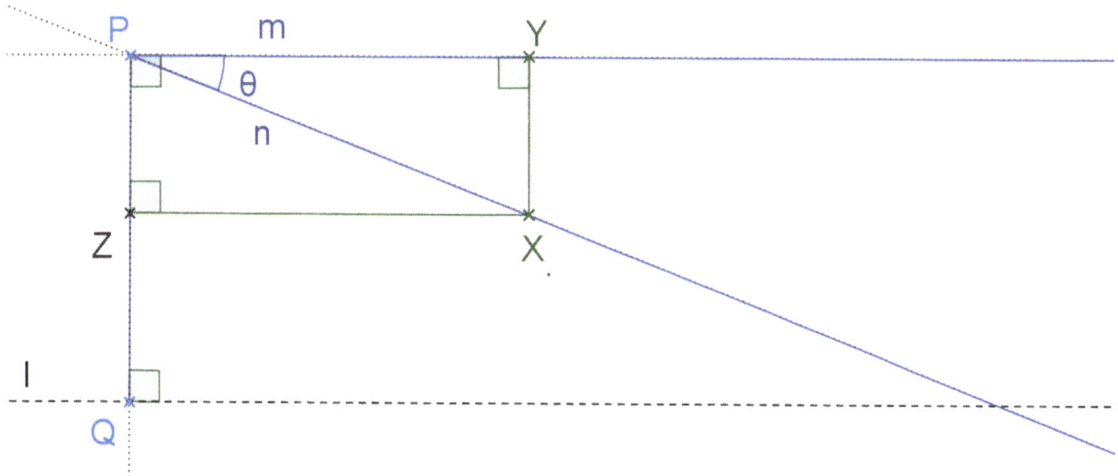

Figure II.11.17: For the angle between two parallels m and n to line l, intersecting at P, the distance from a point on one side to the other side of angle $\angle(m,n)$ is always smaller than the perpendicular PQ.

The perpendicular from point X onto the line m has foot point Y. Thus we have obtained the Lambert quadrilateral $\square ZXYP$. In a semi-elliptic plane, its fourth angle $\chi = \angle ZXY$ is obtuse. From Lemma II.11.1 we conclude

$$PZ > XY$$

Thus we have obtained that $PQ > PZ > XY$ holds for any point X on one of the rays from P on line n. Hence the distance from any point of line n to line m is strictly smaller than PQ, and Aristole's angle unboundedness axiom does not hold. □

Proposition II.11.7. *In a semi-Euclidean as well as in a semi-hyperbolic Hilbert plane, Archimedes' Axiom implies Aristole's Axiom.*

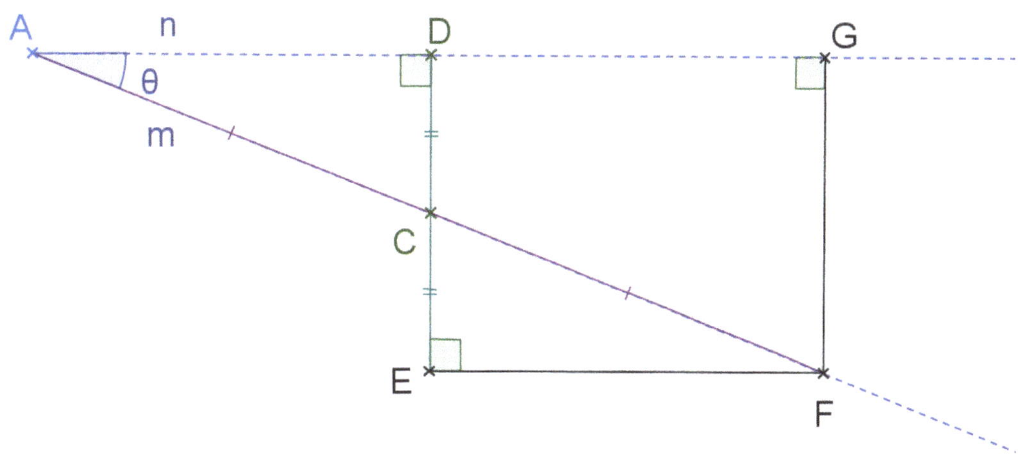

Figure II.11.18: Doubling the distance from a point on one side to the other side of angle θ.

Proof. Let the acute angle $\theta = \angle(m,n)$ be given. We choose any point C on side $m = \overrightarrow{AC}$, and drop the perpendicular onto n to get the foot point D. In order to double the segment CD, we construct point E such that C is the midpoint of segment DE and point F on ray m such that C is the midpoint of segment AF. The triangles

$$\triangle ACD \cong \triangle FCE$$

are congruent by SAS-congruence, because of the vertical angles at point C and two pairs of congruent sides $AC \cong CF$ and $DC \cong CE$.

Because of the triangle congruence, $\angle CDA \cong \angle CEF = \angle DEF$ are both right angles. We drop the perpendicular from point F onto the other side n and get foot point G.

Now $\square DEFG$ is a Lambert quadrilateral, with an acute or right angle $\angle EFG$. By Lemma II.11.1 either side adjacent to the acute angle is longer than the respectively opposite side. Hence we conclude

$$FG \geq ED = 2 \cdot CD$$

We need to assure that doubling the original segment CD is sufficient to get another segment XY longer than any arbitrarily segment PQ. To this end, we use the Archimedean Axiom. Let $FG := C_1D_1$. By induction, we can construct segments

$$C_nD_n \geq 2^n \cdot CD > n \cdot CD$$

for all natural numbers n.

Given any segment PQ, the Archimedean axiom tells there exists a natural number n such that

$$n \cdot CD > PQ$$

Define $XY := C_nD_n$ for such a number n. Both inequalities together imply

$$XY > n \cdot CD > PQ$$

as required for Aristole's angle unboundedness axiom to hold. $\square$

From Proposition II.11.6, the first Legendre Theorem and the Uniformity Theorem we can recapitulate:

Corollary 29. *In a semi-elliptic Hilbert plane, neither Aristole's Axiom nor Archimedes Axiom does hold.*

Together with Proposition II.11.7, we get the remarkable result—for which it would be nice to have a direct proof:

Corollary 30. *In every Hilbert plane, the Archimedean Axiom implies Aristole's Axiom.*

Too, from Proposition II.11.7 and Proposition II.11.5, we get once more the third Legendre Theorem—the direct proof of which we have given earlier.

Corollary 31. *A semi-Euclidean plane for which the Archimedean axiom holds is Pythagorean.*

II.12 Area in Neutral Geometry

II.12.1 Equidecomposable and equicomplementable figures

Definition II.12.1. A *rectilinear figure*, or simply *figure*, is a finite union of non-overlapping triangles. To express that the figure P is the union of non-overlapping triangles T_i for $i = 1 \ldots n$, we write

$$P = \biguplus_{i=1}^{n} T_i$$

We call the union $\biguplus T_i$ a *dissection* of the figure P.

Two triangles (or figures) are *non-overlapping* if their intersection is at most a finite union of points and segments.

Definition II.12.2. Two rectilinear figures P and Q are *equidecomposable* means they are finite unions of non-overlapping, pairwise congruent triangles. Too, we say that the figures P and Q can be *dissected* into congruent triangles, or simply, *have congruent dissections.*

In this case, we write

$$P = \biguplus_{i=1}^{n} T_i \quad \text{and} \quad Q = \biguplus_{i=1}^{n} S_i \quad \text{with} \quad T_i \cong S_i \quad \text{for} \quad i = 1 \ldots n$$

Definition II.12.3. Two rectilinear figures P and Q are *equicomplementable* means there exist figures R and S such that

(1) R and S are equidecomposable

(2) P and R are non-overlapping

(3) Q and S are non-overlapping

(4) $P \uplus R$ and $Q \uplus S$ are equidecomposable

Remark. As a short hand, I write

$P \overset{\boxtimes}{\sim} Q$ for figures P and Q being equidecomposable.

$P \triangleq Q$ to indicate that the figures P and Q are equicomplementable.

Remark. More in everyday language, one may explain as follows: Two figures are equicomplementable means that one can obtain equidecomposable figures from them, by adding to each one an appropriately chosen extra figure. These two extra figures need to be equidecomposable and may not overlap with the two originally given figures.

Remark. In German, Hilbert uses the words "zerlegungsgleich" for equidecomposable, and "ergänzungsgleich" for equicomplementable.

Lemma II.12.1. *Any two dissections of the same figure have a common refinement.*

Proof.

Question. What kinds of polygons occur by intersecting two triangles. How can they be dissected into triangles.

Answer. The intersection of two overlapping triangles is a convex polygon with at most six vertices. In the cases of four, five and six vertices, it can be dissected into triangles by one, two, or three diagonals, respectively.

Question. Explain how to construct a common refinement of two arbitrary dissections of a figure P.

Answer. Given are the two dissections $P = \biguplus_{i=1}^{n} T_i$ and $P = \biguplus_{j=1}^{m} S_j$ of the same figure. A common refinement is easily obtained by dissecting all overlapping intersection $S_j \cap T_i$ into triangles, since these are convex polygons with at most six vertices.

□

Theorem II.12.1. *Any two congruent figures are equidecomposable. Any two equidecomposable figures are equicomplementable.*

All four notions—equality, congruence, being equidecomposable as well as being equicomplementable—are equivalence relations among figures. Indeed these are four different equivalence relations.

Problem II.12.1. *Prove that two figures which are both equidecomposable to a third figure are equidecomposable. You need to use that any two dissections of the same figure have a common refinement, as shown in Lemma II.12.1 above.*

Problem II.12.2. *Prove that two figures which are both equicomplementable to a third figure are equicomplementable. You can use that the relation of being "equidecomposable" is a equivalence relation.*

Remark. Euclid just uses in all four cases the same word "equal", and tells in his common notions that two things equal to a third are equal. That is a good postulate, once there is a clear cut notion of equality. But to prove the Theorem II.12.1 above, we had to deal with a well-defined different equivalence relation, not with equality!

Given is a triangle $\triangle ABC$, a side AB of which has been chosen as base. I call the line through the midpoints of the two remaining sides a *midline*.

Proposition II.12.1 (A neutral version of Euclid I.37). *Two triangles with the same base and the same midline are equicomplementable.*

Two triangles with the same base, and the midpoints of the four remaining sides lying on one line, are equicomplementable.

Remark. Astonishing enough, proposition II.12.1 is even valid in neutral geometry. On the other hand, Euclid's original proposition I.37 is not valid in hyperbolic geometry.

Proof. The proof exploits the correspondence between a triangle and a Saccheri quadrilateral given in Hilbert's Proposition 39. Complete details are given in Proposition II.9.6 from the section about Legendre's geometry.

For a given triangle $\triangle ABC$ and base AB, we obtain the corresponding Saccheri quadrilateral $\square FGBA$ by the construction given in the figure on page 301. We have shown by SAA congruence that

$$\triangle AFD \cong \triangle CHD \quad \text{and} \quad \triangle BGE \cong \triangle CHE$$

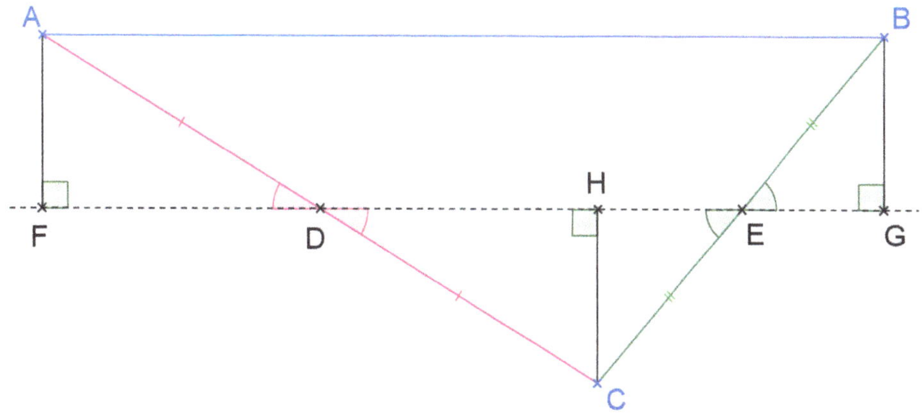

Figure II.12.1: To every triangle corresponds a Saccheri quadrilateral.

Problem II.12.3. *Prove that the segment DE which the triangle cuts out of the middle line has length half the base FG of the corresponding Saccheri quadrilateral.*

Question. Take the case of an acute triangle and convince yourself that the triangle $\triangle ABC$ and the Saccheri quadrilateral $\square FGBA$ are equidecomposable.

We can leave a short answer to this question to the reader. The Saccheri quadrilateral depends only on the base AB and the midline DE of the triangle. Hence we obtain the same Saccheri quadrilateral from another triangle with the same base and same midline. Using the positive answer to the question above, by transitivity (see Theorem II.12.1) we conclude that two acute triangles with the same base and midline are equidecomposable.

In the case of an obtuse triangle, it can happen that the triangle $\triangle ABC$ and the Saccheri quadrilateral $\square FGBA$ are only equicomplementable. The details are explained below. By transitivity, now used for the relation of equicomplementarity, we conclude that any two triangles with the same base and midline are equicomplementable. $\square$

Question. Take the case of an obtuse triangle, as given in the figure on page 302. Explain why the triangle $\triangle ABC$ and the Saccheri quadrilateral $\square FGBA$ are equicomplementable. Let the numbers denote non-overlapping regions.

Answer. Triangle congruences yield

$$1 \uplus 5 \cong 4 \uplus 6 \quad \text{and} \quad 3 \uplus 5 \cong 6$$

The Saccheri quadrilateral and the triangle

$$\square ABGF \overset{\boxtimes}{\sim} 1 \uplus 2 \quad \text{and} \quad \triangle ABC \overset{\boxtimes}{\sim} 2 \uplus 3 \uplus 4$$

are complemented by the triangular region 5 as additional figure. We conclude

$$\begin{aligned} \square ABGF \uplus 5 &\overset{\boxtimes}{\sim} \left(1 \uplus 2\right) \uplus 5 \overset{\boxtimes}{\sim} 2 \uplus \left(1 \uplus 5\right) \overset{\boxtimes}{\sim} 2 \uplus \left(4 \uplus 6\right) \\ &\overset{\boxtimes}{\sim} \left(2 \uplus 4\right) \uplus 6 \overset{\boxtimes}{\sim} \left(2 \uplus 4\right) \uplus \left(3 \uplus 5\right) \overset{\boxtimes}{\sim} \left(2 \uplus 4 \uplus 3\right) \uplus 5 \\ &\overset{\boxtimes}{\sim} \triangle ABC \uplus 5 \end{aligned}$$

Hence the Saccheri quadrilateral and the triangle are equicomplementable.

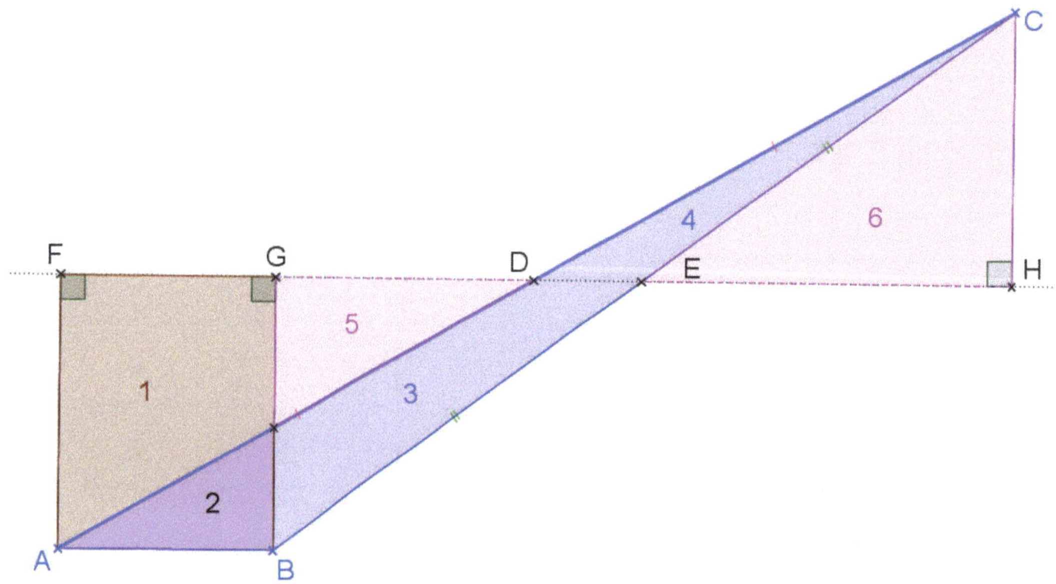

Figure II.12.2: Even an obtuse triangle is equicomplementable to its corresponding Saccheri quadrilateral.

II.12.2 The winding number

In every ordered incidence plane, it is possible to define an orientation. The orientation is fixed, as soon as one has agreed which are the left and right half plane for one fixed ray. We denote the congruence class of right angles by R.

Definition II.12.4 (Winding number function). Given a triangle $\triangle ABC$ with positive (counterclockwise) orientation, we define the *winding number function* $\langle ABC\rangle(X)$ around any point X to be

$$\langle ABC\rangle(X) = \begin{cases} 4R & \text{if } X \text{ lies inside } \triangle ABC. \\ 2R & \text{if } X \neq A, B, C \text{ lies on a side of } \triangle ABC. \\ \alpha & \text{if } X = A. \\ \beta & \text{if } X = B. \\ \gamma & \text{if } X = C. \\ 0 & \text{if } X \text{ lies outside the } \triangle ABC. \end{cases}$$

This function is greater or equal zero. The winding number function for a triangle $\triangle ABC$ with negative (clockwise) orientation is the negative of the above. If the three points A, B, C lie on a line, the winding number function is identical zero.

Lemma II.12.2. *The winding number function $\langle ABC\rangle$ is identically zero if and only if the three points A, B, C lie on a line.*

The winding number functions satisfy $\langle OAB\rangle + \langle OBC\rangle = \langle OAC\rangle$ if and only if point B lies between the points A and C.

Proposition II.12.2. *For any point $X \neq O, A, B, C$ the winding number functions satisfy the permutation property*

$$\langle ABC\rangle = \langle BCA\rangle = \langle CAB\rangle = -\langle BAC\rangle = -\langle ACB\rangle = -\langle CBA\rangle \tag{II.12.1}$$

and the cyclic additivity properties:

$$\langle ABC\rangle = \langle OAB\rangle + \langle OBC\rangle + \langle OCA\rangle \tag{II.12.2}$$

Problem II.12.4. *Check that*

$$\langle ABC\rangle = \langle PAB\rangle + \langle PBC\rangle + \langle PCA\rangle$$

still holds for an arbitrary vertex $P \neq O$.

II.12.3 Area of rectilinear figures

As well known from everyday life, it is more practical to measure the area of figures, instead of only comparing different figures. As a first step in the discussion of area, we state the properties which need to hold for a useful notion of area. Secondly, we give a construction of the area function and thus prove its existence.

Definition II.12.5. The *area* or *measure* is a function assigning to any rectilinear figure P a number area(P) such that the following properties hold:

(1) area$(T) > 0$ for all triangles T

(2) $T \cong S$ implies area$(T) =$ area(S) for any two triangles

(3) area$(P \uplus Q) =$ area$(P) +$ area(Q) for any non-overlapping figures P and Q.

More generally, the value area(P) can be an element of any ordered Abelian group.

Step 1. As the first step, we define the area for any triangle $\triangle ABC$. The is done quite differently for Euclidean geometry on the one side and for semi-hyperbolic or semi-elliptic geometry on the other side.

In any Pythagorean plane the area of a triangle is defined to be half of base times height.

$$\text{area}(\triangle ABC) = \frac{1}{2}ah_a = \frac{1}{2}bh_b = \frac{1}{2}ch_c$$

In any semi-hyperbolic plane the area of a triangle is defined to be the deviation of its angle sum from two right angles

$$\delta(ABC) = 2R - \alpha - \beta - \gamma$$

Recall that $\delta(ABC)$ is the *defect* of the triangle, according to definition II.9.4.

In any semi-elliptic plane the area of a triangle is defined to be the excess of its angle sum over two right angles $\alpha + \beta + \gamma - 2R$

Remark. We see from this definition that the values of area are in different Abelian groups depending on the geometry and are justified in different ways.

- The former definition satisfies the positivity requirement (1) because, as guaranteed by Theorem IV.19.1, the segment lengths in any Pythagorean plane are an *ordered field.* Moreover we have to use the similarly of triangles with opposite orientation.

- The latter two definitions satisfy the positivity requirement (1) because of the Uniformity Theorem 17.

Step 2. We define the left- and right half-planes for any ray, and the clockwise or counterclockwise orientation for any triangle. Note that the orientation depends on the order of the vertices of the triangle. We define the signed area $[ABC]$ of $\triangle ABC$ as

$$[ABC] = \begin{cases} +\text{area}(\triangle ABC) & \text{if } \triangle ABC \text{ has counterclockwise orientation} \\ -\text{area}(\triangle ABC) & \text{if } \triangle ABC \text{ has clockwise orientation} \end{cases}$$

Lemma II.12.3. *For any base point O, the signed triangle areas satisfy*

$$[ABC] = [OAB] + [OBC] + [OCA] \tag{II.12.3}$$

Lemma II.12.4. $[PAB] + [PBC] + [PCA] = 0$ *if and only the three points A, B and C lie on a line.*

Lemma II.12.5. *If the points X and Y lie on the same side of line EF, the two signed areas $[XEF]$ and $[YEF]$ have the same signs. If the points X and Y lie on opposite sides of line EF, the two signed areas $[XEF]$ and $[YEF]$ have the opposite signs.*

Step 3. We define the area of a figure as the sum of the areas of the triangles in any dissection. We show that such a definition cannot lead to a contradiction.

Lemma II.12.6 (Main Additivity Lemma). *If any triangle is dissected into finitely many non-overlapping triangles, its area is the sum of the triangles of the dissection.*

$$T = \biguplus_{i=1}^{n} T_i \quad \textit{implies} \quad \text{area}(T) = \sum_{i=1}^{n} \text{area}(T_i)$$

Proof. We give the triangle $T = \triangle ABC$ and all triangles of the dissection the positive orientation. Any reference point O is chosen. We use the formula (II.12.3) for each term of the sum $\sum_{i=1}^{n} \text{area}(T_i)$. There occur contributions $[OEF]$ of two different types:

- from a side of some triangle T_i in the dissection, which does not entirely lie on one of the sides AB, BC or CA of the big triangle T

- from a side of some triangle T_i lying entirely on a side of the triangle T.

Assume that for side EF from the dissection, the first case applies. There exist two triangles in the dissection having side EF, say $T_1 = \triangle EFX$ and $T_2 = \triangle FEY$, lying in the two half-planes of EF. Hence they do not overlap. They have the same orientation, which we have assumed to be positive. The total area of the quadrilateral formed by the two triangles is

$$\begin{aligned}\text{area}(\square EYFX) &= \text{area}(\triangle EFX) + \text{area}(\triangle FEY) = [EFX] + [FEY] \\ &= [OEF] + [OFX] + [OXE] + [OFE] + [OEY] + [OYF] \\ &= [OFX] + [OXE] + [OEY] + [OYF]\end{aligned}$$

We see that the two contributions from the common side EF cancel. The same cancelation occurs for all interior sides of the dissection.

The contributions from sides of some triangle T_i lying entirely on one of the sides of the big triangle T add up to

$$[OAB] + [OBC] + [OCA] = \text{area}(\triangle ABC)$$

□

Theorem II.12.2. *Assume that the area for triangles has the properties*

(1) $\text{area}(T) > 0$ *for all triangles* T

(2) $T \cong S$ *implies* $\text{area}(T) = \text{area}(S)$ *for any two triangles*

(3t) *If two non-overlapping triangles* P *and* Q *are joined to a larger triangle, this has the* $\text{area}(P \uplus Q) = \text{area}(P) + \text{area}(Q)$.

Then the area of any rectilinear figure is well defined. Indeed, one gets the same sum

$$\sum_{i=1}^{n} \text{area}(T_i) =: \text{area}(P)$$

independent of the dissection $P = \biguplus T_i$ *one has used.*

Furthermore, $\text{area}(P \uplus Q) = \text{area}(P) + \text{area}(Q)$ *holds for any non-overlapping figures* P *and* Q.

Proof of Theorem II.12.2. Given are the two dissections $P = \biguplus_{i=1}^{n} T_i$ and $P = \biguplus_{j=1}^{m} S_j$ of the same figure. Let R_k with $k = 1 \dots K \le 3nm$ be a common refinement. By applying the additivity lemma to each T_i, and each S_j, we obtain

$$\sum_{i=1}^{n} \text{area}(T_i) = \sum_{k=1}^{K} \text{area}(R_k) = \sum_{j=1}^{m} \text{area}(S_j)$$

Hence any two dissections yield the same value of sum of area for the given figure, which is thus well defined.

The proof of the second item can be left to the reader. □

Corollary 32. *The interior domain* R *of a simple closed polygon* $P_1, P_2, \dots P_n, P_{n+1} = P_1$ *has the area*

$$\text{area}(R) = \sum_{i=1}^{n} [OP_iP_{i+1}]$$

We have put $P_{n+1} = P_1$, and assumed positive orientation.

We can dissect the interior region R into triangles, and let all triangles have positive orientation. Then the circumference curves of these triangles add up to the positively oriented polygon surrounding them all.

Theorem II.12.3. (a) *Any two equidecomposable figures have the same area.*

(b) *Any two equicomplementable figures have the same area.*

Proof. Let P and Q be two equidecomposable figures. By definition, they have dissections

$$P = \biguplus_{i=1}^{n} T_i \quad \text{and} \quad Q = \biguplus_{i=1}^{n} S_i \quad \text{such that}$$
$$T_i \cong S_i \quad \text{for} \quad i = 1 \dots n$$

From the definition of area, we get

$$\text{area}(P) = \sum_{i=1}^{n} \text{area}(T_i) = \sum_{i=1}^{n} \text{area}(S_i) = \text{area}(Q)$$

as to be shown. Now let P and Q be two equicomplementable figures. By definition, there exist figures R and S such that

(1) R and S are equidecomposable

(2) the non-overlapping unions $P \uplus R$ and $Q \uplus S$ are equidecomposable

Hence part (a) and the additivity from theorem II.12.2 imply

$$\begin{aligned} \text{area}(R) &= \text{area}(S) \\ \text{area}(P \uplus R) &= \text{area}(Q \uplus S) \\ \text{area}(P) + \text{area}(R) &= \text{area}(Q) + \text{area}(S) \end{aligned}$$

Since substraction is defined in the Abelian group from which the value of area are taken, we get $\text{area}(P) = \text{area}(Q)$, as to be shown. □

Proposition II.12.3. *If the figures P_i are equidecomposable to the figures Q_i for $i = 1 \dots n$, and neither the figures P_i nor the figures Q_i overlap, then the union $\biguplus_{i=1}^{n} P_i$ is equidecomposable to the union $\biguplus_{i=1}^{n} Q_i$.*

The corresponding statement holds in the case of equicomplementable figures.

Proposition II.12.4 (The postulate of de Zolt and Stolz). *Let $Q \subset P$ be two figures such that there exists a triangle T inside P which does not overlap with the smaller figure Q. Then the figures P and Q are not equicomplementable.*

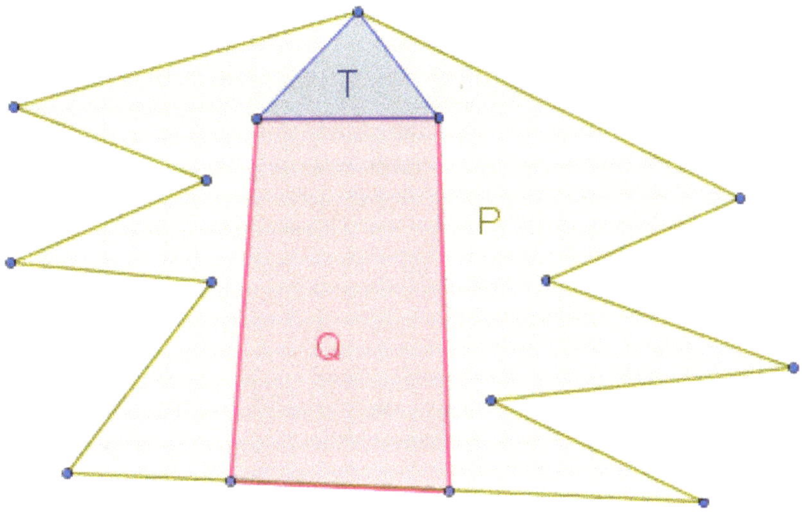

Figure II.12.3: deZolts postulate tells that P and Q are not equicomplementable.

Proof. From dissections of the figures P and Q and $Q \uplus T$, we construct a common refinement R_k such that

$$Q = \biguplus_{k=1}^{K} R_k \,, \quad T = \biguplus_{k=K+1}^{L} R_k \,, \quad P = \biguplus_{k=1}^{M} R_k$$

with $K < L \leq M$. By the additivity of area (Theorem II.12.2), we get

$$\text{area}(Q) = \sum_{k=1}^{K} \text{area}(R_k) < \sum_{k=1}^{M} \text{area}(R_k) = \text{area}(P)$$

Hence $\text{area}(Q) \neq \text{area}(P)$. By part (b) of Theorem II.12.3, the figures P and Q are not equicomplementable. □

II.12.4 A standard equicomplementable form for a figure

There remains the question—are any two figures with the same area equicomplementable? To prove this converse of Theorem II.12.3, one needs a standard equicomplementable form for any given figure. In Euclidean geometry, a rectangle with one side unit length, the other side the given area, is a convenient standard form. In hyperbolic geometry, no rectangles exist. Moreover, the area of Saccheri quadrilaterals with a unit base turns out to be bounded above. A way out is to use for the purpose of comparison a *finite set of non-overlapping congruent triangles.* We present this approach in a way valid for any Hilbert plane for which the circle-line intersection property holds.

Lemma II.12.7. *To any given triangle and a segment longer than its shortest side, there exists an equicomplementable triangle which has the given segment as one of its sides.*

Proof. Let the given triangle $\triangle ABC$ have side $AB < c'$ where c' is the given segment. We bisect sides AB and AC and draw the middle line DE connecting the midpoints D and E, respectively. Let D' be an intersection point of the middle line with the circle around B of radius $c'/2$. We extend the ray $\overrightarrow{BD'}$ to get the point A' such that D' is the midpoint of segment BA'. The triangle $\triangle A'BC$ and the given triangle have the common base BC and the common middle line DE. The triangles $\triangle ABC \triangleq \triangle A'BC$ are equicomplementable by Proposition II.12.1, and obviously the triangle $\triangle A'BC$ has the side $A'B$ of length c' as required. $\square$

Lemma II.12.8. *Any triangle is equicomplementable to an isosceles triangle with the same base and the same midline.*

Proof. For a given triangle $\triangle ABC$ and side AB, we obtain the corresponding Saccheri quadrilateral $\square FGBA$ by the construction given in the figure on page 301. Let DE be the middle line and p be the perpendicular bisector of side AB. The top of the isosceles triangle $\triangle ABC'$ is the point on p which has the same distance from line DE as the points A and B, but lies on the other side. The two triangles $\triangle ABC \triangleq \triangle ABC'$ have the common side AB and the same midline DE. By Proposition II.12.1 they are equicomplementable. $\square$

Lemma II.12.9. *For any two given triangles T_1 and T_2 there exists a pair of non-overlapping congruent triangles $T' \cong T''$ such that the unions*

$$T_1 \uplus T_2 \triangleq T' \uplus T''$$

are equicomplementable.

Proof. If T_i are two congruent equilateral triangles, we are ready. Suppose this is not the case and that triangle T_2 has a side a' longer than one side of T_1. By Lemma II.12.7 there exists a triangle T_1' equicomplementable to T_1 and having the side a'. We have obtained two equicomplementable triangles $T_1' = \triangle A'BC$ and $T_2 = \triangle A_2BC$ which we can put into the two opposite half planes of their common side $a' = BC$.

By Lemma II.12.8 there exist isosceles triangles $T^* = \triangle A^*BC$ and $T^{**} = \triangle A^{**}BC$ on opposite sides of their common base BC such that $T^* \triangleq T_1'$ and $T^{**} \triangleq T_2$. The kite $\square A^*BCA^{**}$ has be constructed as the non-overlapping union of T^* with T^{**} but clearly

$$\square A^*BCA^{**} = T^* \uplus T^{**} = \triangle BA^*A^{**} \uplus \triangle CA^*A^{**}$$

and the latter are two congruent triangles. We put $T' := \triangle BA^*A^{**}$ and $T'' := \triangle CA^*A^{**}$, which are congruent triangles and are ready. $\square$

Lemma II.12.10. *Any given figure is equicomplementable to a disjoint union $2^nT'$ of 2^n congruent non-overlapping triangles T'.*

Proof. By further subdivisions, any figure P is equicomplementable to a union of 2^n non-overlapping triangles $T_1, T_2, \dots T_{2^n}$. Applying the Lemma 2^{n-1} times we construct triangles $T_i' \cong T_i''$ for $i = 1 \dots 2^{n-1}$ such that

$$\biguplus_{i=1}^{2^n} T_i \triangleq \biguplus_{i=1}^{2^{n-1}} \left(T_i' \uplus T_i''\right)$$

In an inductive process for $s = 1 \dots n$, we get a union of 2^n non-overlapping triangles which

- together are still equicomplementable to the originally given figure P;
- consists of 2^{n-s} non-overlapping subsets,
- each of which is the non-overlapping union of 2^s congruent triangles.

In the end, we see that the given figure P is equicomplementable to a disjoint union of 2^n congruent non-overlapping triangles. □

Lemma II.12.11. *For any two given triangles*[38] *S and T there exists isosceles triangles P and Q with a common base such that $S \triangleq P$ and $T \triangleq Q$.*

Proof. If S and T are two congruent equilateral triangles, we are ready. Suppose this is not the case and that triangle T has a side a' longer than one side of S. By Lemma II.12.7 there exists an equicomplementable triangle $S' \triangleq S$ having the side a'. We have obtained two triangles S' and T with the common side $a' = BC$. By Lemma II.12.8 there exist isosceles triangles $P \triangleq S'$ and $Q \triangleq T$, all on the common side a'. □

Proposition II.12.5 (Figures with same area are equicomplementable). *Assume in the given Hilbert plane that*

- *the circle-line intersection property holds;*
- *an* area *has been defined in accordance to Definition II.12.5. The area takes values in an* ordered *and hence free Abelean group such that*

  (1) area$(T) > 0$ *for all triangles T*

  (2) $T \cong S$ *implies* area$(T) =$ area(S) *for any two triangles*

  (3) area$(P \uplus Q) =$ area$(P) +$ area(Q) *for any non-overlapping figures P and Q.*

Then any two figures P and Q for which area$(P) =$ area(Q) *are equicomplementable.*

Remark. An additive group is called *free* if $ng = 0$ implies $g = 0$ for any integer $n \neq 0$ and any group element g. An ordered Abelean group is always free.

Proof. By Lemma II.12.10, the given figure F is equicomplementable to a disjoint union $2^n S'$ of 2^n congruent non-overlapping triangles S' and the given figure G is equicomplementable to a disjoint union $2^n T'$ of 2^n congruent non-overlapping triangles T'. The integer n can be chosen to be equal for both figures, by means of further subdivisions. By assumption, both figures have the same area. Hence

$$2^n \text{area}(T) = \text{area}(F) = \text{area}(G) = 2^n \text{area}(S)$$

Since the area takes values in an ordered and hence *free* Abelean group, we conclude that area$(T) =$ area(S).

By Lemma II.12.11 there exists isosceles triangles P and Q with a common base such that $S \triangleq P$ and $T \triangleq Q$. Hence

$$\text{area}(P) = \text{area}(S) = \text{area}(T) = \text{area}(Q)$$

[38]they need not have the same content

We put the isosceles triangles $P = \triangle XBC$ and $Q = \triangle YBC$ on the same side of their common base BC. Let M be the midpoint of BC. We may assume that the altitude of $XM \geq YM$.

Because of the assumption that the area of any triangle is positive, it is impossible that $X \neq Y$. Indeed in that case

$$\triangle XBC = \triangle YBC \uplus \triangle XYB \uplus \triangle XYC$$

would imply that area$(P) >$ area(Q).

Now since $X = Y$, we see that actually $P = \triangle P_t BC = \triangle Q_t BC = Q$. Wee that the originally given figures are equicomplementable since

$$S \triangleq P = Q \triangleq T$$
$$F \triangleq \uplus 2^n S \triangleq \uplus 2^n T \triangleq G$$

□

II.12.5 The role of the Archimedean axiom

Problem II.12.5. *We assume that the Archimedean axiom holds. Explain why any triangle T is not only equicomplementable, but even equidecomposable to a corresponding Saccheri quadrilateral.*

Solution of Problem II.12.5. The assertion holds for every acute triangle. For an obtuse triangle, we reverse the process used in the first Legendre Theorem. Thus we obtain a sequence of triangles by bisection of the longest side and having all the same base. They have a common middle line. According to Problem II.12.3 and the figure on page 301 , the middle line cuts these triangles in segments DE which have the length half the base FG of the Saccheri quadrilateral. Moreover these segment all lie on one line.. By the Archimedean axiom, it takes only a finite number of steps to obtain a triangle which cuts the midline in two point between or on the base of the Saccheri quadrilateral and hence is equidecomposable to the Saccheri quadrilateral. □

Lemma II.12.12. *We assume that circle-line intersection property and the Archimedean axiom hold. To any given triangle and a segment longer than its shortest side, there exists an equidecomposable triangle which has the given segment as one of its sides.*

Lemma II.12.13. *We assume that circle-line intersection property and the Archimedean axiom hold. Then any two equicomplementable triangles are even equidecomposable.*

Theorem II.12.4. *Assume that the Archimedean axiom* (V.1) *holds, any two equicomplementable figures are even equidecomposable.*

We now turn the case that the Archimedean axiom is not satisfied. We refer to the figure on page 302 with the equicomplementable Saccheri quadrilateral and obtuse triangle

$$\square AFBG \triangleq \triangle ABC$$

Note that AG is the diameter of the Saccheri quadrilateral. Let M be the midpoint of its base FG.

Lemma II.12.14. *Suppose these two figures have congruent dissections into t triangles. Then*

$$|HM| \leq \frac{3t+4}{2}|AG|$$

Proof. Let C be the sum of the circumferences of all t triangles of the dissection.

$$|AB| + |BC| + |AB| \leq C \leq 3t\,|AG|$$

For the circumference of triangle $\triangle ABC$, we need to get a the lower bound from the triangle inequality.

$$2|HM| = |HF| + |HG| \leq 4|FA| + |AB| + |BC| \leq 4|AG| + 3t\,|AG|$$

□

Lemma II.12.15. *Suppose that the Saccheri quadrilateral and obtuse triangle from the figure on page 302 are even equidecomposable*

$$\square AFBG \overset{\boxtimes}{\sim} \triangle ABC$$

and that

$$|HM| \geq N \cdot |AG|$$

for some large integer N. Then the congruent dissections contain at least

$$t \geq \frac{2N-4}{3}$$

triangles.

Proof. The assumption and Lemma II.12.14 imply

$$N \cdot |AG| \leq |HM| \leq \frac{3t+4}{2} \cdot |AG|$$

and solving for t yields the estimate to be shown. □

Proposition II.12.6. *Assume the Archimedean axiom* (V.1) *does* <u>*not*</u> *hold. Then there exist a triangle and a Saccheri quadrilateral that are equicomplementable but* <u>*not*</u> *equidecomposable.*

The logical relations between the different concepts of content gained in this section are summed up in the following diagram:

$$\begin{array}{ccc}
P \text{ and } Q \text{ are equidecomposable} & \xrightarrow{\text{clear}} & P \text{ and } Q \text{ are equicomplementable} \\
\big\downarrow \text{Th. II.12.3} & & \text{Th. II.12.3} \big\downarrow \\
\text{area}(P) & = & \text{area}(Q) \\
\big\downarrow \text{Archimedes and Th. V.25.4} & & \text{Th. V.25.2} \big\downarrow \\
P \text{ and } Q \text{ are equidecomposable} & \xleftarrow[\text{Archimedean axiom}]{} & P \text{ and } Q \text{ are equicomplementable}
\end{array}$$

Corollary 33. *Assuming the Archimedean axiom* (V.1) *holds, any two equicomplementable figures are even equidecomposable. Assuming the Archimedean axiom* (V.1) *does* <u>*not*</u> *hold, there exist two equicomplementable figures that are* <u>*not*</u> *equidecomposable.*

II.13 Neutral Triangle Geometry

II.13.1 Introduction from neutral geometry

Before we start. These exercises deal with some interesting questions from triangle geometry. I could not resist the temptation to discuss neutral and Euclidean triangle geometry separately. Recall that in neutral geometry, only the axioms of incidence, order, congruence are assumed. The attempts of Farkas Bolyai (the father) as well as Legendre to prove the parallel axiom—starting just with neutral geometry—were in vain. But reviewing them from the standpoint of today, they brilliantly show in which tricky and surprising way the parallel axiom is linked to triangle geometry.

Some of the theorems about a triangle remain valid in hyperbolic geometry, others need to be weakened and modified. Therefore I have tried to prove as much as possible in neutral geometry. In a later section, I specialize to Euclidean geometry. Finally, hyperbolic geometry is treated within Klein's and Poincaré's models.

From the section on congruence, recall definition II.7.3 of a triangle, and Euler's conventional notation: in triangle $\triangle ABC$, let $a = BC$, $b = AC$, and $c = AB$ be the sides and $\alpha := \angle BAC$, $\beta := \angle ABC$, and $\gamma := \angle ACB$ be the angles.

Too, we need to recall the characterization of a perpendicular bisector.

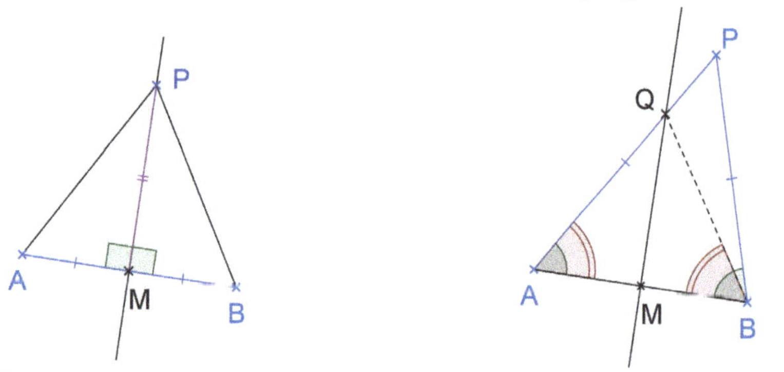

Figure II.13.1: A point lies on the perpendicular bisector if and only if its distances from both endpoints are congruent.

Proposition II.13.1 (Characterization of the perpendicular bisector). *A point P lies on the perpendicular bisector p of a segment AB if and only if it has congruent distances to both endpoints of the segment.*
Point P lies on the same side of the perpendicular bisector p of a segment AB as point B if and only if $|PB| < |PA|$.

Proof. Assume point P lies on the perpendicular bisector. Let M be the midpoint of segment AB. The congruence

$$\triangle AMP \cong \triangle BMP$$

follows by SAS congruence. Indeed, $PM \cong PM$ since a segment is congruent to itself, and $AM \cong BM$ as well as $\angle AMP \cong \angle BMP \cong 90°$ since PM is the perpendicular bisector. Hence $AP \cong BP$.

The converse is a consequence of the second statement. Indeed a point P with congruent distances $AP \cong BP$ from points A and B can neither lie on the same side of bisector p as A nor on the same side as B. Hence the point P lies on the bisector p.

To check the second statement, we assume that point P lies on the same side of the bisector p as point B. Under this assumption, segment AP intersects the bisector—as follows from Pasch's axiom, applied to triangle $\triangle ABP$ and the bisector p. Let Q be the intersection point of p and segment AP. Because Q lies on the bisector, the first part of the proof implies that the triangle $\triangle ABQ$ is isosceles. We apply the triangle inequality for $\triangle BQP$ and conclude

$$|PB| < |PQ| + |QB| = |PQ| + |QA| = |PA|$$

as to be shown. □

II.13.2 The circum-circle

The construction of the circum-circle of a triangle depends on the perpendicular bisectors of its sides.

Proposition II.13.2 (Conditional circum-circle—neutral version). *For any triangle in a Hilbert plane, the following three statements are equivalent:*

(a) *The perpendicular bisectors of two sides of a triangle intersect.*

(b) *The triangle has a circum-circle.*

(c) *The bisectors of all three sides intersect in one point.*

Proof. We show that **(c)** implies **(a)**, **(a)** implies **(b)**, and finally **(b)** implies **(c)**. The first claim is obvious.

*Reason for "***(a)** *implies* **(b)***".* Let O be the intersection point of the perpendicular bisector p_b of segment AC, and the bisector of segment BC, which is called p_a. By Proposition II.13.1 one concludes

$$OA \cong OC \quad \text{and} \quad OB \cong OC$$

Hence transitivity implies $OA \cong OB$, and point O is the center of the circum-circle. □

*Reason for "***(b)** *implies* **(c)***".* Let point O be the center of the circum-circle. Thus O has congruent distances to all three vertices A, B and C. By Proposition II.13.1, congruent distances from A and B imply that center O lies on the bisector p_c. Similarly, congruent distances from B and C imply that center O lies on the bisector p_a, and congruent distances from A and C imply that center O lies on the bisector p_b. Hence center O is the intersection point of all three bisectors. □

Obviously (c) implies (a). Hence we conclude that all three assumptions are equivalent. Note that this means that either all three are true, or all three are false. □

Let a, b, c be any three different mutually parallel lines in an ordered incidence plane. For their mutual placement, the following two cases may occur:

(a) there exists one line among them such that the two other lines are in opposite half planes of the first line;

(b) as seen from any one of the three lines, the two others lie in the same half plane.

We shall see that actually both cases are possible. In some situations it is possible to decide which one of cases (a) or (b) occurs. For easier communication, I have invented the following terms.

Definition II.13.1 (Three parallel lines forming stripes). I say that three different mutually parallel lines in an ordered incidence plane *are forming stripes* iff case (a) occurs,— there exists one line among them such that the two other lines are in opposite half planes of the first line.

Definition II.13.2 (Three parallel lines in triangular shape). I say that three different mutually parallel lines in an ordered incidence plane *are in triangular shape* iff case (b) occurs,—as seen from any one of the three lines, the two others lie in the same half plane.

Lemma II.13.1. *In an ordered* affine *plane, any three different mutually parallel lines are forming stripes. In other words, there do not exist three parallel lines in triangular shape.*

Proof. Let a, b, c be any three different mutually parallel lines in an ordered affine plane. Take any point A on line a, and point B on line b. The line AB intersects line a at the unique point A. By Proclus' lemmaI.3.3, the line AB intersects the parallel $c \parallel a$, too. Let C be the intersection point. By the three-point Theorem II.6.2 (Hilbert's Proposition 4), we know that among any three different points lying on a line, there exists exactly one lying between the two other points. We may assume the case $A * B * C$ to occur, without loss of generality. Then points A and C, and by plane separation the entire lines a and c lie in opposite half -planes of the line b. □

We just mention the following facts about the enclosing line in Hilbert's hyperbolic geometry. The topics in treated in volume II of this book.

Proposition II.13.3. *Given any angle, there exists a unique line the opposite rays of which are limiting parallels to the two sides of the angles.*

A bid more generally, we easily conclude:

Proposition II.13.4 (Hilbert III.3). *For any two rays which are not limiting parallel to each other, there exists a unique line the opposite rays of which are limiting parallels to the two given rays.*

We call this line the *enclosing line* of the angle, or the two given rays. Educated in this way, one finds the relevant counterexamples.

Remark. It is very easy to make the mistake to conjecture that lemma II.13.1 would hold more generally in any ordered incidence plane. But this is not true. Counterexamples can be constructed in Hilbert's hyperbolic geometry as well as in the Poincaré disk model of hyperbolic geometry.

- Any improper triangle with three improper vertices, has as sides three parallel lines which are in triangular shape according to definition II.13.2.

- In Hilbert's hyperbolic geometry, three line in triangular shape may be obtained as follows. One draws any three different rays a, b, c from a common vertex P. By proposition II.13.3 from the section on Hilbert's hyperbolic geometry, there exists an inclosing line for each one of the angles $\angle(a, b)$, $\angle(b, c)$ and $\angle(c, a)$. In case that two among the rays are opposite, the inclosing line is replaced by the line on these opposite rays. These three enclosing lines are parallels, and they form a triangular shape.

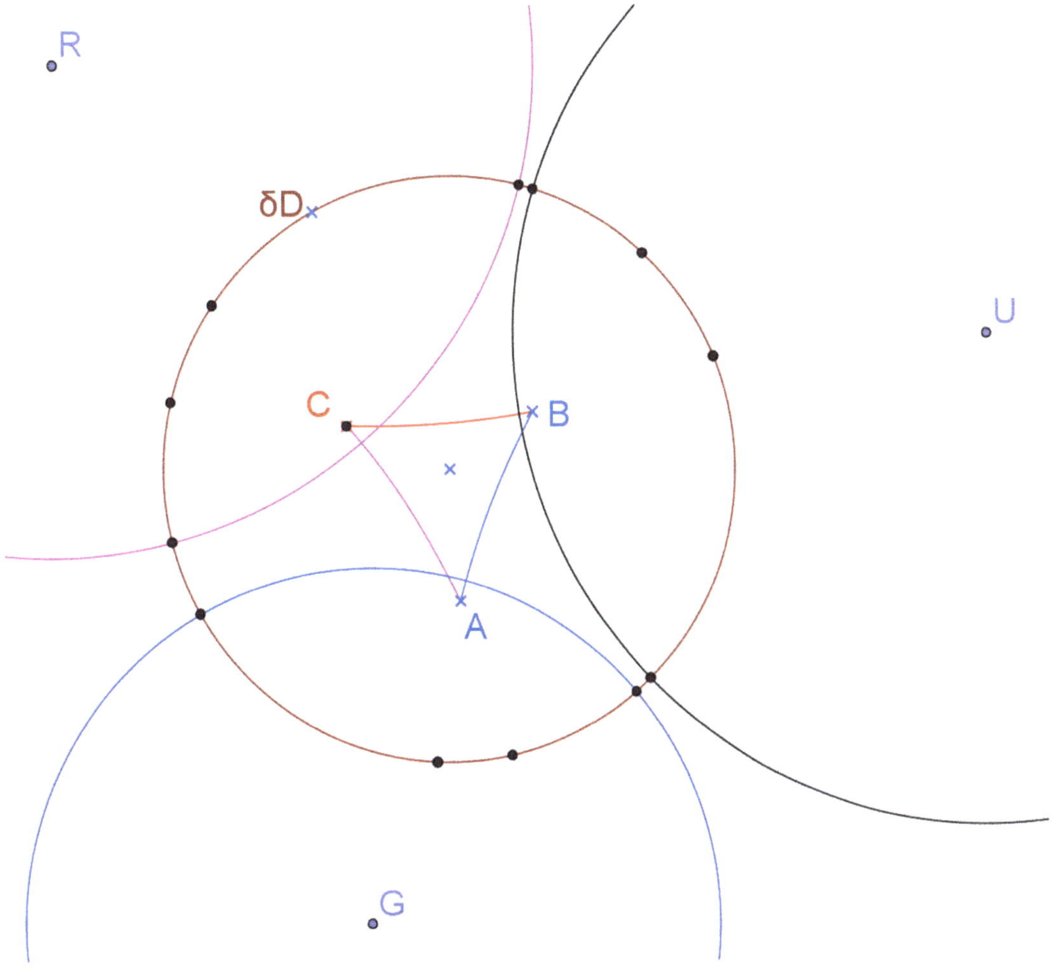

Figure II.13.2: Strange perpendiculars to the three sides of a triangle.

- Another easy example is constructed from a hexagon with six right angles. It is known that such hexagons exist in the hyperbolic plane. Any three nonadjacent sides of such a hexagon lie on three parallel lines which are in triangular shape. The remaining three sides of the hexagon are their common perpendiculars.

- The figure on page 315 gives a most astonishing example. In the hyperbolic plane exist triangles, and perpendiculars erected on their three sides such that all three perpendiculars turn out to be parallel to each other,— and even form a triangular shape according to my definition II.13.2.

Proposition II.13.5 (About triangles without circum-circle)**.** *In any Hilbert plane is given the triangle $\triangle ABC$. Let M_a, M_b, M_c be the midpoints of sides a, b, c, and let m_a, m_b, m_c be their perpendicular bisectors. After a permutation of the vertices, we may assume $c \leq a \leq b$.*

Under the assumption that the bisectors do not intersect neither in the interior of the triangle nor at point M_b, one concludes that $a < b$, and gets the following situation illustrated in the figure

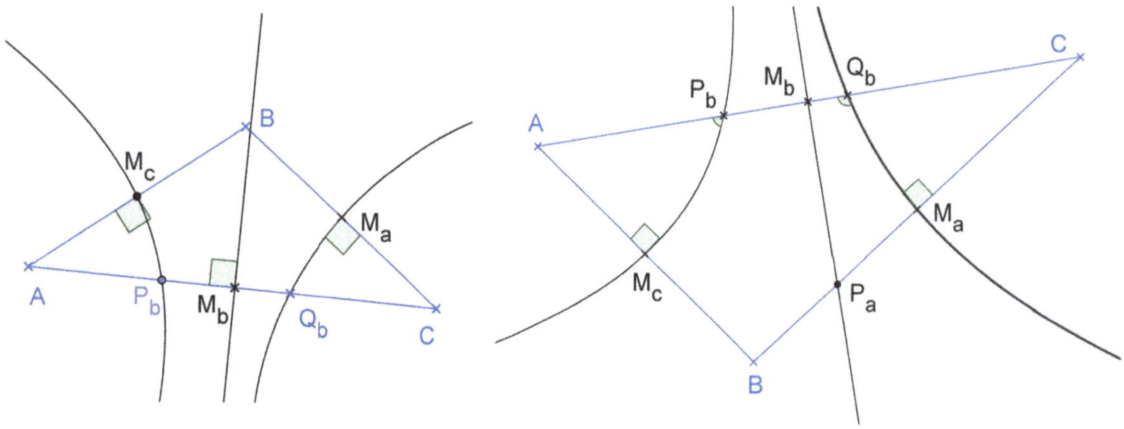

Figure II.13.3: In hyperbolic geometry, the perpendicular bisectors can be parallel as shown—but this figure is impossible in Euclidean geometry.

on page 316:

(II.13.1) $$m_a = M_aQ_b\,,\ m_b = M_bP_a\,,\ m_c = M_cP_b$$

(II.13.2) $$B * P_a * M_a * C \quad \textit{if } c < a. \textit{ The triangle is not isosceles.}$$

(II.13.3) $$B = P_a * M_a * C \quad \textit{if } c = a. \textit{ The triangle is isosceles.}$$

(II.13.4) $$A * P_b * M_b * Q_b$$

Proof. By Pasch's axiom, the bisector m_b of the longest side AC either intersects a further side of triangle $\triangle ABC$, say at point P_a;—or the opposite vertex B lies on the bisector m_b and $P_a = B$. By corollary 9, a point lies on the perpendicular bisector if and only if its distances to the endpoints of the segment are congruent. Hence $P_a = B$ if and only if $c = a$ and the triangle is isosceles. I leave it to the reader to complete the reasonings in this special case.

We now assume the case $c < a$ to occur. From proposition II.13.1 follows that vertices A and B of the shortest side c lie on the same side of bisector m_b, but vertices B and C lie on different sides of m_b. By plane separation, the bisector m_b intersects the side BC at some point P_a. In the right triangle $\triangle P_aM_bC$, the hypothenuse is the longest side. Hence

$$CM_a = a/2 \le b/2 = CM_b < CP_a \quad \text{and hence } C * M_a * P_a * B$$

We use Pasch's axiom for triangle $\triangle P_aM_bC$ to conclude that the bisector m_a intersects either side P_aM_b or side M_bC. In the first case, the bisectors m_b and m_a intersect either in the interior of the triangle or at point M_b,—which cases are ruled out by assumption. Hence bisector m_a intersects side M_bC at some interior point Q_b. We have obtained the order relations

$$A * M_b * Q_b * C$$

Next we rule out the case $b = AC = BC = a$. In that case the bisector m_c of side AB would go through vertex C. The bisectors $m_a = M_aQ_b$ and $m_b = M_bP_a$ would have points in both half-planes of m_c. Hence the three bisectors intersect in the interior of the triangle $\triangle ABC$,—which case we ruled out by assumption.

Since $b = AC > BC = a$, from proposition II.13.1 follows that points A and C lie on different sides of the bisector m_c. Hence m_c intersects the sides AC in some point P_b. From the assumption that the bisectors m_c and m_b do not intersect, we obtain $AP_b < AM_b$, and hence

$$A * P_b * M_b * Q_b * C$$

The segments $Q_bM_a \subset m_a$ and $P_bM_c \subset m_c$ lie in different half planes of bisector m_b. The right triangle $\triangle Q_bM_aC$ has an acute angle at vertex Q_b. The right triangle $\triangle P_bM_cA$ has an acute angle at vertex P_b. Hence the bisectors m_a and m_c may

- either intersect in the the half-plane $\mathcal{H}_b$ of the longest side b opposite to vertex B;
- or all three bisectors are parallel forming stripes.

□

Corollary 34. *In neutral geometry, the following five cases are all possible.*

(a) *The three bisectors intersect in the interior of the triangle.*

(b) *The three bisectors intersect at the midpoint M_b of the longest side b.*

(c) *The three bisectors intersect at a point lying in the half-plane $\mathcal{H}_b$ of the longest side b opposite to vertex B.*

(d) *The three bisectors have a common perpendicular, which lies in the half-plane $\mathcal{H}_b$.*

(e) *Any two of the bisectors are asymptotically parallel. In that case, the rays of all three bisectors converge in the half-plane $\mathcal{H}_b$.*

In the cases (d) *and* (e), *the bisectors all mutually parallel and are forming stripes, according to definition II.13.1. The triangle has* <u>*no circum-circle*</u>.

Corollary 35. *For any triangle in a Pythagorean plane, only cases* (a), (b) *or* (c) *from corollary 34 can occur. Hence every triangle has a circum-circle.*

Proof. In a Pythagorean plane, we may use proposition II.7.44, in short "parallels imply congruent z-angles". As shown above, the three angular bisectors do not have congruent z-angle with the line of the longest side. In a Pythagorean plane, this implies that all three side bisectors pairwise intersect. Thus cases (d) and (e) are excluded. □

Example II.13.1 (**Not every triangle needs to have a circum-circle**). Indeed, for the construction of a counterexample, we use Saccheri quadrilaterals with a baseline far away. From hyperbolic geometry, one needs only the feature that underbarno rectangle exists.

The construction starts with baseline l, and three points X, Y and Z lying on l. Next one sets up two adjacent Saccheri quadrilateral $\square YXAB$ and $\square YZCB$ with common side YB. Hence we get three congruent segments $YB \cong XA \cong ZC$, and right angles at vertices X, Y and Z.

Question. Why do the three points A, B and C not lie on a line?

Answer. Note that $M_a \neq M_b$, otherwise $A = B = C$ and $X = Y = Z$. Now assume towards a contradiction that points A, B and C lie on a line. The three midpoints would lie on that line, too. Hence the quadrilateral $\square EDM_aM_b$ would be a rectangle, contrary to the assumption that no rectangle exists.

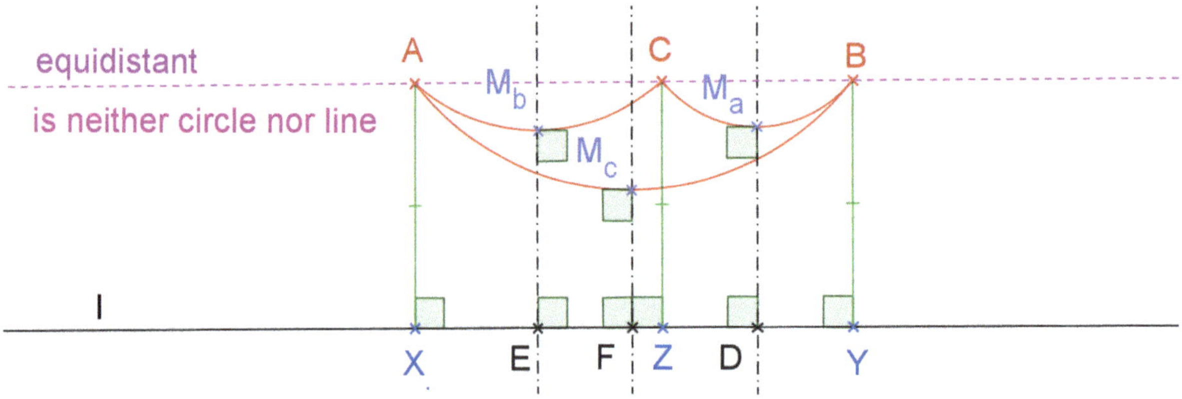

Figure II.13.4: The perpendicular bisectors of the sides of triangle $\triangle ABC$ are all parallel. The triangle has no circum-circle—true, but hard to believe!

The line $l = XYZ$ is the common perpendicular of all three perpendicular bisectors of triangle $\triangle ABC$. Hence all three perpendicular bisectors are parallel. By Proposition II.13.2, the triangle has no circum-circle. But indeed, one realizes a different new feature:

Definition II.13.3 (Equidistance line). Given is a baseline l and a distance AX. The set of all points with distance from a baseline l congruent to AX, and lying on one side of this line, are called an *equidistance line.*

In the example just constructed, all three distances $YB \cong XA \cong ZC$ are congruent, and the three vertices A, B and C lie on one side of line $l = XYZ$. Hence the three vertices of the triangle lie on an equidistance line. One is thus lead to an analog to Proposition II.13.2:

Proposition II.13.6 (Conditional equidistance line). *For any triangle, the following three statements are equivalent:*

(a) *The perpendicular bisectors of two sides have a common perpendicular.*

(b) *The three vertices lie on an equidistance line.*

(c) *The bisectors of all three sides have a common perpendicular.*

Proof of Proposition II.13.6. This proof is very similar to the well known proof of Proposition II.13.2. The missing link is Proposition II.13.7, which is stated and used now. We postpone its proof to the next paragraph.

Proposition II.13.7. *A line is perpendicular to the perpendicular bisector of a segment if and only if the two endpoints of the segment have the same distance to the line, and lie on the same side of it.*

*Reason for "***(a)** *implies* **(b)***" from Proposition II.13.6.* Assume that the perpendicular bisectors of sides AB and BC have the common perpendicular l. Because l and p_c intersect perpendicularly, Proposition II.13.7 yields that vertices A and B have the same distance from l, and lie on the same side of l. By the same reasoning, Proposition II.13.7 yields that the vertices B and C have the same distance from l, and lie on the same side of l, because l and p_a intersect perpendicularly. Hence all three vertices A, B and C lie on an equidistance line. □

Reason for "(b) implies (c)". We assume that all three vertices A, B and C have congruent distances $AX \cong BY \cong CZ$ to baseline l, and lie on the same side of l. Here X, Y and Z are the foot points of the perpendiculars dropped from the vertices A, B and C onto line l.

By Proposition II.13.7, equal distances $AX \cong BY$ implies that the baseline $l = XY$ is perpendicular to the bisector p_c. Similarly, equal distances $BY \cong CZ$ imply $l = YZ$ is perpendicular to the bisector p_a, and $CZ \cong AX$ implies that $l = XZ$ is perpendicular to p_b. Hence their common baseline $l = XYZ$ is perpendicular to all three bisectors. □

Obviously (c) implies (a). Hence we conclude that all three assumptions are equivalent. Once again, this means that either all three are true, or all three are false. □

Remark. We have just seen that l is a common perpendicular to p_a, p_b and p_c. Now one can look at the figure in a different way: Each one of the three sides of the triangle $\triangle ABC$ has as a common perpendicular with l— p_c is the common perpendicular of AB and l, p_a is the common perpendicular of BC and l, finally p_b is the common perpendicular of AC and l.

Hence not only all three sides of the triangle, but even their *extensions* lie on one side of line l.

II.13.3 Interlude about points of congruent distance from a line

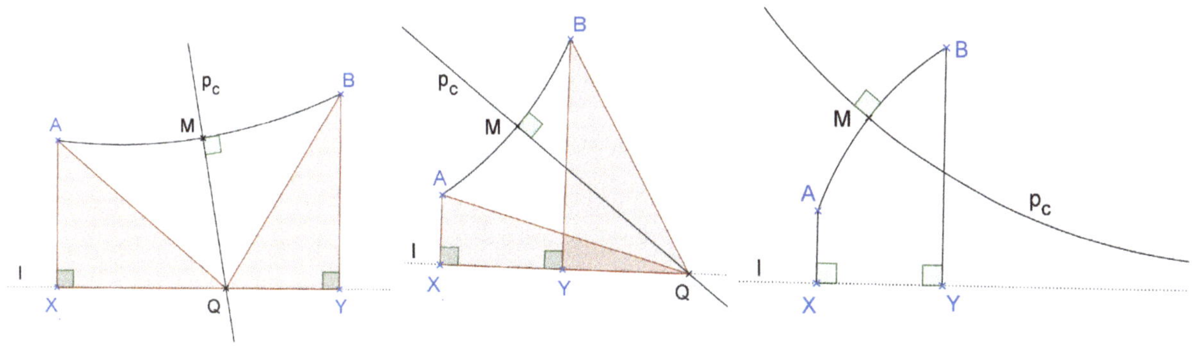

Figure II.13.5: The generic situation for Proposition II.13.7, for which the following statement is proved: segments AX and BY are congruent if and only if lines l and p_c are perpendicular.

Proof of Proposition II.13.7. Let M be the midpoint of segment AB and p_c be its perpendicular bisector. From both points A and B we drop the perpendiculars onto line l. Let X and Y be the foot points.

At first, assume that the lines l and p_c intersect perpendicularly, and let Q be their intersection point. Reversing the steps explained in the proof of Hilbert's Proposition 36, now the first step is to obtain the triangle congruence

(flier) $$\triangle AMQ \cong \triangle BMQ$$

This is done via SAS congruence. As a second step, we get the congruence

(wedge) $$\triangle AQX \cong \triangle BQY$$

by SAA congruence. Hence $AX \cong BY$, and $\square XYBA$ is a Saccheri quadrilateral.

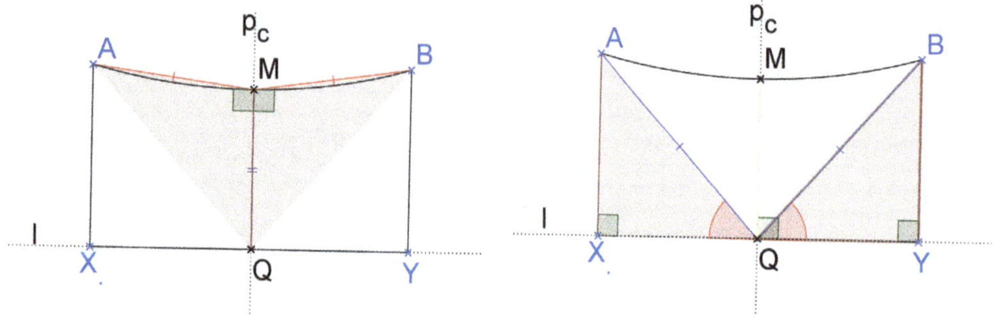

Figure II.13.6: Constructing the Saccheri quadrilateral from its top: if the lines p_c and l are perpendicular to each other, then the segments AX and BY are congruent.

To prove the converse, we assume that the two endpoints A and B have the same distance to line l and lie on the same side of l. Because of $AX \cong BY$, we get the Saccheri quadrilateral $\square XYBA$. As in the proof of Hilbert's Proposition 36, it is shown that the perpendicular bisector q of segment XY intersects segment AB, too—and indeed in its midpoint M.

Question. How does one get these properties of bisector q?

To repeat the details: one begins by checking the congruence (wedge), which is obtained via SAS-congruence. The second step is to get congruence (flier), again by SAS-congruence.

Hence line $q = p_c$ is the perpendicular bisector of both segments AB and XY. Clearly this means that the lines p_c and $l = XY$ are perpendicular. □

II.13.4 The midpoint triangle and its altitudes

Definition II.13.4. Let M_a, M_b and M_c denote the midpoints of the three sides of the triangle. The three segments connecting the vertices to the midpoint of the opposite side are called the *medians* of the triangle. The triangle $\triangle M_a M_b M_c$ is called the *midpoint-triangle.*

Theorem II.13.1. *The line through the midpoints of two sides and the perpendicular bisector of the third side of a triangle are perpendicular.*

Remark. This is a theorem of neutral geometry!

Reason, given in neutral geometry. The theorem follows from Hilbert's Propositions 36 and 39 in the section about Legendre's theorems. We construct the Saccheri quadrilateral $\square AFGB$ corresponding to triangle $\triangle ABC$, To this end, one drops the perpendiculars from the three vertices A, B and C onto line $l = M_a M_b$, and proves the congruence $AF \cong CH \cong BG$. By Hilbert's Proposition 36, the perpendicular bisectors p_c partitions the Saccheri quadrilateral into two Lambert quadrilaterals. Hence p_c is the symmetry line p of this Saccheri quadrilateral. It is perpendicular to both lines $M_a M_b$ and line AB. □

Corollary 36 ("The altitudes of the midpoint-triangle are the side bisectors, and thus form six right angles."). *The altitudes of the midpoint-triangle are the perpendicular bisectors of the sides of the original triangle. Each of the three bisectors is perpendicular to both one side of the original triangle, and one side of the midpoint triangle, thus forming six right angles.*

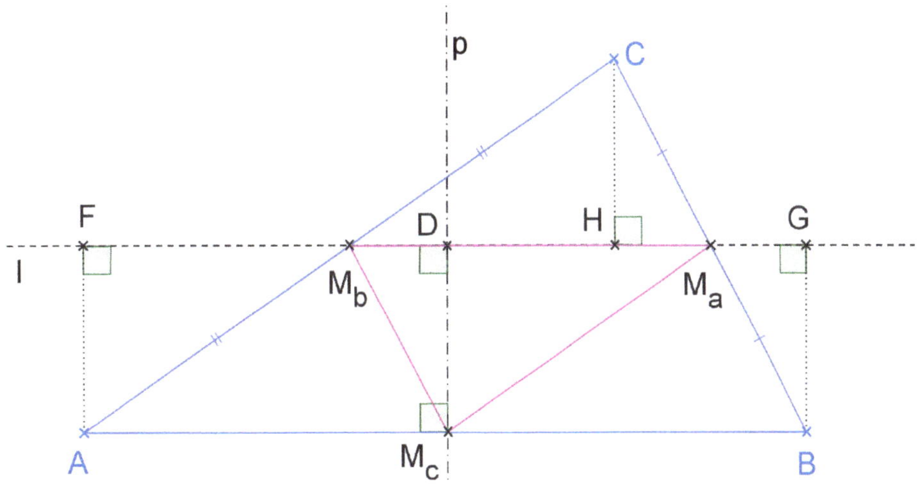

Figure II.13.7: To triangle $\triangle ABC$ corresponds the Saccheri quadrilateral $\square AFGB$.

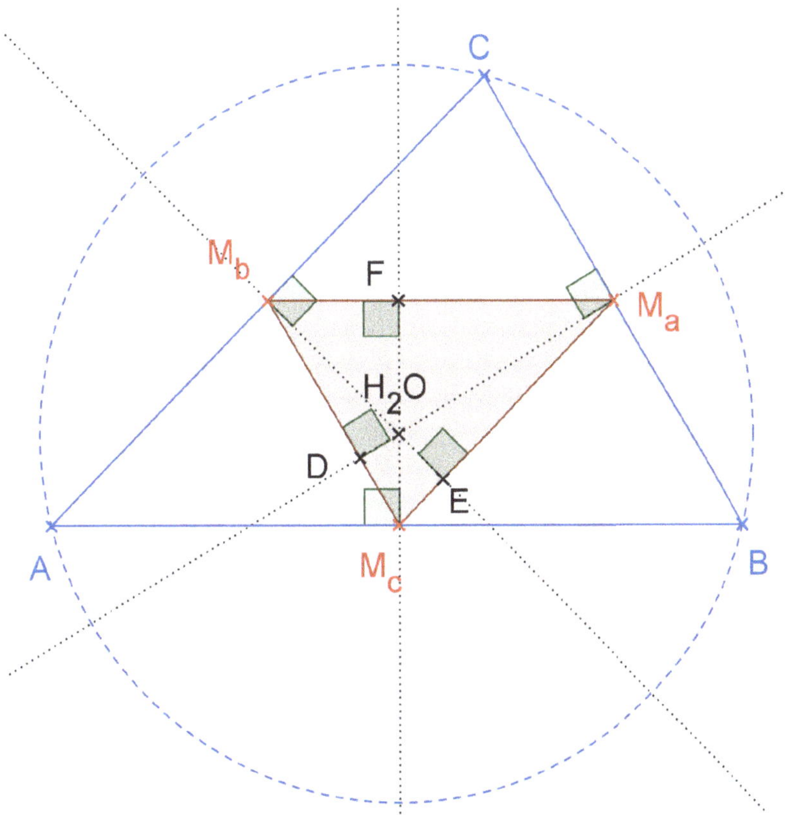

Figure II.13.8: The altitudes of the midpoint triangle are the side bisectors and form six right angles.

Proof. The bisector p_c of side AB, is perpendicular to line M_aM_b, and hence is an altitude of the midpoint triangle $M_cM_aM_b$, too. Similar statements hold for the other two bisectors. $\square$

Proposition II.13.8. *Any two altitudes of an acute triangle insect inside the triangle. Two altitudes of an obtuse triangle may or may not intersect. Any possible intersection point lies outside the triangle, inside the vertical angle to the obtuse angle of the triangle.*

Proof. For an acute triangle, the foot points of the altitude lie on the sides of the triangle, not the

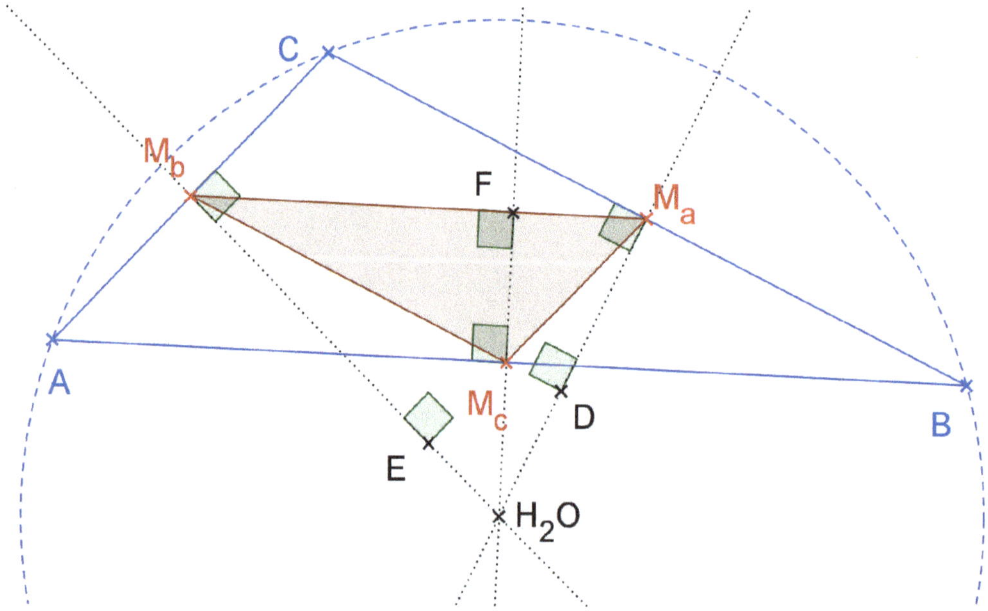

Figure II.13.9: For an obtuse triangle, too, the altitudes of the midpoint triangle are the side bisectors, and form six right angles.

extensions—as follows from the exterior angle theorem. Hence the Crossbar Theorem implies that any two altitude do intersect. The remaining details are left as an exercise. □

Here is stated what we have gathered up to this point. It is little, but more than nothing!

Proposition II.13.9. *For any triangle either one of the following three cases occur:*

"As seen in Euclidean geometry" *The triangle has a circum-circle. The bisectors of all three sides intersect in its center, called circum-center.*

The circum-center of the larger original triangle $\triangle ABC$: is the orthocenter of the midpoint triangle $\triangle M_a M_b M_c$.

(H2O) $$H_2 = O$$

This case always occurs for an acute or right midpoint-triangle. It can but does not need to happen in case of an obtuse midpoint-triangle.

"The genuine hyperbolic case" *The three vertices of the triangle lie on an equidistance line. The bisectors of all three sides have a common perpendicular l. All three extended sides of the triangle are parallel to the baseline l.*

"The borderline case" *The three vertices lie neither on a circle nor an equidistance line. Neither two of the three bisectors intersect, nor do any two of them have a common perpendicular.*

Corollary 37. *In neutral geometry, a triangle $\triangle ABC$ has a circum-circle if and only if two altitudes of the midpoint triangle do intersect. In that case all three altitudes of the midpoint triangle intersect in one point.*

Reason. Suppose now that two altitudes of $\triangle M_a M_b M_c$ do intersect. By Proposition II.13.8, this happens always for an acute or right midpoint-triangle. It can happen, but need not happen in case of an obtuse midpoint-triangle.

Since the two intersecting altitudes are side bisectors for the original triangle, Proposition II.13.2 implies that all three side bisectors intersect in one point. This point is the circum-center of the original triangle and the orthocenter of the midpoint-triangle as stated in formula (H2O).

On the other hand, suppose that all three altitudes of the midpoint-triangle are parallel, but any two of them have a common perpendicular. In that case, we proceed analogously to above: Since the two divergently parallel altitudes are side bisectors for the original triangle, Proposition II.13.6 implies the bisectors of all three sides have a common perpendicular. Furthermore, all three vertices lie on an equidistance line. Thus we arrive at "The genuine hyperbolic case".

The third logical possibility is that all three bisectors are asymptotically parallel—leading to the borderline case. □

Immediate consequences for the altitudes

At this point, it is tempting to look for a more simple result which deal only with the altitudes of the original triangle, without the need to argue via the midpoint triangle. That seems to be astonishingly difficult! Before leaving everything open, I better state my rather awkward partial result:

Proposition II.13.10 (Preliminary and conditional orthocenter—neutral version). *Suppose that the given triangle $\triangle ABC$ is the midpoint-triangle of a larger triangle $\triangle A_0B_0C_0$, and the altitudes of two sides of the triangle $\triangle ABC$ intersect. Then the altitudes of all three sides intersect in one point.*

If the given triangle $\triangle ABC$ is acute, and it is the midpoint-triangle of a larger triangle $\triangle A_0B_0C_0$, then the three altitudes intersect in one point.

Remark. Actually the additional assumption that the given triangle is the midpoint triangle of another (larger) triangle does not need to hold in hyperbolic geometry. Nevertheless the statements about the altitudes of the original triangle remain true! [39]

Proof of Proposition II.13.10. We use the Corollary about the six right angles for triangle $\triangle A_0B_0C_0$. The perpendicular bisectors of triangle $\triangle A_0B_0C_0$ are the altitudes of the midpoint-triangle $\triangle ABC$.

Now Proposition II.13.2 —about the conditional circum-circle— is applied to larger triangle $\triangle A_0B_0C_0$. We conclude that all three perpendicular bisectors of that triangle intersect. But these are just the altitudes of the original given triangle $\triangle ABC$. □

II.13.5 The Hjelmslev Line

Given are two different lines l with the different points $A, B, C, D, \ldots$ and l' with points $A', B', C', D', \ldots$ with congruent distances and same order along these two lines. Let $K, L, M, \ldots$ be the midpoints of the segments $AA', BB', CC', \ldots$.

Theorem II.13.2. *Either one of the following two cases holds:*

[39] The section on hyperbolic geometry contains a proof using Klein's model. A proof in *neutral* geometry is clearly valid a real bottle of wine.

(a) *the exceptional case:* *All midpoints are equal. Furthermore, this point is the midpoint of the common perpendicular of the two lines. They are parallel, and the points $A, B, C, D, \dots$ and $A', B', C', D', \dots$ are ordered in opposite directions along the two lines.*

(b) *the generic case:* *All midpoints are different. They lie on a third line, which is called the* Hjelmslev line.

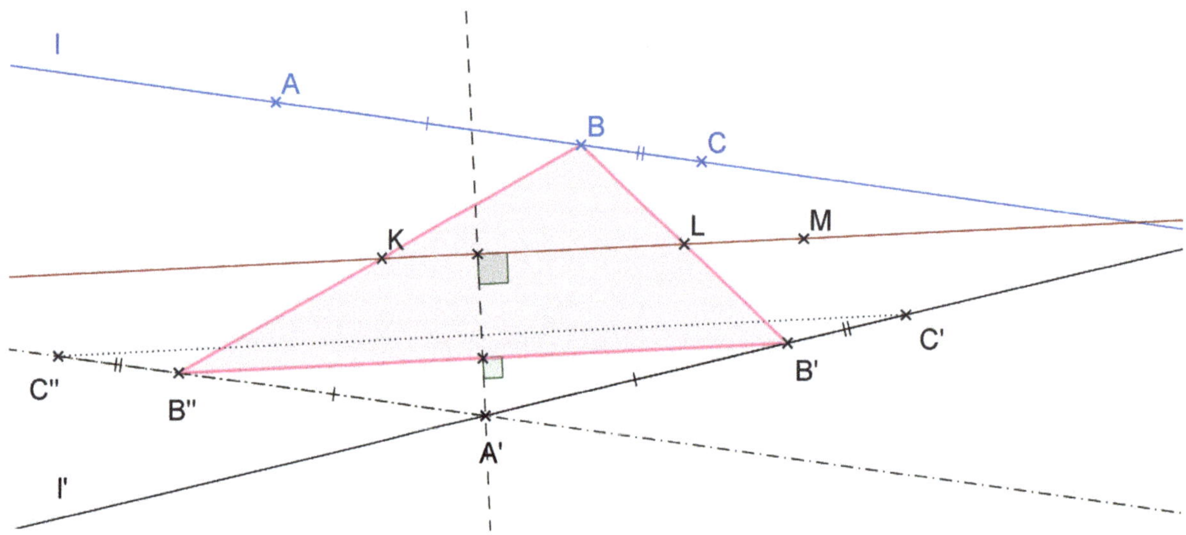

Figure II.13.10: The Hjelmsjev line in the generic case.

Proof. We distinguish the following two cases, from which all others are obtained by renaming:

(a) The mid points K of segment AA' and L of segment BB' are equal.

(b) The mid points of segments AA', BB', CC' are all different.

Problem II.13.1. *Show that in the first case* (a) *all assertions of the theorem hold.*

We now consider case (b). We define point B'' such that K is the midpoint of segment BB''. The assumption $K \neq L$ implies $B' \neq B''$.

Since $AB \cong A'B' \cong A'B''$, the perpendicular bisector p of segment BB'' goes through point $A' = l' \cap l''$ and bisects an angle between the lines $l' = A'B'$ and $l'' := A'B''$. By theorem II.13.1, the line through the midpoints of two sides and the perpendicular bisector of the third side of a triangle are perpendicular. For the triangle $\triangle BB'B''$, we conclude that the lines KL and p are perpendicular. In the special case that points B, B' and B'' lie on a line, we get the same conclusion.

Now we can argue similarly,after replacing B by C and B' by C'. Since $A'C'' \cong A'C'$ and the orders $A' * B' * C'$ and $A' * B'' * C''$, the perpendicular bisector q of segment CC'' is the angle bisector of the same angle $\angle B'A'B'' = \angle C'A'C''$ and hence $p = q$. The line KM is again perpendicular to p. Hence the points K, L and M lie on one line, as to be shown. □

II.13.6 The in-circle and the ex-circles

The construction of the in-circle of a triangle depends on the angular bisectors of its angles. As already stated by definition II.7.12, the *interior angular bisector* of an angle is the ray inside the angle which forms congruent angles with both sides.

Definition II.13.5. The *interior and exterior bisecting lines* of an angle are the two lines which forms congruent angles with either sides of the angle, or the opposite rays.

Question. Show that the bisectors of vertical angles are opposite rays.

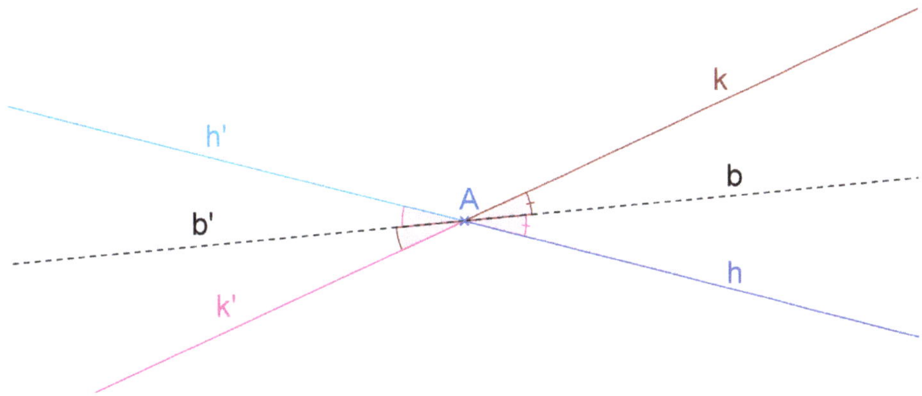

Figure II.13.11: The ray opposite to the bisecting ray b bisects the vertical angle, because all four red angles are congruent.

Solution. Take the vertical angles $\angle(h, k)$ and $\angle(h', k')$. Let ray b be the interior bisector of angle $\angle(h, k)$ and b' be the ray opposite to it. Congruence of vertical angles implies

$$\angle(h, b) \cong \angle(h', b') \quad \text{and} \quad \angle(b, k) \cong \angle(b', k')$$

The definition of the angular bisector yields

$$\angle(h, b) \cong \angle(b, k)$$

From these three formulas, using transitivity and symmetry of angle congruence, one gets

$$\angle(h', b') \cong \angle(h, b) \cong \angle(b, k) \cong \angle(b', k')$$

Hence ray b',which was chosen to be opposite to the bisecting ray b is the angular bisector of the vertical angle $\angle(h', b')$. □

Question. Show that the bisectors of supplementary angles are perpendicular to each other.

Solution. We use the same notation as in the last question, and let ray c be the bisector of the angle $\angle(k, h')$, which is supplementary to angle $\angle(h, k)$. From the last question we get

$$\angle(b, k) \cong \angle(b', k') \cong \angle(h', b')$$

and, by the definition of bisector c, we get

$$\angle(k, c) \cong \angle(c, h')$$

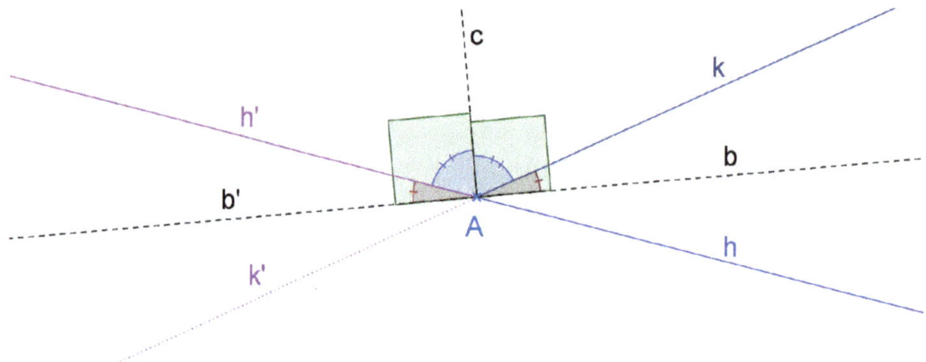

Figure II.13.12: Angle addition of two pairs of congruent angles yields congruent supplementary angles—proving that the bisectors of supplementary angles are perpendicular.

Now we use that sums of congruent angles are congruent. Hence the two formulas yield

$$\angle(b,c) = \angle(b,k) + \angle(k,c) \cong \angle(h',b') + \angle(c,h') \cong \angle(c,b')$$

Thus we see that $\angle(b,c)$ and $\angle(c,b')$ are supplementary congruent angles, and hence they are right angles. □

Here is the characterization of the angular bisectors.

Proposition II.13.11. *A point in the interior of an angle has congruent distances from both sides of the angle if and only if it lies on the interior angular bisector.*

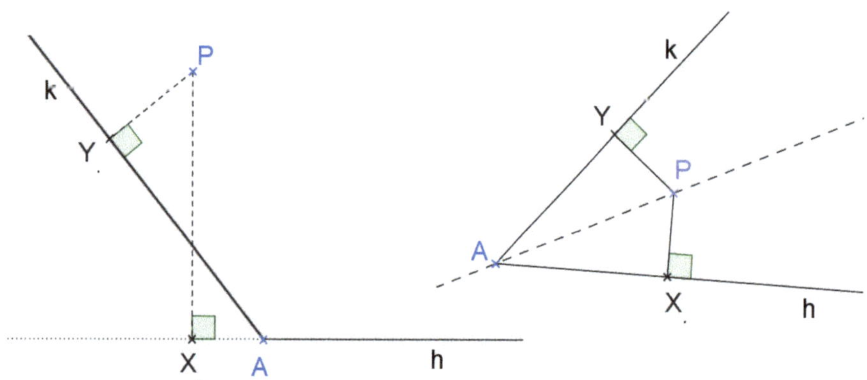

Figure II.13.13: A point P on the interior angular bisector has congruent distances from both sides the angle $\angle(h,k)$.

Proposition II.13.12. *A point has congruent distances from both lines extending the sides of an angle, if and only if it lies on the interior or the exterior bisecting line.*

Remark. Indeed the characterization of the angular bisectors from proposition II.13.11 and II.13.12 fails for a degenerate angle. The assertion of these propositions are wrong for an angle degenerating to zero or two right!

Proof. Given is the angle $\angle(h,k)$ with vertex A, and a point P. The two lines of the sides h and k of the angle $\angle(h,k)$ are assumed to be different, by definition of an angle. Let X and Y be the foot points of the perpendiculars dropped from point P onto the two different lines extending h and k. In the figure on page 326, we give a case were point P neither lies on an angular bisector nor has congruent distances to the sides of the angle—and, in a separate drawing, a case were point P both lies on an angular bisector and has congruent distances to the sides of the angle.

Claim 0. If point P lies on either one of the two lines extending h and k, the assertion of proposition II.13.12 is true.

Reason. For $P \neq A$, neither are the distances from point P to the two lines equal—only one is zero—, nor does point P lie on any one of the bisectors, which lie in the interiors of the four angles with vertex A we have obtained. If $P = A$, all is clear, too. □

We shall now assume that point P does not lie on either one of the lines extending rays h and k. We begin with the case were a triangle congruence characterizes the points P which either lie on the angular bisector or have congruent distances to the sides of the angle.

Claim 1. Assume there exist the triangles $\triangle PAX$ and $\triangle PAY$. In that case,

$$PX \cong PY \quad \text{implies} \quad \angle PAX \cong \angle PAY$$

Reason. The two triangles

$$\triangle PAX \cong \triangle PAY \tag{II.13.5}$$

are congruent by the hypothenuse-leg theorem II.7.32. Hence $\angle PAX \cong \angle PAY$. □

Claim 2. Assume there exist the triangles $\triangle PAX$ and $\triangle PAY$. In that case,

$$\angle PAX \cong \angle PAY \quad \text{implies} \quad PX \cong PY$$

Reason. The two triangles (II.13.5) are congruent by SAA-congruence, given in propositionII.7.54. Hence $PX \cong PY$. □

Claim 3. If the triangles $\triangle PAX$ and $\triangle PAY$ exist, the assertion of proposition II.13.12 is true.

Claim 4. If $A \neq X$, and point P does not lie on the line of ray h, then the triangle $\triangle PAX$ exists. Hence, if $A \neq X$ and $A \neq Y$, the assertion of proposition II.13.12 is true.

Claim 5. Either $A \neq X$ or $A \neq Y$.

Reason. If $A = X = Y$, then both rays h and k would be perpendicular to the segment PA. Hence they would be identical or opposite rays. Hence $A = X = Y$ imply the rays h and k lie on the same line. But this contradicts the definition of an angle, and hence cannot occur. □

What happens in the awkward case $A = X$ and $A \neq Y$? In this case the triangle $\triangle PAX$ is nonexistent (degenerate). But it is easy to deal with the case separately, since the perpendiculars to the sides of an angle cannot be any angular bisector. Let K be any point on the ray k, and H be any point on the ray h.

Claim 6. If $A = X$ and $A \neq Y$, and the rays h and k lie on two different lines, then $R = \angle PAH > \angle PAY$ and $PX = PA > PY$. Similarly, if $A = Y$ and $A \neq X$, then $\angle PAX < R = \angle PAK$ and $PX < PA = PY$. In these cases, point P neither lies on an angular bisector nor has congruent distances to the sides of the angle.
Hence, if $A = X$ or $A = Y$, the assertion of proposition II.13.12 is true.

Reason. Because of claim 5, the assumption $A = Y$ implies $A \neq X$. Hence by claim 4, the right triangle $\triangle PAX$ exists. The right angle across to the hypothenuse is larger than any of the other two angles, with vertices A or X. Hence $\angle PAX < R = \angle PAK$. Its longest side is the hypothenuse PA. □

Claim 7. If the segments $PX \cong PY$ to the foot points are congruent, then point P lies in the interior of the angle $\angle XAY$.

□

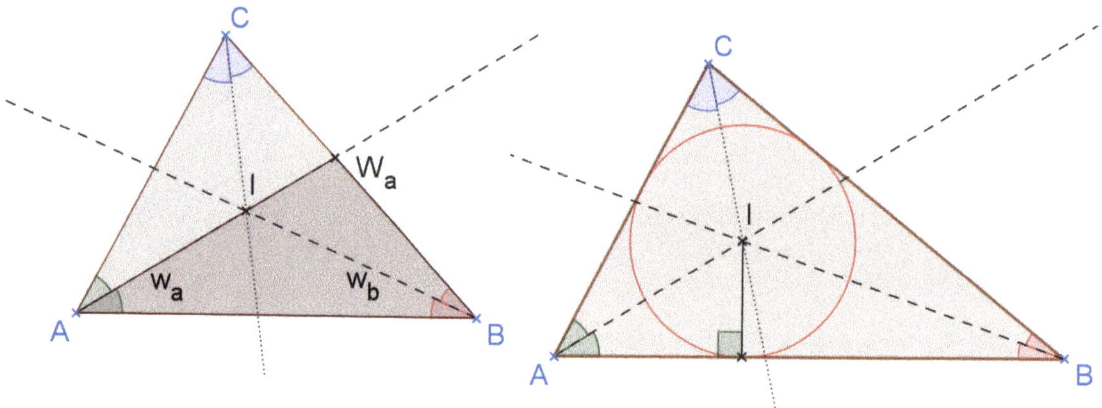

Figure II.13.14: Any two angular bisectors intersect at the center of the in-circle.

Theorem II.13.3 (The in-circle). *All three interior angular bisectors of a triangle intersect in one point. This point has congruent distances from all three sides of the triangle, and hence it is the center of the in-circle.*

Problem II.13.2 (The in-circle). *We want to construct the in-circle of a triangle. Explain the steps of the reasoning and provide drawings. Use the proposition II.13.11 given above, and still other basic facts.*

(a) *Explain why any two interior angle bisectors of a triangle intersect.*

(b) *Explain why this point has congruent distances from all three sides of the triangle, and why it is the center of the in-circle.*

(c) *Explain why all three interior angular bisectors of a triangle intersect in one point.*

(d) *Explain how one can draw the in-circle of a triangle.*

Answer. Here is some rather elaborate recollection:

(a) **Why any two interior angle bisectors of a triangle intersect:** By the Crossbar Theorem, an interior angular bisector intersects the opposite side. Let W_a be the intersection point of the angular bisector of angle $\angle BAC$ with side BC.

Once more, by the Crossbar Theorem, the angular bisector w_b of angle $\angle ABC$ intersects segment AW_a, since points A and W_a lie on the two sides of this angle. The bisector w_b intersects segment AW_a of bisector w_a, say in point I.

(b) **Point I has congruent distances from all three sides:** Since point I lies one the bisector of angle $\angle BAC$, it has congruent distances from the sides AC and AB. Since point I lies one the bisector of angle $\angle ABC$, it has congruent distances from the sides BC and BA. By transitivity, it has congruent distances from all three sides.

(b) **Point I is the center of the in-circle:** Let X be the footpoint of the perpendicular dropped from point I onto the side BC. This side is tangent to the circle around I through X, since the tangent of a circle is perpendicular to the radius IX at the touching point. Indeed, we have already obtained a circle to which <u>all three sides</u> of the triangle are tangent. Let Y, Z be the footpoints of the of the perpendiculars dropped from point I onto the other two sides CA and AB. Since $IX \cong IY \cong IZ$, the same circle goes through Y and Z. Again, the sides CA and AB are tangent to this circle since they are perpendicular to IY and IZ, respectively.

(c) **All three interior angular bisectors of a triangle intersect in one point:** Since point I has congruent distances from the sides AC and BC, the point lies on either one of the bisectors of the angles at C—which we did not even use in the construction. Since point I lies in the interior of the angle $\angle BCA$, it lies on the interior angular bisector of this angle.

(d) **How to draw the in-circle of a triangle:** As seen above, one needs to construct the angular bisectors of any two of the three angles of the triangle. One drops the perpendicular from their intersection point I to any side of the triangle. Finally, one draws the circle around I through the foot point.

The exterior bisectors are bisecting the exterior angles of the triangle. They are the perpendiculars to the interior angular bisectors, erected at the vertices. They are used in the construction of the ex-circles.

Proposition II.13.13 (Conditional ex-circle—neutral version). *If any two of the exterior bisectors at vertices A and B, and the bisector of the third angle $\angle BCA$ intersect, then all these three bisectors intersect in one point. In that case, the triangle has an ex-circle touching side AB from outside, and the extensions of the two other sides.*

Reason. This is an easy consequence of Proposition II.13.12. The details are left to the reader. □

II.13.7 The Hjelmslev quadrilateral in neutral geometry

Definition II.13.6. A quadrilateral with two right angles at opposite vertices is called a *Hjelmslev quadrilateral.*

In a Hjelmslev quadrilateral, we draw the diagonal between the two right angles and drop the perpendiculars from the two other vertices. Their occur some remarkable congruences.

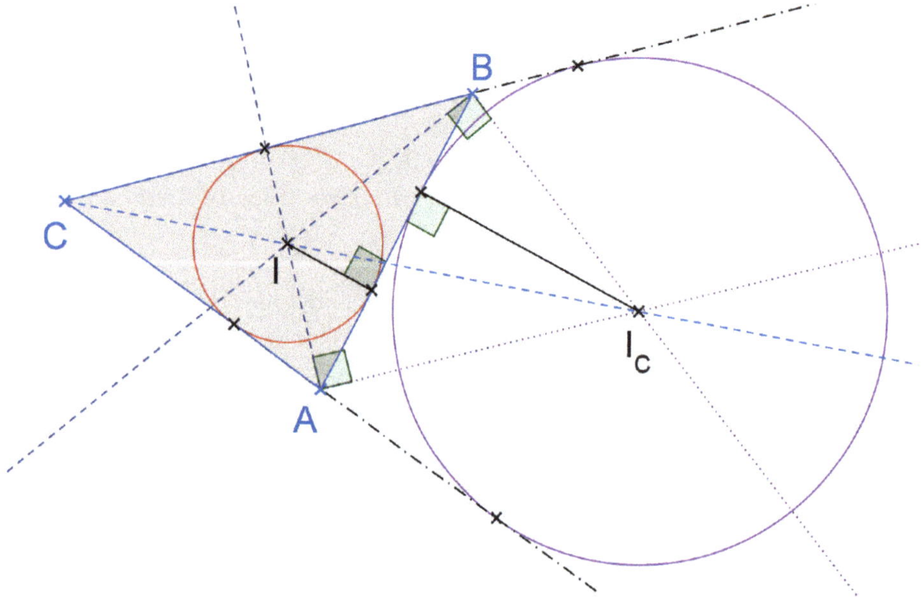

Figure II.13.15: The in-circle and one ex-circle.

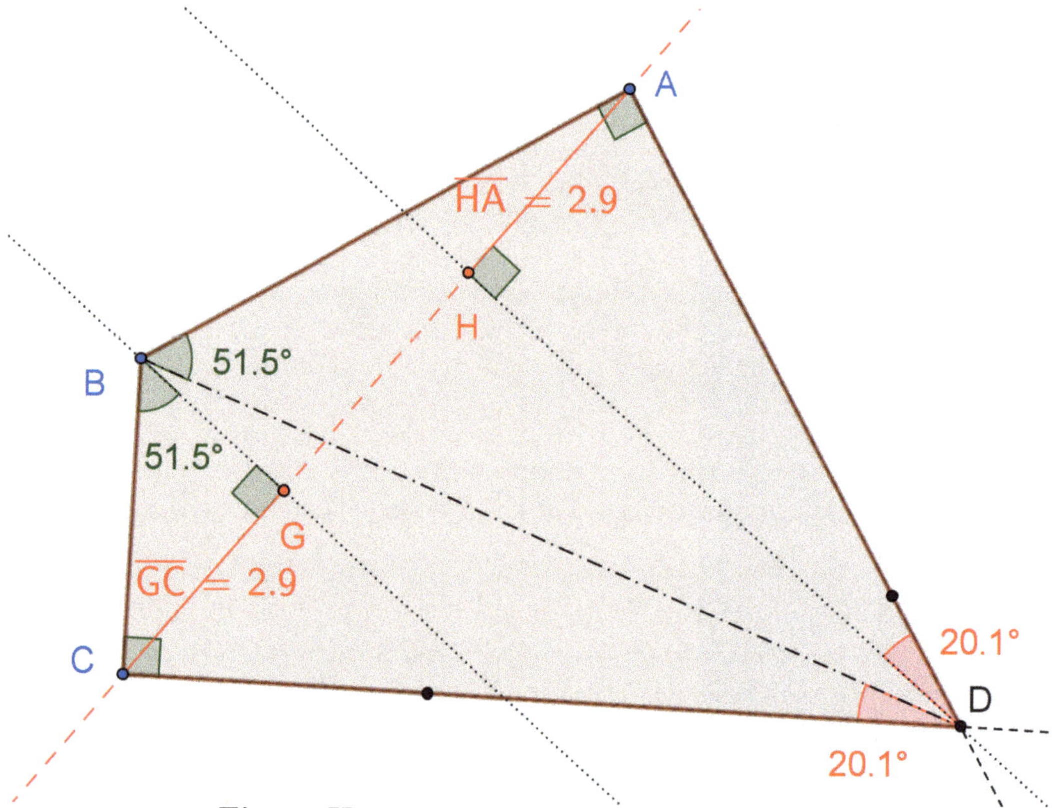

Figure II.13.16: The Theorem of Hjelmslev.

Proposition II.13.14 (Hjelmslev's Theorem). *In a Hjelmslev quadrilateral,*

(a) *at each of the two vertices with arbitrary angle, there is a pair of congruent angles between the adjacent sides, and the second diagonal and the perpendicular dropped onto the first diagonal, respectively;*

(b) *there is a pair of congruent segments on the diagonal between the right vertices, measured between these vertices and the foot points of the perpendiculars.*

For the proof in neutral geometry, we need some basic lemmas about reflections. *These are facts of neutral geometry.*

Definition II.13.7 (Rotation). The composition of two reflections across intersecting lines is a *rotation*. The intersection point O of the reflection lines is the *center* of the rotation. If point P is mapped to point P', the angle $\angle POP'$ is independent of the point P. It is called the *angle of rotation*.

Lemma II.13.2. *Let four lines p, q, r, s intersect in one point. The compositions of reflections across these lines satisfy $p \circ q = r \circ s$ if and only if $\angle(q, p) = \angle(s, r)$ for the directed angles. The angle of rotation is twice the angle between the two reflection lines.*

Definition II.13.8 (Point reflection). The *reflection across the center of reflection O* maps any point P into a point P', such that the center of reflection O is the midpoint of segment PP'.

Lemma II.13.3. *Two reflections commute if and only if the reflection lines are either identical or perpendicular to each other. The composition of two reflections across perpendicular lines depends only on the intersection point of the two lines. It is the reflection across their intersection point.*

Definition II.13.9 (Translation). The composition of two reflections across lines with a common perpendicular t is a *translation along t*.

Lemma II.13.4. *Let four lines p, q, r, s have the common perpendicular t, and let P, Q, R, S be the intersection points. The compositions of reflections across these lines satisfy $p \circ q = r \circ s$ if and only if $\overline{QP} = \overline{SR}$ for the directed segments.*

Problem II.13.3. *Prove the segment congruence of Hjelmslev's Theorem in neutral geometry. Assume that the angle congruence has already been shown. Use compositions of reflections across the lines marked in the figure on page 332 and check that*

$$c \circ g = h \circ a$$

Proof of the segment congruence in neutral geometry.

$$\begin{aligned} e \circ c \circ g &= c_1 \circ b \circ g && \text{by Lemma II.13.3 at point } C \\ &= c_1 \circ f \circ a_1 && \text{by Lemma II.13.2 at point } B \\ &= h \circ d \circ a_1 && \text{by Lemma II.13.2 at point } D \\ &= h \circ e \circ a && \text{by Lemma II.13.3 at point } A \\ &= e \circ h \circ a && \text{by Lemma II.13.3 at point } H \end{aligned}$$

Hence $c \circ g = h \circ a$, and finally Lemma II.13.4 yields $\overline{CG} = \overline{HA}$, as to be shown. □

Proposition II.13.15 (Translation and Saccheri quadrilaterals). *Let point A be mapped to point A' by a translation along the line t, and let F and F' be the foot points of the perpendiculars dropped from the points A and A' onto the line t of translation.*

The quadrilateral $\square FAA'F'$ is a Saccheri quadrilateral. Hence the segments AA' and FF' are parallel.

The base segment $\overline{FF'}$ is double the segment $\overline{QP}$ between the intersection points of the lines of reflection with their common perpendicular, independently of the point A.

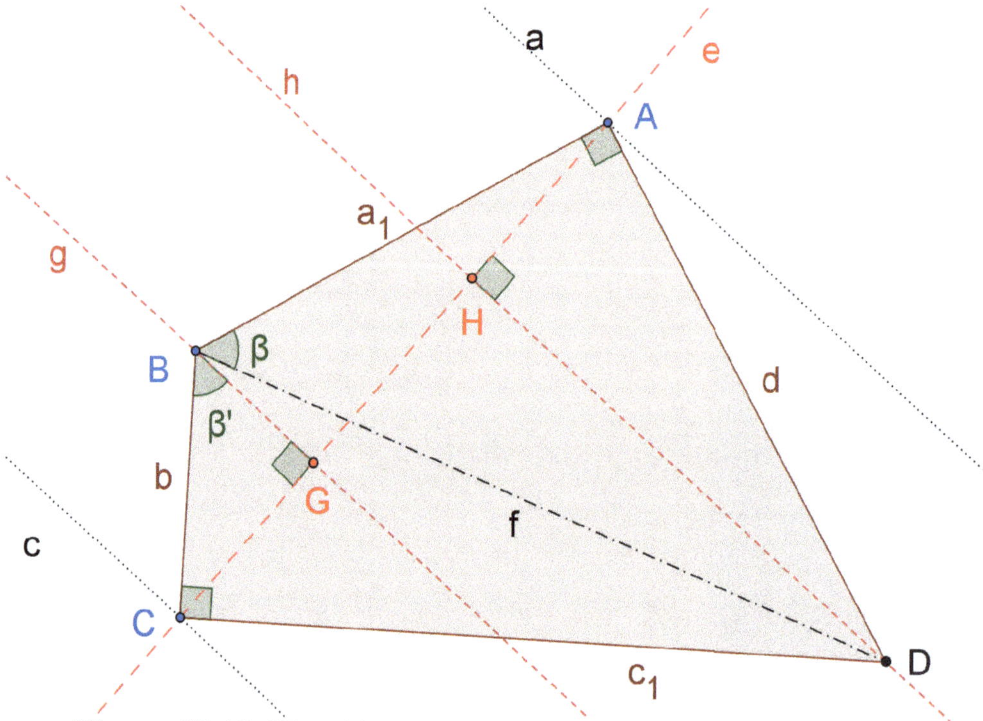

Figure II.13.17: The segment congruence in neutral geometry.

Remark. In Euclidean geometry, the segment $\overline{AA'}$ from any point A to its image A' is double the segment $\overline{QP}$ between the intersection points of the lines of reflection with their common perpendicular, independently of the point A. Note that this is not true in neutral geometry!

Indeed, in hyperbolic geometry, the top segment AA' is longer than the base segment FF', for any point A not on the line of translation.

To prove the entire Theorem of Hjelmslev in neutral geometry, one needs the following more elaborate result.

Definition II.13.10 (Translation-reflection). The composition of two reflections across a line p and a point Q is called a *translation-reflection*. The perpendicular to the line to the point is baseline along which the translation-reflection acts.

Theorem II.13.4 (A translation-reflection specifies both the translation line and distance). *Let two lines p, r and two points Q, S be given. The compositions of reflections across these lines and points satisfy*

$$p \circ Q = r \circ S$$

if and only if the following three conditions hold:

there exists a common perpendicular to lines p and r through the points Q and S;

the respective distances from point Q to the foot point P on p, and from point S to the foot point R on r are congruent;

The two segments $\overline{QP} = \overline{SR}$ have the same orientation on the common perpendicular.

Similarly,

$$p \circ Q = S \circ r$$

holds if and only if the two segments $\overline{QP} = \overline{RS}$ are congruent and have opposite orientations on a common perpendicular of the lines p and r.

Question. What does it mean in terms of order relations or rays that two segments PQ and RS on one line have the same orientation?

Answer. Two segments QP and SR on one line have the same orientation if and only if of the two rays $\overrightarrow{QP}$ and $\overrightarrow{SR}$ one is a subset of the other.

Problem II.13.4. *Prove Hjelmslev's Theorem in neutral geometry. In the Hjelmslev configuration, you need at first to define lines g' and h' by angle congruences.*

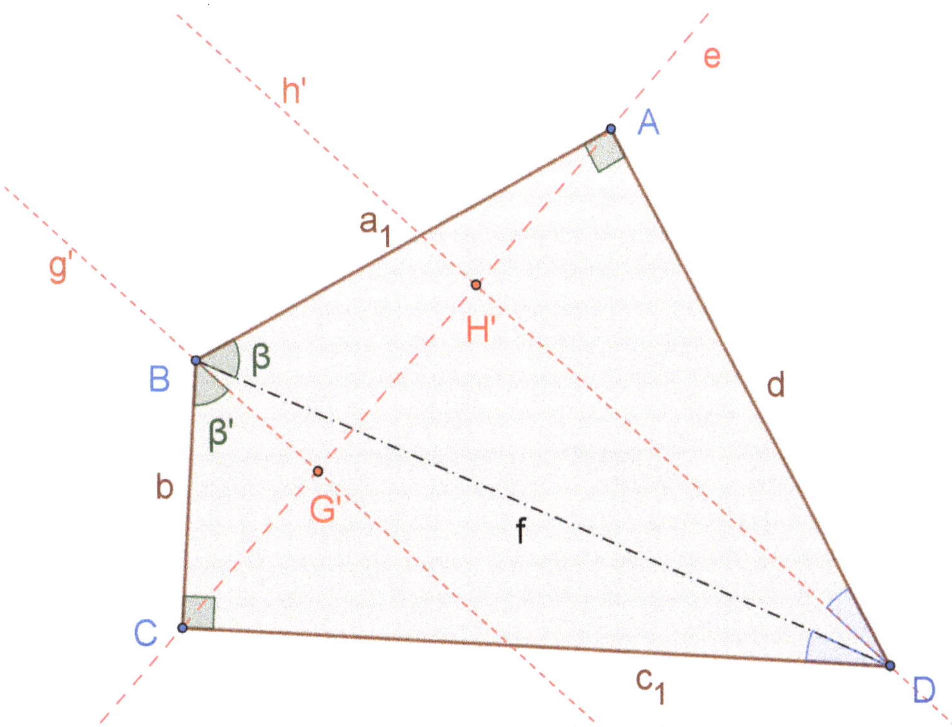

Figure II.13.18: Begin defining the lines g' and h' by the angle congruences.

Proof of the Hjelmslev Theorem in neutral geometry. Begin defining the lines g' and h' by the angle congruences

$$\angle(b, g') \cong \angle(f, a_1)$$
$$\angle(h', d) \cong \angle(c_1, f)$$

for the directed angles. Quite similar above, we check

$$\begin{aligned} C \circ g' &= c_1 \circ b \circ g' && \text{by Lemma II.13.3 at point } C \\ &= c_1 \circ f \circ a_1 && \text{by assumption} \\ &= h' \circ d \circ a_1 && \text{by assumption} \\ &= h' \circ A && \text{by Lemma II.13.3 at point } A \end{aligned}$$

By the translation-reflection theorem II.13.4, he translation-reflections $C \circ g'$ and $h' \circ A$ have the same base line and distance. In other words, line AC is the common perpendicular of lines g' and h'. Hence these lines are the same as the perpendiculars: $g = g'$ and $h = h'$. Furthermore, $CG' \cong H'A$ as directed segments. □

II.13.8 Limitations of neutral triangle geometry

You have for sure realized that the propositions above do not give the complete and beautiful results from Euclidean geometry—instead they contain awkward conditions and unexpected cases.

This state of affairs has good reasons. Indeed many of the more complete theorems of triangle geometry are equivalent in some cases to the Euclidean parallel axiom, or in other cases to the angle sum of a triangle being two right angles. The following theorem is confirming this claim.

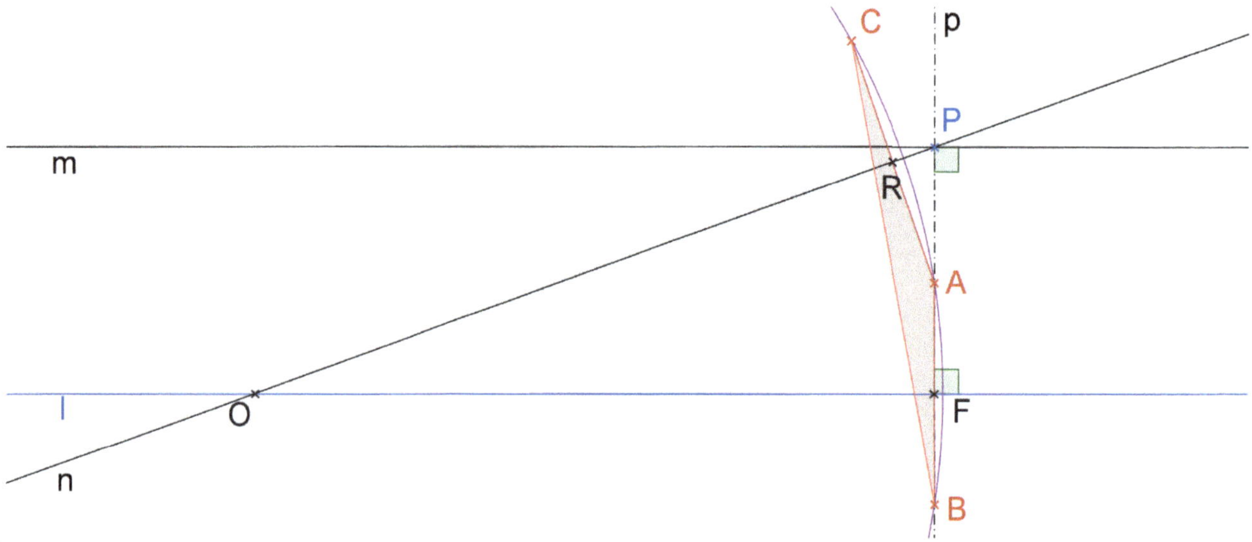

Figure II.13.19: Triangle $\triangle ABC$ has a circum-circle, hence line n intersects line l in its center O—parallels are unique.

Theorem II.13.5 (Farkas Bolyai). *Assume only the axioms of incidence, order, congruence—leading to neutral geometry. If every triangle has a circum-circle, then the Euclidean parallel axiom holds.*

Proof. As explained in Proposition II.7.43 in the section on Legendre's theorems, existence of a parallel can be proved in neutral geometry. Indeed, one parallel is conveniently constructed as "double perpendicular".

We start with this construction from Proposition II.7.43. Given is any line l and any point P not on l. One drops the perpendicular from point P onto line l and denotes the foot point by F. Next, one erects at point P the perpendicular to line PF. Thus, one gets the "double perpendicular" m.

Let A be any point on the segment PF, and B be its image of reflection by line l. By definition of reflection, the foot point F is the midpoint of segment AB. The four points P, F, A and B lie on the common perpendicular p of the two lines l and m.

We want to show the line m is the unique parallel to line l through point P. Assume that line n is a parallel to line l through point P, too. Now let C be the image of reflecting point A by line n. By definition of reflection, this means that segment AC and line n intersect perpendicularly at the midpoint R. If point C lies on line $AB = p$, then $R = P$ and $n = m$, because both are the perpendicular to p erected at P.

But, if one assumes $n \neq m$ is really a second different parallel to line l through point P, one can show that lines n and l intersect, contradictory to being parallel.

Indeed, by the argument above we see that point C does not lie on line AB. Hence the three point A, B and C form a triangle. It is assumed that every triangle has a circum-circle. Let O be

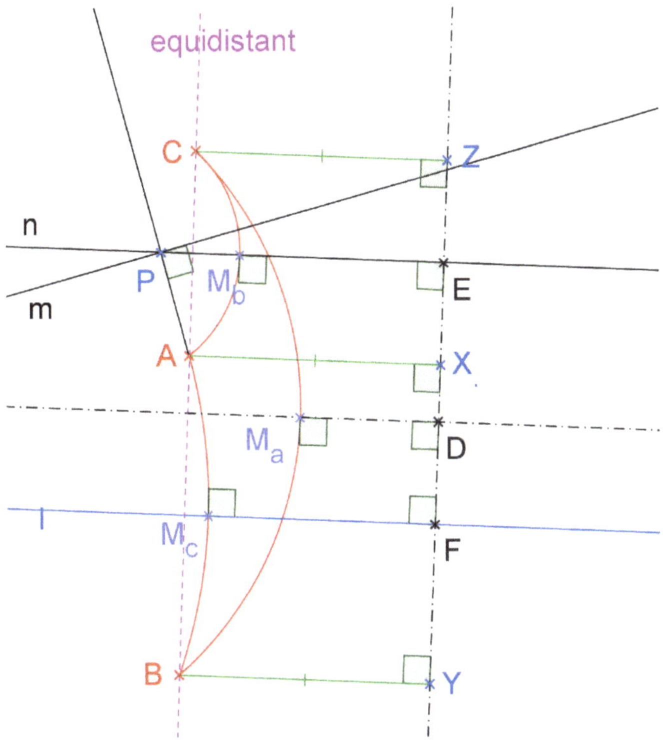

Figure II.13.20: The perpendicular bisectors of the sides of triangle $\triangle ABC$ are all parallel. The triangle has no circum-circle—true, but hard to believe! There exists two parallels m and n to line l through point P.

the center of the circum-circle of triangle $\triangle ABC$. Because O has same distance from points A and B, it lies on the perpendicular bisector of segment AB, which is line l. Similarly, because O has same distance from points A and C, it lies on the perpendicular bisector of segment AC, which is line n. Hence the two lines l and n intersect at point O, and are not parallel. Thus a second parallel to line l through point P does not exist. □

Question. Why are the two lines l and n different?

Answer. Line n goes through point P, but line l does not go through point P.

Corollary 38. *In any Hilbert plane are equivalent*

(i) *for every line l and for every point P lying not on l, there exists a unique parallel m to l through point P;*

(ii) *every triangle has a circum-circle.*

*Reason for "***(i)** *implies* **(ii)***".* Assumption (i) states the Euclidean parallel postulate, resulting in a Pythagorean plane. In corollary 35 is shown that in a Pythagorean plane, every triangle has a circum-circle. □

*Reason for "***(ii)** *implies* **(i)***" .* This is the theorem II.13.5 by Farkas Bolyai. □

Here is another simple proposition that implies semi-Euclidean geometry.

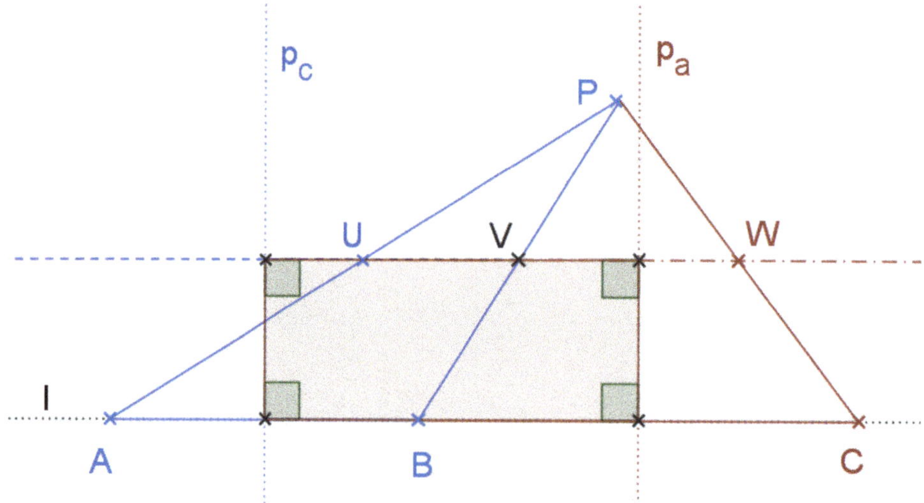

Figure II.13.21: If the three midpoints lie on a line, a rectangle exists.

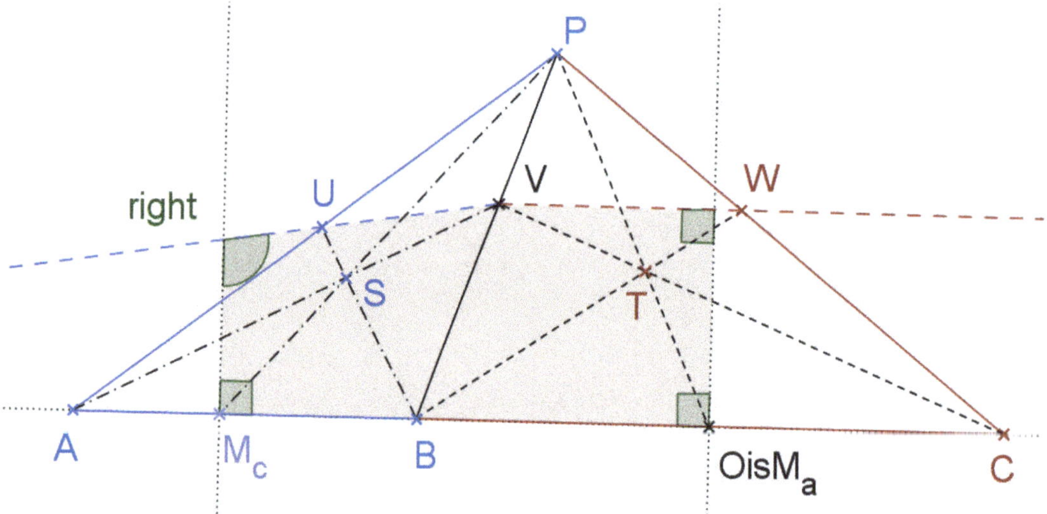

Figure II.13.22: How the hyperbolic case appears in Klein's model.

Proposition II.13.16 (Three midpoints). *Given are three points A, B, C on a line l and a point P not on line l. If the three midpoints U, V, W of segments AP, BP and CP lie on a line, then a rectangle exists. Conversely, if a rectangle exists, then the three midpoints lie on one line.*

Question. Provide a drawing. Explain how one gets a rectangle, in neutral geometry. Use Theorem II.13.1 and once more, the material of my package on Legendre's geometry, especially Proposition 39, to explain your reasoning.

Proof. The perpendicular bisector p_c of segment AB intersects both lines l and UV perpendicularly. This follows from Theorem II.13.1, applied to $\triangle ABP$. By the same reasoning, the perpendicular bisector p_a of segment BC intersects both lines l and VW perpendicularly. This time, one applies Theorem II.13.1 to $\triangle BCP$.

If the three midpoints U, V, W lie on one line, then this line—together with lines l, and the two bisectors p_c and p_a form a rectangle.

Conversely, use that a rectangle exists. By the second Legendre Theorem, the angle sum in every triangle is $2R$, in every quadrilateral $4R$, in every pentagon $6R$. If the three midpoints

U, V, W would not lie on one line, there would exist a pentagon with angle sum different from $6R$, which is impossible. Hence the three midpoints U, V, W lie on one line. □

II.14 A Simplified Axiomatic System of Geometry, my Suggestion

II.14.1 A simplified axiomatization of geometry

Introduction

The axiomatic system of Hilbert shows exceptional insight. On the other hand, it is unnatural in many respects. My simplification restricts attention to two dimensions, introduces circles in the natural classical way, and continuity in the way directly related to proofs. My goal is to give a logical analysis of the natural way to do geometry, and still keep up the level of rigor introduced by Hilbert. But I do not try to minimize the axioms as strictly as Hilbert. Especially the axioms dealing with the order of three points on a line, the SAS congruence, and the angle sum can be stated equivalently, but making less assumptions.

0. Undefined elements and relations

Elements:

- A class of undefined objects called *points*, denoted by $A, B, C, \ldots$.
- A class of undefined objects called *lines*, denoted by $a, b, c, \ldots$.
- A class of undefined objects called *circles*, denoted by $\mathcal{C}, \mathcal{D}, \ldots$.

Relations:

- *Incidence* (being *incident, lying on, containing*)
- *Order* (lying *between*) (for points on a line)
- *Congruence* (for segments)
- *Center* (for circle)

I. Axioms of incidence

I.1 For two points A and B there exists a line that contains both points.

I.2 For two different points A and B there exists no more than one line that contains both points.

I.3a Any line contains at least two points.

I.3b There exist at least three points that do not lie on a line.

Remark. I did separate axiom I.3 into I.3a and I.3b, in order to stress there is no direct logical connection between the two sentences intended.

II. Axioms of order

II.1 If a point B lies between a point A and a point C, then the points A, B, C are three distinct points of a line, and B lies between C and A.

Definition II.14.1 (Segment). Let A and B be two distinct points. The *segment* AB is the set consisting of the points A and B and all points lying between A and B. The points A and B are called the *endpoints* of the segment, they are assumed to be different. The points between the endpoints are called the *interior points* of the segment, and the remaining points on the line AB are called the *exterior* points of the segment.

Definition II.14.2 (Ray). Given two distinct points A and B, the *ray* $\overrightarrow{AB}$ is the set consisting of the points A and B, the points inside the segment AB, and all points P on the line AB such that point B lies between vertex A and P. The point A is called the *vertex* of the ray.

II.2* Of any three different points on a line exactly one lies between the two others.

II.3* For two points A and B, the ray $\overrightarrow{AB}$ contains a point C not lying inside the segment AB.

Definition II.14.3 (Triangle). We define a *triangle* to be union of the three segments AB, BC and CA. The three points A, B and C are assumed not to lie on a line. These three points are the *vertices*, and the segments BC, AC, and AB are the *sides* of the triangle.

II.4* (strong Pasch' Axiom) Given is a triangle $\triangle ABC$ and a line a which does not meet any of its vertices A, B, C. If the line a passes through a point of the side AB, it also passes through a point of either one of the two other sides AC or BC, and intersects exactly two sides of the triangle.

II.5 (Convention for orientation) There exist three points A, B and C such that point C lies in the left half-plane of ray $\overrightarrow{AB}$, and the triangle $\triangle ABC$ is oriented counterclockwise.

III. Axioms for circles

III.1 For any center O and point $A \neq O$, there exists a unique circle with center O through the point A.

III.2 A circle $\mathcal{C}$ and a ray emanating from its center O intersect at exactly one point.

Definition II.14.4 (Interior and exterior of a circle). A point A lies *interior* of a circle if it lies between the center O and the point P where the ray $\overrightarrow{OA}$ cuts the circle.

A point B lies *exterior* of a circle if it lies outside the segment OQ between the center O and the point Q where the ray $\overrightarrow{OA}$ cuts the circle.

III.3 (Circle-circle intersection property) Let $\mathcal{C}$ and $\mathcal{D}$ be two circles. Let P, Q be two points on circle $\mathcal{D}$. Assume that point P lies inside circle $\mathcal{C}$ and Q lies outside circle $\mathcal{C}$. Then the two circles $\mathcal{C}$ and $\mathcal{D}$ intersect each other.

III.4 Under the assumptions of (III.3), the two circles intersect in exactly two points, which lie on different sides of the line connecting their centers.

IVa. Axioms of segment congruence

IV.1 If a segment $A'B'$ and a segment $A''B''$ are congruent to the same segment AB, then segment $A'B'$ is also congruent to segment $A''B''$.

IV.1b For any two different points A and B, the segments AB and BA are congruent.

IV.2 (The sums of congruent segments are congruent.) On a line, let AB and BC be two segments which except for B have no point in common. Furthermore, on the same or another line a', let $A'B'$ and $B'C'$ be two segments which except for B' also have no point in common. In that case,

$$\text{if} \quad AB \cong A'B' \quad \text{and} \quad BC \cong B'C' \,, \quad \text{then} \quad AC \cong A'C'$$

IV.3a For a circle $\mathcal{C}$ with center O and two rays with vertex O which cut the circle at points P and Q, the segments OP and OQ are congruent.

IVb. SAS-axiom for triangle congruence

Definition II.14.5 (Angle). An *angle* is the union of two rays with common vertex not lying on one line.

We introduce a segment of *unit length.*

Definition II.14.6 (Congruences of angles). The two angles $\angle POQ$ and $\angle P'O'Q'$ are called *congruent* if and only if there are four units segments $OP, OQ, O'P', O'Q'$ from the vertices and two congruent segments PQ and $P'Q'$ between the sides of the angles.

IV.4* Two triangles ABC and $A'B'C'$ are congruent if they have a pair of congruent angles and the two pairs of adjacent sides are pairwise congruent. In detail:

$$\begin{array}{ll} \text{The congruences} & AB \cong A'B' \,, \; AC \cong A'C' \,, \; \angle BAC \cong \angle B'A'C' \\ \text{imply the congruences} & BC \cong B'C' \,, \; \angle ABC \cong \angle A'B'C' \,, \; \angle BCA \cong \angle B'A'C' \end{array}$$

V. Axioms of continuity

V.1 (Axiom of Archimedes) If AB and CD are any segments, then there exists a natural number n such that n segments congruent to CD constructed contiguously from A, along a ray from A through B, will pass beyond B.

V.2 (Cantor's principle of boxed intervals) Every sequence of boxed intervals contains a common point. In detail:

For a sequence of segments A_iB_i such that

$$A_i * A_{i+1} * B_{i+1} * B_i \quad \text{for all} \quad i = 1, 2, 3, \ldots$$

there exists a point X^* such that

$$A_i * X^* * B_i \quad \text{for all} \quad i = 1, 2, 3, \ldots$$

VI. Axioms of parallelism

VI.1 (The angle sum of a triangle is two right angles) The sum of the three interior angles of a triangle is two right angles.

VI.2 (Aristotle's Angle Unboundedness Axiom) For any acute angle θ and any segment PQ, there exists a point X on one side of the angle such that the perpendicular XY dropped onto the other side of the angle is longer than the given segment: $XY > PQ$.

Remark. Based on Proclus commentaries to Euclid, Greenberg has suggested the angle unboundedness axiom. This axiom goes back to Aristotle's book I of the treatise *De Caelo* ("On the heavens").

Remark. As shown in Corollary 30, in every Hilbert plane, the Archimedean Axiom implies Aristotle's Axiom.

Problem II.14.1. *Explain, how these axioms imply that each segment is congruent to itself.*

Problem II.14.2. *Explain why congruence of segments is an equivalence relation.*

Problem II.14.3. *Prove that the center of a circle is unique.*

Problem II.14.4. *Prove that every angle is congruent to itself, and to the angle with its sides switched. Explain why congruence of angles is an equivalence relation.*

II.14.2 Fundamental constructions with Euclidean tools

Definition II.14.7 (Traditional Euclidean tools). Constructions with *traditional Euclidean tools* are constructions using only straightedge and compass, done in a finite number of steps of the following types:

1. drawing a line through two different points
2. drawing a circle with a given center through a given point
3. intersecting two lines
4. intersecting a circle and a line
5. intersecting two circles
6. choosing an arbitrary point
7. choosing a point on a given line
8. choosing a point on a given circle
9. deciding whether two points are equal or different

Remark. For the enumeration and count of a construction, we do not count producing the given pieces, neither pieces that were only drawn to facilitate the understanding. But producing the resulting pieces is counted.

Problem II.14.5. *Explain how an angle is bisected with traditional Euclidean tools. Enumerate and count the steps needed, according to definition II.14.7, and provide a drawing.*

Answer. Let $\angle(h,k)$ with vertex A be given.

1. We choose a point $B \neq A$ on the ray h.
2. Draw circle(A,B).
3. Find the intersection point $2 \cap k$.
4. Draw circle$(3,A)$.
5. Draw circle(B,A).
6. Find the intersection point $4 \cap 5$ in the interior of the given angle $\angle(h,k)$.
7. Draw the ray $\overrightarrow{A,6}$. This is the angle bisector to be constructed.

We count that seven steps are needed for the construction.

Problem II.14.6. *Given is a line l and a point lying on the line. Explain how the perpendicular is erected with traditional Euclidean tools.*

Answer. We draw any circle around the given point P and let A and B be the intersection points of line l with this circle. Next we draw two circles, with center A through point B, and with center B through point A. They intersect in two points, let C be one of them. The line PC is the perpendicular to line l at point P to be constructed. We get an extra check of accuracy, since this line goes through the other intersection point of the two circles, too.

Problem II.14.7. *Given is a line l and a point P not lying on the line. Explain how to construct the reflection point P' across the line with traditional Euclidean tools, without using line-circle intersection as a means of construction.*

Answer. We choose any two different points $A \neq B$ on the given line l and draw the circles with center A through point B, and with center B through point A. They intersect in two points: the given point P and a point $P' \neq P$, which is the reflection of point P across the line l to be constructed.

Problem II.14.8. *Given is a line l and a point not lying on the line. Explain how the perpendicular is dropped with traditional Euclidean tools. Avoid line-circle intersection as a means of construction.*

Answer. We choose any two different points $A \neq B$ on the given line l and draw the circles with center A through point B, and with center B through point A. They intersect in two points: the given point P and the reflection point Q. The line PQ is the perpendicular to the line l through point P to be constructed.

Remark. The two points may lie on the same or different sides of the perpendicular, the construction works for any choice of the points $A \neq B$. The two circles may have different radius. In any case, they indeed intersect at two points.

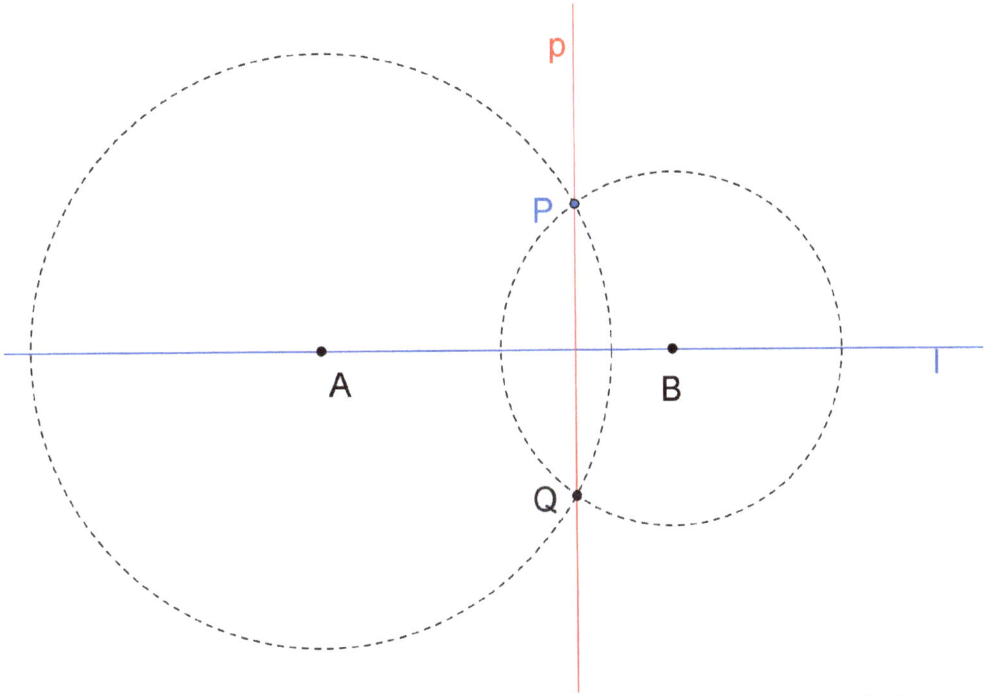

Figure II.14.1: Construction of the perpendicular with just two circles and their intersection.

Problem II.14.9. *Explain that two equilateral triangles the common side of which is any given segment, can be constructed with traditional Euclidean tools. How can the two triangles be distinguished.*

Answer. Let the segment AB be given. We draw the circles with center A through point B, and with center B through point A. They intersect in two points C and D. The triangles $\triangle ABC$ and $\triangle ABD$ are the two equilateral triangles with sides AB to be constructed. They lie on different sides of line AB.

Problem II.14.10. *Explain how the perpendicular bisector of a segment is constructed with traditional Euclidean tools.*

Answer. Let the segment AB be given. We draw the circles with center A through point B, and with center B through point A. They intersect in two points C and D. The line CD is the perpendicular bisector segment AB to be constructed.

Problem II.14.11. *Given is a line l, a point A not lying on the line, and a point B lying on the line. Construct a circle through points A and B that is tangent to the line. Provide a drawing and explain the major steps.*

Answer. We erect the perpendicular at point B onto the given line l. Next we construct the perpendicular bisector of segment AB. The intersection of these two lines is the center O of the circle through A and B, tangent to line l, to be constructed.

II.14.3 Euclidean tools are at least as strong as Hilbert tools

Theorem II.14.1 (The traditional Euclidean tools can do at least all constructions possible with Hilbert tools). *We assume only the axioms of incidence, order, circles, and congruence*

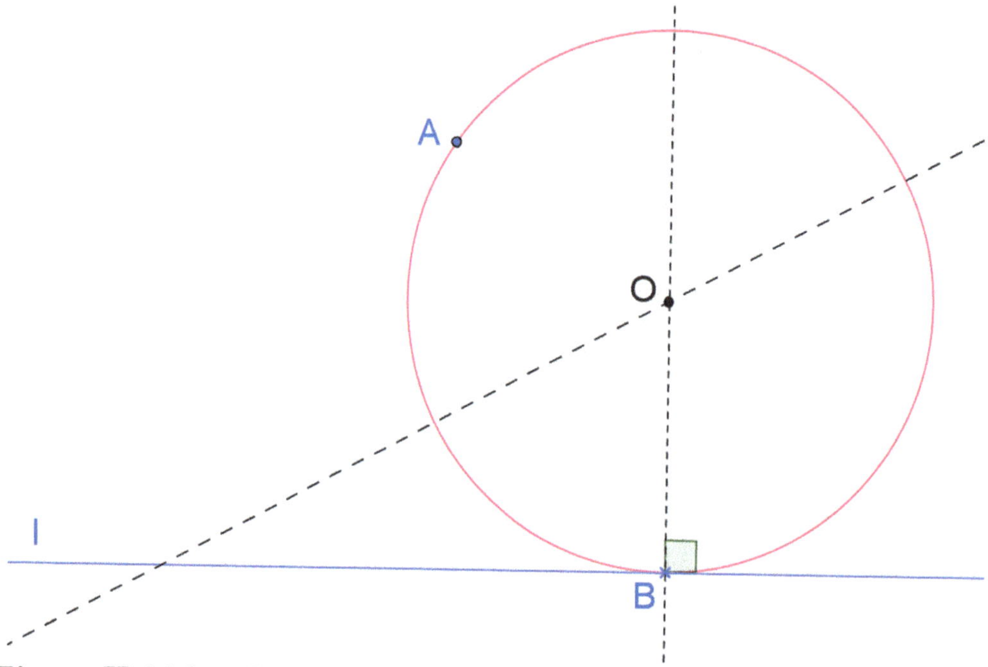

Figure II.14.2: Construction of the circle through A and B tangent to line l.

as stated above. Under these assumptions, it is also possible to transfer any segment as well as any angle, uniquely in the way postulated by Hilbert's axioms (III.1) *and* (III.4)*. In other words, traditional straightedge and compass are stronger tools than Hilbert tools.*

Remark. The axioms of continuity and parallelism are not needed in theorem II.14.1. Too, we have shown that Euclid's collapsible compass can emulate a non-collapsible compass.

Main Theorem 19. *The axiom system given above implies Hilbert's axioms for two-dimensional geometry. On the other hand, the converse is not true. Indeed, it is a more restrictive system.*

We now elaborate the steps needed to check Theorem II.14.1.

Proposition II.14.1 (Transfer of a segment). *Given is a segment AB and a ray a' with vertex A'. It is always possible to find, with traditional Euclidean tools, a point B' on this ray such that the segment AB is congruent to the segment $A'B'$.*

Construction II.14.1 (Transfer a segment). *Construct an equilateral triangle $\triangle AA'C$, getting point C as an intersection of two circles with center A through point A', and center A' through point A.*

A segment $AB_1 \cong AB$ on the ray $\overrightarrow{CA}$ can be obtained from the intersection point B_1 of this ray with the circle around A through point B. A second circle around C through point B_1 enable one to get segment $CB_2 \cong CB_1$ on the ray $\overrightarrow{CA'}$, from the intersection point B_2 of ray and circle. Finally, the segment $A'B' \cong A'B_2$ on the given ray a' is obtained from the intersection point B' of ray a' with the circle around A' through point B_2.

Reason for validity of the construction. Using segment addition or subtraction, it is easy to check that $AB_1 \cong A'B_2$, since by construction both $CB_1 \cong CB_2$ and $CA \cong CA'$. Hence transitivity of congruence implies $AB \cong AB_1 \cong AB_2 \cong AB'$. □

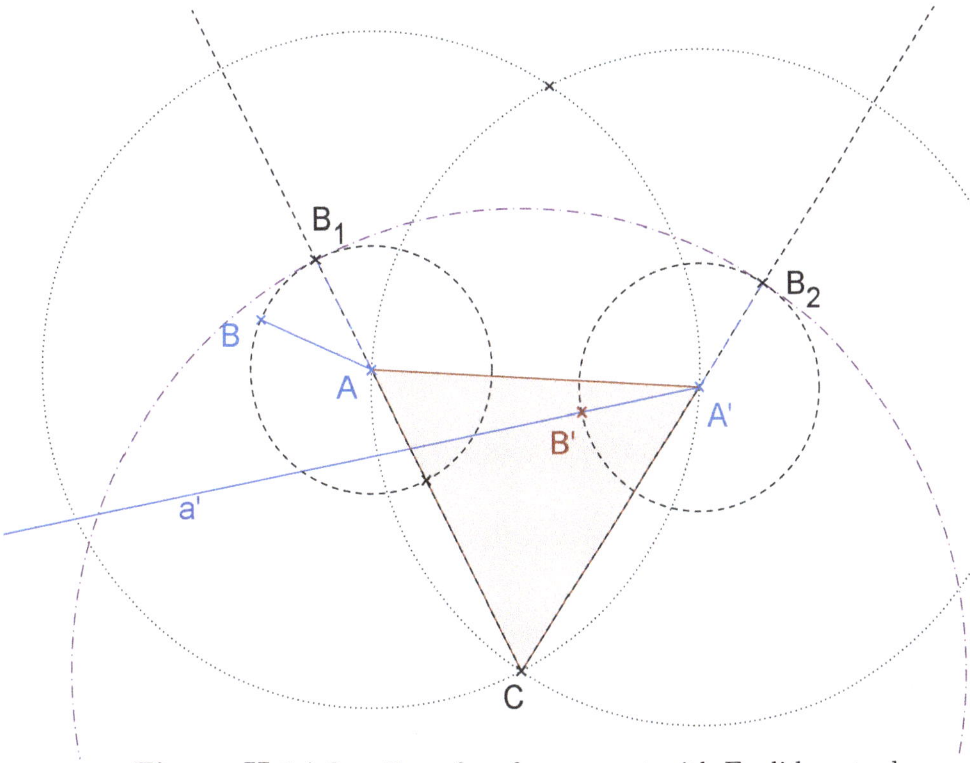

Figure II.14.3: Transfer of a segment with Euclidean tools.

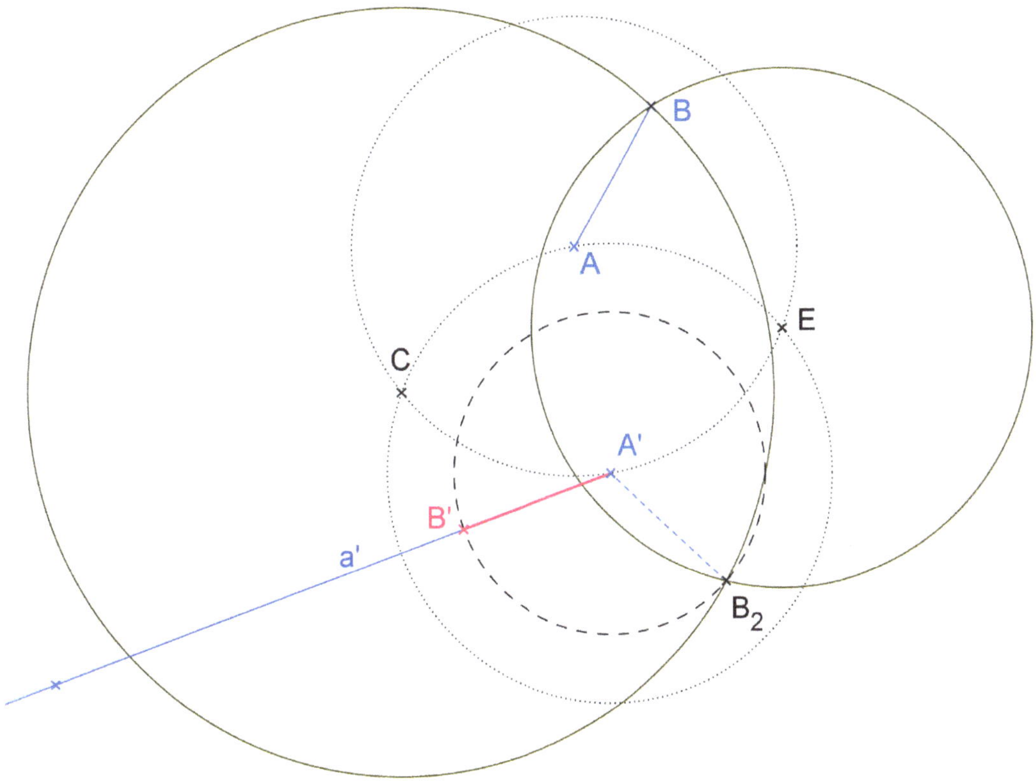

Figure II.14.4: Alternative construction to transfer of a segment with Euclidean tools.

Problem II.14.12. *An alternative construction using more circles, but less lines is shown in the* *figure on page 345. Describe this construction and give the reason for its validity.*

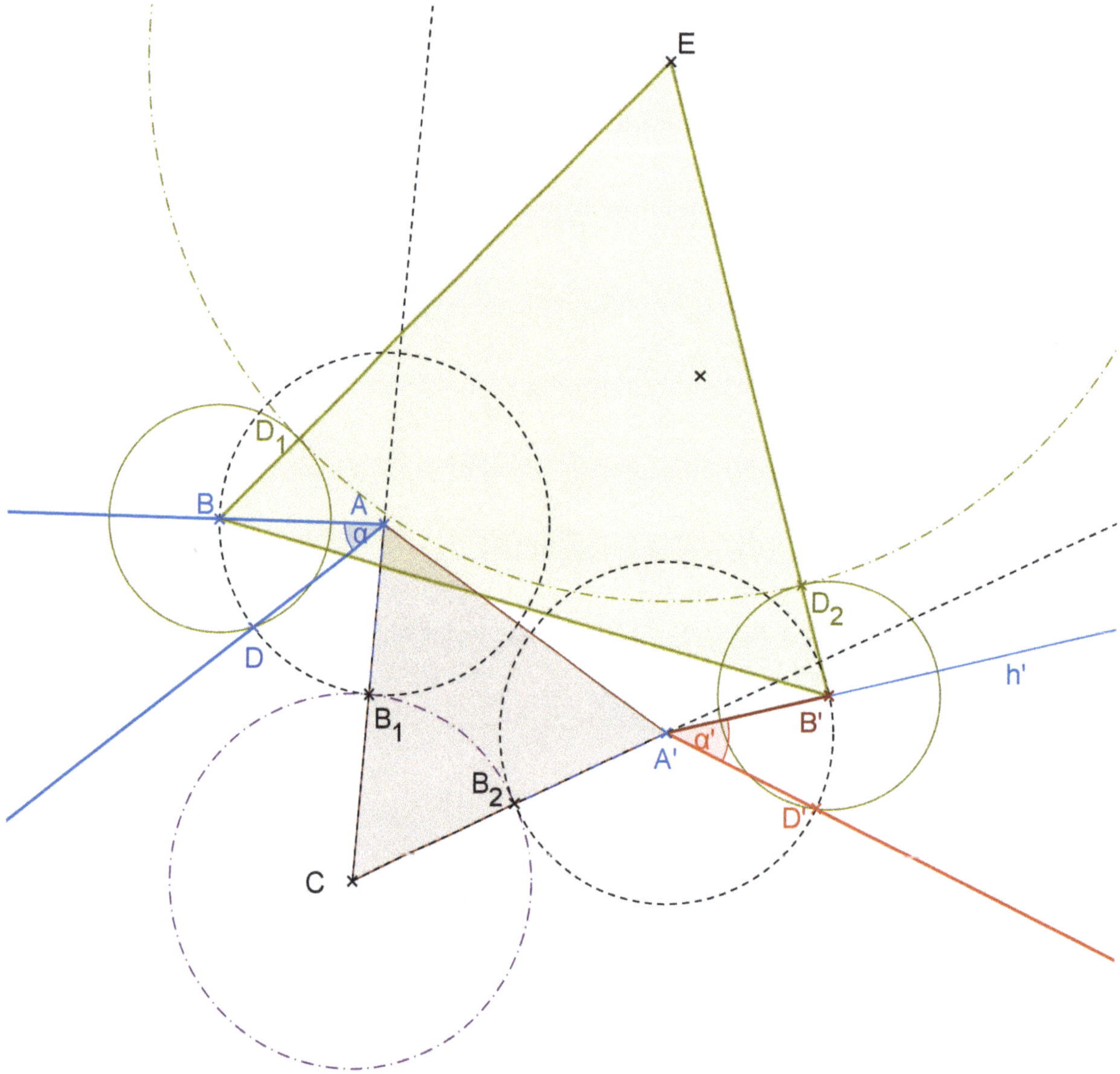

Figure II.14.5: Transfer of an angle with Euclidean tools.

Proposition II.14.2 (Transfer of an angle). *Given is an angle $\angle(h,k)$, a ray h' that emanates from any point A', and a half plane of that ray. Then there exists a unique ray k' such that the angle $\angle(h,k)$ is congruent to the angle $\angle(h',k')$ and at the same time all interior points of the angle $\angle(h',k')$ lie on the given side of h'.*

Construction II.14.2 (Transferring an angle). *We draw a circle around the vertex A of the given angle and get congruent segments $AB \cong AD$ on its sides h and k. We transfer segment AB onto the ray h' by means of the construction II.14.1 above and get the segment $A'B' \cong AB$.*

Finally we transfer segment BD to point B', and get a circle around B' of radius $B'D_2 \cong BD$. This step uses construction II.14.1 once more, and hence produces a second equilateral triangle $\triangle BB'E$.

We construct the intersection points D' and D'' of the circles around A' of radius $A'B' \cong AB$, and around B' of radius $B'D_2 \cong BD$. One chooses among points D' and D'' the one lying in the half plane of h' as required and gets the angle to be constructed: either $\angle B'A'D' \cong \angle BAD$, or $\angle B'A'D'' \cong \angle BAD$.

The intersections of circles and segments need not be postulated, since we can recall:

Proposition II.14.3. *A segment between a point inside a circle and a point outside of this circle intersects the circle in exactly one point.*

Proof. By Proposition II.10.15, the circle-circle intersection property implies the line-circle intersection property. By Proposition II.10.12, the line-circle intersection property implies the segment AB between the point A inside and the point B outside of the circle intersects the circle in exactly one point.

We recall some details, for the convenience of the reader. Assume that the circle-circle intersection property holds. Let $l = AB$ be the line between the two given points A inside, and B outside the circle $\mathcal{C}$. The ray $\overrightarrow{OB}$ intersects the circle in a point Q inside the segment OB, as postulate by axiom. (IV.2) and given by definition II.14.4 of the exterior of a circle.

A special situation occurs in case that the center O of the circle lies on the line l. In this case, the intersection point $Q = X$ of the ray $\overrightarrow{OB}$ with the circle is already the required intersection of line and circle. We now exclude that special case and assume that the center O does not lie on line l.

The ray $\overrightarrow{OA}$ from the center intersects the circle in a point P outside the segment OA (since $O \neq A$). Recall that reflection images of any point across a line can be constructed with the intersection of two circles. Let O' and P', Q' be the reflection images of points O and P, Q across the axis $l = AB$. We construct the circle $\mathcal{C}'$ with center O' through point P'. Points P' and Q' both lie on the reflected circle $\mathcal{C}'$. Thus we obtain the figure already shown on page 277. Line l is the perpendicular bisector of the segment OO'. Points O' and P lie on the same side of line l, since points O and O' lie on different sides of line l but points O and P lie on the same side. Because of proposition II.13.1 we know that any point X lies on the same side of line l as point O if and only if $|OX| < |O'X|$.

Because points P and O' lie on the same side of line l, we get $|OP'| = |O'P| < |OP|$. Hence point P' lies inside circle $\mathcal{C}$. Because points O and O' lie on different sides of line l and points O and Q lie on the same side, the points Q and O' lie on different sides of line l and $|OQ| < |O'Q| = |OQ'|$. Hence point Q' lies outside circle $\mathcal{C}$.

Because point P' lies inside, but point Q' lies outside the circle $\mathcal{C}$, and both points lie on the reflected circle $\mathcal{C}'$, the circle-circle intersection property implies that the two circles intersect in some point X. Since it has the same distance from both centers O and O', by proposition II.13.1 point X lies on the axis l. Thus we have obtained an intersection point of the given circle $\mathcal{C}$ and line AB. □

Remark. In this derivation of the line-circle intersection property, we had to distinguish the cases that the given line goes through the center or not O. This is an instance were the item *"deciding whether two points are equal or different"* is actually needed.

In the special case of a line through the center, we need to invoke the circle axiom (IV.2) directly, since the reflection across the line does not produce a different second circle. In case of a line passing closely to the center, it is still very impractical to use circle-circle intersection because of the small

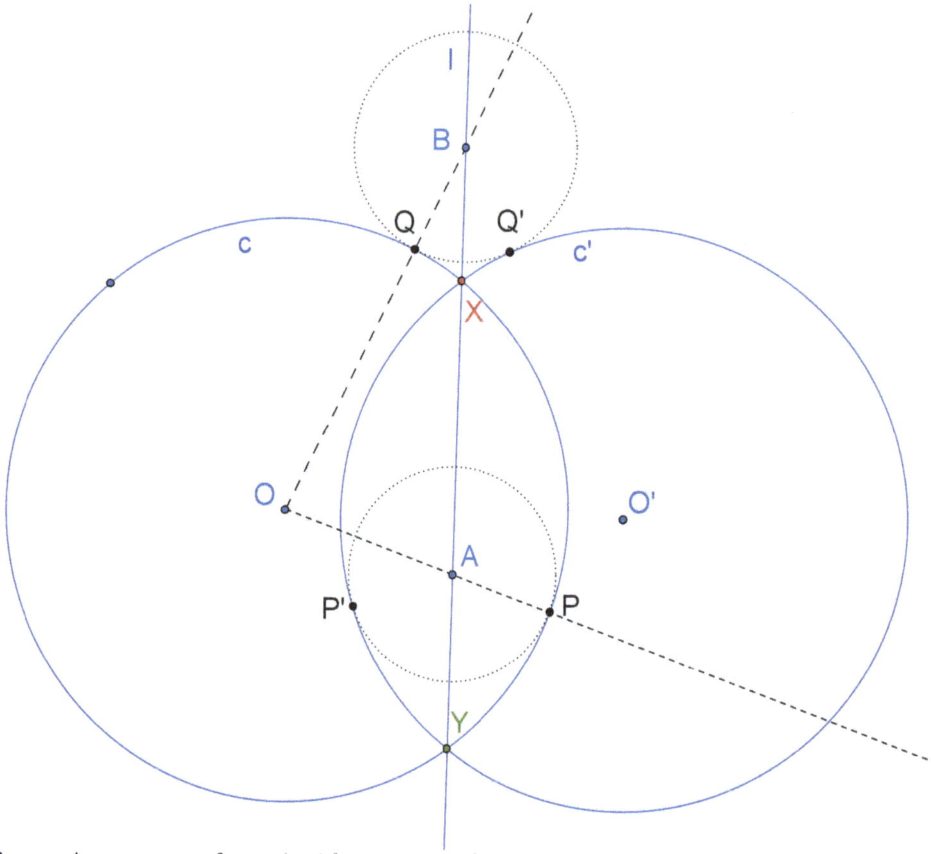

Figure II.14.6: A segment from inside to outside a circle intersects the circle—proved from the circle-circle intersection.

angle at the intersection of the two circles, and the direct application of the circle-line intersection leads to a more accurate construction. Because of that, the line-circle intersection is a practical tool useful to improve the accuracy of constructions.

Too, we have shown in Proposition II.10.13:

Proposition II.14.4. *If a circle has points on both sides of a line, then it intersects the line in two points.*

Problem II.14.13. *Given is a line l and a point not lying on the line. Explain how the perpendicular is dropped using a circle-line intersection.*

Problem II.14.14. *The figure on page II.14.3 shows a variant construction for the transfer of an angle. Enumerate and count the number of elementary steps according to definition II.14.7 needed to perform the construction. Which additional assumption is made in this construction.*

Solution. 1. Draw circle(A, B).

2. Draw circle(B, A).

3. Find an intersection points X of $1 \cap 2$.

4. Find the second intersection point Y of $1 \cap 2$.

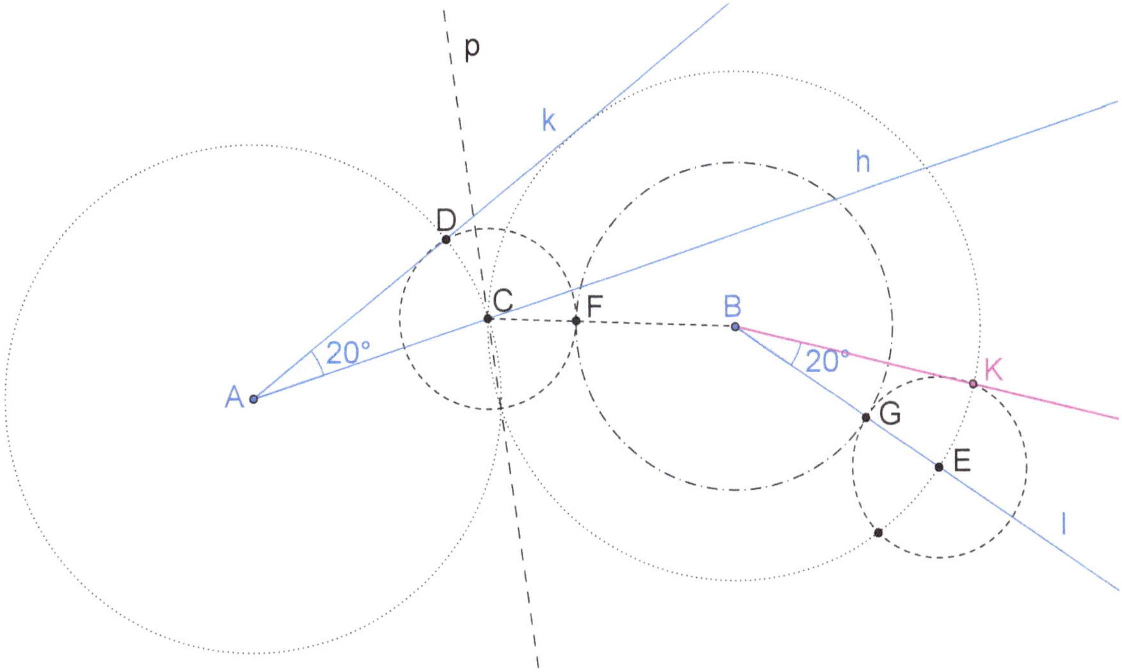

Figure II.14.7: Transfer 1 of an angle.

5. Draw the line $p = XY$. This is the perpendicular bisector of segment AB.
6. Find the intersection points $C = h \cap p$.
7. Draw circle(A, C).
8. Find the intersection $D = k \cap 7$.
9. Draw ray $\overrightarrow{CB}$.
10. Draw circle(C, D).
11. Find the intersection point $F = 9 \cap 10$.
12. Draw circle(B, F).
13. Find the intersection point $G = l \cap 12$.
14. Draw circle(B, C).
15. Find the intersection point $E = l \cap 14$.
16. Draw circle(E, G).
17. Find an intersection point of $14 \cap 16$.
18. If this point does not lie in the half plane as required, find the second intersection point K of $14 \cap 16$.

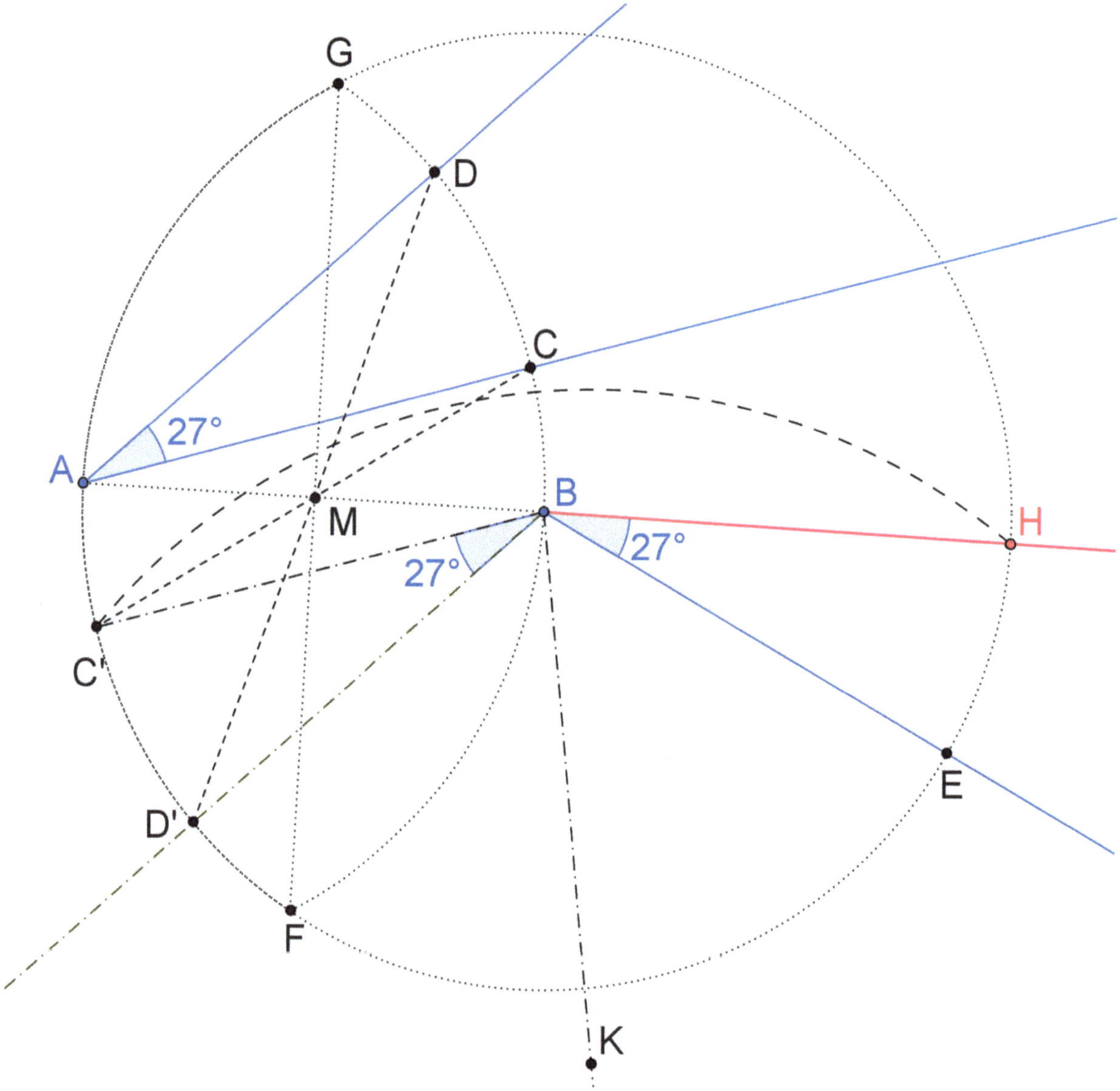

Figure II.14.8: Transfer 2 of an angle.

19. Draw the ray $\overrightarrow{BK}$.

The angle $\angle EBK$ is the transfer of the given angle $\angle(h, k)$ onto the given ray $l = \overrightarrow{BE}$ to be constructed. We see that 18 or 19 steps are needed—provided that intersection point C does exist. □

Remark. We see that the construction made the additional assumption that one side of the given angle intersects the perpendicular bisector of the vertices of the given angle and the given ray. In Euclidean geometry this assumption is not too restrictive, because it always holds for one side of either the given angle or its vertical angle. In hyperbolic geometry, the restriction is more severe.

Problem II.14.15. *Consider transfer 2 of an angle shown in the figure on page II.14.3. Enumerate and count the number of elementary steps according to definition II.14.7 needed to perform the construction.*

Solution. 1. Draw circle(A, B).

2. Draw circle(B, A).
3. Find an intersection points F of $1 \cap 2$.
4. Find the second intersection point G of $1 \cap 2$.
5. Draw the line FG.
6. Draw line AB.
7. Find the intersection point $M = 5 \cap 6$.
8. Find the intersection point $C = h \cap 1$.
9. Find the intersection point $D = k \cap 1$.
10. Draw ray $\overrightarrow{CM}$.
11. Find the intersection point $C' = 2 \cap 10$.
12. Draw ray $\overrightarrow{DM}$.
13. Find the intersection point $D' = 2 \cap 12$.
14. Draw circle(D', B).
15. Find the intersection point $E = l \cap 2$.
16. Draw circle(E, B).
17. Find the intersection point $K = 14 \cap 16$.
18. Draw circle(K, C').
19. Find an intersection point of $H = 2 \cap 18$.
20. If this point H does not lie in the half plane as required, draw circle(E, H).
21. Find the new intersection point $H := 2 \cap 20$.
22. Draw ray $\overrightarrow{BH}$.

The angle $\angle EBH$ is the transfer of the given angle $\angle(h, k)$ onto the given ray $l = \overrightarrow{BE}$ to be constructed. We see that 20 or 22 steps are needed. This construction does work always. □

II.14.4 Birkhoff's set of postulates

In 1932, G. D. Birkhoff created a set of four postulates of Euclidean geometry sometimes referred to as Birkhoff's axioms. These postulates are all based on basic geometry that can be confirmed experimentally with a scale and protractor. Since the postulates build upon the real numbers, the approach is similar to a model-based introduction to Euclidean geometry.

Birkhoff's axiom system was utilized in the secondary-school text *Basic Geometry.* Birkhoff's axioms were also modified by the *School Mathematics Study Group* to provide a new standard for teaching high school geometry, known as SMSG axioms.

(I) Postulate of Line Measure. A set of points $\{A, B, \dots\}$ on any line can be put into a one to one correspondence with the real numbers $\{a, b, \dots\}$ such that $|b - a| = d(A, B)$ for all points A and B.

(II) Point-Line Postulate. There is one and only one line l that contains any two given distinct points P and Q.

(III) Postulate of Angle Measure. A set of rays $\{l, m, n, \dots\}$ through any point O can be put into one to one correspondence with the real numbers $(\mod 2\pi)$ such that for any points $A \neq O$ on line l and $B \neq O$ on line m, the difference $a_m - a_l (\mod 2\pi)$ of the numbers associated with the lines l and m is the angle $\angle AOB$.

Furthermore, if the point B on m varies continuously in a line r not containing the vertex O, the number a_m varies continuously, too.

(IV) Postulate of Similarity. Given are two triangles $\triangle ABC$ and $\triangle A'B'C'$ for which we assume

$$d(A', B') = k \cdot d(A, B)\,,\ d(A', C') = k \cdot d(A, C) \text{ and } \angle B'A'C' = \pm\angle BAC$$

with some constant $k > 0$. Then

$$d(B', C') = k \cdot d(B, C)\,,\ \angle C'B'A' = \pm\angle CBA \text{ and } \angle A'C'B' = \pm\angle ACB$$

II.15 A Synopsis of Euclid

Only the very important items are explained below.

II.15.1 Book I

Definitions

1. A *point* is that which has no part.

2. A *line* is length without breadth.

4. A *straight line* lies evenly with its points.

8. A *plane angle* is the inclination of two lines.

10. When two adjacent angles are equal, they are *right angles.*

15. A *circle* is a line all of whose points are equidistant from a point.

20. A triangle with two equal sides is *isosceles.*

23. *Parallel* straight lines are lines in the same plane that do not meet, no matter how far extended in either direction.

Postulates

1. To draw a line through two points.

2. To extend a given line.

3. To draw a circle with given center through a given point.

4. All right angles are equal.

5. If a line crossing two other lines makes the sum of the interior angles on the same side less than two right angles, then these two lines will meet on that side when extended far enough.

Common Notions

1. Things equal to the same thing are equal to each other.

2. Equals added to equals are equal.

3. Equals subtracted from equals are equal.

4. Things which coincide are equal.

5. The whole is greater than the part.

Propositions

1. To construct an equilateral triangle on a given segment.

2. To draw a segment equal to a given segment at a given point.

3. To cut off a smaller segment from a larger segment.

4. Side-angle-side (SAS) congruence for triangles.

5. The base angles of an isosceles triangle are equal.

6. If the base angles are equal, the triangle is isosceles.

7. It is not possible to put two triangles with equal sides on the same side of a segment.

8. Side-side-side (SSS) congruence for triangles.

9. To bisect an angle.

10. To bisect a segment.

11. To construct a perpendicular to a line at a given point on the line.

12. To drop a perpendicular from a point to a line not containing the point.

13. A line standing on another line makes angles with sum equal to two right angles.

15. Vertical angles are equal.

16. The exterior angle of a triangle is greater than either opposite interior angle.

17. Any two angles of a triangle have a sum less than two right angles.

18. If one side of a triangle is greater than another, then the angle opposite to the greater side is greater than the angle opposite to the smaller side.

19. If one angle of a triangle is greater than another, then the side opposite to the greater angle is greater than the side opposite to the smaller angle.

20. The sum of any two sides of a triangle is greater than the third side.

22. To construct a triangle from all three sides given— provided the sum of any two sides is greater than the remaining side.

23. To produce a given angle at a given point and side.

24. For two triangles with two pairs of equal sides, the triangle with the greater angle between them has the greater opposite side.

25. For two triangles with two pairs of equal sides, the triangle with the greater third side has the greater angle between the two matched sides.

26. Angle-side-angle (ASA) and angle-angle-side (AAS) congruence for triangles.

27. If the two alternate interior angles between a transversal and two lines are equal, then these two lines are parallel.

28. If the exterior angle between a transversal and the first line and the opposite interior angle between the transversal and the second line have sum two right angles, then the two lines are parallel.

28a. If the two interior angles between a transversal and two lines have sum two right angles, then the two lines are parallel.

29. A transversal of two parallel lines makes equal alternate interior angles with the two parallels.

30. Lines parallel to the same line are parallel.

31. To draw a line parallel to a given line through a given point.

32. The sum of angles of a triangle is two right angles.

32a. An exterior angle equals the sum of the two opposite interior angles of a triangle.

33. The lines joining the endpoints of two parallel segments are parallel and the segments cut out by the first two parallels are congruent.

34. The opposite sides and angles of a parallelogram are equal.

35. Parallelograms on the same base and in the same parallels have equal areas.

36. Parallelograms on equal bases and in the same parallels have equal areas.

37. Triangles on the same base and in the same parallels have equal areas.

38. Triangles on equal bases and in the same parallels have equal areas.

39. Triangles with equal areas on the same base lie in the same parallels.

40. Triangles with equal areas on equal bases lie in the same parallels.

41. The area of a parallelogram is twice the area of a triangle on the same base and in the same parallels.

42. To construct a parallelogram with a given angle and area equal to the area of a given triangle.

43. The triangles on opposite sides of a diagonal of a parallelogram are equal.

44. To construct a parallelogram with given side and angle and area equal to the area of a given triangle.

45. To construct a parallelogram with a given angle and area equal to the area of a given figure.

46. To construct a square on a given segment.

47. (Theorem of Pythagoras) The square on the hypothenuse is equal in area to the sum of the squares on the legs of the triangle.

48. If the sum of the areas of squares on two sides of a triangle is equal to the area of the square on the third side, then the triangle is right.

II.15.2 Book II. Propositions

1. The rectangle contained by two lines is the sum of the rectangles contained by one and the segments of the other.

4. The area of the square on the whole segment is equal to the sum of the areas on the two parts of the segment plus twice the area of the rectangle on these two parts.

5. The area of the square on the first longer part of a segment is equal to the area of the rectangle with the difference of the parts and the entire segment as sides, plus the area of the square on the second part. [40]

6. The area of a rectangle with one side a longer segment and an added piece, the second side the added piece plus the area of the square on half of the longer segment add up to the area of the square on the half plus the added piece. [41]

11. To cut a longer segment so that the rectangle on the whole and the part is equal to the square on the other part. [42]

14. To construct a square with area equal to that of a given figure.

II.15.3 Book III. Propositions

1. To find the center of a circle.

2. The segment joining two points of a circle lies inside the circle.

5. If two circles intersect, they do not have the same center.

6. If two circles are tangent, they do not have the same center.

10. Two circles can intersect in at most two points.

11. 12. If two circles are tangent, their centers lie in a line with the point of tangency.

16. The line perpendicular to a diameter at its endpoint is tangent to a circle, and the "angle between the tangent line and the circle" [43] is less than any "rectilineal angle".

17. To draw a tangent to a circle from a point outside the circle.

18. A tangent line to a circle is perpendicular to the radius at the point of tangency.

19. The perpendicular to a tangent line at the point of tangency will pass through the center of a circle.

20. The angle at the center is twice the angle at a point of the circumference subtending a given arc of a circle.

---

[40] $p^2 = [p - (c - p)]c + (c - p)^2$

[41] $(c + p)p + \left(\frac{c}{2}\right)^2 = \left(\frac{c}{2} + p\right)^2$

[42] Solve $cx = (c - x)^2$ for the part x.

[43] The notion in quotes is not well defined by today's standard.

21. Two angles from points of a circle subtending the same arc are equal.

22. The opposite angles of a quadrilateral in a circle have sum equal to two right angles.

31. The angle in a semicircle is a right angle.

32. The angle between a tangent line and a chord of a circle is equal to the angle on the arc cut off by the chord.

35. If two chords cut each other inside the circle, the rectangle on the segments on one chord is equal in area to the rectangle on the segments on the other chord.

36. From a point outside a circle, let a tangent and a secant be drawn. The square on the segment on the tangent is equal in area to the rectangle formed by the two segments from the point outside of the circle to the endpoints of the secant.

37. From a point outside a circle are drawn two lines cutting or touching the circle. If the square on the segment on one line is equal in area to the rectangle formed by the two segments from the point outside to the endpoints of the secant on the other line, then the first line is tangent to the circle.

II.15.4 Book IV. Propositions

1. To inscribe a given segment in a circle.

2. To inscribe a triangle, equiangular to a given triangle, in a circle.

3. To circumscribe a triangle, equiangular to a given triangle, around a circle.

4. To inscribe a circle in a triangle.

5. To circumscribe a circle around a triangle.

10. To construct an isosceles triangle whose base angles are twice the vertex angle.

11. To inscribe a regular pentagon in a circle.

12. To circumscribe a regular pentagon around a circle.

15. To inscribe a regular hexagon in a circle.

16. To inscribe a regular 15-sided polygon in a circle.

II.15.5 Book V

Definitions

4. Magnitude are said to *have a ratio* if either one, being multiplied, an exceed the other.

5. Four magnitudes a, b, c, d are in the same ratio $a : b$ equal to $c : d$, if and only if for any whole numbers m and n either one of these three cases occurs:

$$\text{both} \quad ma > nb \text{ and } mc > nd,$$
$$\text{both} \quad ma = nb \text{ and } mc = nd,$$
$$\text{both} \quad ma < nb \text{ and } mc < nd.$$

II.15.6 Book VI. Propositions

1. The areas of triangles of the same height are in the same ratio as their bases.

2. A line is parallel to the base of a triangle if and only if it cuts the two other sides in the same ratio—proportionately.

3. A line from the vertex of a triangle to the opposite side bisects the angle if and only if it cuts the opposite side in proportion to the remaining sides of the triangle.

4. The sides of equiangular triangles are proportional.

5. If the sides of two triangles are proportional, their corresponding angles are pairwise equal.

6. If two triangles have one pair of equal angles, and the sides containing these pair are proportional, then the triangles are similar.

8. The altitude from the right angle of a right triangle divides it into two triangles similar to each other and to the whole triangle.

12. To find a fourth proportional for three given segments.

13. To find a mean proportional between two given segments.

16. Four segments are proportional if and only if the rectangle on the extremes is equal in area to the rectangle on the means.

30. To cut a segment in extreme and mean ratio. [44]

31. Any figure on the hypothenuse of a right triangle has area equal to the sum of the areas of similar figures on the legs of the triangle.

II.15.7 Book X

1. Given two unequal quantities, if one subtracts from the greater one the smaller one, and repeats this process enough times, there will remain a quantity lesser than the smaller of the two original quantities.

117. [45] The diagonal of a square is incommensurable with its side.

[44] Solve $c : x = x : p$ for x.
[45] not in Heath, but in Commandino

II.15.8 Book XI

Definitions

25. A *cube* is a polyhedron made of six equal squares.

26. An *octahedron* is a polyhedron made of eight equal equilateral triangles.

27. An *icosahedron* is a polyhedron made of twenty equal equilateral triangles.

28. An *dodecahedron* is a polyhedron made of twelve equal regular pentagons.

Propositions

21. The plane angles in a solid angle make a sum less than four right angles.

28. A parallelepiped is bisected by its diagonal plane.

29. 30. Parallelepipeds on the same base and of the same height are equal in volumes.

31. Parallelepipeds on equal bases and of the same height are equal in volumes.

II.15.9 Book XII. Propositions

2. The areas of circles are in the same ratio as the squares on their diameters.

3. A pyramid is divided into two pyramids and two prisms.

5. Pyramids of the same height on triangular bases have volumes in the ratio of the areas of the bases.

6. A prism with triangular bases is divided into three triangular pyramids with equal volumes.

II.15.10 Book XIII. Propositions

7. If at least three angles of an equilateral pentagon are equal, the pentagon is regular.

10. In a circle, the square on the side of the inscribed pentagon is equal in area to the square on the side of the inscribed hexagon plus the square on the inscribed decagon. [46]

13. To inscribe a tetrahedron in a sphere.

14. To inscribe a octahedron in a sphere.

15. To inscribe a cube in a sphere.

16. To inscribe an icosahedron in a sphere.

17. To inscribe a dodecahedron in a sphere.

18. (Postscript) Besides these five figures there is no other contained by equal regular polygons.

Problem II.15.1. *Which ones of these sentences are easily misinterpreted, unclear or even wrong?*

---

[46] $a_5^2 = R^2 + a_{10}^2$

II.16 The History of Translations and Editions of Euclid's Elements

"Jüngling, merke dir in Zeiten,
Wo sich Geist und Sinn erhöht:
Dass die Muse zu g e l e i t e n ,
Doch zu l e i t e n nicht versteht."

Johann Wolfgang Goethe

The highly acclaimed *Elements* is simply a huge collection—divided into 13 books—of 465 propositions from plane and solid geometry. Today, it is generally agreed that relatively few of these theorems were Euclid's own invention. Rather, by compiling, editing, and perfecting the known body of Greek mathematics, he created a superbly organized treatise. The *Elements* were so successful and revered that they thoroughly obliterated all preceding works of its kind. They have become a lasting contribution to mankind.

It is a misfortune that no copy of Euclid's *Elements* has been found that actually dates from the author's own time. Hence his writings have had to be reconstructed from numerous recensions, commentaries, and remarks by other writers. Proclus makes it clear that the *Elements* were highly valued in Greece and refers to numerous commentaries on it. Among the most important must have been those by Heron (100 B.C. - A.D.100), Porphyry and Pappus (third century A.D.).

Modern editions of the work are based upon a revision that was prepared by the Greek commentator Theon of Alexandria (end of the fourth century A.D). He lived almost 700 years after the time of Euclid. Until 1808, Theon's revision was the oldest edition of the *Elements* known. In 1808, Napoleon ordered valuable manuscripts to be taken from Italian libraries and send to Paris. At that occasion, Francois Peyrard found, in the Vatican library, a tenth-century copy of an edition of Euclid's *Elements* that predidates Theon's recension. The historians J.L.Heiberg and Thomas L. Heath have used principally this manuscript. A study of this older edition and a careful sifting of citations and remarks made by early commentators indicate that the introductory material of Euclid's original treatise undoubtedly underwent some editing in the subsequent revisions. But the propositions and their proofs have remained as Euclid wrote them, except for minor additions and deletions.

There are also Arabic translations of Greek works and Arabic commentaries, presumably based on Greek manuscripts no longer available. The first really satisfying Arabic translation of the *Elements* was done by Tâbit ibn Qorra (826-901). These, too, have been used to decide what was in Euclid's original. But the Arabic translations and revisions are on the whole inferior to the Greek manuscripts.

The first complete Latin translation of the *Elements* were not made from the Greek but from Arabic. In the eighth century, a number of Byzantine manuscripts of Greek works were translated by the Arabians. In 1120, the English scholar Adelard of Barth (ca. 1075-1160) made a Latin translation of the *Elements* from one of the older Arabian translations. Other Latin translations were made from the Arabic by Gherardo of Cremona (1114-1187) and, 150 years after Adelard, by Johannes Campanus.

Euclid's *Elements* were widely read, translated and changed by people all over the world. From 1601 to 1607, Mattes Ricci (1552-1610) from Italy and Hsü Kuang-ching (1562-1634) translated the first six books into Chinese. This played a significant role in the subsequent development of mathematics in China.

The first printed edition of the *Elements* was made at Venice in 1482 and contained Campanus' translation. This very rare book was beautifully executed. It was the first mathematical book of any consequence to be printed. In 1574, Clavius (1537-1612) published an edition that is valuble for its extensive scholia. An important Latin translation from the Greek was made by Commandino in 1572. This translation served as a basis for many subsequent translations, including the very influential work by Robert Simson. From the latter, many English editions were derived. The first complete English translation of the *Elements* was the monumental Billingsley translation issued in 1570.

Nicolaus Mercator (ca. 1620-1687) published an edition, along with his works in trigonometry, astronomy and other topics. Adrien-Marie Legendre (1752-1833) is known for his very popular *Éléments de géométrie*, in which he attempted a pedagogical improvement of Euclid's *Elements* by considerably rearranging and simplifying many of the propositions. He used some algebra not in the original. The high school versions most widely used during our century are patterned on Legendre's modification of Euclid's work.

Euclid's *Elements* is to geometry what the Bible is to Christianity. No work, except the Bible, has been more widely used, edited, or studied than Euclid's *Elements*. Over one thousand editions of Euclid's *Elements* have appeared since the first one printed in 1482. For more than two millennia, this work has dominated all teaching of geometry. Abraham Lincoln is credited with learning logic by studying Euclid's *Elements*, as his biographer Carl Sandburg tells. In his autobiography, Bertrand Russell (1872-1970) penned this remarkable recollection: "At the age of eleven, I began Euclid, with my brother as tutor. This was one of the great events of my life, as dazzling as first love." Probably no work has exercised a greater influence on scientific thinking.

Part III

The First Steps in Euclidean Geometry

III.17 Some Euclidean Geometry of Circles

Some interesting properties of circles and rectangles in Euclidean geometry are investigated in this section. The material is from book III of Euclid, and includes furthermore the converse of Thales' theorem, and the basic properties of rectangles. Recall that in Euclidean geometry, the parallel axiom and its consequences are now assumed to hold.

III.17.1 Thales' Theorem

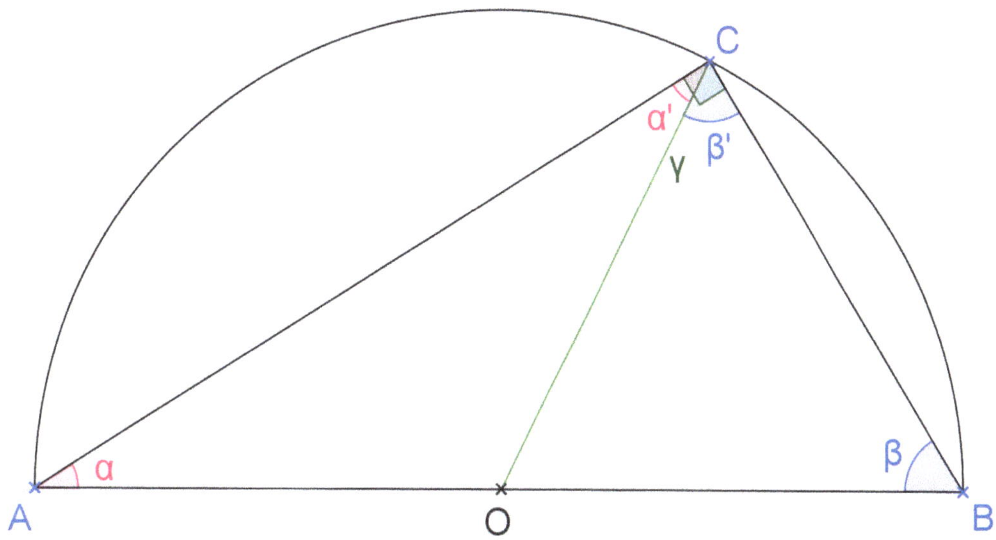

Figure III.17.1: Thales Theorem.

Theorem III.17.1 (Thales' Theorem). *"The angle in a semicircle is a right angle." More precisely stated: If an angle has its vertex C on a circle, and its sides cut the circle at the two endpoints A and B of a diameter, then angle $\gamma = \angle ACB$ is a right angle.*

Thales's lived ca. 624-546 B.C., in Miletus, a Greek island along the coast of Asia Minor. These dates are known rather precisely, because, as reported by Herodotus, he predicted a solar eclipse, which has been determined by modern methods to have occurred on May 28th of 585 B.C.

Tradition names Thales of Miletus as the first Greek philosopher, mathematician and scientist. He is known for his theoretical as well as practical understanding of geometry. Most important, he is credited with introducing the concept of logical proof for abstract propositions.

Traditions surrounds him with legends. Herodotos mentioned him as having predicted a solar eclipse, which put an end to fighting between the Lydians and the Medes.

Aristotle tells this story about him:

> Once Thales somehow deduced that there would be a great harvest of olives in the coming year; so, having a little money, he gave deposits for the use of all the olive presses in Chios and Miletus, which he hired at a low price because no one bid against him. When the harvest-time came, and many were wanted all at once and of a sudden, he let them out at any rate he pleased, and made a quantity of money.

Plutarch tells the following story:

> Solon who visited Thales asked him the reason which kept him single. Thales answered that he did not like the idea of having to worry about children. Nevertheless, several years later Thales anxious for family adopted his nephew Cybisthus.

Thales went to Egypt and studied with the priests. While he was in Egypt, he was able to determine the height of a pyramid by measuring the length of its shadow at the moment when the length of his own shadow was equal to his height. Thales is said to have proved some simple theorems of geometry, as well as the not so obvious theorem about the right angle in a semicircle. As stressed by David Park [25], the story raises an important point, whether or not Thales really invented the proof:

> Babylonians and Egyptians had a number of mathematical tricks. For example, Babylonians knew this proposition, as well as the Pythagorean theorem a thousand years before Thales and Pythagoras found them. If they were known, they must have been proved, but there is no sign that anyone thought the proofs were important enough to preserve. Whoever set the process of proof at the center of the stage is the founder of all the mathematics since then, and if it is not Thales it was someone who lived not long afterwards.

Proof of Thales' Theorem in modern manners. Draw $\triangle ABC$ and, as an extra for the proof, the line from the center O to the vertex C. The base angles of an isosceles triangle are equal by Euclid I.5 . In modern parlance, we say:

`The base angles of an isosceles triangle are congruent.`

We use that fact at first for $\triangle AOC$. Hence

$$\alpha = \angle OAC \cong \angle OCA$$

Secondly, we use Euclid I.5 for $\triangle COB$. Hence

$$\beta = \angle OBC \cong \angle OCB$$

By angle addition at vertex C

$$\gamma = \angle ACB = \alpha + \beta \tag{III.17.1}$$

Next we use Euclid I.32, which tells us:

`The sum  of the angles in a triangle is two right angles.`

Because α, β, γ are just the angles in $\triangle ABC$, we conclude that

$$\alpha + \beta + \gamma = 2R \tag{III.17.2}$$

I have used the letter R to denote a right angle. Pulling formulas (III.17.1) and (IV.20.3) together yields $\gamma + \gamma = \alpha + \beta + \gamma = 2R$, and hence $\gamma = R$, as to be shown. □

Problem III.17.1 (A special triangle). *Use the equilateral triangle and Thales' Theorem to construct a triangle with angles of* $30°, 60°$ *and* $90°$.

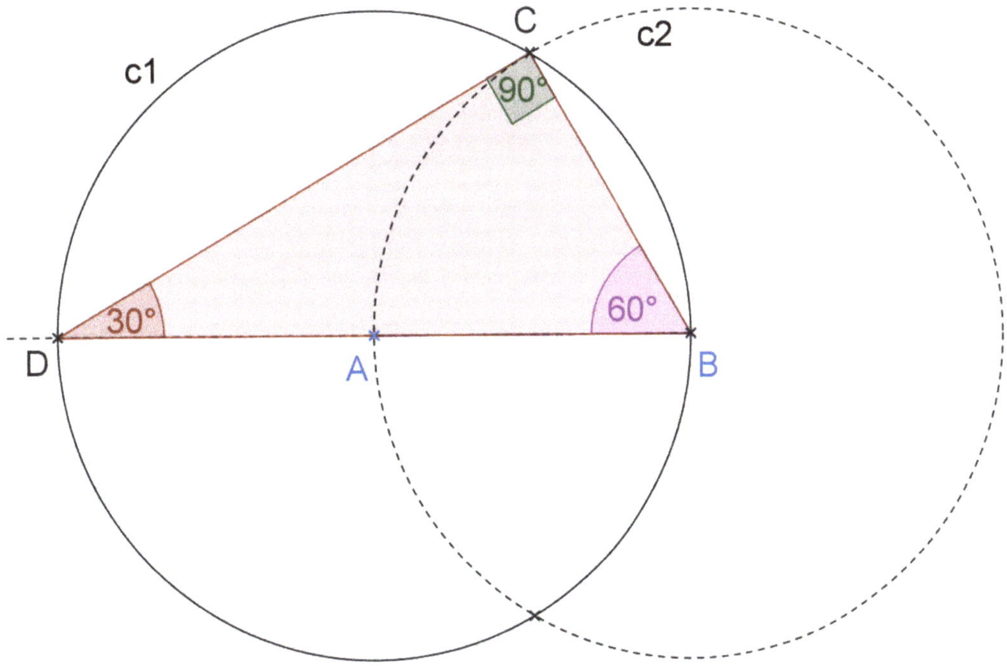

Figure III.17.2: Construction of a triangle with angles of $30°, 60°$ and $90°$.

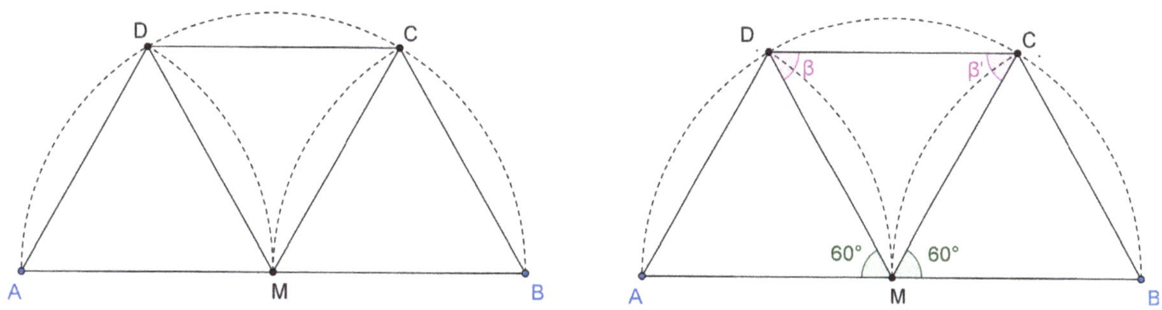

Figure III.17.3: Three triangles.

Answer. On an arbitrary segment AB, an equilateral triangle is erected. This can be done as in Euclid I.1 by finding the intersection point C of a circle around A through B with a circle around B through A. We extend the segment AB on the side opposite to B, and get the second intersection point D of the extension with the circle around A. The triangle $\triangle DBC$ has the angles $30°, 60°$ and $90°$ at its vertices D, B and C.

Problem III.17.2. *Why are all three triangles in the figure on page 367 equilateral. To give a valid reason, use only the following facts:*

- *The base angles of a triangle are congruent if and only if the triangle is isosceles.*
- *The sum of the angles in a triangle is* $180°$.

Proof. By construction, the right-hand triangle $\triangle AMD$ and the left-hand triangle $\triangle MBC$ are both equilateral. Both have three congruent angles. Because of the angle sum, their angles are all 60°. Angle addition at the vertex M implies $\angle CMD = 180° - 60° - 60° = 60°$. Moreover, the middle triangle $\triangle MCD$ is isosceles by construction. Hence the two base angles marked β are congruent. Finally the angle sum of the middle triangle implies $60° + 2\beta = 180°$, and hence $\beta = 60°$. We see that the middle triangle has three angles of 60°, too. Hence all its three sides are congruent, and it is equilateral, too. □

Problem III.17.3. *Construct a right triangle with hypothenuse $c = 5$ and one leg $a = 3$. Use a construction based on Thales' theorem and describe your construction.*

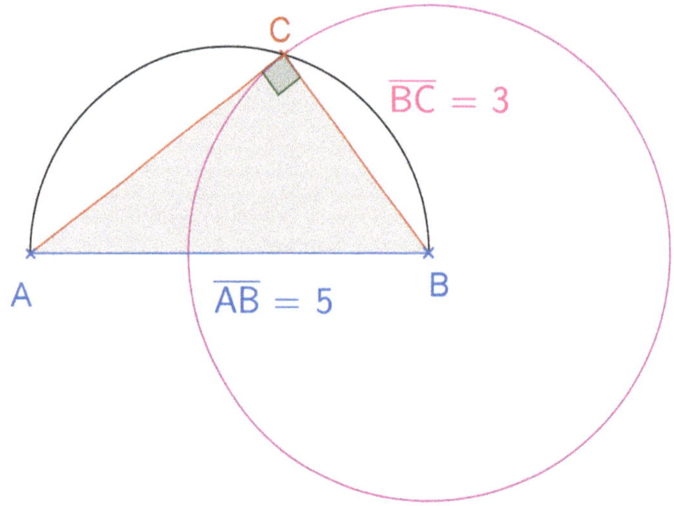

Figure III.17.4: Construction of a right triangle with hypothenuse $c = 5$ and leg $a = 3$.

Answer. We draw a segment of the length $|AB| = c = 5$ as given. Draw a semicircle with diameter AB. Draw a circle around B with radius 3. The semicircle and the circle intersect at point C. The triangle $\triangle ABC$ is a right triangle. Hypothenuse $|AB| = 5$, and leg $|BC| = 3$ have the lengths as required.

Problem III.17.4. *Construct a right triangle with projections $p = 3$ and $q = 4$ of the legs onto the hypothenuse. Use a construction based on Thales' theorem and describe your construction.*

Answer. We draw segments of the lengths as given, $|AF| = q = 4$ and $|FB| = p = 3$, adjacent to each other on one line. Erect the perpendicular on line AB at point F. Draw a semicircle with diameter AB. The semicircle and the perpendicular intersect at point C. The triangle $\triangle ABC$ is a right triangle with hypothenuse AB, and the projections $q = AF$ and $FB = p$ have the lengths as required.

Problem III.17.5. *Use Thales' theorem for a Euclidean construction of the perpendicular p to a given line l through a given point P lying on l. Do and describe the construction.*

Answer. One chooses an arbitrary point O not on line l, and draws a circle c around it through point P. This circle intersects the given line l at point P, and a second point, which is called A. Next one draws the line OA. It intersects the circle c at A and a second point, called B. Finally, the line BP is the perpendicular to be erected on line l at point P.

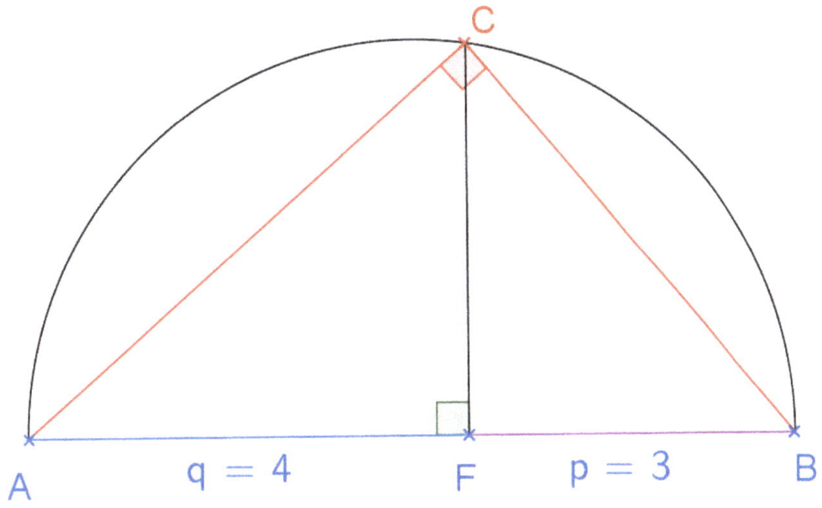

Figure III.17.5: Construction of a right triangle with projections $p = 3, q = 4$.

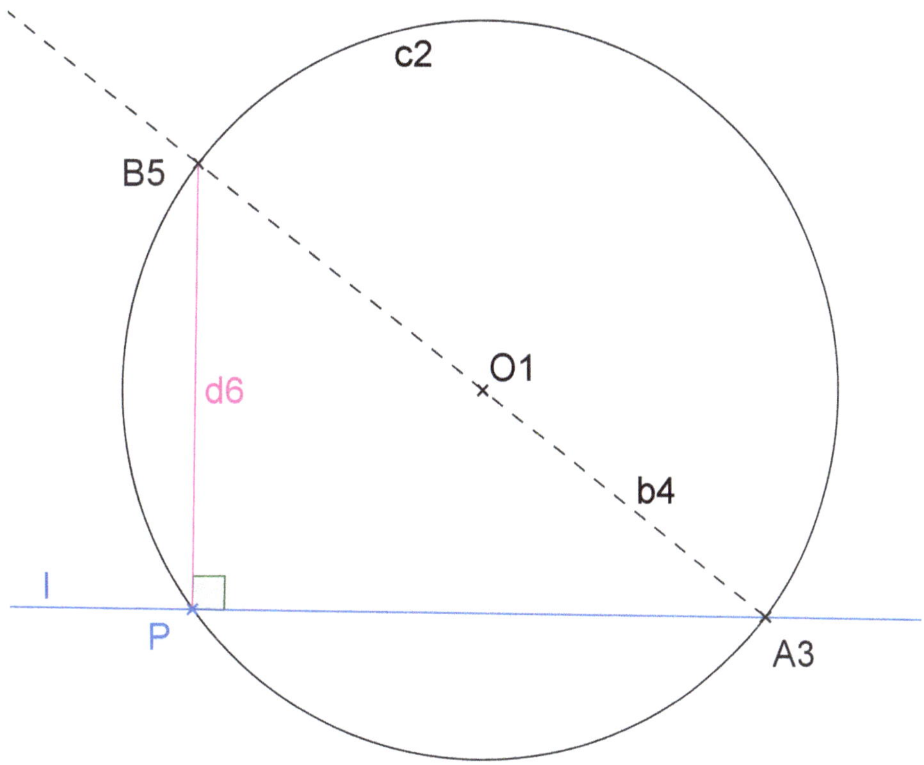

Figure III.17.6: Erect the perpendicular via Thales' Theorem. The numbers indicate the order of the steps.

Problem III.17.6. *Use Thales' Theorem to drop the perpendicular from a given point P not lying on line l onto the given line l.*

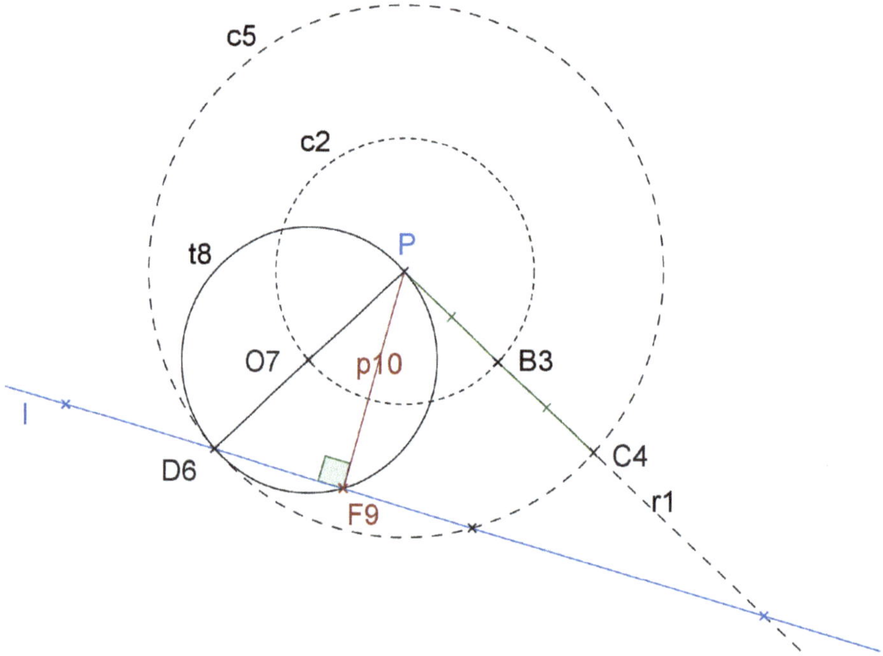

Figure III.17.7: Drop the perpendicular via Thales' Theorem. The numbers indicate the order of the steps.

Answer. The construction is indeed a bid awkward—and done more for fun than of practical value.

Theorem III.17.2 (A strengthening of Thales' Theorem). *Given is a triangle $\triangle ABC$, and a circle with its side AB as diameter.*

(i) *If the third vertex C of the triangle lies inside the circle, the angle at vertex C is obtuse.*

(ii) *If the third vertex C of the triangle lies outside the circle, the angle at vertex C is acute.*

(iii) *If the third vertex C of the triangle lies on the circle, the angle at vertex C is right.*

Corollary 39 (Converse Thales' Theorem). *If the diameter of a circle is one side of a right triangle, then its third vertex lies on this circle, too.*

Theorem III.17.3 (Converse Thales' Theorem). *The vertex of a right angle the sides of which cut a circle at the endpoints of a diameter, lies on this circle.*

Taking Thales' Theorem III.17.1 and the converse Theorem III.17.3 together, we can state:

Corollary 40 (Strong Thales' Theorem). *An angle the sides of which cut a circle at the endpoints of a diameter is a right angle if and only if its vertex lies on this circle.*

Proof of Theorem III.17.2. Following Euler's convention, let $\alpha = \angle BAC$, $\beta = \angle ABC$ and $\gamma = \angle ACB$ denote the angles at the vertices A, B and C, respectively. Let O be the midpoint of the triangle side AB. In the figures on page 371 and page 372, there are given drawings of the $\triangle ABC$ and the circle for the two cases:

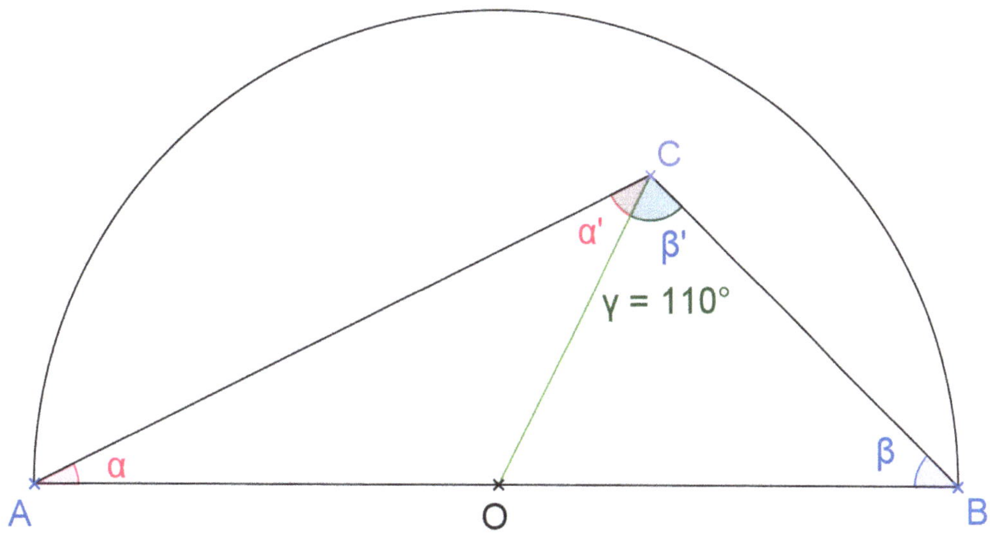

Figure III.17.8: A strengthening of Thales' Theorem, with vertex C inside the circle

(i) $OC < OA$ or

(ii) $OC > OA$.

We use Euclid I.18:

> `If one side of a triangle is greater than another, then the angle opposite to the greater side is greater than the angle opposite to the smaller side.`

First consider the case that (i) $OC < OA$ holds. Using Euclid I.18 for $\triangle AOC$, we conclude that

$$\alpha < \alpha' = \angle OCA \tag{III.17.3}$$

Now $OA \cong OB$, since O is the midpoint of the hypothenuse AB. Hence, because of (i), $OC < OB$ holds, too. Using Euclid I.18 once more, this time for triangle $\triangle BOC$, we conclude that

$$\beta < \beta' = \angle OCB \tag{III.17.4}$$

Adding (III.17.3) and (III.17.4) yields $\alpha+\beta < \alpha'+\beta'$. Angle addition at vertex C yields $\alpha'+\beta' = \gamma$. Now we use that the sum of the angles in a triangle is two right angles, as stated in Euclid I.32. Hence

$$2R = \alpha + \beta + \gamma < \alpha' + \beta' + \gamma = 2\gamma \tag{III.17.5}$$

and hence $\gamma > R$. Hence the triangle is obtuse, as to be shown.

By a similar reasoning, one shows that in case (ii), the assumption $OC > OA$ implies that the angle γ is acute. □

Problem III.17.7. *Provide a drawing for case* (i). *Provide two drawings for case* (ii), *one where the triangle* $\triangle ABC$ *is acute, and, as a catch, a second one where the* $\triangle ABC$ *is obtuse, nevertheless. (The triangle is obtuse because of a different obtuse angle.)*

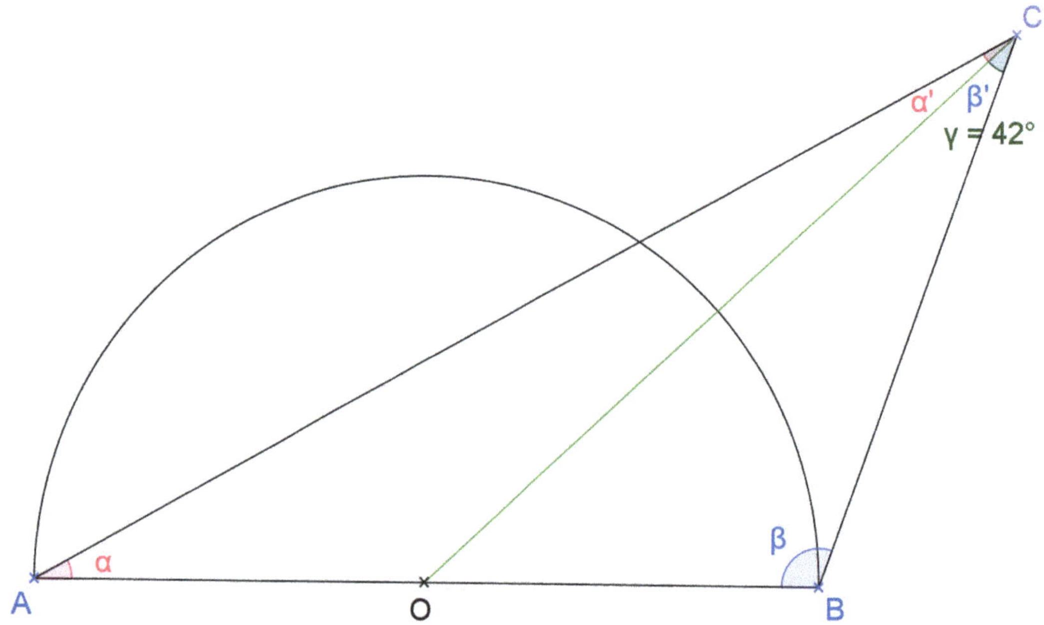

Figure III.17.9: A strengthening of Thales' Theorem, with vertex C outside the circle

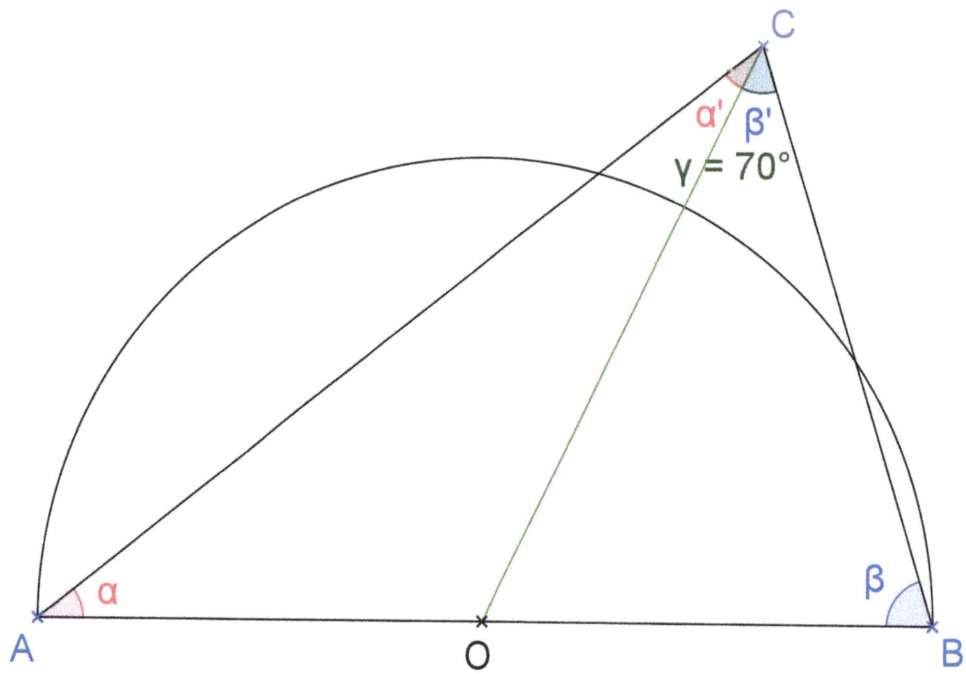

Figure III.17.10: A strengthening of Thales' Theorem, still another case with vertex C outside the circle

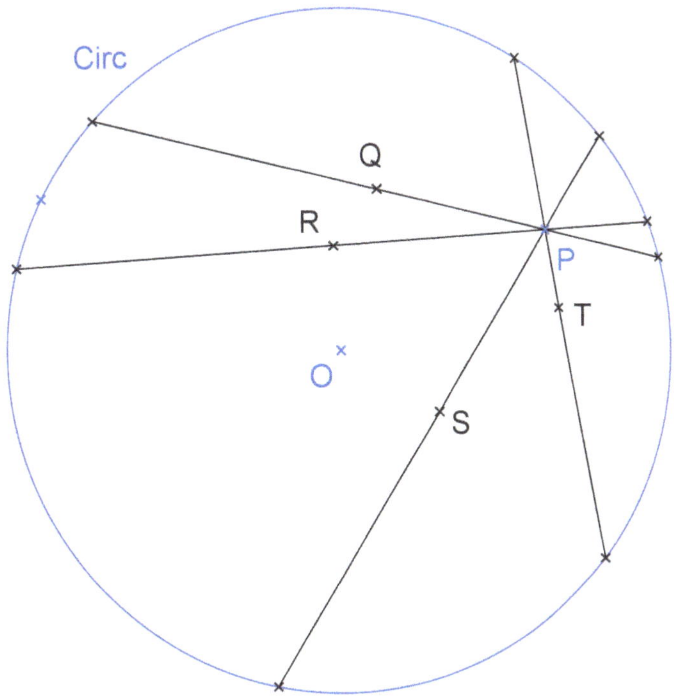

Figure III.17.11: Where lie the midpoints of all chords through point P?

Problem III.17.8 (The midpoints of chords). *Given is a circle $\mathcal{C}$ with center O, and a point P inside $\mathcal{C}$. Describe the location of the midpoints of all chords through point P. Give a reason based on Thales' Theorem or its converse.*

Answer. The midpoints of all chords in the given circle $\mathcal{C}$ through the point P lie on a circle with diameter OP. Here is the reason: The midpoints are the foot points F of the perpendiculars dropped from the center O of the circle onto the respective chords. By the converse of Thales' Theorem, all the vertices F of the right angles with sides through the two points O and P lie on Thales' circle with diameter OP.

Corollary 41 (My strongest Thales' Theorem). *Given is any circle and any angle. Any two of the following statements* (i) (ii) (iii) *imply the third one.*

(i) *The two sides of the angle cut the circle at the endpoints of a diameter.*

(ii) *The vertex of the angle lies on the circle.*

(iii) *The angle is a right angle.*

Problem III.17.9. *Give explanations how to obtain a complete proof of the three parts of the strong Thales' Theorem 41, using all material from this section.*

Answer. The theorem has three parts:

"(i) and (ii) ⇒ (iii)": This is simply the original Thales' Theorem III.17.1.

"(i) and (iii) ⇒ (ii)": This is the Converse Thales' Theorem III.17.3.

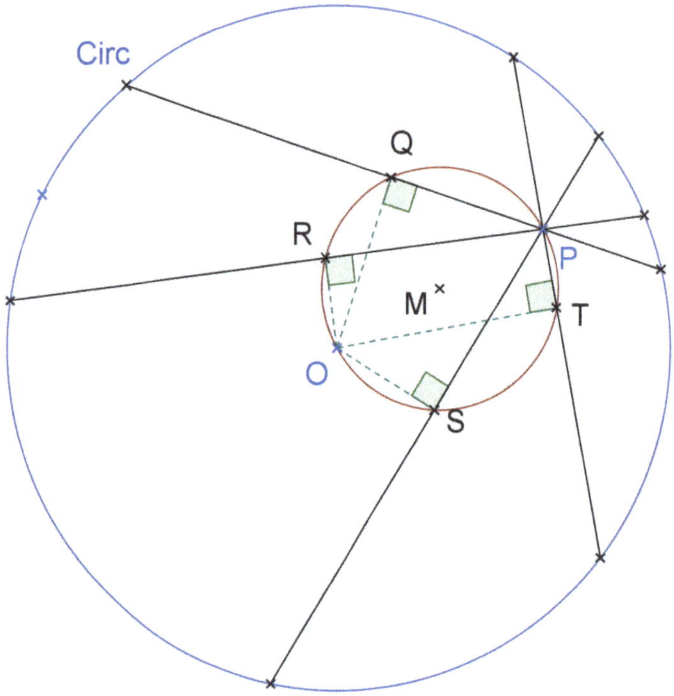

Figure III.17.12: The midpoints of chords lie on Thales' circle with diameter OP.

"(ii) and (iii) ⇒ (i)": This is consequence of Euclid III.20 (Theorem III.17.4) and Euclid III.22 (Theorem III.17.6). By Euclid III.20, the central angle $\angle AOB$ is twice the circumference angle with vertex on the long arc. By Euclid III.22, the circumference angles on the long and the short arc add up to two right angles.

Assume that A, B and C are any three points on a circle and $\angle ACB$ is a right angle. Let D be the other endpoint of diameter CD. By Euclid III.22, the angle $\angle ADB$ is a right angle, too.

Either C or D lies on the long arc AB, or AB is a diameter. Assume that C lies on the long arc AB. By Euclid III.20, we see that the central angle $\angle AOB$ is two right angles. If D lies on the long arc, we get the same conclusion that the central angle $\angle AOB$ is two right angles.. Hence AB is a diameter, anyway.

III.17.2 Rectangles and the converse Thales Theorem

As already stated in the section on Legendre's Theorems, a *rectangle* is defined to be a quadrilateral with four right angles. The segments connecting the opposite vertices are called the *diagonals* of the rectangle. Obviously, a diagonal bisects a rectangle into two right triangles. They turn out to be congruent. Conversely, one can built a rectangle from two congruent right triangles. At first, we use this idea for an independent proof of the converse of Thales' Theorem 39. Secondly, the same idea helps to prove the rather obvious, but important properties of a rectangle.

Independent proof of the converse of Thales' Theorem 39. Let $\triangle ABC$ be the given right triangle, with the right angle at vertex C. The idea of this independent proof is constructing a rectangle from the right triangle.

Step 1: We transfer the angle $\alpha = \angle BAC$ to vertex B in order to produce the congruent angle $\angle ABD = \alpha$ on the side of hypothenuse AB opposite to vertex C. Furthermore, we transfer the segment AC to obtain the congruent segment $BD \cong AC$ on the newly produced ray.

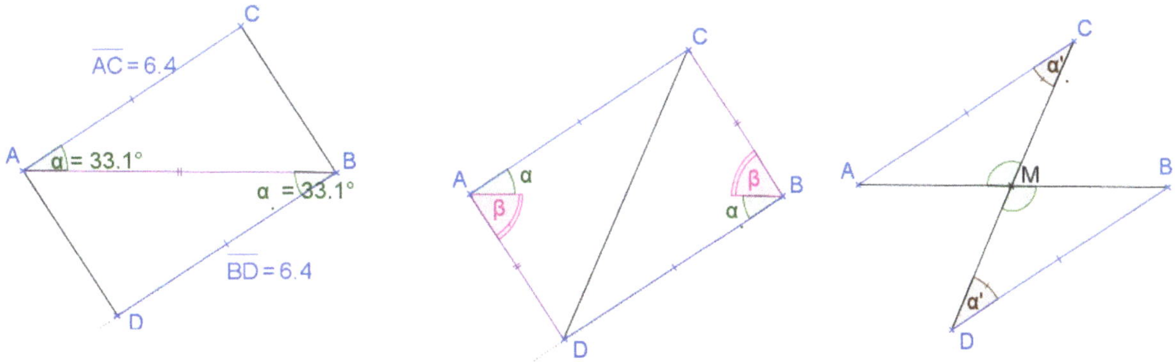

Figure III.17.13: (up-down) $\triangle ABC$ congruent $\triangle BAD$ (left-right) $\triangle CAD$ congruent $\triangle DBC$ (scissors) $\triangle AMC$ congruent $\triangle BMD$

Question. Show the congruence

$$\text{(up-down)} \qquad \triangle ABC \cong \triangle BAD$$

Which congruence theorem do you use?

Answer. This follows by SAS congruence. Indeed, the two triangles have the common side AB, the two sides $AC \cong BD$ are congruent, and the angles $\angle BAC \cong \angle ABD$ are congruent, both by construction.

Step 2: The quadrilateral $\square ADBC$ obtained from the two triangles has remarkable symmetry. It is now bisected along its other diagonal CD to obtain a second pair of congruent triangles

$$\text{(left-right)} \qquad \triangle CAD \cong \triangle DBC$$

Question. Explain how this congruence is shown.

Answer. This follows by SAS congruence. Because of the first congruence (up-down), we have two pairs of congruent sides: $AC \cong BD$ by construction, and $BC \cong AD$ as a consequence of step 1. Furthermore, there are two pairs of congruent z-angles: the angles $\angle BAC \cong \angle ABD = \alpha$ are congruent by construction. The angles $\angle ABC \cong \angle BAD = \beta$ are congruent as a consequence of (up-down). Hence angle addition yields $\angle CAD \cong \angle DBC$. Finally, we get the claim (left-right) via SAS congruence.

The congruence (left-right) now yields two further pairs of congruent z-angles:

$$\angle ACD \cong \angle BDC = \alpha' \quad \text{and} \quad \angle CDA \cong \angle DCB = \beta'$$

Step 3: Since points C and D lie on different sides of line AB, the segment CD intersects this line at midpoint M.

Question. Use the congruence

(scissors) $$\triangle AMC \cong \triangle BMD$$

in order to show that the diagonals AB and CD bisect each other at their common midpoint M. By which theorem is this congruence proved?

Answer. The congruence (scissors) is obtained with the SAA congruence theorem. To use this theorem, we need the congruent sides $AC \cong BD$, the adjacent pair of congruent z-angles α', and the vertical angles at vertex M. Hence one gets the congruent triangles (scissors). Finally, we get the segment congruences

$$AM \cong BM = m' \quad \text{and} \quad CM \cong DM = m$$

Clearly point M lies between C and D, because these two points are on different sides of line AB by construction. But it is quite hard to confirm that M lies between A and B. Indeed, the congruence $AM \cong BM$ implies that point M lies between A and B. [47]

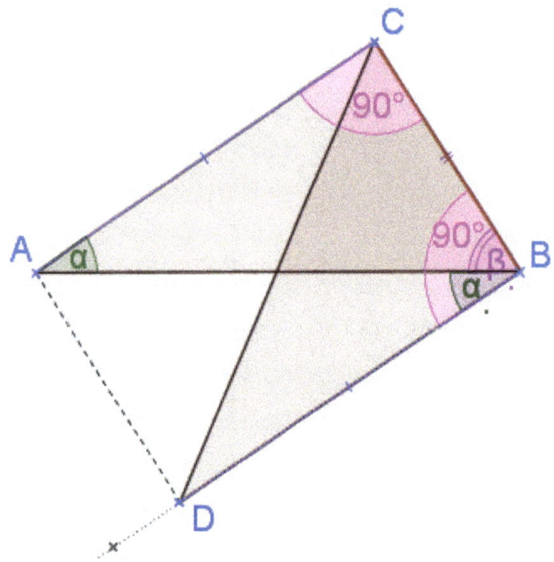

Figure III.17.14: The congruence of $\triangle BCA$ to $\triangle CBD$ implies that the diagonals of a rectangle are congruent

Step 4: We now get to the part of the reasoning which is valid only in Euclidean geometry. Because the angle sum in triangle $\triangle ABC$ is two right, and the angle at vertex C is assumed to be a right angle, we conclude that $\alpha + \beta = R$. Similarly, one confirms that the quadrilateral $\square ADBC$ has a right angle at vertex B. Hence it is a rectangle.

The right angle at vertex A yields still another pair of congruent triangles

(overlap) $$\triangle ABC \cong \triangle DCB$$

Question. Which theorem do you use to confirm this claim?

[47] See also figure on page 203 which is taken from the millenium edition of "Grundlagen der Geometrie", page 26.

Answer. This follows by SAS congruence. Indeed, the two triangles have the common side $BC = CB$, the two sides $AC \cong DB$ are congruent by construction, and the angles $\angle BCA$ and $\angle DBC$ are both right. Indeed, $\angle BCA$ was assumed to be a right angle, whereas

$$\angle DBC = \angle DBA + \angle ABC \cong \alpha + \beta = R$$

because of congruent z-angles α by construction, and the angle sum.

As a consequence of the congruence (overlap), we get $AB \cong CD$. Thus we have shown that the diagonals of the quadrilateral $\square ADBC$ are congruent.

Step 5: This is the final step to get the converse Thales Theorem. Because the diagonals

$$AB \cong CD$$

are congruent, and point M bisects them both:

$$AM \cong MB \quad \text{and} \quad CM \cong MD$$

Hence all four segments from point M to the vertices A, B, C and D are congruent. Hence vertex C lies on a semicircle with diameter AB, as to be shown. □

Remark. As a variant for the final step 5, we compare the distance $AM \cong BM = m'$ with $CM \cong DM = m$.

Assume that $m' > m$. Across the longer side of a triangle lies the larger angle. Using this fact for triangle $\triangle AMC$, we conclude that $\alpha' > \alpha$. Similarly, using triangle $\triangle BMC$, we get $\beta' > \beta$. Hence by addition, we get $R = \alpha' + \beta' > \alpha + \beta$. Hence the angle sum in the right triangle $\triangle ABC$ would be less than two right angles, which is false in Euclidean geometry.

Similarly, the assumption $m' < m$ leads to a contradiction, too. The only remaining possibility is $m = m'$. Hence vertex C lies on a semicircle with diameter AB and center M, as to be shown.

Problem III.17.10. *Which steps are still valid in hyperbolic geometry. Compare these quantities: m with m', α with α', and β with β'.*

Answer. Steps 1,2 and 3 are still valid. We still get $\alpha' + \beta' = 90°$, because the construction was started with a right triangle $\triangle ABC$ with right angle at vertex C. But, because the angle sum of a triangle is less than two right angles, one gets $\alpha + \beta < 90°$. Since $m < m'$ occurs if and only if $\alpha < \alpha'$ and $\beta < \beta'$, these inequalities are all three true. The figure on page 378 indicates the differing angles and segments using different colors. Remarkably enough, one can take the proof a step further without appeal to Euclidean geometry: The perpendiculars dropped from M onto the four sides of the quadrilateral $\square ADBC$ lie on two lines, but these two lines are not perpendicular to each other.

We shall now recapitulate the same ideas, in order to derive the basic properties of a rectangle.

Corollary 42 (The basic properties of the rectangle). *Given is a rectangle. This is a quadrilateral, about which it is* only *assumed that it has four right angles.*

(i) *A diagonal partitions the rectangle into two congruent triangles.*

(ii) *The opposite sides of a rectangle are congruent.*

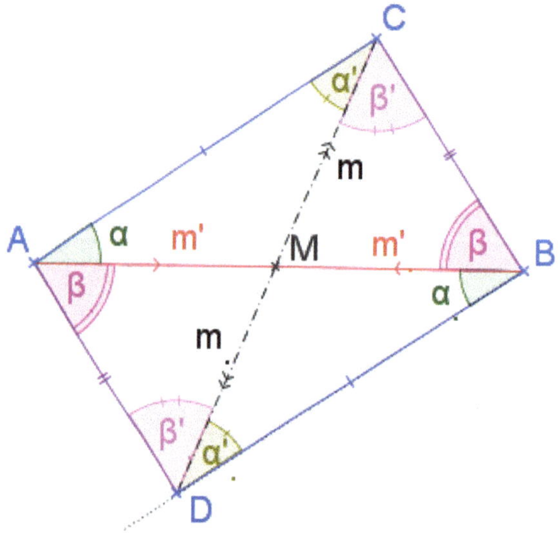

Figure III.17.15: What steps 1,2 and 3 yield in hyperbolic geometry

(iii) *The opposite sides of a rectangle are parallel.*

(iv) *The two diagonals of a rectangle bisect each other.*

(v) *The diagonals of a rectangle are congruent.*

(vi) *The diagonals of a rectangle intersect at the center of its circum circle.*

Proof. By definition, a rectangle is just a quadrilateral with four right angles—that is <u>all</u> what is assumed. The diagonal AB bisects the rectangle $\square ADBC$ into two right triangles $\triangle ABC$ and $\triangle ABD$. Again, let $\alpha = \angle CAB$ and $\beta = \angle ABC$ be the angles of the first of these triangles at vertices A and B, respectively.

We now use the fact that the angle sum in a triangle is two right angles. [48] Hence $\alpha+\beta+R = 2R$ and $\alpha+\beta = R$. The given rectangle has a right angle at vertex C, hence angle subtraction at vertex A now confirms that the second lower triangle $\triangle ADB$ has the $\angle BAD = R - \alpha = \beta$ at vertex A.

Thus we have obtained a pair of congruent z-angles $\beta = \angle ABC \cong \angle BAD$. Similarly, we get a second pair of congruent z-angles $\alpha = \angle BAC \cong \angle ABD$. With the ASA congruence theorem, we can now confirm the two triangles $\triangle ABC \cong \triangle BAD$ are congruent. Finally, the construction done above can be used once more to rebuild the rectangle from these two congruent right triangles.

We see that a diagonal partitions the rectangle into two congruent right triangles, and the opposite sides of the rectangle are congruent. Too, they are parallel because of the congruent z-angles.

The two diagonals of a rectangle are congruent because of

(overlap) $$\triangle BCA \cong \triangle DAC$$

Let M be the intersection point of the diagonals. The congruence (overlap) implies that the triangle $\triangle AMC$ has congruent base angles α. Hence the converse isosceles theorem implies $MA \cong MC$.

[48] Actually, while proving the second Legendre Theorem, we have shown that the existence of rectangle already implies this fact about the angle sum.

Similarly, one can show that the triangle $\triangle BMC$ has congruent base angles β and the converse isosceles theorem implies $MC \cong MB$. Hence the two diagonals of a rectangle bisect each other. and their intersection point is the center of its circum circle. □

Problem III.17.11. *In Euclidean geometry, one defines a parallelogram to be a quadrilateral, the opposite sides of which are parallel. Which parts of the Corollary 42 are true for all parallelograms, in Euclidean geometry.*

Answer. Items (i) true (iv) are true for any parallelogram, but items (v) and (vi) are not true in general.

III.17.3 Construction of tangents to a circle

Problem III.17.12 (Tangents to a circle). *Given is a circle $\mathcal{C}$, with center O and a point P outside of $\mathcal{C}$. Construct the tangents from point P to the circle $\mathcal{C}$. Actually do and describe the construction!*

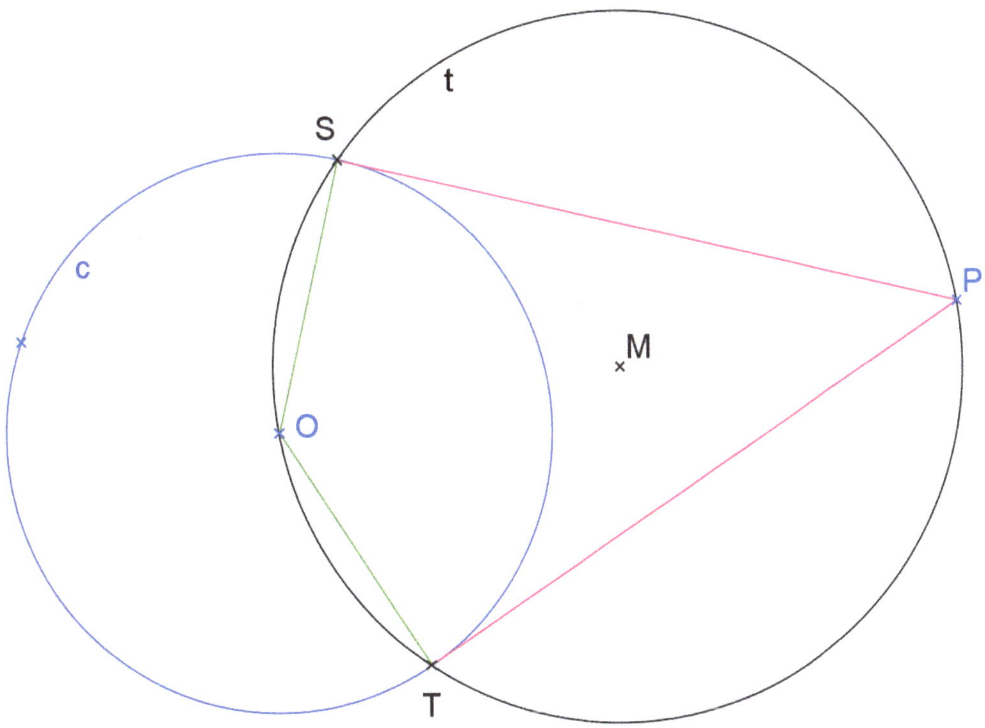

Figure III.17.16: Tangents to a circle

Construction III.17.1 (Tangents to a circle). *One begins by constructing a second circle $\mathcal{T}$ with diameter OP. (I call this circle the Thales circle over the segment OP). The Thales circle intersects the given circle in two points T and S . The lines PT and PS are the two tangents from P to circle $\mathcal{C}$.*

Validity of the Construction. By Thales theorem, the angle $\angle PTO$ is a right angle, because it is an angle in the semicircle over diameter OP. Since point T lies on the circle $\mathcal{C}$, too, the segment OT is a radius of that circle. By Euclid III. 16,

`The line perpendicular to a diameter is tangent to a circle.`

Since TP is perpendicular to the radius OT, and hence to a diameter, it is a tangent of circle $\mathcal{C}$. □

III.17.4 A bid of philosophy

If a Mathematician learns such a nice theorem, as Thales' theorem is for sure, what does he want to do with it? [49] How does mathematics and other clever people benefit from it, after putting all issues of priority aside? Here are some general considerations:

Generalize the theorem. This can mean either getting along with less assumptions, or just putting the given assumptions into a more general context.

Simplify the statement of the theorem.

Sharpen the conclusions. Expressing the conclusions in another way and simplifying the statement can both help to sharpen them.

Ask whether a converse holds. There can be more than one version for the converse, depending one how the theorem is formalized. Too, in case the converse fails to be true, one can ask for similar statements of which the converse does hold.

Consider special cases. They can look surprisingly different—for example, some assumption may turn out to be too obvious to be stated explicitly.

Find applications. Are there constructions or algorithms which follow from the theorem.

Simplify the proof of the theorem.

Build a theory. Find the natural place for the theorem in a larger context.

Problem III.17.13. *We want to construct the tangents through a given point outside to a given circle. Which one of construction* (III.17.1) *and construction* (II.10.1) *—explained in the section on neutral geometry of circles and continuity—remains valid in hyperbolic geometry? Explain why.*

Answer. Construction (III.17.1) is no longer valid in hyperbolic geometry. On the other hand, construction (II.10.1) remains valid in hyperbolic geometry.

Construction (III.17.1) is not valid in hyperbolic geometry, because the angle sum of a triangle is less than two right, and hence Thales' theorem does not hold in hyperbolic geometry.

On the other hand, construction (II.10.1) depends only on SAS-congruence, and the fact that tangent and radius being perpendicular. These statements hold in hyperbolic geometry, too.

Problem III.17.14. *Once more: given is a circle and a point outside of it. We have to construct the tangent from the point to the circle.*

One could simply put the straightedge through this point and move it around until its line just touches the circle. What can one say on the one side in favor, and what on the other side against this "lazy-boy" construction? Gather all arguments coming to your mind.

---

[49] Well, in case he has discovered the theorem himself, one should publish it—with the possibility in mind that somebody else has already discovered something similar before.

Answer. Arguments in favor. The "lazy-boy construction" is quick and easy. The accuracy with which the tangent is obtained is as good as for either one of the traditional constructions—as long as one wants to obtain only the tangent and not the touching point.

After the fact—once it is known that the construction of the tangent is possible with the traditional tools—it turns out that the "lazy-boy construction" does not allow constructions for other problems than those solvable with the traditional tools. The "lazy-boy construction" is just an additional tool helping to do things simpler, in the same sense as the geometrical triangle does that.

Too, the "lazy-boy construction" would still work in hyperbolic geometry. □

Arguments against the easy construction. The "lazy-boy construction" does not allow to obtain the touching point with any reasonable accuracy. One can of course find the touching point by an additional step—dropping the perpendicular from the center of the circle onto the tangent. But in that way, the number of steps needed is nearly the same as for the traditional construction of the tangent.

Before establishing the fact that construction of the tangent is possible with the traditional tools, the only way to formally justify this procedure is by requiring a new axiom— on the same footing as it has been suggested for origamy, or the two marked ruler, or any other extended tools of construction.

Question. Is the axiom

> "The tangent from a point to a circle can be obtained by appropriate placement of the straightedge (lazy-boy construction)."

equivalent to the circle-line intersection property, or to the circle-circle intersection property? Or is it a weaker axiom?

Question. How often does one really want new axioms?

Because of such questions, the "lazy-boy construction" cannot simply replace the traditional construction, nor does it make the traditional construction and its proof superfluous. □

My last word. In the end, I think one can use this simple way to obtain the tangent, if it is not necessary to have the touching point. □

III.17.5 Common tangents of two circles

Problem III.17.15 (Common tangents of two circles). *How many common tangents do two circles have. Informally draw all different cases, with* $0, 1, 2, 3, 4$ *common tangents. Describe how they arise.*

Answer. For any two different circles, there are five possibilities regarding their common tangents:

(0) One circle lies inside the other. They have no common tangents.

(1) One circle touches the other from inside. There is one common tangent, located at this touching point.

(2) The two circles intersect in two points. They have two common tangents, which lie symmetrically to the axis connecting the two centers.

(3) The two circles touch each other from outside. They have three common tangents.

(4) The two circles lie outside of each other. They have four common tangents. These are two pairs lying symmetrically to the axis connecting the two centers.

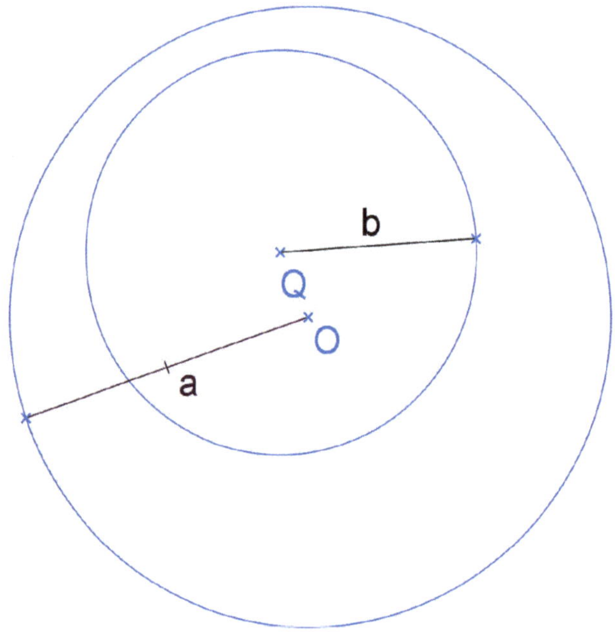

Figure III.17.17: Two circles with no common tangents.

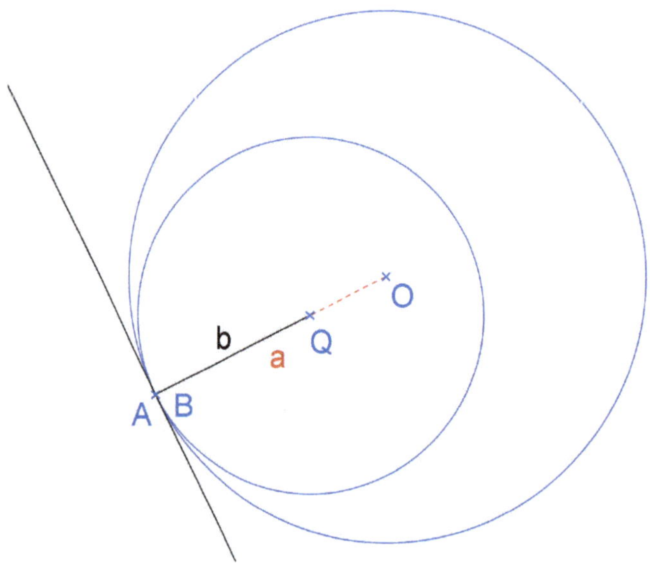

Figure III.17.18: Two circles with one common tangent.

Remark. Let the circles have center O and radius a, and center Q and radius b. These cases correspond to:

(0) $|OQ| < |b - a|$: One circle lies inside the other.

(1) $0 < |OQ| = |b - a|$: One circle touches the other from inside.

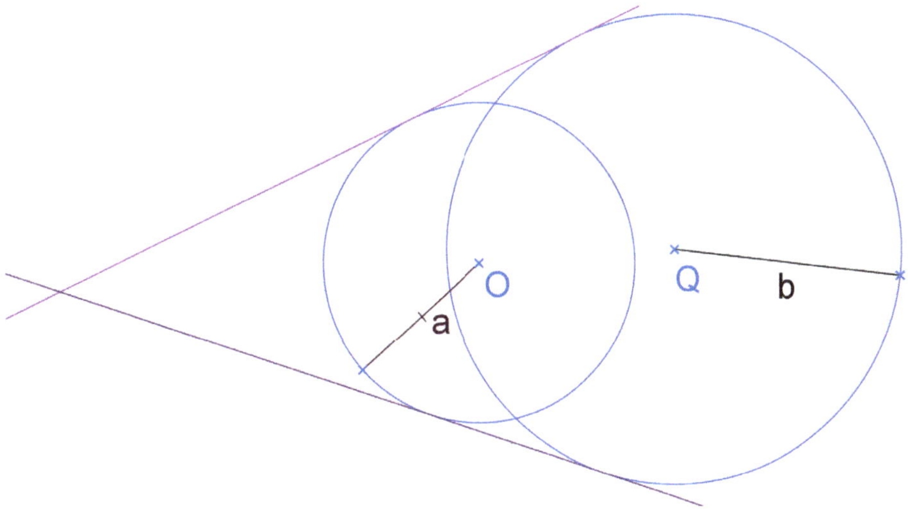

Figure III.17.19: Two circles with two common tangents.

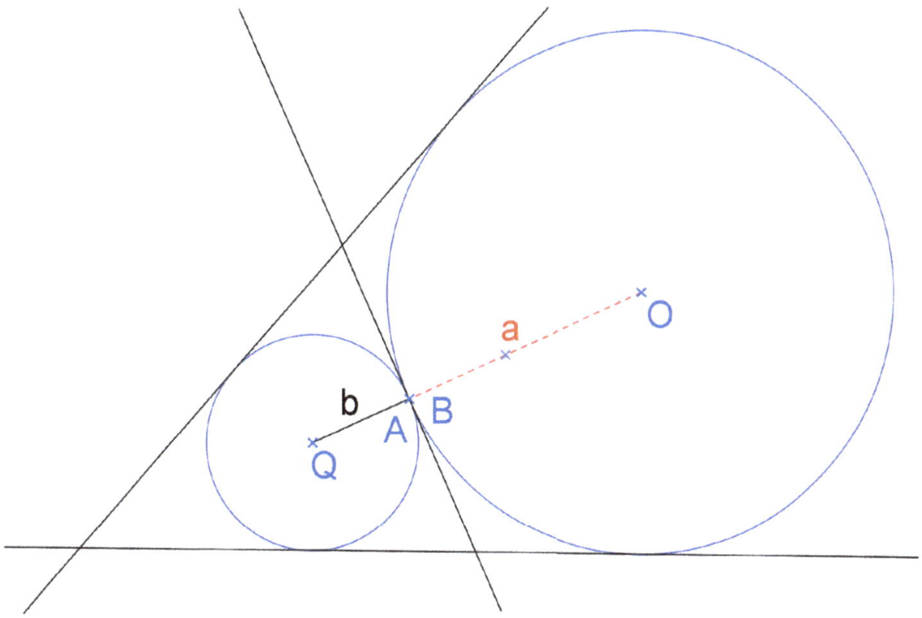

Figure III.17.20: Two circles with three common tangents.

(2) $|b - a| < |OQ| < a + b$: The two circles intersect in two points.

(3) $|b - a| < |OQ| = a + b$: The two circles touch each other from outside.

(4) $a + b < |OQ|$: The two circles lie outside of each other.

(∞) $0 = |OQ| = |b - a|$: The two circles are equal to each other.

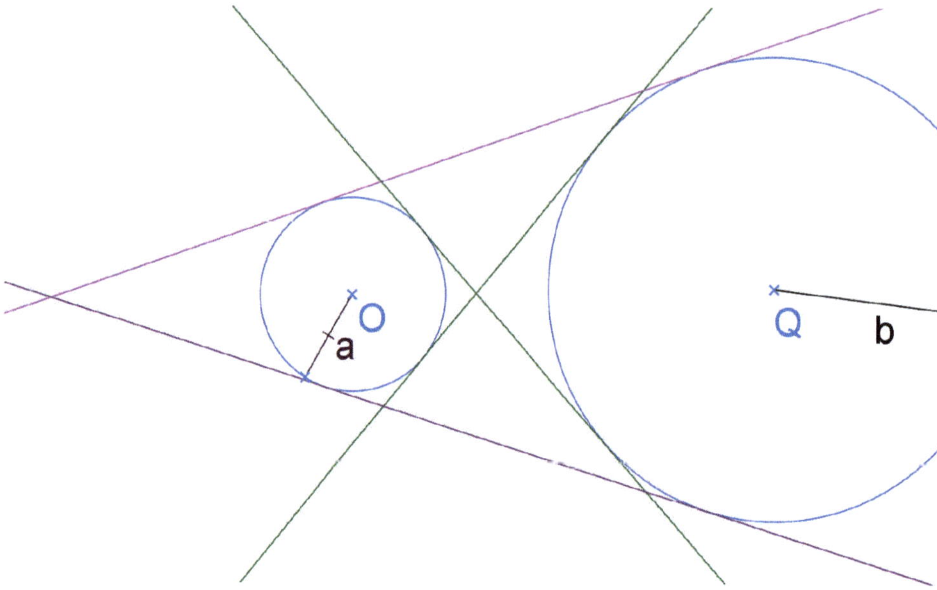

Figure III.17.21: Two circles with four common tangents.

Problem III.17.16 (Construction of the common tangents of two circles). *Given are two circles, with center O and radius a, and center Q and radius b. They lie outside of each other. Explain how to construct, with Euclidean tools, the common tangents of the two circles. Actually do and describe the construction!*

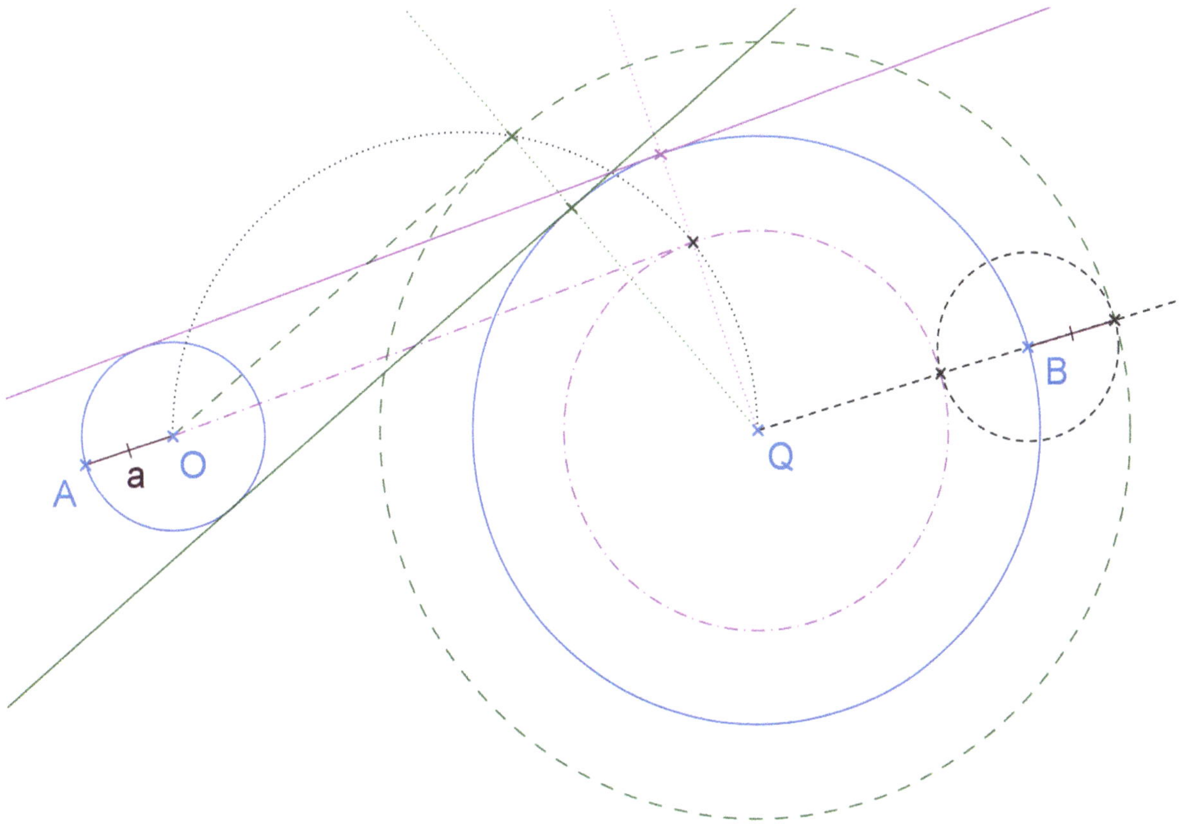

Figure III.17.22: Construction of two of the four common tangents for two circles lying outside each other.

The drawing on page 385 constructs just two common tangents, using construction III.17.2. The other two are their mirror images by line OQ connecting the centers.

Construction III.17.2 (Common tangents of two circles). *One constructs two new circles around one of the given centers, say Q, choosing as their radii the sum $a+b$ and difference $|a-b|$ of the radii a and b of two given circles. Next one constructs the tangents from the second center O to these two new circles. Too, one needs the radii where these tangents touch. The common tangents are produced by parallel shifts of these tangents, to both sides, by a distance given by the radius of the original circle around O.*

Shifting the tangents to the circle of radius $a+b$ yields the inner common tangents, whereas shifting the tangents to the circle of radius $|a-b|$ results in the outer common tangents.

A totally different method to construct the common tangents is described in the section on similar triangles.

III.17.6 Angles in a circle

Given is a circle with center O and a chord AB that is not a diameter. Recall Definition II.10.3 and Proposition II.10.5: a point S on the circle which lies in the interior of the central angle $\angle AOB$ lies on the short arc, and in the half planenn of AB opposite to the center O. A point L on the circle which lies in the exterior of the central angle lies on the long arc of chord AB and in the same half planen of AB as the center O.

Definition III.17.1. We call the angle with vertex at the center, the *central angle*, and the angle with vertex on the opposite arc, the *circumference angle* of the given arc.

Before giving detailed proofs, I state Euclid's important results:

Theorem III.17.4 (Euclid III.20). *The central angle is twice the circumference angle with vertex on the long arc.*

Theorem III.17.5 (Euclid III.21). *If two angles inscribed in a circle subtend the same arc, and if their vertices lie both on the same side of the chord, they are congruent.*

Theorem III.17.6 (Equivalent to Euclid III.22). *The circumference angles on the long and the short arc add up to two right angles.*

The supplement of the circumference angle with vertex on the short angle is congruent to the circumference angle on the long arc.

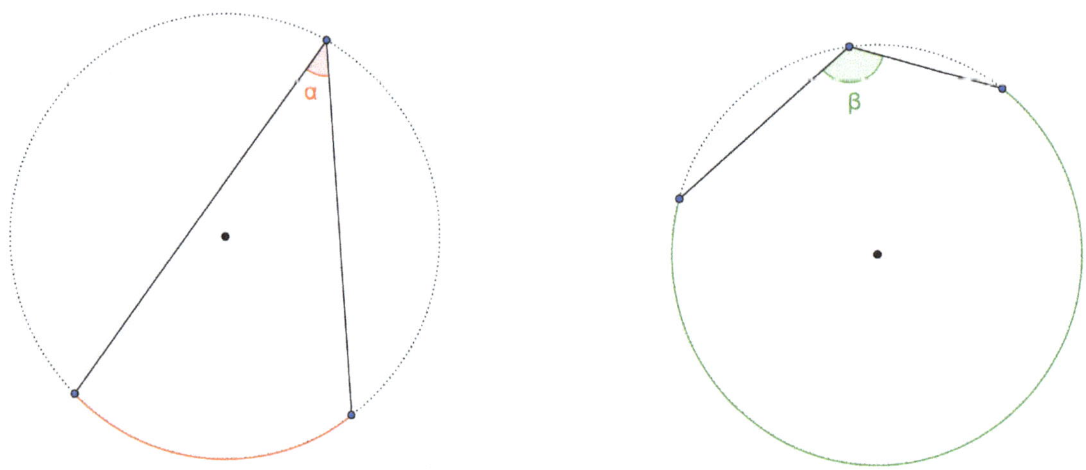

Figure III.17.23: The circumference angle of a short arc is acute, of a long one is obtuse.

Corollary 43. *The circumference angle of the short arc is acute, the circumference angle of the long arc is obtuse, and only the circumference angle of a semicircle is a right angle.*

Remark. Because of Euclid III.22, and the congruence of supplementary angles stated in Proposition II.7.15 [Theorem 14 of Hilbert], we conclude that Euclid III.21 holds both for the long and the short arc.

Definition III.17.2 (Similar arcs). We say that the arcs in two circles are similar if and only if they are either

- both short arcs and have congruent central angles;
- both long arcs and have congruent central angles;
- they are both semicircles.

Corollary 44. *The circumference angles of any congruent or similar arcs are congruent.*

Proof. This statement is justified by Euclid III.21 and Euclid III.22. (see theorem III.17.5 and III.17.6) □

Usually, we first consider the case of an arc AB less or equal half of the circle. It may be, historically Euclid wanted to stick to this situation. With the everyday presence of oscillations in our technical world, it is more natural to allow angles in the entire range from $0°$ to $360°$, and even beyond a whole turn.

As an angle in the usual sense, the central angle $\angle AOB$ is assigned to the *short* arc. By allowing *overobtuse* angles, we assign to the long arc the central angle $360° - \angle AOB$. The long arc is more than a half circle, and hence in this case, the central angle is greater than $180°$.

Furthermore, we want to include the special case that the arc AB is equal to a half circle, the chord AB gets a diameter. In that case, Euclid III.20 becomes Thales' theorem.

Finally, we agree to consider the circumference angle with vertex on the long arc to be the circumference angle of the short arc. Similarly, the circumference angle with vertex on the short arc is considered to be the circumference angle of the long arc. With these agreements put down, we can state Euclid III.20 in a short and more general form:

Corollary 45. *The central angle is twice the circumference angle of the same arc. This holds both for any arc on the circle, being either long or short.*

We are now ready to proceed to the proofs.

Problem III.17.17. *Provide a drawing with appropriate notation. Mark your arc AB in some color, then choose your third point C outside that arc. Mark the circumference angle $\gamma = \angle ACB$ in the same color. Too, mark the central angle $\omega = \angle AOB$. Avoid that the center of the circle lies on any of the chords involved. Prove the claim for the situation occurring in your drawing. The simpler version of the proof uses base angles of isosceles triangles (Euclid I.5), and the exterior angle of a triangle (Euclid I.32). (There are several possibilities, with angle addition or subtraction, but it is enough to do the proof for the situation in your drawing. The other cases are all quite similar.)*

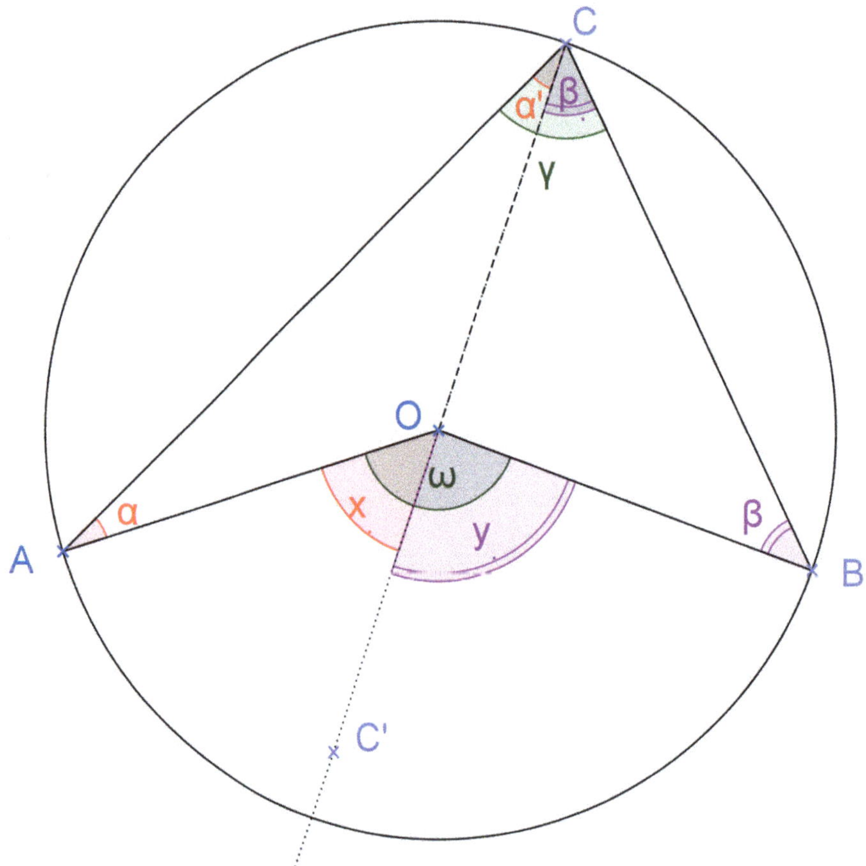

Figure III.17.24: Central and circumference angle of a circular arc

Reason for Euclid III.20. We deal with the central angle $\omega = \angle AOB$ and the circumference angle $\gamma = \angle ACB$ of the arc $\overset{\frown}{AB}$. We use the two isosceles triangles $\triangle BOC$ and $\triangle COA$. Each one of them has a pair of congruent base angles, called α and β, respectively. One extends the segment CO beyond the center O and let C' be any point on this extension. Now we see the exterior angle $x = \angle AOC'$ of triangle $\triangle AOC$ equals 2α. Indeed, we use the congruence of the base angles of the isosceles triangle $\triangle AOC$ together with Euclid I.32:

> The exterior angle of a triangle is the sum of the two nonadjacent interior angles.

Similarly, we see that exterior angle $y = \angle BOC'$ of triangle $\triangle COB$ is 2β.

For the case drawn in the figure on page 388, the center O lies *inside* triangle $\triangle ABC$. Hence the two isosceles triangles do not overlap. In that case, angle addition is used at the vertices C and O. At vertex C, one gets $\gamma = \alpha + \beta$. At vertex O, one gets $\omega = x + y$. Together with the equations

$$x = 2\alpha \quad \text{and} \quad y = 2\beta$$

one gets

$$\omega = x + y = 2\alpha + 2\beta = 2\gamma \tag{III.17.6}$$

as to be shown. In the second figure on page 390, the circumference angle is $\gamma = \alpha - \beta$ and the central angle is $\omega = x - y$. Since points A and O lie on different sides of line BC, central and circumference angle are both obtained as differences. □

The next two problems deal with Euclid III.22. At first, I do a special case, and the other problem completes the proof of Euclid III.22. A quadrilateral the four vertices of which lie on a circle is called a *circular quadrilateral.*

Problem III.17.18. *Let $\square ABCD$ be a circular quadrilateral, and assume that two vertices give a diameter BD, and the other two vertices A and C lie on different sides of it. Use Thales' theorem to show that the sum of the angles at B and D is two right angles. Provide a drawing with appropriate notation. The quadrilateral need not be a rectangle!*

Solution. The diameter BD partitions the quadrilateral into two triangles $\triangle BAD$ and $\triangle BCD$. The sum of the angles of the quadrilateral $\square ACBD$ equals the sum of the angle sums of the two triangles. Since the angle sum in a triangle is two right angles, the angle sum of a quadrilateral is $4R$.

By Thales' theorem, these are both right triangles, with right angles at vertices A and C. After subtraction of the two right angles at vertices A and C, the sum of the two remaining angles of the quadrilateral at vertices B and D is $2R$. □

Theorem III.17.7 (Euclid III.22). *The opposite angles of a circular quadrilateral add up to two right angles.*

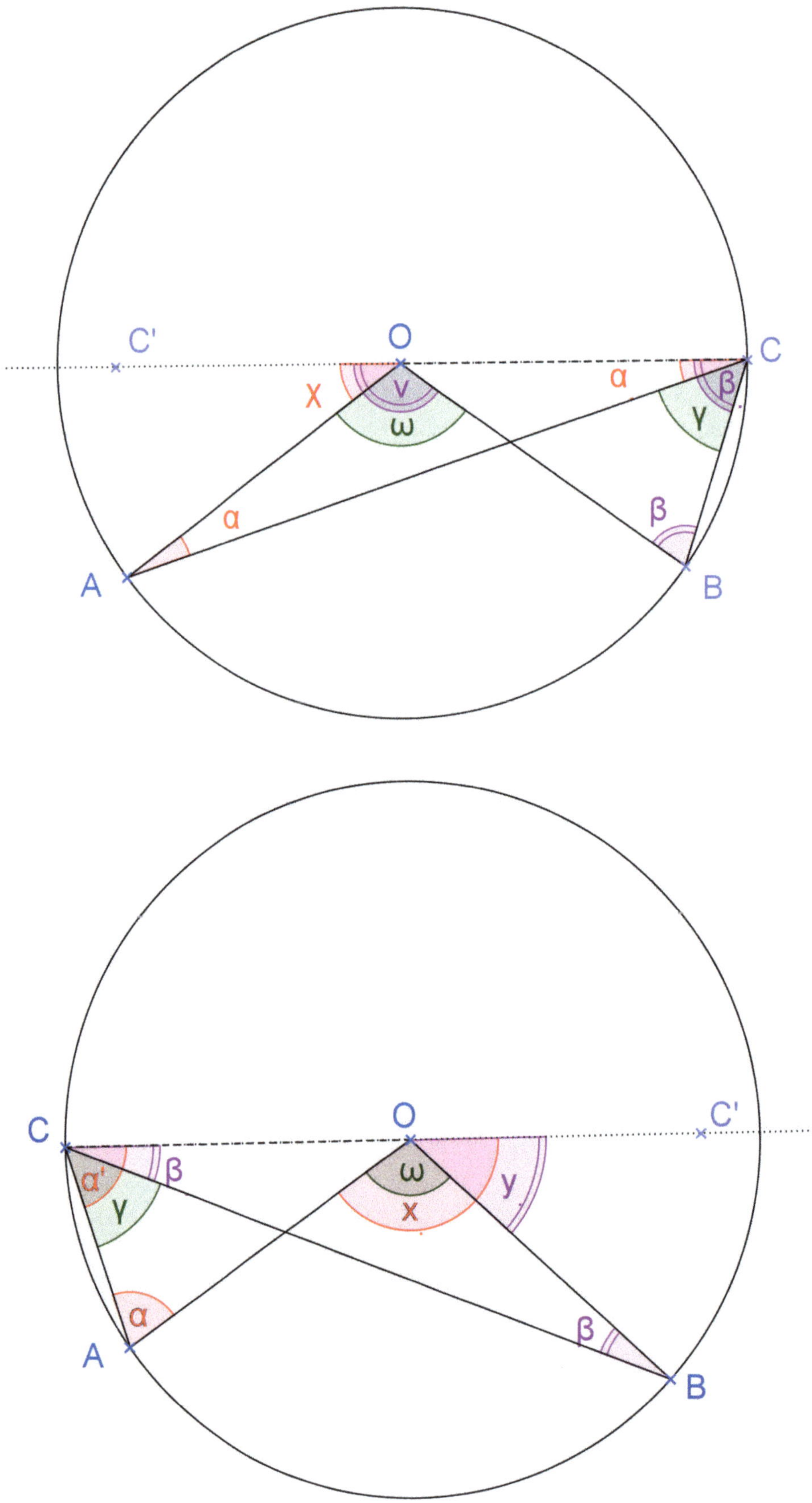

Figure III.17.25: Angles in a circle—still two other cases

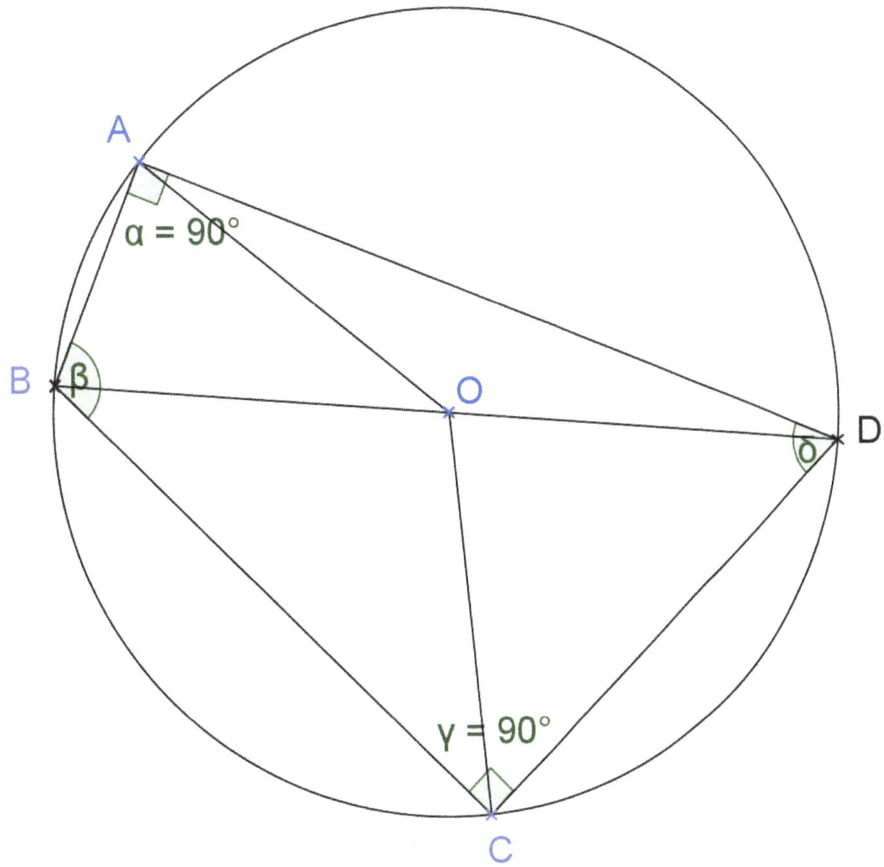

Figure III.17.26: A special quadrilateral

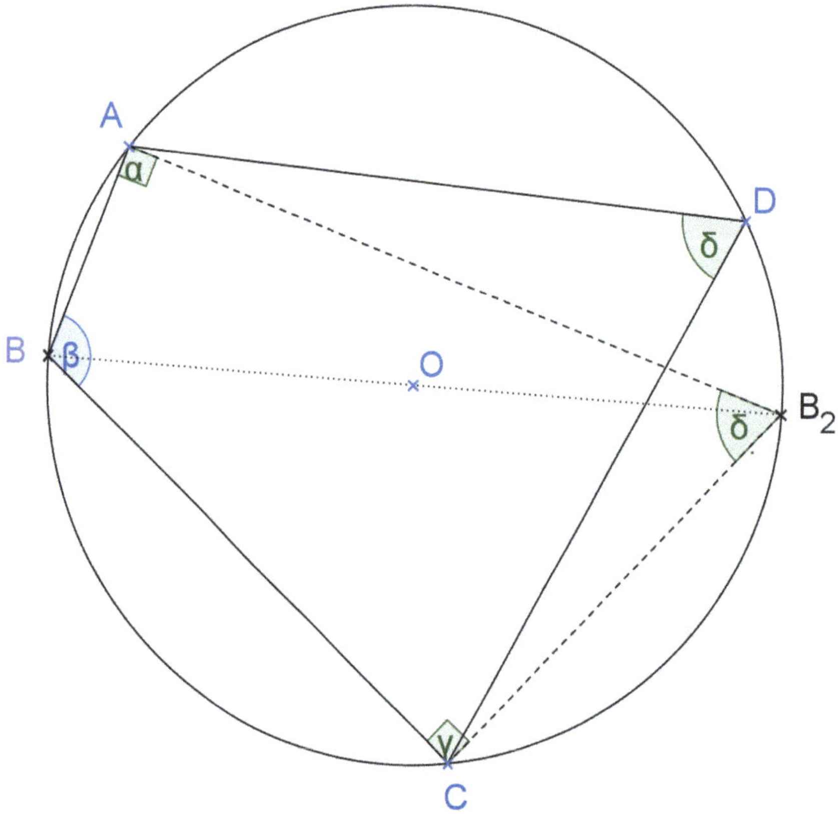

Figure III.17.27: Opposite angles in a quadrilateral with a circum circle

Proof. In the circum circle, either one of the arcs ABC or ADC is less or equal the other one, and hence less or equal a semicircle. We can assume that arc ABC is less or equal a semicircle. Let B_2 be the second endpoint of diameter BB_2. Now B_2 lies on the arc ADC. As shown in the last problem, the angles at opposite vertices B and B_2 add up to two right angles in the quadrilateral $\square ABCB_2$

$$\angle ABC + \angle AB_2C = 2R \tag{III.17.7}$$

By Euclid III.21, the circumference angle of arc ABC is

$$\angle ADC \cong \angle AB_2C \tag{III.17.8}$$

Together, we conclude

$$\angle ABC + \angle ADC \cong \angle ABC + \angle AB_2C = 2R \tag{III.17.9}$$

□

Remark. Here is an argument to get Euclid III.22 at once. We draw the four radial segments OA, OB, OC and OD and partition the quadrilateral $\square ABCD$ into four isosceles triangles. Let p, q, r and s be their respective base angles.

Because of angle addition at each vertex, the sum of the two angles of the quadrilateral $\square ABCD$ at the opposite vertices A and C is

$$\alpha + \gamma = (p + q) + (r + s) = (p + s) + (q + r) = \beta + \delta \tag{III.17.10}$$

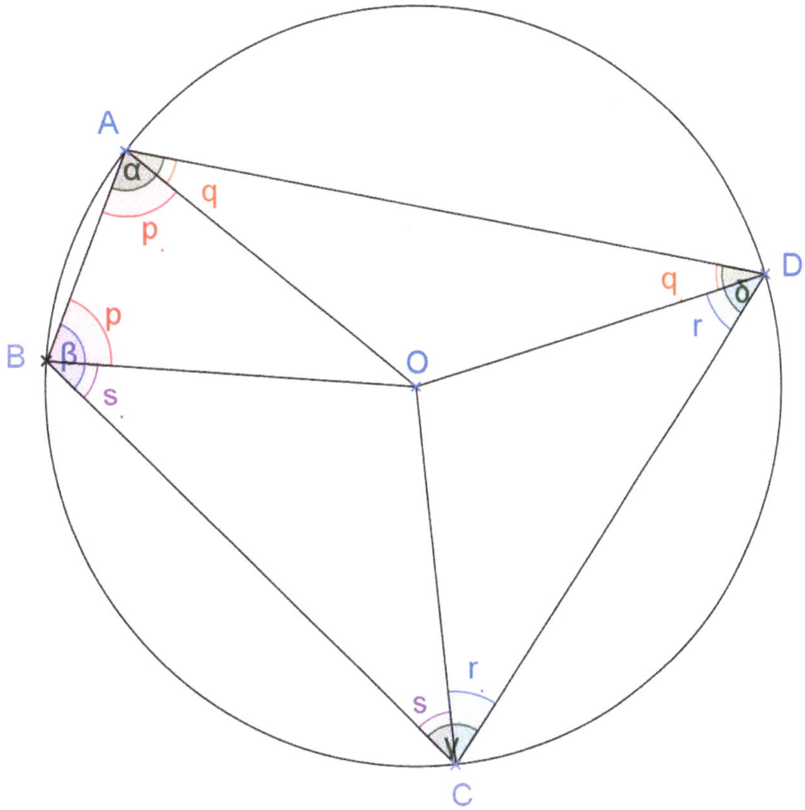

Figure III.17.28: Use isosceles triangles

Since the sum of angles at all four vertices is $\alpha + \beta + \gamma + \delta = 4R$, we conclude

$$\beta + \delta = \alpha + \gamma = 2R \tag{III.17.11}$$

Proposition III.17.1 (Another equivalent formulation of Euclid III.22). *An interior angle of a circular quadrilateral is congruent to the exterior angle at the opposite vertex.*

Proposition III.17.2 (Strengthening of Euclid III.22). *Assume vertices B and D lie on different sides of AC. The opposite angles $\beta + \delta$ of a quadrilateral $\square ABCD$ add up to*

(i) *more than two right angles if and only if vertex D lies inside the circum circle of triangle ABC,*

(ii) *two right angles if and only if vertex D lies on the circum circle,*

(iii) *less than two right angles if and only if vertex D lies outside the circum circle of triangle ABC.*

Problem III.17.19. *Draw a convex quadrilateral $\square ABCD$. The vertices A, B, C, D are named in alphabetic order as they follow each other around the quadrilateral. Draw any instance for these two examples:*

(a) *the opposite angles have sum $\alpha + \gamma = 120°$,*

(b) *the opposite angles have sum $\alpha + \gamma = 270°$.*

Draw a circle through A, B and C and explain how point D is located relative to the circle.

Answer. The angle sum of any quadrilateral is $4R$, and hence $\beta + \delta = 360° - \alpha - \gamma$.

In case (a), one gets $\beta + \delta = 360° - 120° = 240° > 180°$. Hence vertex D lies inside the circum circle of triangle ABC.

In case (b), one gets $\beta + \delta = 360° - 270° = 90° < 180°$, and hence vertex D lies outside the circum circle of triangle ABC.

Given is a circle with center O. Let AB be the chord and T be a point on the perpendicular to the radius at point A. We assume that center O does not lie in the same half plane of the chord AB as point T.

Proposition III.17.3. *The angle $\angle ABT$ is congruent to the circumference angle $\angle ACB$ with vertex C on the long arc.*

Proof. Since we assume that center O does not lie in the same half planen of the chord AB as point T, the angle $\tau = \angle ABT$ is acute or right. We prove the statement first for the circumference angle of an appropriately chosen point D. Let BD be a diameter of the circle. By Thales' theorem, triangle $\triangle ABD$ is a right triangle. Because the angle sum of a triangle is $2R$, its two angles at vertices D and B, add up to a right angle:

$$\beta + \delta = R$$

On the other hand, angle addition at vertex B yields,

$$\beta + \tau = R$$

From these two equation and angle subtraction, we conclude that $\tau = \delta$. Thus the angle $\tau = \angle TAD$ between the perpendicular of the radius at point B and the chord AB is congruent to the circumference angle $\delta = \angle ADB$ of that chord. By Euclid III.21:

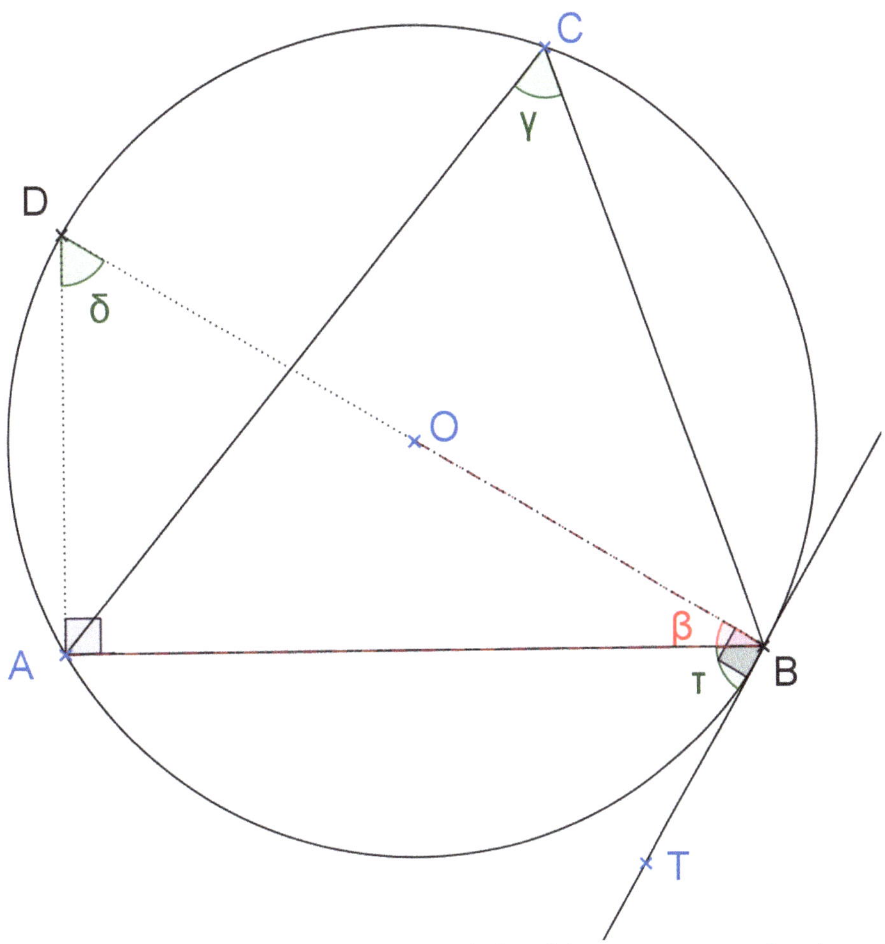

Figure III.17.29: The angle between tangent and chord is congruent to the circumference angle

```
Two angles from points of the circle subtending the same arc
   are congruent.
```

Hence it does not matter that we have chosen a special position for point D. The circumference angle $\gamma = \delta = \tau$ for any point C on the long arc of AB is congruent to the angle τ between the chord and the tangent. □

III.17.7 On the nature of the tangent

Euclid III.16 contains the assertion that the line erected perpendicularly onto a diameter at its end is tangent to the circle. The fact that the tangent is orthogonal to a diameter is stated repeatedly in Euclid III.16, III.18 and III.19—without further clarification how the tangent is defined. Euclid formulates our result proposition III.17.3 in his proposition Euclid III.32:

```
The angle between a tangent line and a chord is congruent
   to the circumference angle of the arc corresponding to the chord.
```

I find this state of affair unsatisfactory. Hence I have defined the tangent to a circle as perpendicular to the radius (see definition II.10.4). Proposition II.10.8 states that a line is tangent to a circle if and only if it has exactly one point in common with the circle. This is just the second obvious property of the tangent, which does not involve limits. We now agree to use the notion of limits. We define

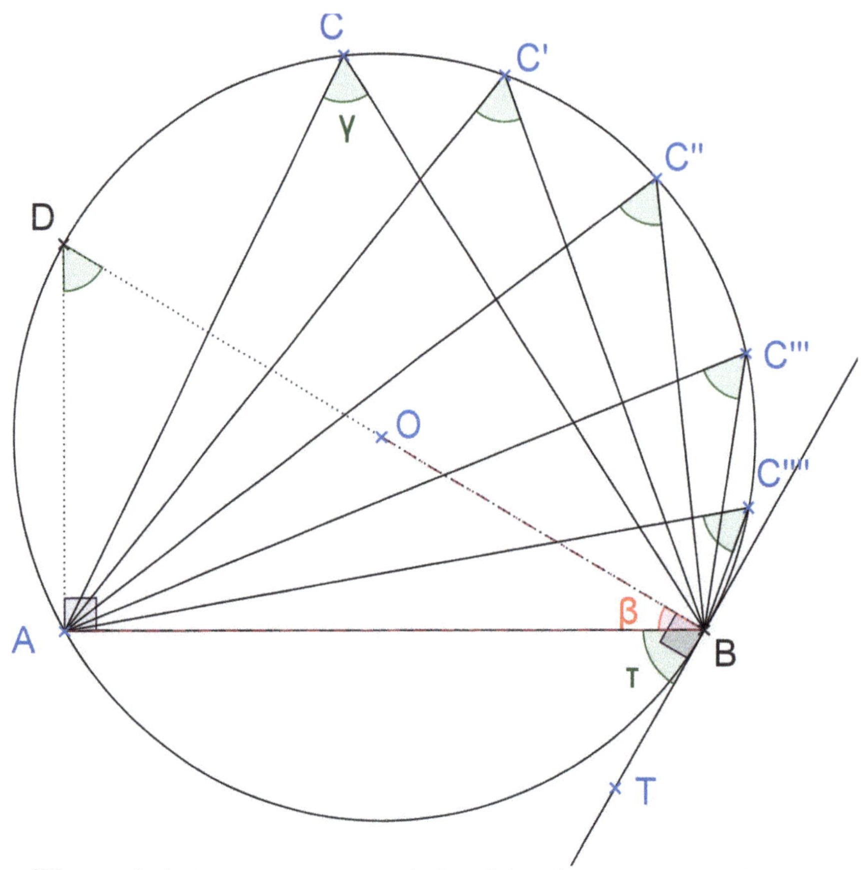

Figure III.17.30: The angle between tangent and chord is a limiting case of the circumference angle

the tangent as the limiting position of a secant line, the two endpoints of which move together at the touching point of the tangent. The circumference angle of the fixed chord AB constantly stays γ. In the limit $C \to B$, one side of the circumference angle $\angle ACB$ becomes the chord AB, and the other side CB becomes the tangent t to the circle at point B. Hence, in the limit we see that the angle between the chord AB and the tangent t at its endpoint B still has the same value γ.

Proposition III.17.4. *We obtain the tangent as limiting position of a secant line, as the length of the secant approaches zero. In this setting, the angle between any chord and the tangent ray at its endpoint, pointing to the side of the short arc, is congruent to the circumference angle with vertex on the long arc.*

Corollary 46 (The tangent as limiting position of a short secant). *The tangent to a circle, defined as the limiting position of a small secant, is perpendicular to the radius at the touching point.*

Proof. Let t be the perpendicular to the radius OB at the touching point B. Let $\tilde{t}$ be the tangent obtained as the limiting position of a small secant with one endpoint B.

By proposition III.17.3, the acute angle between chord AB and t is the circumference angle γ. On the other hand, the remark above and proposition III.17.4 show that the acute angle $\tilde{t}$ between chord AB and tangent as limit is congruent to the circumference angle γ, too. Hence, by the uniqueness of angle transfer, we conclude $t = \tilde{t}$. In other words, the tangent is both the limiting position of a small secant, and the perpendicular to the radius at the touching point. □

Remark. Indeed, we have already obtained the same result even in <u>neutral</u> geometry. The most complete result is indeed Theorem II.10.2.

Remark. Historically, the exact nature of the tangent is not discussed at this point, or any point, by Euclid. Euclid III.16 contains the cryptic statement—I would say, the <u>only</u> cryptic statement in the entire work of Euclid:

```
"The angle between the tangent line and the circle
   is less than any rectilinear angle."
```

III.17.8 The two circle lemma

The following lemma is remarkable by itself, and shall be used again in the proof of Pappus' and Pascal's Theorems.

Lemma III.17.1 (Two Circle Lemma). *If the endpoints of the common chord of two circles lie on two lines, these lines cut the two circles in two further parallel chords.*

Problem III.17.20. *Explain the reason for the two circle lemma. To get a convenient picture, consider the situation drawn in the figure on page 398. The two given circles have the common chord CD. Let AB be a second chord in the left circle. Explain why the second chord PQ in the right circle is parallel to AB.*

Going over the proof, does it matter whether the two given lines are parallel or not?

Answer. We get a circular quadrilateral $\square ABCD$ inside the left circle $\mathcal{C}$, and a second circular quadrilateral $\square CDPQ$ inside the right circle $\mathcal{D}$. As a consequence of Euclid III.22, the opposite angles of a circular quadrilateral sum up to two right angles.

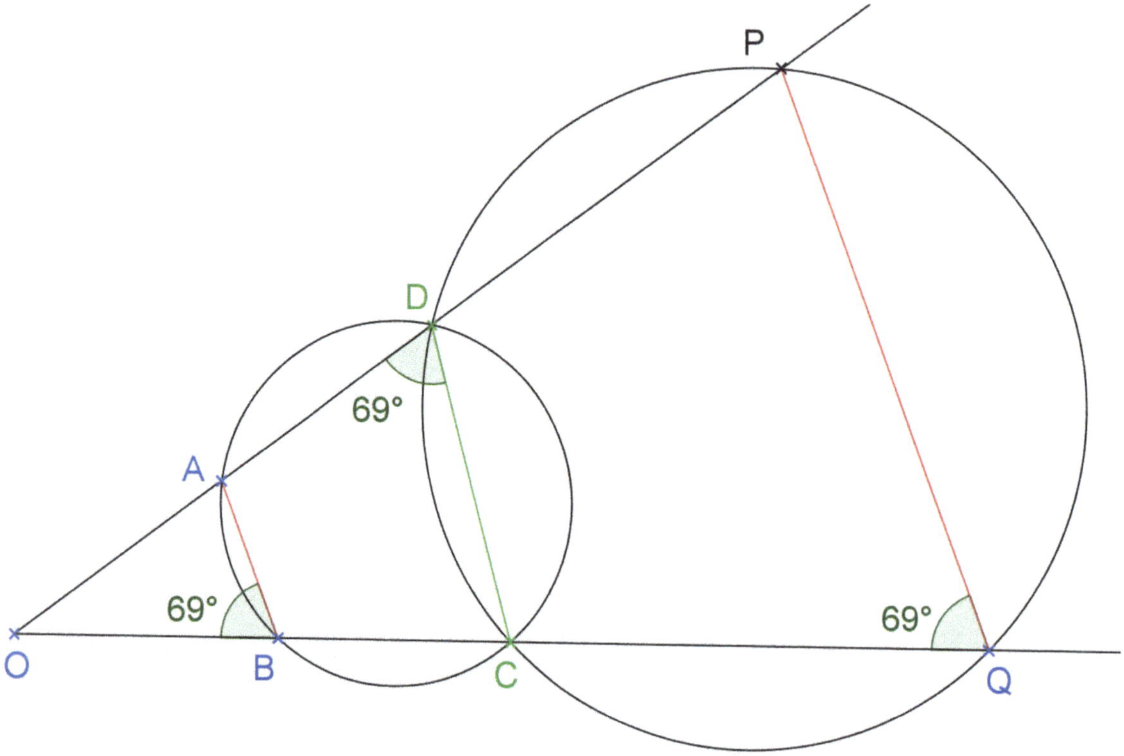

Figure III.17.31: The sides of an angle cutting two circles at the endpoints of their common chord cut them in two further parallel chords.

We check that the segments AB and PQ form congruent angles with one of the given lines. We need to compare the two angles $\angle OBA$ and $\angle OPQ$.

Let O be any point of line BC and O' any point on line AD such that the points mentioned occur on the two lines in the order $O * B * C * Q$, and $O' * A * D * P$. The angle $\angle OBA$ is exterior angle in the circular quadrilateral $\square ABCD$. As a consequence of Euclid III.22, it is congruent to the opposite interior angle:

$$\angle OBA \cong \angle O'DC$$

Now the second angle is exterior angle in the second circular quadrilateral $\square CDPQ$. Hence it is congruent to the opposite interior angle:

$$\angle O'DC \cong \angle OQP$$

From these two congruences, we conclude $\angle OBA \cong \angle OQP$. Hence, by Euclid I.27, the lines AB and PQ are parallel. We see that it does not matter whether the two given lines intersect or not.

Question. Explain how the proof has to be modified in the case drawn in the figure on page 399.

Answer. In this case, the circular quadrilateral has intersecting opposite sides AB and CD. In this case, the interior angles at opposite vertices are congruent. Hence, the exterior angles at opposite vertices are congruent, too.

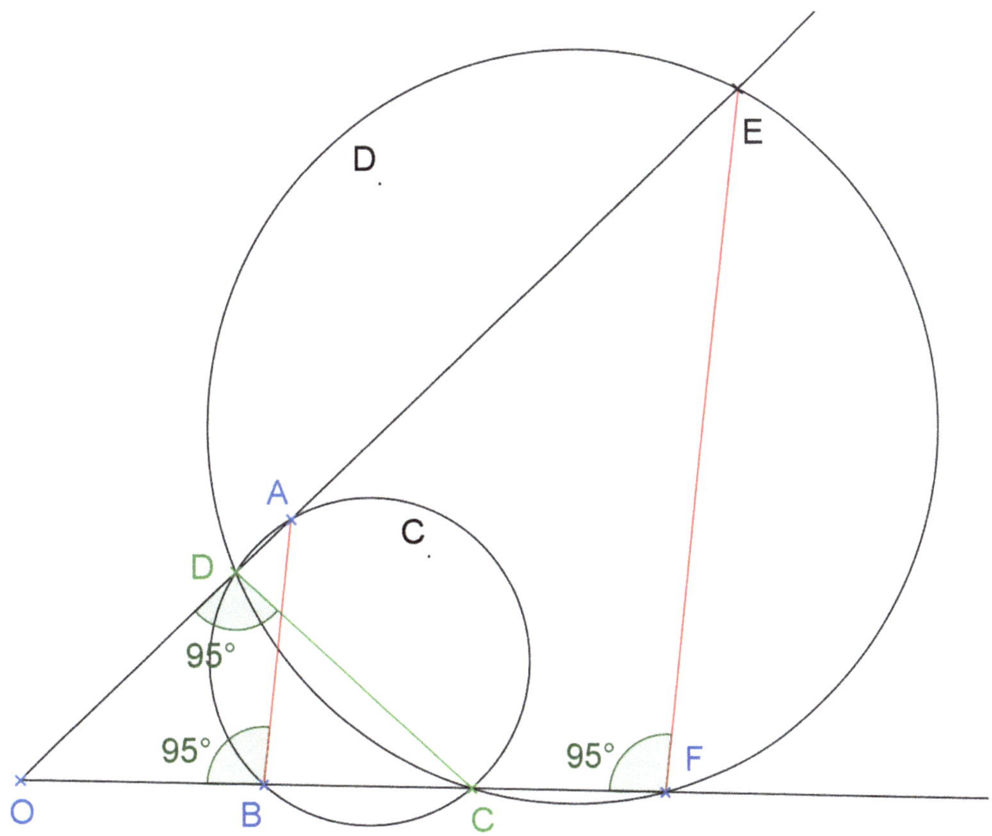

Figure III.17.32: Another example for the two-circle lemma.

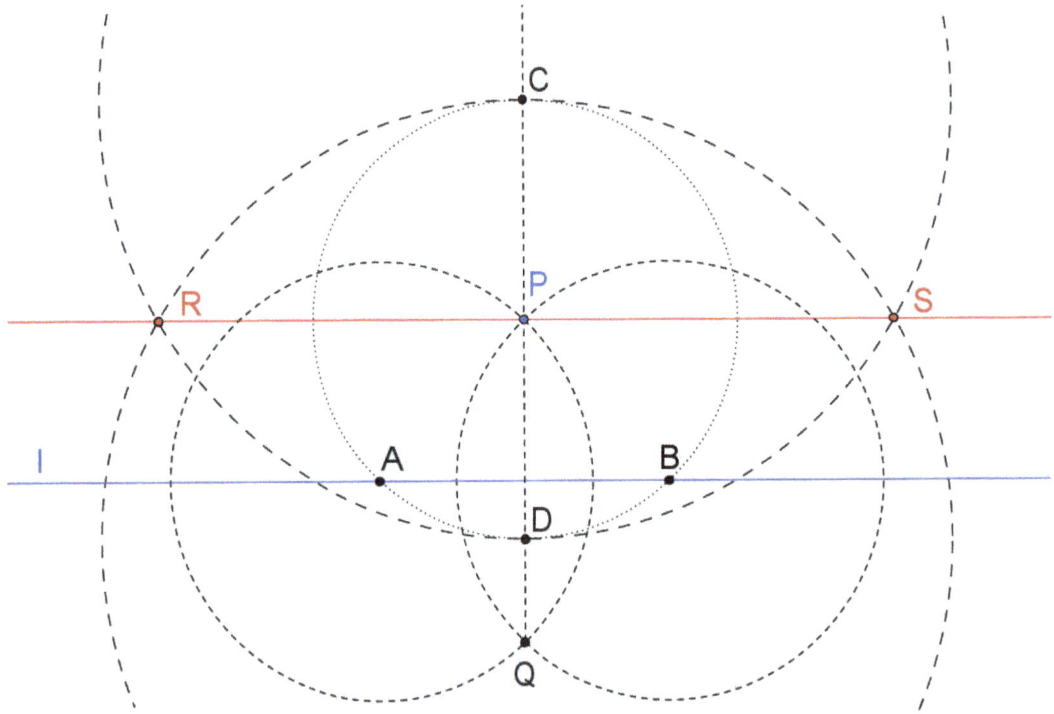

Figure III.18.1: The parallel as double perpendicular.

III.18 Simple Euclidean Geometry

These exercises deal with some interesting questions from Euclidean geometry on the level of Euclid's classic. The parallel axiom and its consequences, and the circle intersection properties are assumed to hold. The Fundamental constructions with Euclidean tools have already been done and described in problems II.14.5 through II.14.10 in the section about *A Simplified Axiomatic system of Geometry—my own Suggestion.*

III.18.1 Five constructions of the parallel

Given is a line l and a point P not lying on the line. We want to construct the parallel to line l through the point P.

Construction III.18.1 (The parallel as double perpendicular). *We draw a circle with center P through point A on the line l, and obtain a second intersection point B. We draw two further circles around A and B, both through point P. They intersect in a second point Q, too, which is the the reflection image of P across the line l. We draw the line $PQ \perp l$ and get the intersection points C and D with the first circle.*

Finally, we need the two circles with center C through point D, as with center D through point C. They intersect in two further points R and S. The line $RS \parallel l$ is the parallel to line l through point P.

Remark. As an extra check, we have obtained indeed three points P, R and S on the parallel.

Problem III.18.1. *Given is a line l and a point P not lying on the line. We want to construct the parallel to line l through the point P. Provide drawing with notations to illustrate the construction III.18.1 given above.*

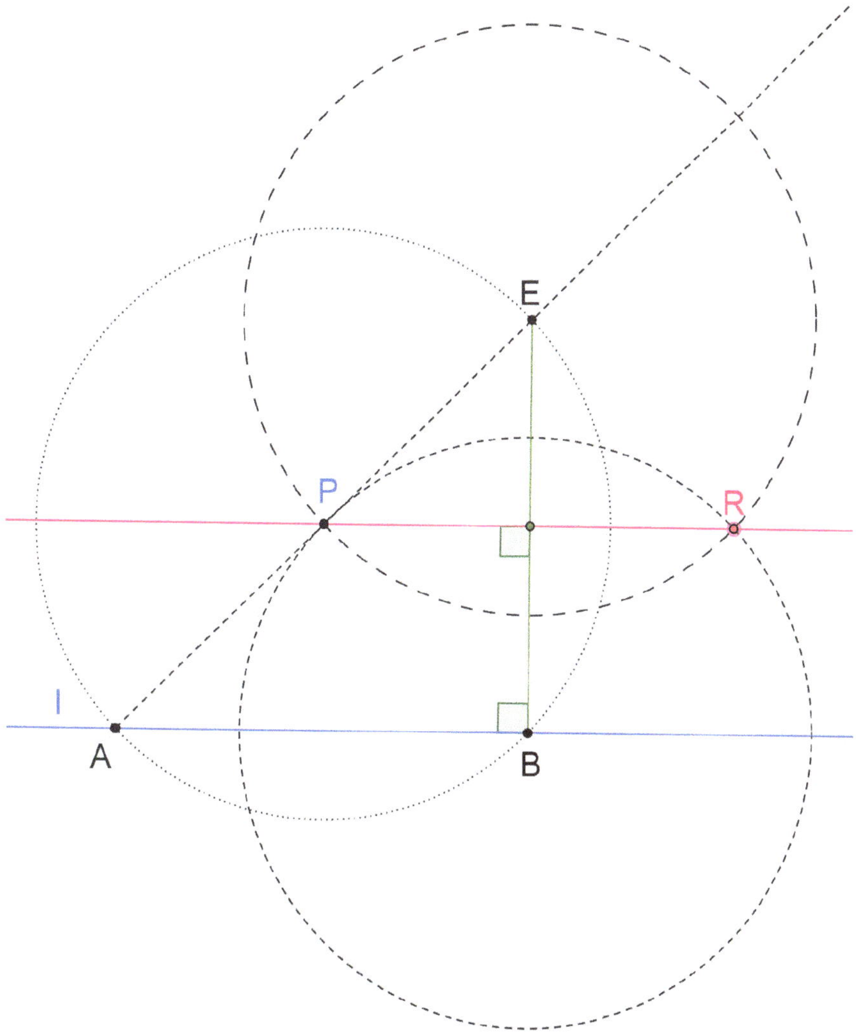

Figure III.18.2: The parallel using Thales' circle.

Remark. As an extra check, we can get the second endpoint F of the diameter BF of the first circle. The two circles with the centers F and A, both through point P. They intersect in a second point S. In this way, we get three points P, R and S on the parallel.

Construction III.18.2 (The parallel using Thales' circle). *We draw a circle with center P through point A on the line l, and obtain a second intersection point B. We draw a further circle around B through point P.*

Let point E be the other endpoint of the diameter AE of the first circle. Finally, we draw two circles with the centers E and B, both through point P. Finally, we draw two circles with the centers E and B, both through point P. They intersect in a second point R, too. The line $RP \parallel l$ is the parallel to line l through point P.

Justification of the construction. The angle $\angle ABE$ is a right angle by Thales' theorem since it is the angle in a semicircle. All three circle drawn have congruent radius $PA \cong PB \cong PE$. Hence the intersection points P and R of the circles around points B and E lie on the perpendicular bisector of segment BE. Hence lines BE and PR are perpendicular. Thus the line given line $l = AB$ and the line PR have a common perpendicular, and hence they are parallel. $\square$

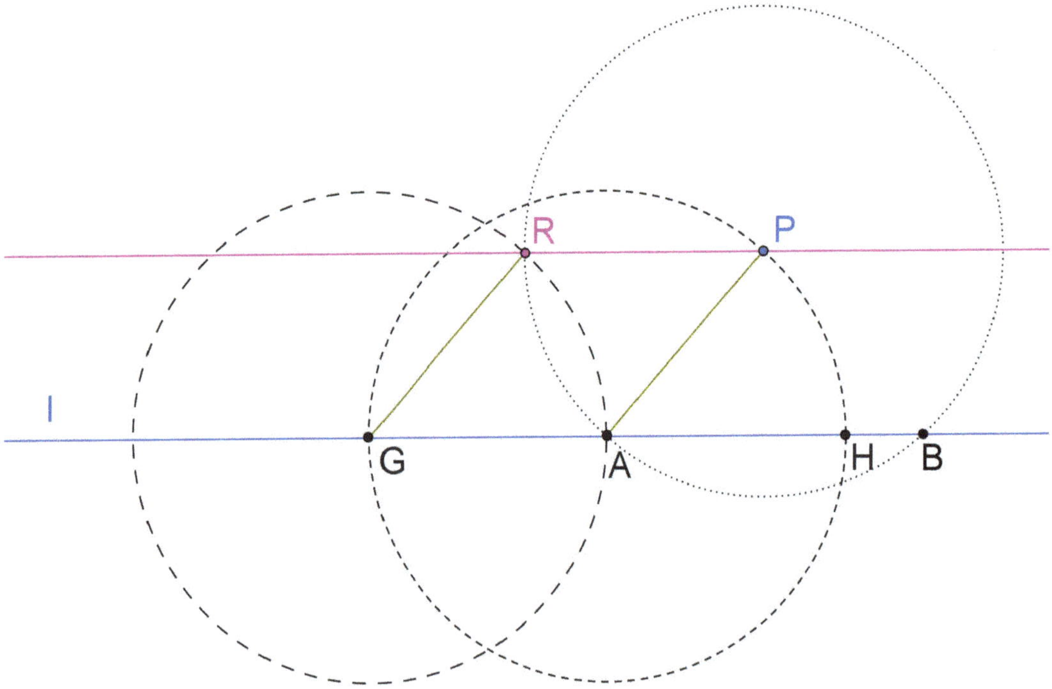

Figure III.18.3: The parallel obtained from a rhombus.

Construction III.18.3 (The parallel obtained from a rhombus). *We draw a circle with center P through point A on the line l. (We obtain a second intersection point B, too.)*

We draw the circle with center A through point P. It intersects the line l in the points G and H. We draw a third circle around G through A. It intersects the very first circle in point R. The line RP is parallel to the given line l since RP and GA are opposite sides of the rhombus $\square GAPR$.

Remark. We can draw a forth circle around H through A. It intersects the very first circle in point S. The line RP is parallel to the given line l since The segments SP and HA are opposite sides a rhombus and hence parallel. In this way, we get three points P, R and S on the parallel, which provides an extra check.

Problem III.18.2. *Given is a line l and a point P not lying on the line. We want to construct the parallel to line l through the point P. Provide drawing with notations to illustrate the construction III.18.3 given above.*

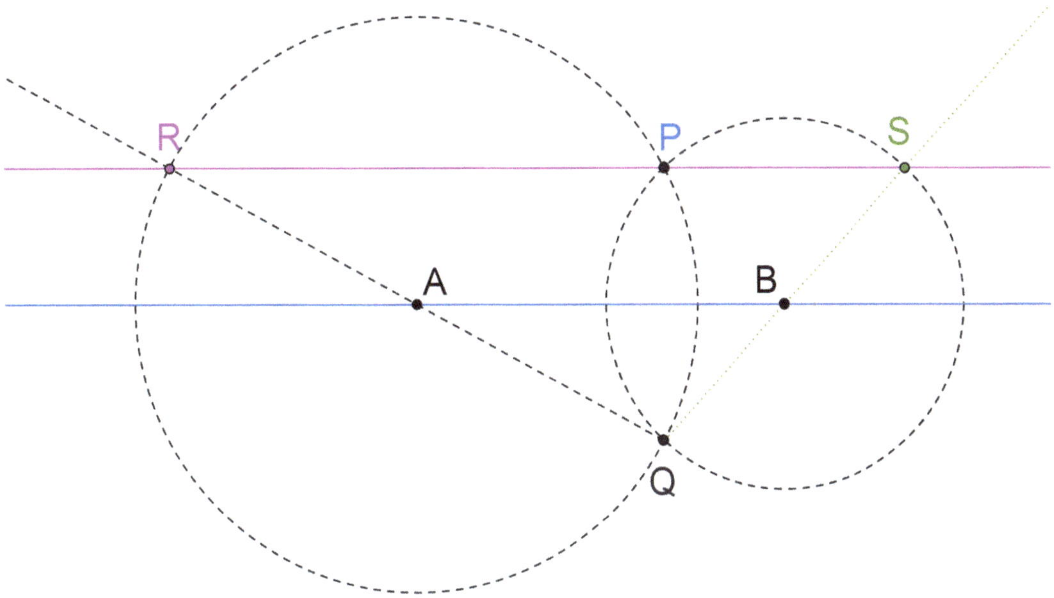

Figure III.18.4: The parallel using similar triangles.

Construction III.18.4 (The parallel obtained from similar triangles). *Choose any two points on the given line and draw circles with these centers through the given point P. They intersect in a second point Q, too, which is the the reflection image of P across the line l.*

The segment PQ intersects the given line in a point O, around which we draw a circle through points P and Q. It intersects the given line l in points A and B. We draw the rays from point Q through points A and B, and circular arcs through these points through point Q. We get the intersection points R and S. The line $RS \parallel l$ is the parallel to line l through point P. All three points P, R and S lie on the parallel, providing a further check.

Problem III.18.3. *Given is a line l and a point P not lying on the line. We want to construct the parallel to line l through the point P. Provide drawing with notations to illustrate the construction III.18.4 given above.*

Construction III.18.5 (The parallel using the harmonic quadrilateral). *We draw a circle with center any point B on the line l, and obtain the diameter AC on this line.*

We choose any point D on the line PC and draw lines DA and DB. Let point E be the intersection of lines AP and DB. Let point F be the intersection of lines EC and DA. The line PF is the parallel to the given line l through point P.

Problem III.18.4. *Given is a line l and a point P not lying on the line. We want to construct the parallel to line l through the point P. Provide drawing with notations to illustrate the construction III.18.5 given above.*

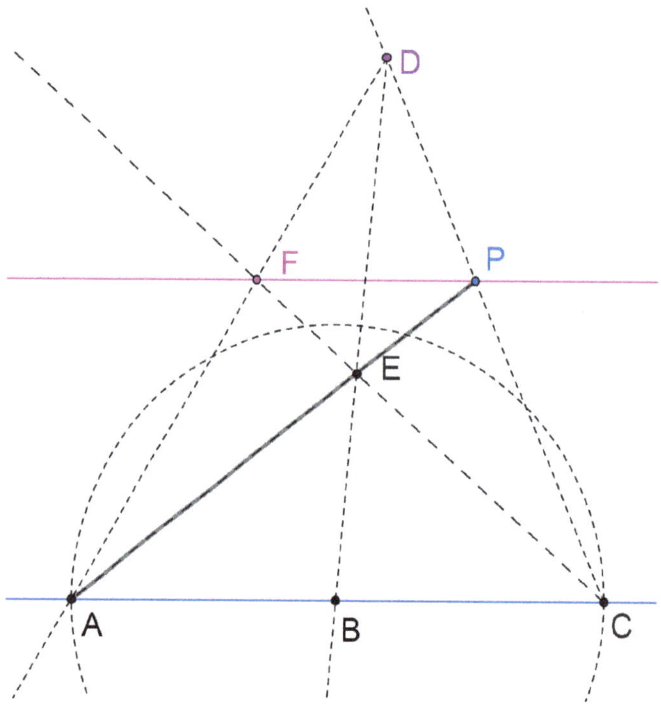

Figure III.18.5: The parallel using the harmonic quadrilateral.

III.18.2 Dividing a segment into any number of congruent parts

Problem III.18.5. *The subdivision of a segment into three congruent parts, can be done as in the figure on page 404. Give a stepwise description of this construction.*

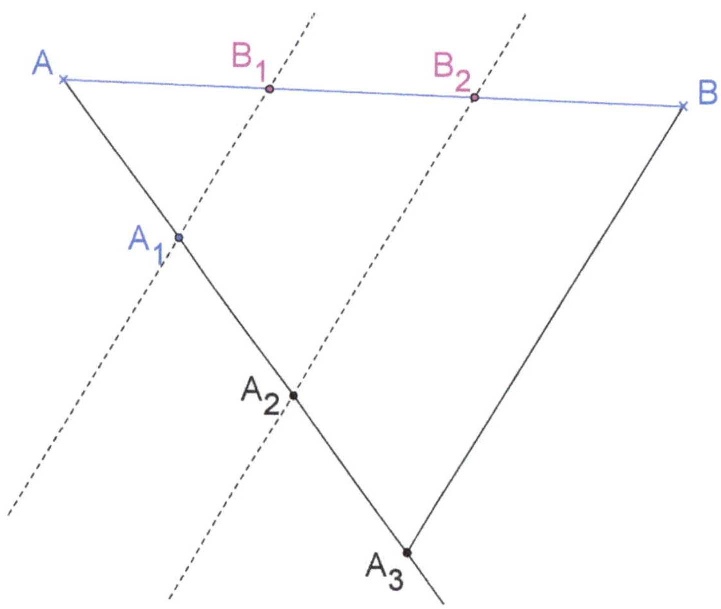

Figure III.18.6: Trisection using parallels.

Construction III.18.6 (Subdivision using parallels). *Given is the segment AB, to be subdivided into n congruent parts. We draw any ray with vertex A and transfer onto it n congruent*

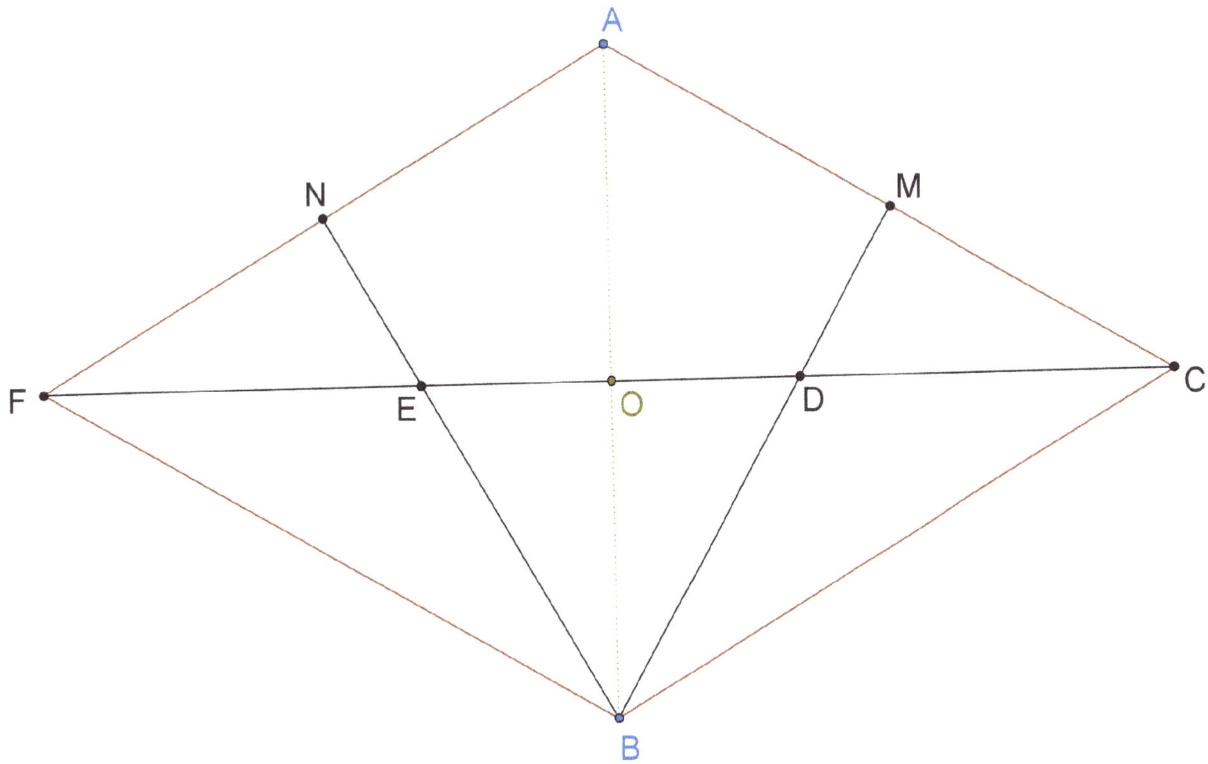

Figure III.18.7: How to trisect a segment but not an angle.

segments $AA_1, A_1A_2, \dots, A_{n-1}A_n$. *We draw the parallels to line* A_nB *through all intermediate points* $A_1 \dots A_{n-1}$. *They intersect the given segment* AB *in the required division points* $B_1 \dots B_{n-1}$ *such that segments*

$$AB_1, B_1B_2, \dots, B_{n-1}B$$

are the n *congruent parts of the given segment* AB.

Remark. The required parallels are easily constructed by at first dropping the perpendicular p from A onto line A_nB, and then dropping perpendiculars onto line p from the intermediate points $A_1 \dots A_{n-1}$.

Take two equilateral triangles with the common side AB. Together they give the rhombus $\square ACBF$ as shown in the figure on page 405. We draw the long diagonal CF. Let M and N be the midpoints of segments AC and AF. The segment BM intersects the long diagonal CF in point D. The segment BN intersects the long diagonal CF in point E. We see and explain the following facts:

- At vertex B we get the angles $\angle CBM = 30°$, $\angle MBN = 60°$ and $\angle NBF = 30°$.
- The points D and E trisect the long diagonal CF.

Construction III.18.7 (Trisection using triangles). *Given is the segment* AB, *to be trisected. One constructs two congruent isosceles* $\triangle ABC$ *and* $\triangle ABD$, *lying on different sides of* AB. *Let* M

and N be the midpoints of segments AC and BC. Let P and Q be the intersections of segment AB with segments DM and DN, respectively.
Claim: *The segments AP, PQ and QB are congruent. Each one measures one third of AB.*

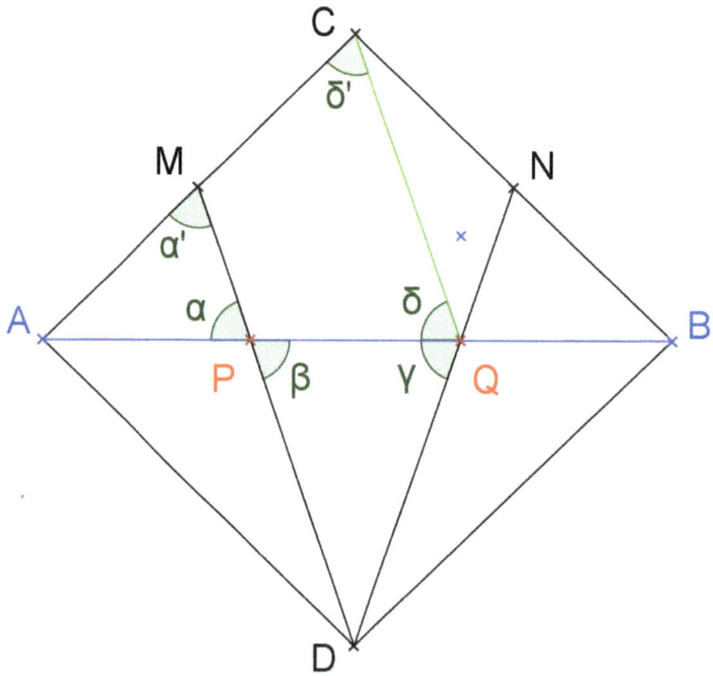

Figure III.18.8: Trisection of a segment

Problem III.18.6. *Complete the following proof of that claim. Indicate is yellow, which part of the argument still works, in green, which part does no longer work in hyperbolic geometry.*

Proof. We show that angles $\alpha = \angle APM, \beta = \angle QPB, \gamma = \angle PQD, \delta = \angle PQC$ are all congruent.

Indeed $\alpha \cong \beta$, because they are vertical angles.

Next $\triangle AMD \cong \triangle BND$, by SAS congruence with angles at vertices A and B and the adjacent segments matched.

Next $\triangle APD \cong \triangle BQD$ by ASA congruence with segments AD and BD and the adjacent angles matched. Hence $\angle APD \cong \angle BQD$, and $\beta = \gamma$, because supplements of congruent angles are congruent.

Next $\triangle APM \cong \triangle BQN$ by SAS congruence with angles at vertices A and B and the adjacent segments matched. Hence especially $AP \cong BQ$.

Next $\triangle AQC \cong \triangle AQD$, by SAS congruence, with angle at vertex A and adjacent segments matched. Hence $\gamma = \delta$.

Those congruences allow to get the claims $\alpha = \delta$ and $AP \cong BQ$. Hence by Euclid's Proposition I.27, the lines MD and CQ are parallel. The reasoning up to this point is valid in neutral geometry. We can carry the argument further in Euclidean geometry. Indeed Euclid I.29 yields $\angle AMP \cong \angle ACQ$. Hence $\triangle AMP$ and $\triangle ACQ$ have all angles pairwise congruent, and are similar. By Euclid VI.4, the sides of equiangular triangles are proportional. Hence we get the proportion

$$\frac{|AC|}{|AM|} = \frac{|AQ|}{|AP|} \tag{III.18.1}$$

Since $MC \cong AM$ and $AC = AM + MC$ by construction, the ratios in (III.18.1) are both equal to 2. Hence $|AQ| = 2|AP|$, too. Now $AQ = AP + PQ$ implies that the segments AP and PQ are congruent. Thus all three segments AP, PQ and QB have been shown to be congruent, and hence all three have $\frac{1}{3}$ of the length of AB. □

Problem III.18.7 (Using only equilateral triangles). *Draw only two circles of radius AB, and then complete the construction just with a straightedge.*

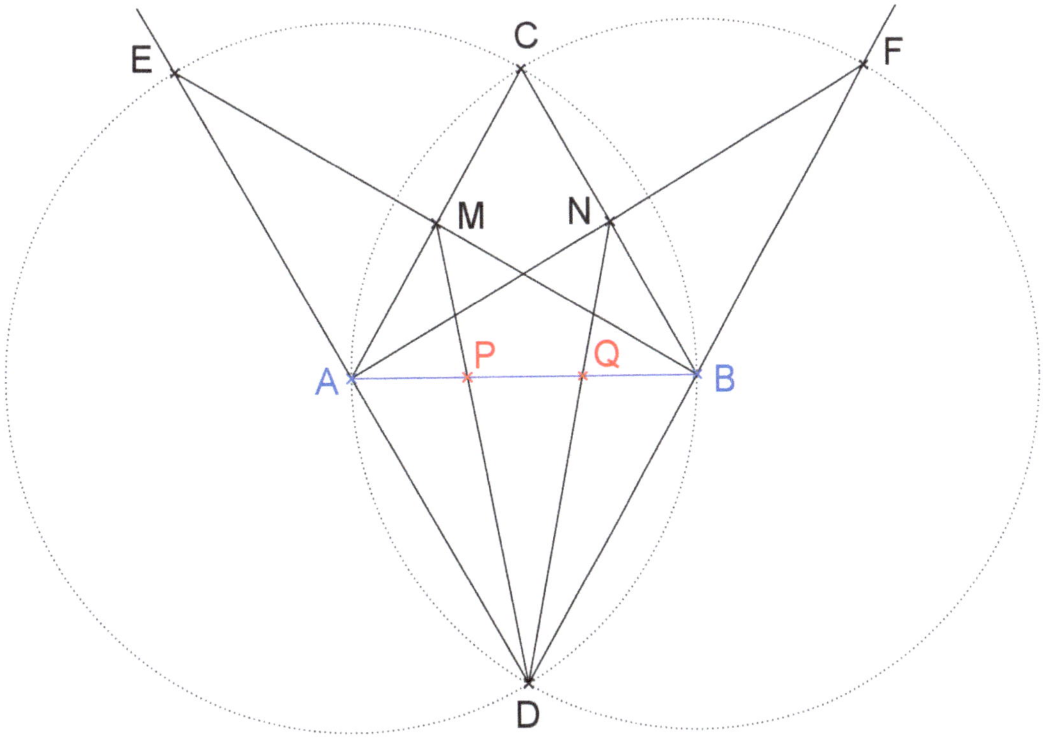

Figure III.18.9: Completed trisection with equilateral triangles

Construction III.18.8. *Begin by constructing two congruent equilateral $\triangle ABC$ and $\triangle ABD$, lying on different sides of AB.*

Line DA intersects the left circle in a second point E. Line DB intersects the right circle in a second point F. Drawn segment BE, which intersects AC in the midpoint M. Draw segment AF, which intersects BC in the midpoint N. Draw segments MD and ND. They intersect the given segment AB in points P and Q.

Claim: *The segments AP, PQ and QB are congruent. Each one measures one third of AB.*

Problem III.18.8. *For clever students: In hyperbolic geometry, does the middle segment get shorter, or longer than the other two segments?*

Answer. In hyperbolic geometry, the middle segment gets shorter than the two other segments. Here is a reasoning for the special case that $\triangle AQC$ is isosceles. In that case $\delta \cong \delta' := \angle ACQ$, because they are base angles of an isosceles triangle.

Next, I use the defect of the angle sum $2R - \alpha - \beta - \gamma$, which is proportional to the area of a triangle. $\triangle APM$ is part of $\triangle AQC$, and hence has smaller area, and smaller defect.

Let $\alpha' := \angle AMP$. The defect of $\triangle AMP$ is less than the defect of the larger $\triangle ACQ$. Comparison of the defects leads to $\alpha + \alpha' > \delta + \delta'$ and $\alpha' > \delta'$. Euclid I.19 says:

```
"If one angle of a triangle is greater than another, then the side opposite
   to the angle is greater than the other."
```

Now, we have found that $\angle AMP = \alpha' > \delta' = \delta = \alpha = \angle APM$, and we can use Euclid I.19 in $\triangle APM$. Hence $AP > AM$. Now $AQ \cong AC$ implies $PQ = AQ - AP < AC - AM = MC \cong AM < AP$.

Remark. In hyperbolic geometry, the problem to trisect a given segment cannot be solved with straightedge and compass. Indeed, the problem leads to the Delian problem to construct $\sqrt[3]{2}$, which is impossible to do with straightedge and compass.

III.18.3 Some triangle constructions

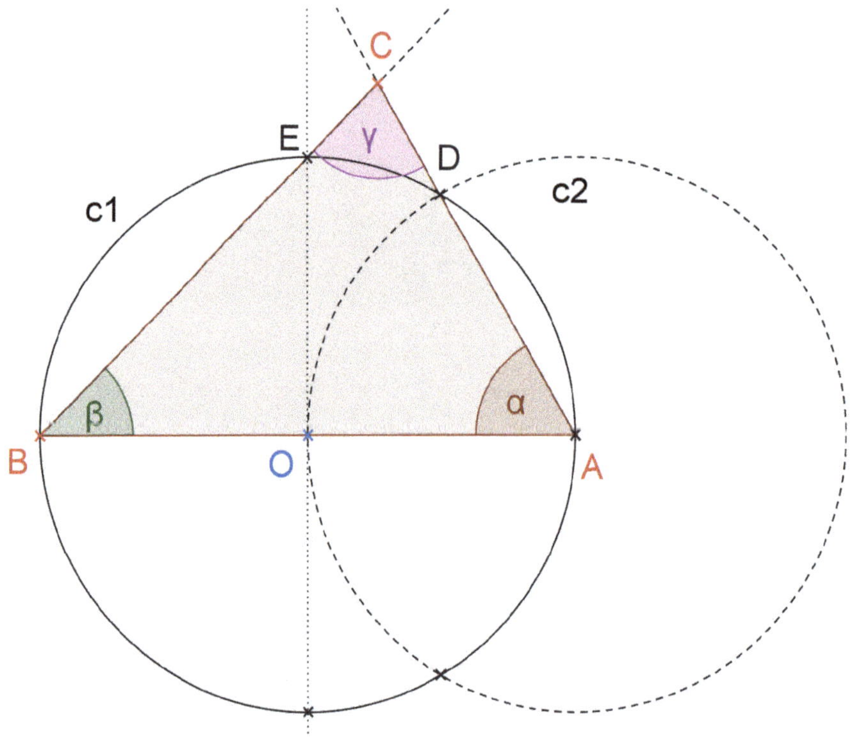

Figure III.18.10: What are the angles of $\triangle ABC$?

Problem III.18.9 (Another special triangle). *Describe the construction done in the figure on page 408. Explain how you determine the three angles of the triangle $\triangle ABC$.*

Answer. On an arbitrary segment OA, an equilateral triangle $\triangle OAD$ is erected. This can be done as described in Euclid I.1. We find the intersection point D of a circle around A through O with a circle around O through A.

Let B be the second endpoint of diameter AOB. At point O, we erect the perpendicular onto this diameter. Let E be the intersection of the perpendicular with the right circle, lying on the same side of AB as point D. The rays $\overrightarrow{AD}$ and $\overrightarrow{BE}$ intersect in point C.

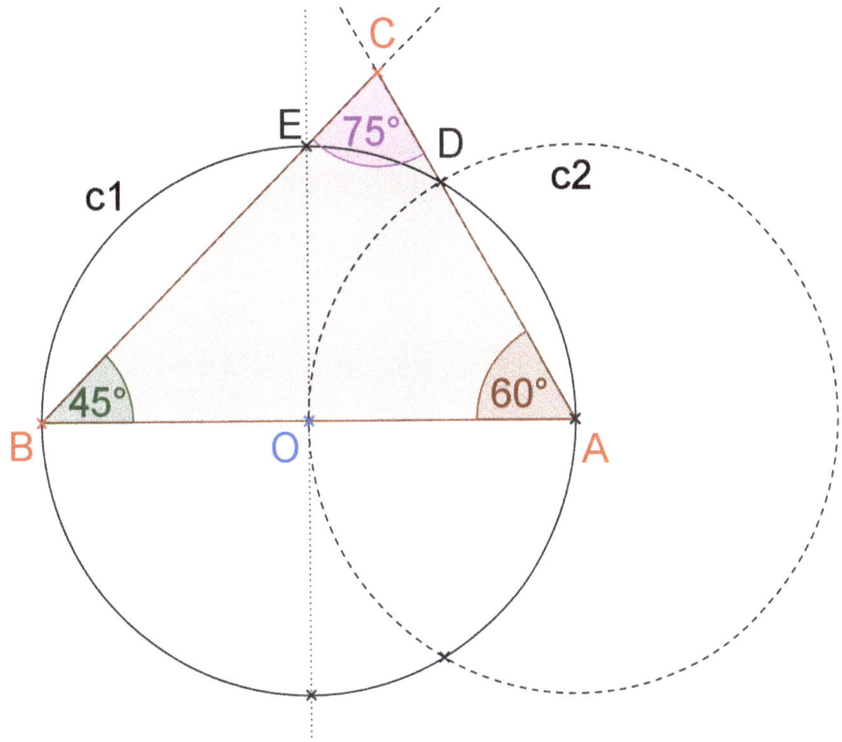

Figure III.18.11: Construction of a triangle with angles of $45°, 60°$ and $75°$.

We thus get a triangle $\triangle ABC$ with the angles $60°, 45°$ and $75°$ at its vertices A, B and C. The angle $\angle CAB = 60°$ is obtained from the equilateral triangle $\triangle OAD$. The angle $\angle ABC = 45°$ is obtained from the right isosceles triangle $\triangle OBE$. Finally, one calculates the third angle $\angle BCA = 180° - 60° - 45°$ by the angle sum of triangle $\triangle ABC$.

Problem III.18.10 (A construction using an altitude). *Using Euclid III.21, construct a* $\triangle ABC$ *with the three following pieces given: side* $c = AB = 6$*, opposite angle* $\gamma = \angle BCA = 30°$*, and altitude* $h_c = 5$*.*
(h_c *is the altitude dropped from vertex* C *onto the opposite side* AB*).*

Question **(a)**. Do the construction and measure your angles α and β.

Answer.

Question **(b)**. Describe the steps for your construction.

Construction III.18.9. *Draw side* $AB = 6$ *and its perpendicular bisector* p*. Let* M *be the midpoint of* AB*. The center* O *of the circum circle lies on the perpendicular bisector. The center angle is double the circumference angle* γ*. Hence* $\angle AOB = 2\gamma = 60°$*, and* $\angle AOM = 30°$*. For the example given, the point* O *is especially easy to find because the* $\triangle AOB$ *is equilateral. Next, we draw the circle around* O *through* A *and* B*. This is the circum circle of* $\triangle ABC$*, on which vertex* C *lies.*

Secondly, vertex C *lies on a parallel* q *to* AB *of distance* $|MD| = h_c = 5$*, because of the given altitude. Hence vertex* C *is an intersection point of this parallel with the circum circle. One may choose any one of these two intersection points.*

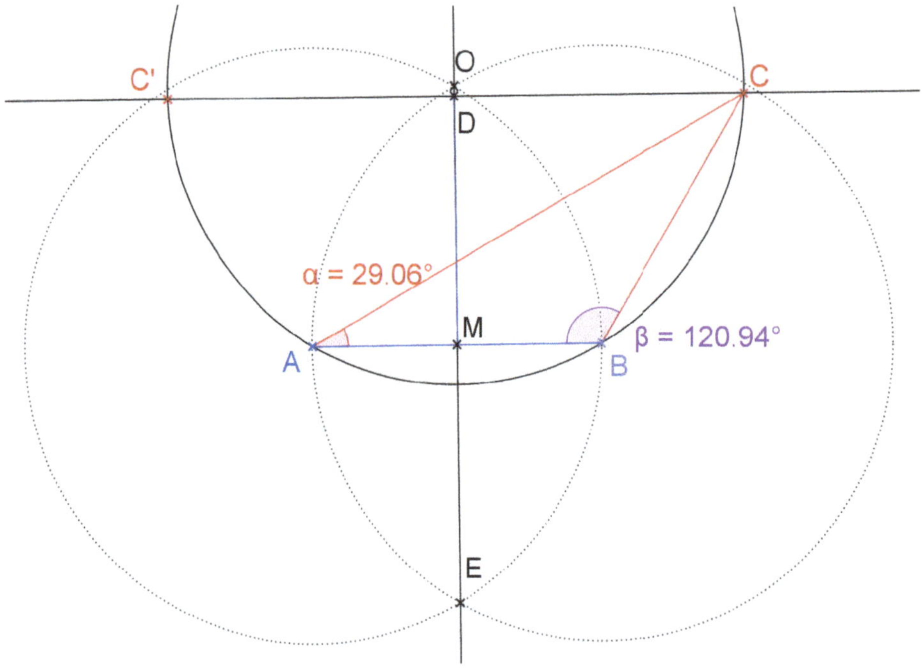

Figure III.18.12: A triangle construction

Remark. One can check that, in Euclidean geometry, the triangle $\triangle ABC$ is <u>not</u> isosceles: Indeed, the isosceles triangle $\triangle ABC'$ with $AB = BC' = 6$ and $\gamma = 30°$ has the altitude

$$h'_a = 6 \cdot \frac{\sqrt{3}}{2} = \sqrt{27} > 5$$

Hence point D lies between M and O, and $\alpha < 30°$.

Problem III.18.11 (A further construction using an altitude). *Using Euclid III.21 as well as the tangent to a circle, construct a triangle $\triangle ABC$ with the three following pieces given: side $c = AB = 6$, opposite angle $\gamma = \angle BCA = 30°$, and altitude $h_a = 4$.*
(h_a is the altitude dropped from vertex A onto the opposite side BC).

Question (**a**). Do the construction and measure your angles α and β.

Answer. The construction is shown in the figure on page 411. The angles are measured to be $\alpha = 108.19°$ and $\beta = 41.81°$.

Question (**b**). Describe the steps for your construction.

Construction III.18.10. *Draw side $AB = 6$ and its perpendicular bisector p. Let M be the midpoint of AB. The center O of the circum circle lies on the perpendicular bisector. The center angle is double the circumference angle γ. Hence $\angle AOB = 2\gamma = 60°$, and $\angle AOM = 30°$. For the example given, the point O is especially easy to find because the $\triangle AOB$ is equilateral. Next, we draw the circle around O through A and B. This is the circum circle $\mathcal{C}$ of $\triangle ABC$, on which vertex C lies.*

Next we draw the circle $\mathcal{A}$ of radius $h_a = 4$ around point A. The triangle side BC lies on the tangent r to this circle from point B. In the figure on page 411, the tangent r has been constructed

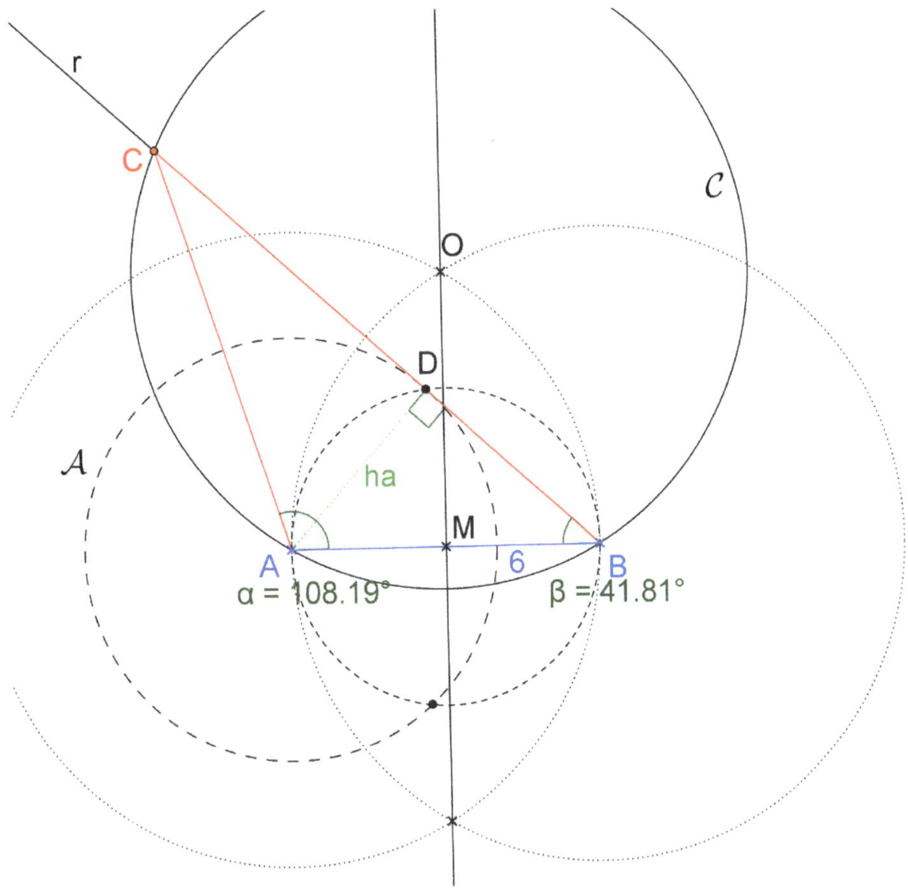

Figure III.18.13: A triangle construction

using the Thales circle with diameter AB. In the figure on page 412, the tangent r has been constructed by Euclid's method.

Vertex C is an intersection point of this tangent r with the circum circle $\mathcal{C}$. In the upper half planennn of AB one gets only one solution.

Problem III.18.12. *Given is a triangle $\triangle ABC$ with its three altitudes. Find in this figure three pairs of congruent angles at the vertices A, B, C.*

Hint: *Draw the circle with a triangle side AB as diameter. Because of the converse Thales' theorem, two foot-points of altitudes lie on this circle. Now use Euclid III.21 to find congruent circumference angles.*

Answer. We draw a semicircle with a triangle side AB as diameter. Because of the converse Thales' theorem, two foot-points D and E of altitudes lie on this circle. By Euclid III.21 , the arc between these two foot points has congruent circumference angles at the vertices A and B. From the triangle sides BC and CA, we get two further pairs of congruent angles.

Problem III.18.13 (A construction using an altitude). *Using Euclid III.21, construct a triangle $\triangle ABC$ with the three following pieces given: side $c = AB = 6$, opposite angle $\gamma = \angle BCA = 60°$, and altitude $h_c = 4$.*
(h_c is the altitude dropped from vertex C onto the opposite side AB).

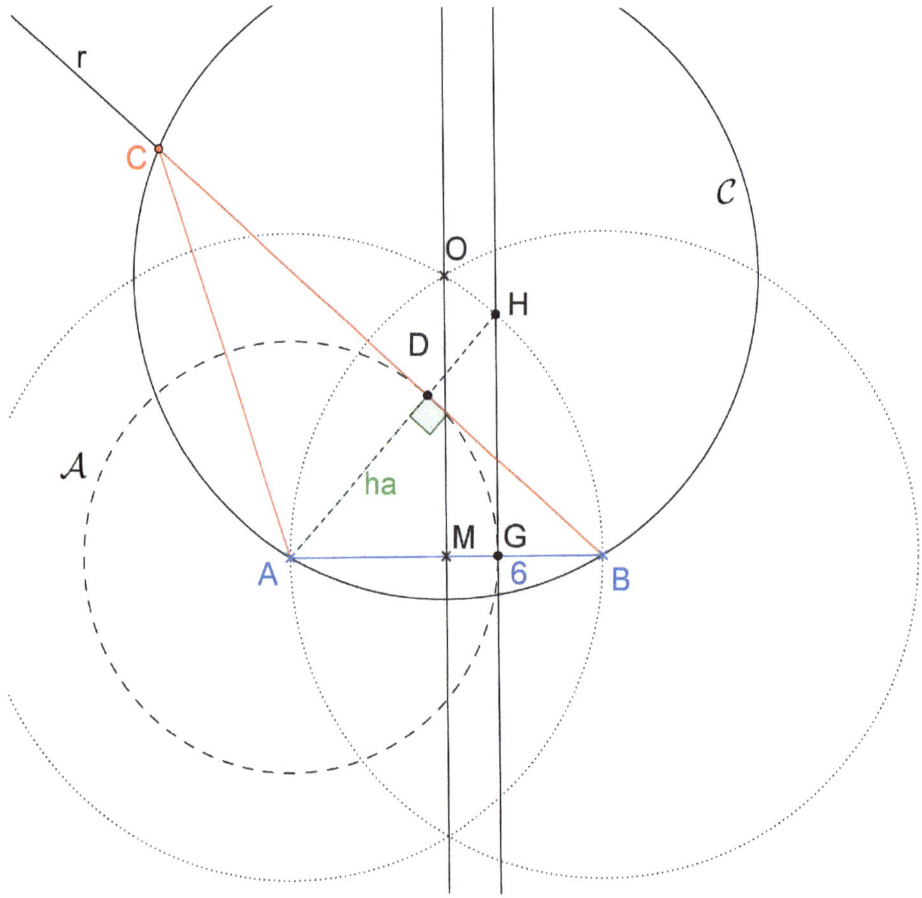

Figure III.18.14: A triangle construction

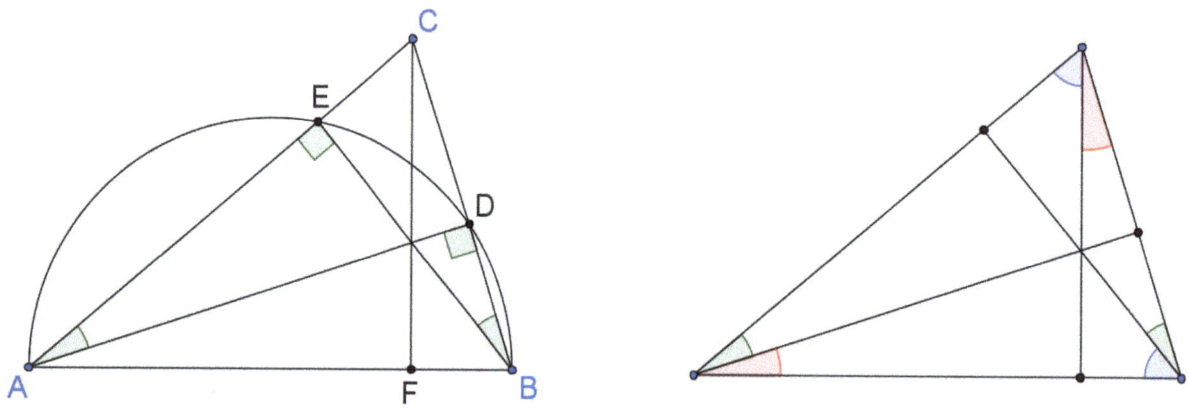

Figure III.18.15: Congruent angles produced by the altitudes of a triangle.

Question (**a**). Do the construction and measure your angles α and β.

Answer. The construction is shown in the figure on page 413. The angles are measured to be $\alpha = 35.49°$ and $\beta = 84.51°$.

Question (**b**). Describe the steps for your construction.

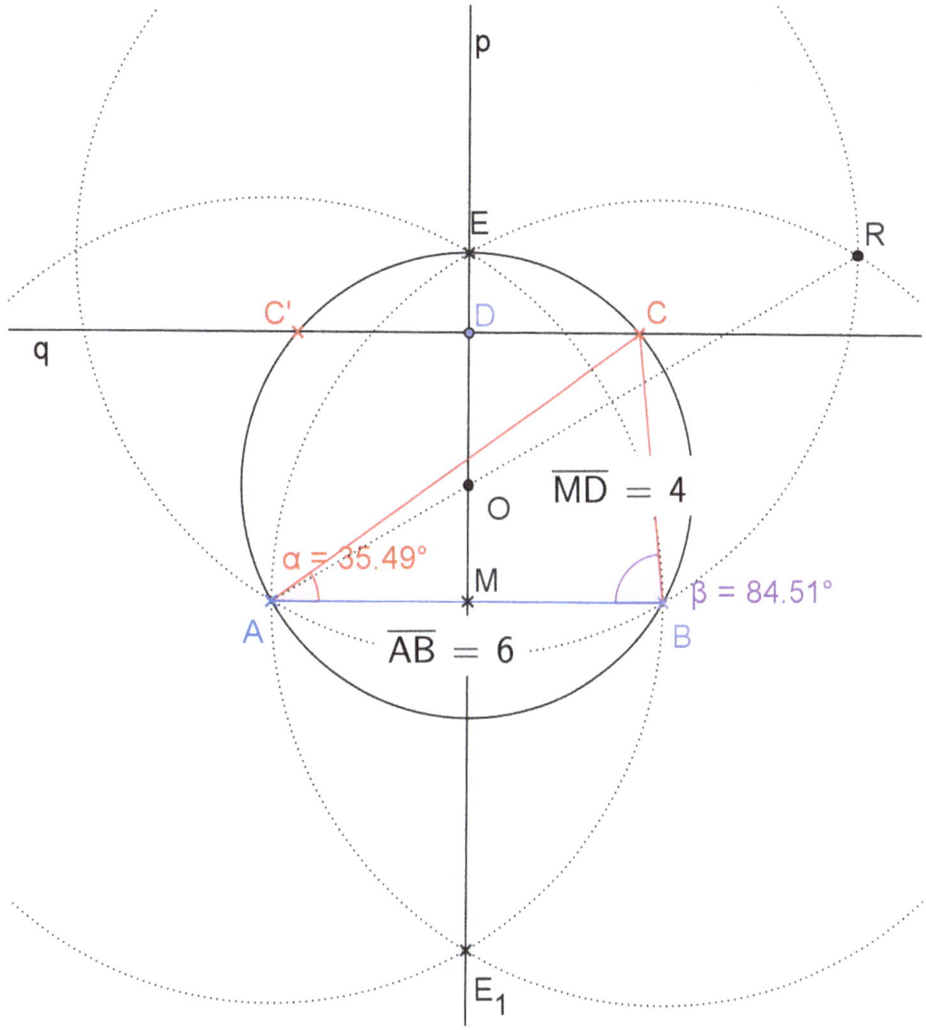

Figure III.18.16: A triangle construction

Answer. Draw side $AB = 6$ and its perpendicular bisector p. Let M be the midpoint of AB. The center O of the circum circle lies on the perpendicular bisector p. Because the value $\gamma = 60°$ is given in the example, we construct at first the equilateral triangle $\triangle AEB$. Next we need the circum circle $\mathcal{C}$ of the equilateral triangle $\triangle ABE$. As follows from Euclid III.21, the vertex C of the triangle $\triangle ABC$ to be constructed lies on the circle $\mathcal{C}$, too .

To get the center of circle $\mathcal{C}$ we draw the circle around E through A and B. It intersects the circle around B through A and E at the points A and R. The line AR is the perpendicular bisector of segment BE, and intersects the perpendicular bisector p of segment AB at point O, which is the center of the circum circle $\mathcal{C}$ of the equilateral triangle $\triangle ABE$. The circum circle $\mathcal{C}$ is now easily drawn.

Secondly, vertex C lies on a parallel q to AB of distance $|MD| = 4$, because of the altitude is given to be $h_c = 4$. The vertex C is an intersection point of this parallel with the circum circle $\mathcal{C}$. In the upper half planenn of AB one gets two solutions. One may choose any one of these two intersection points.

III.18.4 Quadrilaterals

Problem III.18.14. *In Euclidean geometry, find convex quadrilaterals with the interior angles of* $30°, 60°, 120°, 150°$.

(i) *Convince yourself that such quadrilaterals all are either trapezoids or circular quadrilaterals.*

(ii) *We assume additionally that two adjacent sides are congruent. I do not count separately the reflexion symmetric instances, nor congruent or similar instances,—how many different shapes can these quadrilaterals have? Convince yourself that there exist five different shapes for trapezoids, and one shape for a circular quadrilateral.*

(iii) *Give accurate drawings, using a protractor and straightedge, or geogebra.*

Answer. **(i)** If the angles $30°$ and $150°$ are adjacent, one gets a trapezoid, if these two angle are opposite angles, one gets a circular quadrilateral.

(ii) We assume additionally that two adjacent sides are congruent.

- In case the angle of $30°$ includes two congruent sides, the adjacent angle cannot be $60°$. Hence the adjacent angles are $120°$ and $150°$.
- In case the angle of $60°$ includes two congruent sides, the adjacent angle cannot be $30°$. Hence the adjacent angles are $120°$ and $150°$.
- In case the angle of $120°$ includes two congruent sides, the adjacent angle cannot be $30°$. Hence the adjacent angles are $60°$ and $150°$.
- In case the angle of $150°$ includes two congruent sides, all three choices of the adjacent angles among $30°$, $60°$ or $150°$ are possible.

Problem III.18.15. *Given are three points A, B and C. On which curve $\mathcal{D}$ lies the forth vertex of a quadrilateral $\square ABCD$, for which*

(a) *the opposite angles have sum $\alpha + \gamma = 120°$,*

(b) *the opposite angles have sum $\alpha + \gamma = 270°$.*

Again, the vertices A, B, C, D are named in alphabetic order as they follow each other around the quadrilateral.

Actually, the curve $\mathcal{D}$ is a circle. In two instances for the examples (a) *and* (b) *above, construct this circle, and especially its center.*

Answer. Because the angle sum of a quadrilateral is $360°$, the angle at the forth vertex is $\delta = 360° - \alpha - \beta - \gamma$. In case (a), we get $\delta = 360° - \beta - 120°$. This angle can easily be constructed at vertex B since $\beta = \angle ABC$ is given. In the figure on page 416, the equilateral triangle $\triangle BEF$ is used to obtain the angle $\angle ABF = 120°$ and hence the angle $\angle FBC = \delta$. The complementary angle $90° - \delta$ is transferred to the rays $\overrightarrow{AC}$ and $\overrightarrow{CA}$. The intersection of these two rays is the center O of the circle through A and C on which the forth vertex D lies. Indeed, the center angle is $\angle AOC = 2\delta$ and, by Euclid III.21, the circumference angle is $\angle ADC = \delta$ as required.

In case (b), the angle at the forth vertex is $\delta = 360° - \alpha - \beta - \gamma = 90° - \beta$. Hence we transfer angle β to the rays $\overrightarrow{AC}$ and $\overrightarrow{CA}$. The intersection of these two rays is the center O of circle to be constructed.

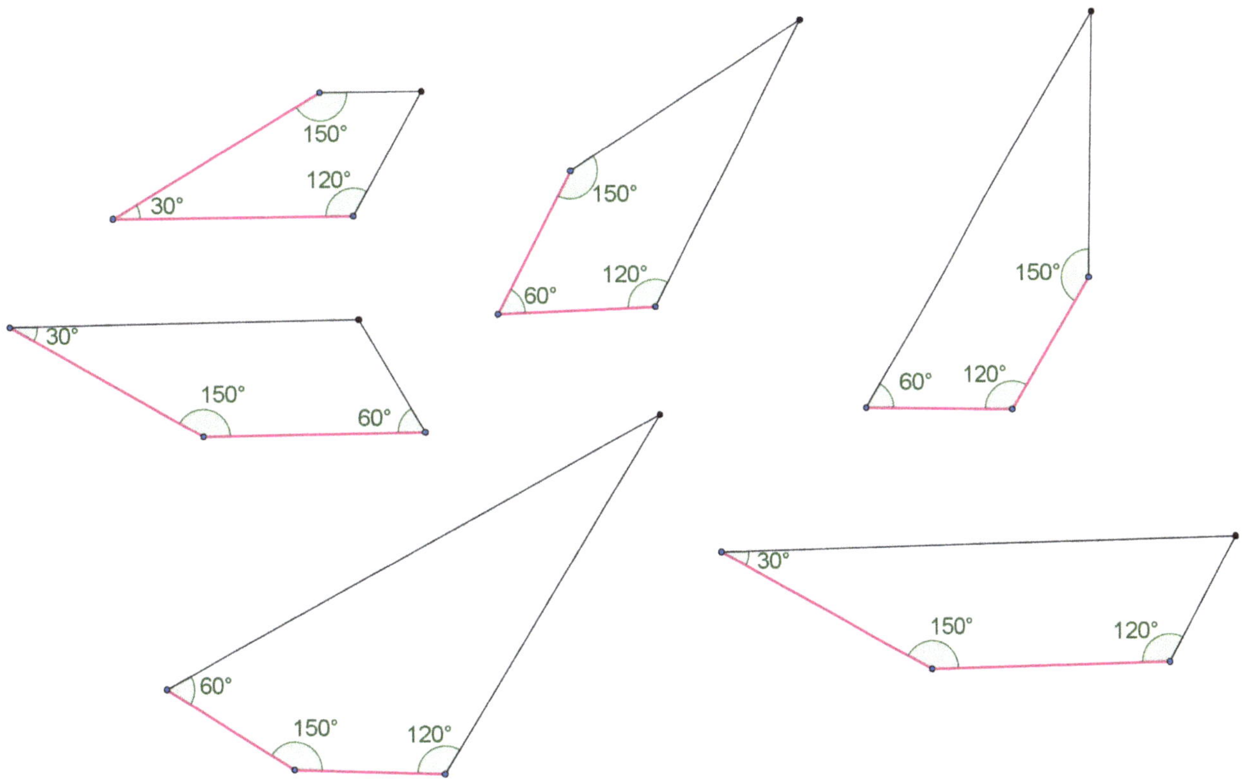

Figure III.18.17: Six quadrilaterals.

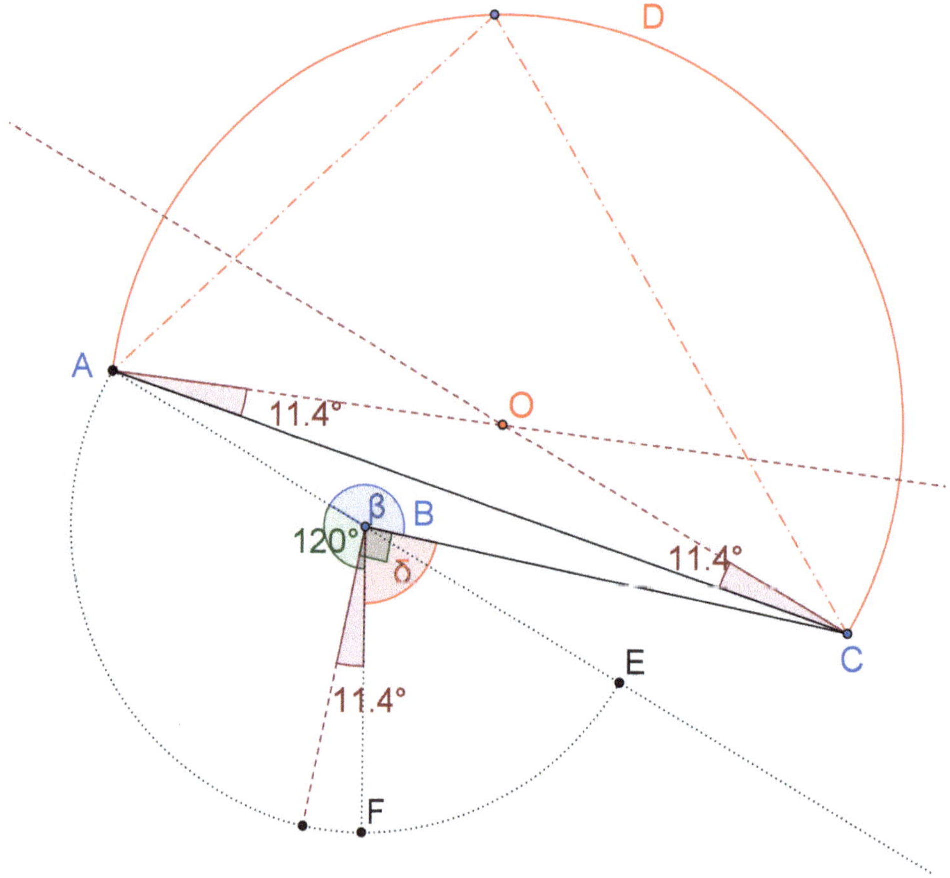

Figure III.18.18: A quadrilateral with $\alpha + \gamma = 120°$ and given vertices A, B, C.

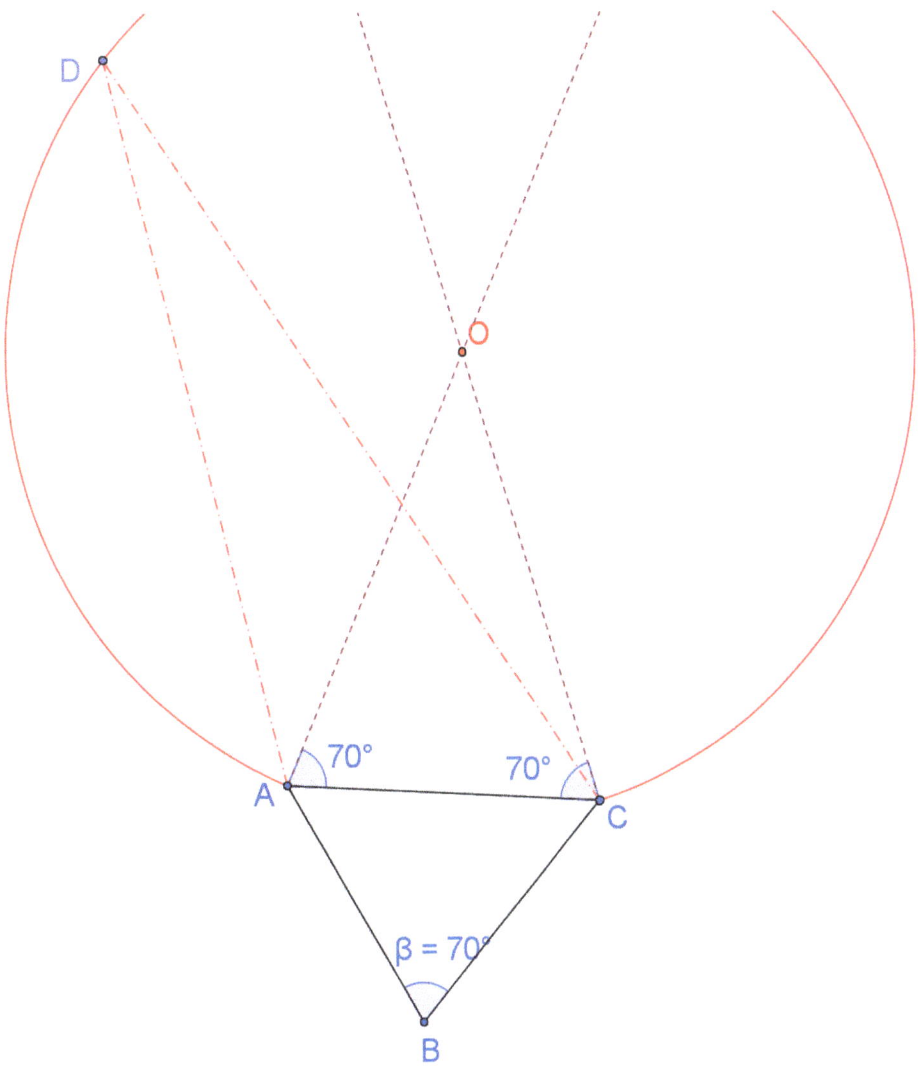

Figure III.18.19: A quadrilateral with $\alpha + \gamma = 270°$ and given vertices A, B, C.

Part IV

From Coordinates to Axioms, and Back

IV.19 Arithmetic of Segments—Hilbert's Road from Geometry to Algebra

In this section, we explain Hilbert's procedure to construct an arithmetic of segments, also called *Streckenrechnung.* Hilbert constructs a field from a given model of geometry and thus opens up a general road from geometry to algebra.

The approach is explained in Hilbert's foundations. It works for an arbitrary affine plane for which the theorem of Pappus holds. Since existence and uniqueness of the parallel is assumed, one can take advantage of similar triangles in the construction of the field of segment arithmetic. Neither one of the axioms of continuity shall be assumed. Depending on the given model of geometry, we obtain a variety of different fields, rather than the real numbers exclusively.

One can even generalize the approach, and assume only the weaker theorem of Desargues instead of Pappus. In this case, one obtains on the side of algebra only a skew field—thus only the associative law still holds, but the commutative law does no longer hold. These important results have already been stated in the Main Theorems 5 and 4. In the context of projective geometry, the commutative law turns out to be equivalent to the theorem of Pappus and the associative law to the theorem of Desargues.

As an alternative, we may avoid the theorem of Pappus and get the segment arithmetic more directly from angles in a circle via Euclid's III.21. This approach is used in Hartshorne's book and presented here in an own variant. It works in every Pythagorean plane and breaks any connection with ideas from projective geometry. Recall the definitions:

Definition (I.2.2 Pythagorean plane). A *Pythagorean plane* is a Hilbert plane for which the Euclidean parallel postulate holds.

Definition IV.19.1 (Pythagorean field). A *Pythagorean field* is an ordered field with the property

(Pyth) If $a, b \in \mathbf{F}$, then $\sqrt{a^2 + b^2} \in \mathbf{F}$.

Main Theorem 20. *Any affine plane for which the Theorem of Pappus holds is isomorphic to the Cartesian plane over a field. Conversely, the Theorem of Pappus holds in a Cartesian plane over a field.*

The first result is elaborated in this section in some detail. The second statement repeats part of the Main Theorem 4, and can be checked directly with some rather tedious calculations which we leave out. Furthermore, this section gives the detailed direct proof of the following result:

Theorem IV.19.1. *Any Pythagorean plane is isomorphic to the Cartesian plane* $\mathbf{F}^2$ *over its field* $\mathbf{F}$ *of segment lengths. Furthermore, this is a Pythagorean field.*

IV.19.1 Fields

Definition IV.19.2 (Field). A *field* is a system of undefined elements (symbols, numbers), two binary relation symbols $+$ and $\cdot$ and two constants 0 and 1 with the following properties:

(1) (i) For any two numbers a and b exists the *sum* $a + b$.

(ii) $(a+b)+c = a+(b+c)$ for any a, b, c.

(iii) $a+b = b+a$ for any a, b.

(iv) $a+0 = a$ for any a.

(v) for each number a there exists a number $-a$ such that $a+(-a) = 0$.

(2) (i) For any two numbers a and b exists the *product* $a \cdot b$.

(ii) $(a \cdot b) \cdot c = a \cdot (b \cdot c)$ for any a, b, c.

(iv) $a \cdot 1 = 1 \cdot a = a$ for any a.

(v) for each number $a \neq 0$ there exists a number a^{-1} such that $a \cdot a^{-1} = a^{-1} \cdot a = 1$.

(3) (i) $0 \neq 1$.

(ii) $(a+b) \cdot c = a \cdot c + b \cdot c$ for any a, b, c.

(iii) $c \cdot (a+b) = c \cdot a + c \cdot b$ for any a, b, c.

(4) $a \cdot b = b \cdot a$ for any a, b.

A system of elements satisfying only (1)(2)(3) is called a *skew field.*

Any model for the field axioms is simply called a field, too. In this context, we denote by $\mathbf{F}$ the set of elements. Clearly $0, 1 \in \mathbf{F}$.

IV.19.2 Construction of the field of segment arithmetic

To relate Euclidean geometry to algebra, we need to construct a field from the congruence classes of segments.

One can define the sum of any two segments by putting them together along a line. Especially, if point B lies between A and C, we define $AC = AB + BC$. More generally, the sum of any two segments is defined by Definition II.7.2. Because of Hilbert's postulate III.3, the addition of segments defines even an addition for the congruence classes of segments. As already shown in Proposition II.7.7 in the section about congruence of segments, angles and triangles, we get an ordered Abelian group from these congruence classes, their negatives and the zero element.

In the theorem 10 about the measurement of segments, we have already used a *unit segment.* To this unit segment, say OE, we assign the length 1. The construction of the length for any segment as an (finite or infinite) binary fraction done in the measurement theorem 10 depends on the Archimedean axiom.

Taking the Archimedean axiom for granted, we could simply define the product and the quotient of segment via the product and quotient of the corresponding binary fractions. Although this "old segment arithmetic" is very practical (and usually tacitly assumed in high school), it depends essentially on the Archimedean axiom, and the possibility to approximate irrational numbers by rational numbers. Only to <u>some</u> very special segments, a finite binary or proper fraction can be assigned as length. Hence the old segment arithmetic is not suitable for a general constructive definition of the multiplication of the field.

The essential step for the "new segment arithmetic" claimed by Theorem IV.19.1 is the correct definition of the product of two segments. This is done using the legs of equiangular right triangles. Thus all dependence on the axioms of continuity can be avoided.

Remark. Hilbert even shows in detail how the assumption of any axioms of congruence can be avoided, if instead the theorem of Desargues is assumed. We can leave these considerations aside, and define the sum of segments still via Definition II.7.2.

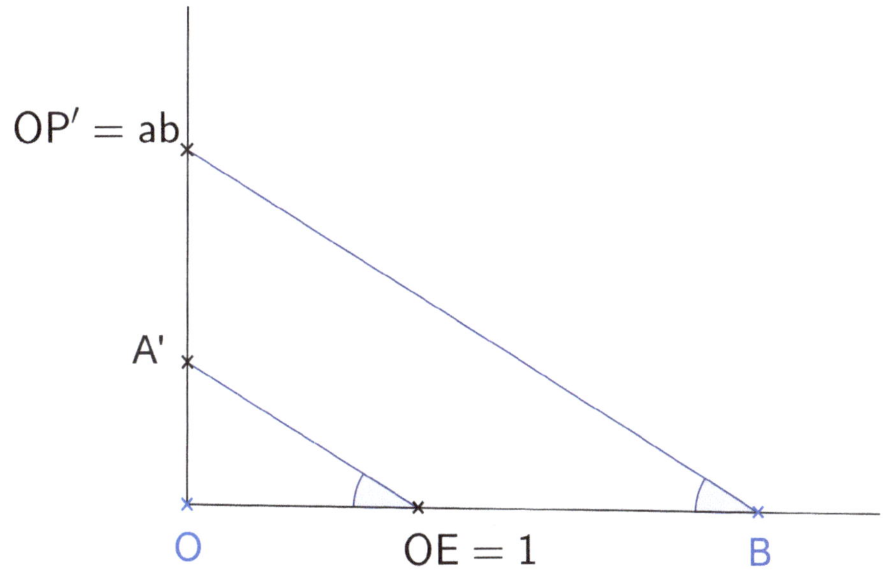

Figure IV.19.1: Construction of the product ab for given segments $|OA'| = a$, $|OB| = b$ and $|OE| = 1$.

To be remembered. For the definition of the product ab, we use two similar right triangles. The first triangle has the legs the <u>first factor</u> a and 1, the second one has as its legs ab and the <u>second factor</u> b. □

Look at the figure on page 423. On the perpendicular ray through O, we have put the segment with prescribed length

$$|OA'| = a$$

On the horizontal ray emanating from vertex O to the right, there are put points O, E and B such that the segments measured from the origin O have the prescribed lengths:

$$|OE| = 1\,, \ |OB| = b$$

Now we draw the parallel to line EA' through point B. Why does this parallel does intersect the vertical line? Indeed, by Proclus' Lemma I.3.3, a line intersecting one of two parallel lines intersects the other one, too. Since the vertical line intersects line EA', it intersects its parallel through B, too. Hence we get the unique intersection point P' and <u>define</u> the product ab as length of the segment OP':

$$ab := |OP'|$$

Indeed, we have used the similar right triangles $\triangle OEA'$ and $\triangle OBP'$.

Question. How does the order of the factors a and b become relevant for the construction?

Answer. The first factor $a = |OA'|$ and the unit segment $|OE| = 1$ appear *together as legs of one triangle* $\triangle OEA'$. The second factor b determine the segment OB' through one end of which one draws the parallels $EA' \parallel BP'$.

Indeed, the product of segments a and b turns out to have the property that the rectangle with sides a and b has the same area as the rectangle with sides 1 and ab.

Remark. The definition of area does not automatically imply commutativity. On the contrary, the usual simple definition of the area of the rectangle as the product of its sides is only possible because of commutativity. [50]

The first obvious question is whether our definition given above makes the product commutative. Furthermore, we need to go on and check whether the remaining axioms required for a field are satisfied, and whether we get an ordered field.

Commutative law

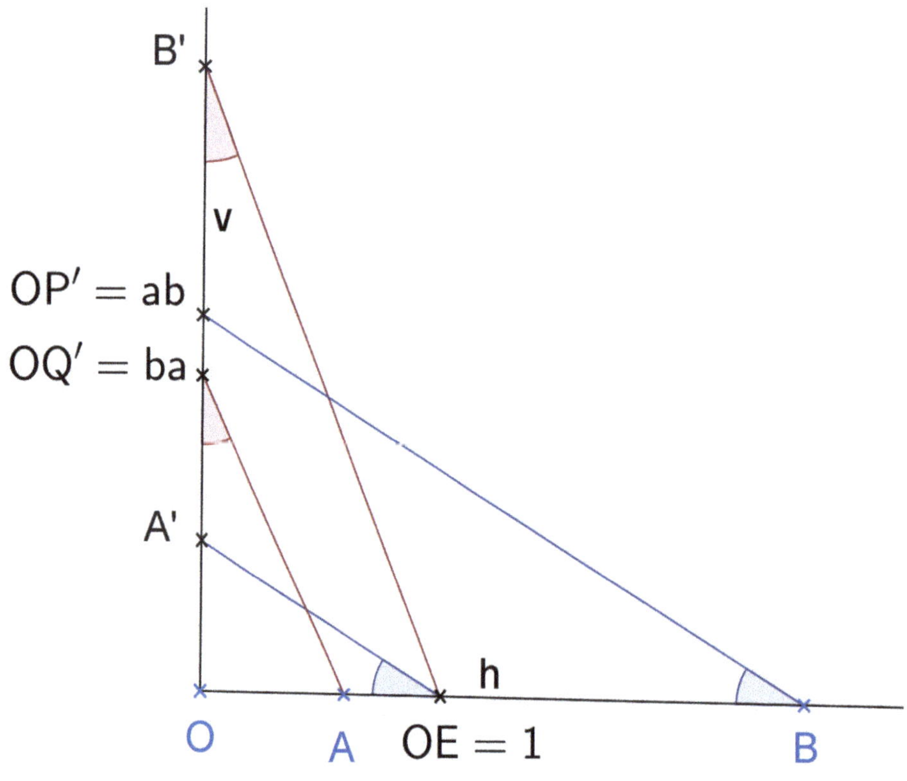

Figure IV.19.2: Construction of the products ab and ba, with an empirical error exaggerated.

To check the commutative law, we repeat the above construction, with the roles of a and b exchanged. We construct point A on the horizontal line such that

$$|OA| = a$$

and a point on the perpendicular line such that

$$|OB'| = b$$

[50] It would be interesting to find out in a psychological study who really knows that!

Now we draw the parallel to line EB' through point A, and get the intersection point Q'. By definition ba is length of the segment OQ':

$$ba = |OQ'|$$

As you can see from the figure on page 424, it is now really a question from geometry, whether $ab = ba$!

Proposition IV.19.1. *The segment arithmetic is commutative in any affine plane where the Theorem of Pappus holds.*

Proof. The product ab is constructed as in the figure on page 424, by the pair of parallels

$$EA' \parallel BP'$$

Because of the isosceles and equiangular triangles $\triangle OAA'$ and OBB', we get another pair of parallel lines $AA' \parallel BB'$. The hexagon $AP'BB'EA'$ and the horizontal and vertical lines h and v are a Pappus configuration.

Remark. To get the three pairs of opposite sides of the hexagon, it is convenient to shift the points cyclically by three:

$$\begin{array}{cccccc} A & P' & B & B' & E & A' \\ B' & E & A' & A & P' & B \end{array}$$

The three pairs of opposite sides of the hexagon appear underneath each other:

$$\begin{array}{ccc} AP' & P'B & BB' \\ B'E & EA' & A'A \end{array}$$

The Pappus configuration contains the two pairs of parallel opposite sides

$$EA' \parallel BP'$$
$$AA' \parallel BB'$$

Hence, by the theorem of Pappus, there is a third pair of parallels

$$EB' \parallel AP'$$

The product ba is constructed as in the figure on page 424, by the pair of parallels

$$EB' \parallel AQ'$$

Both $EB' \parallel AQ'$ and $EB' \parallel AP'$, and both P' and Q' lie on the vertical line v. The uniqueness of the parallels and the uniqueness of intersection point implies $P' = Q'$. Hence $ab = |OP'| = |OQ'| = ba$, as to be shown. □

Remark. Conversely, the commutative law of multiplication, together with the theorem that equiangular triangles have proportional sides, can be used to get Pappus theorem in the simplified version.

Remark. In this and the following proofs, we consider only the case of positive quantities a, b, c. It is left to the reader to check that the corresponding results hold for positive as well as negative quantities.

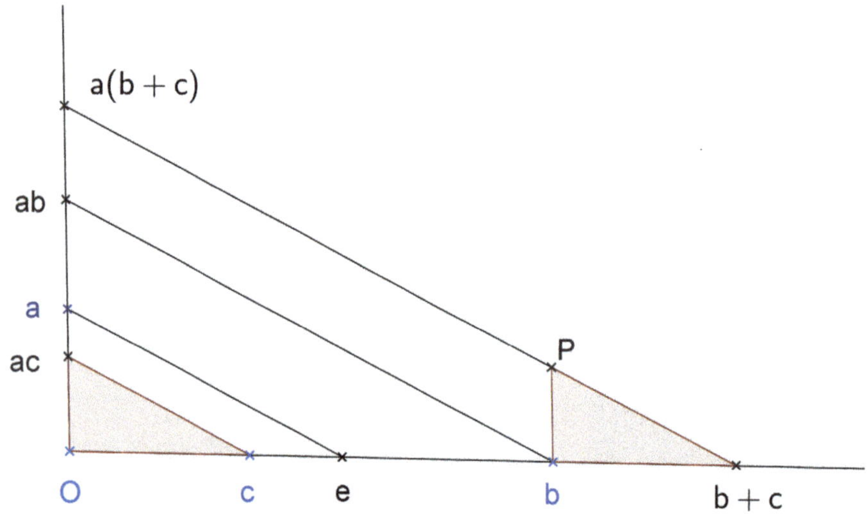

Figure IV.19.3: Proof of the distributive law $ab + ac = a(b + c)$.

The distributive law

Proposition IV.19.2. *The segment arithmetic is distributive in any Pythagorean plane.*

Proof. Given are the segment lengths a, b, c, which we may assume to be positive. We use two perpendicular rays with the common origin O, and transfer the segment a onto one of them,—the vertical one in the drawing; and the two other segments b and c and the unit segment $[O, e]$ onto the other one,—the horizontal one in the drawing.

Next we construct the segment $[b, b + c]$, by transferring the segment $[O, c]$ onto the ray with vertex b on the horizontal axis in order to get segment $[O, b + c]$. The construction of the products ab, ac, and $a(b+c)$ is shown in the figure on page 426. In all three cases, we use parallels to the line EA', where A' is the point on the vertical axis with $a = |OA'|$. These parallels go through points b, c and $(b + c)$ on the horizontal axis, respectively. On the vertical axis, we get the intersection points ab, ac and $a(b + c)$, the distance of which to the origin define the respective products.

Finally, we draw the parallel to the vertical axis through point b, and get intersection point P with the line through points $b + c$ and $a(b + c)$. By ASA congruence, the triangles $\triangle Ob(ab)$ and $\triangle b(b + c)P$ are congruent, since they are equiangular and have one pair of congruent sides. Hence the segments $[O, ac]$ and $[b, P]$ are congruent. Furthermore, the segments $[b, P]$ and $[ab, a(b + c)]$ are congruent since they are opposite sides of a parallelogram. Hence $[O, ac] \cong [ab, a(b + c)]$.

By definition of segment addition, which is now done on the vertical axis, we know that $[O, ac] \cong [ab, ab + ac]$. Hence we finally get $ab + ac = a(b + c)$, as to be shown. □

The associative law

Proposition IV.19.3. *The segment arithmetic is associative in any affine plane where the Theorem of Pappus holds.*

Proof. From given segments a and c on the vertical line, and $e = 1$ and b on the horizontal line, we construct at first segments ab and cb on the vertical line. These two segments are transferred onto

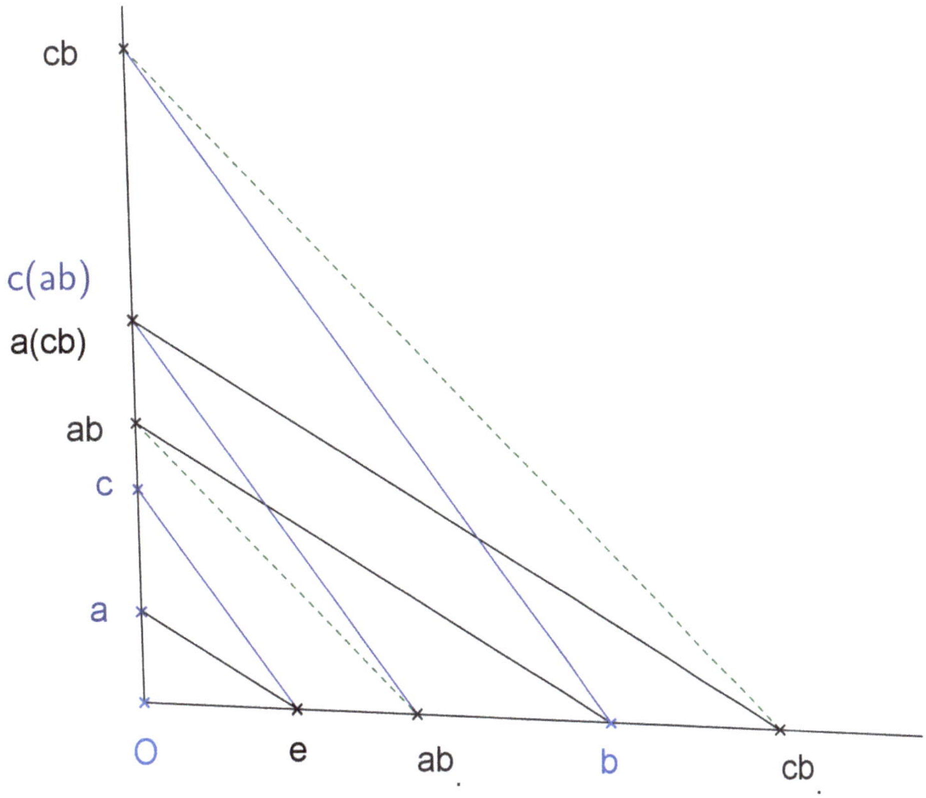

Figure IV.19.4: The proof of the associative law yields at first $c(ab) = a(cb)$.

the horizontal axis, using 45°-lines. Finally, one gets, both on the vertical line, segment $c(ab)$ from the parallels $[e, c] \parallel [ab, c(ab)]$ and segments $a(cb)$ from the parallels $[e, a] \parallel [cb, a(cb)]$.

The three points ab, b and cb on the horizontal axis and the three points ab, $c(ab)$ and cb on the vertical axis give a Pappus configuration with two, and hence three pairs of parallels. We conclude that $a(cb) = c(ab)$. By applying the already proved commutative law twice, we get

$$a(bc) = a(cb) = c(ab) = (ab)c$$

□

A direct proof of commutativity of segment arithmetic

Proposition IV.19.4. *In any Pythagorean plane, the segment arithmetic is commutative.*

Remark. The following proof of commutativity for the segment arithmetic does not presuppose Pappus' theorem. As a prize, we have to restrict our attention to a Pythagorean plane. We use Euclid III.21 about angles in a circle directly. Thus we get a proof that commutativity of segment arithmetic holds in every Pythagorean plane.

Proof. Given are the positive lengths $a > 0$ and $b > 0$, and a unit segment OE. We take the line OE as horizontal axis, the ray $\overrightarrow{OE}$ as the positive x-axis. At point O the perpendicular is erected to obtain the vertical axis.

The reader should look at the figure on page 428 to see how the three points A_-, E'_- and B are obtained: The given positive segment a is transferred to the vertical axis to obtain a point A' such

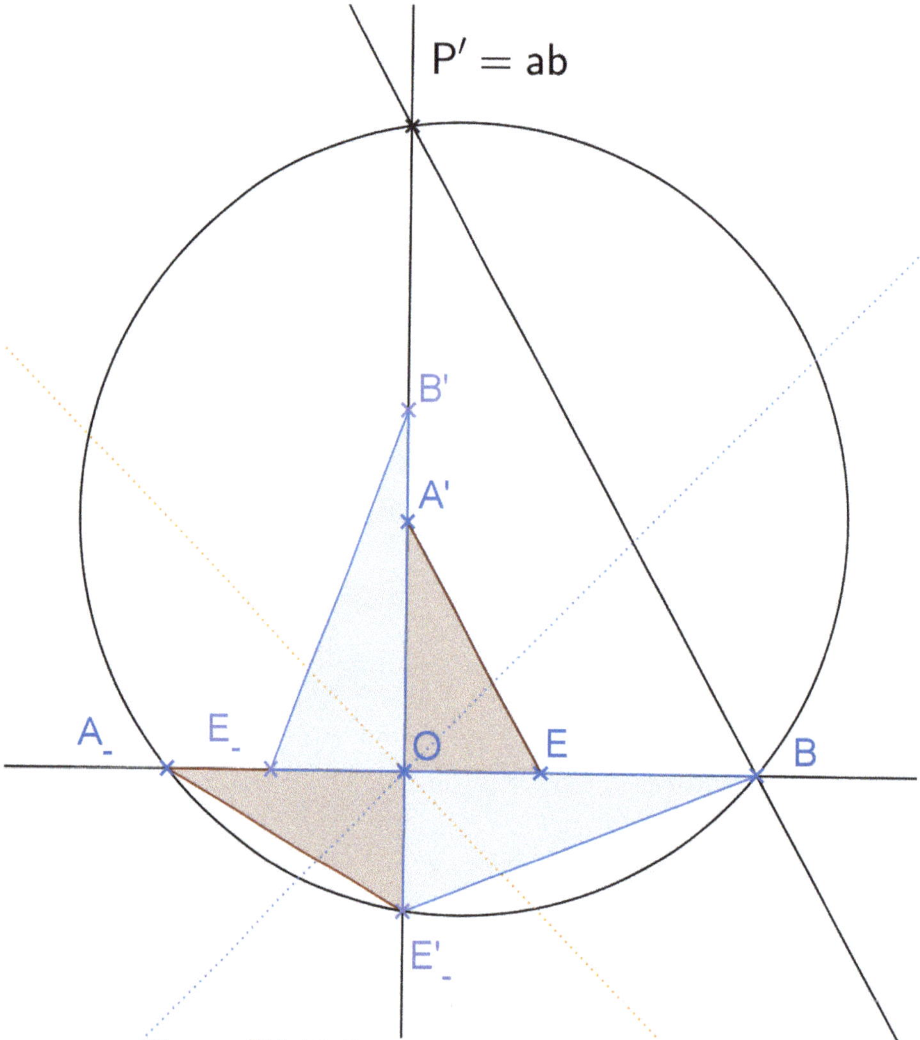

Figure IV.19.5: Prove commutativity $ab = ba$.

that $|OA'| = a$. The same segment a is transferred to the negative horizontal axis to obtain a point A_- such that $A_- * O * E$ and $|OA_-| = a$. The unit segment is transferred to the negative vertical axis to obtain a point E'_- such that $E'_- * O * A'$ and $|OE'_-| = 1$. The segment b is transferred to the positive horizontal axis to obtain a point B such that $|OB| = b$. The three points A_-, E'_- and B do not lie on a line. By theorem 35, in a Pythagorean plane every triangle has a circum-circle. Hence there *exists a circle* through points A_-, E'_- and B. Indeed, this is the only circle used for the proof.

The origin of coordinates O lies in the segment A_-B and hence in the interior of this circle. (Clearly point O need not be the center of the circle.) By construction, the vertical axis intersects the circle at point E'_- and goes through the point O in the interior of the circle. Hence it is easy to prove that the vertical axis intersects the circle at second point P' for which $E'_- * O * P'$. Note that no circle-line postulate is needed to get point P'.

Question. Recall how the product ab has been constructed.

Answer. As explained by the figure 423, the construction of the product ab uses two similar right triangles. The first triangle has the legs the <u>first factor</u> a and 1, the second one has as its legs ab and the <u>second factor</u> b. To get these triangles, the given positive segment a has been transferred to the positive vertical axis to obtain a point A' such that $|OA'| = a$. To obtain the product ab, we draw the parallel to line EA' through point B and get its intersection point Y with the vertical axis. The product is given by the length $|OY| = ab$ of the segment from the origin.

Question. Why is the parallel obtained for free, — and indeed $Y = P'$?

Answer. We reflect the triangle $\triangle OA'E$ across the $-45°$ line to produce the triangle $\triangle OA_-E'_-$. The triangles $\triangle OA'E \cong \triangle OA_-E'_-$ are congruent, as one easily checks by SAS-congruence. Hence the angles $\angle OA'E \cong \angle OA_-E'_-$ are congruent, too.

The angles $\angle BA_-E'_- \cong \angle BP'E'_-$ are both circumference angles of the arc E'_-B, and both lie on the same side of the chord E'_-B. Hence they are congruent by Euclid III.21. Both steps together show the congruence

$$\angle OA'E \cong \angle OA_-E'_- = \angle BA_-E'_- \cong \angle BP'E'_- = \angle OP'B$$

By Euclid I.27, "congruent z-angles imply parallels". (See proposition II.7.42.) In a Pythagorean plane, the parallel is *unique*. Hence the line $P'B$ is the *unique* parallel to line EA' through point B, which is to be used in the construction of the product ab. The intersection point Y with the vertical axis is equal to point P'.

Question. Explain how the product ba has to be constructed, in the left upper quadrant.

Answer. The triangle $\triangle OBE'_-$ is reflected across the $+45°$ line. We obtain the triangle $OB'E_-$, needed to construct the product ba. We draw the line through A'_- parallel to E_-B', and get its intersection point with the vertical axis at point Z with distance $|OZ| = ba$ from the origin.

Question. Explain why, because of these constructions, there appear three congruent angles β.

Answer. The angles $\angle OB'E_- \cong \angle OZA_-$ are congruent because the vertical axis traverses a pair of parallel lines. The angles $\angle OBE'_- \cong \angle OB'E_-$ are congruent because of SAS congruence of the corresponding triangles. We have marked the three congruent angles by β.

Question. Explain why the circle through A_-, E'_- and B—already drawn—goes through the point Z.

Answer. The angles $\angle E'_-BA_- \cong \angle E'_-P'A_-$ are circumference angles of the same arc, and lie both on the same side of its chord E'_-A_-. Hence they are congruent. From

$$\angle E'_-P'A_- \cong \angle E'_-ZA_-$$

we see that point $P' = Z$ lies on this circle.

Since the circle has only one intersection point P' with the vertical ray $\overrightarrow{OA}$, we conclude that $Y = Z = P'$. Hence we have obtained that both ab and ba is this intersection point, we conclude that $ab = ba$. □

Remark. In the drawing from a former version—which was not from Leibniz' "best of all worlds"—the construction did not come out right. In such a case, would Leibniz have trusted the proof or the drawing?

As an additional benefit, we state:

Corollary 47. *The Theorem of Pappus holds in every Pythagorean plane.*

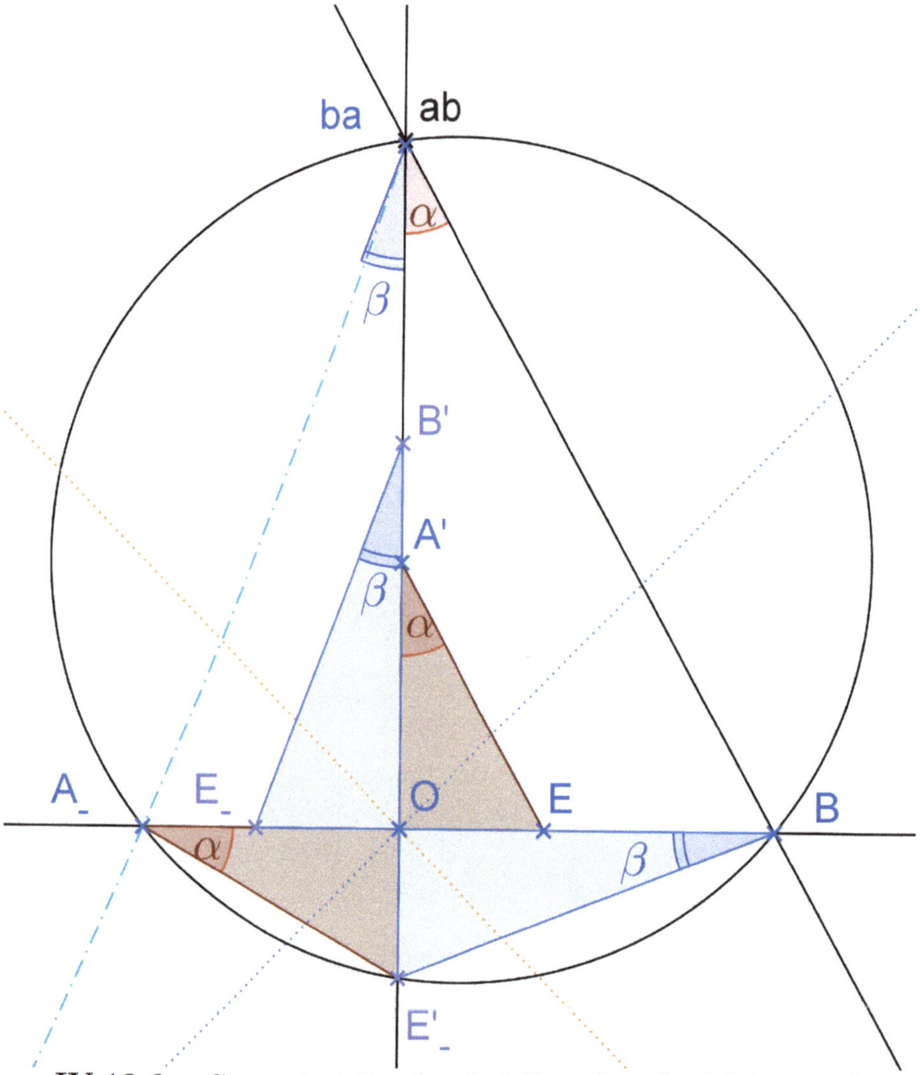

Figure IV.19.6: Commutativity $ab = ba$ follows from both lying on the circle.

Alternative proof of associativity

The given positive segments $|OA| = a > 0$ and $|OB| = b > 0$ are transferred to the vertical axis. The segments $|OE| = 1$ and $|OC| = c > 0$ are transferred to the horizontal axis. The products ac and bc are constructed on the vertical axis.

Problem IV.19.1. *One goes on, almost as above, with E, A, B replaced by C, AC and BC. The angles α and β each appear four time. Explain in detail how the figure on page 431 yields a proof of $a(bc) = b(ac)$ with any three positive $a, b, c > 0$, and thus confirms the associative law.*

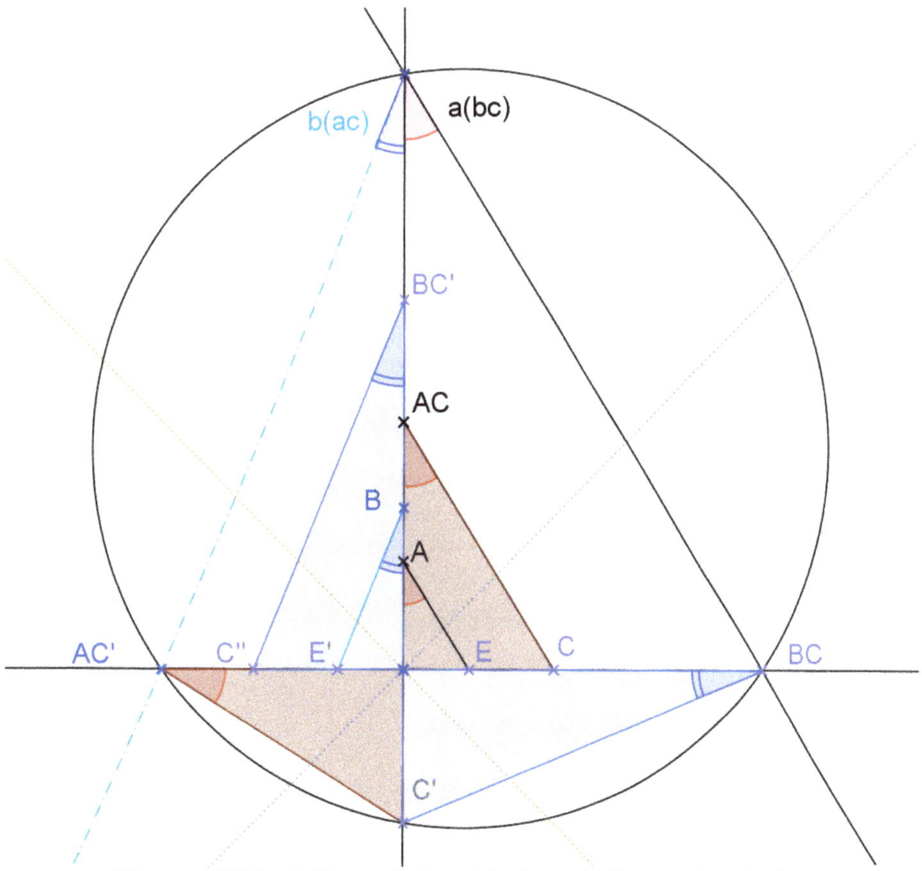

Figure IV.19.7: $a(bc) = b(ac)$ both lie on the circle.

IV.20 Coordinates are Descartes' Road from Algebra to Geometry

The Cartesian plane over a field $\mathbf{F}$ takes as coordinates the elements of an arbitrary finite or infinite field $\mathbf{F}$— in place of the real number coordinates for the real Cartesian plane. We recall from definition I.3.15:

- The "`points`" of the Cartesian plane are ordered pairs (x, y) of elements $x, y \in \mathbf{F}$.
- The "`lines`" of the Cartesian plane of the field $\mathbf{F}$ are equations

  $$ax + by + c = 0$$

  with coefficients a, b, c from the field, of which a and b are not both zero.
- A "`point lies on a line`" if and only if the coordinate pair (x, y) satisfies the equation of the line.

One defines the slope of a line in the usual way. Lines are parallel if and only if they have the same slope. To find the parallel to a given line through a given point, one uses the point slope equation of a straight line. This procedure shows that the Euclidean parallel property holds. Leaving the details aside, we repeat the results:

(Theorem I.3.2). *In a Cartesian plane over any field, there exist a unique line between any two points. There exists a unique parallel to a line through a given point. Hilbert's axioms* (I.1)(I.2)(I.3a)(I.3b) *and the Euclidean parallel postulate hold. Hence a Cartesian plane over a field is an affine plane.*

The Main Theorem 4 implies the converse of the elementary Theorem I.3.2. Together, there is the sharp result:

(Theorem I.4.5). *Any affine plane is isomorphic to a plane over a field if and only if the Theorem of Pappus holds in its projective completion.*

IV.20.1 A Hierarchy of Cartesian planes

Definition IV.20.1 (Ordered field). An *ordered field* is a field with an order relation $<$ having the following properties:

(i) For any two numbers a and b, one of the three relations holds: either $a < b$ or $b < a$ or $a = b$.

(ii) For any two numbers a and b, no more than one of the three relations $a < b$ or $b < a$ or $a = b$ hold.

(iii) $a < b$ and $b < c$ imply $a < c$, for any a, b, c.

(iv) $a < b$ implies $a + c < b + c$, for any a, b, c.

(iv) $a < b$ and $0 < c$ imply $a \cdot c < b \cdot c$ and $c \cdot a < c \cdot b$, for any a, b, c.

Every ordered field contains the rational numbers as a subfield: $\mathbf{Q} \subseteq \mathbf{F}$. Hence every ordered field has the characteristic zero, and especially it is infinite. The points of a plane over an *ordered* field have a natural order relation. It turns out that Hilbert's relevant axioms of order hold, too.

Definition IV.20.2. A point $B = (b_1, b_2)$ lies between a point $A = (a_1, a_2)$ and a point $C = (c_1, c_2)$ if and only if

$$b_i = \lambda a_i + (1 - \lambda) c_i \quad \text{for } i = 1, 2 \text{ with } 0 < \lambda < 1. \tag{IV.20.1}$$

Proposition IV.20.1. *In a Cartesian plane over an* ordered *field, Hilbert's axioms of order* (II.1)(II.2)(II.3)(II.4) *hold.*

Proposition IV.20.2. *Conversely, assume that the axioms of order* (II.1)(II.2)(II.3)(II.4) *hold in any Cartesian plane constructed over some field. Then one can introduce into this field an order compatible with the order of the plane.*

Remark. There exists an affine incidence plane for which the Euclidean parallel property (IV*), the axioms of order (II.1) through (II.4) all hold, but the Theorems of Desargues and Theorem of Pappus are both violated. In 1902, F.R. Moulton [12] has described such an example in his article *"A simple non-desarguesian plane geometry"*. Because of Theorem I.4.5 this is an example of an ordered Euclidean incidence plane that is not a Cartesian plane over a field.

We are now ready to introduce congruence. From a Pythagorean field, we reconstruct a Pythagorean plane as the analytic geometry over the field. This construction works for an arbitrary Pythagorean field, and thus we continue our road from algebra to geometry. The introduction of Cartesian coordinates turns out to be a perfect way to produce many different models for Pythagorean planes. Recall the definitions:

Definition (I.2.2 Pythagorean plane). A *Pythagorean plane* is a Hilbert plane for which the Euclidean parallel postulate holds.

Definition (IV.19.1 Pythagorean field). A *Pythagorean field* is an ordered field with the property

(Pyth) If $a, b \in \mathbf{F}$, then $\sqrt{a^2 + b^2} \in \mathbf{F}$.

This section gives the detailed direct proof of the following result:

Theorem IV.20.1. *The Cartesian plane over a Pythagorean field is a Pythagorean plane.*

The converse result have already been obtained in the section on Hilbert's arithmetic of segments. By Theorem IV.19.1, any Pythagorean plane is isomorphic to the Cartesian plane over its field of segment lengths, which is an Pythagorean field. Taking both results together, we state the important result:

Main Theorem 21 (Fundamental Theorem about Pythagorean Geometry). *The Cartesian plane* $\mathbf{F}^2$ *over a Pythagorean field* $\mathbf{F}$ *is a Pythagorean plane.*

Conversely, any Pythagorean plane is isomorphic to the Cartesian plane over its field of segment arithmetic, which is a Pythagorean field.

Hence there is a bijection between the Pythagorean planes and the Pythagorean fields.

The proof of the first part uses the fact that enough rigid motions exist in a Cartesian plane over a Pythagorean field. For the proof of the second part, one constructs the field with Hilbert's new arithmetic of segments *(Streckenrechnung)* as explained in the section "Arithmetic of Segments—Hilbert's Road from Geometry to Algebra".

Our next step towards more and more familiar geometry is to add some axiom justifying the classical use of ruler and compass.

Definition IV.20.3 (Euclidean plane). A *Euclidean plane* is a Hilbert plane for which both the Euclidean parallel postulate and the circle-circle intersection property hold.

Definition IV.20.4 (Euclidean field). A *Euclidean field* is an ordered field with the following property:

(3) If $a \in \mathbf{F}$ and $a \geq 0$, then $\sqrt{a} \in \mathbf{F}$.

Putting results together we get what I would be willing to name the fundamental theorem of Euclidean geometry:

Main Theorem 22 (Fundamental Theorem about Euclidean Geometry). *The Cartesian plane* $\mathbf{F}^2$ *over a Euclidean field* $\mathbf{F}$ *is a Euclidean plane.*

Conversely, any Euclidean plane is isomorphic to the Cartesian plane over its field of segment arithmetic, which is a Euclidean field.

Hence there is a bijection between the Euclidean planes and the Euclidean fields.

Definition IV.20.5 (real closed field). A field is called *real closed* if every polynomial of odd order has at least one zero.

In a fourth step, we use Archimedes' axiom in order to justify measurements. Finally, we introduce a suitable axiom of completeness.

Definition IV.20.6 (Archimedean plane). A *Archimedean plane* is a Hilbert plane for which the Archimedean Axiom hold.

Definition IV.20.7 (Archimedean field). A field or skew field is called *Archimedean* if for any $a, c > 0$ there exists a natural number n such that $a < n \cdot c$.

Theorem IV.20.2. *The Cartesian plane* $\mathbf{F}^2$ *is Archimedean if and only if the field* $\mathbf{F}$ *is Archimedean.*

Definition IV.20.8 (Real Euclidean plane). A *real Euclidean plane* or simply *real plane* is a Hilbert plane for which both the Euclidean parallel postulate and the Dedekind axiom hold.

Putting results together we get what I would be willing to name the fundamental theorem of geometry:

Main Theorem 23 (Fundamental Theorem of Geometry). *The real Cartesian plane* $\mathbf{R}^2$ *is a complete, Archimedean Euclidean plane.*

Conversely, any complete, Archimedean Euclidean plane is isomorphic to the real Cartesian plane. The real numbers are obtained as its field of segment arithmetic.

We thus get a chain of models of strictly decreasing familiarity:

- The real Euclidean plane is a Euclidean plane.
- Any Euclidean plane is a Pythagorean plane.
- Any Pythagorean plane is a semi-Euclidean plane.
- Any semi-Euclidean plane is a Hilbert plane.
- Any Hilbert plane is an affine plane.
- Any affine plane is an incidence plane.

The converse does not hold in any of these steps.

IV.20.2 About Archimedean fields

Proposition IV.20.3 (Theorem 59 of Hilbert). *Every Archimedean skew field is commutative.*

Proof. We get by an easy induction that

$$n \cdot a = a \cdot n$$

for any natural number n and any a of the skew field. Take two arbitrary elements a and b, for which we need to check the commutative law. It is easy to see that it is enough to consider—towards a contradiction—the case

$$a > 0\,,\ b > 0\,,\ ab - ba > 0 \tag{IV.20.2}$$

From the existence of the inverse, we conclude that there exists an element $c > 0$ such that

$$(a + b + 1) \cdot c = ab - ba \tag{IV.20.3}$$

There exists a number d such that

$$0 < d < 1\,,\ d < c$$

Recall that each nonempty set of natural numbers has a smallest one. Hence it is easy to see that there exist natural numbers $n, m \geq 0$ such that

$$m \cdot d < a \leq (m+1) \cdot d \quad \text{and} \quad n \cdot d < b \leq (n+1) \cdot d$$

Multiplication and subtraction yield

$$\begin{gathered} mn \cdot d^2 < ba < ab \leq mn \cdot d^2 + (m+n+1) \cdot d^2 \\ ab - ba < (m+n+1) \cdot d^2 \\ (m+n+1) \cdot d^2 < (a+b+1) \cdot d < (a+b+1)c \\ ab - ba < (a+b+1)c \end{gathered}$$

contradicting relation (IV.20.3). Hence the assumption (IV.20.2) cannot be true, and the commutative law is confirmed. ☐

IV.20.3 Congruence in a Pythagorean plane

For any two points $A = (a_1, a_2)$ and $B = (b_1, b_2)$, we introduce the distance

$$d(A, B) = \sqrt{(a_1 - b_1)^2 + (a_2 - b_2)^2}$$

and define the congruence of any two segments AB and $A'B'$ by postulating

$$AB \cong A'B' \Leftrightarrow d^2(A, B) = d^2(A', B')$$

From this definition one easily checks that congruence of segment is an equivalence relation and hence Hilbert's axiom (III.2) holds.

Problem IV.20.1 (The rational plane is not Euclidean). *The rational Cartesian plane* $\mathbf{Q}^2$ *is an incidence plane, but not a Hilbert plane. Which are the two among Hilbert's axioms of incidence* (I.1)(I.2)(I.3)*, and the axioms of order, congruence, parallelism, and continuity that do not hold in this case?*

We may assume to be known that $\sqrt{2}$ *is not rational. Explain with a simple counterexample involving* $\sqrt{2}$ *that one of the axioms of congruence does not hold.*

Answer. Only the axiom of congruence (III.1), and the completeness axiom (V.2) are not satisfied in the rational Cartesian plane $\mathbf{Q}^2$.

To see that the axiom of congruence (III.1) is violated, we take the unit square with vertices $(0,0), (1,0), (0,1), (1,1)$ and try to transfer its diagonal onto the positive x-axis, the ray $\overrightarrow{(0,0)(1,0)}$. This is impossible because a segment $(0,0)(\frac{p}{q}, 0)$ with length $\sqrt{2}$ does not exist.

The completeness axiom (V.2) is violated, since it is possible to adjoin additional points into the rational plane in such a way to obtain a Pythagorean plane.

Problem IV.20.2. *Check that Hilbert's axiom* (III.1) *about the transfer of segments holds in a plane over a Pythagorean field.*

Proof. In axiom (III.1) are given arbitrary points A, B and A' and an arbitrary ray with vertex A'. We may suppose this ray is $a' = \overrightarrow{A'C}$ with $C \neq A'$. To check the claim made by the axiom, we need to prove existence of a point B' on this ray for which the segments $AB \cong A'B'$ are congruent.

With the coordinates

$$A = (a_1, a_2) \;,\; B = (b_1, b_2) \;,\; A' = (a'_1, a'_2) \;,\; C = (c_1, c_2)$$

the equation of the ray is

$$x = a'_1 + \lambda(c_1 - a'_1) \quad \text{and} \quad y = a'_2 + \lambda(c_2 - a'_2) \quad \text{with } \lambda \geq 0$$

To get the required congruence $AB \cong A'B'$, we need to satisfy the equation

$$(b_1 - a_1)^2 + (b_2 - a_2)^2 = (x - a'_1)^2 + (y - a'_2)^2$$

Plugging in the equation for the ray yields

$$(b_1 - a_1)^2 + (b_2 - a_2)^2 = \lambda^2 \left[(c_1 - a'_1)^2 + (c_2 - a'_2)^2\right]$$

It is possible to solve for the parameter $\lambda > 0$ since $C \neq A'$. We plug into the equation of the ray and get the coordinates of the required point $B' = (b'_1, b'_2)$ to be

$$b'_1 = a'_1 + \lambda(c_1 - a'_1) \quad \text{and} \quad b'_2 = a'_2 + \lambda(c_2 - a'_2) \quad \text{with } \lambda = \sqrt{\frac{(b_1 - a_1)^2 + (b_2 - a_2)^2}{(c_1 - a'_1)^2 + (c_2 - a'_2)^2}}$$

Hence we have proved existence of the point $B' = (b'_1, b'_2)$. □

Problem IV.20.3 (adding two adjacent segments on a line). *Check that Hilbert's axiom* (III.3) *holds in a plane over a Pythagorean field.*

Proof. In axiom (III.3) are given two segments AB and BC lying on a line with B as the only point in common. In the same manner, two segments $A'B'$ and $B'C'$ are given on a line with only point B' in common. In other words, we assume that point $B = (b_1, b_2)$ lies between point $A = (a_1, a_2)$ and point $C = (c_1, c_2)$, and similarly for the primed items. Hence by definition IV.20.2

$$b_i = \lambda a_i + (1 - \lambda)c_i \quad \text{for } i = 1, 2 \text{ with } 0 < \lambda < 1.$$
$$b'_i = \lambda' a'_i + (1 - \lambda')c'_i \quad \text{for } i = 1, 2 \text{ with } 0 < \lambda' < 1.$$

The axiom (III.3) takes

$$AB \cong A'B' \text{ and } BC \cong B'C$$

as assumption for the given segments. By definition, this means

$$(b_1 - a_1)^2 + (b_2 - a_2)^2 = (b'_1 - a'_1)^2 + (b'_2 - a'_2)^2$$
$$(b_1 - c_1)^2 + (b_2 - c_2)^2 = (b'_1 - c'_1)^2 + (b'_2 - c'_2)^2$$

We plug in equation (IV.20.3) and get

$$(1 - \lambda)^2[(a_1 - c_1)^2 + (a_2 - c_2)^2] = (1 - \lambda')^2[(a'_1 - c'_1)^2 + (a'_2 - c'_2)^2]$$
$$\lambda^2[(a_1 - c_1)^2 + (a_2 - c_2)^2] = \lambda'^2[(a'_1 - c'_1)^2 + (a'_2 - c'_2)^2]$$

If either $A = C$ or $A' = C'$, both equalities hold, and nothing is to be proved. We can assume $A \neq C$ and $A' \neq C'$, and divide the two relations to get

$$\frac{(1 - \lambda)^2}{\lambda^2} = \frac{(1 - \lambda')^2}{\lambda'^2}$$

Since $0 < \lambda < 1$ and $0 < \lambda' < 1$, we obtain via the root

$$\frac{1 - \lambda}{\lambda} = \frac{1 - \lambda'}{\lambda'}$$
$$\lambda = \lambda'$$
$$(a_1 - c_1)^2 + (a_2 - c_2)^2 = (a'_1 - c'_1)^2 + (a'_2 - c'_2)^2$$
$$AC \cong A'C'$$

Hence we have proved

$$\text{if } AB \cong A'B' \text{ and } BC \cong B'C', \text{ then } AC \cong A'C'$$

as was claimed. □

IV.20.4 Transfer of an angle

Recall the definition II.7.14 of an *orientated angle* as an ordered pair (h,k) of two rays h and k with a common vertex . In every ordered incidence plane, it is possible to define and fix an orientation. We establish an orientation in the coordinate plane.

Definition IV.20.9 (Slope of a line). For a line AB through points $A = (a_1, a_2)$ and $B = (b_1, b_2)$, we define the *slope*

$$m = \frac{b_2 - a_2}{b_1 - a_1}$$

Opposite rays on the same line have the same slope. A vertical ray gets the slope $\infty \notin \mathbf{F}$, hence the slope of a line is an element in $\mathbf{F} \cup \{\infty\}$.

Definition IV.20.10. We define the tangent of the orientated angle between two lines p and q with slopes m_p and m_q in $\mathbf{F} \cup \{\infty\}$ by

$$\tan \curvearrowleft (p,q) = \frac{m_q - m_p}{1 + m_p m_q}$$

as an element of $\mathbf{F} \cup \{\infty\}$. The tangent of the orientated angle depends only on the two lines and does not distinguish opposite rays. The tangent of the orientated angle between two rays is defined by the tangent between the lines of the two rays.

Remark. In the case that one of the slopes is $m_p \neq 0$ and the other one $m_q = \infty$, the formula makes sense by using the rule

$$\frac{\infty - m_p}{1 + m_p \cdot \infty} = \frac{1}{m_p}$$

In the case that one of the slopes is $m_p \neq 0$ and $1 + m_p m_q = 0$, the formula makes sense by using the rule $\pm 1/0 = \infty$. Finally there remains the special case of the slopes $m_p = 0$ and $m_q = \infty$, for which we need to agree that $\tan \curvearrowleft (p,q) = \infty$.

Remark. The tangent takes the value infinity if and only if the angle $\angle(p,q)$ is a right angle. This occurs if the slopes are either 0 and ∞, or the denominator $1 + m_p m_q = 0$.

Recall the definition II.7.17 of *congruence of orientated angles*

Definition IV.20.11 (Congruence of angles). Two angles $\angle(h,k)$ and $\angle(h',k')$ are called *congruent* if and only if either

- the orientated angles $\curvearrowleft (h,k)$ and $\curvearrowleft (h',k')$ are congruent, or
- the orientated angles $\curvearrowleft (k,h)$ and $\curvearrowleft (h',k')$ are congruent.

Question. Convince yourself that two orientated angles $\curvearrowleft (h,k)$ and $\curvearrowleft (h',k')$ are congruent if and only if the angles $\angle(h,k)$ and $\angle(h',k')$ are both congruent and they have the same orientation.

Question. Let the angles $\angle(h,k)$ and $\angle(h',k')$ have the same orientation. Convince yourself that $\tan \curvearrowleft (h,k) = \tan \curvearrowleft (h',k')$ if and only if the angles $\angle(h,k)$ and $\angle(h',k')$ are congruent.

Question. Let the angles $\angle(h,k)$ and $\angle(h',k')$ have opposite orientations. Convince yourself that $\tan \curvearrowleft (h,k) = -\tan \curvearrowleft (h',k')$ if and only if the angles $\angle(h,k)$ and $\angle(h',k')$ are congruent.

Lemma IV.20.1. *If an angle is congruent to its supplement, the slopes m_h and m_k of its two sides are either* 0 *and* ∞, *or satisfy* $1 + m_h m_k = 0$.

Proof. Let h' be the ray opposite to h, and assume that the angle $\angle(h,k)$ is congruent to its supplement $\angle(h',k)$. By definition we conclude congruence of the orientated angles $\curvearrowleft (h,k)$ and $\curvearrowleft (k,h')$ since they have the same orientation. Because h and h' have the same slope, we conclude

$$\tan \curvearrowleft (h,k) = \tan \curvearrowleft (k,h')$$
$$\frac{m_k - m_h}{1 + m_k m_h} = \frac{m_h - m_k}{1 + m_k m_h}$$

We conclude that either the slopes $m_h = m_k$ are equal—impossible for a non degenerate angle—or the slopes are 0 and ∞, or $1 + m_h m_k = 0$. ☐

Problem IV.20.4. *Explain why unique angle transfer according to axiom* (III.4) *is possible in the rational plane* $\mathbf{Q}^2$, *and indeed axiom* (III.4) *does hold.*

Indication of the solution. Given is an angle $\angle(h,k)$ and a ray h'. Assume that ray k lies in the left half-plane of ray h, and the angle $\angle(h,k)$ has to be transferred to the left half-plane of h'. In this case, the two orientated angles $\curvearrowleft (h,k)$ and $\curvearrowleft (h',k')$ have the same orientation, and need to be congruent. Hence

$$\tan \curvearrowleft (h,k) = \tan \curvearrowleft (h',k')$$
$$t := \frac{m_k - m_h}{1 + m_k m_h} = \frac{m'_k - m'_h}{1 + m'_k m'_h}$$

We can solve the last equation for the slope m'_k and get

$$m'_k = \frac{m'_h + t}{1 - m'_h t}$$

This determines the slope of the line of k'. Since the ray k' lies in the left half-plane of h', the ray k' has been determined, too.

Now we assume that ray k lies in the left half-plane of ray h, and the angle $\angle(h,k)$ has to be transferred to the right half-plane of h'. In this case, the two orientated angles $\curvearrowleft (k,h)$ and $\curvearrowleft (h',k')$ with h and k switched have the same orientation, and need to be congruent, too. Hence

$$\tan \curvearrowleft (h,k) = -\tan \curvearrowleft (h',k')$$
$$t := \frac{m_k - m_h}{1 + m_k m_h} = -\frac{m'_k - m'_h}{1 + m'_k m'_h}$$

We can solve the last equation for the slope m'_k and get

$$m'_k = \frac{m'_h - t}{1 + m'_h t}$$

This determines the slope of the line of k'. Since the ray k' lies in the right half-plane of h', the ray k' has been determined, too.

Well, I see clear enough that all remaining details can be worked out. We thus have checked that the requested rays h' and h'' have rational slope and hence exist in the rational plane $\mathbf{Q}^2$. ☐

Lemma IV.20.2. *Hilbert's axiom* (III.4) *about the transfer of angles holds in a plane over a Pythagorean field.*

IV.20.5 Enough rigid motions exist in a Pythagorean plane

Definition IV.20.12 (Rigid motion). A *rigid motion* in a Pythagorean plane is a bijection ϕ from the set of points to itself and a bijection of the set of lines to itself which preserves incidence and order, and maps any segment AB or angle $\angle(h,k)$ to a congruent segment $\phi(A)\phi(B) \cong AB$ or angle $\angle(\phi(h),\phi(k)) \cong \angle(h,k)$.

Definition IV.20.13 (Similarity). A *similarity* in a Pythagorean plane is a bijection ϕ from the set of points to itself and a bijection of the set of lines to itself which preserves incidence and order, and maps any angle $\angle(h,k)$ to a congruent angle $\angle(\phi(h),\phi(k)) \cong \angle(h,k)$, and any pair of congruent segments $AB \cong A'B'$ to a pair of congruent segments $\phi(A)\phi(B) \cong \phi(A')\phi(B')$.

Definition IV.20.14 (Preserving versus reversing orientation). A rigid motion ϕ of any ordered affine plane preserves orientation if and only if any triangle $\triangle ABC$ and its image $\triangle\phi(A)\phi(B)\phi(C)$ always have the same orientation.

The rigid motion ϕ reverses orientation if and only if any triangle $\triangle ABC$ and its image $\triangle\phi(A)\phi(B)\phi(C)$ always have the opposite orientation.

Equivalently, we can state that a rigid motion preserves orientation if any ray$\overrightarrow{AB}$ and point C in the left half plane of this ray is mapped to a ray $\overrightarrow{\phi(A)\phi(B)}$ and point $\phi(C)$ in the left half plane of the image ray.

Theorem IV.20.3 (There are enough rigid motions). *In a Pythagorean plane are given are two points P and Q, and rays r_P and r_Q with these vertices. There exist exactly <u>two</u> rigid motions ϕ and ψ, which map P to Q and r_P to r_Q. The first mapping ϕ preserves the orientation, and the second mapping ψ reverses the orientation.*

Lemma IV.20.3. *For any point $A = (a_1, a_2)$, there exists a rigid translation which maps A to O, and which is orientation preserving.*

Lemma IV.20.4. *The reflection P at the origin is a rigid motion which preserves the orientation.*

Lemma IV.20.5. *The rotation by a right angle around the origin is a rigid motion which preserves the orientation.*

Problem IV.20.5. *Check that the linear equations*

$$x' = cx + sy$$
$$y' = -sx + cy$$

with $c^2 + s^2 = 1$ define an isometry in the Pythagorean plane.

Problem IV.20.6. *Given the orientated angle $\angle EOA$ with $E = (1,0), O = (0,0), A = (x,y)$ and $y > 0$. Check that the linear equations*

$$x' = cx + sy$$
$$y' = -sx + cy$$

with $c^2 + s^2 = 1, c > 0, s > 0$ map the orientated angle $\angle EOA$ to a congruent angle.

Answer. The sides $\overrightarrow{OE}$ and $\overrightarrow{OA}$ of the angle $\angle EOA$ have the slopes $0 = m_p$ and $y/x = m_q$. Hence

$$\tan \curvearrowleft (p,q) = \frac{m_q - m_p}{1 + m_p m_q} = \frac{y}{x}$$

Point $E = (1,0)$ is mapped to $E' = (c, -s)$. Hence the ray $\overrightarrow{OE'}$ has the slope

$$m_p' = -\frac{s}{c}$$

The other image ray $\overrightarrow{OA'}$ has the slope

$$m_q' = \frac{y')}{x'} = \frac{-sx + cy}{cx + sy}$$

Hence

$$\tan \curvearrowleft (p',q') = \frac{m_q' - m_p'}{1 + m_p' m_q'} = \frac{(-sx+cy)c + s(cx+sy)}{c(cx+sy) - s(-sx+cy)} = \frac{y}{x}$$

We can also check that both the angle $\angle EOA$ and its image $\angle E'OA'$ have the same orientation. This is easy enough for the special case $A = (0,1)$, from which the general case follows. Hence the angle $\angle EOA$ and its image $\angle E'OA'$ are congruent.

Lemma IV.20.6. *The linear equations*

$$x' = cx + sy$$
$$y' = -sx + cy$$

with $c^2 + s^2 = 1$ define a rigid rotation in the Pythagorean plane.

Lemma IV.20.7. *The reflection I_x across the x-axis is a rigid motion which reverses the orientation.*

Lemma IV.20.8. *Given is a Pythagorean plane. For any ray $\overrightarrow{OA}$ pointing into the right half plane, there exists a rigid rotation which maps this ray into the positive x-axis and preserves orientation.*

Proof. Let $A = (a_1, a_2)$ be the coordinates of A, point $E = (1,0)$ and

$$m = \tan \angle EOA = \frac{a_2}{a_1}$$

be the slope of ray $\overrightarrow{OA}$. In a Pythagorean field, we can get the cosine and sinus of the rotation angle $\angle EOA$ as

$$c = \frac{1}{\sqrt{1+m^2}} = \frac{a_1}{a_1^2 + a_2^2} \text{ and } s = \frac{m}{\sqrt{1+m^2}} = \frac{a_2}{a_1^2 + a_2^2}$$

These quantities exist: Indeed, the square root $\sqrt{1+m^2}$ exists in a Pythagorean field. Moreover, $1 + m^2 \geq 1$ cannot be zero, hence $\sqrt{1+m^2} \neq 0$ and the division by this quantity is possible. The linear equations

$$x' = cx + sy$$
$$y' = -sx + cy$$

can be checked to be the rigid rotation mapping the ray $\overrightarrow{OA}$ to the positive x-axis $\overrightarrow{OE}$. □

Lemma IV.20.9. *Any composition of rigid motions, and the inverse of a rigid motion is again a rigid motion. In other words, the rigid motion are a group. The orientation preserving motions are a subgroup.*

Proof of the Theorem about enough rigid motions. By composition of a translation and a rotation, a ray towards right can be mapped to the positive x-axis. For an arbitrary ray, the same result can be achieved by use of a point reflection or a rotation by a right angle as third factor. We thus obtain an orientation preserving rigid motion with the required property. Composition with a reflection across the x-axis yields an orientation reversing rigid motion with the required property. □

Corollary 48. *Every rigid motion can be realized by a composition a translation, a rotation about the origin, and possibly a reflection across the x-axis.*

Proposition IV.20.4. *In a Pythagorean coordinate plane are given two triangles $\triangle ABC$ and $\triangle A'B'C'$. If the congruences*

$$AB \cong A'B' \,,\ AC \cong A'C' \,,\ \ \angle BAC \cong \angle B'A'C'$$

hold, then the two triangles are congruent. Especially Hilbert's SAS-axiom (III.5) *holds in a plane over a Pythagorean field.*

Proof. I restrict myself to the case that angle $\angle BAC$ is acute. There exists a rigid motion ϕ mapping A to O, the ray $\overrightarrow{AB}$ to the positive x-axis $\overrightarrow{OE}$, and vertex C to a point in the upper half plane. Hence

$$\phi(A) = O = (0,0) \,,\ \phi(B) = (0,c) \quad \text{and} \quad \phi(C) = (x,y) \ \text{ with } \ y > 0$$

The rigidity of the mapping ϕ implies

$$\begin{aligned} c = d(A,B) &= \sqrt{(a_1-b_1)^2 + (a_2-b_2)^2} \,, \\ x^2 + y^2 = d^2(A,C) &= (a_1-c_1)^2 + (a_2-c_2)^2 \quad \text{and} \\ \frac{y}{x} &= \tan \angle BAC \end{aligned}$$

All corresponding claims holds for a mapping ϕ' related to the second triangle $\triangle A'B'C'$. Comparing the relations obtained for both triangles, and using the congruences assumed, we get

$$\begin{aligned} c = d(A,B) &= d(A',B') = c' \\ x^2 + y^2 = d^2(A,C) &= d^2(A',C') = x'^2 + y'^2 \\ \frac{y}{x} = \tan \angle BAC &= \tan \angle B'A'C' = \frac{y'}{x'} \\ y > 0 &\ \text{and}\ y' > 0 \end{aligned}$$

From the assumption that angle $\angle BAC \cong \angle B'A'C'$ is acute, we get $x > 0$ and $y > 0$ and similarly $x' > 0$ and $y' > 0$. Hence we conclude $x = x', y = y'$ and $c = c'$. Hence the two "moved triangles" $\triangle \phi(A)\phi(B)\phi(C)$ with the vertices

$$\phi(A)(0,0) \,,\ \phi(B) = (0,c) \ \text{ and } \ \phi(C) = (x,y)$$

and the corresponding primed one are equal. Hence they are congruent, too! Since a rigid motions maps a triangle to a congruent one, we conclude that the original triangles $\triangle ABC \cong \triangle A'B'C'$ are congruent, too, as to be shown.

In the case of an obtuse angle $\angle BAC \cong \angle B'A'C'$, we get $x > 0$ and $y < 0$ and similarly $x' > 0$ and $y' < 0$. Thus $x = x', y = y'$ follows similarly to above. The proof is even easier for the case that angle $\angle BAC \cong \angle B'A'C'$ is right. □

End of the proof of Theorem IV.20.1. We finish the proof that a Cartesian plane over a Pythagorean field is a Pythagorean plane.

- By Theorem I.3.2, a Cartesian plane over a field is an affine plane. This means that Hilbert's axioms (I.1)(I.2)(I.3a)(I.3b) and the Euclidean parallel property hold.

- By Proposition IV.20.1, in a Cartesian plane over an *ordered* field, Hilbert's axioms of order (II.1)(II.2)(II.3)(II.4) hold.

- In the solution of Problem IV.20.2, we have checked that Hilbert's axiom (III.1) about the transfer of segments holds in a plane over a Pythagorean field.

- From the definition of congruence of segments, one sees that congruence is an equivalence relation, and hence Hilbert's axiom (III.2) holds.

- In the solution of Problem IV.20.3, we have checked that Hilbert's axiom (III.3) about segment addition holds in a plane over a Pythagorean field.

- By Lemma IV.20.2, Hilbert's axiom (III.4) about the transfer of angles holds in a plane over a Pythagorean field.

- By Proposition IV.20.4, Hilbert's SAS-axiom (III.5) holds in a plane over a Pythagorean field.

Thus all five axioms of congruence (III.1) through (III.5) are satisfied in a Cartesian plane over a Pythagorean field. □

IV.20.6 Euclidean fields and intersection properties of circles

Theorem IV.20.4. *The following three properties are equivalent in any Pythagorean coordinate plane* $\mathbf{F}^2$*:*

(1) *The field* $\mathbf{F}$ *is Euclidean.*

(2) *The circle-line intersection property holds in the plane* $\mathbf{F}^2$.

(3) *The circle-circle intersection property holds in the plane* $\mathbf{F}^2$.

Lemma IV.20.10. *A Euclidean field is Pythagorean.*

Lemma IV.20.11 ((1) ⇒ (2)). *The circle-line intersection property holds in the plane over a Euclidean field.*

Proof of (1) ⇒ (2). Given is a circle and a line which contains an interior point of the circle. Let P be the foot point of the perpendicular p dropped from the center of the circle onto the line. Since a Euclidean field is Pythagorean, there exist enough rigid motions in the plane $\mathbf{F}^2$. Hence there exists a rigid motion mapping the ray $\overrightarrow{OP}$ onto the positive x-axis. In the special case $O = P$ any one of the two rays with vertex perpendicular to the given line will be mapped onto the positive x-axis.

Thus we have put the circle with radius $c > 0$ and the line into a Cartesian coordinate system, with coordinates $C = (0, 0)$ for the center and $P = (p, 0)$ for the foot point.

Question. What are the equations for the circle and the line.

Answer.

$$x^2 + y^2 = c^2 \qquad \text{(IV.20.4)}$$
$$x = p$$

Since the line contains a point (p, q) inside the circle, we know that $p^2 \leq p^2 + q^2 < c^2$ and hence $c^2 - p^2 > 0$.

Question. What are the coordinates (x, y) of the intersection points of the circle and the line.

Answer. Plugging $x = p$ into the equation of the circle yields $y^2 = c^2 - p^2$. We have seen above that $c^2 - p^2 > 0$ follows since the line contains a point inside the circle. The field is assumed to be Euclidean, hence we can take the square root of a positive quantity and get the two solutions $y = \pm\sqrt{c^2 - p^2}$. We have obtained two intersection points with the coordinates $(p, \sqrt{c^2 - p^2})$ and $(p, -\sqrt{c^2 - p^2})$.

We have checked that the circle-line intersection property holds in the coordinate plane over a Euclidean field. □

Lemma IV.20.12 **((2) ⇒ (1)).** *If the circle-line intersection property holds for the Pythagorean plane* $\mathbf{F}^2$*, then the field* $\mathbf{F}$ *is Euclidean.*

Proof of (2) ⇒ (1). Given is $f > 0$, a positive element of the field. We construct in the plane $\mathbf{F}^2$ the circle and the line with the equations IV.20.4 with

$$c = \frac{1+f}{2} \quad \text{and} \quad p = \frac{f-1}{2}$$

The assumption $f > 0$ implies $-c < p < c$. Hence the point $(p, 0)$ lies inside the circle. By assumption, the circle-line intersection property holds. Hence circle and line intersect in some point (p, y). From the equation of the circle and the assumption we get

$$y^2 = c^2 - p^2 = f$$

Hence $y = \sqrt{f}$ is a square root of f, the existence of which was to show. □

Lemma IV.20.13 **((1) ⇒ (3)).** *The circle-circle intersection property holds in the plane over a Euclidean field.*

Proof of (1) ⇒ (3). The two circles $\mathcal{C}$ with center C and radius c, and $\mathcal{D}$ with center D and radius d are given. We assume that one of the circles has points both inside and outside the other one.

Question. Which inequalities hold between the two radius $c, d > 0$ and the distance $a = |CD|$ between the two centers. Which case occurs in Proposition II.10.10.

Answer. If $a \geq c + d$, the interiors of the two circles are disjoint, which is case **(i)** of Proposition II.10.10. If $a \leq |c - d|$, the interior of the smaller circle lies entirely in the interior of the larger one. These are the cases **(ii)** and **(iii)** of Proposition II.10.10. These cases are excluded by the assumption that one circle has points both inside and outside the other one. Hence

$$|c - d| < a < c + d$$

These inequalities correspond in Proposition II.10.10 to only remaining possibility **(iv)**.

Since a Euclidean field is Pythagorean, there exist enough rigid motions in the plane $\mathbf{F}^2$. Hence there exists a rigid motion mapping the ray $\overrightarrow{CD}$ between the two centers onto the positive x-axis.

Question. What are the coordinates of the two centers of the circles.

Answer. We have put the two circle in a Cartesian coordinate system, with coordinates of the centers $C = (0, 0)$ and $D = (a, 0)$, where $a = |CD|$ is the distance of the two centers.

Question. What are the equations for the two circles.

Answer.

$$x^2 + y^2 = c^2$$
$$(x - a)^2 + y^2 = d^2$$

Question. What can one conclude from subtracting the two equations about the possible intersection of the two circles?

Answer. Substraction of the two equations yields a <u>linear</u> equation with one unknown x. One can even solve for x.

(IV.20.5) $$x = \frac{a^2 + c^2 - d^2}{2a}$$

This is the equation of a line. Hence the intersection points of the two circles lie on this line l.

Conversely, the intersection points of the line l with either circle lie on the other circle, too. Hence it is now enough to look for the intersection points of circle $\mathcal{C}$ and line l. We may assume that $c \geq d$, and use the abbreviation

$$p = \frac{a^2 + c^2 - d^2}{2a}$$

The point P with coordinates $(p, 0)$ lies on the line l.

Question. Why do the circle $\mathcal{C}$ and the line l intersect?

Answer. The point P lies inside the circle $\mathcal{C}$ since

$$-c < p < c$$

The right inequality can be checked as follows:

$$\frac{a^2 + c^2 - d^2}{2a} < c$$
$$a^2 + c^2 - d^2 < 2ac$$
$$a^2 + c^2 - 2ac < d^2$$
$$(a - c)^2 < d^2$$
$$-d < a - c < d$$
$$c - d < a < c + d$$

The last of these inequalities is true by assumption, and they are all equivalent. Checking that $c + p > 0$ is a quite similar calculation left to the reader. Since the line l has the point P inside circle $\mathcal{C}$, the circle-line intersection property guarantees that they intersect.

Question. What are the coordinates (x, y) of the intersection points of the circle and the line.

Answer. Plugging $x = p$ into the equation of the first circle yields $y^2 = c^2 - p^2$. We have seen above that $c^2 - p^2 = (c+p)(c-p) > 0$. The field is assumed to be Euclidean, hence we can take the square root of a positive quantity y^2 and get the two intersection points with the coordinates $(p, \sqrt{c^2 - p^2})$ and $(p, -\sqrt{c^2 - p^2})$.

Thus we have that the circle-circle intersection property holds for a Euclidean plane. □

End of the proof of Theorem IV.20.4. We know from Proposition II.10.15 that the circle-circle intersection property implies the line-circle intersection property. This holds even more generally in every Hilbert plane. Thus we know that (3) ⇒ (2) for the items of Theorem IV.20.4. We have proved that (2) ⇒ (1)— the circle-line intersection property implies that field is Euclidean— in Lemma IV.20.12 Moreover, Lemma IV.20.13 states that (1) ⇒ (3)—The circle-circle intersection property holds in the plane over a Euclidean field. Thus all three statements (1)(2)(3) of Theorem IV.20.4 are equivalent, as to be shown. □

Problem IV.20.7 ((3) ⇒ (1)). *Give a direct proof of the statement that a Pythagorean plane* $\mathbf{F}^2$ *for which the circle-circle intersection property holds has a Euclidean field* $\mathbf{F}$.

Proof of (3) ⇒ (1). Given is $f > 0$, a positive element of the field. We construct in the plane $\mathbf{F}^2$ two circles with the same radius c and the distance a between there centers. We choose

$$c = \frac{1+f}{2} \quad \text{and} \quad a = |f - 1|$$

From the equations of the two circles

$$\begin{aligned} x^2 + y^2 &= c^2 \\ (x-a)^2 + y^2 &= c^2 \end{aligned}$$

we get the coordinates (x, y) of an intersection point to be $x = \frac{a}{2}$ and

$$y = \sqrt{y^2} = \sqrt{\left(\frac{1+f}{2}\right)^2 - \left(\frac{1-f}{2}\right)^2} = \sqrt{f}$$

Since the intersection point is assumed to exist, we see that the square $\sqrt{f}$ exists for any $f > 0$, as to be shown. □

From the results above, we get a proof of Proposition II.10.16 as promised in the section on the neutral geometry of circles.

Definition (Proposition **II.10.16**)**.** In any Hilbert plane, the line-circle intersection property together with the parallel axiom imply the circle-circle intersection property.

Reason. Assuming the parallel axiom, we may invoke theorem IV.19.1: any Pythagorean plane is isomorphic to the Cartesian plane over its field of segment lengths, which is an Pythagorean field $\mathbf{F}$. Because the circle-line intersection property is assumed, we know from Lemma IV.20.12 above, that the field $\mathbf{F}$ is even Euclidean. Hence by Lemma IV.20.13, the circle-circle intersection holds, too. □

Corollary 49. *If the distance $a = |CD|$ between the centers of two circles $\mathcal{C}$ with center C and radius c, and $\mathcal{D}$ with center D and radius d satisfy the inequalities*

$$|c-d| < a < c+d$$

they intersect in two points.

Corollary 50 (Heron's formula). *In Euclidean geometry, the area of a triangle with the sides a, b and c is*

$$\text{area}(\triangle ABC) = \frac{1}{4}\sqrt{(a+b+c)(a+b-c)(a-b+c)(b+c-a)}$$

Proof. I replace b by d as always above and consider the triangle with the vertices C, D and an intersection point $X = (p, y)$ of the circles $\mathcal{C}$ and $\mathcal{D}$. We get the Euclidean area

$$\text{area}(\triangle CDX) = \frac{ay}{2} = \frac{a}{2}\sqrt{c^2 - p^2}$$

The definition of p implies

$$\begin{aligned} c - p &= \frac{2ac - a^2 - c^2 + d^2}{2a} = \frac{d^2 - (a-c)^2}{2a} = \frac{(d-a+c)(d+a-c)}{2a} \\ c + p &= \frac{2ac + a^2 + c^2 - d^2}{2a} = \frac{(a+c)^2 - d^2}{2a} = \frac{(a+c+d)(a+c-d)}{2a} \end{aligned} \tag{IV.20.6}$$

The coordinates of the intersection points are $(x, y) = (p, \pm\sqrt{c^2 - p^2})$ with

$$\begin{aligned} p &= \frac{a^2 + c^2 - d^2}{2a} \\ c^2 - p^2 &= \frac{(a+c+d)(a+c-d)(a-c+d)(c-a+d)}{4a^2} \end{aligned}$$

] Hence the area of the triangle with sides a, c and d is

$$\text{area}(\triangle CDX) = \frac{a}{2}\sqrt{c^2 - p^2} = \frac{1}{4}\sqrt{(a+c+d)(a+c-d)(a-c+d)(c-a+d)} \tag{IV.20.7}$$

□

Problem IV.20.8. *Confirm that the line point P and the line l are constructible with Hilbert tools and unique parallels, even if no circle intersection properties are assumed.*

Proof. From the expression IV.20.6 for $c - p$ and $c + p$, one gets the proportions

$$\frac{c-p}{a+c+d} = \frac{c+d-a}{2a} \tag{IV.20.8}$$

$$\frac{c+p}{a+c-d} = \frac{a+c+d}{2a} \tag{IV.20.9}$$

From these proportions, construction of either $c-p$ or $c+p$ can be implemented with Hilbert tools, using parallels and similar triangles.

Since the line l has the point P inside circle $\mathcal{C}$, the circle-line intersection property guarantees that they intersect. By the remarks above, there intersection points are the same as the intersection points of the two circles $\mathcal{C}$ and $\mathcal{D}$. □

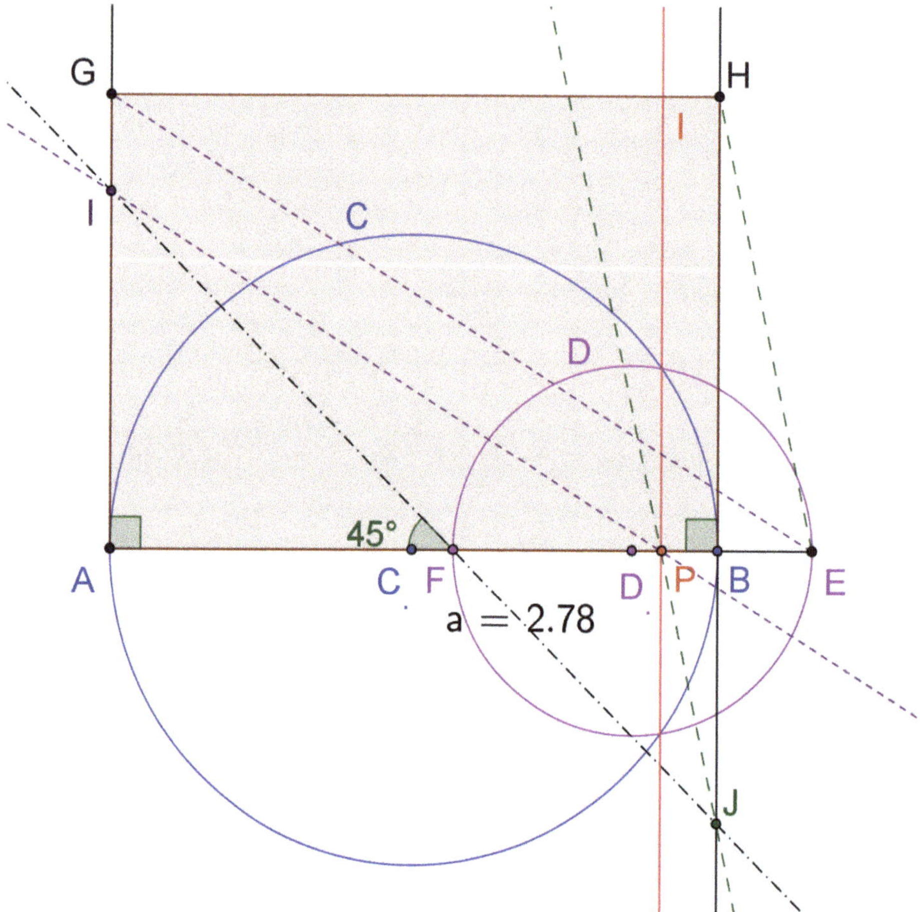

Figure IV.20.1: The intersections $\mathcal{C} \cap \mathcal{D} = \mathcal{C} \cap l = \mathcal{D} \cap l$ are equal—point P and line l can be constructed with Hilbert tools.

Here is a "fairly clean" solution for the construction of point P. Let AB and FE be the diameters of the two given circles $\mathcal{C}$ and $\mathcal{D}$ lying on the line CD.

Question. We assume the order $A * F * B * E$. Calculate the distances $|AF|, |FB|, |BE|$ and $|AE|$.

Answer.

$$\begin{aligned} |AF| &= a + c - d \\ |FB| &= c + d - a \\ |BE| &= a - c + d \\ |AE| &= a + c + d \end{aligned}$$

We construct a rectangle with one side AB, and the perpendicular sides congruent to $2a$. The coordinates of its vertices are

$$A = (-c, 0), B = (c, 0), H = (c, 2a), G = (-c, 2a)$$

We draw a line g with slope -1 through point $F = (a - d, 0)$. It intersects the vertical sides of the rectangle, or their extensions, in the points

$$I = (-c, a + c - d) \quad \text{and} \quad J = (c, -(c + d - a))$$

The parallel to line GE through point I intersects the x-axis in point $P = (p, 0)$, because of the proportion (IV.20.9) and the similar triangles $\triangle API \sim \triangle AEG$. With a similar reasoning, we get that the parallel to line HE through point J intersects the x-axis in the same point $P = (p, 0)$, too.

Part V

Real Progress in Euclidean Geometry

V.21 Similar Triangles

V.21.1 Basic properties from Euclid

Definition V.21.1 (Similar triangles). Two triangles $\triangle ABC$ and $\triangle A'B'C'$ for which the ratios of corresponding sides

$$\frac{a}{b} = \frac{a'}{b'}\,,\ \frac{b}{c} = \frac{b'}{c'}\,,\ \text{and}\quad \frac{c}{a} = \frac{c'}{a'}\,, \tag{V.21.1}$$

are equal, are called *similar*.

Hilbert's—and already Legendre's—view about the theory of similarity is different from Euclid's: In Hilbert's foundations, the results about similar triangles follow from the segment arithmetic. The segment arithmetic, in turn, is justified by Pappus theorem. Following Hilbert's approach, the properties of similar triangles shall be justified on the basis of the segment arithmetic developed in the previous section.

Proposition V.21.1. *For any two similar triangles, there exists a* scaling factor f *by which one needs to multiply the length of a side of the triangle $\triangle ABC$ to obtain the length of the corresponding side of the equiangular triangle $\triangle A'B'C'$.*

Proof. By ordinary calculation with fractions,

$$\frac{a'}{b'} = \frac{a}{b} \quad \text{implies} \quad \frac{a'}{a} = \frac{b'}{b}$$

Similarly we get $\frac{b'}{b} = \frac{c'}{c}$. This common ratio is the scaling factor, which I call f and obtain

$$a' = fa\,,\ b' = fb\,, \quad \text{and} \quad c' = fc$$

□

Definition V.21.2 (Equiangular triangles). Two triangles $\triangle ABC$ and $\triangle A'B'C'$ for which the angles across the corresponding sides are congruent are called *equiangular*.

Theorem V.21.1. *Two triangles are equiangular if and only if they are similar.*

To prove this basic theorem, we begin with right triangles.

The legs of two equiangular right triangles have the same ratio. Given are two equiangular right triangles. We use SAS congruence to get triangles $\triangle OAB$ and $\triangle OA'B'$ congruent to the original ones, which have a common vertex O at the right angle, and the legs $|OB| = a, |OB'| = a'$ and $|OA| = b, |OA'| = b'$ lying on two perpendicular rays from this vertex. For convenience, we take these two rays horizontally to the right and upwards.

We draw the unit segment $|OE| = 1$ on the horizontal ray and the parallel to the hypothenuse AB through point E. We obtain the intersection point Q on the vertical ray. By segment arithmetic, the segment $|OQ| = q$ on the vertical ray satisfies

$$qb = a$$

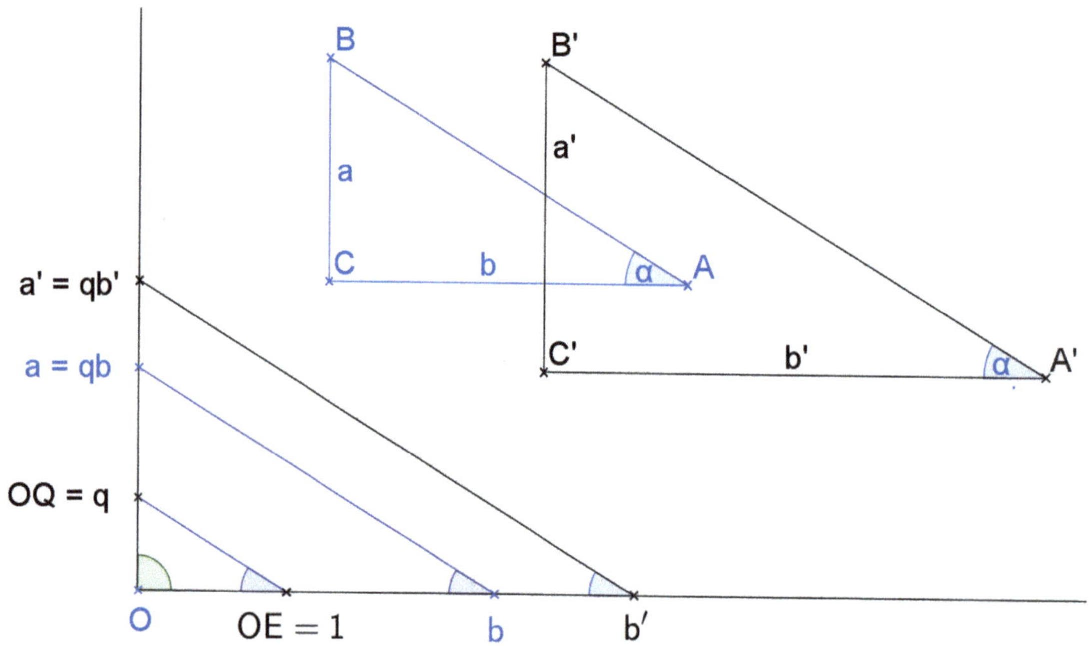

Figure V.21.1: The legs of equiangular right triangles are proportional.

Because the two hypothenuses AB and $A'B'$ have congruent angles α with the horizontal axis, the lines EQ and $A'B'$ are parallel, too. Hence with the same value of q, segment arithmetic now shows that

$$qb' = a'$$

and a simple calculation shows that

$$\frac{a}{b} = q = \frac{a'}{b'}$$

□

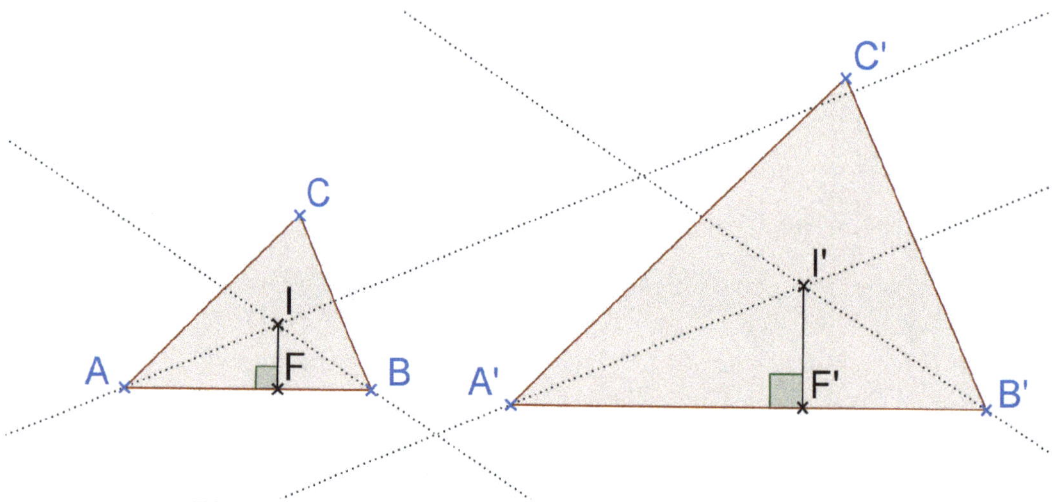

Figure V.21.2: Equiangular triangles are similar.

Since we have shown above that the ratio $\frac{a}{b}$ depends only on the angles of the right triangle, can can define a trigonometric function:

Definition V.21.3 (Tangent function for acute angles). The ratio $\frac{a}{b}$ of the leg a across to the leg b adjacent to the angle α of a right triangle is the *tangent function* of this angle:

$$\tan\alpha = \frac{a}{b}$$

Proposition V.21.2 (Euclid VI.4). *The corresponding sides of equiangular triangles have equal ratios. Equiangular triangles are* similar. *In formulas:*

$$\frac{a}{b} = \frac{a'}{b'}, \ \frac{b}{c} = \frac{b'}{c'} \quad \textit{and} \quad \frac{c}{a} = \frac{c'}{a'}$$

Proof. Given are two equiangular triangles $\triangle ABC$ and $\triangle A'B'C'$. By Theorem II.13.3 from the section about neutral triangle geometry, every triangle has an in-circle.

Its center I is the intersection point of the three interior angular bisectors. We drop the perpendicular from the center I onto the triangle side AB and let F be the foot point. We proceed similarly for the equiangular triangle $\triangle A'B'C'$. Thus we get two pairs of equiangular right triangles: $\triangle AIF$ is equiangular with $\triangle A'I'F'$, and $\triangle BIF$ is equiangular with $\triangle B'I'F'$. We get the proportions for the corresponding legs, which we have to add:

$$\frac{|AF|}{|FI|} = \frac{|A'F'|}{|F'I'|} \quad \text{and} \quad \frac{|FB|}{|FI|} = \frac{|F'B'|}{|F'I'|},$$
$$\text{and hence} \quad \frac{|AB|}{|FI|} = \frac{|A'B'|}{|F'I'|}$$

Now $|FI| = \rho$ and $|F'I'| = \rho'$ are the radii of the in-circles. For the sides $|AB| = c$ and $|A'B'| = c'$, as well as the other corresponding sides we obtain

$$\frac{c}{\rho} = \frac{c'}{\rho'}, \ \frac{a}{\rho} = \frac{a'}{\rho'} \quad \text{and} \quad \frac{b}{\rho} = \frac{b'}{\rho'},$$
$$\text{Hence} \quad \frac{a}{b} = \frac{a'}{b'}, \quad \text{and similarly} \quad \frac{b}{c} = \frac{b'}{c'} \quad \text{and} \quad \frac{c}{a} = \frac{c'}{a'}$$

□

Remark. By the sin Theorem V.23.1 proved below, the ratio of any two sides of a triangle is the ratio of the sinus of the opposite sides:

$$\frac{a}{b} = \frac{\sin\alpha}{\sin\beta}, \ \frac{b}{c} = \frac{\sin\beta}{\sin\gamma}, \quad \text{and} \quad \frac{c}{a} = \frac{\sin\gamma}{\sin\alpha},$$

Hence we see once more why these ratios depend only of the angles of the triangle.

Proposition V.21.3. *For any two equiangular triangles, there exists a scaling factor f by which one needs to multiply the length of a side of the triangle $\triangle ABC$ to obtain the length of the corresponding side of the equiangular triangle $\triangle A'B'C'$:*

$$a' = fa, \ b' = fb, \quad \textit{and} \quad c' = fc$$

Proof. As explained above, the scaling factor is the common ratio $f := \frac{a'}{a} = \frac{b'}{b} = \frac{c'}{c}$. □

Proposition V.21.4 (Euclid VI.5). *Similar triangles are equiangular.*

Proof. Given are two triangles $\triangle ABC$ and $\triangle A'B'C'$ the lengths of the sides of which satisfy

$$\frac{a}{b} = \frac{a'}{b'}, \; \frac{b}{c} = \frac{b'}{c'} \quad \text{and hence} \quad \frac{a}{c} = \frac{a'}{c'}$$

As shown earlier, in a Hilbert plane with unique parallels, there exist equiangular triangles of arbitrary sizes. See the section on the natural axiomatization, Wallis' postulate and Theorem II.11.5.

We transfer the angle $\angle BAC$ to vertex A' and angle $\angle ABC$ to vertex B', each of both with one side along line $A'B'$. We obtain two new rays, both in the same half plane as point C'. The two new rays intersect at a point X. [51] We obtain the equiangular triangles $\triangle ABC$ and $\triangle A'B'X$, with the (possibly new) vertex X,— and both $\triangle A'B'X$ and $\triangle A'B'C'$ lying in the same half plane of line $A'B'$.

By Euclid VI.4 (stated in proposition V.21.2), the equiangular triangles $\triangle ABC$ and $\triangle A'B'X$ are similar. Let $|B'X| = a''$ and $|A'X| = b''$. We have obtained the proportions

$$\begin{aligned} &\frac{b}{c} = \frac{b''}{c'} \quad \text{and} \quad \frac{a}{c} = \frac{a''}{c'} \\ \text{Hence} \quad &\frac{b'}{c'} = \frac{b''}{c'} \quad \text{and} \quad \frac{a'}{c'} = \frac{a''}{c'} \\ \text{Hence} \quad &b' = b'' \quad \text{and} \quad a' = a'' \end{aligned}$$

We see that triangles $\triangle A'B'C'$ and $\triangle A'B'X$ have three pairs of congruent sides. Hence the two triangles are congruent by SSS-congruence Theorem II.7.22. Indeed, they are even equal by the corresponding Lemma for SSS Congruence II.7.2. Hence $X = C'$, and we have already shown that $\triangle ABC$ and $\triangle A'B'C'$ are similar. □

Proposition V.21.5 (Euclid VI.6). *If two triangles have one pair of congruent angles, and the sides containing these pairs are proportional, then the triangles are similar.*

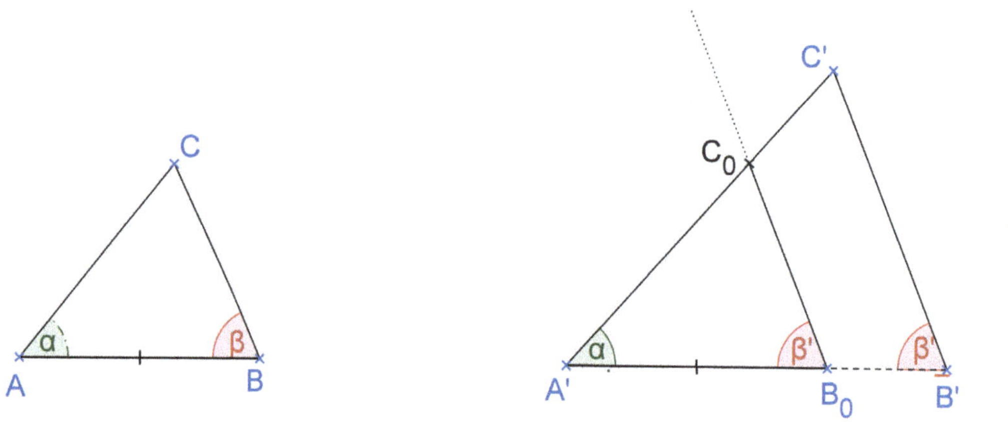

Figure V.21.3: Addendum to the extended ASA congruence.

[51] The existence of the intersection point X is shown Proposition II.11.2, telling equiangular triangles exist in all given sizes.

Proof. We use Euler's notation: angle α lies opposite to side a, angle β lies opposite to side b, and angle γ lies opposite to side c. Given are two triangles $\triangle ABC$ and $\triangle A'B'C'$ with congruent angles $\angle BAC \cong \angle B'A'C'$ and two pairs of adjacent sides b, c and b', c' which have proportional lengths:

$$\alpha = \alpha' \text{ and } \frac{b}{c} = \frac{b'}{c'}$$

Since segment length are comparable, we may assume $AB \leq A'B'$, without loss of generality. There exists a point B_0 in the segment $A'B'$ such that

$$AB \cong A'B_0 \tag{V.21.2}$$

There exists a ray $\overrightarrow{B_0C_0}$ pointing into the half plane of C' and making congruent angles $\beta' = \angle A'B'C' \cong \angle A'B_0C_0$. By congruent z-angles, the lines $B_0C_0 \parallel B'C'$ are parallel. Hence by Pasch's axiom exists the intersection point C_0 in the segment $A'C'$. For a Pythagorean plane, it is assumed the angle sum of any triangle is 180°. Hence the triangles $\triangle A'B_0C_0$ and $\triangle A'B'C'$ are equiangular. By Euclid VI.4 (stated in proposition V.21.2), the equiangular triangles $\triangle A'B_0C_0$ and $\triangle A'B'C'$ are similar. We obtain the proportion

$$\frac{|A'C_0|}{|A'B_0|} = \frac{|A'C'|}{|A'B'|} = \frac{b'}{c'}$$

Hence the assumption

$$\frac{b'}{c'} = \frac{b}{c} = \frac{|AC|}{|AB|}$$

and $A'B_0 \cong AB$ imply $|A'C_0| = |AC|$ and finally

$$AC \cong A'C_0 \tag{V.21.3}$$

By equations (V.21.2) and (V.21.3) and $\angle CAB \cong \angle C_0AB_0$, the SAS congruence implies that the triangles $\triangle ABC$ and $\triangle AB_0C_0$ are congruent.

Since the triangles $\triangle A'B_0C_0$ and $\triangle A'B'C'$ are similar, we finally conclude the triangles $\triangle ABC$ and $\triangle A'B'C'$ are similar, as required. □

V.21.2 Some exercises

Problem V.21.1. *Given are two triangles with one pair of congruent angles, and the sides across these angles are proportional to a pair of sides adjacent to the angles. Are the two triangles always similar. Give examples, counterexamples, a general reason.*

Answer. The two triangles need not always to be similar. Take two triangles for which the assumption holds in the form

$$\gamma = \gamma' , \quad \frac{c}{a} = \frac{c'}{a'}$$

It can happen they are not similar. Actually, this is only possible if the angle $\gamma = \gamma' < 90°$ is acute, and $\frac{c}{a} < 1$. Take the example $\gamma = \gamma' = 30°$ and $\frac{c}{a} = \frac{c'}{a'} = \frac{3}{5}$. There exist two triangles with these pieces which are not congruent, and similar neither. See the figure on page 206. I call this figure a "dangling derrick." The important point is that angle γ and side c are *across* to each other.

Problem V.21.2. *For the two triangles it is assumed*

$$\alpha = \alpha' \, , \quad \frac{b}{c} = \frac{b'}{c'}$$

Are the two triangles always similar. Give examples, counterexamples, a general reason.

Answer. The similarity has just been stated in Proposition V.21.5. The sides b and c the ratio of which is given are adjacent to the given angle α. Hence Euclid VI.6 tells that the two triangles are similar.

Problem V.21.3. *Draw two circles of equal radii, the center of one lying on the circumference of the other one. Draw the line through the two centers. Use some of the intersection points you have just obtained, and draw two triangles with angles* $30°, 30°, 120°$ *of different sizes, in two different colors.*

Let the lengths of sides of the larger triangle be a *and* c*, and the lengths of the sides of the smaller one be* a' *and* c'*. Determine the ratio* $\frac{a}{a'} = \frac{c}{c'}$*.*

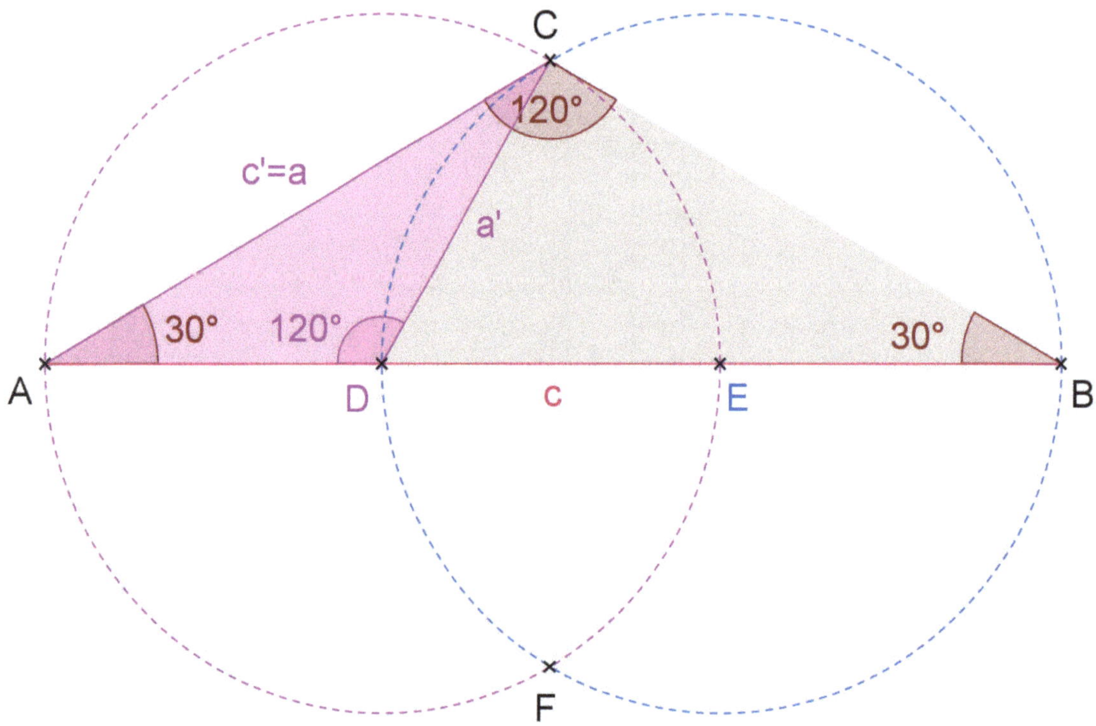

Figure V.21.4: Two isosceles similar triangles with angles of $30°$ and $120°$.

Answer. The line through the two centers A and B intersects the two circles in points D and E, too. In the figure on page 458, these four points have the order $A * D * E * B$. Let C be an intersection point of the two circles. For example, we get two isosceles similar triangles $\triangle ABC \sim \triangle ACD$. Both have the base angles $30°$ and top angle $120°$.

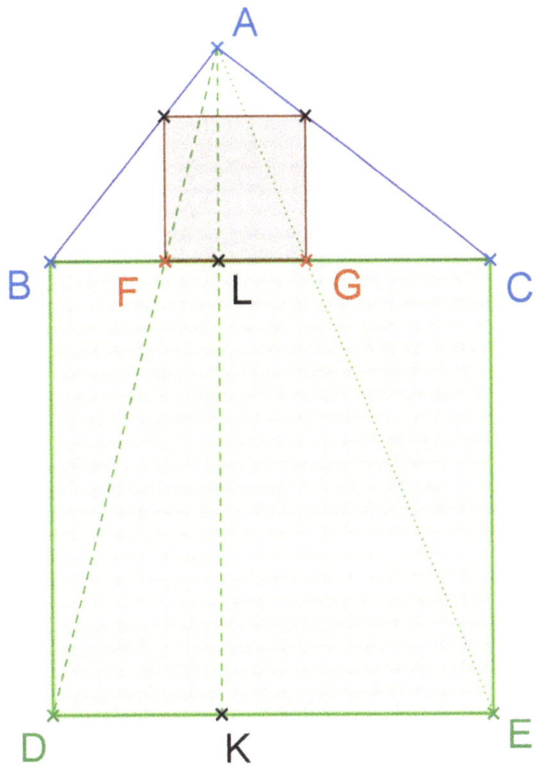

Figure V.21.5: Inscribing a square into a triangle.

We determine the ratio $\frac{a}{a'} = \frac{c}{c'}$ of the sizes of the two triangles. Since $c' = a$ and $c = 3a'$, we get

$$\frac{a}{a'} = \frac{c}{c'} = \frac{3a'}{a}$$

$$\frac{a^2}{a'^2} = 3 \quad \text{and} \quad \frac{a}{a'} = \sqrt{3}$$

Problem V.21.4. *Given is a triangle. Inscribe a square with one side part of a side of the triangle, and all vertices on the sides of the triangle.*

Answer. We erect a square onto one side a of the triangle. We connect the two new vertices of the square to the third vertex A of the triangle. The connecting lines intersect the side a in two points, which are the endpoints of one side of the square to be inscribed. Finally, we draw the inscribed square.

Problem V.21.5. *A square is inscribed with one side lying on the side a of the triangle, and the other two vertices lying on the two remaining sides of the triangle. Determine the side x of the square in terms of the side a and the area F of the triangle.*

Answer. We drop the perpendicular from vertex A onto the opposite side. From the similar triangles $\triangle ADK \sim \triangle AFL$, we get the proportions

$$\frac{|DK|}{|AK|} = \frac{|FL|}{|AL|}$$

Once more one gets proportions form the similar triangles $\triangle AEK \sim \triangle AGL$, and adds

$$\frac{|EK|}{|AK|} = \frac{|GL|}{|AL|}$$
$$\frac{|DE|}{|AK|} = \frac{|FG|}{|AL|}$$
$$\frac{a}{a+h} = \frac{x}{h}$$

were h is the altitude. Hence

$$x = \frac{ah}{a+h} = \frac{2aF}{a^2+2F}$$

Problem V.21.6. *Suppose that the endpoints of one side of the square are midpoints of two sides of the triangle. Show by dissection that the area of the square is half the area of the triangle.*

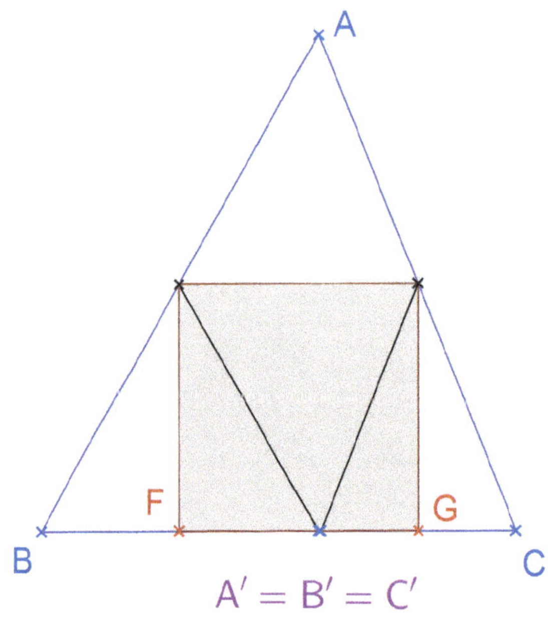

Figure V.21.6: For $h = a$, one can fold the triangle to cover the square twice.

Problem V.21.7. *Is the area of the square from the problem above less, more or equal to half of the area of the triangle. Give a simple folding argument to decide without any calculation.*

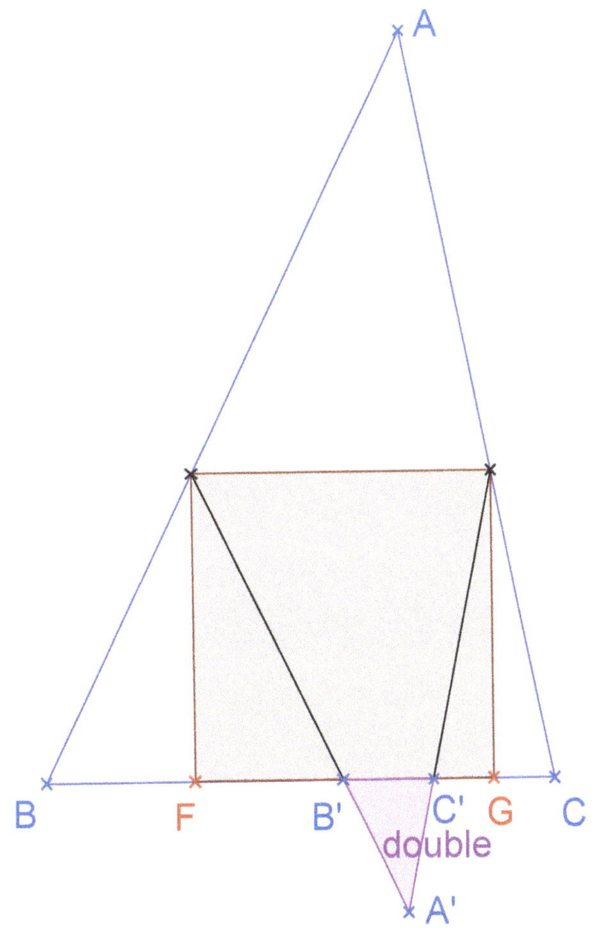

Figure V.21.7: For $a < h$, one can fold the triangle to cover the square twice, and triangle $\triangle A'B'C'$ left over.

Answer. For the special case that the endpoints of one side of the square are midpoints of two sides of the triangle, the area of the square is half of the area of the triangle. In the figure on page 460 you see this borderline case. In all other cases, the area of the square is less than half of the area of the triangle.To see this, we fold the triangle along three sides of the square. We get three reflection images A', B' and C' of the vertices of the triangle. Because of the right angles of the square, the folded sides touch along the lines $A'B'$ and $A'C'$. In the figure on pages 461, you see the case that triangle side $a < h$ is less than the altitude h, and in the figure on page 462, the case that triangle side $a > h$ is longer than the altitude. In the first case, one covers the square and an extra triangle $\triangle A'B'C'$ is left over in the opposite half plane. In the second case, one again covers the square, and the part of the extra triangle $\triangle A'B'C'$ inside the square is covered twice.

Problem V.21.8. *Given are two intersecting lines l and m and a point A different from their intersection point. We have to construct a circle through the point A touching both lines. Describe the construction given in the figure on page 462.*

Answer.

- We draw the angular bisector of the the two given lines, bisecting the angle α in the interior of which the given point A lies.

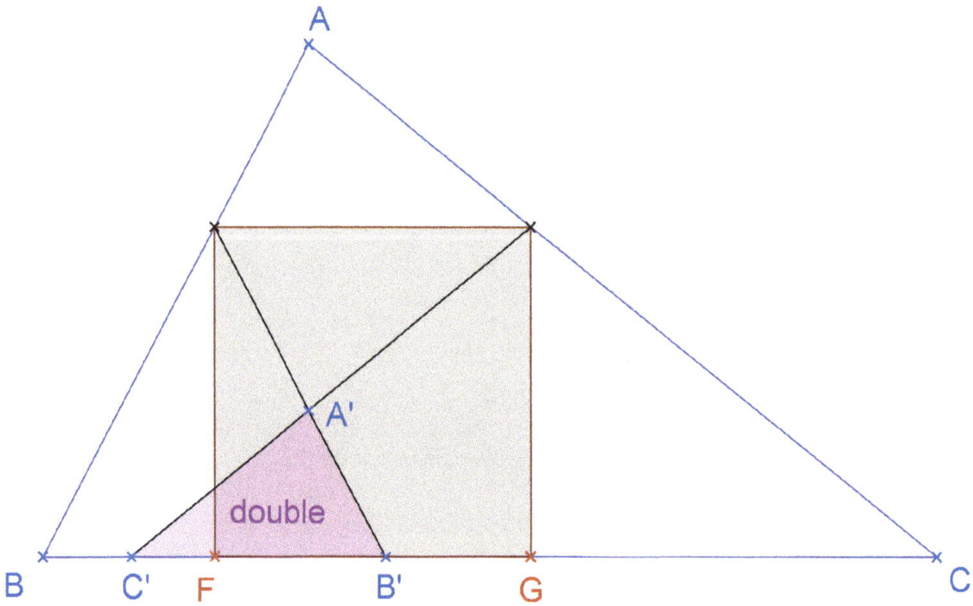

Figure V.21.8: For $a > h$, one can fold the triangle to cover the square twice, and part of triangle $\triangle A'B'C'$ is covered twice.

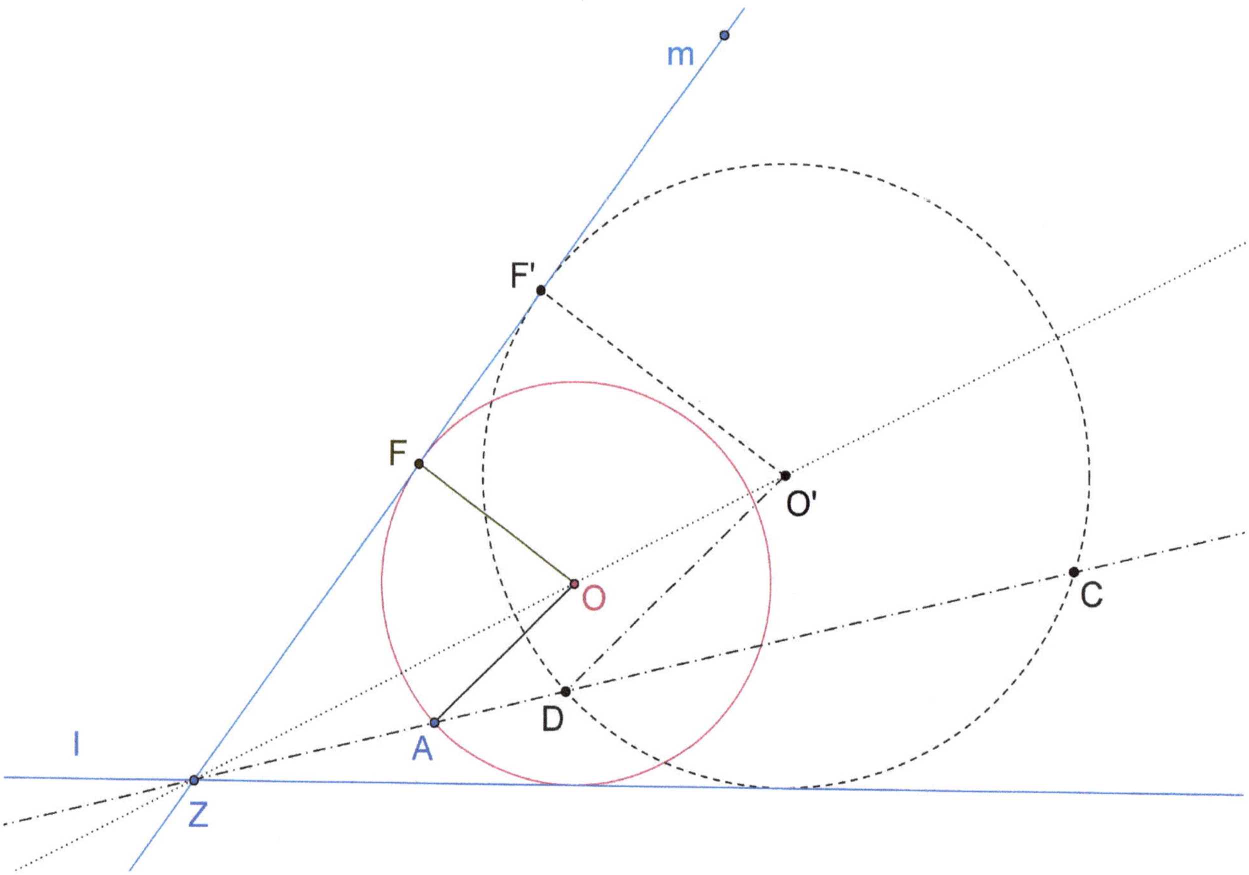

Figure V.21.9: Constructing a circle through a point and touching two lines.

- We choose any point O' on the bisector and draw a circle around it touching both line l and line m.
- We draw the ray emanating from the intersection point Z of the two given lines and pointing into the interior of the angle α. This ray intersects the circle in two points C and D.
- The parallel to line DO' through point A intersects the bisector in point O. The circle around O through point A does touch the two given lines and solves the problem.
- A second solution is obtained by drawing the parallel to line CO' through point A. This line intersects the bisector in the center O_2. The circle around O_2 through point A does touch the two given lines and yields the second solution of the problem, which has not been drawn.

Reason for the construction. The two triangles $\triangle ZAO \sim \triangle ZDO'$ are equiangular because the lines AO and DO' are parallel by construction. Because the lines FO and $F'O'$ are both perpendicular to the same line m and hence parallel, the triangles $\triangle ZFO \sim \triangle ZF'O'$ are equiangular . Since equiangular triangles are similar, we get the proportions

$$\frac{ZO}{AO} = \frac{ZO'}{DO'} \quad \text{and} \quad \frac{ZO}{FO} = \frac{ZO'}{F'O'}$$

Division yields

$$\frac{AO}{FO} = \frac{DO'}{F'O'} = 1$$

Indeed, the latter quotient equals 1 since both points D and F' lie on a circle around O' . Hence the former quotient equals one, too. Hence both points A and F lie on a circle $\mathcal{C}$ around O. This means that the constructed circle $\mathcal{C}$ touches the line m . Since the centers O' and O lie on the angle bisector of the two given lines l and m, the circle $\mathcal{O}$ touches the line l, too. □

Problem V.21.9. *Given are two points A and B and a line l. Construct a circle through the two points touching the line.*

First ideas. It is useful to consider these cases:

(a) One of the points lies on the line.

(b) The lines AB and l are different and parallel.

(c) The lines AB and l intersect perpendicularly.

(d) The lines AB and l intersect, at any other angle.

We have solved case (a) earlier in Problem II.14.11 in the section "A simplified axiomatic system of geometry". I leave cases (a) and (b) to the reader.

Question. What assumption about the two points and the line is needed that the problem can be solvable is cases (c) and (d).

Answer. The two points A and B need to lie on the same side of line l, or one of them on the line. The most interesting case (d) is explained below.

□

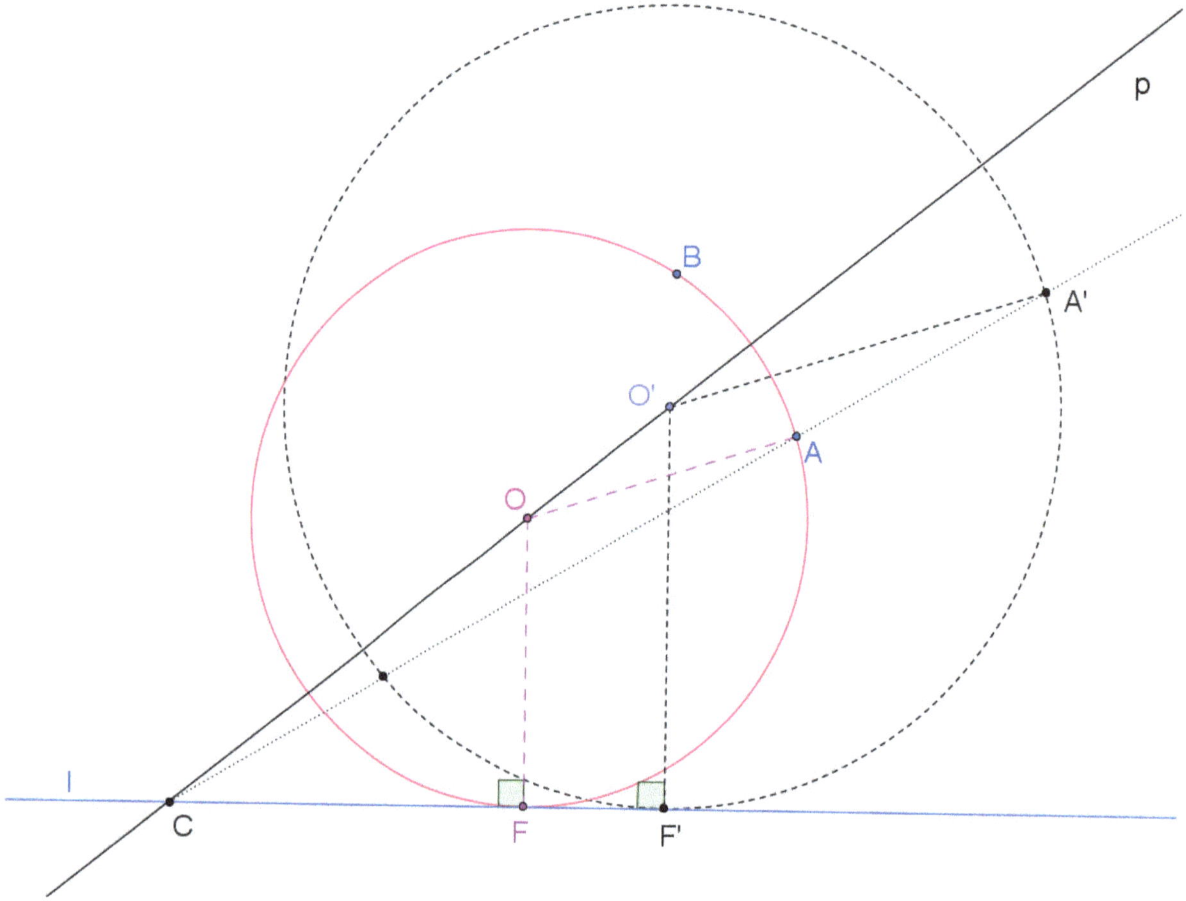

Figure V.21.10: Constructing a circle through two points touching a line.

Problem V.21.10. *Given is a line l, and two points A and B on the same side of this line. It is assumed that the lines AB and l intersect, but not perpendicularly. Construct a circle through the two points touching the line.*

Construction for Problem V.21.10. Let p the perpendicular bisector of segment AB. The lines p and l intersect, say at point C. We use point C as a center for similarities.

We choose any point $O' \neq C$ on the line p which lies on the same side as points A and B. We draw the circle $\mathcal{D}'$ around O' touching the line l, say in point F'. This circle can be mapped by a similarity with center C to the circle to be constructed.

We choose an intersection point A' of $\overrightarrow{CA} \cap \mathcal{D}'$. The parallel of line $A'O'$ intersects the perpendicular bisector p in point O. We construct the circle $\mathcal{D}$ around O touching the line l in point F.

□

Complete justification for Problem V.21.10.

Question. Why do the two rays $\overrightarrow{CA}$ and $\overrightarrow{CB}$ both intersect the circle $\mathcal{D}'$.

Answer. One of the points A and B lies in the interior of the angle $\angle O'CF'$, say point A. By the Crossbar Theorem [52] the segment $O'F'$ and the ray $\overrightarrow{CA}$ intersect, say in point Q.

[52] A segment with endpoints on the two sides of an angle and a ray emanating from its vertex into the interior of the angle intersect. See the section about the axioms of order, Proposition II.6.9

The point Q is in the interior of circle $\mathcal{D}'$, hence the line CA intersects the circle $\mathcal{D}'$. Since all points of a circle lie on the same side of the tangent as the center, even the ray $\overrightarrow{CA}$ intersects the circle $\mathcal{D}'$, indeed in two points. These are rather obvious consequences of the circle-line intersection property, see Proposition II.10.12 and Proposition II.10.6 from the section of neutral geometry of circles and continuity.

The two rays $\overrightarrow{CA}$ and $\overrightarrow{CB}$ are images obtained from each other by reflection across the line p. The circle $\mathcal{D}'$. is symmetric to this line. Hence the ray $\overrightarrow{CB}$ intersects the circle $\mathcal{D}'$, indeed in two further points.

Question. By means of similar triangles, check that circle $\mathcal{D}$ goes through point A.

Answer. The triangles $\triangle COA$ and $\triangle CO'A'$ are equiangular by construction, and hence similar:

$$\frac{|CO|}{|OA|} = \frac{|C'O'|}{|O'A'|}$$

The right triangles $\triangle COF$ and $\triangle CO'F'$ are equiangular by construction, and hence similar:

$$\frac{|CO|}{|OF|} = \frac{|C'O'|}{|O'F'|}$$

Hence

$$\frac{|OA|}{|OF|} = \frac{|O'A'|}{|O'F'|} = 1$$

by calculation, and since A' and F' lie on a circle around O'. Hence A and F lie on a circle around O, as to be shown.

□

V.21.3 Secants in a circle

Theorem V.21.2 (Theorem of chords (Euclid III.35)). *If two chords cut each other inside a circle, the product of the segments on one chord equals the product of the segments on the other chord.*

Assume two chords intersect each other outside a circle. The product of the segments, measured from the point of intersection to the two intersection points with the circle are equal for both chords.

Problem V.21.11. *Provide a drawing, with appropriate notation.*

Proof using proportions. Let AB and CD be two chords of circle γ intersecting at point P inside the circle. We have to check whether $|PA| \cdot |PB| = |PC| \cdot |PD|$. Following Legendre, I am using proportions. Equiangular triangles are, by definition, triangles with pairwise congruent angles at their corresponding vertices. Congruent circumference angles, as shown in Euclid III.21, occur at vertices A and D, as well as at vertices C and D. Furthermore, we get congruent vertical angles at vertex P. The vertices of the equiangular triangles have to be listed in such an order that these congruent angles occur at corresponding vertices. Hence we get the equiangular triangles

$$\triangle PAC \sim \triangle PDB$$

By Euclid VI.4

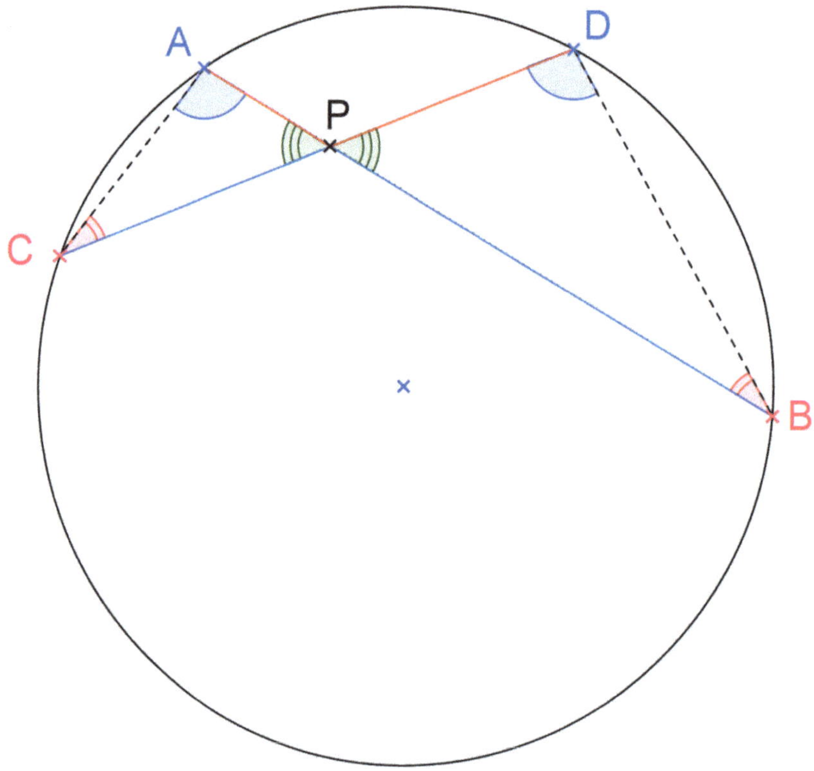

Figure V.21.11: Theorem of chords, case of chords intersecting inside the circle

`The sides of equiangular triangles are proportional.`

Hence the ratios of corresponding sides are the same for two equiangular triangles.

$$\frac{|PA|}{|PC|} = \frac{|PD|}{|PB|} \tag{V.21.4}$$

Now multiplying with the denominators yields

$$|PA| \cdot |PB| = |PC| \cdot |PD|$$

which is just the claim of Euclid III.35.

The proof for the case of the segments intersecting outside the circle is almost identical. But note that the two segments QA and QB now overlap. □

For the case where the extension of the chords intersect outside the circle, one obtains a further result by considering the tangent as a limit of a small secant. In the figures on page 467, we have illustrated the limit $C \to T, D \to T$, where T is the touching point of a tangent drawn from point Q to the circle. Thus one is lead to claim Euclid's next result.

Theorem V.21.3 (Theorem of chord and tangent (Euclid III.36)). *From a point outside a circle, a tangent and a second are drawn. The square of the tangent segment equals the product of the segments on the chord, measured from the point outside to the two intersection points with the circle.*

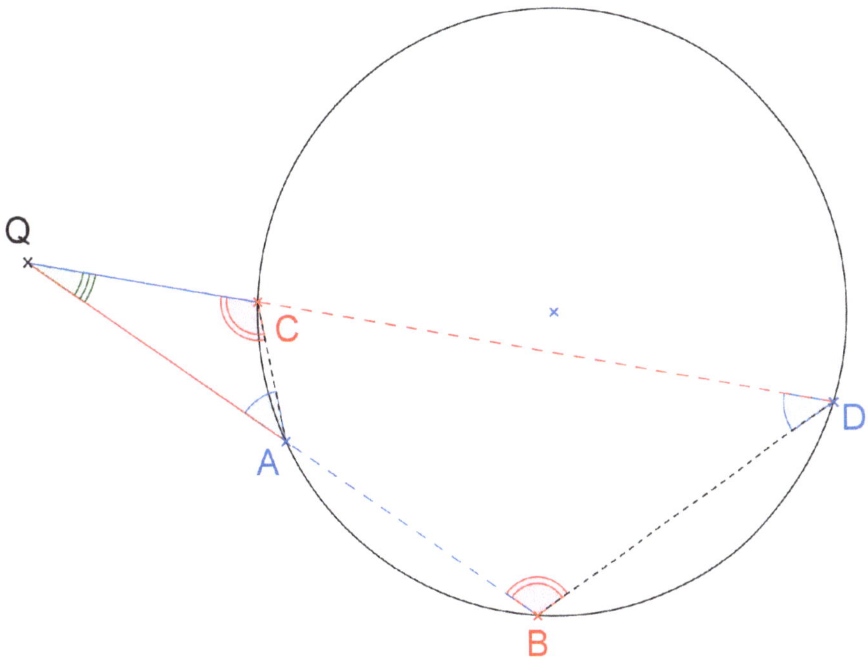

Figure V.21.12: Theorem of chords, case of chords intersecting outside the circle

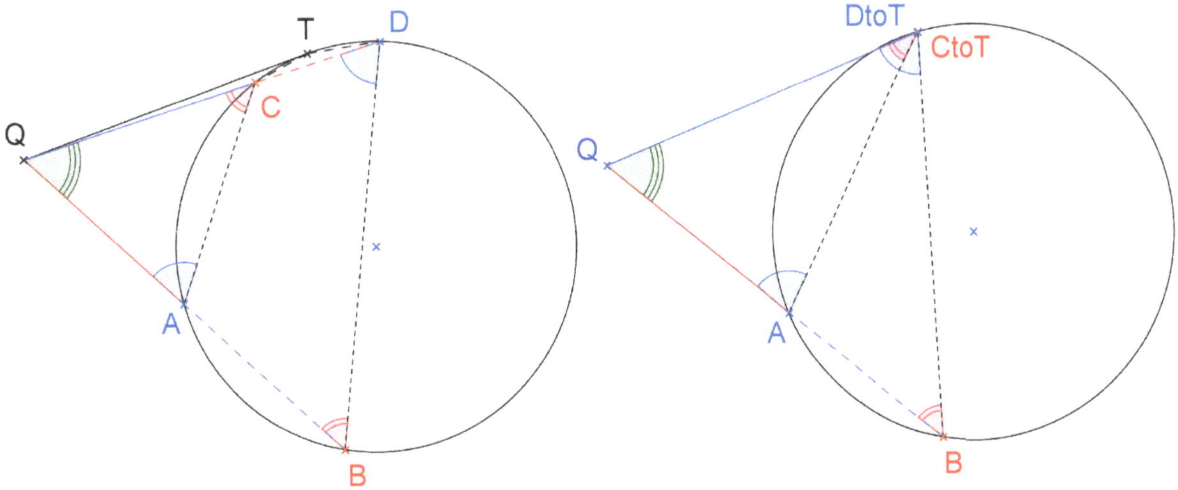

Figure V.21.13: Getting the limit $C \to T, D \to T$

Problem V.21.12. *Provide a drawing, with appropriate notation.*

Independent proof, again using proportions. Let Q be a point outside circle γ, let T be the touching point of the tangent to the circle through Q, and let AB be a chord of the circle the extension of which runs through Q. We assume that B lies between Q and A.

I have to check whether $|QT|^2 = |QA| \cdot |QB|$. To this end, one compares the triangles $\triangle QAT$ and $\triangle QTB$. We use Euclid III.32:

```
The angle between a tangent line and a chord is congruent  to the circumference
   angle of the arc corresponding to the chord.
```

Hence the angles at vertices A and T, for the two triangles, respectively, are congruent. With vertices listed in an order that these congruent angles occur at corresponding vertices, we get equiangular triangles $\triangle QAT \sim \triangle QTB$. By Euclid VI.4:

```
The sides of equiangular triangles are proportional.
```

Hence the ratios of corresponding sides are the same for two equiangular triangles.

$$\frac{|QA|}{|QT|} = \frac{|QT|}{|QB|} \tag{V.21.5}$$

and multiplying with the denominators yields the result. □

V.21.4 The lens equation

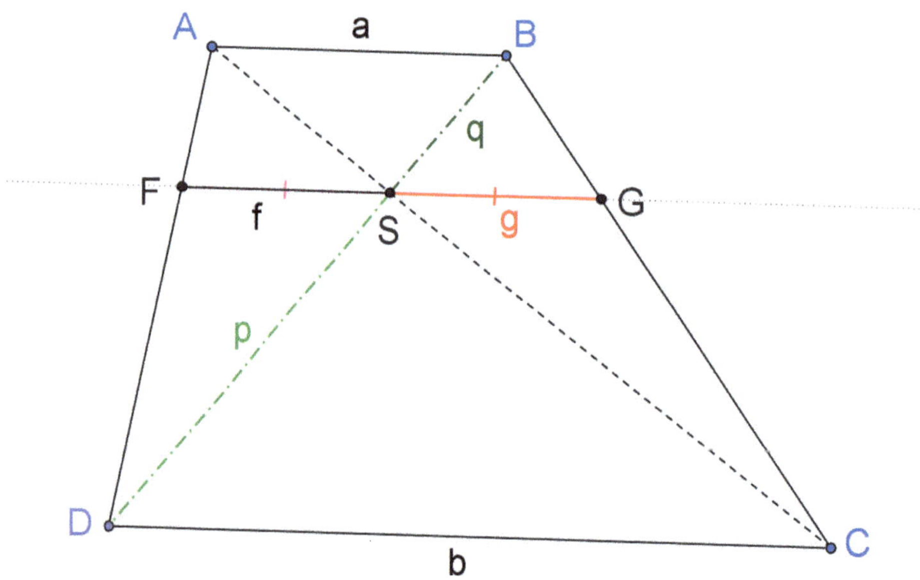

Figure V.21.14: A trapezoid and the lens equation.

Proposition V.21.6. *In a trapezoid $\square ABCD$ is drawn a third parallel to the two parallel sides through the intersection point S of the diagonals. The trapezoid sides cut congruent segments $f = FS$ and $g = SG$ out of the third parallel.*

(i) *The segments $f = FS$ and $g = SG$ cut out on the third parallel are congruent.*

(ii) *The lengths of the parallel sides $a = AB$ and $b = CD$ and the segment $f = FS$ satisfy the lens equation*

$$\frac{1}{a} + \frac{1}{b} = \frac{1}{f} \tag{V.21.6}$$

Proof. We introduce the segments $p = DS$ and $q = SB$. Now we use similar triangles to get three equations among the lengths a, b, f, g, p, q

Question. Use two similar triangles with common vertex D to find a proportion among these quantities.

Answer. The triangles $\triangle DSF$ and $\triangle DBA$ are equiangular and hence similar. One gets the proportion

$$\frac{p}{f} = \frac{p+q}{a} \tag{V.21.7}$$

Question. Use two similar triangles with common vertex B to find a proportion among these quantities.

Answer. The triangles $\triangle BSG$ and $\triangle BDC$ are equiangular and hence similar. One gets the proportion

$$\frac{q}{g} = \frac{p+q}{b} \tag{V.21.8}$$

Question. Use two similar triangles with common vertex S to find a proportion among these quantities.

Answer. The triangles $\triangle SAB$ and $\triangle SCD$ are equiangular and hence similar. One gets the proportion

$$\frac{a}{q} = \frac{b}{p} \tag{V.21.9}$$

We need now to eliminate p and q from these three equation. Dividing equation (V.21.7) by equation (V.21.8) yields

$$\frac{p \cdot g}{f \cdot q} = \frac{b}{a}$$

Together with equation (V.21.9) one gets

$$\frac{p \cdot g}{f \cdot q} = \frac{p}{q}$$

and hence $f = g$ as required.

Question. Use this result now to get the lens equation out of the two equations (V.21.7) and (V.21.8).

Answer. We add these two equations and divides by $p + q$ to obtain the lens equation (V.21.6).

□

The trapezoid from the figure on page 468 appears in the context of geometric optics. Recall that the perpendicular to the plane of the lens through its center is called the *optical axis*. For the light rays through a thin lens and close to the optical axis, one can empirically gets to the following three observations:

(a) Light rays parallel to the optical axis coming in,—say from the left—are bundled at the focus G behind the lens.

(b) There is a second focus F on the other side of the lens at the same distance. Light rays starting at focus F are bend on the other side of the lens into rays parallel to optical axis.

(c) Light rays,, even those coming in from different directions, go straight through the center of the lens.

These assertions are in reality only an approximation, but they are exact enough for a thin lens to be useful. These assumption lead to a trapezoid, which has the unusual form shown in the figure on page 470. The third parallel FG from above turns out to be the optical axis. The light source

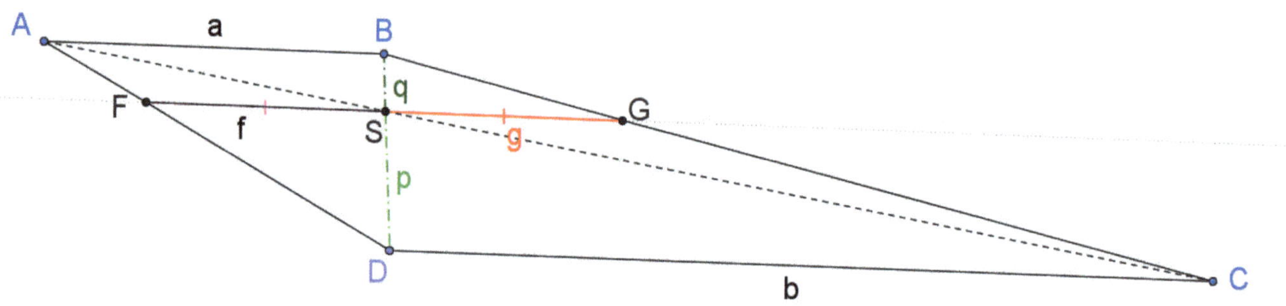

Figure V.21.15: The trapezoid appearing for an optical lens.

is located at vertex A, the lens lies in the plane BD. The light rays emanated point A are focused at vertex C. We can check this to true at least for three rays:

- the ray AD through the left focus F, which is bend to the ray CD parallel to the optical axis;
- the ray AB parallel to the optical axis, which is bend to the ray BC through the right focus G;
- the ray ASC which goes straight through the center S of the lens.

Too, Proposition V.21.6 now confirms the well-known lens equation (V.21.6).

Problem V.21.13. *A candle is standing at distance $a = 3$ from a lens with focus length $f = 2$. At which distance b behind the lens has one to put a screen to get a sharp image of the candle flame. Give a construction and calculate the distance b.*

V.22 The Pythagorean and Related Theorems

V.22.1 The leg and the altitude theorems

Problem V.22.1. *The figure on page 471 explains the common definitions for the lengths a, b, c and p, q, h from a right triangle. Find an example where all six segments have lengths with of integer values.*

Answer. The smallest example, I can find is

$$a = 15\,,\ b = 20\,,\ c = 25 \quad \text{and} \quad p = 9\,,\ q = 16\,,\ h = 12$$

Problem V.22.2. *Explain for which special situation Euclid III.35 implies the theorem about the altitude of a right triangle, usually stated as $h^2 = pq$.*

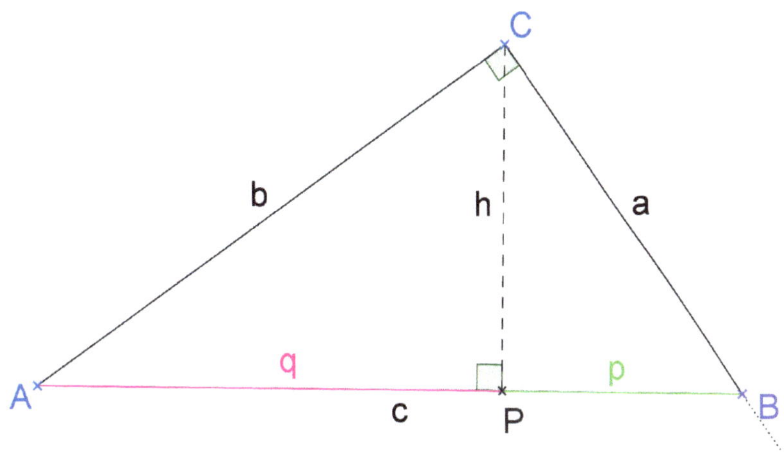

Figure V.22.1: The common notation for a right triangle

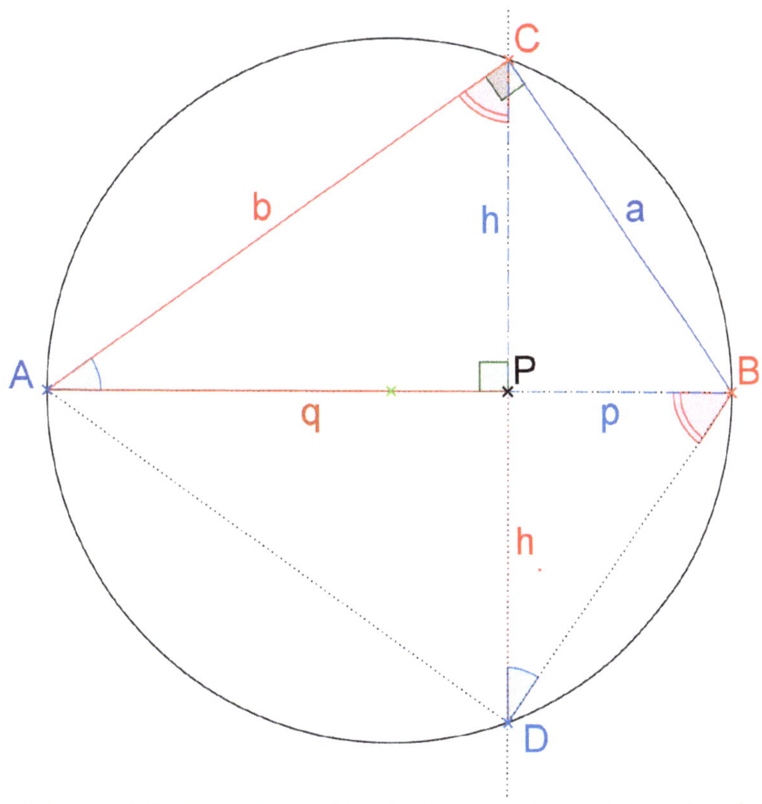

Figure V.22.2: The altitude theorem for a right triangle

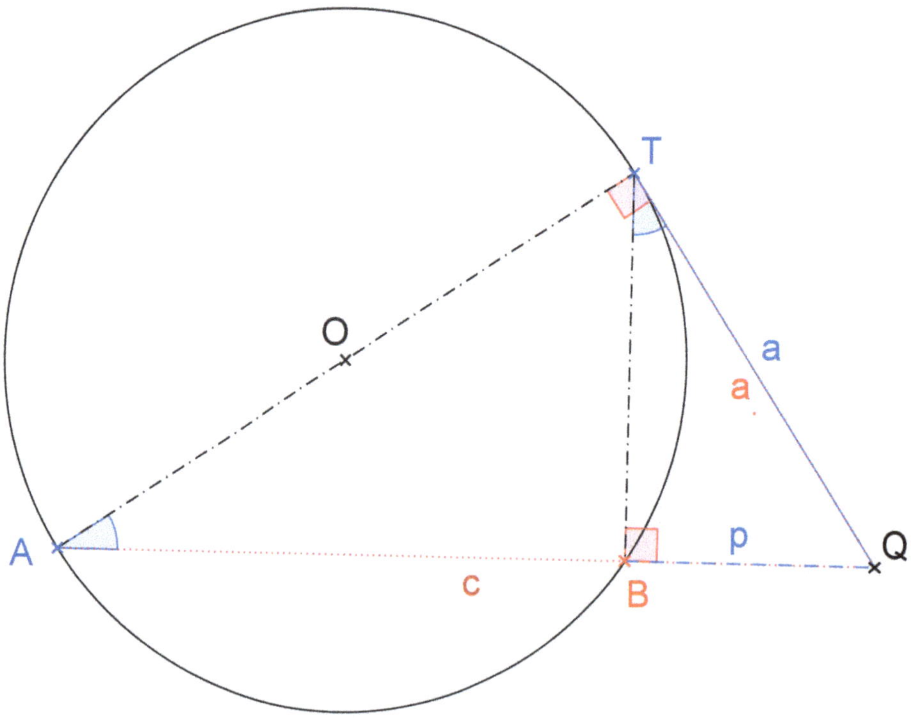

Figure V.22.3: The leg theorem

Answer. Take for segment AB a diameter of circle γ and chose CD to be any segment perpendicular to that diameter. One gets a right $\triangle ABC$, with P as foot point of its altitude. Because of $PC \cong PD$ and $|PC| = |PD|$, Euclid III.35 implies $pq = |PA| \cdot |PB| = |PC| \cdot |PD| = h^2$. □

Problem V.22.3. *Explain how Euclid III.36 implies the leg theorem $a^2 = pc$.*

Answer. In Euclid III.36, as given in theorem V.21.3, we choose points A and T on a diameter of circle γ, and point Q on the tangent to the circle at point T. We see that $\triangle ATQ$ is a right triangle, because tangent and radius are perpendicular to each other. The segment AQ intersects the circle in a second point B. The triangle $\triangle ATQ$ has altitude TB, because Thales' theorem shows that the angle $\angle ABT$ is right. Now Euclid III.36 tells that the square of one leg QT equals the product of the hypothenuse time the projection of that leg onto the hypothenuse. After renaming $T \to C, B \to P, Q \to B, A \to A$, one gets the statement in its usual form $a^2 = pc$. □

V.22.2 The Pythagorean theorem

Problem V.22.4. *Explain how Euclid III.36 implies the Pythagorean Theorem $a^2 + b^2 = c^2$.*

Answer. For the given right triangle $\triangle ABC$, we draw circle with center A through point C. We draw the tangent at point C and see that point B lies on the tangent since tangent and radius are perpendicular to each other.

We draw the ray $\overrightarrow{BA}$. Let points E and D be the endpoints of the diameter on this ray. Now Euclid III.36, as given in theorem V.21.3, tells that

$$|BC|^2 = |BD| \cdot |BE|$$

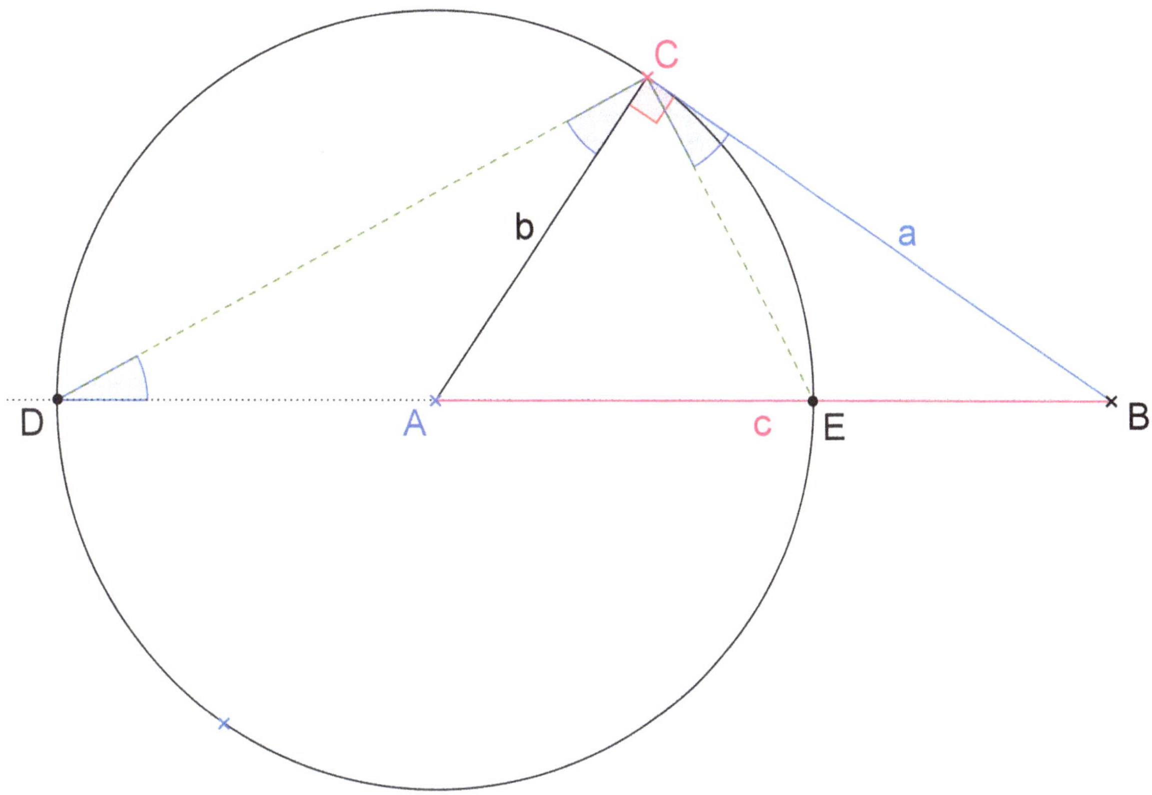

Figure V.22.4: The Pythagorean Theorem

In terms of the sides of the triangle $\triangle ABC$, this shows that

$$a^2 = (c+b)(c-b) = c^2 - b^2 \quad \text{and } a^2 + b^2 = c^2$$

We got the Pythagorean Theorem. □

Remark. In the spirit of Euclid, one has to understand the product of segments occurring in the theorems above as *areas of rectangles or squares.* The equality of their areas can really be shown by obtaining one from the other in a finite sequence of cuts and pastes. In the figure below, I want to visualize this interpretation. Take the case of chords intersecting at point Q outside the circle. We extend the two chords to the other side of their intersection point and transfer the segments QA and QC to these opposite rays to obtain a quadrilateral $\square A'ACC'$ with axial symmetry. It is not hard to check that this quadrilateral has a circum circle and the triangles $\triangle QAC \cong \triangle QA'C'$ are congruent.

The two circle lemma implies that the chords $A'C' \parallel BD$ are parallel. Hence we have obtained a new pair of similar triangles

$$\triangle QA'C' \sim \triangle PDB$$

which have the center of similarity Q.

The figure also contains two triangles of equal area, which confirm Euclid's theorem III.35 directly as an *equality among areas.* Since $A'C' \parallel BD$, the triangles

$$|\triangle A'C'B| \doteq |\triangle A'C'D|$$

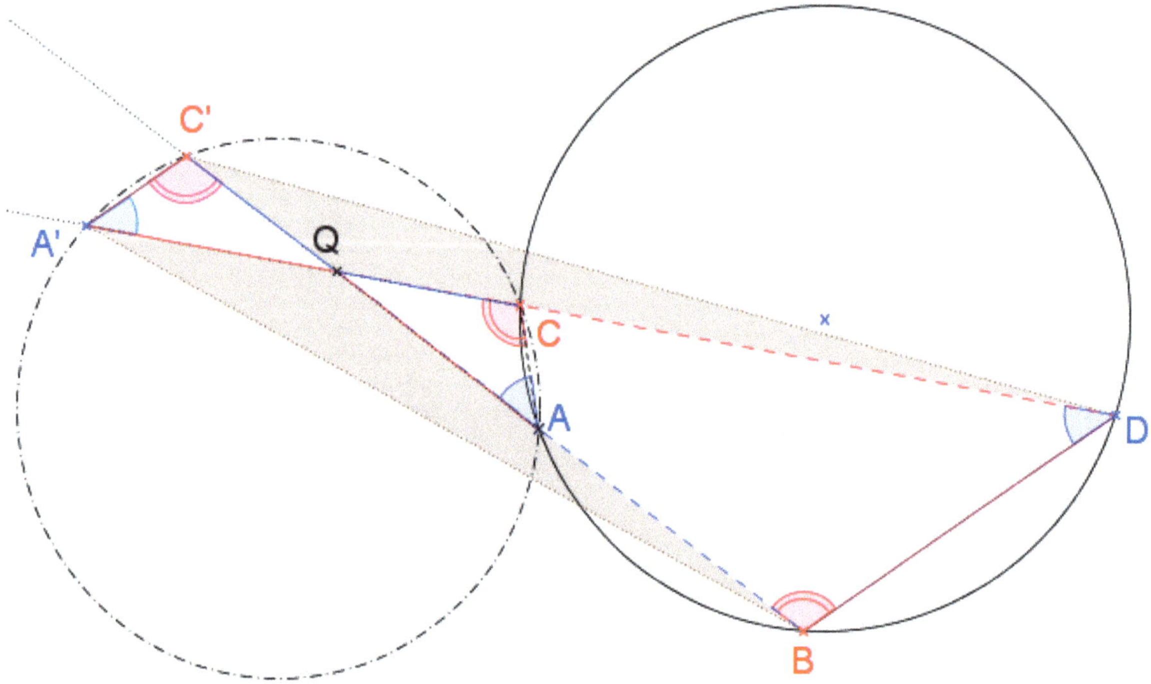

Figure V.22.5: A pair of triangles $\triangle A'QB$ and $\triangle C'QD$ of equal area directly confirms the theorem of chords

have equal base $A'C'$ and equal height, and hence equal areas. Their intersection is the triangle $\triangle QA'C'$. We subtract this triangle from both sides, and obtain the pair of triangles

$$|\triangle A'QB| \doteq |\triangle C'QD|$$

again of equal areas. Now the claim of Euclid III.35—

$$|QA'| \cdot |QB| = |QC'| \cdot |QD|$$

can easily be obtained as an equality among areas.

V.22.3 Some number theory inspired by Pythagoras' theorem

Problem V.22.5. *In the figure on page 471 is shown the common notation for the six segments a, b, c and p, q, h which occur in a right triangle. Any two among the six segments a, b, c and p, q, h determine the other four ones. We assume that the hypothenuse c and the altitude h are given. Find explicit formulas for the legs a and b and the projections p and q in terms of the hypothenuse c and the altitude h. To determine the sign of the roots occurring in the formulas, we assume that $a \geq b$.*

Answer. From the Pythagorean theorem $a^2 + b^2 = c^2$ and the formula $ab = ch$, one obtains $a + b$.

$$(a+b)^2 = a^2 + b^2 + 2ab = c^2 + 2ch$$
$$a + b = \sqrt{c^2 + 2ch}$$

From known product ab and sum $a+b$, one gets a quadratic equation for the unknown a. Thus one determines a and finally b. The sign of the roots is determined by the assumption that $a \geq b$.

$$\begin{aligned} & a^2 - a\sqrt{c^2+2ch} + ch = 0 \\ & a = \frac{\sqrt{c^2+2ch} + \sqrt{c^2+2ch-4ch}}{2} = \frac{\sqrt{c^2+2ch} + \sqrt{c^2-2ch}}{2} \\ & b = \frac{\sqrt{c^2+2ch} - \sqrt{c^2-2ch}}{2} \end{aligned}$$

Now p and q can be calculated by the leg theorem.

$$\begin{aligned} p &= \frac{a^2}{c} = \frac{\left(\sqrt{c^2+2ch} + \sqrt{c^2-2ch}\right)^2}{4c} = \frac{2c^2 + 2\sqrt{(c^2+2ch)(c^2-2ch)}}{4c} \\ &= \frac{2c^2 + 2\sqrt{c^4 - 4c^2h^2}}{4c} = \frac{c + \sqrt{c^2-4h^2}}{2} \quad \text{and} \\ q &= \frac{c - \sqrt{c^2-4h^2}}{2} \end{aligned}$$

Lemma V.22.1. *For any solution of the equation $a^2+b^2=c^2$ with natural numbers a,b,c it is impossible than a and b are both odd.*

Proof. Assume towards a contradiction that a and b are both odd and $a^2+b^2=c^2$. There exists a natural number $n \geq 0$ such that $a = 2n+1$. One checks that $a^2 = (2n+1)^2 = 4(n^2+n)+1 \equiv 1 \mod 4$. Similarly one obtains $b^2 \equiv 1 \mod 4$. Hence $c^2 \equiv 2 \mod 4$ and c is even, but c^2 is not divisible by 4, which is impossible. □

Proposition V.22.1. *All solutions of the equation $a^2+b^2=c^2$ with natural numbers a,b,c, for which b divisible by a* higher *prime power of* 2 *than a, are given by the formulas*

$$a = g(u^2-v^2)\,, \quad b = 2guv\,, \quad c = g(u^2+v^2) \tag{V.22.1}$$

with integer $g \geq 1$ and $u > v \geq 1$. Moreover, if u and v are relatively prime and $u+v$ is odd, then $g = \gcd(a,b,c) = \gcd(a,b) = \gcd(a,c) = \gcd(b,c)$.

Proof. It is enough to deal with the case that $\gcd(a,b,c) = 1$. Hence $\gcd(a,b) = \gcd(a,c) = \gcd(b,c) = 1$ holds, too. By lemma V.22.1 one concludes that b is even, and a and c are odd. Hence $b+c$ and $c-b$ are both odd and $\gcd(c+b, c-b) = 1$. Because of

$$(c+b)(c-b) = c^2 - b^2 = a^2$$

and the uniqueness of the prime decomposition, we conclude that the numbers $c+b$ and $c-b$ are both perfect squares. We conclude

$$c+b = m^2\,, \quad c-b = n^2 \quad \text{and} \quad a = m \cdot n$$

where $m > n$ are both odd natural numbers. Define the natural numbers

$$u := \frac{m+n}{2} \quad \text{and} \quad v := \frac{m-n}{2}$$

Clearly $u > v$ are both natural numbers, they are relatively prime and $u+v=m$ is odd. We easily check that

$$b = \frac{m^2 - n^2}{2} = 2uv$$
$$c = \frac{m^2 + n^2}{2} = u^2 + v^2 \ , \quad \text{and finally}$$
$$a = m \cdot n = u^2 - v^2$$

to confirm that equations (V.22.1) hold with $g = 1$, as to be shown. □

Corollary 51. *The unit circle* $\{(x, y) \ : \ x^2 = y^2 = 1\}$ *contains infinitely many points for which both coordinates are rational.*

Reason. One puts

$$x = \frac{u^2 - v^2}{u^2 + v^2} \quad \text{and} \quad y = \frac{2uv}{u^2 + v^2}$$

with any integers $u, v \in \mathbf{Z}$. □

Remark. Using complex numbers, the equations (V.22.1) can be put together as

$$a + ib = g(u + iv)^2$$

Hence one gets once more

$$c^2 = (a + ib)(a - ib) = g^2(u + iv)^2(u - iv)^2 = g^2(u^2 + v^2)^2$$
$$\text{and} \quad c = g(u^2 + v^2)$$

Corollary 52. *In the ring* $\mathbf{Z} + i\mathbf{Z}$ *of Gaussian integers, the number* $a + ib$ *is a perfect square if and only if* $a^2 + b^2$ *is a perfect square.*

Problem V.22.6. *Find all cases for which all six segments* a, b, c *and* p, q, h *have integer values, and* $\gcd(a, b, c, p, q, h) = 1$. *Give formulas for these six segments in terms of integer solutions of* $a'^2 + b'^2 = c'^2$ *with* $\gcd(a', b') = 1$.

Proof. These integer solutions are

$$\text{(V.22.2)} \qquad \begin{aligned} a = a' \cdot c' \ , \quad b = b' \cdot c' \ , \quad c = c'^2 \ , \\ p = a'^2 \ , \quad q = b'^2 \ , \quad h = a' \cdot b' \end{aligned}$$

Here is a simple reason. From proposition V.22.1 we know that $a = a' \cdot g \ , \quad b = b' \cdot g \ , \quad c = c' \cdot g$ with $g = \gcd(a, c) = \gcd(b, c)$. From $ab = ch$ we get

$$h = \frac{ab}{c} = \frac{(a'b')g}{c'} = a'b' \cdot \frac{g}{c'}$$

which is assumed to be an integer. Now $\gcd(a', c') = 1$ and $\gcd(b', c') = 1$ imply $\gcd(a'b', c') = 1$. But c' divides the product $(a'b') \cdot g$. Hence Euclid's lemma implies than c' divides g. In other words, $\frac{g}{c'}$ is an integer. The remaining lengths are

$$p = \frac{a^2}{c} = a'^2 \cdot \frac{g}{c'} \quad \text{and} \quad q = b'^2 \cdot \frac{g}{c'}$$

We see that both are divisible by the integer $\frac{g}{c'}$, as are the lengths h as well as a and b. Hence the assumption $\gcd(a, b, c, p, q, h) = 1$ implies that $g = c'$. Finally, one obtains the formulas (V.22.2). □

Corollary 53. *In the case that all six segments a, b, c and p, q, h have integer values, and* $\gcd(a, b, c, p, q, h) = 1$*, all segments c, p, q and $c + 2h$ and $c - 2h$ are perfect squares.*

Reason. One see directly from the equations (V.22.2) that c, p, q are perfect squares. Moreover

$$c + 2h = c'^2 + 2a' \cdot b' = a'^2 + b'^2 + 2a' \cdot b' = (a' + b')^2$$
$$\text{and} \quad c - 2h = (a' - b')^2$$

□

V.22.4 Some (sandbox) applications

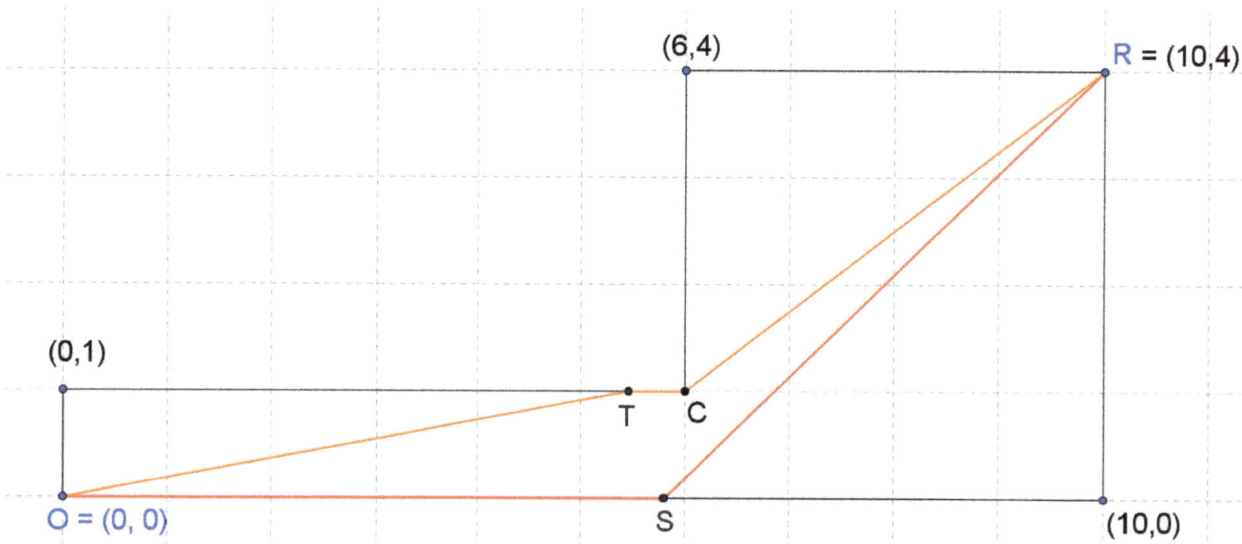

Figure V.22.6: Two equally good runners start at O and R. Do they better meet at point S or point T.

Problem V.22.7. *Two equally fast runners start at the opposite points O and R of the place shown in the figure on page 478. They are allowed to run across the place, but cannot enter any space outside the place. They want to meet on the boundary. Can they meet quicker at point S or at point T.*

(i) *Calculate the distance $|OS|$ for which $|OS| = |SR|$.*

(ii) *Calculate the point T such that $|OT| = |TC| + |CR|$.*

(iii) *Calculate the distance $|OT|$ and decide whether which distance is shorter $|OT|$ or $|OS|$.*

Answer. **(i)** Let $S = (x, 0)$ be the coordinates of the meeting point S. Since the distances $|OS| = |SR|$ are equal, one calculates

$$\begin{aligned} x &= \sqrt{(10-x)^2 + 4^2} \\ x^2 &= 100 - 20x + x^2 + 16 \\ x &= 5.8 \end{aligned}$$

(ii) Let the distance $|OT| = x$. Hence the coordinates of the meeting point are $T = (\sqrt{x^2 - 1}, 1)$. Since the distances $|OT| = |TC| + |CR|$ are equal, $|TC| = 6 - \sqrt{x^2 - 1}$ and $|CR| = \sqrt{3^2 + 4^2} = 5$ gives

$$\begin{aligned} x &= 11 - \sqrt{x^2 - 1} \\ x^2 - 1 &= (11-x)^2 = 121 - 22x + x^2 \\ x &= \frac{122}{22} = 5 + \frac{6}{11} \end{aligned}$$

(iii) We see that point T is where to meet quicker.

Remark. With $T = (t, 1)$

$$\begin{aligned}\sqrt{t^2+1} &= 6 - t + \sqrt{3^2+4^2} = 11 - t\\ t^2 + 1 &= 121 - 22t + t^2\\ t &= 5 + \frac{5}{11}\end{aligned}$$

We get the same distance

$$|OT| = |TC| + |CR| = (6-t) + 5 = 5 + \frac{6}{11}$$

V.22.5 The parallelogram equation

Proposition V.22.2. *The sum of the squares of the four sides of a parallelogram equals the sum of the squares of the diagonals.*

Remark. Any real numbers satisfy

$$(a+b)^2 + (b-a)^2 = 2a^2 + 2b^2$$

The corresponding equation holds for vectors in a space with an inner product. For a parallelogram, the vectors along the sides and diagonals are

$$\begin{aligned}a := \overrightarrow{AB}\ ,\ -a := \overrightarrow{CD}\ ,\\ b := \overrightarrow{BC} = \overrightarrow{AD}\\ a + b = \overrightarrow{AC}\ ,\ b - a = \overrightarrow{BD}\end{aligned}$$

In a vector space with an inner product, the corresponding identity

$$(a+b)^2 + (b-a)^2 = 2a^2 + 2b^2$$

implies that the sum of the squares of the diagonals of a parallelogram equals twice the sum of the squares of two adjacent sides. In this way, we have an analytic proof of the parallelogram equation.

Problem V.22.8 (Parallelogram equation). *The sum of the squares of the diagonals of a parallelogram equals the sum of the squares of its four sides. Prove the parallelogram equation*

$$|AC|^2 + |BD|^2 = 2|AB|^2 + 2|BC|^2$$

using the case and the notation of the figure on page 480.

Proof. For the parallelogram $\square ABCD$, we may assume $a = |AB| > |BC| = b$, and the angle at B to be obtuse. We drop the perpendiculars from vertex C onto the extension of side AB, and from vertex B onto side CD. The foot points are F and G, respectively. Let the rectangle $\square BFCG$ have side lengths $|BG| = |CF| = h$ and $|BF| = |CG| = k$. Apply Pythagoras' Theorem three times:

$$\begin{aligned}|AC|^2 &= |AF|^2 + |FC|^2 = (a+k)^2 + h^2 &&\text{for triangle } \triangle AFC;\\ |BD|^2 &= |BG|^2 + |GD|^2 = h^2 + (a-k)^2 &&\text{for triangle } \triangle BGD;\\ h^2 + k^2 &= |BF|^2 + |FC|^2 = |BC|^2 = b^2 &&\text{for triangle } \triangle BFC.\end{aligned}$$

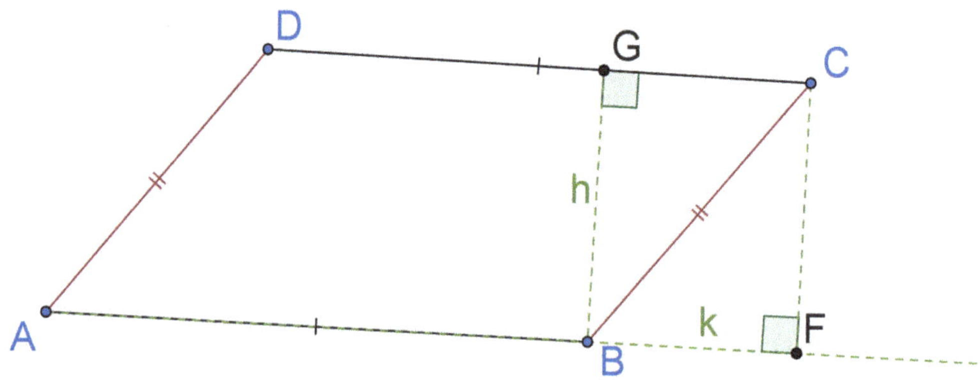

Figure V.22.7: Proof of the parallelogram equation.

Adding the first two equations yields

$$|AC|^2 + |BD|^2 = (a+k)^2 + (a-k)^2 + 2h^2 = 2a^2 + 2k^2 + 2h^2 = 2a^2 + 2b^2$$

We see that the sum of the squares of the diagonals of parallelogram $\square ABCD$ equals the sum of the squares of its four sides. $\square$

V.23 Trigonometry

V.23.1 The law of sines

Definition V.23.1 (sin function for acute and obtuse angles). The ratio $\frac{a}{c}$ of the leg a across the acute angle α to the hypothenuse of a right triangle is the *sin function* of this angle:

$$\sin\alpha = \frac{a}{c}$$

For right and acute angles, one defines

$$\sin(180^\circ - \alpha) = \sin\alpha \quad \text{and} \quad \sin 90^\circ = 1$$

Definition V.23.2 (cosine function for acute and obtuse angles). The ratio $\frac{b}{c}$ of the leg b adjacent the acute angle α to the hypothenuse of a right triangle is the *cosine function* of this angle:

$$\cos\alpha = \frac{b}{c}$$

For right and acute angles, one defines

$$\cos(180^\circ - \alpha) = -\cos\alpha \quad \text{and} \quad \cos 90^\circ = 0$$

Problem V.23.1. *Convince yourself that*

$$\cos\alpha = \sin(90^\circ - \alpha) \tag{V.23.1}$$

$$\tan\alpha = \frac{\sin\alpha}{\cos\alpha} \tag{V.23.2}$$

$$\frac{1}{\tan\alpha} = \tan(90^\circ - \alpha) \tag{V.23.3}$$

$$\tan(180^\circ - \alpha) = -\tan\alpha \tag{V.23.4}$$

Problem V.23.2. *Convince yourself that*

$$\sin^2\alpha + \cos^2\alpha = 1 \tag{V.23.5}$$

is equivalent to the Theorem of Pythagoras

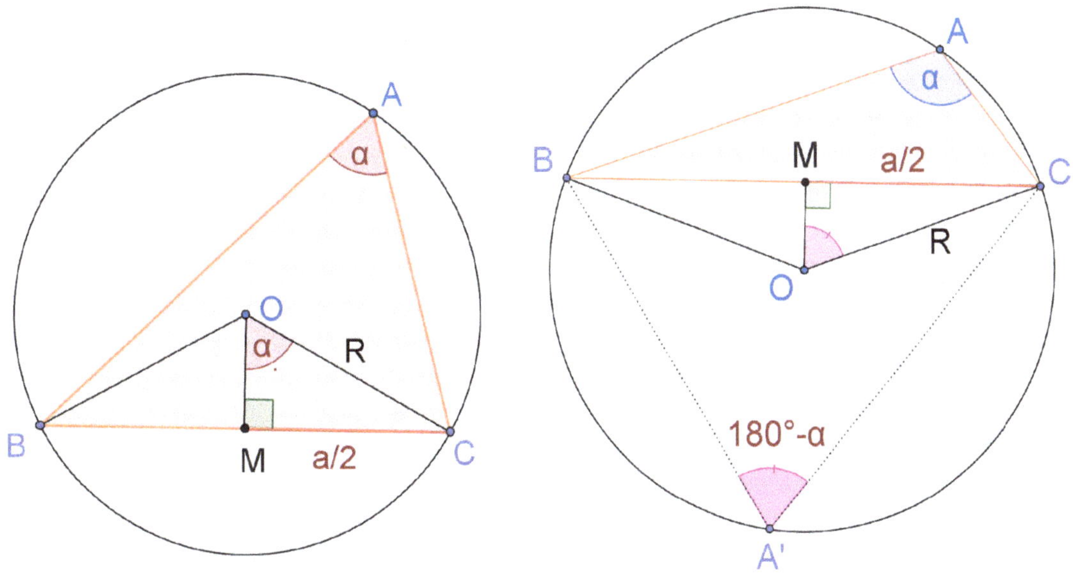

Figure V.23.1: Proof of the sin theorem for acute, and for obtuse angles.

Theorem V.23.1 (The extended sin Theorem). *For any triangle with sides a, b, c and angles α, β, γ, and radius R of the circum-circle*

$$\frac{a}{\sin\alpha} = \frac{b}{\sin\beta} = \frac{c}{\sin\gamma} = 2R$$

Proof. This is an easy consequence of (Euclid III.20) and (Euclid III.21). As stated in theorem III.17.4, the central angle is twice the circumference angle.

We put the circum circle around the given triangle. Let its radius be R. We connect the endpoints of side $a = BC$ with the center O, and drop the perpendicular from O this side. Let M be the foot point. The central angle is $\angle BOC = 2\alpha$, since $\angle BAC = \alpha$. It is bisected by the ray $\overrightarrow{OM}$. The side a is bisected by point M. In the right triangle $\triangle OMC$, the definition of the sin function yields

$$\sin\alpha = \frac{|CM|}{|OC|} = \frac{a}{2R}$$

Thus we obtain the required sin theorem for the acute angles of the triangle.

For the obtuse angle, we need to use (Euclid III.22), telling that the opposite angles of a circular quadrilateral $\square ABCA'$ add up to two right angles. We can take for A' any point on the arc opposite to arc BAC, and obtain $\angle BA'C = 180° - \alpha$. This time the central angle $\angle BOC = 2(180° - \alpha)$. Now this angle is bisected by the ray $\overrightarrow{OM}$ and we obtain

$$\sin(180° - \alpha) = \frac{|CM|}{|OC|} = \frac{a}{2R}$$

Hence, in order to obtain a sin theorem valid for both acute and obtuse triangles, one defines

$$\sin(180° - \alpha) = \sin \alpha$$

$\square$

V.23.2 The law of cosines

Theorem V.23.2 (The cosine theorem). *For any triangle with sides a, b, c and angles α, β, γ*

$$c^2 = a^2 + b^2 - 2ab \cos \gamma$$

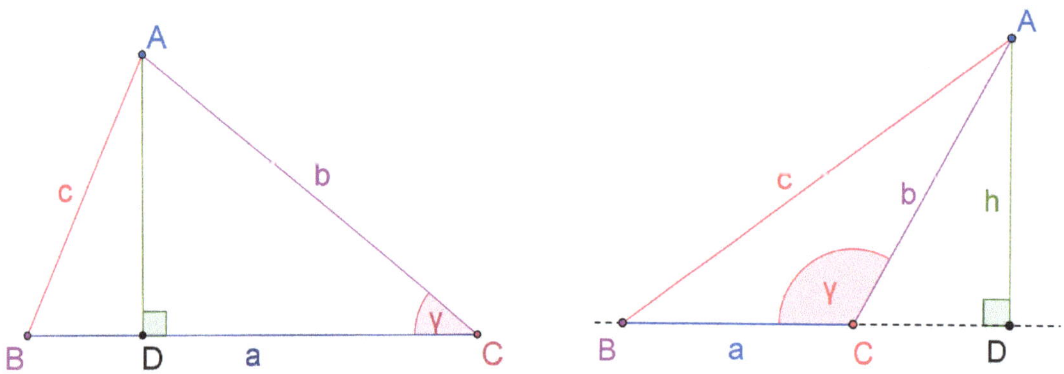

Figure V.23.2: Proof of the cos theorem for acute, and for obtuse angles.

Proof. We drop the perpendicular from vertex A onto the opposite side a. Let D the foot point. The Pythagorean theorem for triangles $\triangle ADC$ and $\triangle ADB$ yields, after subtracting

$$b^2 = h^2 + |DC|^2$$
$$c^2 = h^2 + |DB|^2$$
$$c^2 = b^2 - |DC|^2 + |DB|^2$$

We need now to distinguish the case of an acute and obtuse triangle. Take an acute triangle. By definition of the cosin function $\cos \gamma = \frac{|DC|}{b}$, and hence $|DC| = b \cos \gamma$. From the drawing, we see that $|DC| + |DB| = a$. Hence

$$\begin{aligned}|DB| &= a - |DC| = a - b\cos\gamma \\ |DB|^2 &= a^2 - 2ab\cos\gamma + b^2\cos^2\gamma \\ c^2 &= b^2 - |DC|^2 + |DB|^2 = b^2 - b^2\cos^2\gamma + a^2 - 2ab\cos\gamma + b^2\cos^2\gamma \\ c^2 &= a^2 + b^2 - 2ab\cos\gamma\end{aligned}$$

as required. For the obtuse triangle, we get $|DB| - |DC| = a$. By definition of the cos function $\cos(180° - \gamma) = \frac{|DC|}{b}$, and hence $|DC| = b\cos(180° - \gamma)$. From the drawing, we see that $a = |DB| - |DC|$. Hence

$$\begin{aligned}|DB| &= a + |DC| = a + b\cos(180° - \gamma) \\ |DB|^2 &= a^2 + 2ab\cos(180° - \gamma) + b^2\cos^2(180° - \gamma) \\ c^2 &= b^2 - |DC|^2 + |DB|^2 = b^2 - b^2\cos^2(180° - \gamma) + a^2 + 2ab\cos(180° - \gamma) + b^2\cos^2(180° - \gamma) \\ c^2 &= a^2 + b^2 + 2ab\cos(180° - \gamma)\end{aligned}$$

Hence, in order to obtain a cos theorem valid for both acute and obtuse triangles, one defines

$$\cos(180° - \alpha) = -\cos\alpha$$

□

Corollary 54 (Pythagorean comparison). *For acute angle $\gamma < 90°$, the sides of a triangle satisfy $c^2 < a^2 + b^2$. For obtuse angle $\gamma > 90°$, the sides of a triangle satisfy $c^2 > a^2 + b^2$.*

Corollary 55 (The converse Pythagorean Theorem). *If the sides of a triangle satisfy $c^2 = a^2 + b^2$* in the sense of segment arithmetic, *then the angle across side c is a right angle.*

Problem V.23.3. *For a triangle are given the sides $a = 10$ and $c = 6$, and the angle $\gamma = 30°$ across the latter side. Find angle α, side b and finally angle β. How many non-congruent solutions do you get?*

Answer. The sin Theorem yields

$$\sin\alpha = \frac{a\sin\gamma}{c} = \frac{5}{6}$$

Hence we get two supplementary angles

$$\alpha_1 = 56.44° \quad \text{and} \quad \alpha_2 = 180° - \alpha_1 = 123.56°$$

as possible solutions. The side b is obtained with the cos Theorem:

$$\begin{aligned}c^2 &= a^2 + b^2 - 2ab\cos\gamma \\ b^2 - 10\sqrt{3}b + 64 &= 0 \\ b_{1,2} &= \frac{10\sqrt{3} \pm \sqrt{300 - 256}}{2} = 5\sqrt{3} \pm \sqrt{11}\end{aligned}$$

Question. It is not yet clear how these results combine, and whether there are two or even four solutions. Provide a drawing to decide this question.

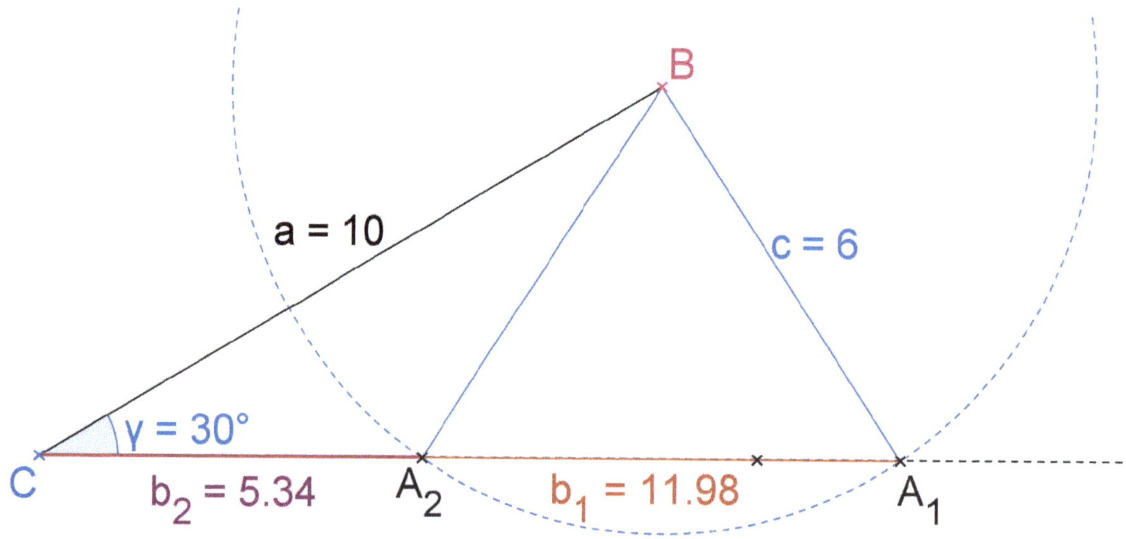

Figure V.23.3: For the given data, there are two non-congruent solutions.

Answer. One sees that the shorter side $b_2 = 5\sqrt{3} - \sqrt{11} = 5.3436$ combines only with the obtuse angle α_2. The longer side $b_1 = 5\sqrt{3} + \sqrt{11} = 11.977$ combines only with the acute angle α_1. Hence one obtains two non-congruent solutions.

Finally, the angle β can be obtained from the angle sum. One gets

$$\beta_1 = 180^\circ - \alpha_1 - \gamma = 93.548^\circ$$
$$\beta_2 = 180^\circ - \alpha_2 - \gamma = 26.45^\circ$$

Question. Obtain the angle β with the sin Theorem. Which problem does arise?

Answer. The sin Theorem yields

$$\sin\beta = \frac{b\sin\gamma}{c} = \frac{5\sqrt{3} \pm \sqrt{11}}{12}$$

One cannot yet know whether the acute or obtuse solution is valid. Indeed $\beta_2 = 26.44$ is acute and $\beta_1 = 93.56$ is obtuse.

Problem V.23.4. *Use the cosine theorem to give another proof of the parallelogram equation from proposition V.22.2.*

Another proof of the parallelogram equation. Refer to the figure on page 204. Given is a parallelogram $\square ACBD$ with two pairs of opposite parallel sides $AC \parallel BD$ and $BC \parallel DA$. Using congruent z-angles and ASA congruence, one sees that the opposite sides are pairwise congruent. Let M be the intersection point of the diagonals. We know already by problem II.7.16 that the diagonals of the parallelogram bisect each other.

We apply the cosine theorem to the triangles $\triangle ACM$ and $\triangle DCM$:

$$|AC|^2 = |AM|^2 + |CM|^2 - 2|AM| \cdot |CM| \cdot \cos\theta$$
$$|AD|^2 = |AM|^2 + |DM|^2 - 2|AM| \cdot |DM| \cdot \cos\eta$$

The angle $\theta = \angle AMC$ and $\eta = \angle DMC$ are supplementary and hence $\cos\theta = -\cos\eta$. We know already by problem II.7.16 that $|CM| = |DM|$. Hence adding the relations yields

$$|AC|^2 + |AD|^2 = 2|AM|^2 + 2|CM|^2$$

Similarly, we prove that

$$|BC|^2 + |BD|^2 = 2|BM|^2 + 2|CM|^2$$

Since the diagonals of the parallelogram bisect each other we know that $2|AM|^2 + 2|BM|^2 = |AB|^2$ and $4|CM|^2 = |CD|^2$. Hence adding the two equations yields

$$|AC|^2 + |AD|^2 + |BC|^2 + |BD|^2 = 2|AB|^2 + 2|CD|^2$$

as to be shown. □

Problem V.23.5. *Get an exact expression for the diameter $2R$ of the circum circle in terms of the three sides a, b, c of a triangle.*

Proof. The cos Theorem yields

$$c^2 = a^2 + b^2 - 2ab\cos\gamma$$
$$\cos\gamma = \frac{a^2 + b^2 - c^2}{2ab}$$

The extended sin theorem gives the diameter of the circum circle

$$\begin{aligned}
2R &= \frac{c}{\sin\gamma} = \frac{c}{\sqrt{1-\cos^2\gamma}} = \frac{2abc}{\sqrt{4a^2b^2 - (2ab\cos\gamma)^2}} \\
&= \frac{2abc}{\sqrt{4a^2b^2 - (a^2+b^2-c^2)^2}} = \frac{2abc}{\sqrt{(2ab + a^2 + b^2 - c^2)}\sqrt{(2ab - a^2 - b^2 + c^2)}} \\
&= \frac{2abc}{\sqrt{((a+b)^2 - c^2)}\sqrt{(c^2 - (a-b)^2)}} \\
&= \frac{2abc}{\sqrt{((a+b+c)(a+b-c)(c-a+b)(c+a-b))}}
\end{aligned}$$

□

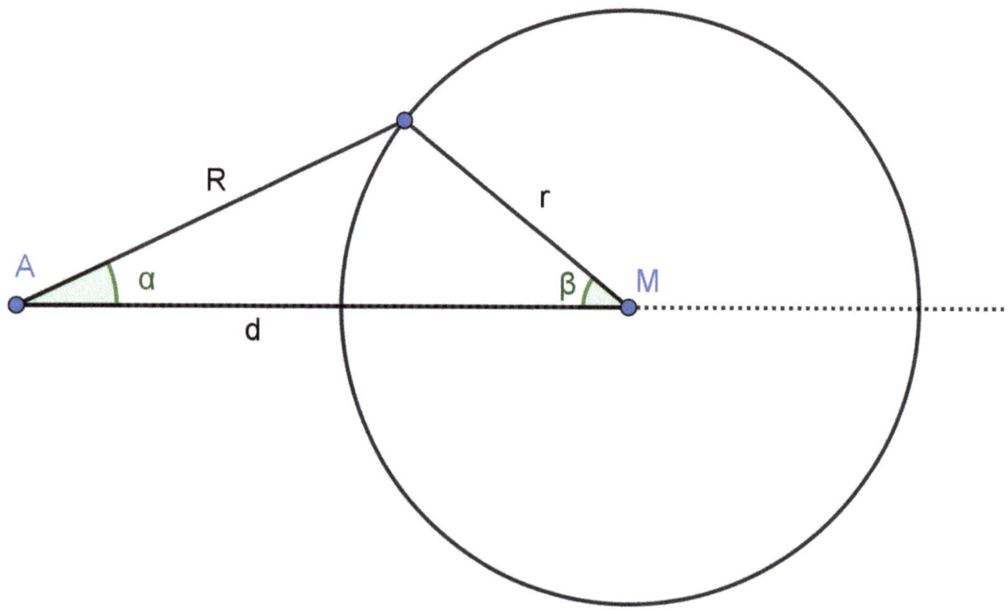

Figure V.23.4: Attraction of a shell to a point mass.

Problem V.23.6. *Calculate the attractive gravitational force between a point mass m and a thin spherical shell of radius r and mass M. The point mass has distance d from the center of the shell.*

Solution. In the figure on page 486, the cosine theorem yields

$$R^2 = r^2 + d^2 - 2rd\cos\beta \quad \text{and} \quad r^2 = R^2 + d^2 - 2Rd\cos\alpha$$

Let μ be the mass density per surface angle of the shell, and hence $M = 4\pi\mu$. Integrated over the azimutal angle, the surface element is $2\pi\sin\beta\,d\beta$ and the range is $0 < \beta < \pi$. The attractive force F has only a component along the line connecting the point mass to the center of the shell.

$$F = m\mu\int_0^\pi \frac{\cos\alpha\cdot 2\pi\sin\beta\,d\beta}{R^2}$$

We may eliminate $\cos\alpha$ and get

$$\begin{aligned} F &= \frac{mM}{8\pi d}\int_0^\pi \frac{2Rd\cos\alpha\cdot 2\pi\sin\beta\,d\beta}{R^3} \\ &= \frac{mM}{4d}\int_0^\pi \frac{(R^2+d^2-r^2)\sin\beta\,d\beta}{R^3} \end{aligned}$$

Variable are R or β which depend on each other by $R^2 = r^2 + d^2 - 2rd\cos\beta$ and differentiating gives

$$2RdR = 2rd\sin\beta d\beta$$

The integral for the force is calculated with independent variable R. The range for the new variable is $|d-r| \le R \le d+r$. In the case that $d > r$, i.e. the point mass *outside the sphere*, one gets

$$\begin{aligned} F &= \frac{mM}{4rd^2} \int_0^\pi \frac{(R^2+d^2-r^2) rd \sin\beta \, d\beta}{R^3} = \frac{mM}{4rd^2} \int_{d-r}^{d+r} \frac{(R^2+d^2-r^2) dR}{R^2} \\ &= \frac{mM}{4rd^2} \left[R - \frac{d^2-r^2}{R} \right]_{d-r}^{d+r} = \frac{mM}{4rd^2} \left[2r - \frac{d^2-r^2}{d+r} + \frac{d^2-r^2}{d-r} \right] \\ &= \frac{mM}{4rd^2} [2r - (d-r) + d + r] = \frac{mM}{d^2} \end{aligned}$$

as Newton found after a painful long time. □

Problem V.23.7. *Explain why for the case of a point mass* inside *the spherical shell, there is no attractive force.*

Solution. In this case holds $d < r$ and the range of integration is $r-d \le R \le d+r$. Hence

$$\begin{aligned} F &= \frac{mM}{4rd^2} \left[R - \frac{d^2-r^2}{R} \right]_{r-d}^{d+r} = \frac{mM}{4rd^2} \left[2d - \frac{d^2-r^2}{d+r} + \frac{d^2-r^2}{r-d} \right] \\ &= \frac{mM}{4rd^2} [2d - (d-r) - (d+r)] = 0 \end{aligned}$$

as Newton found after a painful long time. □

V.23.3 The addition theorem for tangent

Problem V.23.8 (The tangent addition Theorem). *Prove the tangent addition theorem*

$$\tan(x+y) = \frac{\tan x + \tan y}{1 - \tan x \, \tan y} \tag{V.23.6}$$

for any acute or obtuse angles x, y with acute or obtuse sum $x+y$.

Proof. We prove the addition theorem at first for any acute or obtuse angles x, y with acute or obtuse sum $\gamma = x+y$. We assume the perpendicular $|FC| = 1$ has unit length and the two right triangles have angles $x = \angle ACF$ and $y = \angle FCB$. From them together, we get the triangle $\triangle ABC$ as shown in the drawing on page 488.

Question. What are the lengths of the sides of the triangle $\triangle ABC$.

Answer.

$$a = |BC| = \frac{1}{\cos y} \quad , \; b = |CA| = \frac{1}{\cos x} \quad \text{and} \quad c = |AB| = \tan x + \tan y$$

Question. Write the cosine theorem for side c in terms of x and y and solve for $\cos(x+y)$.

Answer.

$$\begin{aligned} c^2 &= a^2 + b^2 - 2ab\cos\gamma \\ (\tan x + \tan y)^2 &= \frac{1}{\cos^2 x} + \frac{1}{\cos^2 y} - \frac{2\cos(x+y)}{\cos x \cos y} \\ \cos(x+y) &= \frac{\cos x \cos y}{2} \left[\frac{1}{\cos^2 x} + \frac{1}{\cos^2 y} - (\tan x + \tan y)^2 \right] \end{aligned}$$

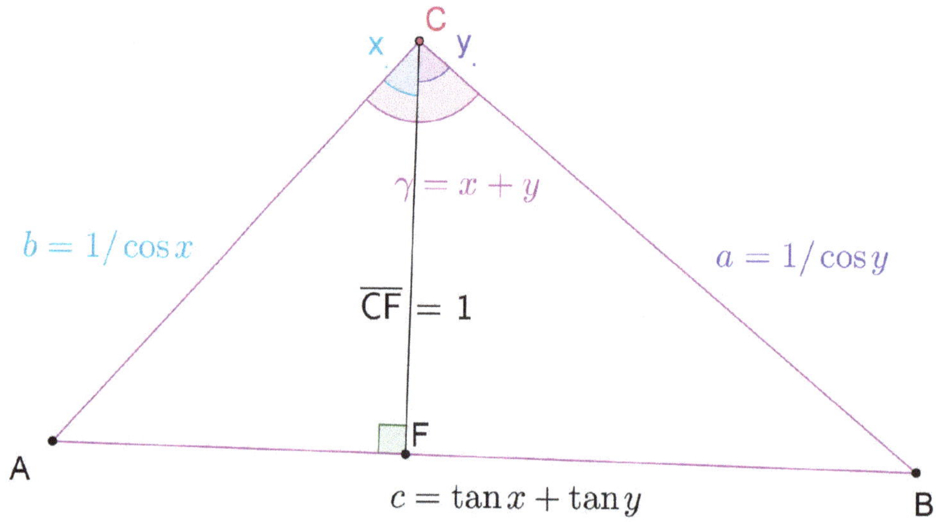

Figure V.23.5: Proof of the tangent addition theorem.

I use the abbreviation $X = \tan x, Y = \tan y$ and $\cos^2 x = 1/(1 + \tan^2 x)$ to eliminate the cosine function.

$$\begin{aligned} \cos(x+y) &= \frac{\cos x \cos y}{2}\left[2 + X^2 + Y^2 - (X+Y)^2\right] = \cos x \cos y\,[1 - XY] \\ 1 + \tan^2(x+y) &= \frac{(1+X^2)(1+Y^2)}{(1-XY)^2} \\ \tan^2(x+y) &= \frac{(1+X^2)(1+Y^2) - (1-XY)^2}{(1-XY)^2} = \frac{X^2 + Y^2 + 2XY}{(1-XY)^2} \\ \tan(x+y) &= \pm\frac{X+Y}{1-XY} \end{aligned}$$

Question. Which sign has $1-XY$ for acute angle $x+y$, which for obtuse angle $x+y$. What happens for right angle $x+y$.

Answer. $1 - XY > 0$ for acute angle $x+y$, and $1 - XY < 0$ for obtuse angle $x+y$. If and only if $x+y$ is a right angle do we get $1 - XY = 0$ since this is the case where the two right triangles $\triangle ACF \sim \triangle CBF$ are similar.

Question. How do we justify

$$\tan(180^\circ - \alpha) = -\tan\alpha \tag{V.23.7}$$

Answer. From the sin and cosine theorems we have obtained

$$\sin(180^\circ - \alpha) = \sin\alpha \ , \ \cos(180^\circ - \alpha) = -\cos\alpha$$

for both obtuse and acute angles. Because of the definition

$$\tan\alpha = \frac{\sin\alpha}{\cos\alpha}$$

which we now <u>extend</u> to obtuse angles, we get equation (V.23.7).

Question. Which sign has $\tan(x+y)$ for acute angle $x+y$, which for obtuse angle $x+y$. What happens for right angle $x+y$.

Answer. $\tan(x+y) > 0$ for acute angle $x+y$, and $\tan(x+y) < 0$ for obtuse angle $x+y$. If $x+y$ is a right angle, we get $\tan(x+y) = \infty$ because of $\sin 90° = 1$ and $\cos 90° = 0$. [53]

From these considerations of the signs, we see that the same formula

$$\tan(x+y) = \frac{\tan x + \tan y}{1 - \tan x \, \tan y} \tag{V.23.6}$$

holds for any acute or obtuse angles x, y with acute or obtuse sum $x+y$. □

V.23.4 Unwound angles

Unwound angles occur in the study of the trigonometric functions, physical oscillations, the theory of hyperbolic area and other contexts. The unwound angles have the structure of an ordered Abelian group. This group actually depends on the Hilbert plane considered.

For a precise definition let $\mathcal{A}cute$ denote the set of the congruence classes for the zero and the acute angles. Let R denote the congruence class of right angles.

Definition V.23.3. The *group of unwound angles* is the Cartesian product $\mathbf{Z} \times \mathcal{A}cute$ with the operation of addition defined by setting

$$(m,\alpha) + (n,\beta) = \begin{cases} (m+n, \alpha+\beta) & \text{if } \alpha+\beta < R \\ (m+n+1, \alpha+\beta-R) & \text{if } \alpha+\beta \geq R \end{cases} \tag{V.23.8}$$

Proposition V.23.1. *In any Hilbert plane, the unwound angles have the structure of an ordered Abelian group. This group actually depends on the Hilbert plane considered.*

There is a natural homomorphism from the group of unwound angles onto the group of rotations around the origin. The kernel of this homomorphism is

$$4\mathbf{Z} \times \{0\} = \{(4n, 0) \, : \, n \in \mathbf{Z}\}$$

in other words the angles which are integer multiples of $360°$.

V.23.5 Extension to unwound angles

Up to now we have defined the sin, cosine and tangent functions for acute and obtuse angles, including the degenerate cases of $0°$ and $180°$.

For the tangent function, we need to allow the infinite value $\tan 90° = \infty$. By allowing ∞ as a value, we extend the tangent function to all unwound angles.

Problem V.23.9. *Convince yourself that*

$$\tan(90° + x) = -\frac{1}{\tan x} \tag{V.23.9}$$

holds for $0° \leq x \leq 90°$.

[53] The meromorphic nature of the tangent function makes a definition meaningful.

Answer. The addition theorem (V.23.6) for the tangent function implies

$$\tan(45^\circ + x) = \frac{1 + \tan x}{1 - \tan x}$$

holds for $0^\circ \leq x \leq 135^\circ$. Iterating the relation implies

$$\tan(90^\circ + x) = \frac{1 + \tan(45^\circ + x)}{1 - \tan(45^\circ + x)} = \frac{1 - \tan x + (1 + \tan x)}{1 - \tan x - (1 + \tan x)} = -\frac{1}{\tan x}$$

□

The tangent function is defined for all x by requiring the relation (V.23.9) for all x.

Proposition V.23.2. *The sin and cosine functions can be extended to all unwound angles by setting*

$$\cos(180^\circ + x) = -\cos x\ , \quad \sin(180^\circ + x) = -\sin x$$

In that way, $\cos x$ *becomes an even function with period* 360°*, and* $\sin x$ *becomes an odd function with period* 360°*. By allowing* ∞ *as a value, we extend the tangent function to all unwound angles: In this way,* $\tan x$ *becomes an odd function with period* 180°*.*

Lemma V.23.1. *We use the imaginary unit* i *which satisfies* $i^2 + 1 = 0$*.*

$$\cos 2x + i \sin 2x = \frac{(1 + i\tan x)^2}{1 + \tan^2 x} \tag{V.23.10}$$

$$\cos 2x = \frac{2\tan x}{1 + \tan^2 x} \tag{V.23.11}$$

$$\sin 2x = \frac{1 - \tan^2 x}{1 + \tan^2 x} \tag{V.23.12}$$

holds for all unwound angles x*. With the obvious interpretation of the fractions with the value* ∞ *for the tangent function, all values are included.*

Reason. Convince yourself that

$$\cos 2x + i \sin 2x = \frac{1 + i\tan 2x}{\sqrt{1 + \tan^2 2x}} \quad \text{for} \quad -90^\circ < 2x < 90^\circ \tag{V.23.13}$$

can be checked from the definition. But the positive sign of the root enforces the restriction of the domain. We can use the addition theorem of the tangent and conclude that

$$\tan 2x = \frac{2\tan x}{1 - \tan^2 x} \quad \text{for} \quad -180^\circ < 2x < 180^\circ \tag{V.23.14}$$

Together the equations (V.23.13) and (V.23.14) imply

$$\cos 2x + i \sin 2x = \frac{1 + i\tan 2x}{\sqrt{1 + \tan^2 2x}} = \frac{1 - \tan^2 x + 2i\tan x}{\sqrt{(1 - \tan^2 x)^2 + 4\tan^2 x}} = \frac{(1 + i\tan x)^2}{1 + \tan^2 x}$$

for $-90° < 2x < 90°$. But we can extend the domain of validity of this equation since it does not contain the root any longer. From the way the extensions of the functions have been defined we get

$$\cos 2(90° + x) + i \sin 2(90° + x) = -\cos 2x - i \sin 2x$$
$$= -\frac{(1 + i \tan x)^2}{1 + \tan^2 x} = \frac{(\tan x - i)^2}{\tan^2 x + 1} = \frac{(1 - i/\tan x)^2}{1 + 1/\tan^2 x} = \frac{(1 + i \tan(90° + x))^2}{1 + \tan^2(90° + x)}$$

Hence equation (V.23.10) still is valid for $-90° < 2x < 180°$. The formula is extended to $-180° < 2x < 180°$ by the symmetry, and then to all x by the periodicity properties of the trigonometric functions. □

Proposition V.23.3. *The extended cosine and sin function satisfy the addition theorem*

$$\cos(x + y) + i \sin(x + y) = (\cos x + i \sin x)(\cos y + i \sin y) \tag{V.23.15}$$
$$\cos(x + y) = \cos x \cos y - \sin x \sin y \tag{V.23.16}$$
$$\sin(x + y) = \cos x \sin y + \sin x \cos y \tag{V.23.17}$$

The extended tangent function satisfied the addition theorem

$$\tan(x + y) = \frac{\tan x + \tan y}{1 - \tan x \tan y} \tag{V.23.6}$$

Reason. The addition theorem for the tangent function is extended via formula (V.23.9). From the original formula valid in the domain $0 \le x, y, x + y \le 180°$ one gets

$$\tan(90° + x + y) = -\frac{1}{\tan(x + y)} = \frac{\tan x \tan y - 1}{\tan x + \tan y}$$
$$\frac{\tan(90° + x) + \tan y \tan x}{1 - \tan(90° + x) \tan y \tan x} = \frac{-1 + \tan x \tan y}{\tan x + \tan y}$$

which are equal. With a similar step for y, one extends the domain of validity of formula (V.23.6) to the domain $0 \le x, y \le 180°$. Now periodicity implies that the formula holds for all x, y.

The addition theorems for sin and cosine follow now via formula (V.23.10):

$$\cos 2(x + y) + i \sin 2(x + y) = \frac{[1 + i \tan(x + y)]^2}{1 + \tan^2(x + y)} = \frac{[1 - \tan x \tan y + i(\tan x + \tan y)]^2}{(1 - \tan x \tan y)^2 + \tan^2(x + y)}$$
$$= \frac{[(1 + i \tan x)(1 + i \tan y)]^2}{(1 + \tan^2 x)(1 + \tan^2 y)} = \frac{(1 + i \tan x)^2}{1 + \tan^2 x} \cdot \frac{(1 + i \tan y)^2}{1 + \tan^2 y}$$
$$= (\cos x + i \sin x)(\cos y + i \sin y)$$

□

V.23.6 Mollweid's formulas

Given any triangle $\triangle ABC$, the radius of the circum-circle is denoted by R and the radius of the in-circle by ρ. The radii of the ex-circles touching sides a,b or c, respectively, are denoted by ρ_a, ρ_b and ρ_c.

Proposition V.23.4 (The sin Theorem and the in-circle). *For any triangle with sides a, b, c and angles α, β, γ across them*

(V.23.18)
$$\frac{a}{\sin\alpha} = \frac{b}{\sin\beta} = \frac{c}{\sin\gamma} = 2R$$

(V.23.19)
$$2R = \frac{\rho}{2\sin\frac{\alpha}{2}\sin\frac{\beta}{2}\sin\frac{\gamma}{2}} = \frac{\rho_a}{2\sin\frac{\alpha}{2}\cos\frac{\beta}{2}\cos\frac{\gamma}{2}} = \frac{\rho_b}{2\cos\frac{\alpha}{2}\sin\frac{\beta}{2}\cos\frac{\gamma}{2}} = \frac{\rho_c}{2\cos\frac{\alpha}{2}\cos\frac{\beta}{2}\sin\frac{\gamma}{2}}$$

Proof. As in the proof of proposition V.21.2, one gets

$$\cot\frac{\alpha}{2} = \frac{|AF|}{|FI|} \quad \text{and} \quad \cot\frac{\beta}{2} = \frac{|FB|}{|FI|} \quad \text{and hence} \quad \frac{c}{\rho} = \frac{|AB|}{|FI|} = \cot\frac{\alpha}{2} + \cot\frac{\beta}{2}$$

Now one uses the addition theorem and $\alpha + \beta + \gamma = 180°$ to get

$$\frac{\rho}{c} = \frac{\sin\frac{\alpha}{2}\sin\frac{\beta}{2}}{\cos\frac{\alpha}{2}\sin\frac{\beta}{2} + \sin\frac{\alpha}{2}\cos\frac{\beta}{2}} = \frac{\sin\frac{\alpha}{2}\sin\frac{\beta}{2}}{\sin\frac{\alpha+\beta}{2}} = \frac{2\sin\frac{\alpha}{2}\sin\frac{\beta}{2}\sin\frac{\gamma}{2}}{\sin\gamma}$$

To deal with the ex-circles, one derives at first the formula

$$\frac{c}{\rho_a} = \cot\frac{\alpha}{2} - \tan\frac{\beta}{2}$$

and proceeds quite similarly. □

Theorem V.23.3 (Mollweid's formulas and the tangent theorem). *Sides and angles of any triangle satisfy*

(V.23.20)
$$\frac{a-b}{c} = \frac{\sin(\alpha-\beta)/2}{\sin(\alpha+\beta)/2}$$

(V.23.21)
$$\frac{a+b}{c} = \frac{\cos(\alpha-\beta)/2}{\cos(\alpha+\beta)/2}$$

(V.23.22)
$$\frac{a+b}{a-b} = \frac{\tan(\alpha+\beta)/2}{\tan(\alpha-\beta)/2}$$

Reason for formula (V.23.20). Use several addition theorems, the angle sum and the cosine theorem three times:

$$\begin{aligned}\frac{\sin(\alpha-\beta)/2}{\sin(\alpha+\beta)/2} &= \frac{2\,[\sin(\alpha-\beta)/2]\cdot[\sin(\alpha+\beta)/2]}{2\sin^2(\alpha+\beta)/2} = \frac{\cos\beta - \cos\alpha}{2\cos^2(\gamma/2)} = \frac{\cos\beta-\cos\alpha}{1+\cos\gamma}\\ &= \frac{b(2ac\cos\beta) - a(2bc\cos\alpha)}{2abc + c(2ab\cos\gamma)} = \frac{b(a^2+c^2-b^2) - a(b^2+c^2-a^2)}{2abc + c(a^2+b^2-c^2)}\\ &= \frac{(b-a)(c^2-(a+b)^2)}{c((a+b)^2-c^2)} = \frac{a-b}{c}\end{aligned}$$

□

Reason for formula (V.23.21). Use the addition theorem, the angle sum, the sin theorem twice and then the cosine theorem:

$$\frac{\sin(\alpha-\beta)}{\sin(\alpha+\beta)} = \frac{\sin\alpha\cos\beta-\cos\alpha\sin\beta}{\sin\gamma} = \frac{a\cos\beta-b\cos\alpha}{c} = \frac{2ac\cos\beta-2bc\cos\alpha}{2c^2}$$
$$= \frac{a^2+c^2-b^2-b^2+c^2+a^2}{2c^2} = \frac{a^2-b^2}{c^2}$$

Now use the half-angle formula and divide by formula (V.23.20):

$$\frac{2\,[\sin(\alpha-\beta)/2]\cdot[\cos(\alpha-\beta)/2]}{2\,[\sin(\alpha+\beta)/2]\cdot[\cos(\alpha+\beta)/2]} = \frac{(a-b)(a+b)}{c^2}$$
$$\frac{\cos(\alpha-\beta)/2}{\cos(\alpha+\beta)/2} = \frac{a+b}{c}$$

□

Formula (V.23.22). —results from dividing formula (V.23.20) by formula (V.23.21) □

V.24 Measurement of the Circle

V.24.1 Approximate measurement of a circular arc

In the interior of the angle $\theta = \angle AOB$, we draw a circular arc of radius $r = 1$. We want to get lower and upper bounds for the length of this arc. Indeed, we want to construct a *convergent sequence* of more and more precise lower and upper bounds.

As a first lower bound, one can simply use the length $a_1 = |AB|$ of the chord. To get an upper bound, we draw the tangents to the circle at the endpoints A and B of the arc. They intersect at point C. The length of the union of the segment AC and CB give the first upper bound $b_1 = |AC| + |CB|$.

Question. Why is $\overrightarrow{OC}$ the bisector of angle $\angle AOB$.

Question. Get the formulas for a_1 and b_1. From which right triangles are they obtained.

Answer. Since $|OA| = |OB| = 1$, we get from right triangles $\triangle OMA$ and $\triangle OAC$

$$a_1 = 2\sin\frac{\theta}{2} \quad\text{and}\quad b_1 = 2\tan\frac{\theta}{2}$$

To get the second, and better lower and upper bounds, we draw the bisector OC of the given angle $\angle AOB$, and mark the point D where it intersects the circular arc $\overset{\frown}{AB}$. The construction of the first lower and upper bounds is now repeated for the two arcs $\overset{\frown}{AD}$ and $\overset{\frown}{DB}$. Since we have doubled the polygons for half the angle, we obtain

$$a_2 = 4\sin\frac{\theta}{4} \quad\text{and}\quad b_2 = 4\tan\frac{\theta}{4}$$

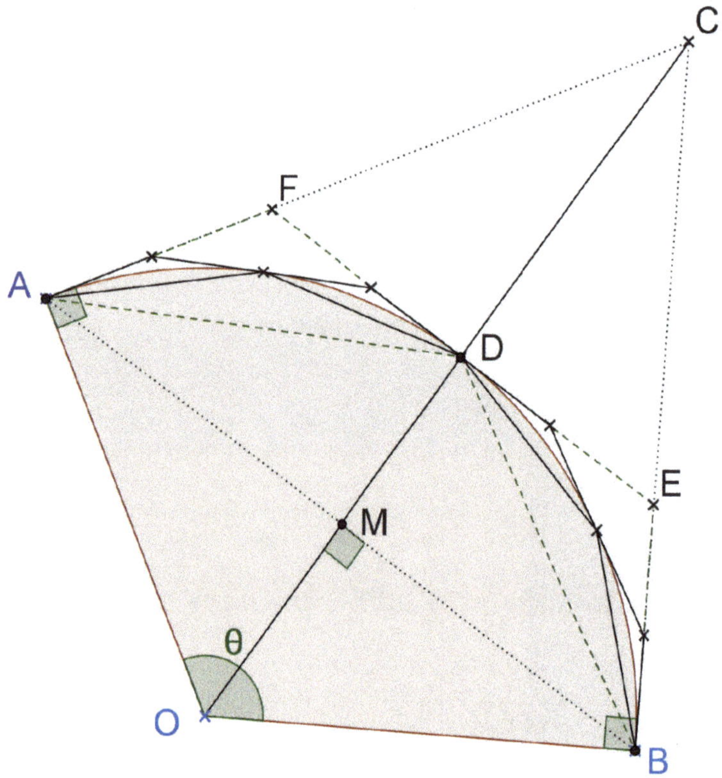

Figure V.24.1: Archimedes' approximation of a circular arc

Proposition V.24.1. *For a short arc θ of the unit circle, the lengths a_n of the inscribed polygon with n sides, and the lengths the b_n of the circumscribed polygon with 2^n sides are calculated recursively: The initial values are*

$$a_1 = 2\sin\frac{\theta}{2} \quad \text{and} \quad b_1 = 2\tan\frac{\theta}{2}$$

and the two sequences a_n and b_n with $n = 1, 2, \ldots$ satisfy the recursion

$$a_{n+1} := 2^{n+1}\sqrt{\frac{b_n - a_n}{2\,b_n}} \quad , \quad b_{n+1} := 2^{n+1}\sqrt{\frac{b_n - a_n}{b_n + a_n}} \tag{V.24.1}$$

Proof.
Question. Give the trigonometric formulas for a_n and b_n.

$$a_n = 2^n \sin\frac{\theta}{2^n} \quad \text{and} \quad b_n = 2^n \tan\frac{\theta}{2^n} \tag{V.24.2}$$

To check the recursion, we use the trigonometric formulas

$$\cos\frac{\alpha}{2} = \sqrt{\frac{1+\cos\alpha}{2}} \quad , \quad \sin\frac{\alpha}{2} = \sqrt{\frac{1-\cos\alpha}{2}} \quad , \quad \tan\frac{\alpha}{2} = \sqrt{\frac{1-\cos\alpha}{1+\cos\alpha}}$$

which hold for $\alpha \in [0, \pi]$ with positive roots. □

Problem V.24.1. *Convince yourself that lengths of the polygons which approximate the circular arc from inside and outside satisfy*

$$a_1 < a_2 < \cdots < a_n < a_{n+1} < \cdots < b_{n+1} < b_n < \cdots < b_2 < b_1 \tag{V.24.3}$$

Proof.

Question. In which triangles does the triangle inequality yield $a_1 < b_1$, $a_1 < a_2$ and $b_2 < b_1$.

Answer. From the triangle inequality in triangle $\triangle ABC$, we obtain $a_1 < b_1$. From the triangle inequality in triangle $\triangle ABD$, we obtain $a_1 < a_2$. From the triangle inequality in triangle $\triangle EFC$, we obtain $b_2/2 = |EF| < |EC| + |CF| = b_1 - (b_2/2)$ and hence $b_2 < b_1$.

Finally, we get $a_2 < b_2$ and hence together $a_1 < a_2 < b_2 < b_1$. Repeated bisections yields two sequences of polygons which approximate the circular arc from inside and outside. Their lengths satisfy

$$a_1 < a_2 < \cdots < a_n < a_{n+1} < \cdots < b_{n+1} < b_n < \cdots < b_2 < b_1$$

□

For the angle $\theta = 60°$ one gets the initial values

$$a_1 := 1 \quad \text{and} \quad b_1 := \frac{2\sqrt{3}}{3}$$

for the sequences with limit

$$\lim_{n\to\infty} a_n = \lim_{n\to\infty} b_n = \frac{\pi}{3}$$

For a calculation of π all a_n and b_n need to be multiplied with three.

Problem V.24.2.

$$A_1 := 3 \quad \text{and} \quad B_1 := 2\sqrt{3}$$

$$A_{n+1} := 3 \cdot 2^{n+1}\sqrt{\frac{B_n - A_n}{2B_n}}\,, \quad B_{n+1} := 3 \cdot 2^{n+1}\sqrt{\frac{B_n - A_n}{A_n + B_n}}$$

Compute numerically and report the values of A_n and B_n for $n = 1, 2, \ldots, 15$. Find a conjecture about the trend for the errors $B_n - \pi$ and $\pi - A_n$.

Which value of n gives the best result for π. In which interval can you say π lies for sure?

Answer. As a trend for the errors, one sees $\pi - A_n \cong (B_n - \pi)/2$ and $B_{n+1} - \pi \cong (\pi - A_n)/2$. The trend breaks down for $n = 11$, where the rounding errors makes it impossible to further improve the result. The result for $n = 10$ is still reliable. Multiplication with $3 \cdot 2^{10} = 3\,072$ makes the last three digits random—but one still gets that $3.141592 < \pi < 3.141594$.

n $3 \cdot 2^n$	B_n A_n	$B_n - \pi$ $\pi - A_n$
1	3.464101615	.3225089615
6	3	.1415926536
2	3.215390309	.0737976556
12	3.105828541	.0357641124
3	3.159659942	.0180672885
24	3.132628613	.0089640403
4	3.146086215	.0044935615
48	3.139350203	.0022424506
5	3.1427146	.0011219461
96	3.141031951	.0005607026872
6	3.14187305	.0002803964226
192	3.141452472	.0001401812718
7	3.141662747	.0000700935935
384	3.141557608	.0000350455515
8	3.141610177	.000017523088
768	3.141583892	.0000087613684
9	3.141597036	.0000043824205
1 536	3.141590465	.0000021886731
10	3.141593778	.0000011239655
3 072	3.141592135	.0000005188065
11	3.141592836	.0000001824518
6 144	3.141592425	.000000228241
12	3.141592234	−.0000004197604
12 288	3.141592131	.0000005224336
13	3.141592458	−.0000001956592
24 576	3.141592432	.0000002213274
14	3.141586207	−.000006446526
49 152	3.141586201	.0000064529797
15	3.141601567	.000008913278
98 304	3.141601565	−.0000089116739

Lemma V.24.1. *The difference of the upper and lower bounds is estimated from above*

$$b_n - a_n < \frac{a_n}{4^{n-1}\, a_1}(b_1 - a_1) < \frac{4b_1(b_1 - a_1)}{4^n\, a_1} \tag{V.24.4}$$

Problem V.24.3. *Explain how to get the estimate* (V.24.4).

Reason.

Question. Check the following formula:

$$a_n^{-2} - b_n^{-2} = 4^{-n}$$

Question. We define the quotients $q_n := \frac{b_n}{a_n}$. Use the trigonometric formula $\cos^2 \frac{\alpha}{2} = \frac{1+\cos\alpha}{2}$ to check the recursion formula

$$q_{n+1}^2 = \frac{2q_n}{1+q_n}$$

Question. Prove that $0 \le q_{n+1} - 1 < (q_n - 1)/4$ for all $n \ge 1$.

Answer. Since $0 < a_n < b_n$ division implies $q_n > 1$ for all $n \ge 1$.

$$q_{n+1}^2 - 1 = \frac{2q_n}{q_n+1} - 1 = \frac{q_n - 1}{q_n + 1}$$
$$q_{n+1} - 1 = \frac{q_n - 1}{(q_n+1)(q_{n+1}+1)} < \frac{q_n - 1}{4}$$

The estimate of q_n implies

$$b_{n+1} - a_{n+1} = a_{n+1}(q_{n+1} - 1) < \frac{a_{n+1}}{4}(q_n - 1) = \frac{a_{n+1}}{4\,a_n}(b_n - a_n)$$

For all $n \ge 1$. Finally one gets the result by induction. □

Problem V.24.4. *Use the estimate* (V.24.4) *and explain which continuity axioms are needed to conclude that both sequences b_n and a_n converge to the same limit.*

Answer. As far as the continuity axioms are concerned, one needs to assume the Archimedean axiom (V.1) and Cantor's principle of boxed intervals (V.2). I have stated these axioms of continuity II.14.1 among a simplified axiomatization of geometry II.14.1.

Because of the estimates (V.24.3), Cantor's principle of boxed intervals implies there exists a length x^* such that

(V.24.5) $$a_1 < a_2 < \cdots < a_n < x^* < b_n < \cdots < b_2 < b_1 \quad \text{for all natural } n.$$

Moreover, the estimate (V.24.4) together with the Archimedean axiom (V.1) imply that the number x^* is unique. Indeed, assume that both x^* and x satisfy the estimate (V.24.5). Hence

$$|x^* - x| \le b_n - a_n < \frac{4b_1(b_1 - a_1)}{4^n\,a_1} = K \cdot 4^{-n}$$

hold for all natural n. This is only possible if $x = x^*$.

Reason. One proves by induction that $N < 2^N \le 4^N$ holds for all natural N. Assume towards a contradiction that $x^* \ne x$. By the Archimedean axiom there would exist a natural number N such that $K < N|x^* - x|$. Hence $|x^* - x| < N|x^* - x|4^{-n}$ and $4^n < N$ would hold for all natural n. This is false even just for $n = N$. □

Remark. The axioms of continuity used above are my own favorites since they can be used directly in the proofs. Moreover they are the axioms of continuity contained in the very first edition of Hilbert's Foundations of Geometry. [54]

[54] In later editions of Hilbert's foundations, his (rather convoluted) axiom of completeness is postulated.

V.24.2 Arc length and area of a circular sector

Definition V.24.1 (Pi). The arc length of a semicircle of radius one is defined to be the number π.

Definition V.24.2 (Measurement of angle by arc length). The arc length of a circular arc of radius one and central angle θ is defined as the *arc measurement* of the central angle.

Proposition V.24.2. *Assume that the Archimedean Axiom holds. The measurement of angles by degrees and arc length are proportional. They are related by*

$$\theta[arc] = \frac{\pi}{180^\circ}\,\theta[^\circ] \quad \text{and} \quad \theta[^\circ] = \frac{180^\circ}{\pi}\,\theta[arc] \tag{V.24.6}$$

Measurement of angle by arc length. The *arc length* of the circular arc with radius one and central angle $\theta = \angle AOB$ is obtained from the common limit of the sequences b_n and a_n

$$\theta[arc] = \lim_{n\to\infty} a_n = \lim_{n\to\infty} b_n$$

Consider at first the special case that the angle $\theta = p2^{-n} \cdot 90^\circ$ with integer p and integer $n \geq 0$. In other words, one begins with a right angle, and uses repeated bisection and takes any integer multiples of the angles one has thus obtained. In all these cases, the simple proportionality given by equation (V.24.6) follows directly from the definition of the number π.

The general case is obtained using the process for measurement of angles explained in theorem 11. Such a measurement process presupposes the Archimedean property for angles, stated in theorem II.8.1. Indeed any angle may be approximated, with arbitrary good accuracy by a sequence of angles $\theta = p2^{-n} \cdot 90^\circ$. Hence the proportionality given by equation (V.24.6) remains valid for any arbitrary angle. □

Theorem V.24.1 (Circumference and area of a circular arc). *A circular arc with radius r and central angle θ, to be measured in arc length, has the* arc length $\theta\, r$ *and the* area $\theta\,\frac{r^2}{2}$.

Theorem V.24.2 (Circumference and area of a circle). *The circle of radius r has the circumference $2\pi r$ and the area πr^2.*

Corollary 56 ("pizza theorem"). *The area of a circle is the product of the radius with half of the circumference.*

Problem V.24.5. *Explain the nickname "pizza theorem".*

The arc length of a circular sector. The approximate measurement of the arc length of a circular arc of radius r is performed as explained in the previous section. One obtains the arclength from the common limit

$$\lim_{n\to\infty} a_n \cdot r = \lim_{n\to\infty} b_n \cdot r = \theta[arc] \cdot r$$

because the measurement of angles by arc length from definition V.24.2. □

Lemma V.24.2. *The area of a regular polygon is the product of the radius ρ of the in-circle with half perimeter s. The area of a regular polygon circumscribed around a circular sector equals the radius times half the length of the polygon.*

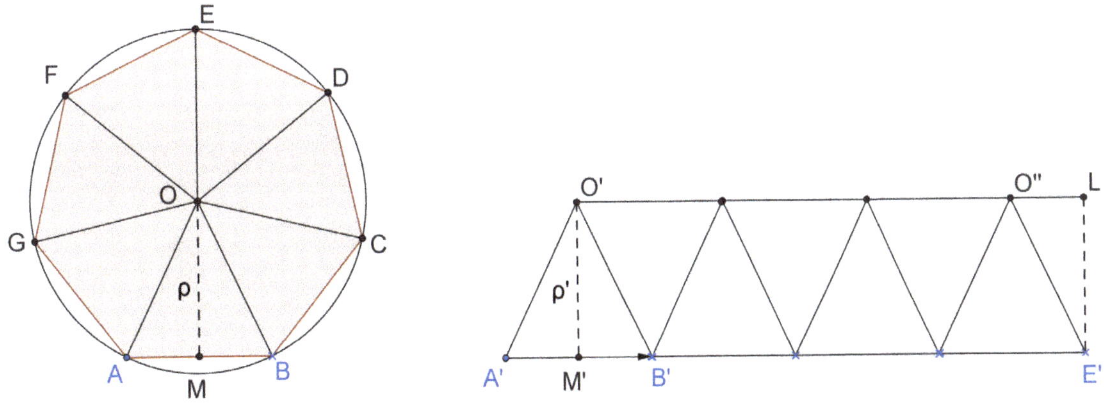

Figure V.24.2: Measuring the area of a regular polygon.

Proof. The reasoning is explained for a 7-gon by the figure on page 499. The 7-gon is partitioned into 7 congruent isosceles triangles $\triangle ABO$ through $\triangle GAO$, where point O is the center of the circum-circle. These triangle are rearranged into a trapezoid $\square A'E'O''O'$. One gets a trapezoid from an n-gon with an odd number n of sides, and one gets a parallelogram from an n-gon with an even number of sides. In both cases, in order to obtain a rectangle, one still needs to bisect the first triangle $\triangle ABO$ into two right triangles, and shift one of them to the opposite side $E'O''$. The rectangle $\square M'E'LO'$ that one has obtained, has the side lengths $|M'O'| = \rho$ which is the radius of the in-circle of the original polygon, and $|M'E'| = s$ which is half of the perimeter of the original polygon. □

The area of a circular sector. To obtain upper- and lower bounds for the area of a circular sector, one uses the areas of the regular polygons circumscribed and inscribed into the circular sector.

By lemma V.24.2, the area of a regular polygon circumscribed around a circular sector equals the radius r times half the length of the polygon. Hence one gets the area

$$r \cdot \frac{b_n r}{2}$$

as an upper bound of the area of the circular sector.

To obtain a lower bound, we use the area of a regular polygon inscribed into the circular sector. The arc length of the inscribed polygon is $a_n \cdot r$. The radius of the circle inscribed into the inscribed polygon is $\frac{a_n}{b_n} \cdot r$. Hence one gets the area

$$\frac{a_n r}{b_n} \cdot \frac{a_n r}{2}$$

as an lower bound of the area of the circular sector. Using the formulas (V.24.2) and the identity $\sin\alpha = 2\sin\frac{\alpha}{2}\cos\frac{\alpha}{2}$, one can simplify

$$\begin{aligned}\frac{a_n r}{b_n} \cdot \frac{a_n r}{2} = \frac{a_n^2}{b_n} \cdot \frac{r^2}{2} &= 2^n \sin\frac{\theta}{2^n}\cos\frac{\theta}{2^n} \cdot \frac{r^2}{2} \\ = 2^{n-1}\sin\frac{\theta}{2^{n-1}} \cdot \frac{r^2}{2} &= \frac{a_{n-1} r^2}{2}\end{aligned}$$

Altogether, we have obtained the bounds

$$\frac{a_{n-1} r^2}{2} < \text{area of circular sector} < \frac{b_n r^2}{2}$$

and in the limit $n \to \infty$, we obtain

$$\text{area of circular sector} = \lim_{n\to\infty} a_n \cdot \frac{r^2}{2} = \lim_{n\to\infty} b_n \cdot \frac{r^2}{2} = \frac{\theta r^2}{2}$$

with the central angle θ measured in arc length. □

Remark. A particular simple "proof" of the pizza theorem is as follows:

> We know that for a regular n-gon, the area is one-half the product of the radius r of the in-circle to the circumference. "Now image that we make the number of sides of the polygon *infinite*. Each triangle has height r and an *infinitesimal* base dr. Since the polygon fills the circle, their areas are equal. Also, the perimeter of the polygon equals the circumference of the circle. Thus the formula in the pizza theorem follows from the corresponding formula for the regular polygons with any number n of sides."

This idea is usually attributed to Johannes Kepler, but similar arguments were already used by Democritus and other early Greek mathematicians.

Because of the ancient Greeks "horror of the infinite" it should come as no surprise that no important mathematician in ancient times,—even those who devised such proofs,—did considered them to be rigorous. The same can be said of all mathematicians since then, including contemporary ones. Currently, we make the proof acceptable as done above: We give lower and upper estimates for the circumference and area. We use a "squeezing argument" which takes advantage of the fact that the lower and upper estimates have the same limit for $n \to \infty$. In this way the result is obtained from a limit $n \to \infty$ for which the number of sides of the polygon approach infinity. To be even more exact, one has to point out,—as we have done above,—which axioms of continuity are involved to confirm existence and uniqueness of such a limit.

The Greeks did not use these terms. But they had their *method of exhaustion* that was quite similar to the modern rigorous use of limits. This method was first clearly articulated by Eudoxus around 370 B.C. and extensively used by Archimedes.

However, without the modern calculus, the method of exhaustion gives no indications how to discover formulas for areas and volumes in the first place. It just needed the genius of Archimedes for every such result! But the mystery about the way Archimedes worked was (at least partly) resolved when a document by Archimedes called "The Method" was discovered around 1903. This document describes an ingenuous procedure in which one or more solids could be mentally sliced into infinitesimally thin pieces. By achieving a balance piece by piece, and using the basic physical law of levers, Archimedes was able to find the volume of the sphere, and similar results.

This *method of equilibrium* is one of the cleverest integration tricks ever devised. As yet, Archimedes did not even mention,—let alone use,—the method in any of his other works, because he did not consider it to be rigorous.

V.25 Area in Euclidean Geometry

V.25.1 Equidecomposable and equicomplementable figures

Problem V.25.1. *Draw two parallelograms on the same base with congruent altitudes, and explain why they are equidecomposable. Clearly draw the dissections you use.*

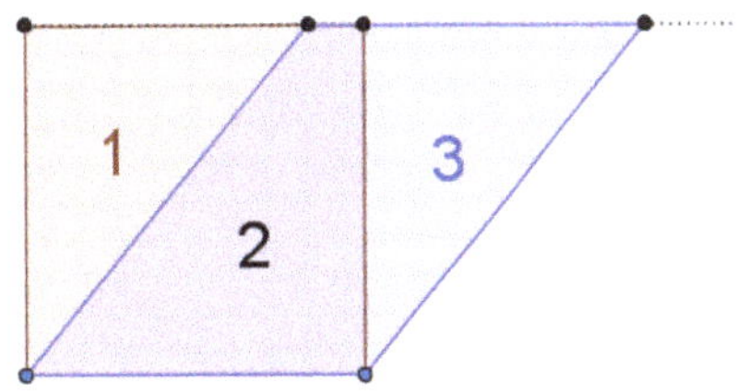

Figure V.25.1: A square and a parallelogram that are easily seen to be equidecomposable.

Answer. In the figure on page 501, the square is decomposed as $1 \uplus 2$. The parallelogram on the same base is decomposed as $2 \uplus 3$. Since the triangles $1 \cong 3$ are congruent, the square and the parallelogram as equidecomposable.

Problem V.25.2. *Draw two parallelograms on the same base with congruent altitudes that are not easily seen to be equidecomposable. Explain why they are equicomplementable. Clearly draw the dissections and mark the additional figures used.*

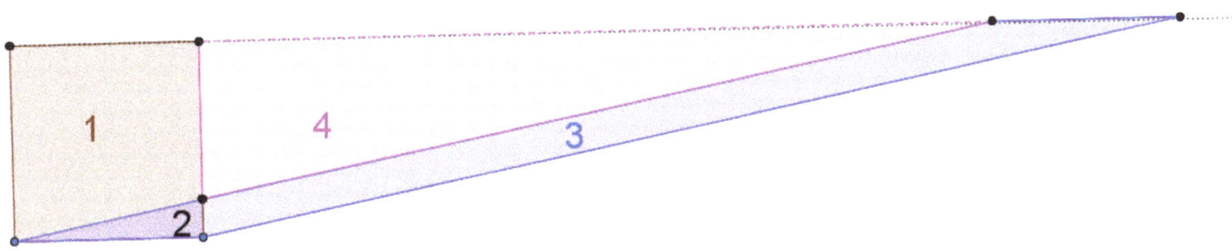

Figure V.25.2: A square and a parallelogram that are only shown to be equicomplementable.

Answer. In the figure on page 501, the square is decomposed as $1 \uplus 2$. The parallelogram on the same base is decomposed as $2 \uplus 3$. But this time 1 and 3 are not congruent.

To see that the square and the parallelogram are equicomplementable, we use the triangle 4 as additional figure. Since the unions

$$1 \uplus 4 \cong 3 \uplus 4$$

are indeed congruent, we conclude

$$\begin{aligned} \text{square } \uplus 4 &\overset{\boxtimes}{\sim} (1 \uplus 2) \uplus 4 \overset{\boxtimes}{\sim} 2 \uplus (1 \uplus 4) \overset{\boxtimes}{\sim} 2 \uplus (3 \uplus 4) \\ &\overset{\boxtimes}{\sim} (2 \uplus 3) \uplus 4 \overset{\boxtimes}{\sim} \text{ parallelogram } \uplus 4 \end{aligned}$$

Hence the square and the parallelogram—with the same base and congruent altitudes—are equicomplementable. Similarly, we prove that any two parallelograms with the same base and congruent altitudes are equicomplementable.

We can now give the exact modern version for the propositions about area in Euclid's book one.

Proposition V.25.1 (Instead of Euclid I.35). *Two parallelograms on the same base with congruent altitudes are equicomplementable.*

Proposition V.25.2 (Instead of Euclid I.36). *Two parallelograms with congruent bases and congruent altitudes are equicomplementable.*

Problem V.25.3. *Explain why two parallelograms with the same base and congruent altitudes are always equicomplementable.*

Proposition V.25.3 (Instead of Euclid I.37). *Two triangles with the same base and congruent altitudes are equicomplementable.*

Remark. Euclid talks about triangles "in the same parallels". Euclid I.37 says literally that `"Triangles on the same base in the same parallels are equal."`

Proposition V.25.4 (Instead of Euclid I.38). *Two triangles with congruent bases and congruent altitudes are equicomplementable.*

Problem V.25.4. *Convince yourself by an informal argument that two triangles with the same base and congruent altitudes are equicomplementable.*

Problem V.25.5. *Use the neutral version of Euclid I.37 given by proposition II.12.1 above to prove in Euclidean geometry that two triangles with the same base and congruent altitudes are equicomplementable.*

Answer. In Euclidean geometry, but only in Euclidean geometry, two triangles with the same base and congruent altitudes have the same midline. (Work out further details!) Hence by proposition II.12.1 above they are equicomplementable, too.

V.25.2 The Theorem of Pythagoras and related results

As a first step towards the Pythagorean Theorem, Euclid proves the following version of the leg theorem:

Proposition V.25.5 (Euclid's leg theorem). *The square on a leg of a right triangle is equicomplementable to the rectangle one side of which is the hypothenuse and the other side is the projection of this leg onto the hypothenuse.*

Modern proof following Euclid. For a given right triangle $\triangle ABC$, we have constructed the square $\square ACED$ over the leg AC and the square $\square ABIH$ over the hypothenuse AB. We have dropped the altitude from vertex C onto the hypothenuse and got the projection AF of the leg AC. Finally we have obtained the rectangle $\square AFGH$, one side of which is the hypothenuse and the other side is the projection AF of the leg AC onto the hypothenuse AB.

To show that the square $\square ACDE$ and the rectangle $\square AFGH$ are equicomplementable, we use as intermediate steps the two parallelograms $\square ABJD$ and $\square AHKC$. Here point J can be obtained by transferring the leg $AD \cong BJ$, and point K can be obtained by transferring the altitude $FC \cong GK$.

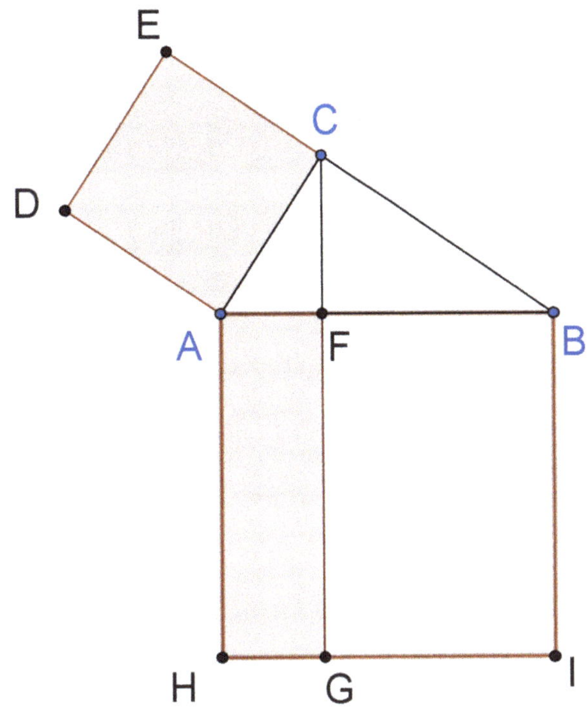

Figure V.25.3: Euclid's leg theorem tells the square over the leg is equicomplementable to the rectangle as indicated.

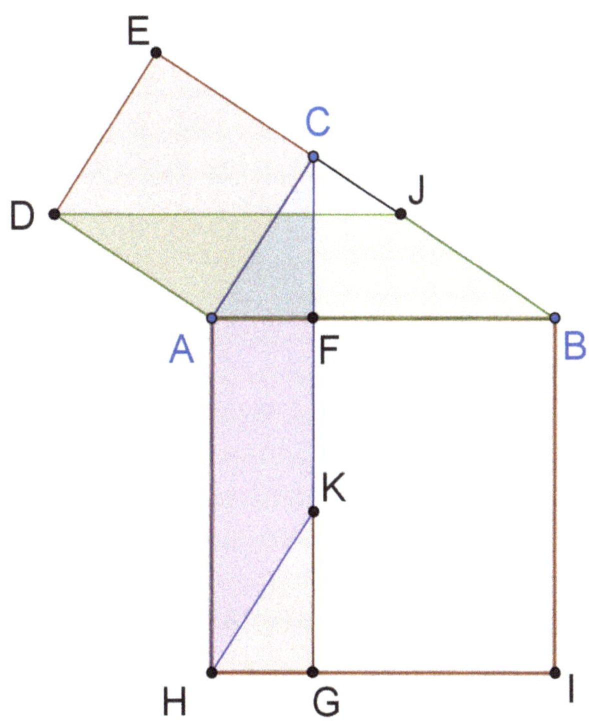

Figure V.25.4: Euclid's proof of the leg theorem.

Question. Convince yourself that both quadrilaterals $\square ABJD$ and $\square AHKC$ are indeed parallelograms.

Question. Convince yourself that these two parallelograms are congruent. How can one easily obtain parallelogram $\square AHKC$ from parallelogram $\square ABJD$.

Question. Why is the square $\square ACED$ equicomplementable to the parallelogram $\square ABJD$.

Answer. These two parallelograms have the same base AD and congruent altitudes. Hence they are equicomplementable by Euclid I.37 (see proposition V.25.3).

Question. Why is the parallelogram $\square ACKH$ equicomplementable to the rectangle $\square AFGH$.

Answer. These two parallelograms have the same base AH and congruent altitudes. Hence they are equicomplementable by Euclid I.37 (see proposition V.25.3).

By these steps, we have obtained the following chain of equicomplementable figures:

$$\square ACED \triangleq \square ABJD \cong \square ACKH \triangleq \square AFGH$$

By Theorem II.12.1, any two congruent figures are equicomplementable. Furthermore, the relation "equicomplementable" is an equivalence relation among figures. Hence we conclude that

$$\square ACED \triangleq \square AFGH$$

as to be shown. □

Proposition V.25.6 (Euclid's (weaker) Pythagorean theorem). *The sum of the squares on the legs of a right triangle is equicomplementable to the square over the hypothenuse.*

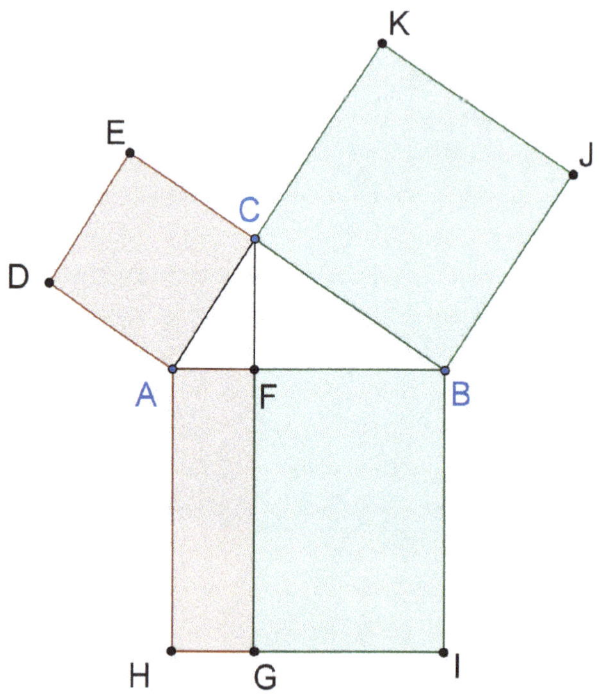

Figure V.25.5: Euclid's proof of the Pythagorean Theorem.

Modern proof following Euclid. We use Euclid's leg theorem V.25.5 twice, for both legs:

$$\square ACED \triangleq \square AFGH \quad \text{and} \quad \square BCKJ \triangleq \square BFGI$$

Since these figures do not overlapping we conclude

$$\square ACED \uplus \square BCKJ \triangleq \square AFGH \uplus \square BFGI \triangleq \square ABIH$$

as to be shown. $\square$

Theorem V.25.1 (The Pythagorean Theorem). *The sum of the squares on the legs of a right triangle is equidecomposable to the square over the hypothenuse.*

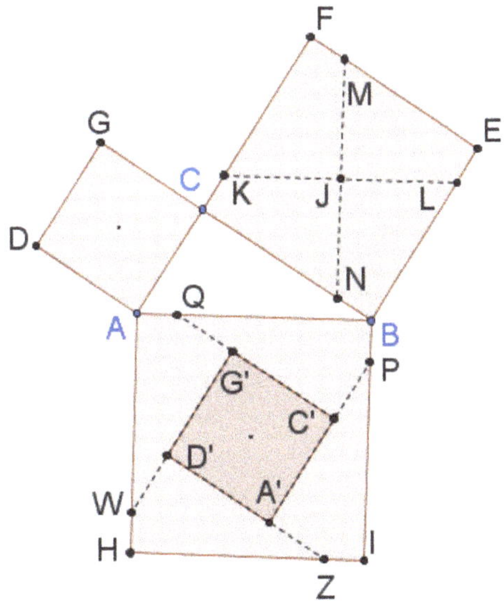

Figure V.25.6: What went wrong?

Problem V.25.6. *We want to prove the Theorem of Pythagoras by dissection. What went wrong in the above figure on page 505 to get an appropriate dissection. Draw a figure with a more useful dissection.*

Problem V.25.7. *Prove the Theorem of Pythagoras using the figure on page 506. For a proof by dissection, we need to*

- *give a construction of this diagram which explains in which order it was obtained.*
- *Based on this construction, we need to prove the congruence of the pieces to which the square on the hypothenuse is dissected to the pieces of which the two squares erected on the legs are made up.*

Proof. Here is a way to do these steps: We construct the center J of the square $\square BCFE$ on the larger leg of the given right triangle $\triangle ABC$. Drop the perpendicular from J onto the hypothenuse AB and erect the "double perpendicular" at point J. The square $\square BCFE$ cuts the segments NM and KL out of these two perpendicular lines.

We need to determine the segment lengths $|CK| = x$ and $|BL| = y$. We see that $y + x = |BL| + |LE| = a$. The parallelogram $\square ABLK$ has two pair of opposite congruent sides. Hence $|KL| = |AB| = c$ and $b + x = |AC| + |CK| = |AK| = |BL| = y$.

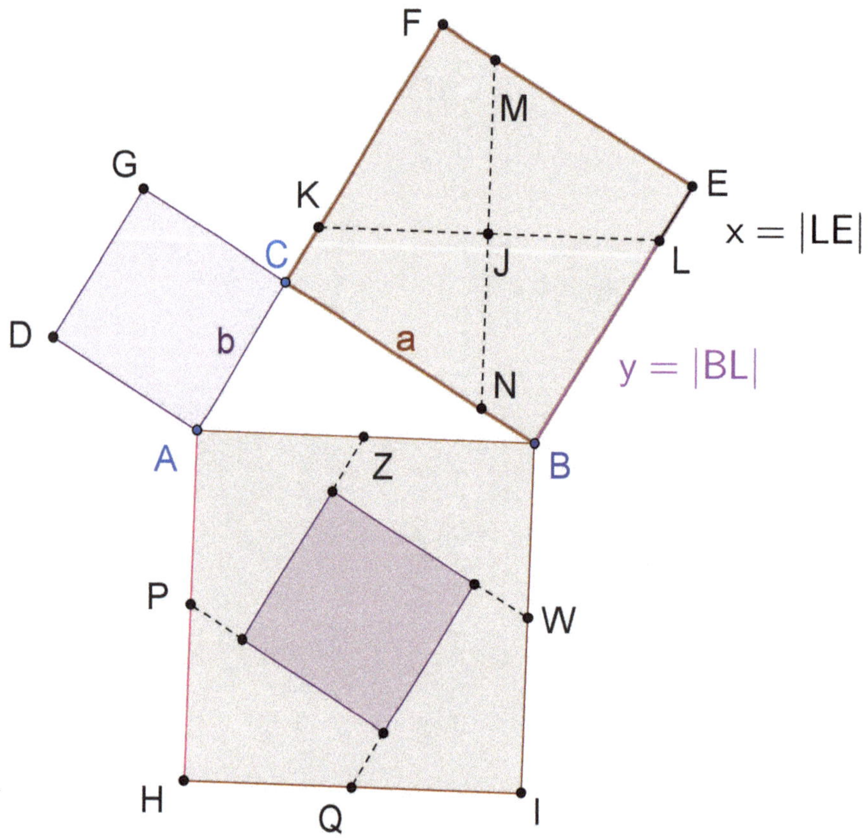

Figure V.25.7: A dissection proof of the Theorem of Pythagoras.

Since $|KJ| = |JL| = \frac{c}{2}$, the four quadrilaterals into which the square $\square BCFE$ is dissected, are congruent. They are transferred by several different parallel shifts to the square $\square ABIH$ on the hypothenuse. We do this in such a manner that the four right angles at vertex J are transferred to the four vertices A, B, I, H. Hence the transferred quadrilaterals have vertices at the midpoints P, Q, W, Z of the sides of the square $\square ABIH$. Furthermore, they have each one vertex A', C', G' or D' lying inside the square.

We get right angles at these four points. Since $A'C' \cong C'G' \cong G'D' \cong D'A' = y - x = b$, we see that the square $\square A'C'G'D'$ is congruent to the square $\square ACGD$ erected over the shorter leg AC.

Hence the square over the shorter leg, and the four congruent quadrilaterals into which the square over the longer leg has been dissected are put together without overlapping to form the square over the hypothenuse. $\square$

Problem V.25.8. *Given is a right triangle $\triangle ABC$ with angle $\alpha = 60°$ and side $a = 2$. Squares are erected on its three sides. As done in the figure on page 506, one can construct dissections of them which prove the Theorem of Pythagoras. What kind of quadrilateral is $\square ALKB$. Determine the segment lengths $x = |LE|$ and $y = |BL|$.*

Answer. The quadrilateral $\square ABLK$ is a parallelogram. Its opposite sides are congruent. Hence $y = |AK| = |AC| + |CK| = |AC| + |LE| = b + x$. Secondly, $a = |BE| = |BL| + |LE| = y + x$.

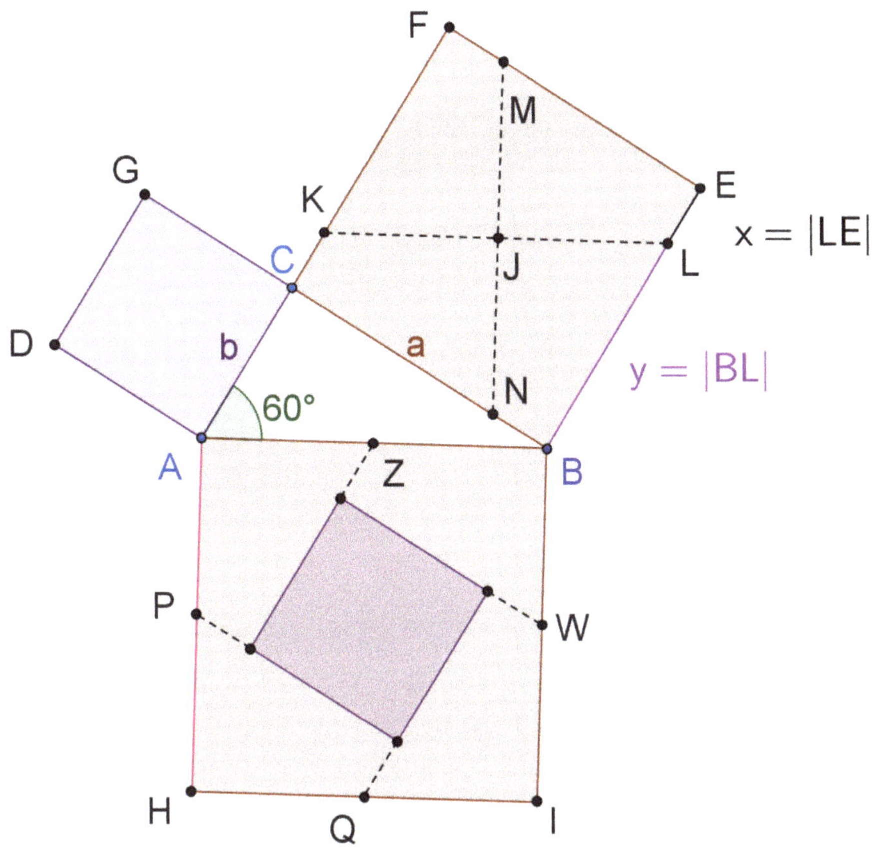

Figure V.25.8: Dissection proof of the Pythagorean theorem in a special case

From the equations

$$x + y = a\,,\ y - x = b \quad \text{one concludes}$$

$$x = \frac{a-b}{2} = 1 - \frac{\sqrt{3}}{3} \quad \text{and} \quad y = \frac{a+b}{2} = 1 + \frac{\sqrt{3}}{3}$$

V.25.3 Dudeney's dissection problem

How can one dissect a unit square into a few polygons and rearrange them into an equilateral triangle? This is a problem from "The Canterbury Puzzles" by H.E. Dudeney [?], first published in 1908. Dudeney gave the solution indicated in the drawing below, which uses just four polygons. A less ingenious solution is given in the figure on page 508. It uses five pieces some of which occur as mirror images.

Problem V.25.9. *We assume the side of the square to be equal to one. Calculate segments a and $x+y$ and the angle α in the accompanying drawing for Dudeney's dissection to confirm its existence.*

Answer. The side a is obtained from the total area $1 = \frac{\sqrt{3}}{4}a^2$ to be

$$a = 2\,(3^{-1/4}) = 2\sqrt{\frac{\sqrt{3}}{3}}$$

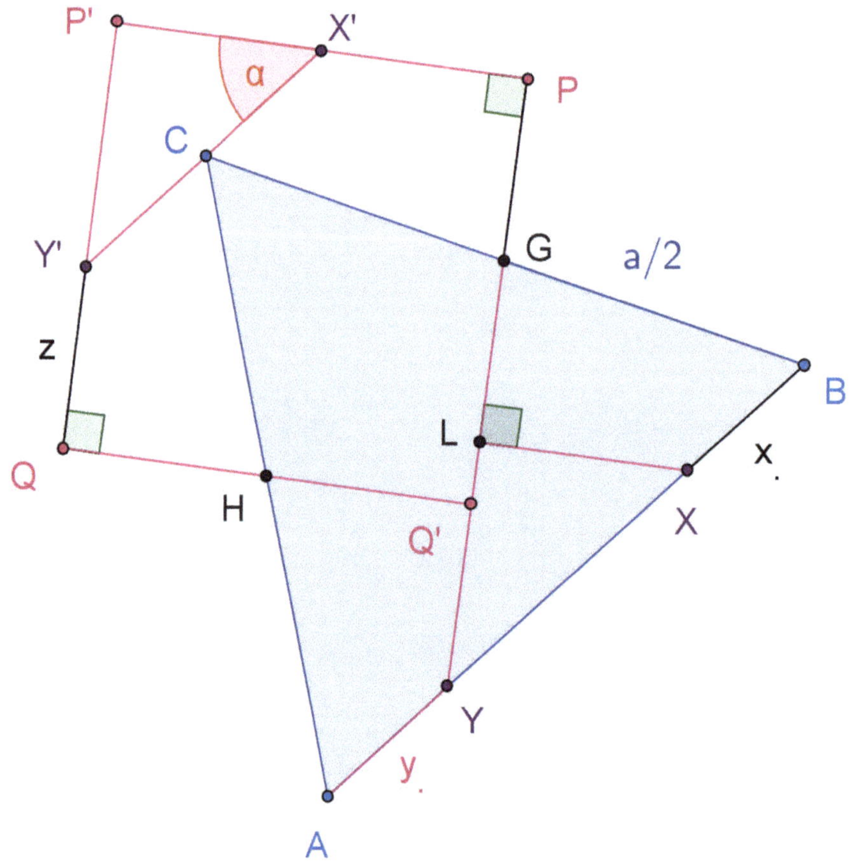

Figure V.25.9: Dudeney's dissection of an equilateral triangle and a square of the same area.

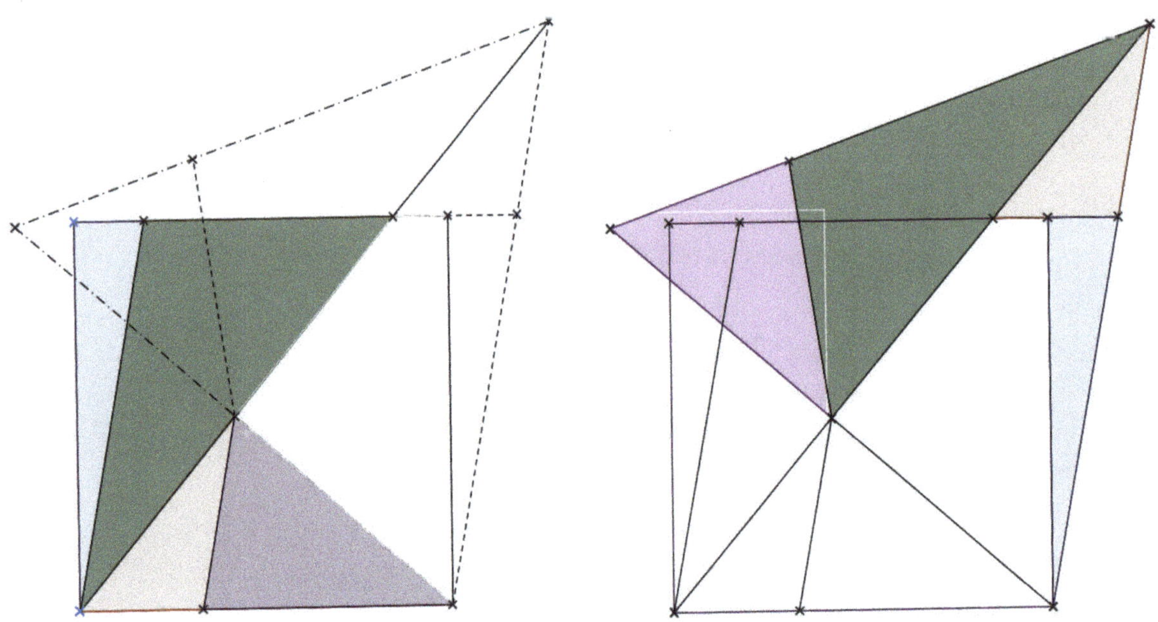

Figure V.25.10: Another dissection of an equilateral triangle and a square of the same area.

The two congruent quadrilaterals on the left side

$$\square HAYQ \cong \square HCY'Q$$

are obtained rotating by 180° about H. Hence $y = |AY| = |CY'|$. The two congruent quadrilaterals on the right side

$$\square GBXL \cong \square GCX'P$$

are obtained rotating by 180° about G. Hence $x = |BX| = |CX'|$. To complete the dissection, the triangles

$$\triangle XLY \cong \triangle X'P'Y'$$

have to be congruent. Hence $a - x - y = |XY| = |X'Y'| = x + y$ and

$$x + y = \frac{a}{2}$$

Since $|P'X'| = |LX| = |X'P| = \frac{1}{2}$, we get from the right triangle $\triangle X'P'Y'$

$$\cos\alpha = \frac{|P'X'|}{|X'Y'|} = \frac{1}{a} = \frac{3^{1/4}}{2} < \frac{\sqrt{2}}{2} \quad \text{and} \quad \alpha \approx 48.85° > 45°$$

Problem V.25.10. *Find four segments of the length $z = |QY'|$. Calculate the length $|Q'L|$ in terms of z.*

Problem V.25.11. *Give a Euclidean construction for Dudeney's dissection beginning with a unit square.*

Answer. The three figures below on page 510, page 510, and page 511 give stages of such a construction.

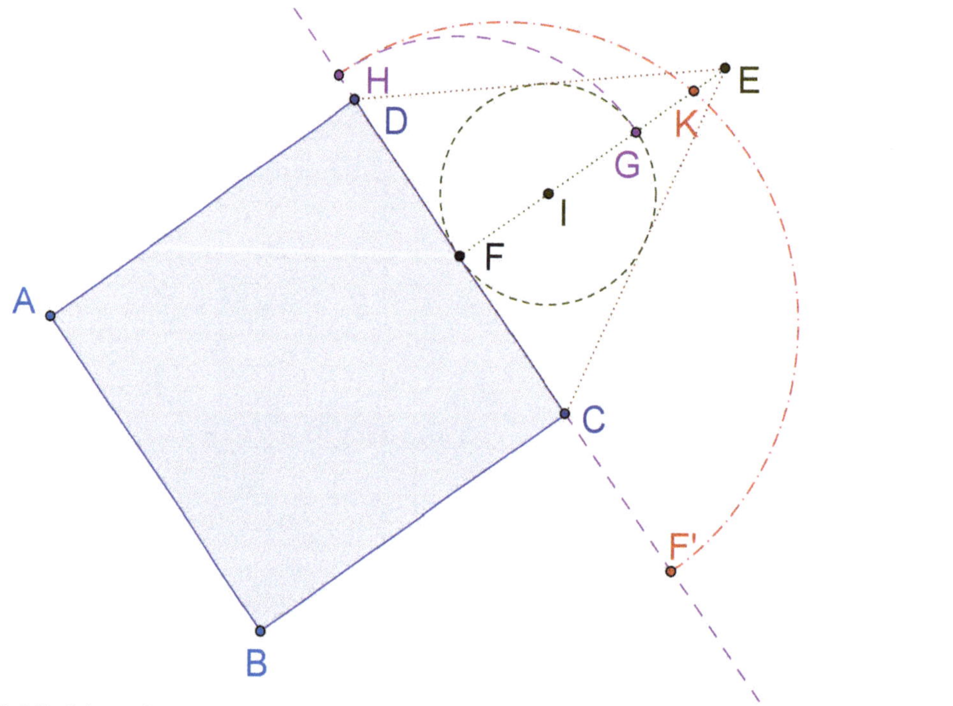

Figure V.25.11: A construction of Dudeney's dissection starting with the square I.

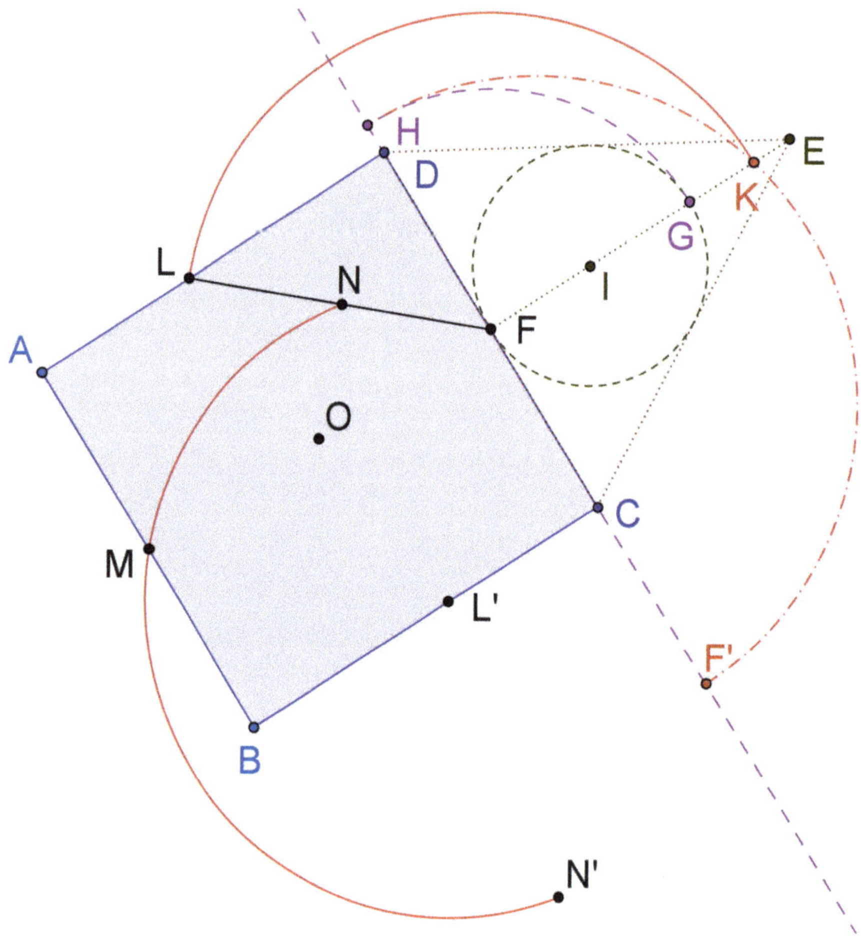

Figure V.25.12: A construction of Dudeney's dissection starting with the square II.

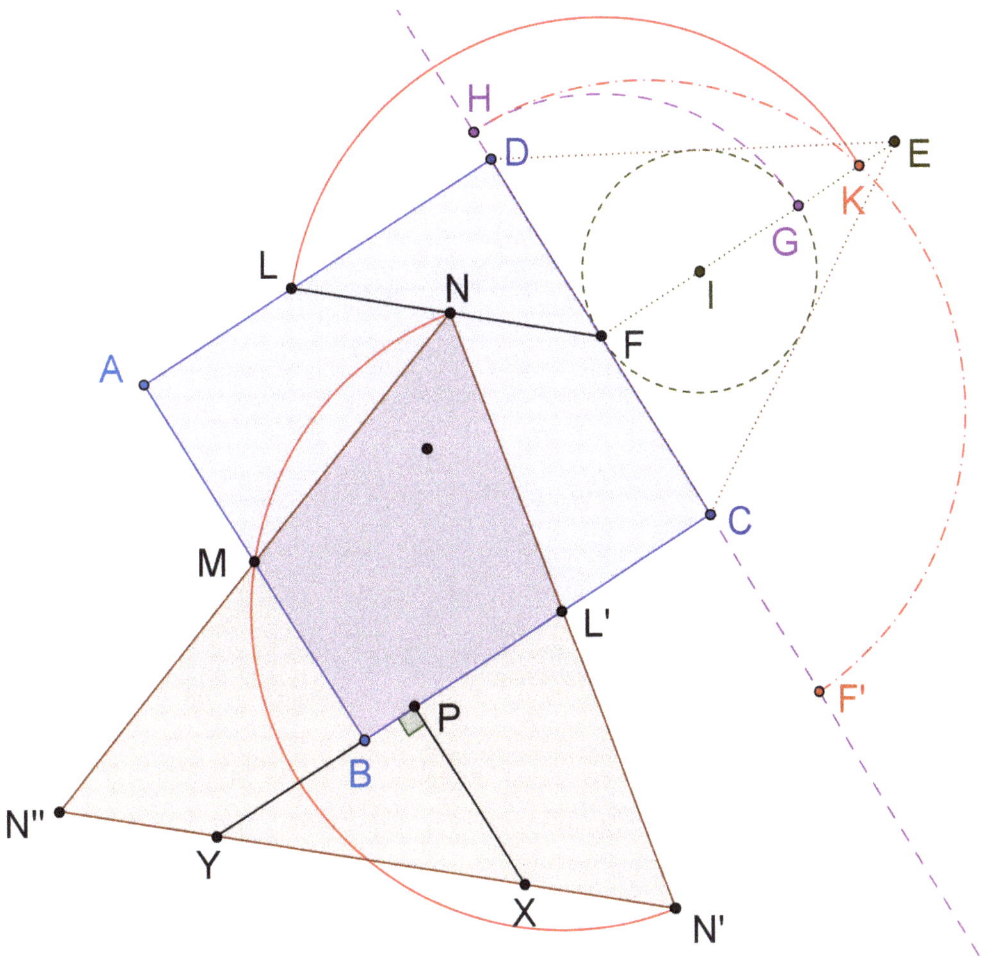

Figure V.25.13: A construction of Dudeney's dissection starting with the square III.

Problem V.25.12. *Calculate the segment x and the ratios $\frac{x}{a}$ and $\frac{y}{a}$. You can use the cosine theorem for the triangle $\triangle YGB$.*

Answer. The sides of the triangle $\triangle YGB$ have the lengths

$$|YG| = 1\,,\ |GB| = \frac{a}{2}\,,\ |BY| = x + \frac{a}{2}$$

Using the $60°$ angle at vertex B, the cosine Theorem yields

$$1^2 = \frac{a^2}{4} + \left(x + \frac{a}{2}\right)^2 - 2\frac{a}{2}\left(x + \frac{a}{2}\right)\cos 60° = x^2 + xa + \frac{a^2}{2} - \frac{a}{2}\left(x + \frac{a}{2}\right)$$

$$x^2 + \frac{ax}{2} + \frac{a^2}{4} - 1 = 0$$

$$x = \frac{-a + \sqrt{16 - 3a^2}}{4}$$

$$\frac{x}{a} = \frac{-1 + \sqrt{4\sqrt{3} - 3}}{4}$$

$$\frac{y}{a} = \frac{3 - \sqrt{4\sqrt{3} - 3}}{4}$$

Hence

$$x \approx .24549 \quad \text{and} \quad y \approx .25451$$

Problem V.25.13. *Give a Euclidean construction for Dudeney's dissection beginning with a given equilateral triangle.*

Answer. The figure on page 513 shows such a construction.

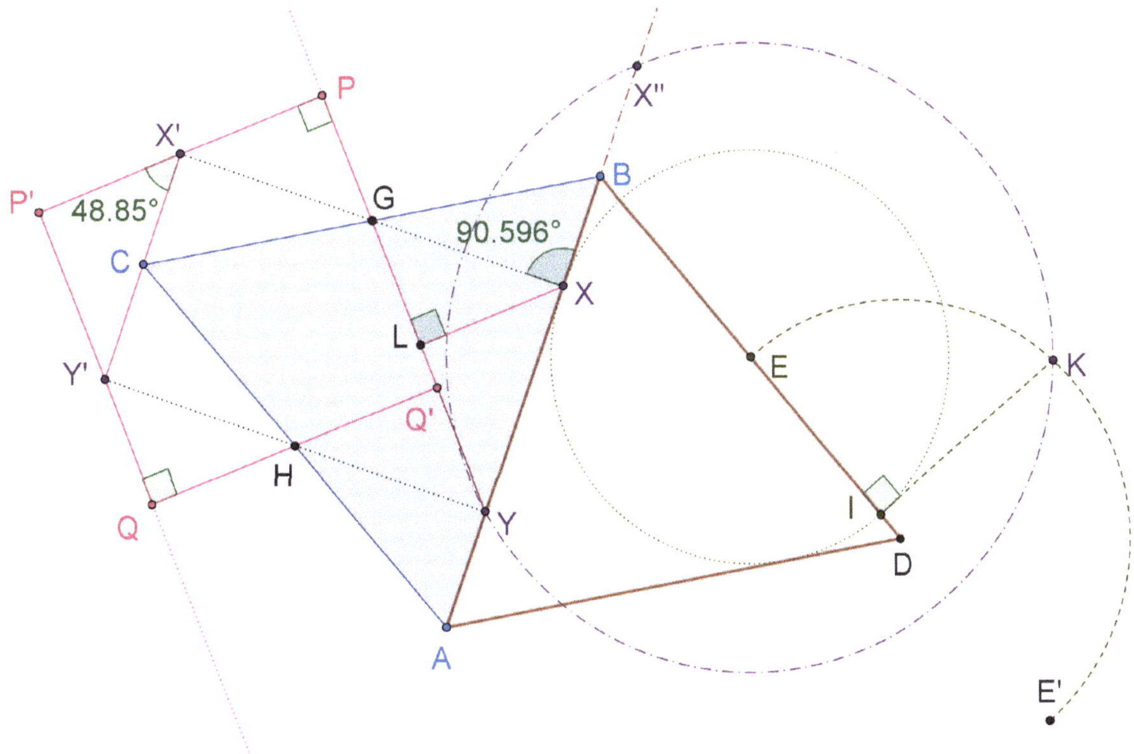

Figure V.25.14: A construction of Dudeney's dissection starting with the triangle.

V.25.4 Euclidean area

As the first step, we define the area for any triangle $\triangle ABC$. In Euclidean geometry, the area of a triangle is defined to be half of base times height.

$$\text{area}(\triangle ABC) = \frac{1}{2}ah_a = \frac{1}{2}bh_b = \frac{1}{2}ch_c$$

Problem V.25.14. *Prove that it does not matter in this definition, which side of the triangle is chosen as base.* [55]

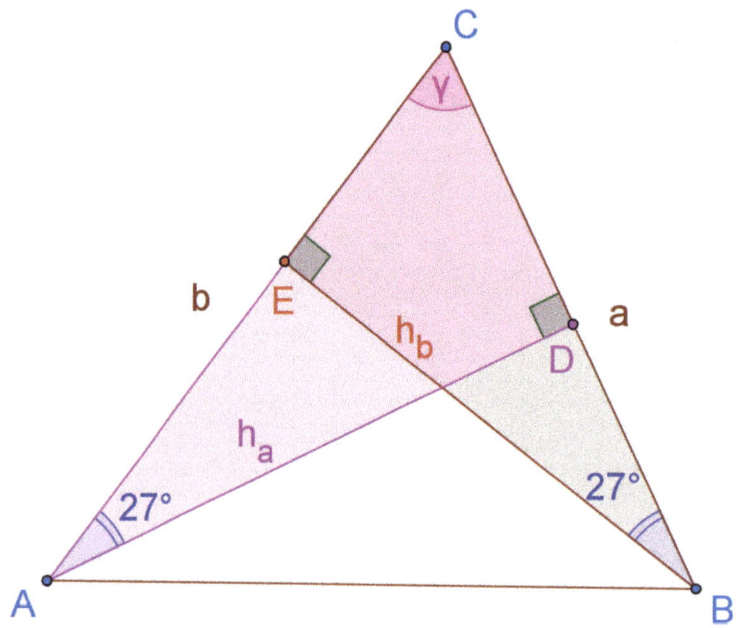

Figure V.25.15: For the calculation of its area, any side of a triangle may be used as its base.

Answer. For the triangle $\triangle ABC$, we can take side BC as base. The corresponding altitude is AD, were D is the foot-point of the perpendicular dropped from vertex A onto side BC.

As a second possibility, we can take side AC as base. The corresponding altitude is BE, were E is the foot-point of the perpendicular dropped from vertex B onto side AC.

The two triangles $\triangle CAD$ and $\triangle CBE$ are equiangular, and hence similar. We get the proportion

$$\frac{h_a}{b} = \frac{|AD|}{|AC|} = \frac{|BE|}{|BC|} = \frac{h_b}{a} = \sin\gamma$$

By multiplication with the denominators, we obtain

$$ah_a = |AD| \cdot |BC| = |BE| \cdot |AC| = bh_b = \sin\gamma \cdot a \cdot b$$

which is both the double area. We see that the area of a triangle well defined to be half of base times height. It does not matter which side one chooses as base.

[55] If one wants to define the area of rectangle as width time length, one get a similar difficulty—which again needs to be resolved by means of similar triangles.

Remark. We see from this definition that the area takes values the field of segment arithmetic. This is indeed an Abelian group, as required. Note that this step uses the similarly of triangles with different orientation.

Proposition V.25.7. *The area of the triangle is given by one half product of two sides time and the sin of the angle between them:*

$$\text{area}(\triangle ABC) = \frac{1}{2}ab\sin\gamma = \frac{1}{2}bc\sin\alpha = \frac{1}{2}ca\sin\beta$$

Proposition V.25.8 (Instead of Euclid I.39). *Two equicomplementable triangles with the same base have congruent altitudes.*

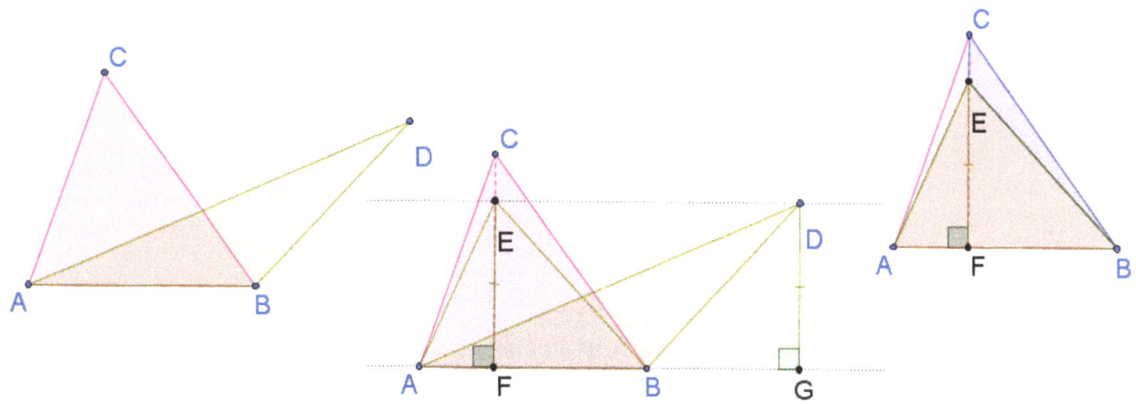

Figure V.25.16: Two triangles with same base and different altitudes are not equicomplementable.

Proof inspired by Euclid. We assume that $\triangle ABC$ and $\triangle ABD$ are equicomplementable. Let their altitudes be $CF \perp AB$ and $DG \perp AB$. To check whether they are congruent, we transfer GD onto the ray $\overrightarrow{FC}$ and get the congruent segment $FE \cong GD$. The triangles $\triangle ABD$ and $\triangle ABE$ have the same base and congruent altitudes. By the proposition V.25.3, which Euclid has already proved, they are equicomplementable. Hence by theorem II.12.1, the triangles $\triangle ABC$ and $\triangle ABE$ are equicomplementable. Too, their altitudes dropped from C and E have the same foot point F, and lie on the same line.

The question is now, whether $C = E$. If not, there are two possible cases left. Either E lies between F and C, or C lies between F and E. It is enough to rule out the first case, the argument in the second case is similar.

Assume—towards a contradiction—that $F * E * C$. In that case the triangle $\triangle ABE$ is a subset of triangle $\triangle ABC$, to which it is equicomplementable. Too, the triangle $\triangle ECB$ lies inside $\triangle ABC$, but does not overlap the smaller triangle $\triangle ABE$. Hence, by deZolt's postulate II.12.4, the triangles $\triangle ABE$ and $\triangle ABC$ cannot be equicomplementable. This contradiction rules out the case $F*E*C$. The case $F * C * E$ can be ruled out similarly.

Hence $E = C$, and the originally given triangles $\triangle ABC$ and $\triangle ABD$ have congruent altitudes $CF = EF \cong DG \perp AB$, as to be shown. □

Remark. Euclid does not succeed to prove deZolt's postulate in his treatment of area. But he did see that an additional assumptions is needed to get besides the simpler Euclid I.37 its converse Euclid I.39. To this end, he suggests an even stronger very general postulate about any magnitudes:

"The whole is more than its parts".

Such a general principle is clearly not justifiable. Indeed, it is not valid if dissection constructed from infinite sets,—obtained with the axiom of choice,—are allowed. This is exemplified by the Banach-Tarski paradox.

As we have seen from the proof above, deZolt's postulate would have been a workable version for a new axiom that Euclid could have used to complete his theory of area.

Short simple proof of Proposition V.25.8. Let $\triangle ABC$ and $\triangle ABD$ be equicomplementable. By theorem II.12.3, they have same area. The area is given by half the product of base times height. Let $CF \perp AB$ and $DG \perp AB$ be their altitudes. Hence we know that

$$\frac{1}{2}|AB| \cdot |CF| = \frac{1}{2}|AB| \cdot |DG|$$

Since segment arithmetic is a field, we can multiply both sides by 2 and divide by $|AB| \neq 0$ and get

$$|CF| = |DG|$$

from which follows $CF \cong DG$, as to be shown. □

Already in the segment arithmetic, it was necessary to introduce a *unit segment*. Areas are naturally measured by taking the square over the unit segment as unit for area.

Lemma V.25.1. *A right triangle T with legs of lengths a and b is equicomplementable to a right triangle with legs 1 and ab, and to a rectangle with legs 1 and* area(T).

Problem V.25.15. *Prove lemma V.25.1.*

Proposition V.25.9. *Every figure P is equicomplementable to a rectangle one side of which is the unit segment. The second side of this rectangle has the length* area(P).

Proof. The given figure P can be dissected into triangles. Each of them is either a right triangle or can be dissected into two right triangles. Hence the figure P can be dissected into finitely many right triangles T_i for $i = 1...n$. Each of them is equicomplementable to a rectangle R_i with sides 1 and area(T_i). By proposition II.12.3,

$$P = \biguplus_{i=1}^{n} T_i \triangleq \biguplus_{i=1}^{n} R_i = R$$

were the latter is a rectangle R with sides 1 and

$$\sum_{i=1}^{n} \text{area}(T_i) = \text{area}(P)$$

as to be shown. □

Theorem V.25.2. *Any two figures with same area are equicomplementable.*

Proof. Let P and Q be two figures with same area area(P) = area(Q). By proposition V.25.9, figure $P \triangleq R$, where R is a rectangle, one side of which is the unit segment, and the other side has the length area(P). Similarly, figure $Q \triangleq R'$, where R' is a rectangle with lengths of sides 1 and area(Q). But two rectangles with the same lengths of sides are congruent, and hence clearly $R \doteq R'$. By theorem II.12.1, "equicomplementable" is an equivalence relation. Hence

$$P \triangleq R \doteq R' \triangleq Q$$

implies the originally given figures P and Q are equicomplementable, as to be shown. □

Corollary 57. *Any two figures are equicomplementable if and only if they have equal areas.*

V.25.5 The role of the Archimedean axiom

Theorem V.25.3. *Assuming the Archimedean axiom* (V.1) *holds, any two figures with same area are even equidecomposable.*

Problem V.25.16. *We assume that the Archimedean axiom holds. Explain why a right triangle T is legs of lengths a and b is even equidecomposable to a rectangle with legs* 1 *and* area(T).

Proof of Theorem V.25.4. As explained in the problem V.25.16 above, with the assumption of the Archimedean axiom, a right triangle T with legs of lengths a and b and a rectangle with legs 1 and area(T) are shown even to be equidecomposable.

In the same way as in proposition V.25.9, we show that any figure P is even equidecomposable to a rectangle one side of which is the unit segment. The second side of this rectangle has the length area(P).

Finally, as in the proof of theorem V.25.2 that any two figures P and Q with same area are equidecomposable. (We do not use the Archimedean axiom in the two last steps, but the reasoning is done now with the stronger equivalence relation "equidecomposable".) □

Problem V.25.17. *Between two parallels $AB \parallel CD$ of distance* 5 *units, there are given two parallelograms $\square ABCD$ and $\square ABEF$. Let the lengths of side be* $|AB| = 1$, $|AC| = 10$ *and* $|AE| = 20$.

Construct dissections into triangles to shown that the two parallelograms are equidecomposable. How many triangles do you need?

Problem V.25.18. *Let n be a natural number. What is the number t of triangles which are needed to get congruent dissections of a square of side n and a rectangle of sides* 1 *and* n^2*?*

(a) *Prove that at least $\ulcorner \sqrt{2}\, n \urcorner$ triangles are needed.*

(b) *Give a dissection using* $2n$ *triangles.*

Can one improve the lower or upper bound?

Answer. Suppose we have congruent dissections of the square of side n and rectangle of sides 1 and n^2 into t triangles. Let L be the length of the longest side of these triangles.

The two opposite long sides of the rectangle are dissected into sides of different triangles. Since each triangle fits into the square of side n, the longest side L can have length at most $\sqrt{2}\, n$. Hence

$$2n^2 \leq t \cdot L \leq t\sqrt{2}\, n$$

which implies $t \geq \ulcorner \sqrt{2}\, n \urcorner$ as claimed.

Proposition V.25.10. *Assume the Archimedean axiom* (V.1) *does* _not_ *hold. Then there exist a square and a rectangle that are equicomplementable but* _not_ *equidecomposable.*

Proof. We assume the Archimedean axiom (V.1) does not hold. Hence there exist segments OB and OA such that

$$OA > n \cdot OB \quad \text{for all natural numbers } n$$

Now OB takes the role of the unit segment. We put the right triangle $\triangle OAB$ and the right isosceles triangle $\triangle OAA'$ on the same side of their common leg OA. We draw the parallel to the hypothenuse AB through point A' and get an intersection point P with the ray $\overrightarrow{OA}$. In this way, we have constructed a segment OP such that

$$|OP| \cdot |OB| = |OA|^2$$

The right triangle $\triangle OBP$ and the right isosceles triangle $\triangle OAA'$ are equicomplementable. By doubling these figures, we get a rectangle with sides OB and OP, as well as a square with side OA, which, too, are equicomplementable.

We claim these two figures are _not_ equidecomposable. Suppose—towards a contradiction—we have congruent dissections of the square and the rectangle into t triangles. Let L be the length of the longest side of these triangles.

The two opposite long sides of the rectangle together have total length $2|OP|$. They are dissected into sides of different triangles. Since each triangle fits into the square, the longest side L can have length at most $\sqrt{2}\,|OA|$. Hence

$$\begin{aligned} 2|OP| &\le t \cdot L \le t\sqrt{2}\,|OA| \\ \sqrt{2}|OP| \cdot |OA| &\le t|OA|^2 = t|OP| \cdot |OB| \\ \sqrt{2}|OA| &\le t|OB| \end{aligned}$$

On the other hand, we have assumed $OA > n \cdot OB$ for all natural numbers n, and hence with $n := t$ we get

$$t|OB| < |OA|$$

Thus we arrive at the contradiction $\sqrt{2}|OA| < |OA|$. This confirms the claim the square and the rectangle are _not_ equidecomposable. ☐

The logical relations between the different concepts of content gained in this section are summed up in the following diagram:

$$\begin{array}{ccc} P \text{ and } Q \text{ are equidecomposable} & \xrightarrow{\text{clear}} & P \text{ and } Q \text{ are equicomplementable} \\ \big\downarrow \text{Th. II.12.3} & & \text{Th. II.12.3} \big\downarrow \\ \text{area}(P) & = & \text{area}(Q) \\ \big\downarrow \text{Archimedes and Th. V.25.4} & & \text{Th. V.25.2} \big\downarrow \\ P \text{ and } Q \text{ are equidecomposable} & \xleftarrow[\text{Archimedean axiom}]{} & P \text{ and } Q \text{ are equicomplementable} \end{array}$$

Corollary 58. *Assuming the Archimedean axiom* (V.1) *holds, any two equicomplementable figures are even equidecomposable. Assuming the Archimedean axiom* (V.1) *does* _not_ *hold, there exist two equicomplementable figures that are* _not_ *equidecomposable.*

Problem V.25.19. *Convince yourself once more that two parallelograms with the same base and congruent altitudes are always equicomplementable. Why can one not claim they are always equidecomposable?*

V.25.6 Some uniqueness results for justification

Given is a Pythagorean plane. Let UQ denote a square the sides of unit length. Does assigning the area 1 to this figure determine the area function uniquely?

Proposition V.25.11. *Take any Pythagorean plane. Let G be any ordered Abelian group. Suppose that μ is an measure function having the following properties:*

(1) *the measure $\mu(P) \geq 0$ of any figure P is a nonnegative value in the ordered Abelian group G;*

(2) *$\mu(T) = \mu(S)$ for any two congruent triangles $T \cong S$;*

(3) *$\mu(P \uplus Q) = \mu(P) + \mu(Q)$ for any non-overlapping figures P and Q.*

Then there exists an order preserving homomorphism $\phi : F \mapsto G$ from field F of segment arithmetic to the group G such that $\mu = \phi \circ \text{area}$. In other word, each measure is a homomorphic image of the area:

$$\mu(P) = \phi(\text{area}(P)) \quad \textit{for all figures } P \tag{V.25.1}$$

and we get the diagram

$$\begin{array}{ccc} & \textit{figure } P & \\ \text{area} \downarrow & & \downarrow \textit{measure } \mu \\ \text{area}(P) & \xrightarrow[\phi]{} & \textit{measure } \mu(P) \end{array}$$

Proposition V.25.12. *Under the additional assumption that*

(1*) *the measure $\mu(T) > 0$ is strictly positive for any triangle T.*

the mapping ϕ is injective, hence an endomorphism.

Proof of Proposition V.25.11. By the same reasoning as used in the proof of Theorem II.12.3, we derive from the assumptions (2) and (3) the following more general properties of the measure μ:

(4) Any two equidecomposable figures $P \overset{\boxtimes}{\sim} Q$ have the same measure $\mu(P) = \mu(Q)$.

(5) Any two equicomplementable figures $P \triangleq Q$ have the same measure $\mu(P) = \mu(Q)$.

We go on to construct the homomorphism $\phi : F \mapsto G$ with the property (V.25.1). For any segment length r, we define $\phi(r) = \mu(R)$ to be the measure of the rectangle R, one side of which is the unit segment and the second side has the length r.

Given is any figure P. By Proposition V.25.9, the figure P is equicomplementable to a rectangle $R \triangleq P$, one side of which is the unit segment and the second side has the length area(P). Hence we get both area$(R) =$ area(P), as well as $\mu(R) = \mu(P)$ from property (5). In the end that means

$$\mu(P) = \mu(R) = \phi(r) = \phi(\text{area}(R)) = \phi(\text{area}(P))$$

holds for any figure P, as required.

It remains to check that the mapping ϕ preserves the addition and the order. Take two rectangles R with sides $r \times 1$, and S with sides $s \times 1$. We may assume they are non overlapping and have a common side of unit length. In that case $R \uplus S$ is a rectangle with sides $r+s$ and 1, and thus

$$\phi(r+s) = \mu(P \uplus R) = \mu(R) + \mu(S) = \phi(r) + \phi(s)$$

confirms the additivity. Now assume that $r \leq s$, and both R and S lie on the same side of their common side of unit length. In this case $R \subseteq S$, and hence

$$\phi(r) = \mu(R) \leq \mu(S) = \phi(s)$$

confirms that ϕ preserves the order. □

Proof of Proposition V.25.12. In order to check whether the homomorphism ϕ is injective, we suppose that r and s are two segment lengths and $\phi(r) = \phi(s)$. The rectangles R with sides $r \times 1$ and S with sides $s \times 1$ have the same measure since $\mu(R) = \phi(r) = \phi(s) = \mu(S)$. Since any two segments are comparable, we may assume $r \leq s$ and hence $R \subseteq S$.

We use the equality $\mu(R) = \mu(S)$ to rule out the R is strictly less that S. In that case, the difference $S \setminus R$ would be the disjoint union of two congruent right triangles. By assumption (1*), they would have a positive measure and hence additivity would imply $\mu(R) < \mu(S)$.

Thus we have seen that $\phi(r) = \phi(s)$ implies $\mu(R) = \mu(S)$, secondly $R = S$ and finally $r = s$, as to be checked for injectivity of ϕ. □

Theorem V.25.4. *In a Pythagorean and Archimedean plane, any measure function μ is equal to the common known* area, *provided it has the following properties:*

(0) *the measure $\mu(UQ) = 1$ for a unit square;*

(1) *the measure $\mu(P) \geq 0$ of any figure P is a nonnegative value in the segment arithmetic;*

(2) *$\mu(T) = \mu(S)$ for any two congruent triangles $T \cong S$;*

(3) *$\mu(P \uplus Q) = \mu(P) + \mu(Q)$ for any non-overlapping figures P and Q.*

It is interesting to note that this uniqueness proof uses really Eudoxus idea to get equality from rational upper and lower bounds. This principle appear in Euclid's book V as a definition for the equality of ratios of magnitudes. We reformulate this result as a proposition about Archimedean fields.

Proposition V.25.13 (Eudoxus principle). *Given are any four elements $a, b, c, d > 0$ in an Archimedean field. The ratios $a : b$ and $c : d$ are equal, if and only if for any whole numbers m and n either one of these three cases occurs:*

both $na > mb$ *and* $nc > md$,

both $na = mb$ *and* $nc = md$,

both $na < mb$ *and* $nc < md$.

Proof. By the the Archimedean axiom there exists a natural number n such that $(n+1)\cdot a > m\cdot b$. By the basic properties of the natural numbers, there exists a *smallest* such n. In this case both $n\cdot a \le m\cdot b$ as well as $(n+1)\cdot a > m\cdot b$ hold. By assumption, the same inequalities hold for a and b replaced by c and d. Hence

$$n\cdot a \le m\cdot b < (n+1)\cdot a \quad \text{and} \quad n\cdot c \le m\cdot d < (n+1)\cdot c$$

Division implies

$$\frac{n}{m} \le \frac{b}{a} < \frac{n+1}{m} \quad \text{and} \quad \frac{n}{m} \le \frac{c}{d} < \frac{n+1}{m}$$

We have squeezed both fractions b/a and c/d
between the same upper and lower rational bounds. By subtraction we conclude

$$-\frac{1}{m} \le \frac{b}{a} - \frac{c}{d} \le \frac{1}{m}$$

Since m is an arbitrary natural number, and the Archimedean axiom has been assumed to hold, we conclude $\frac{b}{a} = \frac{c}{d}$. This last step also called the *principle of exhaustion.* □

Remark. Eudoxus and Archimedes' principles are used by Euclid in book XII.2 to prove that the *"The areas of circles are in the same ratio as the squares on their diameters"*. Too, they are essential for Euclid's theory of volume. Lateron, the same principle appear again in Archimedean proof of the law of the lever, and many times more in mathematical physics.

Proof of Proposition V.25.4 . By Proposition V.25.11, there exists a homomorphism $\phi : F \mapsto F$ from the field of segment arithmetic to itself such that

$$\mu(P) = \phi(\text{area}(P)) \quad \text{for all figures } P \tag{V.25.1}$$

Moreover normalization for the unit square implies

$$\phi(1) = \phi(\text{area}(UQ)) = \mu(1) = 1$$

The additivity of the area and the measure of any n non overlapping congruent rectangles R of dimensions $r \times 1$ imply, by means of an easy induction on n that

$$\begin{aligned} \phi(n\cdot r) &= \phi(\text{area}(n\cdot R)) = \mu(n\cdot R) \\ &= n\cdot\mu(R) = n\cdot\phi(\text{area}(R)) = n\cdot\phi(r) \end{aligned}$$

For any natural n and m we conclude

$$\begin{aligned} m\cdot\phi\left(\frac{r}{m}\right) = \phi\left(m\cdot\frac{r}{m}\right) &= \phi(r) \\ \phi\left(\frac{1}{m}\cdot r\right) &= \frac{1}{m}\cdot\phi(r) \\ \phi\left(\frac{n}{m}\cdot r\right) &= \frac{n}{m}\cdot\phi(r) \end{aligned}$$

proving $\phi(qr) = q\,\phi(r)$ for any rational q. We need the corresponding property for any even irrational segment length s.

Given is any segment $s > 0$ and denominator $m \neq 0$, by the Archimedean axiom there exists a natural number n such that $(n+1) \cdot 1 > m \cdot s$. By the basic properties of the natural numbers, there exists a *smallest* such n. In this case, we get

$$n \cdot 1 \leq m \cdot s < (n+1) \cdot 1 \quad \text{and} \quad n \cdot 1 \leq m \cdot \phi(s) < (n+1) \cdot 1$$

The second inequalities hold since $\phi(qr) = q\phi(r)$ for any rational q and $r = 1$, and $\phi(1) = 1$, and finally the homomorphism ϕ preserves the order. We have squeezed both s and $\phi(s)$ between the same upper and lower rational bounds, and conclude by subtraction

$$-\frac{1}{m} \cdot 1 \leq \phi(s) - s \leq \frac{1}{m} \cdot 1$$

Since m is an arbitrary natural number, and the Archimedean axiom has been assumed to hold, we conclude $\phi(s) = s$. Once more, we have used Archimedes' *principle of exhaustion.* Hence it turns out that ϕ is the identity and the only additive positive measure μ is the common area. □

V.25.7 About the volume of polyhedra

Already Gauss brought it to the attention of mathematicians, whether a corresponding justification of the three-dimensional volume based on dissections is possible. In a letter to Gerling in 1844, Gauss says that it is too bad that the equality of volume of two symmetrical, but not congruent figures, can be proved only using the method of exhaustion. Gerling could solve the dissection problem for this example. In his reply, Gerling gives a direct proof that any triangular pyramid can be dissected into twelve pieces that are congruent to those of a dissection of its mirror image. Gauss responds that this is a nice result. But it is still unfortunate that the proof of Euclid's proposition (XII.5)

"Pyramids of the same height on triangular bases are in the same ratio as their bases"

needs the method of exhaustion. Hilbert conjectured that the method of exhaustion is really necessary. At the International Congress of Mathematicians in Paris 1900, he presented his famous list of the 23 most important problems facing mathematics in the twentieth century. The third problem of this list addresses the present question:

> Give two tetrahedra with equal bases and equal heights, which in no way can be dissected into congruent tetrahedra, and which, even by adding congruent tetrahedra, cannot be complemented into polyhedra, which in turn can be dissected into congruent tetrahedra.

In the same year, Max Dehn has given such a counterexample. Max Dehn proves that it is not possible to dissect a regular tetrahedron into a finite number of pieces that can be reassembled into a cube, even after possibly adding on other figures that are equivalent by dissection. The important consequence is that the method of exhaustion is really necessary in Euclid's proof of (XII.5).

V.25.8 Heron's formula for the area of a triangle

Problem V.25.20. *Derive Heron's formula*

$$\text{area}(\triangle ABC) = \frac{1}{4}\sqrt{(a+b+c)(a+b-c)(a-b+c)(b+c-a)}$$

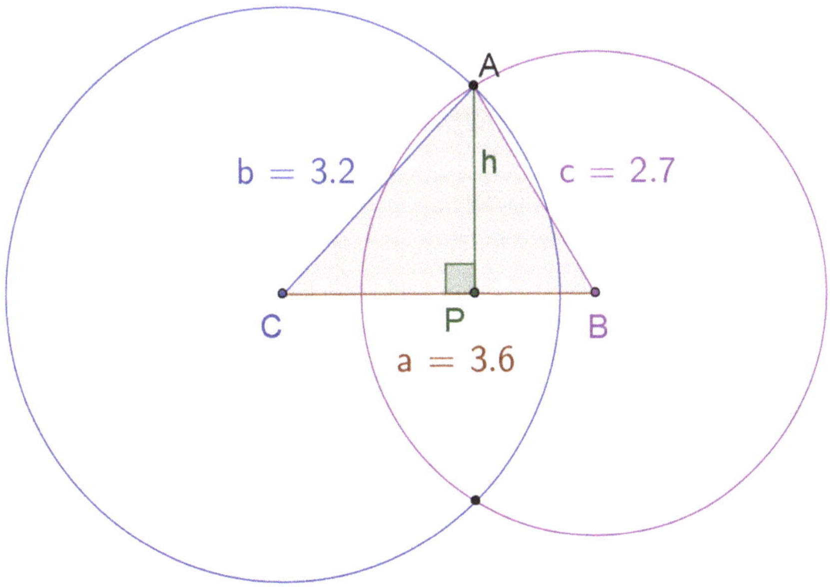

Figure V.25.17: Herons' formula for the area of triangle.

for the area a triangle in terms of its three sides a, b and c.

Answer. Take the triangle $\triangle ABC$ with vertices at the centers of the two circles and their intersection point A. We begin with the basic fact that the area of a triangle is half the product of base and altitude. We use the base CB and the altitude AP and get

$$\text{Area } \triangle ABC = \frac{1}{2}|CB| \cdot |AP|$$

Let $|CP| = p$ and $|BP| = q$. The height is calculated with the Pythagorean theorem in the right triangle $\triangle CPA$

$$h^2 = |AP|^2 = |CA|^2 - |CP|^2 = b^2 - p^2 = (b-p)(b+p)$$

Similarly, the Pythagorean theorem in the right triangle $\triangle BPA$ yields

$$h^2 = |AP|^2 = |BA|^2 - |BP|^2 = c^2 - q^2$$

Subtracting yields

$$\begin{aligned} b^2 - c^2 &= p^2 - q^2 = p^2 - (a-p)^2 \\ a^2 + b^2 - c^2 &= a^2 + p^2 - (a-p)^2 = 2ap \\ 2a(b+p) &= (a+b)^2 - c^2 = (a+b+c)(a+b-c) \\ 2a(b-p) &= -(a-b)^2 + c^2 = (-a+b+c)(a-b+c) \end{aligned}$$

Hence we get

$$\begin{aligned} \text{Area } \triangle ABC &= \frac{1}{2}|CB| \cdot |AP| = \frac{1}{4}\sqrt{2a(b+p)}\,\sqrt{2a(b-p)} \\ &= \frac{1}{4}\sqrt{(a+b+c)(a+b-c)(-a+b+c)(a-b+c)} \end{aligned}$$

V.25.9 Algebraic relations for the pieces of a triangle

Problem V.25.21. *Let r be the radius of the in-circle of triangle $\triangle ABC$. Prove the area is*

$$\text{area}(\triangle ABC) = \frac{(a+b+c)\,r}{2}$$

Answer. It is explained in an earlier problem II.13.3 in the context of neutral triangle geometry, how to construct the in-circle of a triangle. The perpendiculars are dropped from its center I onto

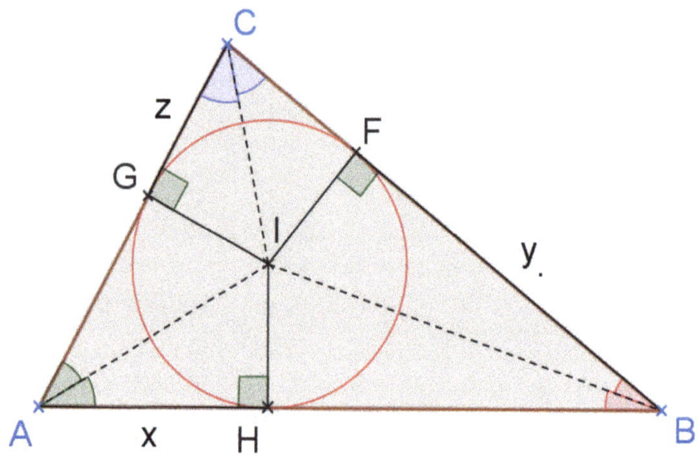

Figure V.25.18: An arbitrary triangle is partitioned into six right triangles.

the three sides. The foot points F, G, H are the touching points of the in-circle to the three sides. The six segments from the center of the in-circle to the vertices and the touching points partition the given triangle into six right triangles. All of them have the radius of the in-circle as one of their legs. The other legs have lengths $x = |AH|$, $y = |BF|$, and $z = |CG|$, each one occurring twice for two congruent triangles. Since the circumference is $a + b + c = 2x + 2y + 2z$, the total area of the given triangle $\triangle ABC$ is

$$\text{area}(\triangle ABC) = 2 \cdot \frac{x\,r}{2} + 2 \cdot \frac{y\,r}{2} + 2 \cdot \frac{z\,r}{2} = (x+y+z)\,r = \frac{(a+b+c)\,r}{2}$$

Problem V.25.22. *Let R is the radius of the circum-circle of triangle $\triangle ABC$. Prove the area is*

$$\text{area}(\triangle ABC) = \frac{abc}{4R}$$

Answer. As already stated in the Proposition V.25.7, the area of a triangle is one half product of two sides time and the sin of the angle between them:

$$\text{area}(\triangle ABC) = \frac{1}{2} ab \sin\gamma$$

On the other hand, the extended sin theorem tells that

$$\frac{c}{\sin\gamma} = 2R$$

Hence $\sin\gamma = \frac{c}{2R}$. Plugging into the first formula yields the required formula

$$\text{area}(\triangle ABC) = \frac{abc}{4R}$$

Problem V.25.23. *Can the three sides of a triangle be determined from the radius of the circum-circle, the radius of the in-circle and the circumference of the triangle?*

Answer. Yes, this is possible, but one needs to solve a third order equation. Hence the problem cannot be solved by a Euclidean construction— one needs to use Cardanos' formula.

We assume as given: the radius r of the in-circle, the radius R of the circum-circle, and the circumference $2s$. From the two previous problems, we get

$$\text{area}(\triangle ABC) = \frac{abc}{4R} = s\,r$$

Hence we can calculate the sum $a + b + c = 2s$ and the product $abc = 4Rrs$. We define the polynomial

$$P(x) := (x-a)(x-b)(x-c) = x^3 - (a+b+c)x^2 + (ab+ac+bc)x - abc$$

To get all coefficients of this polynomial, we have still to use Heron's formula, given in Problem V.25.20 below in the form

$$\text{area}(\triangle ABC) = \frac{1}{4}\sqrt{(a+b+c)(a+b-c)(a-b+c)(b+c-a)}$$

This can also be written with the semi-perimeter s as

$$\text{area}(\triangle ABC) = \sqrt{s(s-a)(s-b)(s-c)}$$

Hence we get one further value of the polynomial P:

$$P(s) = \frac{\text{area}(\triangle ABC)^2}{s} = rs^2$$

One now can calculate the polynomial and obtain

$$P(x) = x^3 - 2sx^2 + (rs + s^2 + 4Rr)x - 4Rrs$$

From this information, it is possible to calculate the three sides, for example by Cardanos' formula or by some numerical method.

Bibliography

[1] Michael Artin, *Algebra*, second ed., Prentice Hall, 2011.

[2] R. Baldus and F. Löbell, *Nichteuklidische Geometrie, Hyperbolische Geometrie der Ebene*, Sammlung Göschen 970, Berlin, 1964.

[3] Arthur Baragar, *Constructions using a compass and twice-notched straightedge*, The American Mathematical Monthly **109** (2002), 151–164.

[4] Paul Bernays, *Betrachtungen über das Vollständigkeitsaxiom und verwandter Axiome*, Math. Zeitschrift **63** (1955), 219–292.

[5] Edward B. Burger and Robert Tubbs, *Making Transcendence Transparent*, Springer, 2004.

[6] Barry A. Cipra, *Computer search solves an old math problem*, Science **242** (1988), 1507–1508.

[7] H.S.M. Coxeter and S.L. Greitzer, *Geometry Revisited*, ninth printing ed., New Mathematical Library 19, The Mathematical Asociation of America, 1967.

[8] H.E. Dudeney, *The Canterbury Puzzles, and other curios problems*, Nelson, London, 1929.

[9] E.H.Moore, Trans. Math. Soc. (1902).

[10] Solomon Feferman Anita B. Feferman, *Alfred Tarski: Life and Logic*, Cambridge University Press, 2004.

[11] G. Feigl, *Über die elementaren Anordnungssätze der Geometrie*, Jahresbericht der Deutschen Math.-Vereinigung **33** (1924).

[12] F.R.Moulton, *A simple non-desarguesian plane geometry*, Trans. Math. Soc. (1902).

[13] Chaim Goodman-Strauss, *Compass and straightedge in the Poincaré disk*, The American Mathematical Monthly **108** (2001), 38–49.

[14] Marvin J. Greenberg, *Euclidean and Non-Euclidean Geometry*, fourth ed., W.H. Freeman and Company, New York, 2008.

[15] ______, *Old and new results in the foundations of elementary plane Euclidean and Non-Euclidean geometries*, The American Mathematical Monthly **117** (2010), 198–219.

[16] Robin Hartshorne, *Geometry: Euclid and Beyond*, second printing, Springer, 2002.

[17] ______, *Non-Euclidean iii.36*, The American Mathematical Monthly **110** (2003), 495–502.

[18] David Hilbert, *Foundations of Geometry*, 2nd English ed., Open Court, La Salle, 1971.

[19] ———, *Grundlagen der Geometrie*, 14. Auflage ed., B.G.Teubner, Stuttgart, Leibzig, 1999.

[20] Ross Honsberger, *Episodes in nineteenth and twentieth century Euclidean geometry*, New Mathematical Library, vol. 37, The Mathematical Association of America, 2005.

[21] Roger A. Johnson, *Advanced Euclidean Geometry*, second printing, Dover Publications, Inc., Mineola, New York, 2007.

[22] Wilbur Richard Knorr, *The Ancient Tradition of Geometric Problems*, Dover Publications, 1945/1993.

[23] J. Kürschàk, *Das Streckenabtragen*, Math. Ann. **55** (1902).

[24] Stefan Mykytiuk and Abe Shenitzer, *Four significant axiomatic systems and some of the issues associated with them*, The American Mathematical Monthly **102** (1995), 62–67.

[25] David Park, *The Grand Contraption— the world in myth, number, and chance*, second printing, Princeton University Press, 2005.

[26] Abraham Pais, *Subtle is the Lord—the science and life of Albert Einstein*, Oxford University Press, 1982.

[27] Kenneth H. Rosen (ed.), *Handbook of Discrete and Computational Mathematics*, CRC Press, 2000.

[28] A. Rosenthal, Math. Ann. **71**.

[29] Alexander Ryba and Joseph Stern, *Equimodular polynomials and the tritangency theorems of Euler, Feuerbach and Guinand*, The American Mathematical Monthly **118** (2011), 217–228.

[30] James R. Smart, *Modern Geometries*, fifth ed., Brooks/Cole Publishing Company, 1997.

[31] John Stillwell, *The Four Pillars of Geometry*, Springer, 2005.

[32] H. Tverberg, *A proof of the Jordan Curve Theorem*, Bull. Lond. Math. Soc. **12** (1980), 34–38.

[33] B.L. van der Waerden, *Science Awakening I*, Kluwer Academic Publishers, 1954.

[34] O. Veblen and J.H. Maclagan-Wedderburn, *Non-desarguesian and non-pascalian geometries*, Transactions of the American Mathematical Society **8** (1907), 379–388.

[35] Robert S. Wolf, *A Tour through Mathematical Logic*, the carus mathematical monographs 30, Mathematical Association of America, 2005.

Index

www.ingramcontent.com/pod-product-compliance
Lightning Source LLC
LaVergne TN
LVHW071344210926
848486LV00003B/229